# Computational Imaging and Vision



This comprehensive book series embraces state-of-the-art expository works and advanced research monographs on any aspect of this interdisciplinary field.

Topics covered by the series fall in the following four main categories:

- Imaging Systems and Image Processing
- Computer Vision and Image Understanding
- Visualization
- Applications of Imaging Technologies

Only monographs or multi-authored books that have a distinct subject area, that is where each chapter has been invited in order to fulfill this purpose, will be considered for the series.

Volume 38

For other titles published in this series, go to
www.springer.com/series/5754

# The Theory of the Moiré Phenomenon

## Volume I: Periodic Layers

Second Edition

by

Isaac Amidror
*Peripheral Systems Laboratory,*
*Ecole Polytechnique Fédérale de Lausanne (EPFL), Switzerland*

Springer

Isaac Amidror
Department Informatique
Ecole Polytechnique Fédérale de Lausanne
Lab. Systemes Peripheriques
1015 Lausanne
Switzerland

ISSN 1381-6446
ISBN 978-1-84882-180-4 ISBN 978-1-84882-181-1 (eBook)
DOI 10.1007/978-1-84882-181-1

British Library Cataloguing in Publication Data
A catalogue record for this book is available from the British Library

Library of Congress Control Number: 2009921442

Mathematics Subject Classification (2000): 42-xx, 78-xx, 68U10



Printed on acid-free paper

Springer Science+Business Media
springer.com

*To my parents*

*No one admires Fourier more than I do. It is the only entertaining mathematical work I ever saw. Its lucidity has always been admired. But it was more than lucid. It was luminous. Its light showed a crowd of followers the way to a heap of new physical problems.*

**Oliver Heaviside** [Heaviside71 p. 32]

**Front cover image:** A heart-shaped moiré which is generated in the off-centered superposition of two circular gratings with slightly different radial periods. See Problem 11-8 and Fig. 11.4(c).

**Back cover images:** Interesting moiré effects in the superposition of two bell-shaped curvilinear gratings. See Figs. 10.34(c),(d).

# Contents

# Appendices

# Preface to the Second Edition

Since the first edition of this book was published several new developments have been made in the field of the moiré theory. The most important of these concern new results that have recently been obtained on moiré effects between correlated aperiodic (or random) structures, a subject that was completely absent in the first edition, and which appears now for the first time in a second, separate volume.

This also explains the change in the title of the present volume, which now includes the subtitle "Volume I: Periodic Layers". This subtitle has been added to clearly distinguish the present volume from its new companion, which is subtitled "Volume II: Aperiodic Layers". It should be noted, however, that the new subtitle of the present volume may be somewhat misleading, since this book also treats (in Chapters 10 and 11) moiré effects between *repetitive* layers, which are, in fact, geometric transformations of periodic layers, that are generally no longer periodic in themselves. The most suitable subtitle for the present volume would therefore have been "Periodic or Repetitive Layers", but in the end we have decided on the shorter version.

Although this revised edition maintains the general structure of the original book, it also includes some important improvements. It provides additional topics that were not explicitly treated in the first edition, such as the hybrid (1,-1)-moiré effects with 2D intensity profiles (now in Sec. C.14 of Appendix C), the moiré effects between hexagonal screens (now in Sec. C.15 of Appendix C) or the extension of the indicial equations method to the case of 2D screens (in Sec. 11.2). The present edition of the book also includes several new figures and some new or revised problems. New references have been added throughout the book, and all the Internet references have been verified and updated. And finally, cross-references have been added wherever appropriate to the second volume, and in particular to those of its appendices which may be of interest to readers of the present book. Note, however, that the two volumes are basically independent of each other. Each volume thus contains its own Glossary, List of notations and symbols, References and Index.

In preparing this second edition, we have also taken the opportunity to correct errors and typos that crept into the original edition of the book. However, some errors may have been overlooked, and some may have been inadvertently added in this new edition. Such errors, when detected, will be listed along with their corrections in the Internet site of the book, and we therefore encourage readers to inform us of any errors they may find.

The material in this book is based on the author's personal research at the Swiss Federal Institute of Technology of Lausanne (EPFL: *Ecole Polytechnique Fédérale de Lausanne*), and on his Ph.D. thesis (thesis No. 1341: "Analysis of Moiré Patterns in Multi-Layer Superpositions") which won the best EPFL thesis award in 1995.

This work would not have been possible without the support and the excellent research environment provided by the EPFL. In particular, the author wishes to express his gratitude to Prof. Roger D. Hersch, the head of the Peripheral Systems Laboratory of the EPFL, for his encouragement throughout the different stages of this project. Many thanks are also due to the publishers for their helpfulness and availability throughout the publishing cycle.

# From the Preface to the First Edition

Who has not noticed, on one occasion or another, those intriguing geometric patterns which appear at the intersection of repetitive structures such as two far picket fences on a hill, the railings on both sides of a bridge, superposed layers of fabric, or folds of a nylon curtain? This fascinating phenomenon, known as the *moiré effect*, has found useful applications in several fields of science and technology, such as metrology, strain analysis or even document authentication and anti-counterfeiting. However, in other situations moiré patterns may have an unwanted, adverse effect. This is the case in the printing world, and, in particular, in the field of colour reproduction: moiré patterns which may be caused by the dot-screens used for colour printing may severely deteriorate the image quality and turn into a real printer's nightmare.

The starting point of the work on which this book is based was, indeed, in the research of moiré phenomena in the context of the colour printing process. The initial aim of this research was to understand the nature and the causes of the superposition moiré patterns between regular screens in order to find how to avoid, or at least minimize, their adverse effect on colour printing. This interesting research led us, after all, to a much more far-reaching mathematical understanding of the moiré phenomenon, whose interest stands in its own right, independently of any particular application. Based on these results, the present book offers a profound insight into the moiré phenomenon and a solid theoretical basis for its full understanding. In addition to the question of moiré minimization between regular screens, the book covers many interesting and important subjects such as the navigation in the moiré parameter space, the intensity profile forms of the moiré, its singular states, its periodic or almost-periodic properties, the phase of the superposed layers and of each of the eventual moirés, the relations between macro- and micro-structures in the superposition, polychromatic moirés between colour layers, etc. All this is done in the most general way for any number of superposed layers having any desired forms (line-gratings, dot-screens with any dot shape, etc.). The main aim of this book is, therefore, to present all this material in the form of a single, unified and coherent text, starting from the basics of the theory, but also going in depth into recent research results and showing the new insight they offer in the understanding of the moiré phenomenon.

Fourier-based tools are but a natural choice when dealing with periodic phenomena; and, indeed, our approach is largely based on the Fourier theory. We consider each of the superposed layers as a function (*reflectance* or *transmittance* function) having values in the range between 0 and 1. We study the original layers, their superpositions, and their moiré effects by analyzing their properties both in the image domain and in the spectral, frequency domain using the Fourier theory. Further results are obtained by investigating the spectrum using concepts from geometry of numbers and linear algebra, and by interpreting the corresponding image-domain properties by means of the theories of periodic and almost-periodic functions. However, no prior knowledge of these fields of mathematics is assumed, and the required background is fully introduced in the text (in Chapter 5 and in Appendices A and B, respectively). The only prerequisite mathematical background is limited to undergraduate mathematics and an elementary familiarity with the Fourier theory (Fourier series, Fourier transforms, convolutions, Dirac impulses, etc.).

This book presents a comprehensive approach that provides a full explanation of the various phenomena which occur in the superposition, both in the image and in the spectral domains. This includes not only a quantitative and qualitative analysis of the moiré effect, but also the synthesis of moiré effects having any desired geometric forms and intensity profiles. In the first chapters we present the basic theory which covers the most fundamental case, namely: the superposition of monochrome, periodic layers. In later chapters of the book we extend the theory to the even more fascinating cases of polychromatic moirés and moirés between repetitive, non-periodic layers. Throughout the whole text we favour a pictorial, intuitive approach supported by mathematics, and the discussion is accompanied by a large number of figures and illustrative examples, some of which are visually striking and even spectacular.

This book is intended for students, scientists, and engineers wishing to widen their knowledge of the moiré effect; on the other hand it also offers a beautiful demonstration of the Fourier theory and its relationship with other fields of mathematics and science. Teachers and students of imaging science will find moiré phenomena to be an excellent didactic tool for illustrating the Fourier theory and its practical applications in one or more dimensions (Fourier transforms, Fourier series, convolutions, etc.). People interested in the various moiré applications and moiré-based technologies will find in this book a theoretical explanation of the moiré phenomenon and its properties. Readers interested in mathematics will find in the book a novel approach combining Fourier theory and geometry of numbers; physicists and crystallographers may be interested in the intricate relationship between the macro- and microstructures in the superposition and their relation to the theories of periodic and almost-periodic functions; and colour scientists and students will find in the polychromatic moirés a vivid demonstration of the additive and subtractive principles of colour theory. Finally, the occasional reader will enjoy the beauty of the effects demonstrated throughout this book, and — it is our hope — may be tempted to learn more about their nature and their properties.

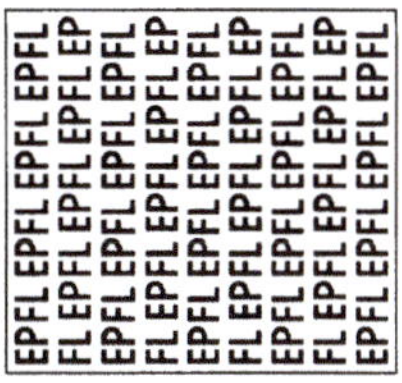

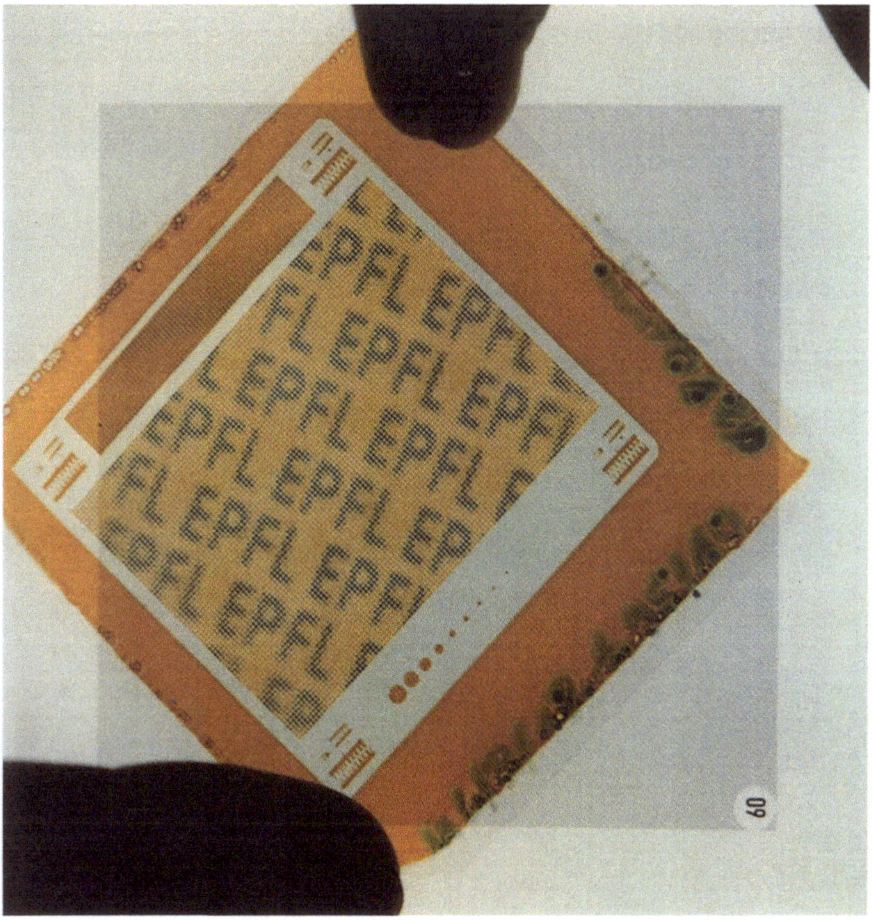

**Plate 1:** A (1,0,-1,0)-moiré generated by overlaying a periodic microlens array on a periodic dot-screen consisting of microscopic “EPFL”-shaped dots (the top inset shows a largely magnified detail of this dot-screen). The microlens array used here was produced by IMT, Neuchâtel, Switzerland; its lenslets are yellow, and the substrate between them is transparent. See Sec. 4.4 and Problem 4-12.

**Plate 2:** A (1,-1)-moiré in the superposition of two line-gratings of various colours. In all cases the periods and the angles of the gratings (and hence, of the moiré bands, too) are identical, and only the internal colouring of the period varies from case to case. The superpositions shown here consist of: (a) Two black gratings. (b) A black grating on a red grating. (c) A black grating on a green grating. (d) A green grating on a red grating. (e) A black grating on a grating whose period consists of R,G stripes. (f) Two gratings whose periods consist of R,G stripes. (g) A black grating on a grating whose period consists of R,G,B stripes (like in a Trinitron TV screen). (h) A grating whose period consists of R,G,B stripes on a grating whose period consists of B,G,R stripes (i.e., in inverse colour order). (i) Two gratings whose periods consist of R,G,B stripes. The microstructures of cases (h) and (i) were first described in [Bryngdahl81]. See Sec. 9.5 in Chapter 9.

**Plate 3:** A polychromatic (1,0,-1,0)-moiré in the superposition of a colour screen having red triangular dots with a green background, and a black-and-white screen having black circular dots. Both screens have identical frequencies but gradually increasing dots, as shown in the magnified bands to both sides of the superposition. Note that the top of the black-and-white screen consists of tiny "pinholes" on a black background; similarly, the bottom of the coloured screen contains green tiny "pinholes" on a red background. The moiré effect consists of red triangles on a green background (at the top of the superposed area), which gradually turn into red circles on a yellow background (at the bottom of the superposed area). See Sec. 9.6 in Chapter 9.

**Plate 4:** A polychromatic (1,-1)-moiré that is obtained by superposing a black grating on a colour grating whose period consists of C,M,Y stripes (see problem 9-5 in Chapter 9). The details of the two original gratings can be best perceived in the non-superposed areas.

# Chapter 1

# Introduction

## 1.1 The moiré effect

The moiré effect is a well known phenomenon which occurs when repetitive structures (such as screens, grids or gratings) are superposed or viewed against each other. It consists of a new pattern of alternating dark and bright areas which is clearly observed at the superposition, although it does not appear in any of the original structures.[1]

The moiré effect occurs due to an interaction between the overlaid structures. It results from the geometric distribution of dark and bright areas in the superposition: areas where dark elements of the original structures fall on top of each other appear brighter than areas in which dark elements fall between each other and fill the spaces better (see Fig. 1.1).

Because of its extreme sensitivity to the slightest displacements, variations, or distortions in the overlaid structures the moiré phenomenon has found a vast number of applications in many different fields. For example, in strain analysis moirés are used for the detection of slight deflections or object deformations, and in metrology moirés are used in the measurement of very small angles, displacements or movements [Patorski93; Kafri89; Shepherd79; Takasaki70; Durelli70; Theocaris69]. Among the numerous applications of the moiré one can mention fields as far apart as optical alignment [King72], crystallography [Oster63 p. 58], and document anti-counterfeiting [Renesse05 pp. 146–161]. Moiré effects have been used also in art [Oster65; Witschi86; Durelli70 pp. xiii–xxxviii], and even just for fun, enjoying their various intriguing shapes.

However, in other situations moiré patterns prove to be an undesired nuisance, and many efforts may be required to avoid or to eliminate them. This is the case, for example, in the printing world, and in particular in the field of colour image reproduction, where moiré patterns may appear between the dot-screens used for colour printing and severely corrupt the resulting image.

Clearly, mastering the moiré theory is essential for the proper use and control of moiré-based techniques, as well as for the elimination of unwanted moirés. It is the aim of this book, therefore, to provide the reader with a full theoretical understanding of the moiré phenomenon.

---

[1] The term *moiré* comes from the French, where it originally referred to watered silk, a glossy cloth with wavy, alternating patterns which change form as the wearer moves, and which is obtained by a special technique of pressing two watered layers of cloth together. Note that the term moiré does *not* refer to a presumed French physicist who studied moiré patterns, as has sometimes been stated (either mistakenly or humorously; see, for example, [Coudray91] and [Weber73]). Therefore the term moiré should not be written with a capital letter.

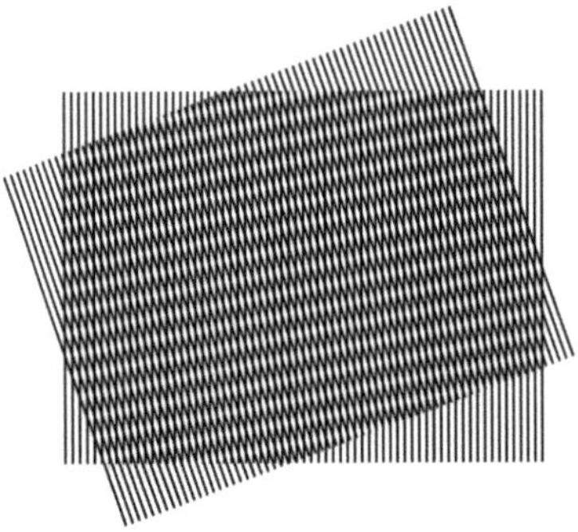

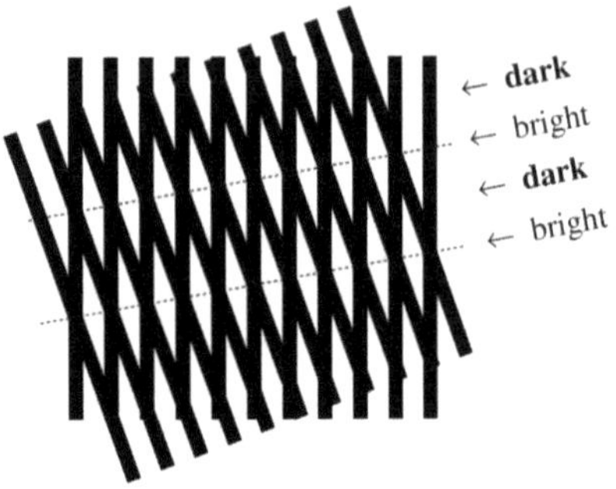


(a) (b)

**Figure 1.1:** (a) Alternating dark and bright areas which form the moiré effect in the superposition of two identical, mutually rotated line-gratings. (b) Enlarged view.

## 1.2 A brief historical background

The moiré phenomenon has been known for a long time; it was already used by the Chinese in ancient times for creating an effect of dynamic patterns in silk cloth. However, modern scientific research into the moiré phenomenon and its application started only in the second half of the 19th century with pioneering works such as [Rayleigh74] and [Righi87]. During almost a full century since then the theoretical analysis of moiré phenomena has been based on purely geometric or algebraic approaches (see Sec. 2.1). Based on these approaches many special purpose mathematical developments have been devised for the needs of specific applications such as strain analysis, metrology, etc. More recently several new approaches have been proposed for studying moiré phenomena, based, respectively, on non-standard analysis [Harthong81], on elementary geometry and potential theory [Firby84], or on algebraic geometry [Kendig80]. It was, however, undoubtedly the Fourier-based approach that most significantly contributed to the theoretic investigation of the moiré phenomenon.

The first significant steps in the introduction of the Fourier theory to the study of moiré phenomena can be traced back to the 1960s and 1970s. This pioneering work can be divided into two distinct stages: First came the use of Fourier series decompositions, purely in the image domain, for representing the original repetitive structures, their superpositions and their moirés (see, for example, [Lohmann67]). Only then were introduced further elements of the Fourier theory, such as the dual role of the image and the spectral domains [Bryngdahl74; Bryngdahl75], and the interpretation of the moiré in spectral terms as an aliasing phenomenon [Legault73]. Since then the Fourier approach has been used occasionally for the needs of some particular applications [Steinbach82; Takeda82; Morimoto88], but no systematic effort has been made to explore the full possibilities it offers. A possible reason for this fact may be that, as we show in this book, any such systematic attempt inevitably leads one to some branches of mathematics that are not very widespread, notably the theory of almost-periodic functions (see Appendix B) and the theory of geometry of numbers (see Chapter 5). The present book offers, for the first time, a full scale theoretical exploration of the moiré phenomenon which is based on the Fourier approach, and it contains several new results, both qualitative and quantitative, which have been obtained thanks to this fruitful approach.

A more detailed historical account on the research of the moiré phenomenon can be found in [Patorski93]. This book also gives a survey of various applications of the moiré effect, and an extensive bibliography on the subject. A collection of key scientific papers, both new and old, on the moiré effect and its applications can be found in [Indebetouw92].

## 1.3 The scope of the present book

The theory of the moiré phenomenon is an interdisciplinary domain whose range of applications is extremely vast. Its various theoretical and practical aspects concern the fields of physics, optics, mechanics, mathematics, image reproduction, colour printing, the human visual system, and numerous other fields. It would be in order, therefore, to clearly delimit here the scope of our present work.

Our main aim in this book is to present the moiré theory in the form of a unified and coherent text, starting from the basics of the theory, but also going in depth into recent research results. Among other topics we will discuss questions such as the minimization of moirés between regular screens, the moiré profile forms, its singular states, its periodic or almost-periodic properties, the phase of the superposed layers and of each of their eventual moirés, the relations between macro- and microstructures in the superposition, polychromatic moirés, moirés between repetitive, non-periodic layers, etc. These questions will be treated in the most general way, for any number of superposed layers having any desired forms (line-gratings, dot-screens with any dot shape, etc.).

It is clear, however, that it was impossible to include all the interesting material related to the moiré phenomenon in the present book. In the following list we enumerate some of the main points which have remained beyond the scope of this volume.

- First of all, we limit ourselves here to the analysis of moiré effects in the superposition of periodic or repetitive layers (like straight or curved line-gratings, dot-screens, etc.). Moiré effects between aperiodic or random layers are treated in *Vol. II* of the present work. Other types of moiré phenomena, such as moirés between almost-periodic or fractal structures (like Penrose tilings [Steinhardt90] or Cantor structures [Zunino03]), temporal moirés [Yule67 p. 330], etc., are not directly addressed in the present book, although they can be considered as natural extensions of the theory presented here.

- We do not consider here effects such as light scattering, light diffraction through the gratings, or any other physical questions concerning the nature of light (coherent/

incoherent) and its influence on the moiré [Patorski93 Chapter 4]. In particular, we will always assume that the line spacing in each grating is coarse enough for diffraction effects to be ignored [Durelli70 pp. 16, 35–42; Ebbeni70 p. 338; Theocaris75 p. 280].

- We do not consider here, either, the discrete nature of gratings and screen elements which are produced on digital devices such as laser printers, high-resolution filmsetters, etc., and the influence of this discrete nature on the moiré (this question is discussed in [Réveillès91 pp. 176–183]). The jagged aspect of discrete lines or dots is considered here as a real-world constraint, and we try to avoid it (or at least to reduce its influence) by producing our samples on appropriate devices with high enough resolutions (normally, at least 600 dots per inch).

- We suppose here that the different layers are superposed *in contact* (see [Post94 p. 90]), for example by overprinting, and we ignore the possible effects of the distance between the layers on the resulting moiré patterns, such as parallax-related phenomena [Huck03; Huck04] or the Talbot effect [Latimer93; Post94 pp. 76–78; Kafri90 pp. 102–103]; see also Sec. 1.8.6 in [Durelli70].

- We do not treat explicitly kinematic aspects of the moiré patterns, such as the speed of movement in the superposed layers and the speed of the resulting evolution in the moiré patterns (see, for example, Problems 7-6 and 7-7 at the end of Chapter 7). But although the kinematic aspects of the moiré theory are not explicitly developed here, they can be obtained in a rather straightforward manner by introducing the notion of time, and by considering shifts, rotations or any other layer transformations as functions of this new parameter.

- We also intentionally content ourselves here with a simplified model of the human visual system (see Sec. 2.2), and we avoid going any further into the complex questions related to the modelization of the human visual system and its performance in an inhomogeneous environment (like the perception of a moiré pattern on the irregular background of a screen superposition). More details about human vision and its modelization can be found, for example, in [Wandell95] and [Daly92].

- Finally, we usually prefer a pictorial, intuitive approach supported by mathematics over a rigorous mathematical treatment. In many cases we give informal demonstrations rather than formal proofs, or defer detailed derivations to an appendix.

It should be noted that although we occasionally use questions related to image reproduction to illustrate our discussion, this book has not been written with any specific moiré application in mind. In fact, our principal aim is to present the theoretical aspects of the moiré phenomenon in a general, application-independent way. Consequently a full discussion on the various applications of the moiré remains beyond the scope of the book; this material can be found in other books such as [Patorski93] or [Post94]. However, we felt that presenting the moiré theory without giving at least some flavour of its numerous applications would not serve the interest of the reader. Therefore, as a reasonable compromise, we have included among the problems at the end of each chapter some of the main applications of the theory being covered, along with additional references for the benefit of the interested readers. This should give the reader a general idea about the vast range of applications that the moiré effect has found in various different fields.

## 1.4 Overview of the following chapters

Chapter 2 lays the foundations for the entire book. This chapter presents the Fourier spectral approach that is the basis for our investigation of the moiré effect, and shows, step by step and in a systematic way, how this approach explains the moiré phenomenon between superposed layers: Starting with the simplest case, the superposition of cosinusoidal gratings, it gradually proceeds through the cases of binary gratings and square grids to the superposition of dot-screens. It also presents the notational system that we use for the identification, classification and labeling of the moiré effects, and introduces several fundamental notions such as the order of a moiré, singular moiré states, etc.

Chapter 3 presents the problem of moiré minimization, namely: the question of finding stable moiré-free combinations of superposed screens. In this chapter we focus on the moiré phenomenon from a different point of view: we introduce the moiré parameter space, and show how changes in the parameters of the superposed layers vary the moiré patterns in the superposition. This leads us to an algorithm for moiré minimization which provides stable moiré-free screen combinations that can be used, for example, for colour printing. Other methods for fighting unwanted moirés are also briefly reviewed (see, in particular, the problem section at the end of the chapter).

In Chapter 4 we show how, by considering not only the impulse locations in the spectrum but also their amplitudes, the Fourier-based approach provides a full quantitative analysis of the moiré intensity profiles, in addition to the qualitative geometric analysis of the moiré patterns which is already offered by the earlier classical approaches. We analyze the profile forms and intensity levels of moirés of any order which are obtained in the superposition of any periodic layers (gratings, dot-screens, etc.), and we show how they can be derived analytically from the original superposed structures, either in the spectral domain or directly in the image domain. We show how this analysis method can fully explain the surprising profile forms of the moiré patterns that are generated in the superposition of screens with any desired dot shapes, and how it can be used to synthesize moiré effects with any desired intensity profiles.

In Chapter 5 we set up a new algebraic formulation that will help us better understand the structure of the spectrum-support of the superposition, based on concepts from the theory of geometry of numbers and on linear algebra. In this discussion we completely ignore the impulse amplitudes, and we only consider their indices, their geometric locations, and the relations between them. This algebraic abstraction provides a new, important insight into the properties of the spectrum of the layer superposition and its moiré effects. Yet, this chapter can be skipped upon first reading and revisited later, when a deeper understanding is required.

In Chapter 6 we reintroduce the impulses on top of the spectrum-support, and we investigate the properties of the impulse amplitudes that are associated with the algebraic structures discussed in Chapter 5. Through the Fourier theory we see how both the structure and the amplitude properties of the spectral domain are related to properties of the layer superposition and their moirés in the image domain. In particular, we show the fundamental relationship between the Fourier expression of the layer superposition and the algebraic structure of the spectrum support.

In Chapter 7 we introduce the notion of phase, and we investigate what happens to the layer superposition (and particularly to the moiré effects) when the superposed layers are shifted on top of each other while keeping their angles and frequencies unchanged.

In Chapter 8 we focus our attention to the microstructure which occurs in the superposition and its relationship with the macro-moirés. We will see, in particular, that any moiré effect in the superposition is generated, microscopically speaking, by a repetitive alternation between zones of different microstructure in the superposition; when observed from a distance, this microstructure alternation is perceived as a repetitive gray level alternation in the superposition, i.e., as a macroscopic, visible moiré pattern.

In Chapter 9 we extend our Fourier-based approach to polychromatic moirés in the superposition of any coloured periodic layers. This will be done by considering the full colour spectrum of each point in any of the superposed layers; we will be dealing, therefore, with both colour spectra and Fourier spectra simultaneously. This extension of our theory will allow a full qualitative and quantitative analysis of moirés in colour, and it will enable us to synthesize moirés of any desired colours.

In Chapter 10 we further extend the scope of our Fourier-based approach, this time, to the superposition of repetitive, non-periodic layers such as curvilinear gratings or curved screens. We will see that although the Fourier spectrum in such cases are no longer purely impulsive, the fundamental principles of the theory remain valid in these cases, too. In particular, we will obtain the *fundamental moiré theorem*, which is a generalization of the results that were obtained in Chapter 4. We will see also how this approach can be used to synthesize moiré effects having any desired geometric layout and any intensity profile.

Finally, in Chapter 11 we briefly review some of the most widely used classical methods of moiré analysis, which are not directly based on the Fourier approach. We show that these alternative methods are, in fact, encompassed by the spectral approach, so that the results they can provide are only partial to the full information which can be obtained by the spectral approach. Nevertheless, these methods remain very useful in many real-world applications in which the use of the full scale spectral approach may prove to be impractical.

The main body of the book is accompanied by several appendices:

In Appendix A we review the main properties of 1D and 2D periodic functions both in the image and in the spectral domains. We also introduce the notion of step-vectors (in contrast to period-vectors), and the vector notations for 2D Fourier series that are extensively used in our work (notably in Chapters 6–10).

In Appendix B we review the main properties of 1D and 2D almost-periodic functions, both in the image and in the spectral domains. This appendix serves us as an introduction to the mathematical theory of almost periodic functions, which is not usually covered by standard textbooks on the Fourier theory.

In Appendix C we group together various issues, including the derivations of several results that we preferred, for different reasons, not to include in the main text of our work.

And finally, in Appendix D we provide a glossary of the most important terms that have been used in the present book.

The organization of Chapters 1–11 is as follows:

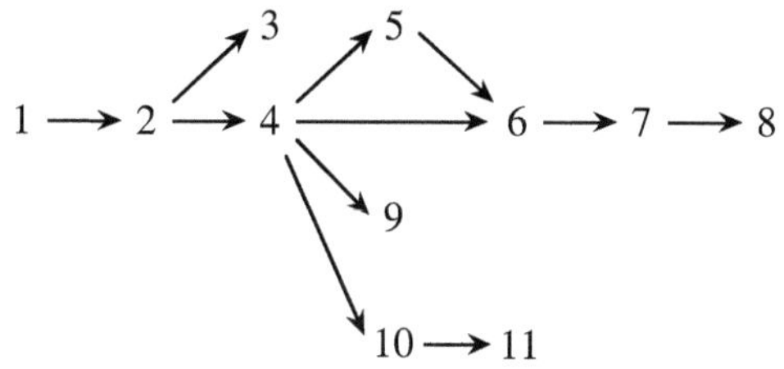


Although reading the chapters in their sequential order is recommended, any reader may choose to concentrate on one (or more) of the branches in this organization chart, according to his own needs and preferences.

## 1.5 About the exercises and the moiré demonstration samples

At the end of each chapter we provide a section containing a number of problems and exercises. Many of these problems are not merely routine exercises, but really intriguing and sometimes even challenging problems. Their aim is not only to aid the assimilation of the material covered by the chapter, but also to develop new insights beyond it. As already mentioned, these problems also include examples of real-world applications of the theory discussed in the chapter, along with references to existing publications on these applications (books, scientific papers, patents, etc.). We therefore highly encourage readers to dedicate some time for reviewing these exercises.

Since moiré effects are best appreciated by a hands-on experience, some of the key figures of this book have been also provided in the form of *PostScript*® programs [Adobe90], which can be printed on transparencies using any standard desktop laser printer. These PostScript programs and the instructions for using them can be found in the Internet site of this book, at the address:

`http://lspwww.epfl.ch/books/moire/`

By printing these demonstration samples the reader will obtain a kit of transparencies offering a vivid illustration of the moiré effects and their dynamic behaviour in the superposition. This demonstration set will allow the interested reader to make his own experiments by varying different parameters (angles, frequencies, etc.) in order to better understand their effects on the resulting moirés. This will not only be a valuable aid for the understanding of the material, but certainly also a source of amusement and fun.

* * *

Finally, a word about our notations. Throughout this book we adopt the following notational conventions:

Sec. 3.2 — Section 2 of Chapter 3.

Sec. A.2 — Section 2 of Appendix A.

Fig. 3.2 — Figure 2 of Chapter 3.

Fig. A.2 — Figure 2 of Appendix A.

(3.2) — Equation or formula 2 of Chapter 3.

(A.2) — Equation or formula 2 of Appendix A.

Similar conventions are also used for enumerating tables, examples, propositions, remarks, etc.; for instance, Example 3.2 is the second example of Chapter 3.

Whenever reference is made to the second volume of this work, we use the abbreviation "*Vol. II*". When referring to a section, a figure or an equation in *Vol. II*, we simply add the prefix "*II*" to the specified number; for example, Sec. *II*.3.2 refers to Sec. 2 of Chapter 3 in *Vol. II*, Fig. *II*.3.2 means Fig. 2 of Chapter 3 in *Vol. II*, and Eq. (*II*.A.2) refers to the second equation in Appendix A of *Vol. II*.

The mathematical symbols and notations used in the present volume are listed at the end; a glossary of the main terms is provided in Appendix D.

# Chapter 2

# Background and basic notions

## 2.1 Introduction

Several mathematical approaches can be used to explore the moiré phenomenon. The classical geometric approach [Nishijima64; Tollenaar64; Yule67] is based on a geometric study of the properties of the superposed layers, their periods and their angles. By considering relations between triangles, parallelograms, or other geometric entities generated between the superposed layers, this method leads to formulas that can predict, under certain limitations, the geometric properties of the moiré patterns. Another widely used classical approach is the indicial equations method (see Sec. 11.2); this is a pure algebraic approach, based on the equations of each family of lines in the superposition, which also yields the same basic formulas [Oster64]. A more recent approach, introduced in [Harthong81], analyzes the moiré phenomenon using the theory of non-standard analysis. This approach can also provide the intensity levels of the moiré in question, in addition to its basic geometric properties.

However, the best adapted approach for investigating phenomena in the superposition of periodic structures is the spectral approach, which is based on the Fourier theory. This approach, whose first applications to the study of moiré phenomena appeared in the 1960s and 1970s (see Sec. 1.2), is the basis of our work, and it will be largely developed in the present book. Unlike the previous methods, this approach enables us to analyze properties not only in the original layers and in their superposition but also in their spectral representations, and thus it offers a more profound insight into the problem and provides indispensable tools for exploring it. We will discuss the advantages that the spectral approach offers in the study of moiré phenomena at the end of this chapter (Sec. 2.14), after having introduced the basic notions of the theory.

The present chapter lays the foundations for the entire book. In Sec. 2.2 we present the background and the basic concepts of the spectral approach, and we determine the image types with which we will be concerned in our work. Then we proceed in the following sections by showing step by step, in a didactic way, how our approach explains the various moiré phenomena between superposed layers. We start in Secs. 2.3–2.4 with the simplest case, the superposition of cosinusoidal gratings, and then we gradually proceed to the more interesting cases involving binary gratings, grids and dot-screens. On our way we will also introduce some fundamental terms and notions of the theory, such as first-order and higher-order moirés, singular moirés, stable and unstable moiré-free superpositions, etc. The problems at the end of the chapter include some of the main applications of the moiré effect in various fields of science and technology, along with additional references for the interested readers.

## 2.2 The spectral approach; images and their spectra

The spectral approach is based on the duality between functions or images in the spatial image domain and their spectra in the spatial frequency domain, through the Fourier transform. A key property of the Fourier transform is its ability to allow one to examine a function or an image from the perspective of both the space and frequency domains. By allowing us to analyze properties not only in the original image itself but also in its spectral representation this approach combines the best from both worlds, namely: it accumulates the advantages offered by the analysis in each of the two domains.[1]

In this book we will be concerned with bidimensional (2D) structures in the continuous $x,y$ plane, that we will call *images*, and their 2D spectra in the continuous $u,v$ plane which are obtained by the 2D Fourier transform.[2] In fact, we will restrict ourselves only to some particular types of 2D images, such as line-gratings or dot-screens, which are liable to generate moiré effects when superposed. In this section we will review the basic properties of the image types with which we are concerned, and the implications of these properties both in the image and in the spectral domains.

First, let us mention that we will mainly deal here with moiré effects between monochrome, black and white images; the extension of our discussion to the fully polychromatic case will be delayed until Chapter 9. In the monochrome case each image can be represented in the image domain by a *reflectance* function, which assigns to any point $(x,y)$ of the image a value between 0 and 1 representing its light reflectance: 0 for black (i.e., no reflected light), 1 for white (i.e., full light reflectance), and intermediate values for in-between shades. In the case of transparencies, the reflectance function is replaced by a *transmittance* function which is defined in a similar way: it gives 0 for black (i.e., no light transmittance through the transparency), 1 for white (or rather *transparent*, i.e., full light transmittance through the transparency), or any intermediate value between them. A superposition of such images can be obtained by means of overprinting, or by laying printed transparencies on top of each other. Since the superposition of black and any other shade always gives here black, this suggests a *multiplicative* model for the superposition of monochrome images. Thus, when $m$ monochrome images are superposed, the reflectance of the resulting image (also called the *joint reflectance*) is given by the *product* of the reflectance functions of the individual images:

$$r(x,y) = r_1(x,y) \cdot \ldots \cdot r_m(x,y) \tag{2.1}$$

---

[1] It should be emphasized that since the Fourier transform is reversible, no information is gained or lost by its application. It only reveals certain image features which were present but not explicitly apparent before the image was transformed.

[2] Note that throughout this book we adopt the Fourier transform conventions that are commonly used in optics (see [Bracewell86 p. 241] or [Gaskill78 p. 128]); thus, the Fourier transform of a function $f(x,y)$ and its inverse are given by:

$$F(u,v) = \int_{-\infty}^{\infty}\int_{-\infty}^{\infty} f(x,y)\, e^{-i2\pi(ux+vy)}\, dx\, dy, \qquad f(x,y) = \int_{-\infty}^{\infty}\int_{-\infty}^{\infty} F(u,v)\, e^{i2\pi(ux+vy)}\, dx\, dy.$$

For alternative definitions used in literature and the relationships between them see [Bracewell86 pp. 7 and 17] or [Gaskill78 pp. 181–183].

The same rule applies also to the superposition of monochrome transparencies, in which case $r_i(x,y)$ and $r(x,y)$ simply represent transmittance rather than reflectance functions. Now, according to the convolution theorem [Bracewell86 p. 244], the Fourier transform of a function product is the convolution of the Fourier transforms of the individual functions. Therefore, if we denote the Fourier transform of each function by the respective capital letter and the 2D convolution by $**$, the spectrum of the superposition is given by:

$$R(u,v) = R_1(u,v) ** \ldots ** R_m(u,v) \tag{2.2}$$

**Remark 2.1:** It should be noted, however, that the multiplicative model is not the only possible superposition rule, and in other situations different superposition rules can be appropriate. For example, when images are superposed by making multiple exposures on a positive photographic film (assuming that we do not exceed the linear part of the film's response [Shamir73 p. 85]), intensities at each point are summed up, which implies an *additive* rule of superposition. In another example, when images are superposed by making multiple exposures on a negative photographic film (again, assuming a linear response) an *inverse additive* rule can be appropriate. More exotic superposition rules (involving, for example, various Boolean operations etc.) can be artificially generated by computer, even if they do not correspond to any physical reality. The interested reader may find examples which illustrate various superposition rules in references like [Bryngdahl76], [Asundi93] or Chapter 3 of [Patorski93]. Note that different superposition rules in the image domain will have different spectrum composition rules in the spectral domain, which are determined by properties of the Fourier transform. For example, in the case of the additive superposition rule, where Eq. (2.1) is replaced by:

$$r(x,y) = r_1(x,y) + \ldots + r_m(x,y) \tag{2.3}$$

the spectrum of the superposition is no longer the spectrum-convolution given by Eq. (2.2), but rather the *sum* of the individual spectra:

$$R(u,v) = R_1(u,v) + \ldots + R_m(u,v) \tag{2.4}$$

As we will see in Remark 2.3 at the end of Sec. 2.3 below, this case is less interesting from the point of view of moiré generation. ■

Second, until Chapter 10 we will be basically interested in *periodic* images, such as line-gratings or dot-screens, and in their superpositions. This implies that the spectrum of the image on the $u,v$ plane is not smooth but rather consists of impulses, which represent the frequencies in the Fourier series decomposition of the periodic image [Bracewell86 p. 204].[3] A strong impulse in the spectrum indicates a pronounced periodic component in the original image at the frequency and direction of that impulse.

Each impulse in the 2D spectrum is characterized by three main properties: its *label* (which is its index in the Fourier series development); its *geometric location* (or *impulse*

[3] A short survey of the spectral Fourier representation of periodic functions is also provided in Appendix A.

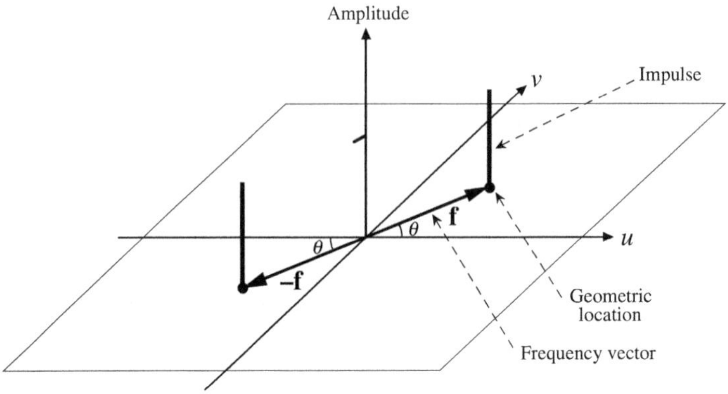


**Figure 2.1:** The *geometric location* and *amplitude* of impulses in the 2D spectrum. To each impulse is attached its *frequency vector*, which points to the geometric location of the impulse in the $u,v$ spectrum plane.

*location*); and its *amplitude* (see Fig. 2.1). To the geometric location of any impulse is attached a *frequency vector* **f** in the spectrum plane, which connects the spectrum origin to the geometric location of the impulse. This vector can be expressed either by its polar coordinates $(f,\theta)$, where $\theta$ is the direction of the impulse and $f$ is its distance from the origin (i.e., its frequency in that direction), or by its Cartesian coordinates $(u,v)$, where $u$ and $v$ are the horizontal and vertical components of the frequency. In terms of the original image, the *geometric location* of an impulse in the spectrum determines the frequency $f$ and the direction $\theta$ of the corresponding periodic component in the image, and the *amplitude* of the impulse represents the intensity of that periodic component in the image.[4] Note, however, that the impulse which is located on the spectrum origin is rather unique in its properties and requires some particular attention: It represents the zero frequency, which corresponds in the image domain to the constant component of the image, and its amplitude corresponds to the intensity of this constant component.[5] This particular impulse is traditionally called the *DC impulse* (because it represents in electrical transmission theory the direct current component, i.e., the constant term in the frequency decomposition of an electric wave), and we will maintain here this convention.

The periodic images with which we will be dealing will normally be of a symmetric nature (gratings, grids, etc.). For the sake of simplicity we also assume, unless otherwise mentioned, that the given images are not shifted, but indeed centered symmetrically about

[4] It should be stressed that the direction $\theta$ of the impulse, i.e., the direction of the corresponding periodic component in the image, is *perpendicular* to the corrugations of the periodic component (see, for example, gratings (a) and (b) and their spectra (d) and (e) in Fig. 2.2).

[5] In fact, as shown in Appendix C.2, the amplitude of the DC impulse represents the average intensity level of the image (which is, in our case, a number between 0 and 1, since our images can only take values between 0 and 1).

the origin. As a result, we will normally deal with images (and image superpositions) which are *real-valued* and *symmetric*, and whose spectra are consequently also real-valued and symmetric [Bracewell86 pp. 14–15]. This means that each impulse in the spectrum (except for the DC impulse at the origin) is always accompanied by a twin impulse of an identical amplitude, which is symmetrically located at the other side of the origin as in Fig. 2.1 (their frequency vectors being **f** and –**f**). Note, however, that if the original image is not symmetric about the origin (but, of course, still real-valued), the amplitudes of the twin impulses at **f** and –**f** are complex conjugates; in this case the amplitude of each impulse in the spectrum (except for the DC impulse) may also have a non-zero imaginary component. We will return to such cases in more detail in Chapter 7, where we will discuss the superposition of non-centered or shifted images. It is important to understand, however, that even in such cases each frequency **f** of the image is still represented in the spectrum by a pair of impulses, whose geometric locations are **f** and –**f**. [6]

However, the question of whether or not an impulse pair in the spectrum represents a *visible* periodic component in the image strongly depends on properties of the human visual system. The fact that the eye cannot distinguish fine details above a certain frequency (i.e., below a certain period) suggests that the human visual system model includes a low-pass filtering stage. This is a bidimensional bell-shaped filter whose form is anisotropic (since it appears that the eye is less sensitive to small details in diagonal directions such as 45° [Ulichney88 pp. 79–84]).[7] However, for the sake of simplicity this low-pass filter can be approximated by the *visibility circle*, a circular step-function around the spectrum origin whose radius represents the *cutoff frequency* (i.e., the threshold frequency beyond which fine detail is no longer detected by the eye). Obviously, its radius depends on several factors such as the contrast of the observed details, the viewing distance, light conditions, etc. If the frequencies of the image details are beyond the border of the visibility circle in the spectrum, the eye can no longer see them; but if a strong enough impulse in the spectrum of the image superposition falls inside the visibility circle, then a moiré effect becomes visible in the superposed image. (In fact, the visibility circle has a hole in its center, since very low frequencies cannot be seen, either.)

Another possible property of our images (although it is not necessarily a requirement) comes from the fact that most printing devices are only bilevel, namely: they are only capable of printing solid ink or leaving the paper unprinted, but they cannot produce intermediate ink tones. (This is also true for most colour printing devices, where each of the printed primary colours is bilevel.) In such devices the *visual impression* of intermediate tone levels is usually obtained by means of the halftoning technique, i.e., by breaking the continuous-tone image into small dots whose size depends on the tone level (see Sec. 3.2). Therefore, in most practical cases the reflectance function of a printed

---

[6] For the sake of completeness we mention here that this conjugate symmetry property in the spectrum only breaks up in the case of complex-valued images. For example, a single impulse at the point $(u,v)$ in the spectrum corresponds to the complex-valued function $p(x,y) = e^{-2\pi i(ux+vy)}$ in the image domain. We will rarely be concerned with such cases, since all physically realizable images are purely real.

[7] For a more detailed account on the human visual system and its properties the reader is referred to specialized references on this subject such as [Cornsweet70], [Wandell95] or Chapter 34 in [Boff86].

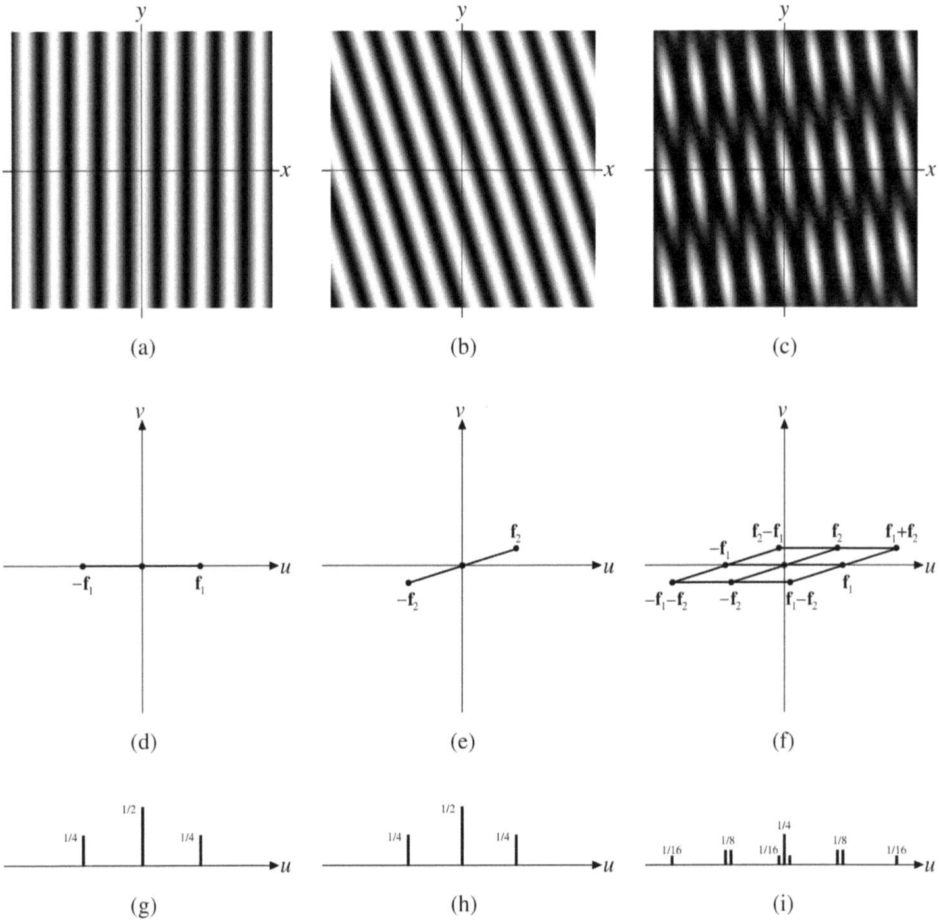


**Figure 2.2:** First row: cosinusoidal gratings (a) and (b) and their superposition (c) in the image domain. Second row: top view of the respective spectra (d), (e) and their convolution (f). Black dots in the spectra indicate the geometric locations of the impulses; the line segments connecting them have been added only in order to clarify the geometric relations. (g), (h), (i): Side view of the same spectra, showing the impulse amplitudes. Note the two new impulse pairs which have appeared in the spectrum convolution (f); the isolated contributions of these two impulse pairs to the superposition (c) are shown in (j) and (k): (j) is the periodic component contributed by the new impulse pair which is located at the difference frequencies $\mathbf{f}_1-\mathbf{f}_2$ and $\mathbf{f}_2-\mathbf{f}_1$, and (k) is the periodic component contributed by the new impulse pair at the sum frequencies, $\mathbf{f}_1+\mathbf{f}_2$ and $-\mathbf{f}_1-\mathbf{f}_2$.

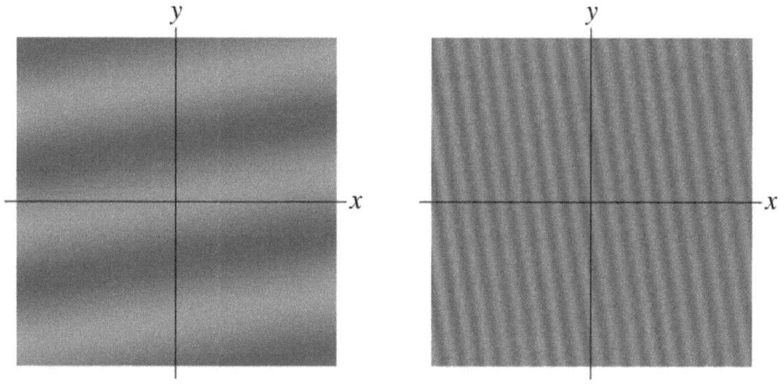


**Figure 2.2:** (*continued*).

image is *binary*, taking only values 0 and 1 (signifying the existence or absence of ink on the white paper, respectively). However, in order to present the ideas in a more didactic way we will first review in the following sections the fundamental case of continuous tone cosinusoidal gratings and their superpositions, and only then we will proceed to cases involving binary gratings, grids and dot-screens.

## 2.3 Superposition of two cosinusoidal gratings

Let us first consider the case of gratings with a cosinusoidal intensity profile. Since reflectance functions always take values between 0 and 1, the cosinusoidal reflectance function has the form of a "raised" cosinusoidal wave (see Fig. 2.2(a)):

$$r_1(x,y) = \tfrac{1}{2}\cos(2\pi f_1 x) + \tfrac{1}{2} \tag{2.5}$$

We will henceforth call such a "raised" cosinusoidal wave a *cosinusoidal grating*, without explicitly mentioning each time its "raised" nature. This periodic function has a frequency of $f_1$ cycles per unit, i.e., its period is $T_1 = 1/f_1$ units (in the $x$ direction, $\theta_1 = 0$). Similarly, the reflectance function of a cosinusoidal grating with a frequency of $f_2$ which is rotated by angle $\theta_2$ (Fig. 2.2(b)) is given by:

$$r_2(x,y) = \tfrac{1}{2}\cos(2\pi f_2[x\cos\theta_2 + y\sin\theta_2]) + \tfrac{1}{2} \tag{2.6}$$

The 2D Fourier transform of each of these reflectance functions consists of exactly three impulses (see Fig. 2.2(d),(e)). In fact, it is the sum of the Fourier transform of the cosinusoidal term, which consists of two symmetric impulses of amplitude $\tfrac{1}{4}$ located at a distance of $f_i = 1/T_i$ from the origin in the direction $\theta_i$, plus the Fourier transform of the additional constant $\tfrac{1}{2}$, which is an impulse of amplitude $\tfrac{1}{2}$ at the origin (the DC impulse).

According to Eqs. (2.1) and (2.2) the spectrum of the superposition of $r_1(x,y)$ and $r_2(x,y)$ is the Fourier transform of their product, i.e., the convolution of their individual spectra $R_1(u,v)$ and $R_2(u,v)$. Since in our case each of these spectra consists of 3 impulses, their convolution consists of 9 impulses (see Fig. 2.2(f)). Carrying out this convolution graphically by the "move and multiply" method [Bracewell86 pp. 29–30; Rosenfeld82 pp. 13–14] we see that the convolution is 0 throughout the $u,v$ plane, except at the points where the impulses of the moving copy of $R_2(u,v)$ fall whenever the origin of $R_2(u,v)$ is placed on top of an impulse of $R_1(u,v)$. This means that the *geometric location* of the impulses in the convolution can be found simply by placing on top of each impulse of $R_1(u,v)$ a centered copy of $R_2(u,v)$ (or vice versa, since convolution is commutative). The *amplitude* of each impulse thus received is the product of the amplitudes of the two impulses involved: the impulse in the first spectrum on top of which the moving spectrum is centered, and the impulse in the moving spectrum which then determines the location of the impulse in question. If a newly generated impulse falls on top of an already existing impulse, their amplitudes are summed. The amplitudes received in our case are $\frac{1}{4}$ for the DC impulse, $\frac{1}{8}$ for the two impulse pairs of the original cosines, and $\frac{1}{16}$ for the two new impulse pairs generated by the convolution (see Fig. 2.2(g)–(i)).

The spectrum resulting from the convolution contains all the impulse pairs of the original spectra (only their amplitudes have been modified, but not their geometric locations). However, in addition to the original impulses, two new impulse pairs which did not exist in any of the original spectra have appeared in the convolution (see Fig. 2.2(f)). The geometric locations of these new impulse pairs are determined by the vectorial sum and the vectorial difference of the frequency vectors of the original impulses, namely: $\mathbf{f}_1+\mathbf{f}_2$, $-\mathbf{f}_1-\mathbf{f}_2$, and $\mathbf{f}_1-\mathbf{f}_2$, $\mathbf{f}_2-\mathbf{f}_1$.

Since each impulse pair in the spectrum reflects a periodic component with the corresponding frequency and angle in the image domain, these two new impulse pairs suggest that the superposition of the two original images includes two new periodic components which did not exist in either of the original images. And indeed, two new periodic components are present in the superposed image, Fig. 2.2(c): The more obvious one has the frequency and the direction of the difference vector, $\mathbf{f}_1-\mathbf{f}_2$ (see Fig. 2.2(j)); the other one, with the frequency and the direction of the sum vector, $\mathbf{f}_1+\mathbf{f}_2$, contributes to the superposed image the fine, high-frequency details, but its isolated contribution, shown in Fig. 2.2(k), is not easily discerned by the eye in the superposition. The first periodic component is more visible than the other since its frequency is lower, i.e., its period is larger. While the frequency of the vector sum is larger than the frequencies of each of the individual vectors, the frequency of the vector difference may be significantly smaller than either of the original frequencies. Consequently, the periodic component in the superposed image which corresponds to the vector difference can have a significantly larger period, and therefore be much more visible, than the cosines of the original images.[8] This prominent periodic component is, in fact, the moiré effect that we see in the superposition

[8] Note that the situation is inversed when the angle between $\mathbf{f}_2$ and $\mathbf{f}_1$ is obtuse, since in this case the length of $\mathbf{f}_1+\mathbf{f}_2$ is smaller than the length of $\mathbf{f}_1-\mathbf{f}_2$.

of the two original images (Fig. 2.2(c)). The other periodic component corresponds to another moiré effect, that is not currently visible since its frequency vector $\mathbf{f}_1+\mathbf{f}_2$ is located beyond the visibility circle.[9]

**Remark 2.2:** Note that the impulse pairs at the frequencies of the original cosinusoidal gratings ($\pm\mathbf{f}_1$ and $\pm\mathbf{f}_2$) appear in the spectrum of the superposition only thanks to the "raised" nature of the cosines (2.5) and (2.6) which represent the reflectance of our cosinusoidal gratings. It is precisely the additive constant $\frac{1}{2}$ that contributes the DC impulse to each of the original spectra (Fig. 2.2(g),(h)), and it is only thanks to these DC impulses that the original frequency impulses appear in the spectrum convolution (Fig. 2.2(f)). For when two pure, unraised cosine functions are multiplied, the spectrum of their product (i.e., the spectrum convolution) contains only the new impulse pairs at the sum and difference frequencies, but the original impulse pairs at the frequencies of $\pm\mathbf{f}_1$ and $\pm\mathbf{f}_2$ are lost. This is exactly what happens in communication theory in the case of amplitude modulation (AM) when both the carrier and the modulator are cosinusoidal signals (see, for example, [Taub86 pp. 115–116]). This can be also verified by means of the following simple trigonometric identities, that give the product of pure or raised cosines, respectively:

$$\cos\alpha\cos\beta = \tfrac{1}{2}\cos(\alpha-\beta) + \tfrac{1}{2}\cos(\alpha+\beta)$$

$$(\tfrac{1}{2}\cos\alpha + \tfrac{1}{2})(\tfrac{1}{2}\cos\beta + \tfrac{1}{2}) = \tfrac{1}{4} + \tfrac{1}{4}\cos\alpha + \tfrac{1}{4}\cos\beta + \tfrac{1}{8}\cos(\alpha-\beta) + \tfrac{1}{8}\cos(\alpha+\beta) \quad \blacksquare$$

**Remark 2.3:** It is also interesting to note that the new impulse pairs at the sum and difference frequencies (i.e., the moiré impulses) appear in the spectrum of the superposition (Fig. 2.2(f)) only thanks to the non-linearity of the superposition rule being used, in our case: the multiplicative rule.[10] When the original cosinusoidal functions are superposed using the additive superposition rule (see Remark 2.1), which is, of course, a linear operation, it follows from the linearity of the Fourier transform that $\mathcal{F}[r(x,y)] = \mathcal{F}[r_1(x,y)] + \mathcal{F}[r_2(x,y)]$. Therefore, the spectrum of the superposition only contains the DC impulse and the impulses at the original cosine frequencies $\pm\mathbf{f}_1$ and $\pm\mathbf{f}_2$, but no impulses are generated at new frequencies such as $\mathbf{f}_1-\mathbf{f}_2$ or $\mathbf{f}_1+\mathbf{f}_2$. This means, back in the image domain, that in cases where the additive superposition rule applies no moiré effects appear in the superposition. In such cases, only when a non-linear operation is applied to the cosine sum (for example: by recording it on a film with a non-linear response, or by printing it with a printer having a non-linear reproduction curve) can new frequencies of the form $k_1\mathbf{f}_1 + k_2\mathbf{f}_2$ be generated in the spectrum, and moiré effects may become visible in

[9] The moiré which corresponds to the frequency difference is often called in the literature a *difference moiré* or a *subtractive moiré*, whilst the moiré which corresponds to the frequency sum is called an *additive moiré*. We will usually prefer not to use these terms in order to avoid any possible confusion with the superposition rules (*additive superposition*, etc.; see Remark 2.1).

[10] A superposition operation $S[f,g]$ which takes a pair of functions $f(x)$ and $g(x)$ and returns their superposition according to a given mathematical rule is called *linear* if for any functions $f_i(x)$ and $g_i(x)$ and constants $a$ and $b$ we have $S[af_1+bf_2, g] = aS[f_1,g] + bS[f_2,g]$ and $S[f, ag_1+bg_2] = aS[f,g_1] + bS[f,g_2]$. In our case the additive superposition operation $S_1[f,g] = f+g$ is linear, but the multiplicative superposition operation $S_2[f,g] = fg$ is not. Note that if we apply to $S_1[f,g]$ a non-linearity, as in $S_3[f,g] = (S_1[f,g])^2 = (f+g)^2$, the resulting superposition operation $S_3$ is non-linear.

the superposition. For a more detailed discussion on this subject the interested reader is referred to [Post94 pp. 91–92], [Bryngdahl76 p. 88] and particularly to [Eschbach88]. ■

## 2.4 Superposition of three or more cosinusoidal gratings

If we now superpose a third cosinusoidal grating on top of the first two, the resulting spectrum will be the convolution of all three spectra — i.e., the result of convolving the third spectrum with the convolution of the first two spectra. The geometric location and the amplitude of the impulses in the resulting spectrum can be determined graphically in the same manner as above: a centered copy of the new 3-impulse spectrum of the third cosinusoidal grating is placed on top of each of the 9 impulses of the 2-layer convolution, thus generating 9 additional impulse pairs in the combined spectrum (Fig. 2.3). The amplitude of each of the impulses of the copied spectrum is scaled by the amplitude of the impulse on top of which the copied spectrum has been placed. If any of the newly generated impulses falls inside the visibility circle, a new periodic component (or moiré effect) can be seen in the image.

Generalizing this to the superposition of $m$ cosinusoidal gratings, we see that the final convolution contains the frequency vectors of each of the original images as well as all the new frequency vectors obtained in each successive convolution. This means that the final convolution contains all the frequency vectors which can be obtained as a vectorial sum of 1, 2, ... or $m$ frequency vectors, one (or none) from each original spectrum. If we consider the DC impulse of each spectrum as having a zero frequency vector, it can be said that *each* of the individual spectra contributes *one* of its frequency vectors to every vectorial sum in the final convolution. In other words, the frequency vector $\mathbf{f}$ of any individual impulse in the final convolution (i.e., in the spectrum of the superposition of the $m$ images) is a vectorial sum of $m$ frequency vectors $\mathbf{f}'_i$, where $\mathbf{f}'_i$ is one of the 3 frequency vectors contained in the spectrum of the $i$-th image ($\mathbf{f}_i$, $-\mathbf{f}_i$, or $\mathbf{0}$):

$$\mathbf{f} = \mathbf{f}'_1 + \ldots + \mathbf{f}'_m \tag{2.7}$$

If the polar coordinates of the vectors $\mathbf{f}'_i$ are $(f_i,\theta_i)$, i.e., if $f_i$ is the frequency of the $i$-th original cosinusoidal image and $\theta_i$ is the angle that it forms with the positive horizontal axis, then the two Cartesian components of the above vectorial sum $\mathbf{f}$ can be written as:

$$u = f_1 \cos\theta_1 + \ldots + f_m \cos\theta_m$$
$$v = f_1 \sin\theta_1 + \ldots + f_m \sin\theta_m$$

and the frequency, the period and the angle of the impulse in question are given by the length and the direction of the sum vector $\mathbf{f}$:

$$f = \sqrt{u^2 + v^2} \qquad T = 1/f \qquad \varphi = \arctan(v/u) \tag{2.8}$$

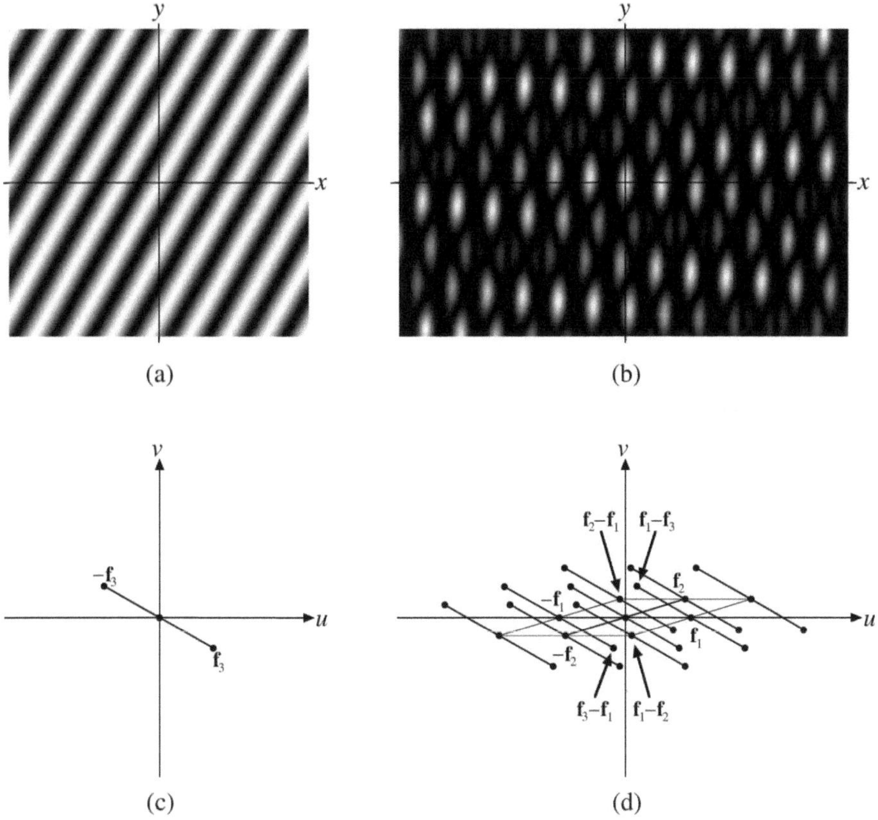


**Figure 2.3:** First row: (a) a third cosinusoidal grating and (b) its superposition with Fig. 2.2(c), in the image domain. Second row: a top view of their respective spectra (c) and (d) in the spectral domain. The strongest visible moirés in (b), which are best seen from a distance of about 3 meters, belong to the two impulse pairs marked by arrows in the spectrum (d).

The way in which any frequency vector **f** in the spectrum of the superposition is obtained from frequency vectors $\mathbf{f}_i$ in the individual spectra of the original images can be illustrated graphically using the geometric rules of vector addition, as shown in Figs. 2.8 and 2.11. Such *vector diagrams* can be drawn for any frequency vector **f** in the spectrum of the superposed image, but they are of particular interest for those impulses which correspond to moirés in the superposed image. The vector diagrams provide in the spectral domain a clear pictorial explanation of the nature of any moiré in question.

If the impulse whose frequency vector is **f** falls inside the visibility circle and represents a visible moiré in the superposition of the $m$ original images, the above formulas (2.8) express the frequency, the period and the angle of this moiré. Note that, as shown in Sec. C.1 of Appendix C, in the special case of $m = 2$, where a moiré effect occurs due to the vectorial sum of the frequency vectors $\mathbf{f}_1$ and $-\mathbf{f}_2$ (see Fig. 2.2), these formulas are reduced to the familiar geometrically obtained formulas of the period and angle of the moiré effect between two gratings [Righi87; Nishijima64]:[11]

$$T_M = \frac{T_1 T_2}{\sqrt{T_1^2 + T_2^2 - 2T_1T_2\cos\alpha}} \qquad \varphi_M = \arctan\left(\frac{T_2\sin\theta_1 - T_1\sin\theta_2}{T_2\cos\theta_1 - T_1\cos\theta_2}\right) \tag{2.9}$$

where $T_1$ and $T_2$ are the periods of the two original gratings and $\alpha$ is the angle difference between them, $\theta_2$–$\theta_1$. (Note, however, that these formulas are only valid when $T_1 \approx T_2$; the reason for this restriction will become clear at the end of Sec. 2.6.) In the particular case where $T_1 = T_2$ this is further simplified into the formulas:

$$T_M = \frac{T}{2|\sin(\alpha/2)|} \qquad \varphi_M = \begin{cases} \tfrac{1}{2}(\theta_1+\theta_2) - 90^\circ & \alpha > 0 \\ \tfrac{1}{2}(\theta_1+\theta_2) + 90^\circ & \alpha < 0 \end{cases} \tag{2.10}$$

Note that in this case the direction of the moiré is perpendicular to the bisector of the angle formed between the grating directions. In another interesting particular case where $T_1 \neq T_2$ but $\theta_1 = \theta_2$ (namely, the superposition of two parallel gratings) Eqs. (2.9) are reduced into the following equally well-known formulas (see also Remark C.9 in Sec. C.15.7), and the resulting moiré is parallel to the original gratings:

$$T_M = \frac{T_1T_2}{|T_2 - T_1|} \quad \text{(i.e., } f_M = |f_1 - f_2|\text{)}, \qquad \varphi_M = \begin{cases} \theta & T_2 > T_1 \\ \theta + 180^\circ & T_2 < T_1 \end{cases} \tag{2.11}$$

Eqs. (2.7), (2.8) and their derived formulas (2.9)–(2.11) give the *geometric properties* of an impulse in the spectrum of the superposition (and of the periodic component or moiré that it represents in the image domain), namely, the period and the direction. The *amplitude* of any individual impulse, which represents the strength of the corresponding periodic component in the image, is a product of the amplitudes of the $m$ impulses from which it has been obtained in the convolution, one from each of the $m$ spectra:

$$a = a_1 \cdot \ldots \cdot a_m \tag{2.12}$$

Note, however, that if two or more impulses in the convolution happen to fall on top of each other exactly in the same location, their individual amplitudes are summed.

As we can see from Eqs. (2.7) and (2.12), the convolution of impulsive spectra can be considered as an operation in which frequency vectors of the original spectra are added vectorially, whereas the corresponding impulse amplitudes are multiplied. These rules follow from the properties of convolution, and they can be readily verified by the "move and multiply" method. Note that if all the convolved spectra are real-valued and symmetric

[11] Note that the moiré angle formulas found in literature may vary according to the angle conventions being used.

about the origin (see Sec. 2.2), the resulting spectrum is also real-valued and symmetric, and contains for each impulse at location **f** an identical twin impulse at –**f**. If, however, the original images are not symmetric, the amplitudes of each such impulse pair in the convolution are complex conjugates, and the impulse amplitudes in the spectrum may also have imaginary components.

## 2.5 Binary square waves and their spectra

Let $r(x)$ be a one-dimensional binary (0, 1 valued) periodic square wave, i.e., a sequence of square pulses. We will denote the period of this function by $T$ and its *opening* (i.e., the width of its *white* square pulses) by $\tau$; see Fig. 2.4. We may also assume here that the square wave is symmetrically centered about the origin, so that both the original image and its spectrum are real-valued and symmetric. The square wave $r(x)$ can be expressed by:

$$r(x) = \begin{cases} 1 & |x - nT| < \tau/2 \\ 0 & |x - nT| > \tau/2 \end{cases} \qquad \forall n \in \mathbb{Z} \tag{2.13}$$

or, equivalently, by:

$$r(x) = \sum_{n=-\infty}^{\infty} \operatorname{rect}\left(\frac{x - nT}{\tau}\right)$$

where:

$$\operatorname{rect}(x) = \begin{cases} 1 & |x| < \frac{1}{2} \\ 0 & |x| > \frac{1}{2} \end{cases}$$

According to the Fourier theory the periodic square wave $r(x)$ can be expressed, by means of its Fourier series expansion, as an infinite series of weighted sine and cosine functions at the fundamental frequency of $1/T$ and all its harmonics. As explained in Sec. A.2 of Appendix A, the general expansion (or decomposition) of a one-dimensional periodic function $p(x)$ into a two-sided Fourier series is given by:

$$p(x) = \sum_{n=-\infty}^{\infty} a_n \cos(2\pi nx/T) + \sum_{n=-\infty}^{\infty} b_n \sin(2\pi nx/T) \tag{2.14}$$

where the Fourier series coefficients are:

$$a_n = \frac{1}{T}\int_T p(x) \cos(2\pi nx/T)\, dx \qquad b_n = \frac{1}{T}\int_T p(x) \sin(2\pi nx/T)\, dx \tag{2.15}$$

Moreover, if $p(x)$ is symmetric there are no sine components, and $b_n = 0$ for all $n$.

The fact that we only deal with periodic functions whose values are bounded between 0 and 1 affects also the possible range of the Fourier series coefficients. In fact, we have:

**Proposition 2.1:** If the values of the periodic function $p(x)$ are bounded between 0 and 1, then all its Fourier series coefficients (impulse amplitudes) have absolute values between 0

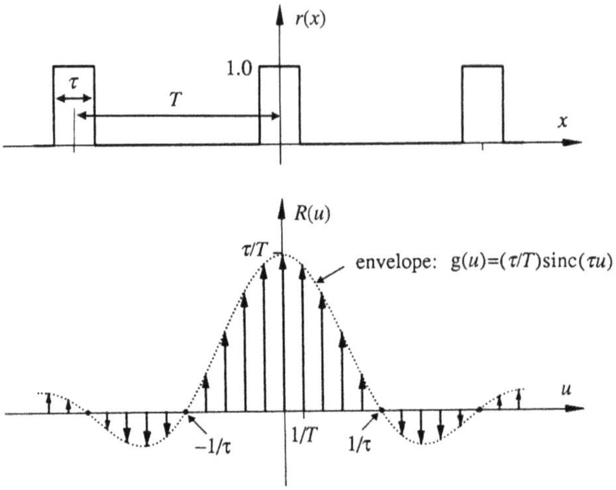


**Figure 2.4:** A symmetric square wave with period $T$ and opening $\tau$ and its Fourier transform. The dotted line indicates the envelope of the impulse train. The opening ratio in this case is $\tau/T = \frac{1}{5}$, and therefore every fifth impulse in the spectrum has a zero amplitude.

and 1. More accurately, they satisfy: $0 \le a_0 \le 1$, and for any $n \ne 0$: $|a_n| \le 1/\pi$, $|b_n| \le 1/\pi$ (see Sec. C.2 in Appendix C). Furthermore, it is true for any convergent Fourier series that $a_n$ and $b_n$ tend to 0 as $n \to \pm\infty$; and moreover, if $p(x)$ is $k+1$ times continuously differentiable, then $a_n$ and $b_n$ tend to 0 faster than $1/n^k$ [Cartwright90 pp. 64–65]. ■

As an example, in the case of our symmetric binary square wave (see Fig. 2.4) the Fourier coefficients (2.15) are:

$$\begin{aligned} a_0 &= \frac{\tau}{T} \\ a_n &= \frac{1}{n\pi}\sin\left(\frac{\pi n \tau}{T}\right) = \frac{\tau}{T}\operatorname{sinc}\left(\frac{n\tau}{T}\right) \qquad (2.16) \\ b_n &= 0 \end{aligned}$$

(the *sine* components here are all null owing to the symmetry of the square wave; note also that $a_0$ is a particular case of $a_n$ since sinc(0) = 1). The fact that the square wave can be expressed as a constant term $a_0$ plus an infinite sum of cosine functions implies that the Fourier transform of the square wave contains a DC impulse whose amplitude is $a_0$, plus an infinite series of impulse pairs (the $n$-th harmonics) that are located at the frequencies $\pm n/T$, and whose amplitudes are given by the cosine coefficients $a_n$ (Eq. (2.16)). And indeed, the spectrum of the square wave $r(x)$ is given, according to Eq. (A.8), by:

$$R(u) = \sum_{n=-\infty}^{\infty} a_n\, \delta(u - n/T) \qquad (2.17)$$

where $\delta(u)$ is the impulse symbol [Bracewell86, Chapter 5]. This is an impulse train (or a *comb*) which samples the continuous function ("envelope") $g(u) = \frac{\tau}{T}\text{sinc}(\tau u)$ at the frequency $u = 1/T$ and all its harmonics, $u = n/T$ (Fig. 2.4). The amplitude of the impulses oscillates and fades out symmetrically in both directions from the center.

Note that the period $T$ of the square wave determines the interval $1/T$ between each two successive impulses of the comb in the spectrum, while the opening $\tau$ $(0 \leq \tau \leq T)$ determines the length of the lobes in the envelope sinc function. The height of the envelope at the origin, i.e., the amplitude of the DC impulse, is determined by the *opening ratio* of the square wave, $\tau/T$ $(0 \leq \tau/T \leq 1)$.

An additional observation from the formula of the Fourier coefficients $a_n$ in the case of a square wave (Eq. (2.16)) is that if the opening ratio $\tau/T$ of a given square wave is rational, i.e., if it can be expressed as a ratio $l/k$ between two integers, then for any $n$ that is a multiple of $k$ the impulse amplitude $a_n$ is zero. For instance, if $\tau/T = \frac{1}{2}$ then every even impulse in the comb has a zero amplitude, and if $\tau/T = \frac{1}{4}$ or $\tau/T = \frac{3}{4}$ then every fourth impulse in the comb is zero.

Let us mention here one more result that will be used later: If $R(u)$ is the Fourier transform of a periodic wave $r(x)$, then the Fourier transform of the "negative" wave $1 - r(x)$ is given by:

$$\mathcal{F}[1 - r(x)] = \mathcal{F}[1] - \mathcal{F}[r(x)] = \delta(u) - R(u) \qquad (2.18)$$

Therefore, if $a_n$ are the impulse amplitudes in the spectrum of $r(x)$, $R(u) = \sum a_n \delta(u - n/T)$, then the impulse amplitudes in the spectrum of $1 - r(x)$ are: $d_0 = 1 - a_0$, $d_n = -a_n$. The impulse locations, on their part, remain unchanged.

In the sections which follow we will proceed to the case of 2D binary images. As shown in Appendix A, the general expansion of a 2-fold periodic function $p(x,y)$ with periods $T_x$ and $T_y$ in the $x$ and $y$ directions into a 2D Fourier series is given by Eq. (A.9), with the coefficients (A.10). If $p(x,y)$ is symmetric, its 2D Fourier series only contains cosine terms; each of these terms, multiplied by a coefficient $a_{m,n}$, specifies the contribution to the image $p(x,y)$ of the cosinusoidal periodic component in the direction and frequency of the $(m,n)$-th Fourier harmonic. As shown in Appendix A, the spectrum of $p(x,y)$ is, in the general case, an impulse-nailbed which contains for each of the $(m,n)$-th harmonics an impulse of amplitude $a_{m,n}$. The frequency vector of each impulse of the nailbed indicates the direction and the frequency of the corresponding periodic component in the image.

## 2.6 Superposition of binary gratings; higher order moirés

In this section we discuss the 2D case of line-gratings and their superpositions. As our main example we will consider the case of binary square-wave gratings, in which the reflectance function $r(x,y)$ is a binary square wave. Note that since a 2D line-grating is a 1-fold periodic function, which is constant perpendicularly to its main direction, its 2D

Fourier series representation (and therefore also its 2D spectrum) are in fact of 1D nature; see Sec. A.3.3 in Appendix A. For the sake of convenience we will assume here that $r(x,y)$ is symmetric about the origin. The reflectance function of such a line-grating with period $T_1$ in the $x$ direction (see Fig. 2.5(a)) is given by the two-sided Fourier series:

$$r_1(x,y) = \sum_{n=-\infty}^{\infty} a^{(1)}{}_n \cos(2\pi nx/T_1) \qquad (2.19)$$

If $r_1(x,y)$ is a binary square-wave grating, its Fourier coefficients $a^{(1)}{}_n$ are given by Eq. (2.16), where $T$ and $\tau$ equal $T_1$ and $\tau_1$, respectively. Similarly, the reflectance function of a grating with period $T_2$ which is rotated by angle $\theta_2$ (Fig. 2.5(b)) is given by:

$$r_2(x,y) = \sum_{n=-\infty}^{\infty} a^{(2)}{}_n \cos(2\pi n[x\cos\theta_2 + y\sin\theta_2]/T_2) \qquad (2.20)$$

where $a^{(2)}{}_n$ are the corresponding Fourier coefficients (in the case of a square-wave grating: the same coefficients as above, but with $T_2$ and $\tau_2$).

The Fourier transform $R_1(u,v)$ of the reflectance function $r_1(x,y)$ is a symmetric 1D impulse comb on the $u$ axis (Fig. 2.5(d)); the intervals between the impulses are $1/T_1$ and their amplitudes are $a^{(1)}{}_n$. Note, however, that depending on the function $r_1(x,y)$, some (or even most) of the amplitudes $a^{(1)}{}_n$ may be 0. Similarly, the Fourier transform $R_2(u,v)$ of the reflectance function $r_2(x,y)$ is a symmetric 1D impulse comb lying on a straight line through the origin of the $u,v$ plane whose orientation is given by $\theta_2$; its impulse intervals equal $1/T_2$ and its impulse amplitudes are $a^{(2)}{}_n$ (Fig. 2.5(e)). Note that in the case of square-wave gratings the impulse combs have the envelope shapes of $g_1(u) = \frac{\tau_1}{T_1}\text{sinc}(\tau_1 u)$ and $g_2(u) = \frac{\tau_2}{T_2}\text{sinc}(\tau_2 u)$, respectively, like in Fig. 2.4.

Let us now consider the superposition of the two line-gratings, $r(x,y) = r_1(x,y)r_2(x,y)$. Its spectrum $R(u,v)$ is, according to the convolution theorem (Eq. (2.2)), the convolution of spectra $R_1(u,v)$ and $R_2(u,v)$. This convolution can be carried out graphically by the "move and multiply" method as in the case of two cosinusoidal gratings, the only difference being that in our present case each of the individual spectra contains an infinite number of impulses (a *comb*) rather than only 3 impulses. The result of the convolution is an infinite oblique nailbed on the $u,v$ plane, which is obtained by placing a centered, parallel copy of the comb $R_2(u,v)$ on top of each impulse of the comb $R_1(u,v)$, or vice versa (see Fig. 2.5(f), and compare with the case of cosinusoidal gratings shown in Fig. 2.2(f)). More precisely, the results of this convolution can be described as follows:

(a) The *impulse location* (frequency-vector) of the $(m,n)$-th impulse of the convolution in the $u,v$ plane is the vectorial sum of the frequency-vector of the $m$-th impulse in the first comb and the frequency-vector of the $n$-th impulse in the second comb.

(b) The *amplitude* of the $(m,n)$-th impulse is the product of the amplitudes of the $m$-th impulse in the first comb and the $n$-th impulse in the second comb:[12]

$$a_{m,n} = a^{(1)}{}_m \, a^{(2)}{}_n \qquad (2.21)$$

[12] Note that in the following expressions we tacitly use the impulse indexing notation that will be formally introduced in Sec. 2.7.

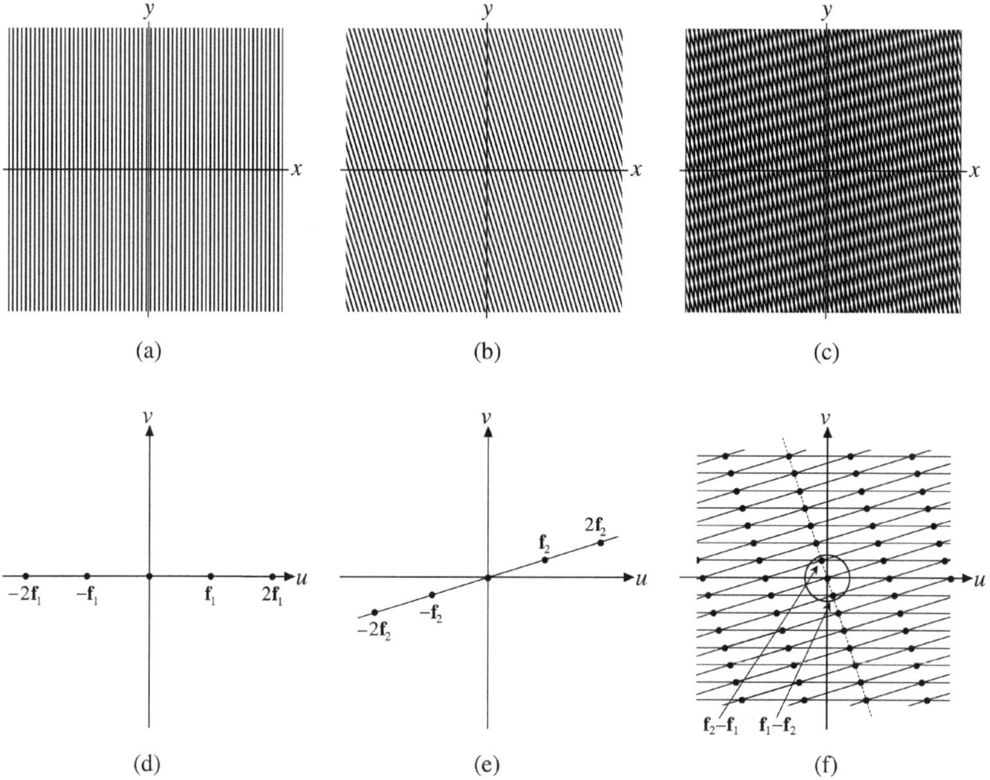


**Figure 2.5:** Binary gratings (a) and (b) and their superposition (c) in the image domain; their respective spectra are the infinite combs shown in (d) and (e) and their convolution (f). (The scale of the *x,y* axes in the image domain is reduced with respect to Figs. 2.2(a)–(c) for the sake of clarity). The circle in the center of spectrum (f) represents the visibility circle (for a viewing distance where the original gratings are no longer visible). It contains the impulse pair whose frequency vectors are $\mathbf{f}_1-\mathbf{f}_2$ and $\mathbf{f}_2-\mathbf{f}_1$; this is the fundamental impulse pair of the moiré seen in (c). The dotted line in (f) shows the infinite comb of impulses which represents the moiré.

In our example of symmetric square-wave gratings this gives, according to Eq. (2.16):

$$\begin{aligned} a_{m,n} &= \frac{1}{mn\pi^2}\sin\left(\frac{\pi m\tau_1}{T_1}\right)\sin\left(\frac{\pi n\tau_2}{T_2}\right) \\ &= \frac{\tau_1\tau_2}{T_1T_2}\,\mathrm{sinc}\left(\frac{m\tau_1}{T_1}\right)\mathrm{sinc}\left(\frac{n\tau_2}{T_2}\right) \end{aligned} \tag{2.22}$$

As in the cosinusoidal case, we see that here, too, the superposition of gratings introduces new impulses in the spectrum. If any of these impulses fall inside the visibility circle, as in Fig. 2.5(f), this indicates that in the superposed image there exists a visible periodic component (i.e., a moiré effect) at the corresponding direction and frequency.[13]

As a consequence of points (a) and (b) above we obtain the following results for the superposition of square-wave gratings:

(1) Angle changes or spatial scalings in any of the superposed gratings (being transformations that preserve the $\tau/T$ ratios of the original gratings) only influence the geometric location of the impulses in the spectrum convolution, but not the individual impulse amplitudes. In the image domain this means that rotation and spatial scale operations on the original gratings only influence the angle and period of each moiré, but not its amplitude.[14]

(2) Varying the opening ratio $\tau/T$ of any of the superposed square-wave gratings only influences the amplitudes of the impulses in the spectrum convolution, but not their locations. In the image domain this means that only the amplitude and the profile (waveform) of the moiré are affected, but not its angle or period.

These important results can be generalized to the superposition of *any* periodic functions (see Sec. C.3 in Appendix C):

**Proposition 2.2:** The impulse locations and the impulse amplitudes in the spectrum of the superposition are *independent properties*: While the impulse locations only depend on the periods and angles of the superposed layers, the *individual* impulse amplitudes are only affected by the intensity profile of each superposed layer.[15] This also applies, of course, to impulses which fall inside the visibility circle, and hence to the corresponding moiré effects in the image domain: While the moiré periods and angles only depend on the periods and angles of the superposed layers, their intensity profiles (waveforms) within these periods and angles are only affected by the intensity profiles of the superposed layers. ■

The generalization of this result to the superposition of $m$ line gratings is straightforward. In fact, the geometric location of each impulse in the resulting spectrum is a vectorial sum of frequency vectors, one from the spectrum of each of the superposed gratings, while its amplitude is the product of the amplitudes of the original impulses involved. This is very similar to the case of cosinusoidal gratings, except that in the general case the spectrum of each line grating consists of an infinite comb of impulses rather than only 3 impulses, so that the convolution of $m$ spectra gives an infinite nailbed of impulses

---

[13] Note that by convention, a moiré continues to *exist* even when its fundamental impulse exceeds the visibility circle or has a low amplitude, and the moiré is no longer visible.

[14] Note, however, that in some cases angle or period changes may cause impulses in the spectrum convolution to fall on top of each other, in which case their individual amplitudes are summed. Such combined impulses will be called in Chapter 6 *compound impulses*, and it will be shown there that they only occur in *singular states* (see Sec. 2.9 below).

[15] The stress is on *individual* impulse amplitudes, since in cases where compound impulses are generated in the spectrum convolution (see the previous footnote) it is clear that the summed amplitude of a compound impulse only exists at the precise angle and frequency combination in which the compound impulse is generated.

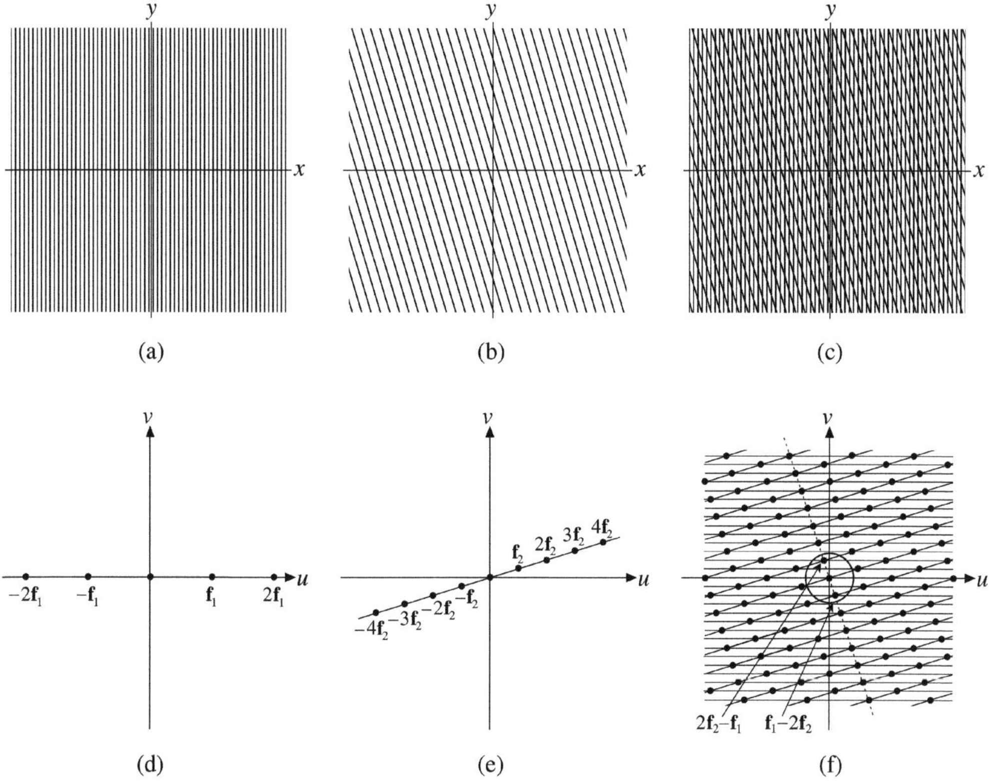


**Figure 2.6:** Binary gratings (a) and (b) as in Fig. 2.5 but with (b) having half the frequency, and their superposition (c); their respective spectra are (d), (e) and (f). The visibility circle in the center of the spectrum (f) contains the impulse pair with frequency vectors $\mathbf{f}_1-2\mathbf{f}_2$ and $2\mathbf{f}_2-\mathbf{f}_1$, which originate from the second harmonic of $\mathbf{f}_2$, and represent the fundamental impulse pair of the moiré. Note that the moiré seen in (c) is a (1,-2)-moiré, but it still has the same angle and frequency as the (1,-1)-moiré of Fig. 2.5, and only its intensity is weaker. (The moiré index notation is explained in Secs. 2.7–2.8).

rather than just a finite number ($3^m$) of impulses. This means that in the general case of line gratings each of the components $\mathbf{f}'_i$ in Eq. (2.7) may originate from any impulse of the comb of the $i$-th spectrum. In fact, if $\mathbf{f}_i$ is the frequency-vector of the fundamental impulse in the $i$-th spectrum, then the frequency-vector of its $k_i$-th harmonic impulse is $k_i\mathbf{f}_i$, and Eqs. (2.7) and (2.12) for any individual impulse in the convolution become:[16]

---

[16] These expressions will be presented in their final form in Proposition 2.3, after introducing the impulse indexing notation in Sec. 2.7.

$$\mathbf{f} = k_1\mathbf{f}_1 + ... + k_m\mathbf{f}_m \tag{2.23}$$

$$a = a^{(1)}{}_{k_1} \cdot ... \cdot a^{(m)}{}_{k_m} \tag{2.24}$$

where $a^{(i)}{}_{k_i}$ is the amplitude of the $k_i$-th impulse of the $i$-th grating. Note that according to Proposition 2.1, each successive multiplication in the product of Eq. (2.24) further scales down the amplitude $a$ of the impulse in question. In the particular case of centered square-wave gratings we obtain for any impulse of the convolution, using Eq. (2.16):

$$\begin{aligned} a &= \frac{1}{\pi^m k_1 ... k_m} \sin\left(\frac{\pi k_1 \tau_1}{T_1}\right) ... \sin\left(\frac{\pi k_m \tau_m}{T_m}\right) \\ &= \frac{\tau_1 ... \tau_m}{T_1 ... T_m} \operatorname{sinc}\left(\frac{k_1 \tau_1}{T_1}\right) ... \operatorname{sinc}\left(\frac{k_m \tau_m}{T_m}\right) \end{aligned} \tag{2.25}$$

As in the cosinusoidal case, if any impulse of the convolution falls inside the visibility circle, then a moiré effect is visible in the superposition.

As we can see, two important differences emerge between the superpositions of cosinusoidal gratings and the superpositions of binary gratings:

(i) First, in the case of binary gratings, each impulse in the visibility circle (like any other impulse in the spectrum) belongs to an infinite comb of impulses, which lies on a straight line through the origin (see Fig. 2.5(f)). This means that each moiré is represented in the spectrum by an *infinite* comb of impulses. The fundamental impulse of this comb (i.e., the first impulse next to the DC) determines the period and the direction of the moiré. If further harmonic impulses of this comb also fall inside the visibility circle, the intensity profile of the moiré is no longer perceived as a pure cosinusoidal function, but rather as a more complex form (a sum of cosines).

(ii) Second, in the case of binary grating superposition the visibility circle may contain impulses which originate from higher harmonic impulses in the spectra of the original gratings. This means that, unlike the cosinusoidal case, moiré effects between binary gratings can be also obtained from higher harmonics of the fundamental grating frequencies. Such moiré effects are called *higher order moirés* [Bryngdahl75]. This is illustrated for the case of two superposed gratings in Fig. 2.6; note that in this example the visible moiré effect is caused by the vectorial sum (or rather difference) of $\mathbf{f}_1$, the fundamental frequency of the first grating, and *twice* $\mathbf{f}_2$, i.e., the second harmonic of the other grating, while the vectorial difference $\mathbf{f}_1-\mathbf{f}_2$ is outside the visibility circle. In the image domain this means that the visible moiré is actually due to the intersection of every *second* line in the first grating with each line of the second grating.[17] Fig. 2.7 shows (both in the image and in the spectral domains) some of the different moirés which may exist between two superposed gratings. In this case both $\mathbf{f}_1-\mathbf{f}_2$ and $\mathbf{f}_1-2\mathbf{f}_2$ are inside the visibility circle, and indeed, the corresponding moirés can be both

[17] Note, however, that such attempts to interpret the different moirés by counting line intersections in the image domain cannot be extended to superpositions involving more than two gratings (see, for example, Fig. 2.8(h)).

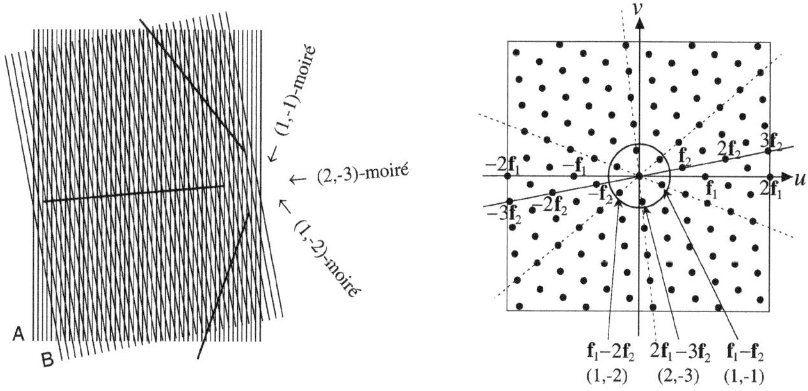


(a) (b)

**Figure 2.7:** (a) Some of the different moirés between two superposed gratings. Note that the different moirés are best seen by holding the figure at grazing angle, rotating it and looking along the indicated lines. (b) The corresponding spectrum. The dotted lines show the impulse combs of the three moirés shown in (a); their directions are perpendicular to the respective moiré bands. (The moiré index notation is explained in Secs. 2.7–2.8).

simultaneously observed in the superposition. Note, however, that although $2\mathbf{f}_1-3\mathbf{f}_2$ is also located inside the visibility circle, meaning that its period length is visible, the corresponding third-order moiré is hardly perceptible in the superposition since its impulse amplitudes are too weak (see Sec. 2.10).

Now that we are already armed with some basic notions of the spectral approach, it may be interesting to return for a moment to the classical, geometrically-obtained formulas (2.9) which provide the period and the angle of the moiré between two superposed gratings. It was mentioned in Sec. 2.4 that these formulas only hold when the grating periods satisfy the condition: $T_1 \approx T_2$. At this stage we are already in a position to understand the reason for this restriction: Formulas (2.9) only give the geometric parameters for the first-order moiré which is caused by the vectorial difference $\mathbf{f}_1-\mathbf{f}_2$ (remember that we obtained these formulas for $m = 2$ gratings from Eq. (2.7), or equivalently from Eq. (2.23) with indices $k_1 = 1$ and $k_2 = -1$). This moiré is visible, of course, only when $\mathbf{f}_1-\mathbf{f}_2 \approx \mathbf{0}$, which implies, indeed, $f_1 \approx f_2$ or $T_1 \approx T_2$. However, in other two-grating superpositions the impulse of $\mathbf{f}_1-\mathbf{f}_2$ may be located outside the visibility circle, while another impulse, for example that of $\mathbf{f}_1-2\mathbf{f}_2$, may be found inside the visibility circle (as in Fig. 2.6(f)). In this case the visible moiré is the second-order moiré which corresponds to the vectorial difference $\mathbf{f}_1-2\mathbf{f}_2$; but formulas (2.9) still predict the angle and the frequency (or period) of the first-order $\mathbf{f}_1-\mathbf{f}_2$ impulse, which is irrelevant now and does not correspond to the visible moiré. The restriction $T_1 \approx T_2$ ensures, therefore, that the dominant moiré in the superposition is, indeed, the first-order moiré belonging to $\mathbf{f}_1 - \mathbf{f}_2$, whose parameters are predicted by formulas (2.9).[18]

## 2.7 The impulse indexing notation

We have seen in Sec. 2.2 that each impulse in the spectrum of a periodic image is characterized by three main properties: its label, its geometric location and its amplitude. This also remains true in the spectrum of any superposition of periodic images, i.e., in the spectrum convolution. In Sec. 2.6 we have seen what are the *geometric location* and the *amplitude* of any impulse in the spectrum of an $m$-grating superposition; the remaining property, the *label* or the *index* of each impulse, is the subject of the present section.

The indexing notation for impulses in the spectrum of a superposition is based on the impulse indices in each of the individual spectra, i.e., on the indices in the Fourier series of each of the individual superposed layers. Consider a superposition of $m$ gratings, and its spectrum — the convolution of the $m$ original spectra. As we have seen, each impulse in the spectrum-convolution is generated during the convolution process by the contribution of *one* impulse from *each* of the $m$ original spectra: its location is given by the sum of their frequency vectors, and its amplitude is given by the product of their amplitudes. This allows us to introduce an indexing method for naming each of the impulses of the spectrum-convolution in a unique, unambiguous way. The general impulse in the spectrum-convolution will be called the $(k_1,...,k_m)$*-impulse*, where $m$ is the number of superposed gratings, and each integer $k_i$ is the index (harmonic), within the comb (the Fourier series) of the $i$-th spectrum, of the impulse that this $i$-th spectrum contributed to the impulse in question in the convolution. $(k_1,...,k_m)$ will be called the *index-vector* (or simply: the *index*) of the impulse, and the highest absolute value in $(k_1,...,k_m)$ will be called the *order* of the impulse. The index-vector of the symmetric twin impulse, which is also present in the spectrum, is $(-k_1,...,-k_m)$.

**Remark 2.4:** Note that the presence of a zero index $k_i = 0$ in the $(k_1,...,k_m)$ index-vector indicates that the $i$-th spectrum contributes its 0-th impulse (namely: the DC impulse, whose frequency is zero). This means that the $i$-th spectrum does not contribute to the *geometric location* of the $(k_1,...,k_m)$-impulse, although it does scale down by its DC value the *amplitude* of that impulse (remember that by Proposition 2.1 the DC amplitude is always a fraction between 0 and 1). In terms of the image domain this means that although the $i$-th grating is present in the superposition, it does not participate in the generation of the periodic component of the $(k_1,...,k_m)$-impulse (or more accurately, it only contributes its average intensity value). In this case the periodic component of the $(k_1,...,k_m)$-impulse

[18] Note that formulas (2.9) can be generalized to the moiré caused by any given $k_1\mathbf{f}_1 + k_2\mathbf{f}_2$ impulse, by deriving them from Eq. (2.23) using the given values of $k_1, k_2$ instead of $k_1 = 1$, $k_2 = -1$. Such a formula has been obtained in [Tollenaar64 p. 626] using the geometric approach; however, without the help of the spectral approach, i.e., without having a panoramic view of the spectrum, we cannot know which integers $k_1, k_2$ correspond to the dominant moiré and should be therefore used in the formula.

already exists in the superposition without the presence of the $i$-th grating, and the addition of the $i$-th grating only decreases to some extent the visibility of this periodic component in the superposition. ■

**Example 2.1:** Consider the two cosinusoidal grating superposition of Fig. 2.2(c) and its spectrum-convolution in Fig. 2.2(f). The impulse whose frequency-vector is $\mathbf{f}_1-\mathbf{f}_2$ is called the (1,-1)-impulse, since it is formed in the convolution by the 1-st impulse in the comb of the first spectrum (d) and the –1-st impulse in the comb of the second spectrum, (e). Table 2.1 below gives for each of the 9 impulses in this spectrum-convolution its frequency-vector, as it appears in Fig. 2.2(f), and its $(k_1,k_2)$ index-vector according to our new notation. ■

| The frequency-vector of the impulse | The $(k_1,k_2)$-index of the impulse |
|---|---|
| $\mathbf{0} = (0\mathbf{f}_1+0\mathbf{f}_2)$ | (0,0) |
| $\mathbf{f}_1 = (1\mathbf{f}_1+0\mathbf{f}_2)$ | (1,0) |
| $-\mathbf{f}_1 = (-1\mathbf{f}_1+0\mathbf{f}_2)$ | (-1,0) |
| $\mathbf{f}_2 = (0\mathbf{f}_1+1\mathbf{f}_2)$ | (0,1) |
| $-\mathbf{f}_2 = (0\mathbf{f}_1-1\mathbf{f}_2)$ | (0,-1) |
| $\mathbf{f}_1+\mathbf{f}_2 = (1\mathbf{f}_1+1\mathbf{f}_2)$ | (1,1) |
| $-\mathbf{f}_1-\mathbf{f}_2 = (-1\mathbf{f}_1-1\mathbf{f}_2)$ | (-1,-1) |
| $\mathbf{f}_1-\mathbf{f}_2 = (1\mathbf{f}_1-1\mathbf{f}_2)$ | (1,-1) |
| $\mathbf{f}_2-\mathbf{f}_1 = (-1\mathbf{f}_1+1\mathbf{f}_2)$ | (-1,1) |

**Table 2.1:** The frequency-vectors and the index-vectors of the impulses in the spectrum-convolution of Fig. 2.2(f).

**Example 2.2:** Consider the three cosinusoidal grating superposition of Fig. 2.3(b), which is obtained by superposing a third cosinusoidal grating on top of the two gratings of Fig. 2.2(c). Here, the index-vector of each impulse in the spectrum-convolution consists of three indices, the third index being the contribution of the spectrum of the third superposed layer. For example, the rightmost impulse in Fig. 2.3(d), whose frequency-vector is $\mathbf{f}_1+\mathbf{f}_2+\mathbf{f}_3$, is called the (1,1,1)-impulse. The impulse whose frequency-vector in Fig. 2.3(d) is $\mathbf{f}_1-\mathbf{f}_2$, which was called in the 2-layer case of Example 2.1 the (1,-1)-impulse, is called now the (1,-1,0)-impulse; note that it is only generated by the first two layers (see Fig. 2.2(c)), and the addition of the third layer does not change its angle or period, but only makes its amplitude weaker (see Fig. 2.3(b) and Remark 2.4). ■

Note that both in Fig. 2.2 and in Fig. 2.3 there appear no impulses whose order is higher than 1, since each of the original combs here contains only three impulses whose indices in the Fourier series are -1, 0 and 1. In Figs. 2.5 and 2.6, however, there exist also higher-order impulses. For instance, the impulse pair inside the visibility circle of Fig. 2.6(f) consists of the (1,-2)-impulse and the (-1,2)-impulse of the spectrum convolution.

**Remark 2.5:** In order to keep our discussion general and for reasons that will become clear later, we will henceforth adopt the convention that *every* grating (or 1-fold periodic function) is represented in the spectral domain by an *infinite* impulse-comb, although some or even most of the comb impulses (Fourier harmonics) may have zero amplitudes, as in the case of cosinusoidal gratings. Consider, for example, the spectra shown in Figs. 2.2(d)–(f); according to our new convention, instead of saying that in these spectra higher-order impulses "do not exist", we will say henceforth that all the higher-order impulses are present, like in Figs. 2.5(d)–(f), but their amplitudes are zero. Although it may seem to be just a difference of wording, this convention will prove to be very useful in Chapter 5. ■

With this convention it follows that the spectrum-convolution of $m$ superposed gratings contains an infinite number of impulses, whose index-vectors $(k_1,...,k_m)$ run through the whole of $\mathbb{Z}^m$. Some or even most of these impulses may, however, have zero amplitudes (as in the case of cosinusoidal gratings).

Using the formal impulse notation presented in this section we can now reformulate Eqs. (2.23), (2.24) as follows:

**Proposition 2.3:** The geometric location of the general $(k_1,...,k_m)$-impulse in the spectrum convolution of an $m$-grating superposition is given by the vectorial sum (or linear combination):

$$\mathbf{f}_{k_1,...,k_m} = k_1\mathbf{f}_1 + ... + k_m\mathbf{f}_m \tag{2.26}$$

and its amplitude is given by the product:

$$a_{k_1,...,k_m} = a^{(1)}{}_{k_1} \cdot \ ... \ \cdot a^{(m)}{}_{k_m} \tag{2.27}$$

where $\mathbf{f}_i$ denotes the frequency vector of the fundamental impulse in the spectrum of the $i$-th grating, and $k_i\mathbf{f}_i$ and $a^{(i)}{}_{k_i}$ are respectively the frequency-vector and the amplitude of the $k_i$-th harmonic impulse in the spectrum of the $i$-th grating. If several impulses in the convolution happen to fall on the same location (i.e., they have identical frequency-vectors), their individual amplitudes are summed, and they form together a *compound impulse* (see Chapter 6). ■

The vectorial sum of Eq. (2.26) can be also written in terms of its Cartesian components. If $f_i$ are the frequencies of the $m$ original gratings and $\theta_i$ are the angles that they form with the positive horizontal axis, then the coordinates $(u,v)$ of the $(k_1,...,k_m)$-impulse in the spectrum-convolution are given by:

$$u_{k_1,...,k_m} = k_1 f_1 \cos\theta_1 + ... + k_m f_m \cos\theta_m$$
$$v_{k_1,...,k_m} = k_1 f_1 \sin\theta_1 + ... + k_m f_m \sin\theta_m \qquad (2.28)$$

and, as in the case of cosinusoidal gratings, they can be inserted into Eqs. (2.8) in order to obtain the frequency, the period and the angle of the impulse in question (and of the moiré it may represent, in case it falls inside the visibility circle).

**Proposition 2.4:** Let $(k_1^{(1)},...,k_m^{(1)})$ and $(k_1^{(2)},...,k_m^{(2)})$ be two impulses in the spectrum-convolution of an $m$-grating superposition, and let $\mathbf{f}_1$ and $\mathbf{f}_2$ be their impulse locations. Then for any integers $i,j$ the impulse $i(k_1^{(1)},...,k_m^{(1)}) + j(k_1^{(2)},...,k_m^{(2)}) = (ik_1^{(1)}+jk_1^{(2)},..., ik_m^{(1)}+jk_m^{(2)})$ also figures in the spectrum-convolution, and its impulse location is $i\mathbf{f}_1+j\mathbf{f}_2$. ■

## 2.8 The notational system for superposition moirés

Using the impulse indexing method that we presented in the previous section, we can introduce now a systematic formalism for the notation of superposition moirés. This formalism provides a useful means for the identification, classification and labeling of the moiré effects.

As we have seen in Sec. 2.2, each impulse pair in the spectrum represents a periodic component in the image domain, whose angle and frequency are determined by the location of the impulse pair. Furthermore, we have seen in Secs. 2.3 and on that the spectrum of a superposition of periodic layers contains: (a) all the impulse pairs from the individual spectra of the original layers (note that only their amplitudes have been modified, but not their geometric locations); and (b), new impulse pairs which did not exist in any of the individual spectra, and whose impulse locations are integer linear combinations of the original impulse locations. These new impulse pairs correspond to periodic components which appear in the superposition due to the interaction between the superposed layers; in other words: they are the spectral evidence to the appearance of the superposition moiré effects in the image domain.

We saw in point (i) of Sec. 2.6 that each $(k_1,...,k_m)$-impulse in the spectrum-convolution belongs to an infinite comb through the spectrum origin; this comb consists of all the impulse pairs $\pm(nk_1,...,nk_m)$ for any integer $n$. Therefore, every $(k_1,...,k_m)$-impulse in the spectrum-convolution whose indices $k_1,...,k_m$ are mutually prime (i.e., their greatest common divisor equals 1) is the fundamental impulse of such a comb. We will call this comb the $(k_1,...,k_m)$-comb; its fundamental impulse is the $(k_1,...,k_m)$-impulse of the spectrum-convolution and its $n$-th harmonic impulse is the $(nk_1,...,nk_m)$-impulse. In terms of impulse locations we may say that the $(k_1,...,k_m)$-comb is *spanned* by the $(k_1,...,k_m)$-impulse. Clearly, the $(k_1,...,k_m)$-comb represents in the image domain a 1-fold periodic structure, whose angle and frequency are determined by the $(k_1,...,k_m)$-impulse. The structure in the $m$-grating superposition which belongs to the $(k_1,...,k_m)$-comb will be called the $(k_1,...,k_m)$-*moiré*.[19] This 1-fold periodic moiré can be visible if at least its fundamental impulse, the $(k_1,...,k_m)$-impulse, falls inside the visibility circle. The fundamental frequency of the $(k_1,...,k_m)$-moiré is given by the frequency of the $(k_1,...,k_m)$-impulse, namely by Eq. (2.26). The order of the $(k_1,...,k_m)$-moiré is defined to be the order of its fundamental impulse, i.e., the highest absolute value in $(k_1,...,k_m)$.

It is interesting to note that this formal definition of the $(k_1,...,k_m)$-moiré covers all the moiré effects of all orders, in an exhaustive and systematic way, and on an equal basis. Hence, with this definition there is no need any longer to introduce a "moiré hierarchy", i.e., to consider first "simple moirés" between the original layers, then "moirés of moirés", etc., as was the case, for example, in [Yule67 pp. 336–337], [Foster72 p. 134] or [Delabastita92 pp. 63–64]. Such "moirés of moirés" are nothing but higher-order moirés, and they are treated in our approach on the same basis as any other moiré.

**Example 2.3:** The visible moiré between the two superposed gratings in Fig. 2.5(c) is the (1,-1)-moiré. Its fundamental impulse is the (1,-1)-impulse in the spectrum-convolution (see Fig. 2.5(f)); its (1,-1)-comb is marked in Fig. 2.5(f) by a dotted line. The moiré seen in Fig. 2.6, whose fundamental impulse is the (1,-2)-impulse, is the (1,-2)-moiré; note that the (1,-1)-moiré in this case is not visible (although it still formally exists!), since its fundamental impulse is outside the visibility circle. ■

It should be noted, however, that although to each $(k_1,...,k_m)$-impulse in the spectrum-convolution there belongs in the image-domain a periodic component which may become visible when this impulse is located inside the visibility circle, some of the convolution impulses span combs which do not represent "valid" but rather "degenerate" moirés:[20]

(1) Any impulse whose $(k_1,...,k_m)$ index-vector contains only one non-zero index, such as (1,0,0,0), (0,0,-1,0), etc., is generated in the spectrum of the $m$-layers superposition by only one of the layers (see Remark 2.4), and hence it does not correspond to a moiré effect. For example: in Fig. 2.2(f), the impulse pairs located at $\mathbf{f}_1$, $-\mathbf{f}_1$ and at $\mathbf{f}_2$, $-\mathbf{f}_2$, whose indices are respectively (1,0), (-1,0) and (0,1), (0,-1), do not represent an interaction between two layers, but rather, the periodic components contributed by the original layers themselves.

(2) A $(k_1,...,k_m)$-impulse in which all the indices $k_1,...,k_m$ have a common divisor bigger than 1 is not the fundamental impulse on its comb, and therefore it does not represent a valid moiré. For example: the impulses (2,-4), (3,-6), etc. are simply higher harmonics within the comb whose fundamental impulse is (1,-2) (see Fig. 2.6(f)). If such an impulse is located inside the visibility circle, it does not represent an independent moiré effect, but just a higher harmonic of the (1,-2)-moiré (whose fundamental impulse, the (1,-2)-impulse, is also located inside the visibility circle, closer to the origin).

---

[19] Note, however, that this notation is only unique up to a sign, since the twin impulse $(-k_1,...,-k_m)$ also spans the same comb.

[20] In fact, the distinction between "valid" and "degenerate" moirés is just a matter of convention; in some circumstances it can be more convenient to consider all of them as periodic components (cosines) in the superposition and to treat them all on an equal basis.

Suppose now that two 1-fold periodic moirés, the $(k_1^{(1)},...,k_m^{(1)})$-moiré and the $(k_1^{(2)},...,k_m^{(2)})$-moiré, whose frequency vectors $\mathbf{f}_1$ and $\mathbf{f}_2$ are linearly independent, are simultaneously visible in the superposition. It is often convenient to consider both of these moirés as a single 2-fold periodic moiré, which is, in fact, the product of the two original moirés (see, for example, Fig. 2.10(a)). The spectrum of this combined moiré is, therefore, the convolution of the spectra (combs) of the individual moirés: it consists of the impulses $i(k_1^{(1)},...,k_m^{(1)}) + j(k_1^{(2)},...,k_m^{(2)})$ for all integers $i,j$ (whose impulse-locations are, according to Proposition 2.4, $i\mathbf{f}_1+j\mathbf{f}_2$). In other words: the combined moiré is not only represented by a union of the two 1D combs of the impulses $(k_1^{(1)},...,k_m^{(1)})$ and $(k_1^{(2)},...,k_m^{(2)})$, but indeed, by the full 2D *impulse-cluster* which they span together. This point is fundamental to the understanding of 2D moirés between line-grids or dot-screens (Secs. 2.11 and 2.12 below), and it will be the key of our discussion in Chapter 4 on the extraction of the moiré intensity profile from the spectrum convolution. Such a 2-fold periodic moiré will be henceforth called the $((k_1^{(1)},...,k_m^{(1)}), (k_1^{(2)},...,k_m^{(2)}))$-moiré.[21, 22]

The same general principle applies also to the case of three or more moirés that are simultaneously visible in the superposition. Note, however, that in this case the spectrum of the combined moiré is not generally a discrete 2D cluster, but rather a 2D cluster which is everywhere dense. This subject will be explained in more detail in Chapter 5.

Note that sometimes, when no confusion may occur, it proves more convenient to use a shorthand notation in which zero indices and negative signs are omitted. For instance, the (2,1,-2,0) or the (2,-1,0,2)-moiré between 4 gratings may be considered as belonging to the same family and be simply called {1,2,2}-moirés. Furthermore, both the 4-grating (1,0,-1,0)-moiré (in which only two of the gratings actively participate) and the 2-grating (1,-1)-moiré may be called in short {1,1}-moirés. As we will see in Chapter 3, the use of this shorthand notation is sometimes more convenient than explicitly specifying the precise moiré indices, or enumerating all the similar variants within the same family of moirés.

## 2.9 Singular moiré states; stable vs. unstable moiré-free superpositions

We have seen that if one or several of the new impulse pairs in the spectrum-convolution fall close to the origin, inside the visibility circle, this implies the existence of one or several moirés with visible periods in the superposed image (see, for example, Figs. 2.5(f),

---

[21] Note that although to each 2D moiré there corresponds in the spectrum a *unique* impulse-cluster, the notation of a 2D moiré is not necessarily unique, since several different pairs of impulse-twins within the visibility circle may span the same 2D cluster. The reason is that a dot-lattice (the support of the impulse-cluster on the $u,v$ plane) does not have a unique basis (see Sec. 5.2.1).

[22] Note also that a $(k_1,...,k_m)$-impulse whose index-vector is an integer linear combination of two index-vectors of a lower order does not correspond to a "valid" but rather to a "degenerate" 2-fold periodic moiré. For example, the (2,0,-1,1)-impulse is simply a higher harmonic impulse in the impulse-cluster spanned by the (1,1,-1,0)-impulse and the (1,-1,0,1)-impulse: (2,0,-1,1) = (1,1,-1,0) + (1,-1,0,1), and therefore it does not correspond to a "valid", independent moiré effect. This remark is, in fact, a 2D generalization of point (2) above.

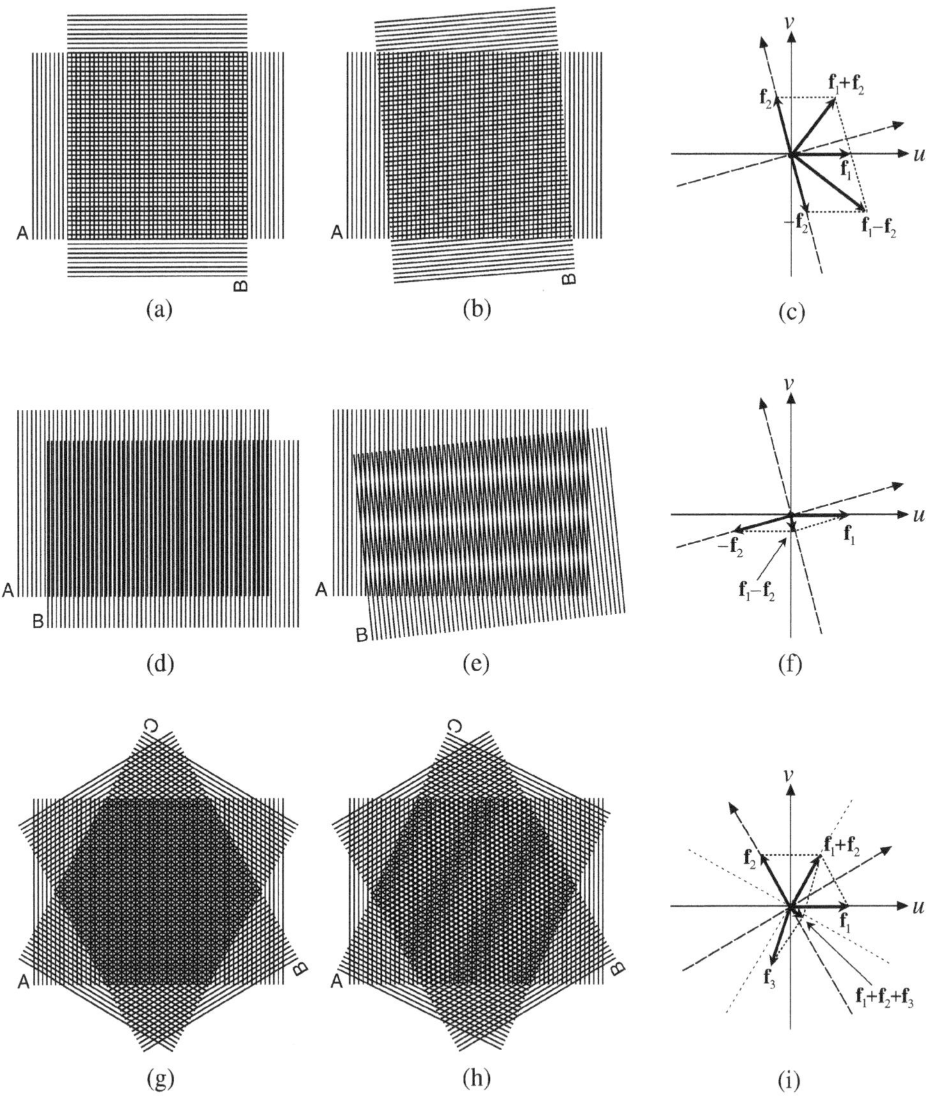


**Figure 2.8:** Examples of stable and unstable (=singular) moiré-free states. First row: (a) the superposition of two identical gratings at an angle difference of 90° gives a stable moiré-free state; small angle or frequency deviations, as in (b), do not cause the appearance of any visible moiré. The spectral interpretation of (b) is shown in the vector diagram (c). Second row: (d) the superposition of two identical gratings at an angle difference of 0° gives an unstable (singular) moiré-free state: a small angle or frequency deviation in any of the layers, like in (e), may cause the reappearance of the moiré with a very significant visible period. The spectral interpretation of (e) is shown in the vector diagram (f). Third row: (g) the superposition of three identical gratings with

*(continued from Fig. 2.8)*

angle differences of 120° gives an unstable (singular) moiré-free state; again, any small angle or frequency deviation may cause the reappearance of this {1,1,1}-moiré, as shown in (h) and in its vector diagram, (i).

2.6(f)). An interesting special case occurs when some of the impulses of the convolution fall *exactly* on top of the DC impulse, at the spectrum origin. This happens, for instance, in the trivial superposition of two identical gratings in match, with an angle difference of 0° or 180°; or, more interestingly, when three identical gratings are superposed with angle differences of 120° between each other (see second and third rows of Fig. 2.8). As can be seen from the vector diagrams, these are limit cases in which the vectorial sum of the frequency vectors is exactly $\mathbf{0}$. This means that the moiré frequency is 0 (i.e., its period is infinitely large), and therefore, as shown in Figs. 2.8(d),(g), the moiré is not visible. This situation is called a *singular moiré state*. But although the moiré effect in a singular state is not visible, this is a very unstable moiré-free state, since any slight deviation in the angle or in the frequency of any of the superposed layers may cause the new impulses in the spectrum-convolution to move slightly off the origin, thus causing the moiré to "come back from infinity" and to have a clearly visible period, as shown in Figs. 2.8(e),(h).

It is important to understand, however, that not all the moiré-free superpositions are singular (and hence unstable). For example, the superposition of two identical gratings at an angle of 90° is indeed moiré-free; however, it is not a singular state, but rather a stable moiré-free state: as shown in the first row of Fig. 2.8, no moiré becomes visible in this superposition even when a small deviation occurs in the angle or in the frequency of any of the layers. The corresponding situation in the spectral domain is clearly illustrated in Fig. 2.8(c), which shows the vector diagram of the superposition of Fig. 2.8(b).

Formally, we say that a singular moiré state occurs whenever a $(k_1,...,k_m)$-impulse (other than (0,...,0)) in the spectrum convolution falls exactly on the spectrum origin, i.e., when the frequency-vectors of the $m$ superposed gratings, $\mathbf{f}_1,...,\mathbf{f}_m$, are such that $\sum_{i=1}^{m} k_i\mathbf{f}_i = \mathbf{0}$. This implies, of course, that all the impulses of the $(k_1,...,k_m)$-moiré comb fall on the spectrum origin. As it can be easily seen, *any* $(k_1,...,k_m)$-impulse in the spectrum convolution can be made singular by sliding the vector sum $\sum_{i=1}^{m} k_i\mathbf{f}_i$ to the spectrum origin, namely: by appropriately modifying the vectors $\mathbf{f}_1,...,\mathbf{f}_m$ (i.e., the frequencies and angles of the superposed layers). When the $(k_1,...,k_m)$-impulse is located exactly on the spectrum origin we say that the corresponding $(k_1,...,k_m)$-moiré has become singular.[23]

A superposition is said to be singular (or in a singular state) if at least one moiré, say, the $(k_1,...,k_m)$-moiré, is singular in this superposition. It may also happen that several different moirés become singular simultaneously in the same superposition. A

[23] Note, however, that singular states are mainly of interest for $(k_1,...,k_m)$-impulses which represent "valid" moirés, i.e., impulses which satisfy conditions (1) and (2) in Sec. 2.8.

superposition in which none of the $(k_1,...,k_m)$-impulses falls on the spectrum origin and hence no singular moiré occurs is said to be *non-singular* or *regular*.

**Remark 2.6:** It should be noted that a singular superposition is not necessarily moiré-free: although the $(k_1,...,k_m)$-moiré itself is not visible in the singular superposition, other impulses may be present at the same time within the visibility circle and cause other moirés to be visible. ■

**Remark 2.7:** Note also that there may exist different degrees of singularity. For example, if the superposition of three of the $m$ superposed layers is already singular, independently of the remaining $m$–3 layers that are superposed on top of them, then we say that the $m$-layer superposition has only a singularity of order 3. This is still a singular case, since for the last $m$–3 layers we may choose $k_i = 0$, so that we still have for the $m$ layers $\sum k_i\mathbf{f}_i = \mathbf{0}$. Note that in a *pure* singular state, i.e., when the order of the singularity is $m$, all the $m$ superposed layers participate in the generation of the singularity, and by removing any of the $m$ superposed layers the singularity is destroyed. This is the case, for example, in the 3-grating superposition shown in the third row of Fig. 2.8. The 4-grating superposition which is obtained by adding a fourth, vertical grating on top of this singular superposition is still singular, but its singularity is only of order 3: the superposition still remains singular when this fourth grating is removed.[24] ■

The distinction between stable and unstable (=singular) moiré-free states is fundamental in the moiré theory. We will return to this subject again in Chapter 3, where we discuss the problem of moiré minimization.

## 2.10 The intensity profile of the moiré and its perceptual contrast

As has been shown above, each 1-fold moiré in the superposition of line gratings is represented in the spectrum by an infinite series of impulses (a comb). The amplitudes of these impulses are the coefficients of the Fourier series development of the function which represents the *intensity profile* of the moiré, i.e., the surface which defines the intensity level of the moiré at any point of the $x,y$ plane. The exact profile shape of the moiré in the image domain can be reconstructed as an infinite Fourier series by inserting the amplitudes $a_n = a_{nk_1,...,nk_m}$ of the moiré comb impulses given by Eq. (2.27) into the Fourier series (2.14) (with $b_n = 0$ for any $n$, since we are dealing with symmetric cases). An approximate profile shape of the moiré can already be obtained from the first few impulse pairs of the comb which fall inside the visibility circle, i.e., in terms of the image domain, as a sum of the first few cosine terms in the Fourier series (plus a constant term $a_0$ due to the DC impulse). A coarse, cosinusoidal shaped approximation of the moiré profile can be obtained from the DC impulse and the fundamental impulse pair, whose frequency vector

[24] As we will see in Remark 5.1 (Sec. 5.6.2), Remark 2.7 can be expressed in terms of the linear dependence over $\mathbb{Z}$ of the frequency vectors $\mathbf{f}_i$ of the superposed layers: if $\mathbf{f}_1,...,\mathbf{f}_m$ are all linearly independent over $\mathbb{Z}$ in the $u,v$ plane, then the superposition is regular; if $\mathbf{f}_1,...,\mathbf{f}_m$ are linearly dependent over $\mathbb{Z}$, then the superposition is singular; and if $\text{rank}_{\mathbb{Z}}(\mathbf{f}_1,...,\mathbf{f}_m) = r < m$, then the singularity is of order $m-r$.

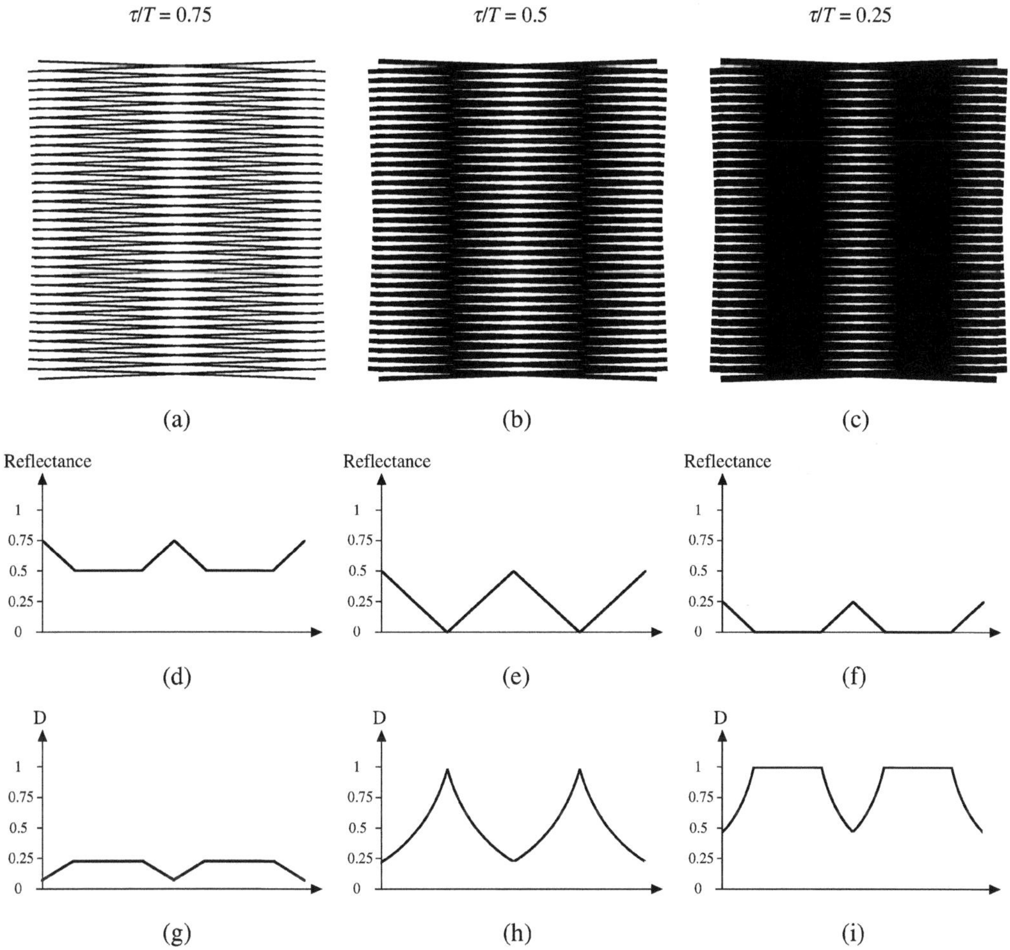


**Figure 2.9:** A (1,-1)-moiré between two identical binary gratings with the same periods and angles, and with opening ratios of: (a) 0.75; (b) 0.5; and (c) 0.25. (d), (e) and (f) show the respective moiré profiles in terms of *reflectance*, as received from the mathematical model. (g), (h) and (i) show the same moiré profiles after their adaptation to the human visual perception, i.e., in terms of *density*.[26]

determines the basic angle and frequency of the moiré. The extraction of the moiré intensity profile will be discussed in detail in Chapter 4.

We have seen in Sec. 2.2 that our mathematical model assigns to each point of the image a reflectance value between 0 and 1, where 0 means black, 1 means white, and intermediate values represent in-between reflectance values. This also applies to the moiré intensity profiles.[25] Note that even when all the original images are binary and only take the values 0 and 1, their moiré profiles still may contain intermediate values. In fact, the value of the moiré intensity profile at each point represents the average ratio of white per unit of area, i.e., the average reflectance at that point. For example, in the case of the (1,-1)-moiré between two gratings, when the width of the black and white lines is identical (i.e., the opening ratio is $\tau/T = 0.5$; see Fig. 2.9(b)), the value of the moiré profile along the center of the dark moiré bands is 0 (no white at all), and this value gradually climbs up to 0.5 at the center of the bright moiré bands (where black and white are equally distributed, and therefore the white ratio is 0.5). This is shown graphically in the reflectance profile of the moiré, in Fig. 2.9(e); similar profiles are shown in Figs. 2.9(d) and (f) for other opening ratios which correspond to the moirés of Figs. 2.9(a) and (c).

However, as we may notice by observation, the difference between the maximum and the minimum reflectance of the moiré profile does not faithfully correspond to the contrast of the moiré as it is perceived by the human eye. For example, in the (1,-1)-moiré between two identical gratings, the difference between the maximum and the minimum reflectance profile values is identical for gratings with opening ratios of $\tau/T = 0.75$ and $\tau/T = 0.25$, while the eye clearly sees a much higher contrast in the second, darker case (compare Figs. 2.9(a),(c) with their reflectance profiles 2.9(d),(f)). The reason for this phenomenon is that the response (or sensitivity) of the human visual system to light intensity is not linear in its nature, but rather closer to logarithmic [Pratt91 pp. 27–29]. Therefore, if we plot the intensities or the moiré profiles logarithmically, i.e., in terms of density rather than in terms of reflectance, we get a more realistic representation of the perceptual contrast of the moiré, which better corresponds to the human perception (see Figs. 2.9(g)–(i)).[26] A still better correspondence can be achieved by replacing the logarithmic approximation of the human visual response by an empiric function which is based on the experimental data obtained from physiological research [Schreiber93 pp. 60–67].

## 2.11 Square grids and their superpositions

A line-grid can be seen as a superposition (i.e., multiplication) of two non-collinear binary gratings, which together form a pattern of identical parallelograms. If the two gratings have identical periods and they are superposed orthogonally to each other, forming together a pattern of identical squares, the resulting line-grid is called a *square grid* or a *regular grid*. In most practical cases we will be mainly interested in square grids that are centered about the origin. Clearly, the spectrum of a square grid is the convolution of two identical but perpendicular combs, and its impulses are located on a square lattice.

---

[25] It should be mentioned that the range [0,1] of reflectance values is only respected by the *precise* profile reconstruction which takes into account all the Fourier terms up to infinity. An approximation using only a finite number of terms, such as the DC plus the first harmonic cosine (×2), may somewhat exceed the range of [0,1].

[26] Note that in Figs. 2.9(g)–(i) we have approximated the unbounded logarithmic function $D = -\log_{10} r$ by $D = -\log_{10}(0.9r + 0.1)$, whose values for the reflectance $r$, $0 \le r \le 1$, vary between 0 (for $r = 1$) and 1 (for $r = 0$).

The amplitude of each impulse in this spectrum is given by Eq. (2.22), with $T_1 = T_2 = T$ and $\tau_1 = \tau_2 = \tau$, where $T$ and $\tau$ are the period and the opening of both gratings. As we can see, this spectrum has the form of an impulse-nailbed which samples the continuous 2D function ("envelope") $g(u,v) = (\frac{\tau}{T})^2 \operatorname{sinc}(\tau u)\operatorname{sinc}(\tau v)$ at the points $(u,v) = (m/T,n/T)$ for all integers $m,n$. The shape of such a spectrum can be seen in Fig. 2.12(f).

The spectrum of a superposition of two square grids is, therefore, the convolution of two such nailbeds. This convolution can be carried out pictorially by placing a centered copy of one of the nailbeds on top of each impulse of the other nailbed (the amplitude of each copied nailbed being scaled down by the amplitude of the impulse on top of which it has been copied). By analogy with the case of grating superposition, where the comb-convolution generates in the spectrum a new impulse-comb centered at the spectrum origin that represents a 1-fold periodic moiré (see Fig. 2.5(f)), in our case the nailbed-convolution generates around the spectrum origin a new 2D impulse-nailbed (or *cluster*), that represents a 2-fold periodic moiré. This will be largely explained and illustrated in the following chapters (see, for example, the spectra in Fig. 4.3). For the moment, we can see the superposition of two square grids as a special case of a four binary grating superposition, each grid being composed of two identical but perpendicular gratings. It is clear, therefore, that each impulse pair in the spectrum of the superposition is accompanied by an identical, perpendicular impulse pair.[27] This means that every moiré effect which occurs between square grids is generated by two perpendicular impulse pairs; in fact, each such 2D moiré is represented in the spectrum convolution by the 2D impulse-cluster that is spanned around the spectrum origin by these two impulse pairs.

This allows us to introduce a simplified notation for the combined 2D moirés between square grids. Since the index of the second fundamental impulse of the combined moiré, $(k_1^{(2)},...,k_m^{(2)})$, is determined by the index of the first impulse, $(k_1^{(1)},...,k_m^{(1)})$, we can denote the combined 2D $\big((k_1^{(1)},...,k_m^{(1)}), (k_1^{(2)},...,k_m^{(2)})\big)$-moiré by the shorthand notation: $(k_1^{(1)},...,k_m^{(1)})$-moiré, implying that the second dimension is contributed by the orthogonal twin impulse.[27] This shorthand notation is very handy in many practical, real-world cases, like in Chapters 3 and 4, where each of the superposed layers is orthogonal and has identical frequencies to both directions. In such cases the second index-vector of the moiré is determined by the first one, so it can be omitted without risking any ambiguities.

**Example 2.4:** The simplest first-order moiré between two superposed square grids is the $\big((1,0,-1,0),(0,1,0,-1)\big)$-moiré, which consists of two perpendicular (1,-1)-moirés (see Fig. 2.10(a)): Its horizontal bands consist of the (1,-1)-moiré generated between the gratings A,C and its vertical bands consist of the (1,-1)-moiré generated between the gratings B,D. Thanks to the orthogonality considerations explained above, this 2-grid moiré can be called, as in Fig. 2.10(a), a (1,0,-1,0)-moiré; the omission of (0,1,0,-1) is permitted since it is simply the orthogonal twin of the (1,0,-1,0)-impulse. Similarly, the 2D $\big((1,2,-2,-1),(-2,1,1,-2)\big)$-moiré between two superposed square grids (see Fig. 2.10(c)) will

[27] If $(k_1,k_2,k_3,k_4)$ and $-(k_1,k_2,k_3,k_4)$ are an impulse pair (in the sense of Fig. 2.1) in the spectrum of two superposed square grids, then their *perpendicular impulse pair* (or *orthogonal twin*) is the impulse pair consisting of the impulses $(-k_2,k_1,-k_4,k_3)$ and $(k_2,-k_1,k_4,-k_3)$. This is further explained in Problem 2-15.

(a) (1,0,-1,0)-moiré: $\alpha = 15°$, $T_1 = T_2 = T_3 = T_4$.

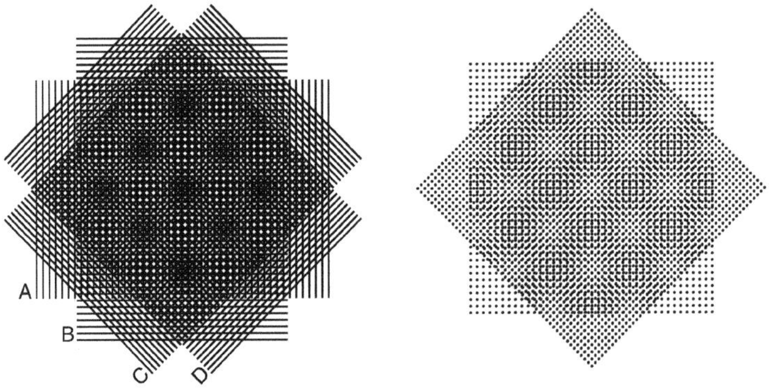


(b) (1,1,-1,0)-moiré: $\alpha = 45°$, $T_1 = T_2 = 1.3T_3 = 1.3T_4$.

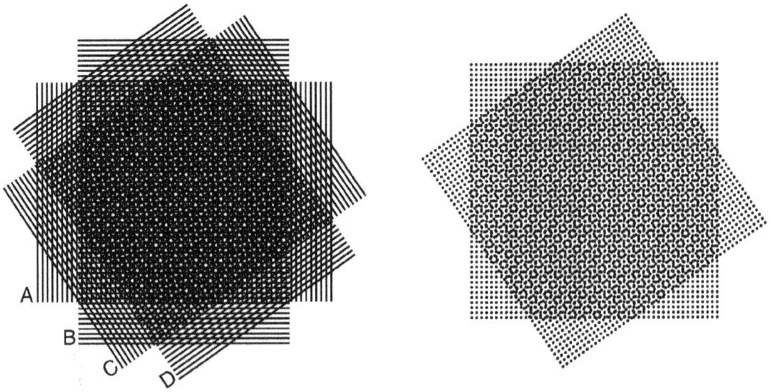


(c) (1,2,-2,-1)-moiré: $\alpha = 34.5°$, $T_1 = T_2 = T_3 = T_4$.

**Figure 2.10:** Three types of moirés between two square grids (left) or two dot-screens (right), which are generated by: (a) two; (b) three; or (c) four of the gratings involved. $T_1$, $T_2$ and $T_3$, $T_4$ are the periods of the first and of the second layers.

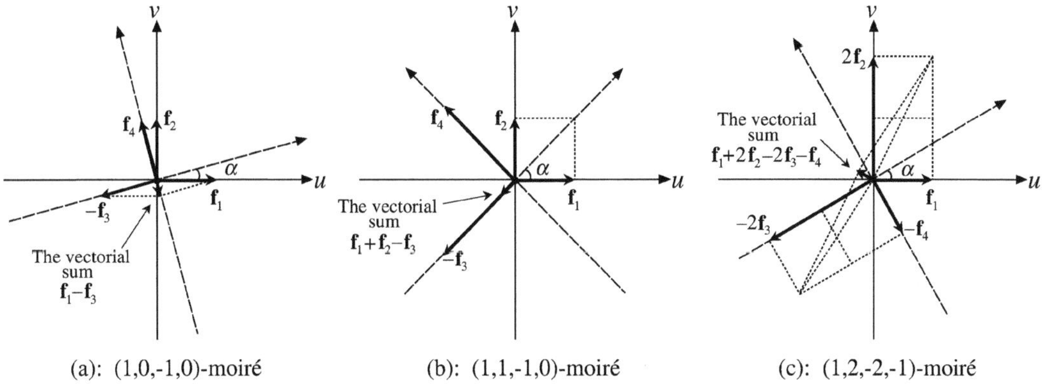


**Figure 2.11:** The spectral interpretation (vector diagrams) of the three types of moirés between two square grids shown in Fig. 2.10, which are generated by two, three or four of the gratings involved. For the sake of clarity, only the frequency vectors of one of the two perpendicularly symmetric moirés are shown in each case. The low frequency vectorial sum which corresponds to the fundamental impulse of the moiré in each of these cases is graphically found by the parallelogram law. Dashed axes belong to the rotated layers.

be called henceforth the (1,2,-2,-1)-moiré; and the 2D ((0,1,-1,0,1,0), (-1,0,0,-1,0,1))-moiré between three superposed square grids will be simply called the (0,1,-1,0,1,0)-moiré, like in Fig. 3.5. An example where such a shorthand notation should not be used is given in Example 5.8 in Sec. 5.7 (see Fig. 5.5): In this case the combined 2D moiré is generated by the (1,1,1) and the (1,-1,0) impulses, which are not obviously determined from one another. ■

Since the superposition of two grids can be seen also as a superposition of 4 gratings, it is clear that the moiré effects which occur between two square grids may originate from 2, 3, or 4 of the gratings involved. This means that the frequency vector of the fundamental moiré impulse in the visibility circle of the spectrum may be a vectorial sum of 2, 3, or 4 frequency vectors. In other words, the index vector $(k_1,k_2,k_3,k_4)$ of the fundamental moiré impulse may consist of 2, 3, or 4 non-zero indices (see Remark 2.4). Examples of such types of moirés between two square grids are shown in Fig. 2.10; their respective vector diagrams are shown in Fig. 2.11. In each diagram, the low frequency vectorial sum of the indicated frequency vectors from the two original layers gives the fundamental frequency vector of the visible moiré effect, and determines the angle and the frequency of the moiré. (The perpendicular twins of the original frequency vectors and of their vectorial sums have been omitted from the vector diagrams for the sake of clarity.)

**Remark 2.8:** Note that there exists a simple practical way to test which of the gratings of the superposed grids actually takes part in the generation of a given moiré, and this even without having to examine the spectrum. One has simply to reconstruct the 2-grid superposition as a 4 grating superposition, using the same angles and frequencies (or periods) as in the original grids, and to print the 4 superposed gratings. (This can be done, for example, by means of an appropriate PostScript program; if necessary, all the grating periods may be scaled up or down by the *same* factor, in order to improve the visibility of the moiré.) Now, all that is left to do is to reprint the grating superposition several more times, each time eliminating one or more of the 4 gratings. If the moiré in question is still visible, it means that the eliminated gratings do not take part in the moiré; if, however, the moiré disappears — it means that the eliminated gratings do participate in the generation of that moiré. This can be also seen in the non-superposed margins of the individual gratings; see the left column in Figs. 2.10(a)–(c). For example, the horizontal bands of the (1,0,-1,0)-moiré in (a) are only generated by the gratings A and C, while the vertical bands of the perpendicular (0,1,0,-1)-moiré are only generated by the gratings B and D. ■

The generalization of this section into the case of three or more superposed square grids is straightforward. The general expressions for the frequency vector and the amplitude of an impulse in the spectrum of the superposition are given by Eqs. (2.26) and (2.27), where $m$ is the number of gratings involved, i.e., twice the number of the square grids.

## 2.12 Dot-screens and their superpositions

A dot-screen is a 2-fold periodic structure which consists of identical and equidistant dots that are ordered in parallel rows along two main directions (axes). A dot-screen is called *regular*[28] if its two axes are perpendicular and the periods in both directions are identical. A dot-screen is in fact a generalization of a square grid, since any square grid can be considered as a special case of a dot-screen having square, white dots on a black background (see Fig. 2.12(e)). A regular screen $r(x,y)$ of white dots on a black background can be considered, as shown in Fig. 2.12, as a convolution of a function $d(x,y)$, which describes a single white dot, with a nailbed of period $T$. Therefore, according to the convolution theorem, the spectrum of such a dot-screen is the product of the continuous function $D(u,v)$ (the Fourier transform of a single dot) and the Fourier transform of the nailbed which is itself a scaled and stretched nailbed. This product is a nailbed which samples the "envelope" function $D(u,v)$ at intervals of $1/T$, scaling its amplitude by $1/T^2$ (see Fig. 2.12). In the case of a square white dot whose side is $\tau$ (as in Fig. 2.12), the single dot is given by $d(x,y) = \mathrm{rect}(x/\tau, y/\tau)$, and therefore the envelope of the spectrum is $D(u,v) = \tau^2 \,\mathrm{sinc}(\tau u)\,\mathrm{sinc}(\tau v)$ [Bracewell86 p. 246]. In the case of a circular dot, the envelope $D(u,v)$ is given by a Bessel function [Bracewell86 p. 248], which is sometimes called the *sombrero function* owing to its circular symmetry. Note that if the shape of the individual dot $d(x,y)$ is not centrosymmetric, or if the screen is not centered on the origin, then the envelope $D(u,v)$ is no longer purely real, and the impulse amplitudes in the nailbed spectrum of the dot-screen may have non-zero imaginary components.

[28] We prefer to avoid the term "square screen" which could be coined from "square grid", in order to avoid confusion with the dot shape of the screen, which may be square, circular, or anything else.

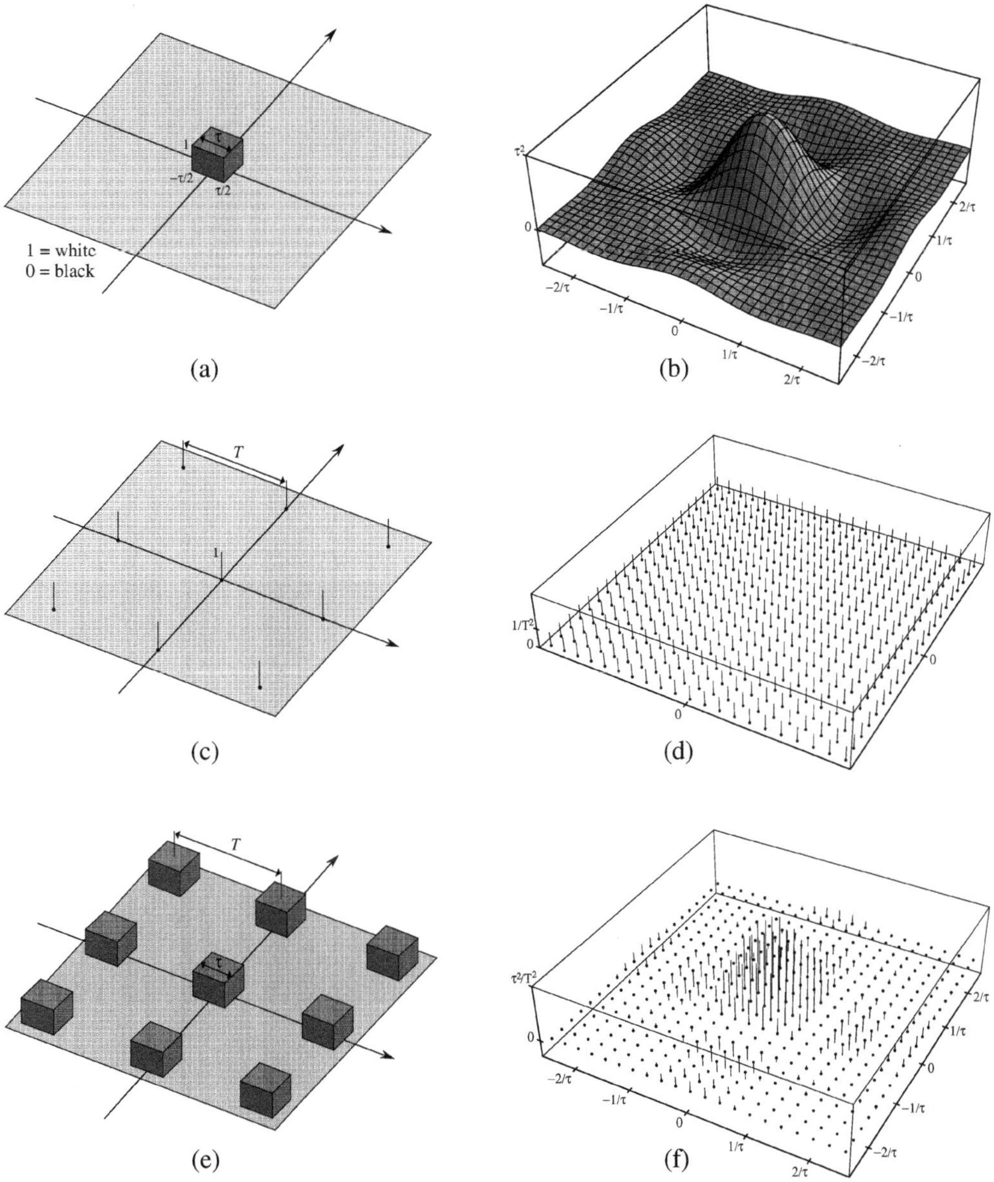


**Figure 2.12:** (a) A square white dot $d(x,y)$ with side $\tau$; (b) its continuous spectrum $D(u,v) = \tau^2 \operatorname{sinc}(\tau u)\operatorname{sinc}(\tau v)$. (c) A nailbed with period $T$ and amplitude 1; (d) its spectrum is a nailbed with period $1/T$ and amplitude $1/T^2$. (e) A screen of square white dots is the convolution of (a) and (c); (f) the spectrum of this screen is the product of (b) and (d): a nailbed that samples the "envelope" (b) at intervals of $1/T$, scaling its amplitude by $1/T^2$. Note that the impulses in (f) represent the coefficients of the 2D Fourier series development of (e).

As we can see, the shape of the individual dot $d(x,y)$ of the periodic screen determines the shape of the envelope function $D(u,v)$ in the spectrum, and therefore it determines the amplitude of each of the impulses in the spectrum of the screen. However, the geometric locations of the impulses in the spectrum of the screen are not influenced by the dot shape, and they are determined only by the nailbed with which the dot is being convolved.

A regular screen of *black dots on white background* can be seen as the "negative" of the above white-on-black screen $r(x,y)$, namely: $1-r(x,y)$. According to the 2D version of Eq. (2.18), if $a_{m,n}$ are the impulse amplitudes in the spectrum of $r(x,y)$, then the impulse amplitudes in the spectrum of $1-r(x,y)$ are: $d_{0,0} = 1-a_{0,0}$, $d_{m,n} = -a_{m,n}$. Here again, only the impulse amplitudes have been influenced, but not their geometric locations.

The spectrum of the superposition of two dot-screens is the convolution of the two bell-shaped nailbed spectra of the individual screens. Just as in the case of grid superpositions, this convolution process generates in the spectrum a new 2D impulse cluster centered about the origin. If sufficiently strong impulses of this new cluster fall inside the visibility circle, then a new visible moiré effect becomes apparent in the superposition. This is very similar to the special case of square grids discussed in Sec. 2.11 (i.e., the case of square white dots on a black background), and in fact the moiré effects generated in both cases have very similar *macroscopic* properties (see Fig. 2.10). Note, however, that the *microscopic* properties of the two cases may look quite different. The microscopic structure in the superposition of dot-screens is characterized by a fine pattern of small dot groups, called *rosettes* [Yule67 p. 339], which can be seen even when no visible moirés are present.

Although dot-screens and square grids are closely related to each other, there exists an important difference between them from the mathematical point of view: Square grids are represented by 2D reflectance functions which are *separable*, i.e., they can be presented as a product of two independent 1D functions; this reflects the fact that each square grid can be seen as a product (superposition) of two line gratings. As a result, the spectrum of a grid is simply given by the convolution of the spectra of its two gratings. However, dot-screens in the general case (e.g., screens with circular dots) are not necessarily separable, and their 2D spectrum can not always be seen as the convolution of two 1D spectra. We will therefore say that the 2D spectrum of a dot-screen, either separable or not, contributes *two* fundamental frequency vectors to the convolution; in the special case of a regular dot-screen, these two frequency vectors are perpendicular and have the same frequency. Hence, just as every line-grid counts for two gratings in the superposition, we may say that every dot-screen counts for two "virtual gratings" in the superposition. For example, in Eqs. (2.26) and (2.27) the case of $m = 4$ may correspond either to the superposition of 4 gratings, or to the superposition of 2 dot-screens, or even to the case of one dot-screen and 2 gratings. In all of these cases we say that the *equivalent grating number* is $m = 4$.

Note that in the case of regular dot-screens the spectrum has the same 90° symmetry as in the case of square grids, and therefore each moiré in their superposition is 2D and has a 90° symmetry. This will permit us to use the same shorthand moiré notations as in the

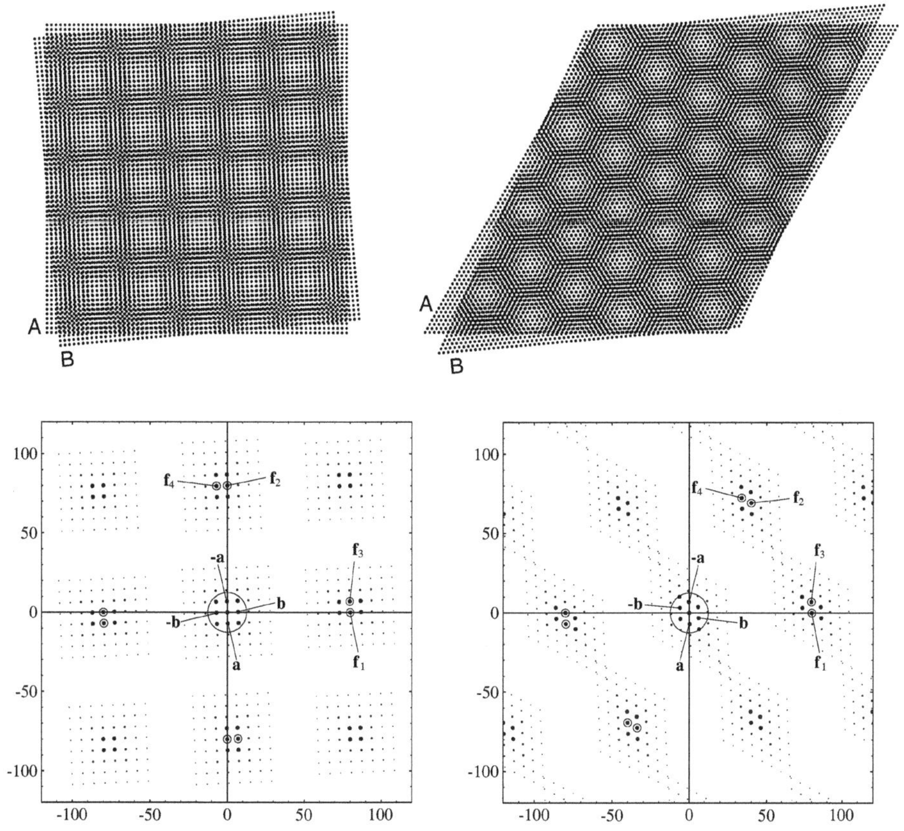


(a) (b)

**Figure 2.13:** A ((1,0,-1,0),(0,1,0,-1))-moiré: (a) in the superposition of two regular dot-screens; and (b) in the superposition of two "hexagonal screens" (that are, in fact, skew-periodic dot-screens; see Sec. A.3.4 in Appendix A). In both cases the angle difference between the superposed screens is 5°. The second row shows the spectra of the two superpositions; only impulses up to the 4-th order are shown. In both (a) and (b) the visible moiré is due to the cluster spanned by the (1,0,-1,0)- and (0,1,0,-1)-impulses, whose geometric locations are $\mathbf{a} = \mathbf{f}_1 - \mathbf{f}_3$ and $\mathbf{b} = \mathbf{f}_2 - \mathbf{f}_4$, respectively. Note the hexagonal (skew-periodic) form of the moiré in case (b).

case of square grids (see Sec. 2.11). This shorthand notation will be used throughout Chapters 3 and 4, which basically deal with regular dot-screens. If, however, the periodic dot-screens are not regular, the moiré effects in their superposition may have other symmetries; an example of this type is shown in Fig. 2.13(b). Note that in cases where the orthogonality between the fundamental impulses of a combined 2D moiré is not obvious from the circumstances, a full-length moiré notation $((k_1^{(1)},...,k_m^{(1)}),\ (k_1^{(2)},...,k_m^{(2)}))$ should be preferred (see Sec. 2.11), in order to avoid any possible confusion.

The test of Remark 2.8 for identifying the nature of a given moiré can be used also in the case of screen superpositions. The only difference is that in case of dot-screens we have first to replace each dot-screen by an *equivalent* line-grid, i.e., a line-grid having the same angles and periods as the original dot-screen, like in Fig. 2.10. This means, in fact, changing the dot shape of the screen, and possibly also taking its negative (since a grid consists of white dots on a black background). But as we have seen above, both changing the dot shape and taking the negative of a dot-screen have no influence on the *angle* and the *period* of the moiré; we may say that they give *geometrically equivalent* moirés. Therefore, even if the dot-screen is not separable, the contribution of each of its two "virtual gratings" (or each of its two frequency vectors in the spectrum) can be still determined by applying Remark 2.8 to an equivalent line-grid (see Fig. 2.10).

The fact that the geometric locations of the impulses in the spectrum of a dot-screen do not change when the shape of the individual dot is modified (or when the negative of the screen is taken) is fundamental for the understanding of moiré effects between superposed screens. It means that the period and the direction of the different moirés in the superposition do not depend on the shape of the individual dots of the superposed screens; changing the dot shape (or even the dot size within the fixed screen period) may only affect the *intensity* and the *profile shape* of each of the moiré effects. This is, in fact, the 2D generalization of Proposition 2.2 (Sec. 2.6); it is illustrated in Figs. 2.14 and 4.1. The influence of the size and shape of the screen dots on the moiré will be discussed in depth in Chapter 4.

## 2.13 Sampling moirés; moirés as aliasing phenomena

In the digital image processing literature moiré effects are often considered, in view of the sampling theory, as *aliasing* phenomena. This point of view is particularly common in the case of sampling moirés, i.e., moirés which occur in the analog to digital conversion of a given image between repetitive patterns in the image (such as the halftone screen of a printed image) and the device sampling lattice (the scanner resolution); see Sec. C.13 in Appendix C. Sampling moirés can be seen as a special case of superposition moirés, in which the first layer $r_1(x,y)$ is the original image to be scanned and the second layer $r_2(x,y)$ is the theoretic impulse nailbed that represents the scanning lattice. Suppose that the image $r_1(x,y)$ is sampled at the lattice points $(mT,nT)$, $m,n \in \mathbb{Z}$. The sampled image can be seen then as the product of $r_1(x,y)$ and the 2D nailbed $r_2(x,y)$ having steps of $T$. Therefore, according to the 2D convolution theorem, the spectrum of the sampled image is a convolution of the spectrum $R_1(u,v)$ with the spectrum $R_2(u,v)$, which is a 2D nailbed having

(a)

(b)

**Figure 2.14:** The influence of the dot size and of the dot shape of the superposed screens on the moiré intensity. Both images show the superposition of two screens, one at 61.7° and with frequency $f_1$ and the other at 38.66° and with frequency $f_2 = 1.153 f_1$. In both images the gray level of the first screen is varied horizontally from 0 to 1 (by modifying the dot size within the fixed screen period), and the gray level of the second screen varies vertically from 0 to 1. This means that each image contains all the possible dot size combinations of the two screens. The only difference between images (a) and (b) is in the shape of the screen dots: the dots are circular in (a), and rhombic (i.e., squares rotated by 45°) in (b). In each of the images two distinct moirés appear: The large, weak square pattern at 10.4°, which belongs to a {1,2,2}-moiré, and the small square pattern at –20.6°, which is a {1,1}-moiré. Note that the intensity of each of the two moirés within each of the images varies with the gray-level combination of the two superposed screens, and reaches its maxima in different gray-level combinations. The influence of the dot shape is demonstrated by the significant difference in the moiré intensities between the two images.

steps of $f = 1/T$.[29] This spectrum convolution gives an infinite set of replicas of $R_1(u,v)$, each of which being centered on an impulse of the nailbed $R_2(u,v)$. Now, according to the 2D sampling theorem [Rosenfeld82 p. 78], if the sampling rate used $f$ is lower than the Nyquist frequency $2f_{max}$ (twice the maximal frequency of $R_1(u,v)$),[30] then overlapping may occur between the replicas of $R_1(u,v)$ in the convolution, meaning that high frequencies of the replicas will fall into the central copy of $R_1(u,v)$ and appear there as

[29] If the scanner's aperture cannot be considered as a theoretical impulse-like pinhole, the nailbed $R_2(u,v)$ simply takes the envelope shape of the Fourier transform of the aperture. This can be pictorially illustrated by Fig. 2.12, where (a) is interpreted as a single aperture, (b) is its continuous spectrum, and (e) is an infinite array of such apertures that represents $r_2(x,y)$. As we can see in (f), the spectrum $R_2(u,v)$ of this aperture-array is a bell-shaped nailbed having the envelope form of (b).

[30] Or in other words, if the spectrum $R_1(u,v)$ of the original function $r_1(x,y)$ contains frequencies that exceed the region $-\frac{1}{2}f < u < \frac{1}{2}f$, $-\frac{1}{2}f < v < \frac{1}{2}f$.

parasite, false low frequencies, known as *aliased frequencies*; this aliasing effect excludes, according to the theorem, the faithful recovery of $r_1(x,y)$ by low-pass filtering of the sampled image.

If the spectrum $R_1(u,v)$ is rather continuous and smoothly fades out, the intruding frequencies from neighbouring replicas of $R_1(u,v)$ only slightly modify the high-frequency range of $R_1(u,v)$ in the spectrum convolution, so that the effect of aliasing in the scanned image is limited to slight distortions, such as the loss of high-frequency details or the appearance of jaggies on sharp edges [Foley90 pp. 132, 617–619]. However, if the original image $r_1(x,y)$ is periodic, or if it contains periodic patterns such as a halftone screen, then its spectrum $R_1(u,v)$ as well as all its replicas in the spectrum convolution contain strong impulses. In this case aliasing may cause in the spectrum convolution an intrusion of strong impulses from neighbouring replicas of $R_1(u,v)$ into the frequency domain of the central copy of $R_1(u,v)$, and even into the visibility circle around the origin. Such impulses are perceived, back in the image domain, as strong moiré effects in the sampled (scanned) image. In other words, the sampling moiré can be considered as an aliasing effect, in which, due to the periodic patterns in the image $r_1(x,y)$ (and hence the impulsive nature of its spectrum $R_1(u,v)$) the aliased frequencies are clearly manifested in the sampled image as parasite, low-frequency periodic patterns. This explanation appears, for example, in [Legault73] and in [Rosenfeld82, Sec. 4.1].

However, the approach we use in the present book to explain moiré phenomena is not based on the sampling theory, and this for several reasons:

(1) The sampling theory only deals with the case of two layers, one of which is a periodic comb or nailbed (representing pinholes with small apertures). However, our approach is much more general, and it covers *all* the moiré effects between *any* given periodic or repetitive layers (see, for example, Figs. 2.2, 2.3, 2.5, 2.6–2.10, as well as the various moirés between geometrically transformed layers shown in Chapter 10).

(2) The classical sampling theorem has been basically conceived for cases where the spectrum $R_1(u,v)$ is continuous. In our case, however, due to the periodic patterns in the image $r_1(x,y)$, the spectrum $R_1(u,v)$ consists of a nailbed of impulses which are separated from each other by wide gaps. This fact completely changes the situation: It opens the way to the use of sampling frequencies *below* the Nyquist frequency of $2f_{max}$ in some particular angle and frequency combinations where all aliased impulses from replicas of $R_1(u,v)$ fall outside the visibility circle of the spectrum convolution (an example of such a case is shown in Fig. 4.3(a)). Hence our approach improves on the classical sampling theorem in that it allows moiré-free scanning even with frequencies that are far below the Nyquist frequency; this will be further explained in Chapter 3 (see in particular Fig. 3.2).

(3) While spectral overlapping indeed prevents a faithful reconstruction of $r_1(x,y)$ from its samples, as stated by the sampling theorem, this is not yet enough to generate visible moiré effects. Visible moirés will only occur if this overlapping causes aliased impulses to fall inside the visibility circle.

## 2.14 Advantages of the spectral approach

The spectral approach whose foundations have been presented in this chapter proves to be very useful in the investigation of superposed periodic layers and their moiré effects. As we have seen, all the results provided by this approach are basically derived from the three different properties which are associated with each impulse in the spectrum: its label (or index), its geometric location in the $u,v$ plane, and its amplitude. Let us conclude this chapter with a summary of the main advantages of the spectral approach in the investigation of the moiré phenomenon.

First, the spectrum of the layer superposition contains all the information about each of the generated moirés: the period and the angle of each moiré are given by the geometric locations of its fundamental impulses, and its intensity profile is determined by the impulse-amplitudes of its fundamental and higher harmonic impulses (the moiré cluster). This enables a full *quantitative* analysis of the intensity levels of each moiré (see Chapter 4), in addition to the *qualitative* geometric analysis of the moiré, which is already offered by the classical approaches. Note that all the information which can be already obtained by the classical methods, namely, the geometric properties of the moiré patterns (periods, angles, etc.), is simply obtained in the spectral approach from the *geometric locations* of the impulses in the spectrum; this point will be further developed in Chapter 11. The additional dimension offered in the spectral approach by the *impulse amplitudes* has no equivalent in the classical methods.

Moreover, the study of the relationship between the *index* and the *geometric location* of the impulses in the spectrum leads to a simple algebraic formalism which is very helpful in the interpretation of the image-domain phenomena related to the superposition. This formalism also provides a useful notational system for the identification, classification and labeling of each of the possible moiré effects in the superposition individually. Thus, each moiré in the $m$-layer superposition has its own "identity" or index notation: the $(k_1,...,k_m)$-moiré. We also use this formalism for the definition of various fundamental terms such as the order of a moiré, a singular moiré state, etc.

Furthermore, since the spectrum of the superposition contains *simultaneously* all the impulses which may represent moiré effects, it provides an overall, panoramic view of all the different moirés of all orders which are simultaneously present in the same layer superposition. This approach also permits us to see how rotations, scalings and other changes in the superposed layers influence the spectrum. This enables us, in particular, to trace in the spectral domain the evolution of each of the moirés, and to identify at any moment which of them is visible, singular, or simply irrelevant (beyond the visibility circle). This also opens the way to a method for finding stable moiré-free combinations of superposed periodic layers. These subjects will be studied in depth in Chapter 3.

And finally, the spectral approach provides an easy explanation for all multiple-layer moirés, including the more complex cases where geometric analysis may become too complicated. In the spectral approach moirés of all orders, and between any number of superposed layers, are all treated on an equal basis, and there is no longer any need to deal first with "simple moirés", then with "moirés of moirés", etc.

## PROBLEMS

2-1. Suppose that two identical "raised" cosinusoidal gratings are superposed with an angle difference $\alpha$.

(a) Draw the spectrum of the superposition for the following cases: $\alpha = 10°$; $\alpha = 45°$; $\alpha = 90°$; $\alpha = 135°$; $\alpha = 170°$. Identify the impulses in each spectrum by their labels.

(b) Based on these spectra what do you expect to see in each of the superpositions in the image domain? In particular, which of the (1,-1)- and the (1,1)-moirés would be visible in each case?

(c) What happens in the spectrum and what happens in the image domain when $\alpha$ reaches 0°? And when it reaches 180°?

(d) For what range of angles $\alpha > 0°$ do we obtain in Eq. (2.10) $T_M > T$?

2-2. Within what range of angle differences $\alpha$ would you expect the "additive" (1,1)-moiré to be visible in the superposition of two identical gratings? Draw the spectrum of one such superposition, and compare with the spectrum of the "subtractive" (1,-1)-moiré in Fig. 2.2(f). Can the (1,-1)-moiré be visible simultaneously with the (1,-1)-moiré? Explain by referring to the spectra of the two cases.

2-3. The derivation of formulas (2.9) for the (1,-1)-moiré between two gratings is given in Appendix C.1. What will be the analogous formulas for the (1,1)-moiré? What will be the analogous formulas for the second-order (1,-2)-moiré (which is shown, for example, in Fig. 2.6)? Can you derive the formulas for the general $(k_1,k_2)$-moiré between two gratings?

2-4. Derive formulas (2.10) as a special case of the (1,-1)-moiré formulas (2.9), where $T_1 = T_2$. (*Hint*: you may use the trigonometric identity $\arctan(-1/x) = 90° + \arctan(x)$). Show that the analogous formulas for the (1,1)-moiré are given by:

$$T_{1,1} = \frac{T}{2\cos(\alpha/2)} \qquad \varphi_{1,1} = \tfrac{1}{2}(\theta_1+\theta_2)$$

2-5. Derive formulas (2.11) as a special case of the (1,-1)-moiré formulas (2.9), where $T_1 \neq T_2$ but $\theta_1 = \theta_2$. What are the analogous formulas for the (1,1)-moiré? And for the (1,-2)-moiré? Can you derive the formulas for the general $(k_1,k_2)$-moiré?

2-6. Fig. 2.2 shows the superposition of two "raised" cosinusoidal gratings with a small angle difference $\alpha$, and the impulse pairs in the corresponding spectra. As shown in the figure, the superposition consists of four cosinusoidal components: the two original cosinusoidal gratings, which are represented in the spectrum by the impulse pairs at the vector frequencies of $\mathbf{f}_1$, $-\mathbf{f}_1$ and $\mathbf{f}_2$, $-\mathbf{f}_2$, and the two new cosinusoidal components belonging to the difference frequency and to the sum frequency, which are represented in the spectrum by the impulse pairs at the vector frequencies of $\mathbf{f}_1-\mathbf{f}_2$, $\mathbf{f}_2-\mathbf{f}_1$ and $\mathbf{f}_1+\mathbf{f}_2$, $-\mathbf{f}_1-\mathbf{f}_2$. Write a computer program (in Mathematica, C, or any other suitable language) which displays the isolated periodic component which corresponds in the image domain to each of the four impulse pairs (see Figs. 2.2(a),(b),(j) and (k)). Now, display the sum of these individual periodic components, weighting them by their respective amplitudes, and verify that this sum gives, indeed, the superposed image (i.e., Fig. 2.2(c)), in accordance with the trigonometric identity:

$$(\tfrac{1}{2}\cos\alpha + \tfrac{1}{2})(\tfrac{1}{2}\cos\beta + \tfrac{1}{2}) = \tfrac{1}{4} + \tfrac{1}{4}\cos\alpha + \tfrac{1}{4}\cos\beta + \tfrac{1}{8}\cos(\alpha-\beta) + \tfrac{1}{8}\cos(\alpha+\beta)$$

(Note that in our case $\alpha = 2\pi f_1[x\cos\theta_1 + y\sin\theta_1]$ and $\beta = 2\pi f_2[x\cos\theta_2 + y\sin\theta_2]$, where $f_i$ and $\theta_i$ are the frequency and the direction of the two gratings, i.e., the polar coordinates of $\mathbf{f}_i$; see Eq. (2.6).)

2-7. *The isolated contribution of the sum frequency.* The new periodic component in the grating superposition of Fig. 2.2 which belongs to the difference frequency (i.e., to the impulse pair located at $\mathbf{f}_1-\mathbf{f}_2$, $\mathbf{f}_2-\mathbf{f}_1$) contributes to the grating superposition the prominent, low-frequency (1,-1)-moiré effect which is clearly visible in the superposition. However, the periodic component belonging to the sum frequency (i.e., to the impulse pair located at $\mathbf{f}_1+\mathbf{f}_2$, $-\mathbf{f}_1-\mathbf{f}_2$) is not visible as a prominent (1,1)-moiré in the superposition, although it does contribute the sum frequency to the superposed image. How can you explain this? Using the computer program of the previous problem, display the weighted sum of the individual periodic components, but this time without the periodic component of the sum frequency. How does the resulting image differ from the image of the grating superposition? The difference between these two images shows the contribution of the periodic component of the sum frequency to the superposed image.

2-8. *"Raised" vs. pure cosinusoidal gratings* (see Remark 2.2). Write a computer program to display the superposition (i.e., the product) of two cosinusoidal gratings with a small angle difference $\alpha$:

(a) When the two original gratings are "raised" and scaled to the intensity range of [0,1], like in Fig. 2.2;

(b) When the two gratings are pure cosines which oscillate in the range of [-1,1].

What are the differences between the superpositions obtained in (a) and (b)? What are the differences between the spectra of these two superpositions?

2-9. *Different layer superposition rules* (see Remarks 2.1 and 2.3). Using the two "raised" cosinusoidal gratings of case (a) in the previous problem, display:

(a) The multiplicative superposition of the two gratings (see Fig. 2.2);

(b) The additive superposition of the same gratings;

(c) The superposition obtained by the rule: $r(x,y) = \max(r_1(x,y),r_2(x,y))$;

(d) The superposition obtained by other superposition rules that you can think of.

What are the differences between the images obtained by the different superposition rules? What are the differences between the spectra of the superpositions in cases (a) and (b)? What can you say about the spectra of the superpositions in cases (c) and (d)?

2-10. *Pseudo (1,-1)-moiré bands in the additive superposition.* Show that the image obtained in case (b) of the previous problem is a 2D amplitude modulation of the cosinusoidal component of the sum frequency by the cosinusoidal component of the difference frequency. Note the washed-out gray bands which appear in case (b) exactly where the dark (1,-1)-moiré bands appear in case (a). Are these washed-out gray bands a real (1,-1)-moiré? Do they really contain a low-frequency component at the difference frequency? Hang the two images obtained in cases (a) and (b) on a wall, and observe them from a sufficient distance, so that the original gratings are no longer discerned by the eye. Do you see a (1,-1)-moiré in the multiplicative superposition? and in the additive superposition? (*Remark*: Remember that printers normally have a significantly non-linear reproduction curve; this may explain the weak traces of a (1,-1)-moiré in the additive superposition, too. But even so, these weak traces will look much weaker than the real (1,-1)-moiré bands which are clearly visible in the multiplicative superposition.)

*Solution*: The bands in the additive superposition only correspond to microstructure variations in the superposition, but unlike in the multiplicative superposition, they do not indicate the presence of macrostructure variations (a real low frequency component): In spite of the microstructure variations, the average value of the additive superposition remains constant throughout. This explains why in the additive case these bands are only visible while the observer can still resolve the individual lines of the original gratings, while in the multiplicative case these bands are clearly visible even from a distance (see, for example, [Post94 pp. 89–92]). Note that while in the multiplicative case the bands in question are indeed represented in the spectrum by a low-frequency impulse pair (at the difference frequency), in the additive case no such impulse pair exists in the spectrum (see Remark 2.3). It is only the human eye-brain system that connects adjacent microstructure details into moiré-like bands that are "seen" in the additive case (see, for example, [Badcock85; Badcock89]).

2-11. *Pseudo (1,-1)-moiré bands in the additive superposition (continued).* Show that the pseudo-moiré bands in the additive superposition, which are due to 2D amplitude modulation (see the previous problem), have the same period and angle as the real (1,-1)-moiré bands in the multiplicative superposition. Is it just a coincidence? Is it true to say that every periodicity in the image domain is represented by a corresponding impulse pair (or impulse comb) in the spectrum? Explain.

2-12. *Higher order pseudo moiré bands.* Pseudo moiré bands which "mimic" the behaviour of higher order moirés may occur in various circumstances, including multiplicative superpositions. The two examples given in this problem are derived from the situation shown in Fig. 2.6 — the multiplicative superposition of two binary square-wave gratings. As you can see in the spectrum of this superposition (Fig. 2.6(f)), the low-frequency bands which are visible in this configuration belong to the second-order (1,-2)-moiré.

(a) Consider the multiplicative superposition of two "raised" cosinusoidal gratings having the same angles and periods as in Fig. 2.6. In order to illustrate the situation, display this superposition using a computer program (for the sake of clarity you may rescale the $x,y$ axes like in Fig. 2.2). Do you see in the superposition diagonal bands which correspond to the (1,-2)-moiré? If such a moiré exists, it must be generated by the impulse pair $\mathbf{f}_1-2\mathbf{f}_2$, $2\mathbf{f}_2-\mathbf{f}_1$ in the spectrum convolution (like in Fig. 2.6(f)); however, as shown in Fig. 2.2(f), in a superposition of "raised" cosinusoidal gratings there exist no second harmonic impulses (their amplitudes are necessarily zero). How do you explain this apparent contradiction? (*Tip*: Remember that the bands you see are partially due to the non-linearity of the reproduction curve of the printer; a close examination of the superposition shows that the bands in question only represent a repetitive variation in the microstructure, but not a variation in the macrostructure).

(b) Consider the multiplicative superposition of two *binary* square-wave gratings having the same angles and periods as in the configuration of Fig. 2.6, but this time both having an opening ratio of $\tau/T = \frac{1}{2}$. In order to illustrate the situation, display this superposition using a computer program. Do you see in the superposition diagonal bands which correspond to the (1,-2)-moiré? If such a moiré exists, it must be generated by the impulse pair $\mathbf{f}_1-2\mathbf{f}_2$, $2\mathbf{f}_2-\mathbf{f}_1$ and the whole comb that it spans in the spectrum convolution (see Fig. 2.6(f)). However, according to Eq. (2.16), which gives the Fourier coefficients $a_n$ in the case of a square wave, it follows that when the opening ratio of a square wave is $\tau/T = \frac{1}{2}$, every even impulse in the comb has a zero amplitude. As a consequence, it follows from Eq. (2.24) that the impulses located at $\mathbf{f}_1-2\mathbf{f}_2$, $2\mathbf{f}_2-\mathbf{f}_1$ in the spectrum convolution, as well as all the other impulses on the comb of the (1,-2)-moiré (see Fig. 2.6(f)), have zero amplitudes. How do you explain this apparent contradiction?

(c) Is it a general rule that for any real moiré effect there may exist, in some circumstances, a "pseudo" moiré counterpart having the same periodicity and orientation?

2-13. *Binary square-wave gratings.* Fig. 2.4 shows a cross-section through a binary square-wave grating with opening ratio $\tau/T = \frac{1}{5}$, and the impulse comb that forms the spectrum of this grating (see also the top view of a binary grating and of its spectrum in Figs. 2.2(a),(d)).

(a) Draw similar figures (like Fig. 2.4) for binary gratings having the same period $T$ but opening ratios $\tau/T$ of 0, $\frac{1}{4}$, $\frac{1}{2}$, $\frac{3}{4}$ and 1 (find the impulse amplitudes using Eq. (2.16)).

(b) Which impulses have zero amplitudes in each case? Will there be impulses with zero amplitudes when the opening ratio $\tau/T$ is irrational?

(c) Do changes in the opening $\tau$ of the grating while keeping the period $T$ constant influence the geometric locations of the impulses in the spectrum? Explain why.

2-14. *Moiré sharpening.* Explain why the moiré bands in the superposition of two binary gratings appear sharpest (their black minima are most narrow and have a sharp center-line) when the two gratings are *complementary*, i.e., when their black-to-white ratios are reciprocal: $\frac{T_1-\tau_1}{\tau_1} = \frac{\tau_2}{T_2-\tau_2}$ [Post94 p. 105]. Note that when $T_1 = T_2$ this condition is simplified into: $\tau_1 + \tau_2 = T$, or equivalently: $\frac{\tau_1}{T} + \frac{\tau_2}{T} = 1$ (which is the form given in [Patorski93 pp. 8–11]). The particular case with $T_1 = T_2$ and $\tau_1 = \tau_2$ is shown in Fig. 2.9.

2-15. *Index-vector orthogonality and frequency-vector orthogonality in the spectrum.* Let $(k_1,k_2,k_3,k_4)$ be an impulse in the spectrum of two superposed line-grids (or dot-screens). Its symmetric twin (in the sense of Fig. 2.1) is the $-(k_1,k_2,k_3,k_4)$-impulse, and together they form an impulse pair. The $(-k_2,k_1,-k_4,k_3)$-impulse is an *orthogonal twin* of these two impulses, since their index-vectors are orthogonal in $\mathbb{Z}^4$ (their scalar product equals zero: $(k_1,k_2,k_3,k_4)\cdot(-k_2,k_1,-k_4,k_3) = -k_1k_2 + k_2k_1 - k_3k_4 + k_4k_3 = 0$). The same is true, of course, for its symmetric twin, the $(k_2,-k_1,k_4,-k_3)$-impulse. Show that if the two superposed grids (or screens) are *regular* then this orthogonality of index-vectors implies also the orthogonality of their impulse locations in the spectrum. In other words: show that if $\mathbf{f}_1 \perp \mathbf{f}_2$, $\mathbf{f}_3 \perp \mathbf{f}_4$ and $|\mathbf{f}_1| = |\mathbf{f}_2|$, $|\mathbf{f}_3| = |\mathbf{f}_4|$ then the frequency vectors $\mathbf{a} = k_1\mathbf{f}_1 + k_2\mathbf{f}_2 + k_3\mathbf{f}_3 + k_4\mathbf{f}_4$ and $\mathbf{b} = -k_2\mathbf{f}_1 + k_1\mathbf{f}_2 - k_4\mathbf{f}_3 + k_3\mathbf{f}_4$ are orthogonal (i.e., $\mathbf{a}\cdot\mathbf{b} = 0$). (Fig. 2.13(b) shows that if the superposed layers are not regular, vectors $\mathbf{a}$ and $\mathbf{b}$ are no longer orthogonal, even though their index-vectors are.)

2-16. Design a superposition of two non-regular screens in which the visible moiré is 1-fold periodic. *Hint*: The orthogonal impulse pair may be located outside the visibility circle.

2-17. *Singular moirés.* Superpose two identical square grids or dot-screens on top of each other, and slowly rotate one of them while keeping the other fixed. Observe the different moiré effects that grow up and then disappear as the rotation goes on; some of them are strong while others are very week and hardly visible. At what angle difference $\alpha$ each of these moirés becomes singular (i.e., momentarily disappears with an infinitely large period)? Can you identify these different $(k_1,k_2,k_3,k_4)$-moirés by drawing the corresponding vector diagrams? (See, for example, Figs. 2.10(a),(c) and their vector diagrams in Figs. 2.11(a),(c).)

2-18. *Singular moirés.* Which of the following superpositions are singular states?

(a) Two identical gratings with an angle difference of 180°/2 = 90°;

(b) Three identical gratings with equal angle differences of 180°/3 = 60°;

(c) Four identical gratings with equal angle differences of 180°/4 = 45°;

(d) Five identical gratings with equal angle differences of 180°/5 = 36°;
(e) Two identical regular dot-screens with an angle difference of 90°/2 = 45°;
(f) Three identical regular dot-screens with equal angle differences of 90°/3 = 30°;
(g) Four identical regular dot-screens with equal angle differences of 90°/4 = 22.5°;
(h) Five identical regular dot-screens with equal angle differences of 90°/5 = 18°.

2-19. *High precision measurement of small angles.* The moiré effect generated in the superposition of two identical gratings is very sensitive to angular deviations. The slightest angular misalignment between the two gratings will cause, according to formula (2.10), moiré bands with a very significant period $T_M$ (note that $T_M$ is inversely proportional to $\sin(\alpha/2)$, which is itself proportional to $\alpha$ when $\alpha$ is a small angle). Based on this phenomenon, can you propose a method for the detection of small angular deviations? Can you also devise a high precision method for measuring small angles? Are there any practical limitations on the smallest angle detectable by such methods? (More details on this technique can be found, for example, in [Morimura75] or [Kumar83].)

2-20. *High precision measurement of small periods.* The moiré effect generated in the superposition of two collinear gratings of periods $T_1 \approx T_2$ is very sensitive to period (or frequency) deviations. The slightest period (or frequency) mismatch between the two gratings will cause, according to formula (2.11), moiré bands with a very significant period $T_M$ (note that $T_M$ is inversely proportional to $|T_2 - T_1|$). Based on this phenomenon, can you propose a method for the detection of small period or frequency deviations? Can you also devise a high precision method for measuring small periods? Are there any practical limitations on the period sizes detectable by such methods? Explain.

2-21. *Grating deformations and moirés.* Let $r_1(x,y) = r_2(x,y) = \frac{1}{2}\cos(2\pi fx) + \frac{1}{2}$ be two identical cosinusoidal gratings. Suppose now that the cosinusoidal grating $r_1(x,y)$ undergoes a slight deformation which is given by $g(x,y)$:

$$r_1(x,y) = \tfrac{1}{2}\cos(2\pi f\,(x - g(x,y))) + \tfrac{1}{2}.$$

What kind of moiré effects do you expect to see in the superposition $r_1(x,y)r_2(x,y)$? (see Sec. 11.3). What would you expect to see in the superposition if the two gratings had a binary square-wave intensity profile rather than a "raised" cosinusoidal profile?

2-22. *Measurement of in-plane deformations; strain analysis.* Let $r_1(x,y) = r_2(x,y)$ be two identical gratings. Grating $r_1(x,y)$ is fixed onto a flat surface of the body to be examined, for example by gluing, by printing or by etching. When a stress is applied to the examined body its surface undergoes an in-plane deformation, i.e., a deformation in $x$ and $y$ but not in $z$. The grating on the body surface undergoes the same deformation and is therefore expressed like $r_1(x,y)$ in the previous problem. Now, if we place the reference grating $r_2(x,y)$ on top of the slightly deformed grating $r_1(x,y)$, a moiré pattern of coarse distorted bands or contours will appear in the superposition. These moiré bands represent *contours of equal displacement* (also called *isothetics* [Kafri90 p. 89]).

(a) Based on this phenomenon, can you propose a method for the detection of microscopic in-plane deformations? Can you also devise a method for measuring such deformations quantitatively? (More on this subject can be found, for example, in [Gåsvik95 Sec. 7.3].)

(b) Strain analysis of bodies under various loads is an important application of in-plane deformation measurement [Post94 p. 1]. Strain, which is defined as the derivative of displacement, is a result of stress and is proportional to it. How would you determine strains in the $x$ and $y$ directions from the moiré contours? (Strain analysis is discussed, for example, in Chapter 5 of [Kafri90] or in [Theocaris69].)

2-23. *Measurement of out-of-plane deformations; moiré topography.* In a technique called *shadow moiré*, moiré fringes are formed between a grating and its own shadow. A flat grating placed over a curved surface (a given object, a human body, etc.) is illuminated under a certain angle of incidence and viewed from another fixed angle. The light falling on the object through the grating projects the grating lines onto the object surface; when the object is viewed through the same grating, a moiré effect is generated between the grating and its shadow on the surface of the object. This moiré effect consists of dark and bright contours which represent *the elevation of the object surface*, like level lines in a topographic map. This method offers, therefore, a good qualitative visualization of the surface shape. Can you derive a mathematical expression permitting to evaluate quantitatively the topography (or the deformations) of the object in question? (More on this technique can be found in books such as [Gåsvik95 Sec. 7.4] and [Post94 pp. 119–124], or in the two original papers which presented this method independently of each other, [Meadows70] and [Takasaki70].)

2-24. *Vibration analysis. Flatness analysis.* How can moiré be used for vibration analysis? (see, for example, [Kafri90 p. 72; Harding83]). And for flatness analysis?

2-25. *Moiré deflectometry.* In the previous applications the distorted grating (or the distorted shadow) is brought as close as possible to physical contact with the reference grating. Moiré deflectometry, however, is based on a different approach, which can be used for both in-plane and out-of-plane distortions (see Chapters 6 and 7 in [Kafri90]). The two gratings are placed apart from each other, and the rays deflected by the sample under test distort the shadow of the front grating which falls on the rear grating. Hence, the deformation of the moiré pattern is caused in this case by the deflection of light rays from the inspected surface. When the two gratings are parallel, the resulting moiré fringes represent *contours of equal deflection angle*, which correspond to the optical properties of the inspected object. What is the role of the distance between the two gratings in this method? How does it influence the quality of the moiré effects? (A detailed discussion on this subject is provided in [Kafri90 pp. 102–108].)

2-26. *Moiré interferometry.* Moiré interferometry combines the concepts and techniques of geometrical moiré and optical interferometry (see Chapter 4 in [Post94]). It is capable of measuring in-plane displacements with a very high sensitivity. In this method a diffraction grating is produced on the specimen, so that when stress is applied to the specimen the grating exactly follows its deformations. Two beams of coherent light illuminate the specimen grating obliquely at angles of $+\alpha$ and $-\alpha$, generating constructive and destructive interferences which play the role of a virtual reference grating. The deformed specimen grating and the virtual reference grating generate together a moiré effect which reveals the strain pattern or the deformations. This moiré effect can be photographed by a camera in order to be analyzed. Note that this method enables the use of gratings with very high frequencies; a frequency of 2400 lines per millimeter (60960 lines per inch) is typical for the reference grating. How does this fact influence the accuracy of the method? (Note that this extreme sensitivity restricts interferometric measurements to low-noise environments, and requires high mechanical stability and vibration-isolated optical benches [Post94 pp. 156–157]).

2-27. *Testing lenses.* Assume that two identical binary gratings are superposed with a small angle difference $\alpha$, so that they generate visible moiré bands. Suppose now that two lenses, one positive and one negative, are placed side by side between the two gratings. Clearly, the positive (convergent) lens expands the bottom grating, whereas the negative (divergent) lens contracts it. Therefore, according to Eq. (2.9), within the lens areas the moiré bands are rotated to opposite directions (see figures in [Nishijima64 p. 4] or

[Oster63 p. 57]). Based on these observations, can you devise a method for measuring the diopter of optical lenses? What happens to the moiré bands when the examined lens contains distortions? How would you use this moiré system to control the quality of lenses? (A short description of the main lens aberrations can be found, for example, in [Kafri90 pp. 175–176].) For further reading see also [Nakano85] and [Konno94].

2-28. *Magnification checking by moiré.* Can you propose a high precision technique for checking the magnification of an optical system, based on the moiré effect between two gratings? (see, for example, [Swift74]).

2-29. *Moiré magnification.* All the moiré applications in the previous problems are based on the magnification property of the moiré phenomenon: depending on the case, the moiré patterns magnified small periods, small angles, or slight distortions in one of the superposed gratings. Discuss and explain the various facets of this general magnification property of the moiré phenomenon. (Further aspects of moiré magnification will be encountered in Chapters 4 and 7.)

2-30. *Moiré fringe multiplication.* Fig. 2.15 illustrates a technique known in moiré applications as moiré fringe multiplication [Post94 pp. 106–108]: By doubling the frequency of one of the superposed gratings (the reference grating), the frequency of the obtained moiré is doubled, too; this allows a more precise analysis of surface deformations etc. in various moiré applications.

(a) Sketch the spectra of the two grating superpositions shown in Fig. 2.15, and explain using them the principle of moiré fringe multiplication.

(b) This phenomenon seems to be in contradiction with Figs. 2.6 and 2.5, in which doubling the frequency of grating (b) does not influence the frequency of the visible moiré in the superposition. How do you explain this contradiction? *Hint*: Compare the spectra of Figs. 2.6 and 2.5 with the spectra that you have drawn in (a).

(c) Can you design a 2-grating superposition in which doubling the frequency of one of the superposed gratings completely changes the frequency and the orientation of the moiré seen in the superposition? *Hint*: Once again, thinking about this question in the spectral domain is much simpler.

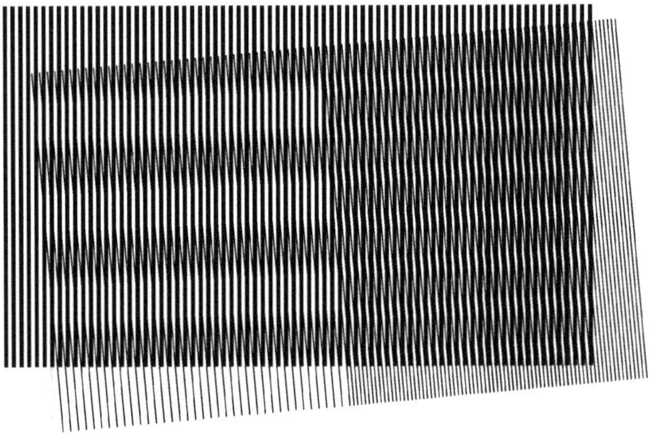

**Figure 2.15:** An example of moiré fringe multiplication: A superposition of two gratings with an angle difference of $\alpha = 5°$ and periods $T_1 = T_2$ is shown in the left side. Doubling the frequency of one of the gratings (see in the right side) doubles the moiré frequency, too.

# Chapter 3

## Moiré minimization

### 3.1 Introduction

As an application of the theoretical concepts introduced thus far, we present in this chapter the interesting problem of moiré minimization, namely, the question of how to find stable moiré-free combinations of several superposed dot-screens. Historically, this question finds its origin in the field of colour printing; and indeed, this context will be used here as our principal setting to illustrate the problem. Simpler cases involving only two superposed screens may also find their use in some other applications, such as the minimization of moirés which occur when a halftoned image is scanned by a digital scanner or sent by fax.

In general, three different strategies can be used to fight undesired moiré patterns: *avoiding moirés*, i.e., taking measures to prevent their generation; *minimizing moirés*, i.e., finding layer combinations which minimize their appearance; and *removing moirés*, i.e., getting rid of them once they already exist. Methods for avoiding moirés include the use of random screens rather than periodic ones [Blatner98 p. 264; Rodriguez94; Ostromoukhov93; Allebach78], the use of analog rather than digital devices (photocopiers, cameras, etc.), or filtering. Removal of already existing moirés can be done by the application of various special-purpose filtering methods. Different combinations of such techniques are used in modern digital scanners to reduce moiré effects which may occur when a halftoned image is scanned-in [Blatner98 pp. 280–281]. In the present chapter, however, we will concentrate on the least known strategy, that of moiré minimization. This strategy is useful in cases where the use of periodic screens or digital devices cannot be avoided, and post-treatment for moiré removal is undesired (being too lossy, too costly, or impractical). The problems at the end of this chapter give some further insights into the different methods for fighting undesired moirés.

We open this chapter in Secs. 3.2–3.3 with a brief description of the moiré problem in the context of halftone screen superpositions in colour printing. We describe the conventional screen combination traditionally used in colour printing, which is an unstable (i.e., singular) moiré-free state, and we explain our goal of finding alternative stable moiré-free combinations, and the difficulties in attaining this goal. Then, based on our spectral approach and on the basic notions presented in Chapter 2, we focus in Sec. 3.4 on the moiré phenomenon from a new point of view: we introduce the moiré parameter space, and show how changes in the parameters of the individual layers vary the moiré patterns in the superposition. This leads us to an algorithm for moiré minimization which provides stable moiré-free screen combinations that may be used, for example, for colour printing. We present the algorithm in Sec. 3.5, and conclude this chapter by a discussion of the results, in Sec. 3.6.

## 3.2 Colour separation and halftoning

The principle of colour image printing is based on the fact that most printable colours can be obtained (or at least approximated) by using only 3 or 4 primary colour inks: cyan (C), magenta (M), yellow (Y), and usually also black (K). The original colour image is first separated, either by photographic or by digital means, into 3 or 4 layers (*colour planes*), one for each of the CMYK primary colours. Then, each of these layers is printed, one on top of the other, using the same technique as in the printing of monochrome, gray-scale images — the *halftoning* technique.

Halftoning is a widely used technique for image printing. Since most existing printing devices are bilevel, they are only capable of printing solid ink or leaving the paper unprinted, but they are unable to produce intermediate ink tones. This is also the case in most colour printing devices, where each of the primary colours is only bilevel. In the halftoning technique, the original continuous-tone image (or each of its CMYK colour planes, in case of a colour image) is broken into tiny dots whose size varies depending on the tone level. When printed, this gives to the eye (looking from a normal viewing distance) an illusion of a full range of intermediate tone levels, although the printing device is only bilevel. For a more detailed introduction to the art of colour printing see, for example, [Hunt87, Chapter 26] or [Kipphan01].

Many different halftoning methods are currently in use in a large variety of printing devices, which range from high performance colour reproduction systems down to cheap desk-top printers. However, most of the high and medium quality image printing (notably for books, magazines, newspapers, advertising publications, posters, etc.) is done using a halftoning method with *clustered-dot elements*. In this method the original continuous-tone image (or each of its 3 or 4 colour planes) is transformed, either by photographic or by digital processing, into a regular screen of equidistant dots where the size of the screen dots varies according to the image tone level, but the frequency and the angle of the screen remain fixed.[1] Each such halftone screen is characterized by three parameters:

1. The *screen angle*, i.e., the direction in which the screen dots are aligned;
2. The *screen frequency*, i.e., the number of screen dots per inch (or per centimeter) in the direction of the screen;
3. The *dot shape* (circular, elliptic, square, diamond-shaped, etc.), or more generally: the *spot function*, a function which defines the way the size and the shape of the screen dots change as they grow from 0% to 100% coverage.

The significance of these parameters for the print quality is discussed for instance in [Blatner98 pp. 207–252]. In general, the screen frequency determines the coarseness of the printed image, and the dot shape influences the smoothness of tone gradations and

[1] There exist also other halftoning methods, such as dispersed-dot halftoning, error-diffusion, dot-frequency modulation, etc., in which the dot elements at the device resolution are not clustered into larger screen dots and do not form a regular screen. For more details concerning the various halftoning methods the reader is referred to [Ulichney88] or to [Kang97, Chapter 9].

may be also used to introduce some special or artistic effects [Blatner98 Chapters 20, 29; Ostromoukhov95]. In black and white printing the screen angle is usually set to 45° in order to make the halftone screen the least obstructive to the eye, since the human eye is less sensitive to small details in diagonal directions [Hunt87 p. 531]. However, in the case of colour printing, when the halftoning technique is applied separately to each of the CMYK colour planes of the image and the resulting 3 or 4 colour screens are printed on top of each other, this superposition of halftone screens may introduce a new serious problem, which did not exist in the single-layer halftoning of gray scale images: an interference in the form of a moiré pattern may appear between the superposed halftone screens of the different colours. Therefore in colour printing the screen angles and frequencies play a new major role, as it will be shown below, in the elimination of superposition moiré effects between the halftone screens of the 3 or 4 primary colours.

The traditionally used screen combination for colour printing consists of three regular screens with identical frequencies, which are oriented at equal angle differences of 30° (or 60°).[2] Usually the screen of the black ink, which is the most prominent colour, is set to 45° for the aforementioned reason, and the cyan and magenta screens are set to 45°±30°, namely, 15° and 75°. These angle differences of 30° between the superposed screens are large enough to make the strong {1,1}-moiré between each pair of layers practically disappear (note that by formula (2.10) the period of this moiré is less than twice the screen period; see Fig. 3.3(c)). However, for the fourth screen, belonging to the yellow ink, there remain no free angles left, and therefore it is placed 15° from two of the preceding screens; normally it is set to 0°. This inevitably generates between the yellow screen and each of its ±15° neighbours a {1,1}-moiré effect (whose period is, according to formula (2.10), almost 4 times the screen period; see Fig. 3.3(b)). But since yellow is a less prominent colour, the moiré it generates is hardly visible [Blatner98 p. 288].

Note, however, that this traditional combination of three screens with identical frequencies and equal angle differences of 30° (or 60°) is in fact a singular case. This singular moiré is simply the 3-layers {1,1,1}-moiré that we have already encountered between three identical gratings with angle differences of 120° (see bottom row in Fig. 2.8), which occurs in our case twice (see Fig. 3.1(b)): once between the screen directions 0°, 120°, 240° (one direction from each of the three screens), and once again, between the perpendicular directions of 90°, 210°, 330° (the second, perpendicular direction from each of the screens). Note that for the sake of convenience we use here for the three screens the angles of 0°, 30° and –30°, i.e., a 45° rotation of the angles 45°, 75° and 15° that are used in practice. The fact that the traditional screen configuration is singular makes it very sensitive to small angle or frequency deviations, as shown in Fig. 3.1(b). But in spite of this significant drawback, this classical screen configuration is widely used in the printing world thanks to its other virtues, notably the smooth and uniform microstructure it generates, which seems to be particularly pleasing to the eye and does not irritate it even

[2] Note that due to the 90° symmetry of regular screens, angle differences of 30° or 60° give identical superpositions. However, if the dot shapes in the individual screens do not have a 90° symmetry (e.g., elliptic dots), small microstructure differences will occur between the two cases.

when it becomes visible, at close viewing distance or under a large magnification (see the small irregular flower-like dot structures, called *rosettes*, in Fig. 3.1(a)).

**Remark 3.1:** It may be asked here, with reason, why shouldn't one prefer to use for the CMYK screens equal angle differences of $90°/4 = 22.5°$ rather than three angle differences of $90°/3 = 30°$ plus a problematic angle difference of 15° for the fourth screen? The answer to this question is two-fold: First, an angle of 22.5° between two identical screens already generates a {1,1}-moiré which is on the limit of visibility (according to formula (2.10) its period is about 2.56 times the screen period). And second, unlike the conventional 3-screen combination which gives a smooth and uniform microstructure that is particularly pleasing to the eye, the 4-screen superposition with angle differences of 22.5° has a less uniform microstructure with a rather unpleasant, granular aspect (see Fig. 8.16). ■

## 3.3 The challenge of moiré minimization in colour printing

Formally speaking, a halftoned image with clustered screen elements is a dot screen consisting of equidistant dots, whose dot size and shape are not constant. The study of moirés between halftoned images is therefore based on the case of regular dot screens that we have already seen in Sec. 2.12.[3] Since the colour separation technique involves overprinting of 3 or 4 halftoned images, one for each of the primary ink colours, undesired moiré patterns may occur between these halftone screens if special care is not taken.[4]

In order to be clearly visible, a moiré between two or more screens should have a large enough period (i.e., its fundamental impulse pair in the spectrum must be located inside the visibility circle), and its perceptual contrast (see Sec. 2.10) should be relatively strong. However, it is important to note that in colour printing not only strong moirés with large periods may be harmful. Experience shows that even weaker moirés with large periods (say, 10 times the screen period or more), or strong moirés with small periods (3-5 times the screen period) can still be visible; the latter may cause a rough linen-like texture and give the printed image a grainy aspect. As a rule of thumb it can be said that pronounced moiré structures which are larger than 1 millimeter are already visible in normal viewing conditions and should therefore be avoided.

The minimization of moirés between superposed screens consists of finding parameter combinations for the superposed layers which give as far as possible moiré-free results. But the task of finding a good moiré-free combination of more than two halftone screens is not an easy one, since many different moirés of various orders may appear at each angle and frequency combination (this is illustrated in Fig. 2.7 between two superposed gratings and in Fig. 2.14 for the case of two superposed dot-screens). Minimizing one of these

[3] Strictly speaking, a halftoned image with clustered dots is not a periodic function, since its dot sizes and forms may vary throughout the image. However, since the screen periods and angles are preserved throughout the halftoned image, the geometric locations of the impulses in the spectrum are not affected, and only their amplitudes are influenced by the dot size and form (see Sec. 10.11).

[4] Other possible sources of undesired moirés in image reproduction are given in Appendix C, Sec. C.13.

moirés (by varying the screen frequencies and angles) does not guarantee the minimization of the others, and may even introduce new moirés which were not visible before. In fact, the task of finding a good screen combination is a trade-off between the contradictory tendencies of the various moirés involved.

It should be also noted that not all the moiré minima are indeed stable solutions. Many interesting moiré-free combinations are in fact singular moiré states (see Sec. 2.9); in such cases any small deviation in the angle or in the frequency of any of the superposed layers may cause the reappearance of the moiré pattern in its full strength. The conventional 3-screen combination with equal frequencies and equal angle distribution is, indeed, such a case. Singular moiré-free states may be used in colour printing only if high accuracy can be guaranteed for the screen angles and frequencies. But moiré-free combinations which are not singular would have the advantage of being more stable and much less sensitive to small angle or frequency deviations in the reproduction and printing process.[5]

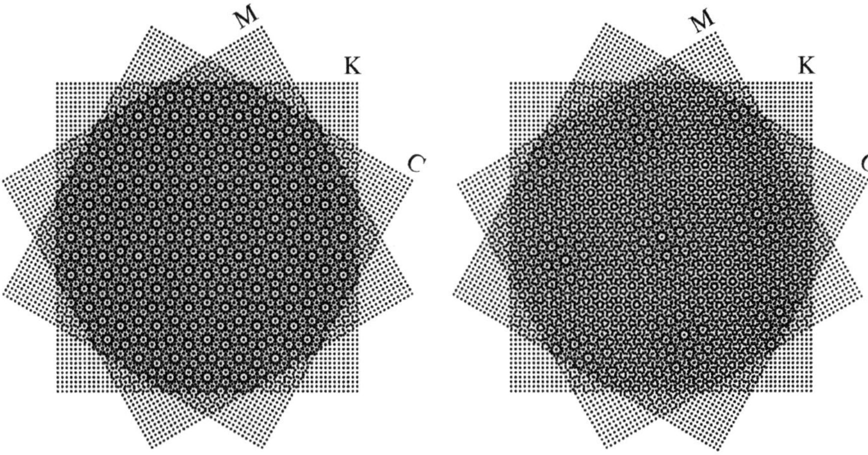


(a) (b)

**Figure 3.1:** The conventional 3-screen combination traditionally used in colour printing consists of 3 screens of identical frequencies with angle differences of 30° (or 60°). (a) The 3-screen superposition exactly at the singular point. (b) Slightly off the singular point, with a small angle and frequency deviation. For the vector diagram of the moiré refer to Fig. 3.5. (Note that in practice the three screens used in colour printing have a much higher frequency, so they are only visible under magnification; they are also rotated by 45° with respect to this figure, so that the black screen, K, is printed at 45°).

[5] Small *angle* deviations between the screens of the different inks may occur in multi-pass printing since a perfect registration of the paper between the different passes requires a very high mechanical precision. Small *frequency* deviations may occur, for example, owed to the dilatation of the printed paper after it has been moistened by the wet ink of the first printed colour.

Since the task of finding such stable moiré free combinations is not an easy one, the only reasonable way to face it is by writing a computer program that scans all the possible screen combinations, and checks all the moiré effects which are generated in each case. This can be done much more efficiently in the spectral domain, by checking the impulse locations and amplitudes in the spectrum convolution for each permitted angle and frequency combination of the screens. However, as we have seen in Sec. 2.12, the intensity of the moirés between halftone screens is significantly influenced by the shape and the size of the screen dots. Moreover, when several moirés are present simultaneously in the same screen superposition, each of them may be dominant in a different combination of gray levels. These facts are clearly illustrated in Fig. 2.14. It is therefore not sufficient to look for good screen combinations by checking the impulse amplitudes only at one predefined gray level, since an innocent moiré in one gray level may prove to be much stronger in another gray level. However, an exhaustive run through a large range of screen parameters and gray levels in order to find the best screen combinations proves to be impractical: First, due to the large amount of calculation; and second, due to the fact that for many of the practically used halftone dots (such as black circles which gradually grow into diamonds at 50% coverage and then gradually turn into white circles) we have no precise analytic expressions for the impulse amplitudes.[6] For these reasons (and also in order to avoid the very complex questions related to the precise modelization of the psycho-physiological behaviour of the human visual system, e.g., the perceptibility of a moiré pattern on the irregular background of a screen superposition), we prefer here a simpler approach for the moiré minimization, which is only based on geometric parameters such as angles and frequencies (although perceptual intensities of the moirés are also taken implicitly into account by experimental measurements). This approach will be explained in the following sections.

## 3.4 Navigation in the moiré parameter space

In Chapter 2 we analyzed the moiré phenomenon between superposed layers only from the *static* point of view. That is, given a fixed combination of angles and frequencies for the individual layers we analyzed the effects of their superposition exactly at the specified parameters, both in the image domain and in the spectral domain. Based on these results, we will consider now the moiré phenomenon from a different angle, namely: how changes in the parameters of the superposed layers affect the moiré phenomena in the superposition, both in the image and in the spectral domains.[7] This point of view will give us a new insight into the behaviour of the moiré phenomenon, and lead us to a method for finding moiré-free screen combinations.

---

[6] Note that a quantitative estimation of the impulse amplitudes by FFT is too coarse for this need, because of the various artifacts inherent to FFT (such as aliasing, leakage, etc.) which may significantly corrupt the impulse amplitudes obtained by FFT.

[7] We will restrict ourselves here only to the *geometric* parameters, i.e., to the influence of varying the *angles* and *frequencies* of the superposed layers on the resulting moirés. The effects of varying the gray levels or the dot shapes in the superposed layers were already discussed above (see Figs. 2.9, 2.14).

### 3.4.1 The case of two superposed screens

Let us start with the simple case of two superposed regular dot-screens. As we already know, any of the convolution impulses in the spectrum which fall within the visibility circle represent a moiré phenomenon between the two screens. Let the frequencies of the two screens be $f_1$ and $f_2$ and the angles they form with the positive horizontal axis be $\theta_1$ and $\theta_2$. Then according to Eq. (2.28) the geometric location of the $(k_1,k_2,k_3,k_4)$-impulse in the spectrum is given by the Cartesian coordinates $(u_{k_1,k_2,k_3,k_4}, v_{k_1,k_2,k_3,k_4})$, where:

$$\begin{aligned} u_{k_1,k_2,k_3,k_4} &= k_1 f_1 \cos\theta_1 + k_2 f_1 \cos(90°+\theta_1) + k_3 f_2 \cos\theta_2 + k_4 f_2 \cos(90°+\theta_2) \\ v_{k_1,k_2,k_3,k_4} &= k_1 f_1 \sin\theta_1 + k_2 f_1 \sin(90°+\theta_1) + k_3 f_2 \sin\theta_2 + k_4 f_2 \sin(90°+\theta_2) \end{aligned} \qquad (3.1)$$

Note that this expression can be also written in vector form:

$$\begin{pmatrix} u \\ v \end{pmatrix}_{k_1,k_2,k_3,k_4} = f_1 \begin{pmatrix} \cos\theta_1 & -\sin\theta_1 \\ \sin\theta_1 & \cos\theta_1 \end{pmatrix} \begin{pmatrix} k_1 \\ k_2 \end{pmatrix} + f_2 \begin{pmatrix} \cos\theta_2 & -\sin\theta_2 \\ \sin\theta_2 & \cos\theta_2 \end{pmatrix} \begin{pmatrix} k_3 \\ k_4 \end{pmatrix}$$

where the two matrices represent the rotations of the two screens by the angles $\theta_1$ and $\theta_2$.

Obviously, if we now let the parameters $f_i$ and $\theta_i$ vary, the geometric location of the $(k_1,k_2,k_3,k_4)$-impulse in the spectrum will vary accordingly. For the sake of simplicity we may assume without loss of generality that the first screen is fixed, with the angle $\theta_1 = 0°$ and the frequency $f_1$, and only the second screen is free to vary. Thus, the number of independent parameters (degrees of freedom) in the superposition of two regular dot-screens is reduced to two. It is convenient to choose them to be the angle difference between the two screens, $\alpha = \theta_2 - \theta_1 = \theta_2$, and their frequency ratio, $q = f_2/f_1$. Expression (3.1), the geometric location of the $(k_1,k_2,k_3,k_4)$-impulse in the spectrum, can be rephrased, therefore, as a function of the two variables $\alpha$ and $q$, as follows:

$$\begin{aligned} u_{k_1,k_2,k_3,k_4} &= k_1 f_1 + q f_1 [k_3 \cos\alpha + k_4 \cos(90°+\alpha)] \\ v_{k_1,k_2,k_3,k_4} &= k_2 f_1 + q f_1 [k_3 \sin\alpha + k_4 \sin(90°+\alpha)] \end{aligned} \qquad (3.2)$$

The distance of this impulse from the spectrum origin (i.e., the frequency of the impulse) is given, according to Eq. (2.8), by:

$$f_{k_1,k_2,k_3,k_4} = \sqrt{u^2_{k_1,k_2,k_3,k_4} + v^2_{k_1,k_2,k_3,k_4}}$$

and its period is: $T_{k_1,k_2,k_3,k_4} = 1/f_{k_1,k_2,k_3,k_4}$

Let us take as an example the (1,1,-1,0)-impulse, which represents the (1,1,-1,0)-moiré of Fig. 2.10(b). As we can see in Fig. 2.11(b), the geometric location of this impulse (denoted in the figure by the vectorial sum) is quite close to the spectrum origin; and in fact, if we rotate and scale the free screen so that $\alpha$ and $q$ vary towards the values of $\alpha = 45°$ and $q = \sqrt{2}$, the impulse will gradually approach the origin, and finally reach it. At this precise point the period of the moiré is infinitely large, but at any values of $\alpha$ and $q$

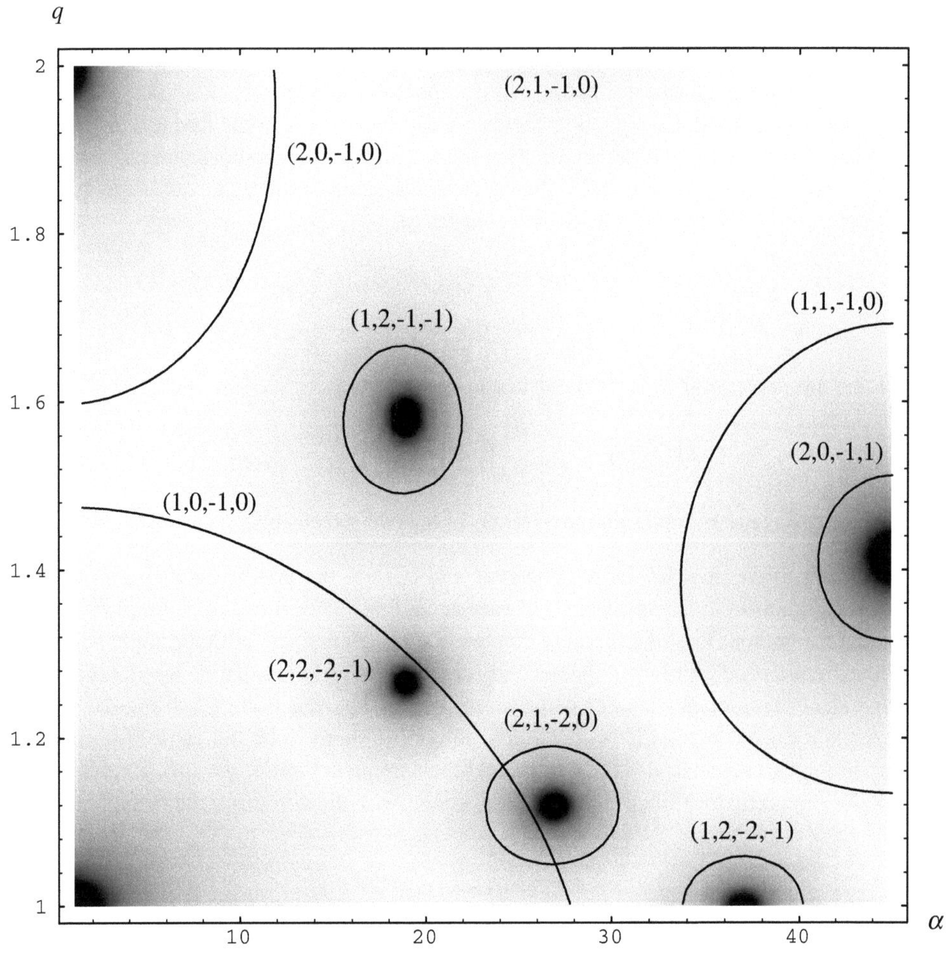


**Figure 3.2:** A panoramic view of the 2D parameter space $(\alpha,q)$ within the range $0° \le \alpha \le 45°$, $1 \le q \le 2$. It shows the most significant moirés between two superposed regular screens in this range, and illustrates how their period varies as a function of the angle difference $\alpha$ and the frequency ratio $q$ between the two screens. Darker shades represent bigger moiré periods. Each moiré is centered at a *singular point* in which its period is infinitely large; this point is surrounded by a spherical zone in which the period of the moiré gradually decreases from the center outwards. The contour line around each moiré delimits the parameter combinations $(\alpha,q)$ for which the fundamental impulse of the moiré is located in the spectrum inside the visibility circle. Note that the (1,0,-1,0)-moiré is clearly more dominant than the others. Note also that the (2,2,-2,-1)-moiré is already too weak to be visible; in practice it is assimilated by the residues of the strong (1,0,-1,0)-moiré.

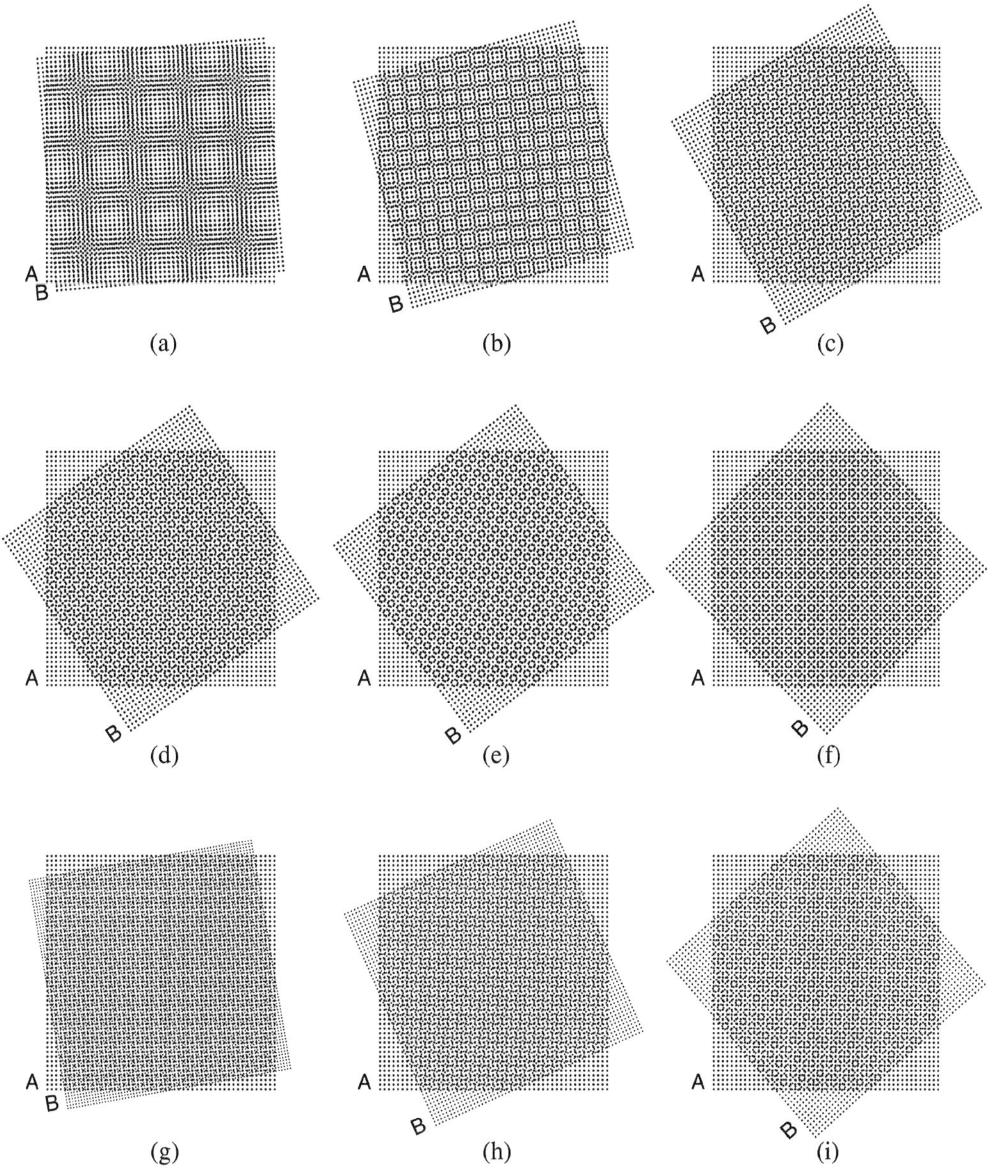


**Figure 3.3:** Two-screen superpositions with various $\alpha,q$ parameters (angle difference and frequency ratio). Each superposition corresponds to one $(\alpha,q)$ point in the 2D parameter space shown in Fig. 3.2. The first and second rows show the superposition of screens with identical periods ($q = 1$), and with angle differences of: (a) $\alpha = 5°$; (b) $\alpha = 15°$; (c) $\alpha = 30°$; (d) $\alpha = 34.5°$; (e) $\alpha = \arctan\frac{3}{4} \approx 36.8699°$; (f) $\alpha = 45°$. Note the appearance of the (1,2,-2,-1)-moiré in (d) and its singular state in (e). The last row illustrates points in which, according to Fig. 3.2, no significant moirés (up to order 2) appear: (g) $\alpha = 10°$, $q = 1.55$; (h) $\alpha = 24°$, $q = 1.25$; (i) $\alpha = 41°$, $q = 1.08$.

around this point the impulse in the spectrum will still be located within the visibility circle, so that the moiré will have a large, visible period. It can be said, therefore, that in the 2D *parameter space* defined by all the possible values of the parameters $\alpha$ and $q$, the point $(\alpha,q) = (45°,1.4142)$ represents a singular state of the (1,1,-1,0)-moiré. Moreover, this is the *only* singular point of the moiré, as a glance at Fig. 2.11(b) shows: Since the first screen is fixed, so is the sum of its two fundamental vectors, $\mathbf{f}_1+\mathbf{f}_2$; and only when $\alpha = 45°$ and $q = \sqrt{2}$ the vector $-\mathbf{f}_3$ of the second (varying) screen cancels it out and brings the vectorial sum (the small arrow) to zero. A similar reasoning shows that every $(k_1,k_2,k_3,k_4)$-moiré between two screens has a unique singular point in the 2D parameter space.

Using Eq. (3.2) we can find the parameters $\alpha$, $q$ of the singular point of any $(k_1,k_2,k_3,k_4)$-moiré, namely: the values $(\alpha,q)$ for which $u_{k_1,k_2,k_3,k_4} = v_{k_1,k_2,k_3,k_4} = 0$. Fig. 3.2 gives a panoramic view of the parameter space $(\alpha,q)$ and the most important moirés (up to the second order) which appear between the two superposed screens. The contour line around each moiré in the figure delimits the parameter combinations $(\alpha,q)$ for which the fundamental impulse of the moiré is located inside the visibility circle.

The results shown in Fig. 3.2 may be useful for finding angle and frequency combinations that minimize moiré effects when a halftoned image is scanned by a digital scanner, or sent by fax. In such cases the two screens involved are the screen of the halftoned image and the sampling-lattice of the digital device in question (the lattice defined by its horizontal and vertical scanning resolution); the angle difference $\alpha$ is measured between the direction of the halftone screen and the scanning direction. A glance at Fig. 3.2 shows the various frequency and angle combinations which are free of significant moiré interferences up to order 2; higher order moirés as well as a wider range of $q$ values may be also added to the figure, if required. This offers a minimization of the scanning moiré without increasing the data quantity due to high-frequency scanning, and without the need for time-consuming post-scan processing. It is also interesting to note that this moiré minimization approach improves on the classical sampling theorem in that it allows moiré-free scanning even with frequencies that are far below the Nyquist frequency, as clearly shown in Fig. 3.2 (see Sec. 2.13).

### 3.4.2 The case of three superposed screens

Having understood how to navigate in the parameter space of moiré effects in the case of two superposed dot-screens, we are now ready to examine the case of three superposed screens, which is the basic configuration for colour printing.

For the sake of clarity, let us adopt here the following notational conventions: we will call the three dot-screens, in descending order of their frequencies, the K-screen, the M-screen and the C-screen (shorthand for: black, magenta and cyan). The fourth, yellow screen will be introduced later, in Sec. 3.5. We suppose without loss of generality that the K-screen is fixed, with the angle $\theta_1 = 0°$ and the frequency of $f_K$, and that only the C- and M-screens are free to vary. The angle between the K- and the M-screens will be denoted by $\alpha$ and the angle between the K- and the C-screens by $\beta$ (see Fig. 3.4).

The case of three superposed screens therefore has four independent parameters: the angles $\alpha$ and $\beta$, and the frequency ratios $q_{\mathrm{MK}} = f_{\mathrm{M}}/f_{\mathrm{K}}$ and $q_{\mathrm{CK}} = f_{\mathrm{C}}/f_{\mathrm{K}}$. The geometric location of the $(k_1,k_2,k_3,k_4,k_5,k_6)$-impulse in the spectrum of the superposition can be expressed according to Eq. (2.28) as a function of these four parameters, as follows:

$$\begin{aligned} u_{k_1,\dots,k_6} &= k_1 f_{\mathrm{K}} + q_{\mathrm{MK}} f_{\mathrm{K}}\,[k_3 \cos\alpha \;+\; k_4 \cos(90°+\alpha)] \\ &\quad + q_{\mathrm{CK}} f_{\mathrm{K}}\,[k_5 \cos(-\beta) + k_6 \cos(90°-\beta)] \\ v_{k_1,\dots,k_6} &= k_2 f_{\mathrm{K}} + q_{\mathrm{MK}} f_{\mathrm{K}}\,[k_3 \sin\alpha \;+\; k_4 \sin(90°+\alpha)] \\ &\quad + q_{\mathrm{CK}} f_{\mathrm{K}}\,[k_5 \sin(-\beta) + k_6 \sin(90°-\beta)] \end{aligned} \qquad (3.3)$$

with $f_{k_1,\dots,k_6} = \sqrt{u^2_{k_1,\dots,k_6} + v^2_{k_1,\dots,k_6}}$ and $T_{k_1,\dots,k_6} = 1/f_{k_1,\dots,k_6}$.

Since the parameter space in this case is 4-dimensional, $(\alpha, \beta, q_{\mathrm{MK}}, q_{\mathrm{CK}})$, a full graphic representation (like Fig. 3.2 for the 2D case) is no longer possible. But, except for the abstraction due to the four dimensions, the situation remains basically similar. The main difference is that unlike in the 2D case, the locus of the singular points of a $(k_1,k_2,k_3,k_4,k_5,k_6)$-moiré in the 4D parameter space is no longer a single point, but rather consists of a 2D manifold (a curved surface) within the 4D space. This is illustrated in Figs. 3.5–3.6.

As the parameters $(\alpha, \beta, q_{\mathrm{MK}}, q_{\mathrm{CK}})$ move away from this singular manifold, the period of the moiré becomes smaller until at a certain distance from the singular manifold the moiré fades out and becomes practically invisible. If we "draw" in the 4D space the locus of all the points $(\alpha, \beta, q_{\mathrm{MK}}, q_{\mathrm{CK}})$ at which the $(k_1,k_2,k_3,k_4,k_5,k_6)$-impulse is located inside the

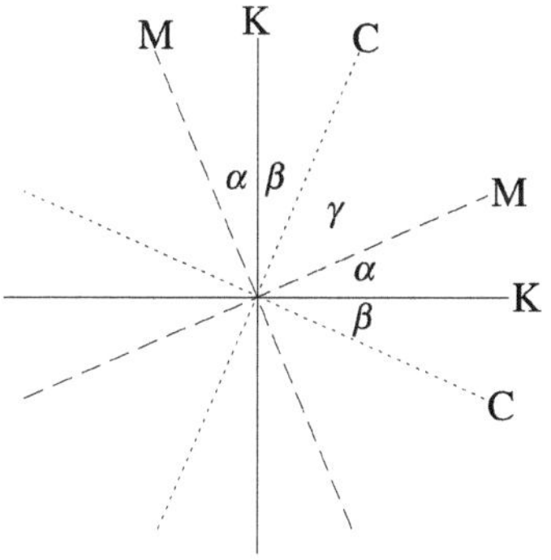


**Figure 3.4:** A schematic view of the three superposed C,M,K screens (each represented by its two main axes) showing the conventional notations used in the text. Note that the angle differences $\alpha$ and $\beta$ are measured with respect to the K-screen; thus, angle $\alpha$ is measured counter-clockwise, while angle $\beta$ is measured clockwise.

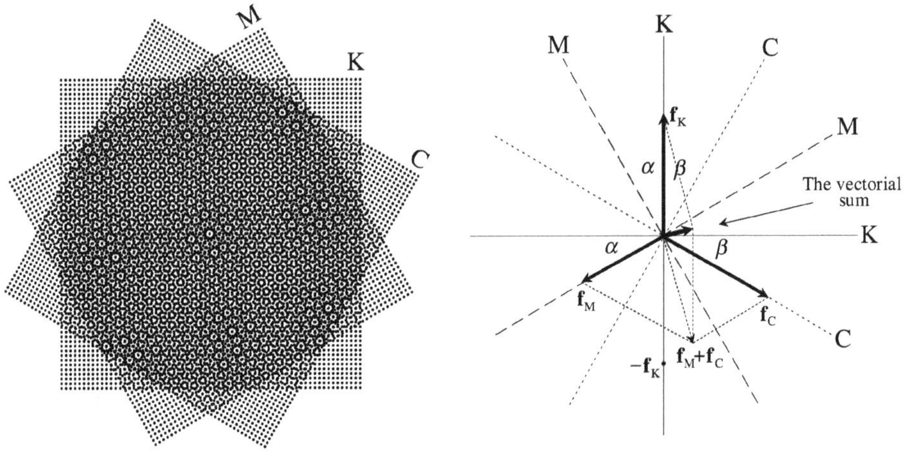


(a) (b)

**Figure 3.5:** A (0,1,-1,0,1,0)-moiré between three screens (a), and its vector diagram (b); for the sake of clarity, the vector diagram belonging to the perpendicularly symmetric (-1,0,0,-1,0,1)-moiré is not shown. By convention it is assumed that the K screen is fixed, and only the M and C screens are free to vary. For any choice of $\alpha$, $\beta$ and $q_{\mathrm{MK}}$ (or $\mathbf{f}_{\mathrm{M}}$) there exists one value of $q_{\mathrm{CK}}$ (i.e., one value of $\mathbf{f}_{\mathrm{C}}$) for which $\mathbf{f}_{\mathrm{M}}+\mathbf{f}_{\mathrm{C}}$ falls on $-\mathbf{f}_{\mathrm{K}}$, thus yielding a singular case (where the vectorial sum is **0**). This means that the locus of all the singular points of this moiré forms a 1D manifold in the 4D space ($\alpha$, $\beta$, $q_{\mathrm{MK}}$, $q_{\mathrm{CK}}$). Note that one of these singular points, the point with $\alpha = \beta = 30°$ and $q_{\mathrm{MK}} = q_{\mathrm{CK}} = 1$, is the case used in traditional colour printing.

visibility circle in the spectrum (i.e., $f_{k_1,\dots,k_6}<$ visibility circle radius), we will obtain around the singular manifold a 4D zone that is analogous to the spherical zone around each singular point in the 2D case of Fig. 3.2. We will call this zone the *forbidden zone* of the moiré in question. In 2D sections taken through the 4D parameter space this forbidden zone may appear as a thick, curved or straight line (Fig. 3.6(a)), or as an elliptic or egg-like shape (Fig. 3.6(b)).

Now, if we "draw" in the 4D space the forbidden zones belonging to all the $(k_1,k_2,k_3,k_4,k_5,k_6)$-impulses in the spectrum which represent perceptible moirés, we will get the following picture: the singular manifolds of the different moirés will appear dispersed in the 4D space, each of them serving as a skeleton which is surrounded by the forbidden zone of the corresponding moiré. These forbidden zones are sometimes intersected, and sometimes they only almost touch in the 4D space. The spaces left *between* these forbidden zones include the parameter combinations for which no disturbing moiré occurs between the three screens.

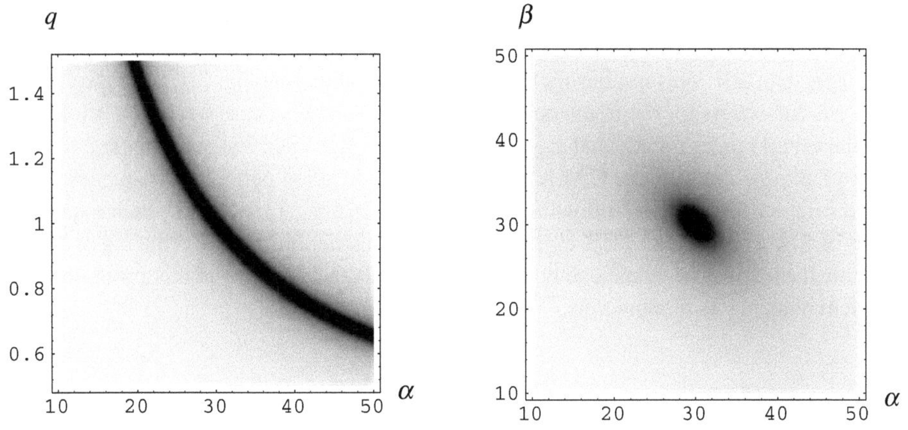


(a) (b)

**Figure 3.6:** Two 2D sections through the 4D parameter space ($\alpha$, $\beta$, $q_{MK}$, $q_{CK}$) showing their intersections with the singular manifold of the (0,1,-1,0,1,0)-moiré and the forbidden zone surrounding it. (a) A section along the diagonal plane $\alpha = \beta$, $q_{MK} = q_{CK}$. (b) A section along the ($\alpha$,$\beta$) plane with $q_{MK} = q_{CK} = 1$. Darker shades represent higher moiré periods. Note that the singular point $\alpha = \beta = 30°$, $q_{MK} = q_{CK} = 1$ is included in this singular manifold.

## 3.5 Finding moiré-free screen combinations for colour printing

The above discussion immediately suggests a practical way of searching for moiré-free combinations for the C,M,K dot-screens in colour printing. In principle, any parameter combination in which no significant moiré is visible is a solution to the problem, including the singular states at the centers of the forbidden zones (unless such a singular point is also covered by a forbidden zone belonging to another moiré). As we have already seen, the traditional combination of $\alpha = \beta = 30°$ and $q_{MK} = q_{CK} = 1$ is an example of such a singular point. But when the accuracy of the printing process is not enough to guarantee the high precision required for the use of a singular screen combination, it may be preferable to use a stable moiré-free solution, i.e., a screen combination which is located in the 4D space at the center of a moiré-free zone, with as high as possible tolerances in each of the four parameters involved. Therefore, rather than searching for singular points (in the center of the forbidden zones), as proposed for example in [Fink92], we will be looking here for stable solutions in the center of the moiré-free zones.

Before we proceed to describe the algorithm itself, a word about the useful ranges of the four parameters $\alpha$, $\beta$, $q_{MK}$, $q_{CK}$ is in order. It is clear from Fig. 3.4 that the angles $\alpha$ and $\beta$ are limited to the range $0° \leq \alpha+\beta \leq 90°$, but since we know experimentally that any angle difference lower than 20° inevitably generates strong {1,1}-moirés (such as in Fig.

2.10(a)) we may limit ourselves only to cases in which all of the three angles, $\alpha$, $\beta$ and $\gamma$, are larger than 20°. This implies the following angle range (see Fig. 3.7): $20° \leq \alpha \leq 50°$, $20° \leq \beta \leq 70° - \alpha$. As for the frequency ratios, since we suppose that $f_C, f_M \leq f_K$ it follows that the useful ranges for $q_{MK}$ and $q_{CK}$ are: $q_{CK}, q_{MK} \leq 1$. But since it is inappropriate to mix within the same color image C,M,K screens of very different periods or frequencies we will restrict ourselves to the following period ranges: $T_K \leq T_M, T_C \leq 1.5T_K$, which means in terms of frequency ratios: $2/3 \leq q_{CK}, q_{MK} \leq 1$. These restrictions help in reducing as far as possible the number of screen combinations to be analyzed by the computer program, and hence in reducing its running time.

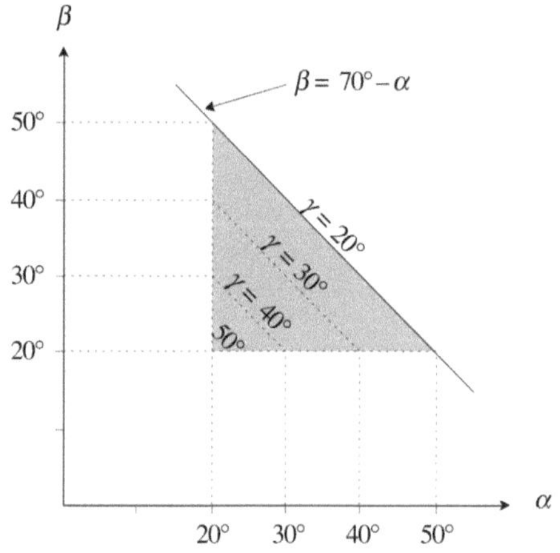


**Figure 3.7:** The admitted range for the angles $\alpha$, $\beta$ and $\gamma$. Any point outside the gray range violates at least one of the conditions $\alpha \geq 20°$, $\beta \geq 20°$, $\gamma = 90° - (\alpha + \beta) \geq 20°$.

The method we adopt here for finding moiré-free screen combinations for colour printing is a straightforward search through all the permitted parameter combinations, within their admitted ranges. For reasons to be explained below the algorithm first searches for good combinations of the C,M,K screens, and only then, for each good solution, it attempts to find appropriate parameters for the Y screen. Here are the main steps of the algorithm:

**Step 1:** Generating a list of the "dangerous" impulses in the spectrum convolution.

The aim of this step is to select from the list of all the impulses which are generated in the spectrum of the superposed C,M,K screens a relatively small subset which contains only the "potentially dangerous" impulses. This will be further explained below. It should be noted that what is needed here is not the actual impulse locations or amplitudes, but only their $(k_1,k_2,k_3,k_4,k_5,k_6)$ index-vectors (for example: (2,0,-1,0,0,1)).

**Step 2:** Scanning the 4D parameter space for moiré-free combinations.

In this stage we run through all combinations of the four parameters $\alpha$, $\beta$, $q_{\mathrm{MK}}$, $q_{\mathrm{CK}}$ (within their admitted ranges, as described above). The step size should be small enough to ensure that no moiré-free zones escape through the mesh, but not too small either in order to avoid excessive running time. For each parameter combination the program runs through the list of "dangerous" impulses and calculates for each of them its actual coordinates in the spectrum, using Eq. (3.3). If any of the impulses fall inside the corresponding visibility circle, the current parameter combination is rejected. Only if all the dangerous impulses fall outside their visibility circle is the parameter combination accepted and registered in a file which accumulates the potentially good moiré-free combinations.

**Step 3:** Finding the best solutions inside the detected moiré-free zones.

Having completed the scanning of all the four parameter combinations in step 2, we are left with a file containing all the potentially good combinations found. As we have seen above, these points are located within the moiré-free zones in the 4D parameter space. In this step we search for the best solutions among these points, according to the tolerances they offer. Then, each of the best solutions found should be experimented with high resolution film superpositions, in order to eliminate cases with visible residual moirés that could not be detected by the program (for instance, higher order moirés).

**Step 4:** Finding for each good C,M,K combination an appropriate Y screen.

After the proposed moiré-free combinations for the C,M,K screens have been tested and verified as good solutions, a similar approach is taken to find for each of them a good Y screen. In this case the values of $\alpha$, $\beta$, $q_{\mathrm{MK}}$ and $q_{\mathrm{CK}}$ are already known, and the varying parameters are the angle $\delta$ and the frequency ratio $q_{\mathrm{YK}}$ of the Y screen. (Obviously, the list of "dangerous impulses" as well as Eq. (3.3) must be readapted to the case of four superposed screens).

The reason for separating the Y screen from the others is two-fold: first, the introduction of a fourth screen in steps 1–3 would significantly increase the number of moirés which occur between the different layers, and make it practically impossible to find good moiré-free combinations between all four layers simultaneously. As a compromise, it is better to find the best possible moiré-free combinations for the C,M,K screens first, even at the expense of having somewhat stronger moirés with the Y screen (which are anyway much less visible due to the non-prominent nature of the Y colour). As a second benefit, the separation of the Y screen from the others reduces from 6 to 4 the number of dimensions to be scanned in step 2; this significantly improves the performance of the program, particularly in terms of running time.

Let us return now in more detail to step 1, and estimate the number of impulses involved. Since each of the C,M,K screens contributes two perpendicular impulse-combs to the convolution, the total number of impulses in the convolution, if only $N$ impulses (harmonics) are considered from each side of the DC, is $(2N+1)^6$. For $N = 2$ this gives $5^6 = 15625$ impulses, and for $N = 3$ the impulse number becomes as high as $7^6 = 117649$. Fortunately, however, the actual number of impulses to be considered can be significantly reduced by means of three different considerations:

• First, according to our practical experience, an impulse generated by a harmonic higher than 2 very rarely contributes a significant visible moiré.[8] It is therefore more efficient to search for good solutions taking into account only two harmonics, and only then to test each of the solutions for possible higher order moirés.

• Second, since we only deal with real images having no imaginary part, each impulse in the spectrum (except for the DC) is accompanied by a twin impulse of an identical amplitude, which is symmetrically located at the other side of the spectrum origin (see Sec. 2.2). A judicious consideration allows us to leave in the impulse list only one impulse from each such pair, thus reducing by half the number of impulses in the list.

• Third, even after the above reductions in the number of impulses, most of the impulses remaining in the list have no practical significance from the moiré point of view since they represent moirés which are far too weak to be perceived, even when they appear at low frequencies (for instance, such are all the impulses of order 2 having more than two indices with the value of ±2). After discarding all these negligible impulses, we are left with a list of only about 2000 "potentially dangerous" impulses, i.e., impulses which may represent, when they fall inside the visibility circle, significant moirés that should be rejected.

However, as we already know, even these 2000 impulses are not all of the same importance. For example (see Figs. 2.10 and 3.3), a (1,0,-1,0,0,0)-moiré is much stronger than a (1,2,-2,-1,0,0)-moiré, and consequently its cutoff frequency (i.e., the frequency at which it becomes invisible to the eye) is higher. In fact, strong moirés may still cause a rough and clearly visible linen-like structure even at very small periods, where the weaker moirés already disappear completely. Therefore the 2000 impulses in the list can be classified into several categories, each category including moiré families with similar relative moiré strength. Each category is assigned its own visibility circle, according to its cutoff frequency. The cutoff frequency for each type of moiré is found experimentally by measurements performed on a light-table, using various high-resolution films with different screen frequencies, dot shapes and gray levels. This can be done by observing the corresponding screen superpositions from a normal reading distance, and slightly rotating the films with respect to each other so as to decrease the period of the moiré in question (i.e., to increase its frequency) until, at a certain moment, the moiré completely disappears. The highest visible frequency of the moiré, just before its disappearance, is

[8] In fact, this is a consequence of Proposition 2.1 and Eq. (2.27).

measured and is taken as its cutoff frequency (allowing for some extra safety margins). In our specific implementation four impulse categories have been used: one for the strongest moirés (of the {1,1} type), a second one for the slightly weaker moirés of the {1,1,1} type, a third category for the medium-strength moirés such as {1,1,2} and {1,2,2}, and the last category for all the weaker cases still in the list. It is interesting to mention that if no classification of the impulses is carried out and all of them use the visibility circle of the strongest moirés, the forbidden zones they generate in the 4D parameter space become larger and cover the entire space, so that no solutions can be found.

Note that these experimental measurements of the cutoff frequency have the advantage of being performed using real halftone screen superpositions, thus taking into account the irregular background and the masking effects which influence the moiré visibility. This is better than using standard data taken from literature, since the standard data have been measured on ideal sinusoidal images, which are free of noise such as rosettes or textures due to other moirés in the background.

## 3.6 Results and discussion

An important advantage of the above moiré minimization method is that, due to the relative simplicity of the calculations involved, it is capable of scanning the full admitted ranges of the four parameters. This provides a global, panoramic view of the entire 4D parameter space, including the forbidden zones inside it and the moiré-free zones between them. Note that all the large enough moiré-free zones are detected by this scan; moiré-free zones which "escape" through the scanning mesh (the sampling points in the 4D space) are inevitably smaller than the scanning step, which means that their tolerance margins are very low so that they do not correspond to our requirements anyway. Such are, for example, all the singular moirés. Obviously, after an important moiré-free zone is discovered, the program may be used again to rescan the interesting zone alone in greater detail ("zoom in").

The results obtained by this method are quite interesting. Several good screen combinations have been found which in experimental tests give a very satisfactory, uniform appearance with no evident presence of moirés or other disturbing structures (some of these solutions correspond to previously known cases listed in [Schoppmeyer85]). An example of such a good solution is shown in Fig. 3.8. Note that each solution in fact represents a whole family of equivalent screen combinations which can be received from each other by various symmetries and transformations such as mirror-images, rotations, scalings, flipping the colour labels between the three screens, etc. The tolerance obtained around a typical good solution is in the order of at least ±0.5° for the angles $\alpha$ and $\beta$ and ±1% for the screen periods or frequencies (these values are indicated for the narrowest dimension of the moiré-free zone; in its more prolonged directions the tolerances may be even higher). This is significantly better than the tolerance required by singular cases such as the traditionally used screen combination of $\alpha = \beta = 30°$ and

$q_{MK} = q_{CK} = 1$, which is ±3 minutes of arc for the angles (DIN standard No. 16547). And moreover, in our good solutions when the specified tolerances are somewhat exceeded, the visual effect only consists of weak, small-period moirés that just start to appear, whereas in singular moiré-free cases the slightest deviations immediately bring back from infinity large-period, highly objectionable moirés. In other words, our solutions are *stable* moiré-free states rather than *unstable* moiré-free states.

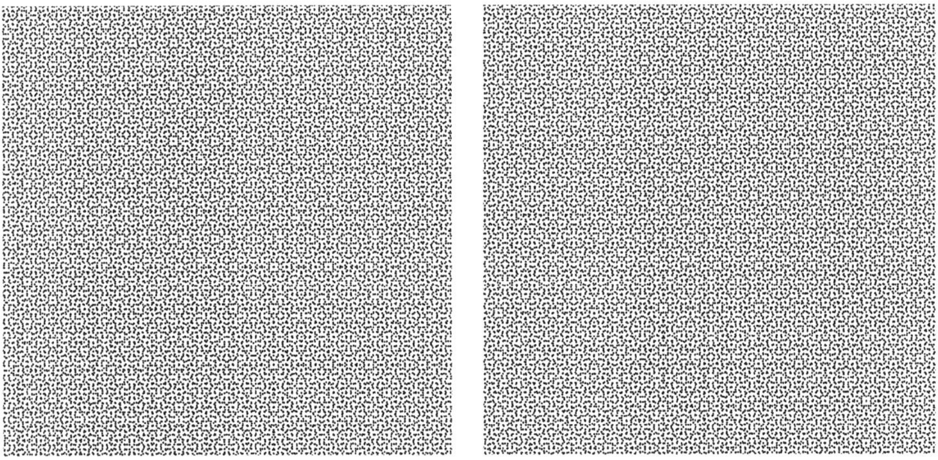

(a) (b)

**Figure 3.8:** (a): An example of a stable moiré-free combination of 3 screens obtained by our algorithm: $\alpha = 23.5°$, $\beta = 24°$, $q_{MK} = 0.833$, $q_{CK} = 0.833$. Small, irregular rosettes, of about the same size as in the conventional 30°-30°-30° singular (unstable) moiré-free case (see Fig. 3.1(a)), can be seen in the superposition, but no significant moirés (such as in Fig. 2.10) are present. (b) A similar moiré-free structure is still maintained slightly off the precise screen combination, within the specified tolerances.

Another interesting observation from our results is that the best solutions in the moiré-free zones, according to our criteria of maximal tolerance, are systematically located on a symmetry axis of the 4D parameter space, such as $\alpha = \beta$, $q_{MK} = q_{CK}$ (or any of its symmetric counterparts, like $\beta = \gamma$, $q_{CK} = q_{CM}$ etc.). But, quite interestingly, these highly symmetric cases were not necessarily the winners during the experimental tests with real film superpositions, and often the cases found most suitable for practical use were slightly *off* the symmetry axes. The reason for this result appears to be the following: in highly symmetrical cases the dots of the three screens form together in the superposition a fine uniform geometrically ordered microstructure which looks to the eye very regular. In such cases even very weak moirés (including moirés arising from higher harmonic orders), which are normally completely negligible on the irregular, noisy background of the screen superposition, may become quite disturbing. On a uniform, geometrically ordered background even the slightest perturbation is very clearly perceived, because the human eye is very sensitive to periodic irregularities on a uniform background.

Therefore, even if the most promising cases obtained by the program are located on a symmetry axis, it might be advisable to prefer a neighbouring case slightly off that axis, in order to break the symmetry of the microstructure generated in the superposition and to make it more disordered.

Finally, as a concluding remark, it may be interesting to mention that in spite of the remarkable advantages of stable moiré-free screen combinations, they have rarely been used in practice in the printing industry. In fact, recent progress in modern colour printing technology (higher precision equipment, stochastic or pseudo-random screening, etc. [Rodriguez94; Widmer92]) reduced their potential importance in this field. But even so, independently of any possible applications, the question of moiré minimization remains an interesting subject in its own right, as the present chapter may have shown.

## PROBLEMS

3-1. *Reducing sampling moirés by high-frequency sampling.* A classical method for reducing sampling moirés when scanning a halftoned image is to increase the sampling rate (the scanning resolution) to at least twice the halftone frequency, and if possible, even beyond that, up to several samples per halftone period [Morimoto90 p. 101; Huang74]. Explain how this method works, and what are its practical limitations.

3-2. *Reducing sampling moirés by pre-filtering.* A standard method for reducing sampling moirés when scanning a halftoned image is to low-pass filter the image before or during sampling [Shu89 p. 807; Morimoto90 p. 101]. Explain how this can be done, and how this method works. *Hint*: The halftoned image can be blurred, scanned out-of-focus (for example by lifting the image above the scanner's glass), or scanned with a scanning aperture larger than the halftone period. (Note that the scanner's sensor averages light coming throughout the aperture and converts it into a single gray-level value between 0–255; this averaging is, in fact, a low-pass filtering process.) What happens if we low-pass filter the image *after* sampling?

3-3. *Inverse halftoning.* Inverse halftoning (also known as *unscreening* or *descreening*) is the process of reconstructing a continuous-tone image from a given halftoned image (see, for example, [Miceli92; Stevenson97; Luo98]). Inverse halftoning is useful for converting between halftone techniques, or to allow the application of various image processing operations which can only be implemented on continuous-tone images, when only the halftoned image is available. Can inverse halftoning be applied to avoid sampling moirés when the image to be scanned is halftoned?

3-4. *Avoiding sampling moirés by random sampling.* It is well known that sampling moiré artifacts can be avoided by random sampling (see, for example, [Ahumada83]).

However, because of various implementation difficulties this method is rarely used. Explain the practical problems in the implementation of this method; can you think of ways to overcome them?

3-5. *Reducing sampling moirés by dual sampling.* Propose a method for reducing sampling moirés that is based on scanning of the halftoned image twice, at different angles $\theta_1$ and $\theta_2$. (*Hint*: Note that between the two scans the image information will simply undergo a rotation of $\theta_2-\theta_1$, while moiré patterns will be significantly altered — scaled, rotated or shifted). Is this a practical way for reducing sampling moirés? Explain. (A similar method has been proposed in [Ohyama86].)

3-6. *Removing existing moirés by post-filtering.* Suppose that you are given an already scanned image which contains an undesired moiré pattern due to uncareful scanning. Propose a filtering method for removing the unwanted moiré. Will a smoothing operation help to remove the moiré?

3-7. *Removing existing moirés by direct manipulation of the spectrum.* Removing of existing moirés (or any other periodic noise) can be also done in the spectral domain. Suppose that you are given an already scanned image which contains an undesired periodic moiré pattern. The FFT of the scanned image contains, therefore, in addition to the information belonging to the original image, a number of strong impulses that represent the periodic moiré effect. These impulses can be wiped off from the FFT by means of any image retouching utility; then, by applying an inverse FFT, one obtains a "clean" version of the image with no pronounced moiré patterns [Russ95 pp. 319–327]. What are the advantages and the shortcomings of this method? (Note that this filtering method attacks directly the moiré patterns, whereas low-pass filtering methods attack the halftone screen of the scanned image.)

3-8. *Minimization of sampling moirés.* How can you use Fig. 3.2 to minimize the sampling moiré when scanning a monochrome halftoned image (assuming that you know the scanner's frequency as well as the halftone angle and frequency)? Is this a practical way for reducing sampling moirés? Will it also work when scanning a colour halftoned image? Explain.

3-9. Compare the advantages and drawbacks of the different anti-moiré methods proposed in the previous problems. (See, for example, [Shu89].)

3-10. Often, the combination of several moiré reducing methods gives better results than any of the methods alone. For example, an often recommended strategy for reducing sampling moirés which appear when scanning a halftoned image consists of the following steps [Matteson04 Secs. 7.2 and 7.3.2]: Rescan the image at twice the resolution; apply a Gaussian blur filter; downsample the image to the intended resolution; and finally, sharpen the image by applying a sharpening filter. A similar combined strategy is often built-in within the scanner's own software, and invoked when the user specifies that a halftoned document is to be scanned; this scanning option is often called *descreening* [Scantips97]. How does this combined method improve on each of the individual methods?

3-11. *Undesired moirés in image reproduction.* Sec. C.13 in Appendix C reviews the different types of unwanted moirés which may occur in image reproduction and their various causes. Suggest for each of these moiré types a practical and well-adapted strategy for fighting it.

3-12. *Using the same angle and frequency for the CMYK screens in colour printing.* It might be thought that the best way to avoid moiré patterns between the CMYK halftone screens in colour printing would be to print all of them at exactly the same angle. Explain why this solution is normally not satisfactory. (See also Problem 9-7.)

3-13. Show that the $(k_1,k_2,k_3,k_4)$-moiré between two regular line-grids or dot-screens is singular when the angle difference and the frequency ratio between them are given by:

$$\alpha = \alpha_1 + \alpha_2 = \arctan\frac{k_2}{k_1} + \arctan\frac{k_4}{k_3}$$

$$q = \frac{f_2}{f_1} = \sqrt{\frac{k_1^2 + k_2^2}{k_3^2 + k_4^2}}$$

(see Fig. 3.9). Using these formulas verify the $(\alpha,q)$ coordinates of the singular points of the various $(k_1,k_2,k_3,k_4)$-moirés shown in Fig. 3.2. Write a computer program that gives the parameters $(\alpha,q)$ for all the $(k_1,k_2,k_3,k_4)$-singular superpositions of two regular grids or screens up to the second order, i.e., with $k_i \in \{-2,-1,0,1,2\}$; compare the results of your program with Fig. 3.2. *Tip*: Make sure that the signs of the $k_i$ parameters are correctly taken care of by the arctan function that you use.

3-14. Using the computer program of the previous problem find the $(\alpha,q)$ coordinates of all the singular points of the third order in the superposition of two regular grids or screens, and add them on top of Fig. 3.3. Many of these points have the unfortunate property of being located precisely in the centers of the moiré-free zones up to order 2. Can you explain why?

3-15. As we can see in Fig. 3.2, the (1,1,-1,0)- and (2,0,-1,1)-moirés have a common singular point. How can you explain this? *Hint*: Does the (2,0,-1,1)-impulse really represent an independent moiré (or a "valid" moiré, using the terms of Sec. 2.8)? Note that (2,0,-1,1) is in fact a higher harmonic impulse in the cluster that is spanned by the impulse (1,1,-1,0) and its orthogonal counterpart (1,-1,0,1). Is it true that a $(k_1,k_2,k_3,k_4)$-impulse and an $(l_1,l_2,l_3,l_4)$-impulse belong to two independent moirés *iff* they have different singular points $(\alpha,q)$?

3-16. *Extensions of Fig. 3.2.* Will the extension of Fig. 3.2 beyond $\alpha = 45°$ or below $q = 1$ offer essentially new information? Which extensions of Fig. 3.2 would you expect to be the most interesting?

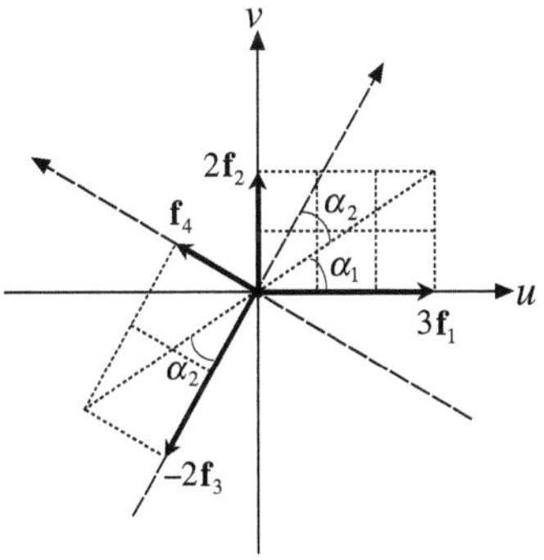


**Figure 3.9:** A vector diagram showing the geometric configuration in the spectral domain when the $(k_1,k_2,k_3,k_4)$-moiré between two regular grids or screens reaches its singular state. The angle and the frequency of layer A are $\theta_1 = 0$ and $f_1 = |\mathbf{f}_1|$, and those of layer B are $\theta_2 = \alpha_1+\alpha_2$ and $f_2 = |\mathbf{f}_2|$. The dashed, rotated axes belong to layer B. The particular case shown in this figure corresponds to the singular state of the third-order (3,2,-2,1)-moiré: $3\mathbf{f}_1 + 2\mathbf{f}_2 - 2\mathbf{f}_3 + \mathbf{f}_4 = \mathbf{0}$.

3-17. *A mysterious singular moiré?* While running through the 4D parameter space $(\alpha,\beta,q_{MK},q_{CK})$ representing the superposition of three dot-screens, a singular 3-layers moiré has been discovered at the point $\alpha = 43°$, $\beta = 23.5°$, $q_{MK} = 1$, $q_{CK} = 0.797$. Analyze this singular moiré and find its $(k_1,...,k_6)$-notation. *Hint*: This is a {1,1,1}-moiré which is located on the same singular manifold as the singular moiré $\alpha = \beta = 30°$, $q_{MK} = q_{CK} = 1$.

3-18. *Moiré-free screen combinations.* Searching for moiré-free screen combinations is most easily performed in the spectral domain, by searching for angle and frequency combinations of nailbeds (the spectra of the dot-screens) where no strong impulses fall inside the visibility circle of the nailbed convolution. Why isn't it a straightforward task to find moiré-free screen combinations using a simple computer program that runs through all the permitted angle and frequency combinations of the screens, calculates the impulse locations in the spectrum convolution, and keeps only those combinations that satisfy the above condition?

3-19. *Moiré-free combinations of non-regular screens.* How would the search for stable moiré-free screen combinations be affected by allowing for non-regular screens with non-orthogonal axes and with different frequencies along these axes? How many dimensions will we have now in our parameter space in the 2-screen case and in the 3-screen case? Would you expect this relaxation of constraints to yield new stable moiré-free combinations that are significantly better than the solutions found for regular screens? Explain.

3-20. *4D-space intuition: 2D sections through the 4D space.* Since the parameter space in the case of three superposed screens is 4-dimensional, a full graphic representation of this space is no longer possible, and we content ourselves with various 2D sections through this space, like in Fig. 3.6. It should be noted, however, that our intuition, which is based on our experience in the 3D world, may be misleading when contemplating the 4D space. As an example, it is perfectly possible for two 2D planes within $\mathbb{R}^4$ to intersect at a single point: Consider the 2D planes A = $\{(x,y,z,w) \mid x = x_0, y = y_0\}$ and B = $\{(x,y,z,w) \mid z = z_0, w = w_0\}$; these planes clearly have a single intersection point in $\mathbb{R}^4$, which is the point $(x_0,y_0,z_0,w_0)$. How is 2D plane A represented in the 2D section through $\mathbb{R}^4$ which is defined by the $x$ and $y$ axes?

3-21. A friend suggests to you to search for 3-screen moiré-free combinations by finding, using Fig. 3.2, simultaneous moiré-free combinations of the screen pairs CK, MK and CM. How would you respond to this suggestion? (*Hint*: Can this method cover all the possible moirés between three screens?)

3-22. An earlier attempt of moiré minimization in a 3-screen superposition [Foster72] consisted of calculating using Eq. (2.9), for each angle and period combination of the screens, the periods of: (a) all the possible "primary moirés" (the moirés generated between dot-rows of two superposed screens); (b) all the "secondary moirés" (the moirés generated between one screen and the diagonals of another screen); and (c) all the "moirés of moirés" which are generated between a primary (or a secondary) moiré and the dot-rows of a screen, between two primary moirés, and between a secondary and a primary moiré. Explain how all these moirés are automatically taken care of by our approach in the spectral domain.

3-23. *Line-screens.* Suppose that the CMYK colour separations for a given image are to be printed using line-screens rather than dot-screens (i.e., using for each of the CMYK layers a line-grating whose line widths are modulated at each point by the respective colour intensity). Which angle and frequency combination would you recommend for the four line-screens?

# Chapter 4

# The moiré profile form and intensity levels

## 4.1 Introduction

We have seen in Chapter 2 that the moiré patterns obtained in the superposition of periodic structures can be described at two different levels. The first, basic level is only concerned with geometric properties within the $x,y$ plane, such as the *periods* and *angles* of the original images and of their moiré patterns. The second level also takes into account the *amplitude* properties, which can be seen as a surface $z = g(x,y)$ on top of the planar 2D description of each of the original structures or their moiré patterns, showing their intensity levels.[1] This 3D description of the moiré is called its *intensity profile* or *intensity surface*. Note that the term *intensity* refers here to the reflectance (or transmittance) values ranging along the $z$ axis between 0 and 1 (see Sec. 2.2). The term *profile* originates from the simple moiré bands that occur in the superposition of two line-gratings, in which a 1D plot (a side view or profile) is enough to describe the intensity levels of the moiré-bands (see Fig. 2.9). However, in more complex cases where the moiré patterns are no longer simple bands, the term *moiré intensity surface* is more appropriate for describing the shape and the intensity variations of the moiré pattern. In the discussions which follow we will use all of these terms interchangeably as synonyms.

In the present chapter we concentrate on the analysis of the profile forms and the intensity levels of moiré patterns which are obtained in the superposition of periodic layers; as shown in Fig. 4.1, such superpositions may result in quite spectacular moiré patterns. Based on the Fourier approach, we will show how the intensity profile of each moiré can be derived analytically from the original superposed structures, either in the spectral domain or in the image domain. This will permit us to determine *quantitatively* the intensity levels of each superposition moiré, in addition to its *qualitative* geometric properties. Moreover, this approach will also allow us to *synthesize* moiré effects with any desired period and intensity profile, by properly designing the layers to be superposed.

We start, in Sec. 4.2, with the simple case of line-grating superpositions. Then, in Sec. 4.3, we generalize the discussion to the superposition of 2-fold periodic structures such as dot-screens. These two sections show, in particular, how the duality between the image and the spectral domains is extended to moiré profiles as well. Then, in Secs. 4.4–4.5 we show how the spectral approach allows us to fully explain and predict the surprising profile forms of the moiré patterns which occur in the superposition of screens with any desired dot shapes, including in cases of higher-order moirés. A further discussion, including an example with hexagonal screens, is provided in Sec. C.15 of Appendix C.

---

[1] In terms of the spectral domain, the first level considerations only take into account impulse locations (or frequency vectors) within the $u,v$ plane, while the second level considerations also take into account the impulse amplitudes.

## 4.2 Extraction of the profile of a moiré between superposed line-gratings

Assume that we are given two line-gratings (see Fig. 2.5); for the sake of simplicity we may assume that both of them are centered with respect to the origin, so that their spectra are purely real. According to Eqs. (2.19), (2.20) we have, therefore:

$$\begin{aligned} r_1(x,y) &= \sum_{n=-\infty}^{\infty} a^{(1)}_n \cos(2\pi n[x\cos\theta_1 + y\sin\theta_1]/T_1) \\ r_2(x,y) &= \sum_{n=-\infty}^{\infty} a^{(2)}_n \cos(2\pi n[x\cos\theta_2 + y\sin\theta_2]/T_2) \end{aligned} \tag{4.1}$$

where $\theta_i$ and $T_i$ are the angle and the period of the respective gratings.

As we have seen in Chapter 2, the spectrum of each of these line-gratings consists of an infinite impulse-comb, in which the amplitude of the $n$-th impulse is given by the coefficient of the $n$-harmonic term in the Fourier series development of that line-grating:

$$\begin{aligned} R_1(u,v) &= \sum_{n=-\infty}^{\infty} a^{(1)}_n \, \delta_{n\mathbf{f}_1}(u,v) \\ R_2(u,v) &= \sum_{n=-\infty}^{\infty} a^{(2)}_n \, \delta_{n\mathbf{f}_2}(u,v) \end{aligned} \tag{4.2}$$

Here, $\delta_{n\mathbf{f}_i}(u,v)$ denotes an impulse located in the spectrum at the frequency-vector $n\mathbf{f}_i = (n/T_i,\ \theta_i)$ (in terms of polar coordinates), or in other words: $\delta_{n\mathbf{f}_i}(u,v) = \delta(u-nu_i, v-nv_i)$, $u_i$ and $v_i$ being the Cartesian coordinates of the frequency-vector $\mathbf{f}_i$.

When we superpose (i.e., multiply) the line-gratings $r_1(x,y)$ and $r_2(x,y)$, the spectrum of the superposition is, according to the convolution theorem, the convolution of the two original combs, $R_1(u,v)**R_2(u,v)$, which gives an oblique nailbed of impulses (see Fig. 2.5(f)). Each moiré which appears in the grating superposition is represented in the spectrum of the superposition by a comb of impulses centered on the origin which is included in the aforementioned nailbed. If a moiré is visible in the superposition, it means that in the spectral domain the fundamental impulse-pair of the moiré comb is located inside the visibility circle, close to the spectrum origin; this impulse-pair determines the period and the direction of the moiré. Now, by extracting from the spectrum-convolution only this infinite moiré comb and taking its inverse Fourier transform, we can reconstruct, back in the image domain, the isolated contribution of the moiré in question to the image superposition; this is the intensity profile (or profile form) of the moiré (see Fig. 4.2). In the present section we will see in detail how the profile of any moiré between superposed line-gratings can be extracted from the spectral domain, or, even more interestingly, directly from the image domain without having to resort to the spectral domain.

Let us denote by $d_n$ the amplitude of the $n$-th impulse of the moiré-comb. If the moiré in question is a $(k_1,k_2)$-moiré, the fundamental impulse of its comb is the $(k_1,k_2)$-impulse in the spectrum-convolution, and the $n$-th impulse of its comb is the $(nk_1,nk_2)$-impulse in the spectrum-convolution. Its amplitude $d_n$ is given by:

$$d_n = a_{nk_1,\, nk_2}$$

(a)

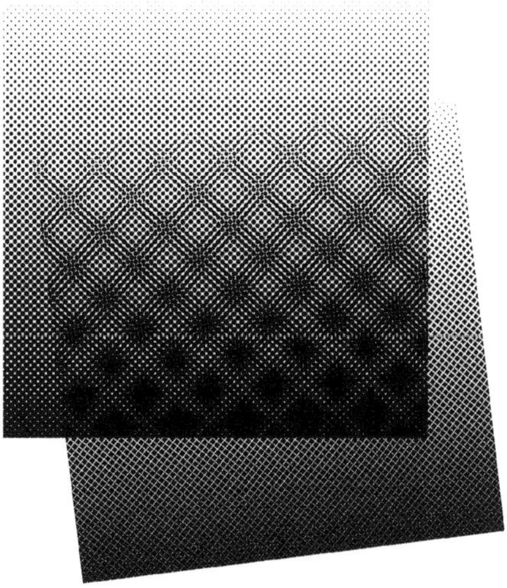

(b)

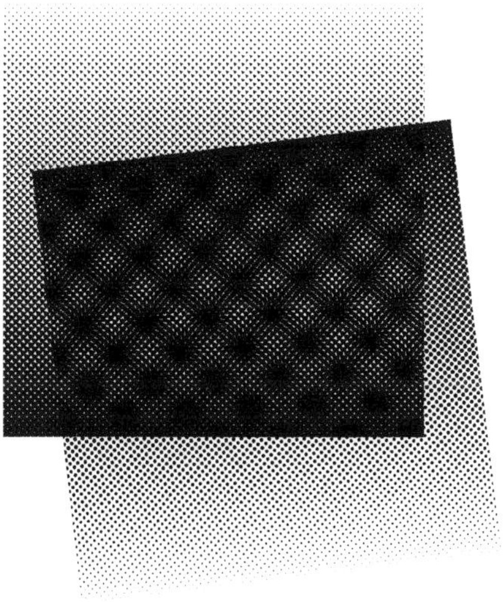

(c)

(d)

**Figure 4.1:** The superposition of dot-screens may yield moiré effects with spectacular profile forms. In all the cases (a)–(d), two binary dot-screens with identical frequencies and gradually increasing dots are superposed with the same angle difference of 7°; this implies that in all of the cases the moiré in question is a (1,0,-1,0)-moiré. (a) Two screens with black circular dots; (b) top screen with black circular dots and bottom screen with black square dots; (c) top screen with black triangular dots and bottom screen with black circular dots; (d) top screen with black square dots and bottom screen with black circular dots. As we can see, the shape and the size of the screen dots only affect the shape and the intensity levels of the moiré profile; but the period and the direction of the moiré remain unchanged (unless the angles and frequencies of the superposed screens are modified).

and according to Eq. (2.27) of Sec. 2.7 we obtain:

$$d_n = a^{(1)}_{nk_1} a^{(2)}_{nk_2}$$

where $a^{(1)}_i$ and $a^{(2)}_i$ are the respective impulse amplitudes from the combs of the first and of the second line-gratings. In other words, we can say:

**Proposition 4.1:** The impulse amplitudes of the moiré comb in the spectrum-convolution are obtained by a simple term-by-term multiplication of the combs of the original superposed gratings (or subcombs thereof, in case of higher order moirés). ■

The $(k_1,k_2)$-moiré extracted from the 2-grating superposition is given, therefore, by the Fourier series:

$$m_{k_1,k_2}(x,y) = \sum_{n=-\infty}^{\infty} d_n \cos(2\pi n[x\cos\varphi_M + y\sin\varphi_M]/T_M) \tag{4.3}$$

where $d_n = a^{(1)}_{nk_1} a^{(2)}_{nk_2}$, and $\varphi_M$ and $T_M$ are the angle and the period of the moiré (Eq. (2.9)).[2]

The spectral representation of this isolated moiré is:

$$M_{k_1,k_2}(u,v) = \sum_{n=-\infty}^{\infty} d_n \, \delta_{n\mathbf{f}_M}(u,v) \tag{4.4}$$

where $\delta_{n\mathbf{f}_M}(u,v)$ denotes an impulse located in the spectrum at the frequency-vector $n\mathbf{f}_M = (n/T_M, \theta_M)$ (in terms of polar coordinates), or in other words: $\delta_{n\mathbf{f}_M}(u,v) = \delta(u-nu_M, v-nv_M)$, $u_M$ and $v_M$ being the Cartesian coordinates of the moiré frequency-vector $\mathbf{f}_M$.

For example, in the case of a (1,-1)-moiré (as in Fig. 2.5(f)) the amplitudes of the moiré-comb impulses are given by $d_n = a_{n,-n} = a^{(1)}_n a^{(2)}_{-n}$ and in the case of the second-order (1,-2)-moiré (see Fig. 2.6) the impulse amplitudes of the moiré comb are given by $d_n = a_{n,-2n} = a^{(1)}_n a^{(2)}_{-2n}$. In each of these cases we know also the exact *locations* of the impulses of the moiré comb, according to Eq. (2.26), and in particular we know the moiré angle $\theta_M$ and the moiré period $T_M$ (see Eq. (2.8)). This means that the spectrum of the isolated moiré in question is fully determined. In order to find, back in the image domain, the intensity profile of this moiré, we can take the inverse Fourier transform of the isolated moiré-comb (Eq. (4.4)). Practically, this can be done either by interpreting the moiré comb as a Fourier series development (Eq. (4.3)), and reconstructing the moiré-profile it represents back in the image domain by summing up the corresponding cosinusoidal functions up to the desired precision; or, more efficiently, by approximating the continuous inverse Fourier transform of the isolated moiré comb by means of the inverse discrete Fourier transform (using FFT).

Proposition 4.1 was already obtained by Patorski *et al.* [Patorski76 pp. 444–446], who also realized that in the particular case of two rectangular line-gratings the product comb (the comb of the moiré) is in fact the Fourier series development of a trapezoidal or

[2] The most general form of the Fourier series development of the $(k_1,k_2)$-moiré, which incorporates the explicit values of $\theta_M$ and $T_M$ and also covers the case of non-symmetric or shifted gratings, will be given in a concise and elegant way in Chapter 6 (see Eq. (6.3)). This will become possible after having introduced the algebraic notations of Chapter 5 and their Fourier interpretation in Chapter 6.

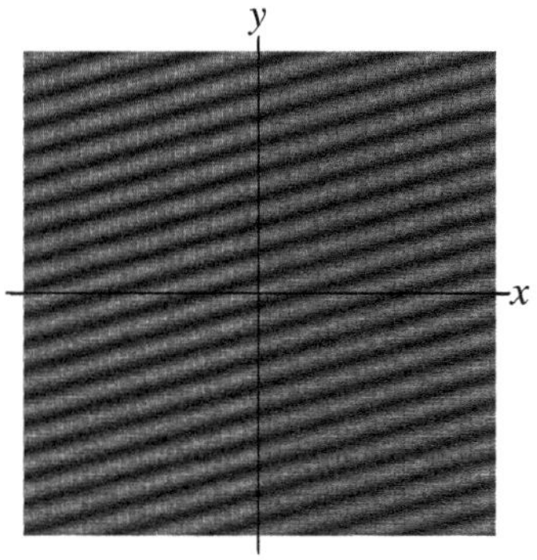


(a)

(*continued from* Fig. 2.5)

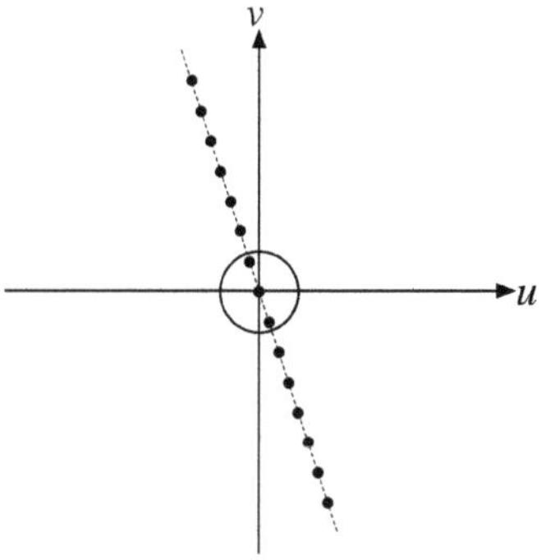


(b)

**Figure 4.2:** Extraction of the (1,-1)-moiré of Fig. 2.5. (b) shows the isolated comb of the (1,-1)-moiré after its extraction from the full spectrum of Fig. 2.5(f). The impulse amplitudes of this comb are the term-by-term products of the respective impulse amplitudes from the combs of Figs. 2.5(d) and 2.5(e). (a) shows the image domain function which corresponds to the spectrum (b). This is the intensity profile of the (1,-1)-moiré shown in Fig. 2.5(c); its crests are triangular or trapezoidal, as shown in Figs. 2.9. Note that although the moiré is visible both in the grating superposition (Fig. 2.5(c)) and in the extracted moiré-profile (Fig. 4.2(a)), the latter does not contain the fine structure of the original gratings but only the isolated form of the extracted moiré, i.e., its isolated contribution to the superposition.

triangular periodic wave. This explains, back in the image domain, the trapezoidal or triangular profile shape of the moiré between two binary gratings (see Figs. 2.9(a)–(f), or Fig. 3 in [Patorski76]). However, this term-by-term multiplication of the original combs (i.e., the term-by-term product of the Fourier series of the two original gratings) can be interpreted in a more general way using the following theorem, which is the equivalent of the convolution theorem in the case of periodic functions [Zygmund68 p. 36; Champeney87 p. 166]; the full importance of this theorem will become clear later in this chapter.

**$T$-convolution theorem:** Let $p_1(x)$ and $p_2(x)$ be periodic functions of period $T$ integrable on a one-period interval $(0,T)$, and let $\{a^{(1)}_n\}$ and $\{a^{(2)}_n\}$ (with $n = 0, \pm1, \pm2, ...$) be their Fourier series coefficients. Then the function:

$$h(x) = \frac{1}{T}\int_T p_1(x-x')\, p_2(x')\, dx' \tag{4.5}$$

(where $\int_T$ means integration over a one-period interval), which is called the *T-convolution* of $p_1$ and $p_2$ and denoted by $p_1 * p_2$, is also periodic with the same period $T$ and has Fourier series coefficients $\{a_n\}$ given by: $a_n = a^{(1)}_n a^{(2)}_n$ for all integers $n$.[3] ■

The $T$-convolution theorem can be rephrased, in a less rigorous but more illustrative way, as follows: If the spectrum of $p_1(x)$ is a comb with fundamental frequency of $1/T$ and impulse amplitudes $\{a^{(1)}_n\}$, and the spectrum of $p_2(x)$ is a comb with the same fundamental frequency and impulse amplitudes $\{a^{(2)}_n\}$, then the spectrum of the $T$-convolution $p_1 * p_2$ is a comb with the same fundamental frequency and with impulse amplitudes of $\{a^{(1)}_n a^{(2)}_n\}$. In other words, the spectrum of the $T$-convolution of the two periodic images is the product of the combs in their respective spectra.

We would now like to apply this theorem to the case where $p_1(x)$ and $p_2(x)$ are our two given line-gratings. Using this theorem, the fact that the comb of the (1,-1)-moiré in the spectral domain is the term-by-term product of the combs of the two original gratings (Proposition 4.1) could be interpreted, back in the image domain, as follows:

The profile of the (1,-1)-moiré generated in the superposition of two line-gratings with identical periods $T$ is the *T-convolution* of the two original line-gratings.[4]

However, there still remains here a certain difficulty. The $T$-convolution theorem requires that $p_1(x)$ and $p_2(x)$ have the same period $T$, and moreover, the resulting $T$-convolution $p_1 * p_2$ also has that same period. This requirement is necessary for the definition of the integral (4.5); or equivalently, from the spectral-domain point of view, this requirement is necessary since the comb multiplication in the spectrum is only meaningful

[3] $T$-convolution (also called *cyclic convolution*) is the periodic analog of the normal convolution with integration limits of $(-\infty,\infty)$. Note that normal convolution cannot be used in the case of periodic functions [see Gaskill78 pp. 157–158]. In general, the normal convolution of a single period of $p_1$ with a single period of $p_2$ is *not* equal to a single period of the $T$-convolution $p_1 * p_2$. Such an equality only occurs in cases in which the normal convolution of the two single periods is not longer than the period $T$; otherwise the outer ends which exceed the boundaries of each convolution period $T$ inevitably penetrate (additively) into the neighbouring periods in the $T$-convolution, thus generating a cyclic wrap-around effect which does not exist in the case of normal convolution. The discrete counterpart of the cyclic convolution is widely used in the discrete Fourier transform theory [Bracewell86 p. 362].

[4] In fact, in the case of (1,-1)-moiré it may be more appropriate to use the term *T-cross-correlation* of $p_1(x)$ and $p_2(x)$, which is defined, following [Gaskill78 p. 172], as $p_1(x) \star p_2(x) = p_1(x) * p_2(-x)$. The reason is that in the case of (1,-1)-moiré we have $d_n = a^{(1)}_n a^{(2)}_{-n}$, which means that the second comb in the term-by-term product is reflected about the origin, and therefore represents in the image domain the reflected image $p_2(-x)$; the resulting moiré-profile is therefore the $T$-cross-correlation of $p_1(x)$ and $p_2(x)$. However, for the sake of consistency in the general case of the $(k_1,...,k_m)$-moiré, where some of the indices are positive and others are negative, we prefer to stick to the terminology of *T-convolution*, understanding that for any negative index in the list the image it represents must be reflected. In the common case where the original images are symmetric about the origin, the two terms coincide.

if the two combs have a common support (i.e., if their impulse locations in the spectrum coincide). However, in line-grating superpositions the original gratings may, of course, have different periods, and moreover, the resulting moiré normally has yet a different period, $T_M$, which is given by Eq. (2.9). What happens then when $p_1(x)$, $p_2(x)$ and $p_1{*}p_2$ have different periods, $T_1$, $T_2$ and $T_M$?

From the spectral-domain point of view this difficulty is settled thanks to the complete independence between the impulse locations and the impulse amplitudes, as formulated by Proposition 2.2 and Eqs. (2.26) and (2.27). The term-by-term multiplication of the combs in the spectrum only yields the *impulse amplitudes* of the resulting moiré-comb, but their actual *geometric locations* in the spectrum are determined, independently of the impulse amplitudes, by the frequencies and the angles of the superposed layers (i.e., by Eq. (2.26), or by its special case, Eq. (2.9)).

This difficulty can be also settled, in a more formal way, directly in the image domain, by the addition of a preliminary stage before the application of the $T$-convolution theorem. Before applying the theorem, the two original gratings must be *normalized*, i.e., stretched (and in the 2D case also rotated) in order that their periods coincide (or equivalently, in terms of the spectral domain, in order that their two combs have a common support). According to well known results in the Fourier theory (see Sec. C.3 in Appendix C) stretching and rotation of the original gratings do not affect the Fourier coefficients (impulse amplitudes) of their combs, but only their impulse locations in the spectrum. Therefore, according to Proposition 4.1, the amplitudes of the moiré comb are not affected by the normalization, either. This normalization therefore allows the theorem to be applied even to line-gratings with periods $T_1 \neq T_2$. Moreover, by selecting the new common period and angle of the two normalized gratings to coincide with the period and angle of the moiré, as determined by Eqs. (2.26) or (2.9), the resulting $T$-convolution obtained by the theorem will fit the actual period and direction of the moiré. We can summarize the above discussion as follows (as an illustration, refer to Figs. 2.5 and 4.2):

**Proposition 4.2:** The profile of the (1,-1)-moiré that is generated in the superposition of two line-gratings with periods $T_1$ and $T_2$ and an angle difference $\alpha$ can be seen from the image-domain point of view as the result of a 2-stage process:
(1) Normalization of the original gratings (by linear stretching- and rotation-transformations) in order to bring each of them to the period and the direction of the moiré.
(2) $T$-convolution of the normalized line-gratings. (This can be done by multiplying their combs in the spectrum and taking the inverse Fourier transform of the product.) ■

Thus, while the period and the orientation of the (1,-1)-moiré bands are determined by Eq. (2.9), their intensity profile is governed by Proposition 4.2. Note that in the particular case where $T_1 = T_2$ and $\theta_1 \approx \theta_2$ the (1,-1)-moiré bands are approximately perpendicular to the original gratings, and their orientation is given by Eq. (2.10).

It is interesting to note that Proposition 4.2, for the special case of $T_1 = T_2$, has been obtained by Harthong [Harthong81 pp. 30–33] using the theory of non-standard analysis.

This proposition can be further generalized to also cover higher-order moirés (as an illustration, refer to the second-order (1,-2)-moiré shown in Fig. 2.6):

**Proposition 4.3:** The profile of the general $(k_1,k_2)$-moiré that is generated in the superposition of two line-gratings with periods $T_1$ and $T_2$ and an angle difference $\alpha$ can be seen from the image-domain point of view as a normalized $T$-convolution of the images belonging to the $k_1$-subcomb of the first grating and to the $k_2$-subcomb of the second grating. In more detail, this can be seen as a 3-stage process:

(1) Extracting the $k_1$-subcomb (i.e., the partial comb which contains only every $k_1$-th impulse) from the comb of the first original line-grating, and similarly, extracting the $k_2$-subcomb from the comb of the second original grating.

(2) Normalization of the two subcombs by linear stretching- and rotation-transformations in order to bring each of them to the period and the direction of the moiré, as they are determined by Eq. (2.26).

(3) $T$-convolution of the *images belonging to the two normalized subcombs.* (This can be done by multiplying the normalized subcombs in the spectrum and taking the inverse Fourier transform of the product.) ■

A more rigorous formulation of Propositions 4.2 and 4.3 will be given in Sec. 10.9.1 using the general exponential Fourier series formulation. The present results will also be extended there to the more general case of moirés between curvilinear gratings.

As for the perceptual contrast of the moiré intensity profiles thus obtained, refer to Sec. 2.10 and to Fig. 2.9 there. In particular, in order to graphically represent a moiré intensity profile which has been calculated mathematically as explained above, one should remember to use a logarithmic intensity presentation, which better approximates the way in which the moiré is actually perceived by the human eye.

In conclusion, we see that thanks to the $T$-convolution theorem the duality between the image and the spectral domains is further extended to include the moiré profiles as well. This enables us to present the extraction of the moiré-profile between two gratings in either of the two domains. From the *spectral* point of view, the profile of any $(k_1,k_2)$-moiré between two superposed (= multiplied) gratings is obtained by extracting from their spectrum-convolution only those impulses which belong to the $(k_1,k_2)$-moiré comb, thus reconstructing back in the image domain only the isolated contribution of this moiré to the image of the superposition. On the other hand, from the point of view of the *image* domain, the profile form of any $(k_1,k_2)$-moiré between two superposed gratings is a normalized $T$-convolution of the images belonging to the $k_1$-subcomb of the first grating and to the $k_2$-subcomb of the second grating.

The importance of the image-domain interpretation of the moiré-profile as a $T$-convolution is not in the actual calculation of the profile, which can be done much more efficiently in the spectral domain (as a term-by-term multiplication followed by an inverse Fourier transform). But as we will see later in this chapter, this image-domain interpretation of the moiré profile sheds a new light on the understanding of the moirés and their profile forms: In essence, it allows us to express the profile form of the resulting moiré effect *in terms of the profile forms of the two original layers.* An interesting variant in which the intensity profile of the (1,-1)-moiré bands carries 2D information is provided in Sec. C.14 of Appendix C.

**Remark 4.1:** In order to avoid any risk of confusion it should be noted that we are dealing here with *two* different convolutions which take place in parallel. First, we have the convolution of the individual combs in the *spectral domain*, which gives the spectrum of the superposition of the original gratings. This convolution corresponds to the multiplication of the original gratings in the image domain. And second, we have the (normalized) $T$-convolution of the original gratings in the *image domain*, which gives the profile of their first-order moiré. This $T$-convolution corresponds in the spectral domain to the term-by-term multiplication of the combs of the original gratings. ■

## 4.3 Extension of the moiré extraction to the 2D case of superposed screens

We have seen in the previous section how the intensity profile of a moiré in the superposition of two line-gratings can be extracted either from the spectrum, or directly from the superposed images. How can this process be generalized to the superposition of 2-fold periodic images such as dot-screens, where the moiré patterns in the superposition are really of a 2D nature (i.e., moiré cells rather than moiré bands)?

Let $r_1(x,y)$ be a 2-fold periodic image (for the sake of simplicity we assume that $r_1(x,y)$ is periodic in two orthogonal directions, $\theta_1$ and $\theta_1 + 90°$, with an identical period $T_1$ in both directions). Its spectrum $R_1(u,v)$ is a nailbed whose impulses are located in the $u,v$ plane on a regular lattice $L_1(u,v)$, rotated by the same angle $\theta_1$ and with period of $1/T_1$; the amplitude of a general $(k_1,k_2)$-impulse in this nailbed is given by the coefficient of the $(k_1,k_2)$-harmonic term in the 2D Fourier series development of the function $r_1(x,y)$.[5]

The lattice $L_1(u,v)$ can be seen as the 2D *support* of the nailbed $R_1(u,v)$ on the plane of the spectrum, i.e., the set of all the nailbed impulse-locations. Its unit points (0,1) and (1,0) are situated in the spectrum at the geometric locations of the two perpendicular fundamental impulses of the nailbed $R_1(u,v)$, whose frequency vectors are $\mathbf{f}_1$ and $\mathbf{f}_2$. Therefore, the location $\mathbf{w}_1$ in the $u,v$ plane of a general point $(k_1,k_2)$ of this lattice is given by a linear combination of $\mathbf{f}_1$ and $\mathbf{f}_2$ with the integer coefficients $k_1$ and $k_2$; and the location $\mathbf{w}_2$ of the perpendicular point $(-k_2,k_1)$ can be also expressed in a similar way:

$$\begin{aligned} \mathbf{w}_1 &= k_1\mathbf{f}_1 + k_2\mathbf{f}_2 \\ \mathbf{w}_2 &= -k_2\mathbf{f}_1 + k_1\mathbf{f}_2 \end{aligned} \tag{4.6}$$

Let $r_2(x,y)$ be a second 2-fold periodic image whose periods in the two orthogonal directions $\theta_2$ and $\theta_2 + 90°$ are $T_2$. Again, its spectrum $R_2(u,v)$ is a nailbed whose support in

[5] Obviously, some (or even most) of the nailbed impulses may have a zero amplitude, as in the case of $f(x,y) = \cos(x) + \cos(y)$, for instance.

the $u,v$ plane is a regular lattice $L_2(u,v)$, rotated by $\theta_2$ and with a period of $1/T_2$. The unit points (0,1) and (1,0) of the lattice $L_2(u,v)$ are situated in the spectrum at the geometric locations of the frequency vectors $\mathbf{f}_3$ and $\mathbf{f}_4$ of the two perpendicular fundamental impulses of the nailbed $R_2(u,v)$. Therefore the location $\mathbf{w}_3$ of a general point $(k_3,k_4)$ of this lattice and the location $\mathbf{w}_4$ of its perpendicular twin $(-k_4,k_3)$ are given by:

$$\begin{aligned} \mathbf{w}_3 &= k_3\mathbf{f}_3 + k_4\mathbf{f}_4 \\ \mathbf{w}_4 &= -k_4\mathbf{f}_3 + k_3\mathbf{f}_4 \end{aligned} \tag{4.7}$$

Assume now that we superpose (i.e., multiply) $r_1(x,y)$ and $r_2(x,y)$. According to the convolution theorem the spectrum of the superposition is the convolution of the nailbeds $R_1(u,v)$ and $R_2(u,v)$; this means, as we have seen in Sec. 2.11, that a centered copy of one of the nailbeds is placed on top of each impulse of the other nailbed (the amplitude of each copied nailbed being scaled down by the amplitude of the impulse on top of which it has been copied). This convolution gives a "forest" of impulses scattered throughout the spectrum (see Fig. 4.3). These impulses are generally not even located on a common lattice, since the product of two periodic functions is generally not periodic, but rather *almost-periodic* (see Appendix B); its spectrum is still impulsive, but its support is no longer a lattice and it may even be everywhere dense.

Fig. 4.3(a) shows the locations of the impulses in the spectrum-convolution in a typical case where no moiré effect is visible in the superposition. Figs. 4.3(b) and 4.3(c), however, show the impulse locations obtained in the spectrum-convolution in typical cases in which the superposition does generate a visible moiré effect, say a $(k_1,k_2,k_3,k_4)$-moiré. As we can see, in these cases the DC impulse at the spectrum origin is closely surrounded by a full cluster of impulses. The cluster impulses closest to the DC, inside the visibility circle, include the $(k_1,k_2,k_3,k_4)$-impulse, which is the fundamental impulse of the moiré in question,[6] and its perpendicular counterpart, the $(-k_2,k_1,-k_4,k_3)$-impulse, which is the fundamental impulse of the same moiré in the perpendicular direction. Naturally, each of these two impulses is also accompanied by its respective symmetrical twin to the opposite side of the origin. The locations (frequency vectors) of these four impulses are marked in Figs. 4.3(b),(c) by: **a**, –**a**, **b** and –**b**. Note that in Fig. 4.3(b) the impulse-cluster belongs to the second-order (1,2,-2,-1)-moiré, while in Fig. 4.3(c) the impulse-cluster belongs to the first-order (1,0,-1,0)-moiré, and it consists of a different set of impulses.

If we look attentively at the impulse-cluster surrounding the DC, we can see that this cluster is in fact a nailbed whose support is the regular lattice which is spanned by **a** and **b**, the geometric locations of the fundamental moiré impulses $(k_1,k_2,k_3,k_4)$ and $(-k_2,k_1,-k_4,k_3)$. This infinite impulse-cluster represents in the spectrum the 2D $(k_1,k_2,k_3,k_4)$-moiré, and its basis vectors **a** and **b** determine the period and the two perpendicular directions of the moiré. This impulse-cluster is in fact the 2D generalization of the moiré-comb that we had in Sec. 4.2 in the case of line-grating superpositions. We will call the infinite

[6] Note that this impulse is generated in the convolution by the $(k_1,k_2)$-impulse in the spectrum $R_1(u,v)$ of the first image and the $(k_3,k_4)$-impulse in the spectrum $R_2(u,v)$ of the second image.

impulse-cluster of the $(k_1,k_2,k_3,k_4)$-moiré the $(k_1,k_2,k_3,k_4)$-*moiré-cluster*, and we will denote it by $M_{k_1,k_2,k_3,k_4}(u,v)$; its explicit form will be given below in Eq. (4.15). If we extract from the spectrum of the superposition only the impulses of this infinite cluster we obtain the spectrum of the isolated $(k_1,k_2,k_3,k_4)$-moiré. To find, back in the image domain, the intensity profile of this moiré, we can interpret the moiré cluster as a 2D Fourier series development, and reconstruct the $(k_1,k_2,k_3,k_4)$-moiré-profile that it represents in the image domain by summing up the corresponding cosinusoidal functions belonging to each impulse pair in the cluster, up to the desired precision. Alternatively, we can also reconstruct the intensity profile of this moiré by taking the inverse 2D Fourier transform of the extracted cluster. Denoting the intensity profile of the $(k_1,k_2,k_3,k_4)$-moiré between the images $r_1(x,y)$ and $r_2(x,y)$ by $m_{k_1,k_2,k_3,k_4}(x,y)$, we therefore have:

$$m_{k_1,k_2,k_3,k_4}(x,y) = \mathcal{F}^{-1}[M_{k_1,k_2,k_3,k_4}(u,v)] \tag{4.8}$$

The intensity profile of the $(k_1,k_2,k_3,k_4)$-moiré is therefore a function $m_{k_1,k_2,k_3,k_4}(x,y)$ in the image domain whose value at each point $(x,y)$ indicates quantitatively the intensity level of the moiré in question, i.e., its particular intensity contribution to the image superposition.[7] Note that although this moiré is visible both in the layer superposition $r_1(x,y)\cdot r_2(x,y)$ and in the extracted moiré intensity profile $m_{k_1,k_2,k_3,k_4}(x,y)$, the latter does not contain the fine structure of the original layers $r_1(x,y)$ and $r_2(x,y)$ but only the isolated form of the extracted $(k_1,k_2,k_3,k_4)$-moiré.[8] As we have seen in Fig. 2.14, a single superposition $r_1(x,y)\cdot r_2(x,y)$ may include several visible moirés simultaneously; but each of these moirés will have a different moiré intensity profile $m_{k_1,k_2,k_3,k_4}(x,y)$ of its own.

It should be noted that the $(k_1,k_2,k_3,k_4)$-moiré-cluster exists in the spectrum-convolution even in cases like Fig. 4.3(a), where no moiré effect is visible in the superposition. In such cases the fundamental impulses $(k_1,k_2,k_3,k_4)$ and $(-k_2,k_1,-k_4,k_3)$ are simply located at a bigger distance from the DC, beyond the visibility circle. Note that at the other extremity, when the $(k_1,k_2,k_3,k_4)$-moiré reaches its singular point and its period becomes infinitely large (i.e., its frequency becomes zero), the entire infinite moiré-cluster which surrounds the spectrum origin collapses down onto the DC impulse.

Let us now find the expressions for the *location*, the *index* and the *amplitude* of each of the impulses of the $(k_1,k_2,k_3,k_4)$-moiré cluster. If **a** is the frequency vector of the $(k_1,k_2,k_3,k_4)$-impulse in the convolution and **b** is the frequency vector of the $(-k_2,k_1,-k_4,k_3)$-impulse, then we have according to Eq. (2.26) of Sec. 2.7:

$$\begin{aligned} \mathbf{a} &= k_1\mathbf{f}_1 + k_2\mathbf{f}_2 + k_3\mathbf{f}_3 + k_4\mathbf{f}_4 \\ \mathbf{b} &= -k_2\mathbf{f}_1 + k_1\mathbf{f}_2 - k_4\mathbf{f}_3 + k_3\mathbf{f}_4 \end{aligned} \tag{4.9}$$

---

[7] The explicit mathematical expression of the function $m_{k_1,k_2,k_3,k_4}(x,y)$ can be given in the form of a 2D Fourier series, which is a 2D extension of Eq. (4.3). However, we prefer to wait for this expression until Chapter 6, where we will be able to give it in a much more concise and elegant way (see Eq. (6.8)), thanks to the algebraic notations of Chapter 5 and their Fourier interpretation in Chapter 6.

[8] This has already been illustrated, in the case of 2-gratings superposition, by the difference between Fig. 2.5(c) (the image superposition) and Fig. 4.2(a) (the extracted intensity profile of the (1,-1)-moiré).

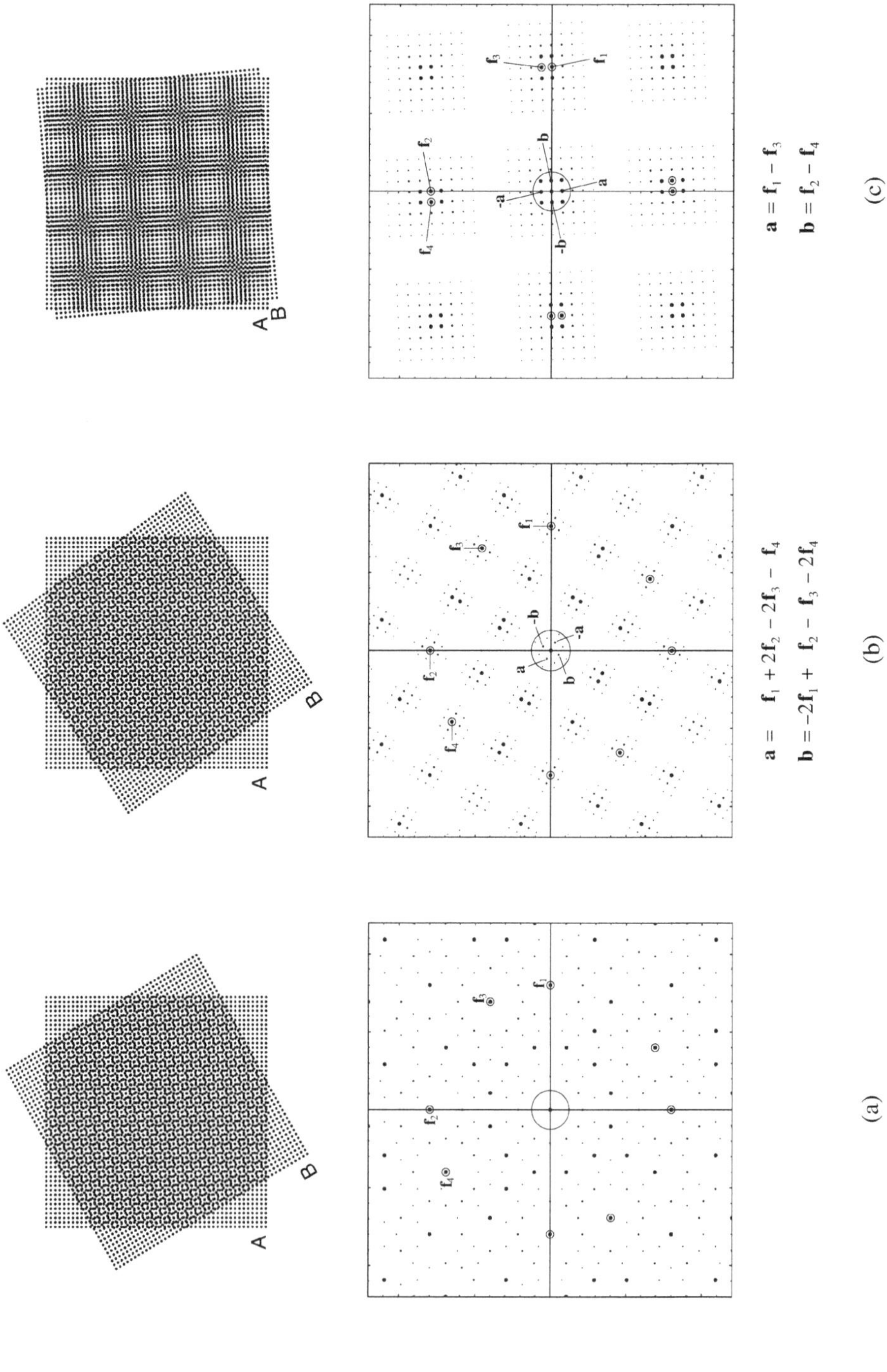
A
B
f1
f2
f3
f4
(a)
a
b
-a
-b
a = f1 + 2f2 − 2f3 − f4
b = −2f1 + f2 − f3 − 2f4
(b)
a = f1 − f3
b = f2 − f4
(c)

**Figure 4.3:** The superposition of two dot-screens with identical frequencies and with an angle difference of: (a) $\alpha = 30°$, (b) $\alpha = 34.5°$, and (c) $\alpha = 5°$, and the corresponding spectra. Only impulse locations are shown in the spectra, but not their amplitudes. Encircled points denote the locations of the fundamental impulses of the two original dot-screens. Large points represent convolution impulses of the first order (i.e., $(k_1,k_2,k_3,k_4)$-impulses with $k_i = 1, 0,$ or $-1$); smaller points represent convolution impulses of higher orders. (Note that only impulses of the first few orders are shown; in reality each impulse-cluster extends in all directions *ad infinitum*.) The circle around the spectrum origin represents the visibility circle. Note that while in (a) no significant impulses are located inside the visibility circle, in (b) the spectrum origin is closely surrounded by the impulse-cluster of the second-order (1,2,-2,-1)-moiré, and in (c) the spectrum origin is closely surrounded by the impulse-cluster of the (1,0,-1,0)-moiré.

According to Proposition 2.4 the *index-vector* of the $(m,n)$-th impulse in the $(k_1,k_2,k_3,k_4)$-moiré cluster is, therefore:

$$m(k_1,k_2,k_3,k_4) + n(-k_2,k_1,-k_4,k_3)$$
$$= (mk_1 - nk_2,\ mk_2 + nk_1,\ mk_3 - nk_4,\ mk_4 + nk_3) \qquad (4.10)$$

and furthermore, the *location* of the $(m,n)$-th impulse within this moiré-cluster is given by the linear combination $m\mathbf{a} + n\mathbf{b}$:

$$m\mathbf{a} + n\mathbf{b} = (mk_1 - nk_2)\mathbf{f}_1 + (mk_2 + nk_1)\mathbf{f}_2 + (mk_3 - nk_4)\mathbf{f}_3 + (mk_4 + nk_3)\mathbf{f}_4 \qquad (4.11)$$

As we can see, the $(k_1,k_2,k_3,k_4)$-moiré cluster is an infinite subset of the full spectrum-convolution which only contains those impulses whose indices are given by Eq. (4.10), for all integers $m,n$.

**Example 4.1:** In the case of the simplest first-order moiré between two dot-screens, the (1,0,-1,0)-moiré (see Fig. 4.3(c)), the index-vector of the $(m,n)$-th impulse in the moiré-cluster is:

$$m(1,0,-1,0) + n(0,1,0,-1) = (m, n, -m, -n)$$

and the location of this impulse in the spectrum is given by:

$$m\mathbf{a} + n\mathbf{b} = m\mathbf{f}_1 + n\mathbf{f}_2 - m\mathbf{f}_3 - n\mathbf{f}_4$$

For instance, the (1,0)-th impulse in the moiré-cluster has the index-vector (1,0,-1,0), and it is located in the spectrum at the point $\mathbf{a} = \mathbf{f}_1 - \mathbf{f}_3$. Similarly, the (0,1)-th impulse in this moiré-cluster has the index-vector (0,1,0,-1), and it is located in the spectrum at the point $\mathbf{b} = \mathbf{f}_2 - \mathbf{f}_4$. ■

Finally, the *amplitude* $d_{m,n}$ of the $(m,n)$-th impulse in the $(k_1,k_2,k_3,k_4)$-moiré cluster is given by:

$$d_{m,n} = a_{mk_1-nk_2,\, mk_2+nk_1,\, mk_3-nk_4,\, mk_4+nk_3} \tag{4.12}$$

and according to Eq. (2.27) in Sec. 2.7 we obtain:

$$d_{m,n} = a^{(1)}{}_{mk_1-nk_2}\, a^{(2)}{}_{mk_2+nk_1}\, a^{(3)}{}_{mk_3-nk_4}\, a^{(4)}{}_{mk_4+nk_3} \tag{4.13}$$

But since we are dealing here with the superposition of two orthogonal layers (dot-screens) rather than with a superposition of four independent layers (gratings), each of the two 2D layers may be inseparable. Consequently, we should rather group the four amplitudes of Eq. (4.13) into pairs, so that each element in the expression corresponds to an impulse amplitude in the nailbed $R_1(u,v)$ or in the nailbed $R_2(u,v)$:

$$d_{m,n} = a^{(1)}{}_{mk_1-nk_2\,,\, mk_2+nk_1}\, a^{(2)}{}_{mk_3-nk_4\,,\, mk_4+nk_3} \tag{4.14}$$

This means that the amplitude $d_{m,n}$ of the $(m,n)$-th impulse in the $(k_1,k_2,k_3,k_4)$-moiré cluster is the product of the amplitudes of its two generating impulses: the $(mk_1 - nk_2, mk_2 + nk_1)$-impulse of the nailbed $R_1(u,v)$ and the $(mk_3 - nk_4, mk_4 + nk_3)$-impulse of the nailbed $R_2(u,v)$. This can be interpreted more illustratively in the following way:

Let us call the *$(k_1,k_2)$-subnailbed of the nailbed* $R_1(u,v)$ the partial nailbed of $R_1(u,v)$ whose fundamental impulses are the $(k_1,k_2)$- and the $(-k_2,k_1)$-impulses of $R_1(u,v)$; its general $(m,n)$-impulse is the $m(k_1,k_2) + n(-k_2,k_1) = (mk_1 - nk_2, mk_2 + nk_1)$-impulse of $R_1(u,v)$. Similarly, let the $(k_3,k_4)$-subnailbed of the nailbed $R_2(u,v)$ be the partial nailbed of $R_2(u,v)$ whose fundamental impulses are the $(k_3,k_4)$- and the $(-k_4,k_3)$-impulses of $R_2(u,v)$; its general $(m,n)$-impulse is the $(mk_3 - nk_4, mk_4 + nk_3)$-impulse of $R_2(u,v)$. It therefore follows from Eq. (4.14) that the amplitude of the $(m,n)$-impulse of the nailbed of the $(k_1,k_2,k_3,k_4)$-moiré in the spectrum-convolution is the product of the $(m,n)$-impulse of the $(k_1,k_2)$-subnailbed of $R_1(u,v)$ and the $(m,n)$-impulse of the $(k_3,k_4)$-subnailbed of $R_2(u,v)$. This means that:

**Proposition 4.4:** (2D generalization of Proposition 4.1): The impulse amplitudes of the $(k_1,k_2,k_3,k_4)$-moiré cluster in the spectrum-convolution are the term-by-term product of the $(k_1,k_2)$-subnailbed of $R_1(u,v)$ and the $(k_3,k_4)$-subnailbed of $R_2(u,v)$. ■

**Example 4.2:** In the case of the simplest first-order moiré between the dot-screens $r_1(x,y)$ and $r_2(x,y)$, the (1,0,-1,0)-moiré (see Fig. 4.3(c)), the amplitudes of the moiré-cluster impulses in the spectrum-convolution are given by $d_{m,n} = a^{(1)}{}_{m,n}\, a^{(2)}{}_{-m,-n}$. This means that in this case the moiré-cluster is simply a term-by-term product of the nailbeds $R_1(u,v)$ and $R_2(-u,-v)$ of the original images $r_1(x,y)$ and $r_2(-x,-y)$. In the case of the second-order (1,2,-2,-1)-moiré (see Fig. 4.3(b)) the amplitudes of the moiré-cluster impulses are $d_{m,n} = a^{(1)}{}_{m-2n\,,\, 2m+n}\, a^{(2)}{}_{-2m+n\,,\, -m-2n}$. ■

Now, since we also know the exact *locations* of the impulses of the moiré-cluster (according to Eq. (4.11)), the spectrum of the isolated moiré in question is fully determined, and it is given analytically by:

$$M_{k_1,k_2,k_3,k_4}(u,v) = \sum_{m=-\infty}^{\infty} \sum_{n=-\infty}^{\infty} d_{m,n}\, \delta_{m\mathbf{a}+n\mathbf{b}}(u,v) \tag{4.15}$$

where $\delta_{\mathbf{f}}(u,v)$ denotes an impulse located at the frequency-vector **f** in the spectrum. This is, indeed, the 2D analog of Eq. (4.4). Therefore, as already mentioned above, we can reconstruct the intensity profile of the moiré, back in the image domain, by formally taking the inverse Fourier transform of the isolated moiré cluster. Practically, this can be done either by interpreting the moiré cluster as a 2D Fourier series, and summing up the corresponding cosinusoidal functions (up to the desired precision);[9] or, more efficiently, by approximating the continuous inverse Fourier transform of the isolated moiré-cluster by means of the inverse 2D discrete Fourier transform (using FFT).

As in the case of grating superposition (Sec. 4.2), the spectral domain term-by-term multiplication of the moiré-clusters can be interpreted directly in the image domain by means of the 2D version of the *T*-convolution theorem:

**2D *T*-convolution theorem:** Let $p_1(x,y)$ and $p_2(x,y)$ be 2-fold periodic functions of period $T_x$, $T_y$ integrable on a one-period interval ($0 \le x \le T_x$, $0 \le y \le T_y$), and let $\{a^{(1)}{}_{m,n}\}$ and $\{a^{(2)}{}_{m,n}\}$ (with $m,n = 0, \pm1, \pm2, ...$) be their 2D Fourier series coefficients. Then the function:

$$h(x,y) = \frac{1}{T_xT_y}\iint_{T_xT_y} p_1(x-x',y-y')\, p_2(x',y')\, dx'dy' \tag{4.16}$$

(where $\iint_{T_xT_y}$ means integration over a one-period interval), which is called the *T-convolution* of $p_1$ and $p_2$ and denoted by $p_1{**}p_2$, is also 2-fold periodic with the same periods $T_x$, $T_y$ and has Fourier series coefficients $\{a_{m,n}\}$ given by: $a_{m,n} = a^{(1)}{}_{m,n}\, a^{(2)}{}_{m,n}$ for all integers $m,n$. ■

From this theorem we obtain the following result, which is the generalization of Proposition 4.3 to the general 2D case:

**Proposition 4.5:** The profile form of the $(k_1,k_2,k_3,k_4)$-moiré in the superposition of $r_1(x,y)$ and $r_2(x,y)$ is a *T*-convolution of the (normalized) images belonging to the $(k_1,k_2)$-subnailbed of $R_1(u,v)$ and the $(k_3,k_4)$-subnailbed of $R_2(u,v)$. Note that before applying the *T*-convolution theorem, the images must be normalized by stretching and rotation transformations, to fit the actual period and angle of the moiré, as determined by Eq. (2.26) (or by the lattice $L_M(u,v)$ of the $(k_1,k_2,k_3,k_4)$-moiré, which is spanned by the fundamental vectors **a** and **b**). As shown in Sec. C.3 of Appendix C, normalizing the periodic images by stretching and rotation does not affect their impulse amplitudes in the spectrum, but only the impulse locations. ■

A more rigorous formulation of Proposition 4.5 will be given in Sec. 10.9.2 using the general exponential Fourier series formulation. The present results will also be extended there to the more general case of moirés between curved screens.

Finally, let us mention that all our results here can be easily generalized to any $(k_1,...,k_m)$-moiré between any number of superposed images in a simple, straightforward way.

---

[9] The explicit Fourier series development of the $(k_1,...,k_m)$-moiré in the most general case, which includes non-symmetric or shifted dot-screens, will be given in a concise and elegant way in Secs. 6.7 and 6.8.

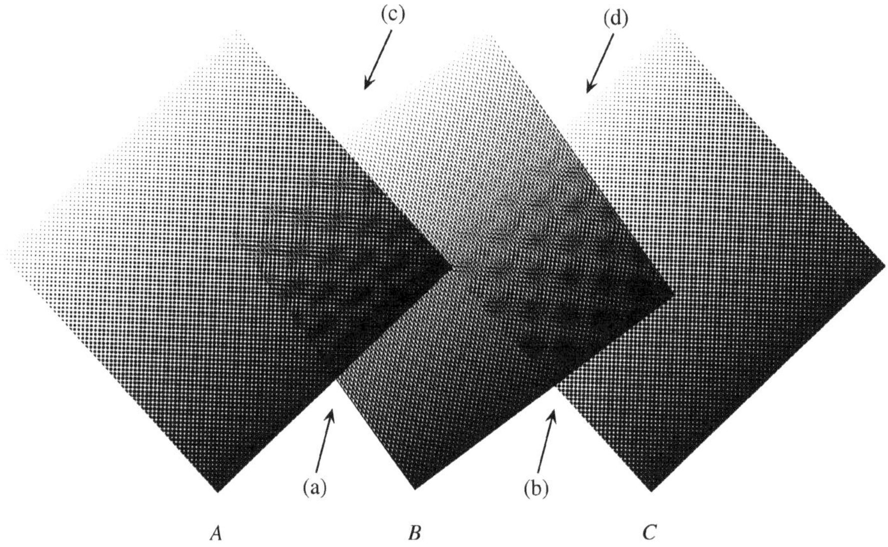


**Figure 4.4:** Demonstration of the magnification and rotation properties of the (1,0,-1,0)-moiré between two dot-screens. Dot-screen *B* consists of black "1"-shaped dots and is superposed with two identical dot-screens of black circular dots, *A* and *C*, forming with each of them an angle difference of 7°. All of the three screens consist of gradually increasing dots with identical frequencies. It can be seen that where one of the two superposed screens is relatively dark and consists of tiny white dots (see (a), (b)), the moiré profile form is essentially a magnified version of the other screen; and where one of the two superposed screens consists of tiny black dots (see (c), (d)), the moiré profile form is essentially a magnified, inverse-video version of the other screen. Note that in both cases the orientation of the "1"-shaped moiré is almost perpendicular to that of the original "1"-shaped dots of screen *B*. Note also the gradual moiré form transitions between (a) and (c) and between (b) and (d), through all the intermediate, blurred stages.

## 4.4 The special case of the (1,0,-1,0)-moiré

In this section we will apply the results that we have obtained above to the special case of the (1,0,-1,0)-moiré. In particular, we will see how these results explain the striking moiré effects observed in superpositions of two dot-screens with identical periods and a small angle difference (like in Fig. 4.1), which are clearly (1,0,-1,0)-moirés. These results will also allow us to synthesize (1,0,-1,0)-moirés with any desired period and intensity profile.

As we have seen in Example 4.2, in the case of the (1,0,-1,0)-moiré the impulse amplitudes of the moiré-cluster are simply a term-by-term product of the nailbeds $R_1(u,v)$ and $R_2(-u,-v)$ themselves: $d_{m,n} = a^{(1)}_{m,n} a^{(2)}_{-m,-n}$. Since the impulse locations of this moiré-

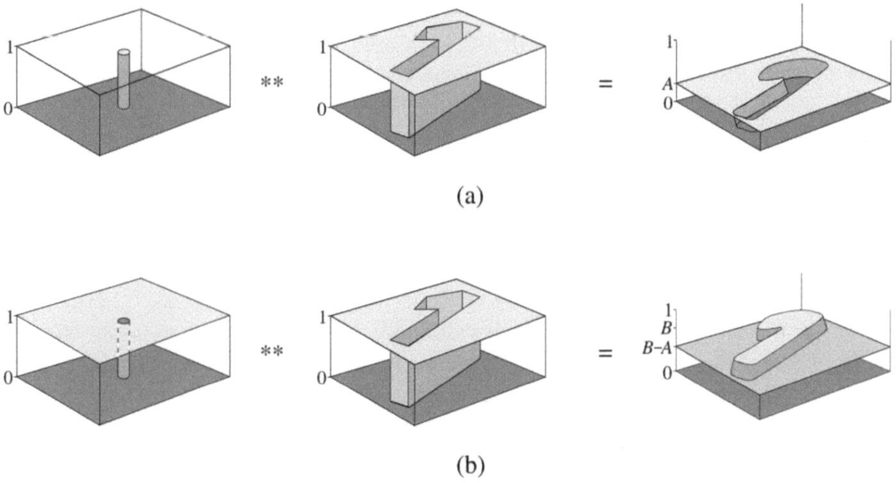


**Figure 4.5:** (a) The $T$-convolution of tiny white dots (from the first screen) with dots of any given shape (from the other screen) gives dots of essentially the same given shape; (b) The $T$-convolution of tiny black dots (from the first screen) with dots of any given shape (from the other screen) gives dots of essentially the same shape, but in inverse video.

cluster are also known, according to Eq. (2.26), we can obtain the intensity profile of the (1,0,-1,0)-moiré by extracting this moiré-cluster from the full spectrum-convolution, and taking its inverse Fourier transform.

However, according to Proposition 4.5 the intensity profile of the (1,0,-1,0)-moiré can be also interpreted directly in the image domain: in this case the moiré intensity profile is simply a $T$-convolution of the original layers $r_1(x,y)$ and $r_2(-x,-y)$ (after they undergo the necessary stretching and rotation transformations to make their periods, or their supporting lattices in the spectrum, coincide). This result has been previously derived by Harthong [Harthong81 p. 69] using non-standard analysis; as we can see, this result is obtained here (for any 2-fold periodic images $r_1(x,y)$, $r_2(x,y)$) as a simple particular case of our general moiré extraction method.

Let us see now how $T$-convolution sheds a new light on the profile form of (1,0,-1,0)-moirés, and explains the striking visual effects observed in superpositions of dot-screens like in Figs. 4.1(a)–(d).

#### 4.4.1 Shape of the intensity profile of the moiré cells

**Case 1:** As we can see in Figs. 4.1(a)–(d), the form of the moiré cells in the superposition is most clear cut and striking where one of the two screens is relatively dark (see for example Fig. 4.4(a) and (b)). This happens because the dark screen includes only tiny white or transparent dots (pinholes), which play in the $T$-convolution the role of very narrow pulses with amplitude 1. As shown in Fig. 4.5(a), the $T$-convolution of such narrow pulses (from one of the screens) and dots of any shape (from the other screen) gives dots of the latter shape, in which the zero values remain at zero, the 1 values are scaled down to the value $A$ (the volume or the area of the narrow white pulse divided by the period area, $T_x{\cdot}T_y$), and the sharp step transitions are replaced by slightly softer ramps. This means that the dot shape received in the normalized moiré-period is practically identical to the dot shape of the second screen, except that its white areas turn darker. This normalized moiré-period is stretched back into the real size of the moiré-period, $T_M$, as it is determined by Eqs. (2.28) and (2.8). Note that in our case the moiré period is determined by the angle difference $\alpha$ alone, since the screen frequencies are fixed; more precisely, according to Eq. (2.10), the moiré period becomes larger as the angle $\alpha$ tends to 0°. This means that the moiré-form in this case is essentially a magnified version of the second screen, where the magnification rate is controlled only by the angle $\alpha$. This interesting magnification property of the moiré effect can be used in certain applications as a "virtual microscope" for visualizing the detailed structure of a given screen. It should be noted, however, that details in the screen to be magnified which are smaller than the pinhole size in the first screen will not be clearly visible in the resulting moiré, since they will be blurred and smoothed-out with their background by the $T$-convolution.

**Case 2:** A similar effect, albeit somewhat less impressive, occurs in the superposition where one of the two screens contains tiny *black* dots (see Fig. 4.4(c),(d)). Tiny black dots on a white or transparent background can be interpreted as "inverse" pulses of 0-amplitude on a background of amplitude 1. As we can see in Fig. 4.5(b), the $T$-convolution of such inverse pulses (from one of the screens) and dots of any shape (from the other

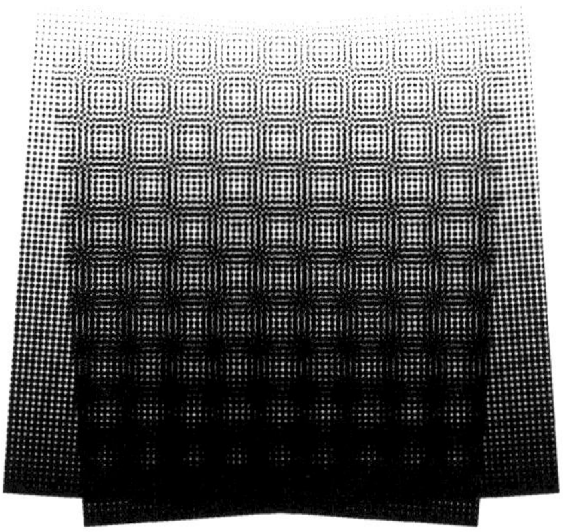

**Figure 4.6:** Two circular black dot-screens which are superposed, unlike in Fig. 4.1(a), with matching gray levels (dot sizes). The moiré profile form in this case is no longer circular as in Fig. 4.1(a), but rather has a squarish form in the darker gray levels.

screen) gives dots of the latter shape, where the zero values are replaced by the value $B$ (the volume under a one-period cell of the second screen divided by the period area $T_x{\cdot}T_y$) and the 1 values are replaced by the value $B-A$ (where $A$ is the volume of the "hole" of the narrow black pulse divided by $T_x{\cdot}T_y$). This means that the dot shape of the normalized moiré-period is similar to the dot shape of the second screen, except that it appears in inverse video and with slightly softer ramps. And indeed, looking at Figs. 4.1 and 4.4, we see that wherever one of the screens in the superposition contains tiny black dots, the moiré appears to be a magnified version of the other screen, but this time in inverse video.

Note that although the amplitude difference (max. value – min. value) in both of the cases above is identical (in both cases it equals $A$), the perceived contrast in the first case appears to the eye much stronger than in the second one. As we have seen in Sec. 2.10 and in Fig. 2.9 there, the reason for this phenomenon is that the response (or sensitivity) of the human visual system to light intensity is not linear in its nature, but rather close to logarithmic [Pratt91 pp. 27–29]. If we plot the intensities or the moiré profiles logarithmically, i.e., in terms of density rather than in terms of reflectance, we get a more realistic representation of the perceptual contrast of the moiré, which corresponds better to human perception (see Figs. 2.9(g)–(i) in Chapter 2).

**Case 3:** When none of the two superposed screens contains tiny dots, either white or black, the profile-form of the resulting moiré is still a magnified version of the *T*-convolution of the two original screens. This *T*-convolution gives, as before, some kind of blending between the two original dot shapes, but this time the resulting shape has a rather blurred or smoothed-out appearance and the moiré looks less attractive to the eye. For example, note in Figs. 4.1(a)–(d) the sharp-cut moiré profile shapes at the bottom and at the top ends of the superposed area (where the white or black tiny dots are located), and the gradual transition between them through intermediate, blurred shapes (where none of the screens contains tiny dots).

Another interesting example of this type occurs when two screens with circular black dots are superposed, unlike in Fig. 4.1(a), with their gray levels (dot sizes) in match (see Fig. 4.6). In this case the resulting moiré profile form is no longer mostly circular, as it was in Fig. 4.1(a), and it rather has a squarish form at the darker gray levels. This reflects the forms obtained by *T*-convolution of two periodic screens with identical, black circular dots: Indeed, these forms tend to become squarish as the circular dots increase, owing to the cyclical wrap-around effect caused at the four boundaries of the period-cell. This can be verified by actually calculating the *T*-convolution. Note that 2D *T*-convolutions of periodic images on the continuous $x,y$ plane (or rather, their discretized approximations) can be easily performed by a computer program using 2D DFT: since the discrete Fourier transform is inherently periodic, it follows that the discrete convolution obtained by using it (i.e., by multiplying the DFT of a one-period cell from each of the original screens and taking the inverse DFT of the product) is also periodic and cyclic [Bracewell86 p. 362]; this is, indeed, the discrete counterpart of *T*-convolution. Fig. 4.7 shows the *T*-convolution obtained in this manner for the case of two identical black circular dots of various sizes;

these results are identical to the moiré profile forms obtained at the corresponding gray levels in Fig. 4.6.

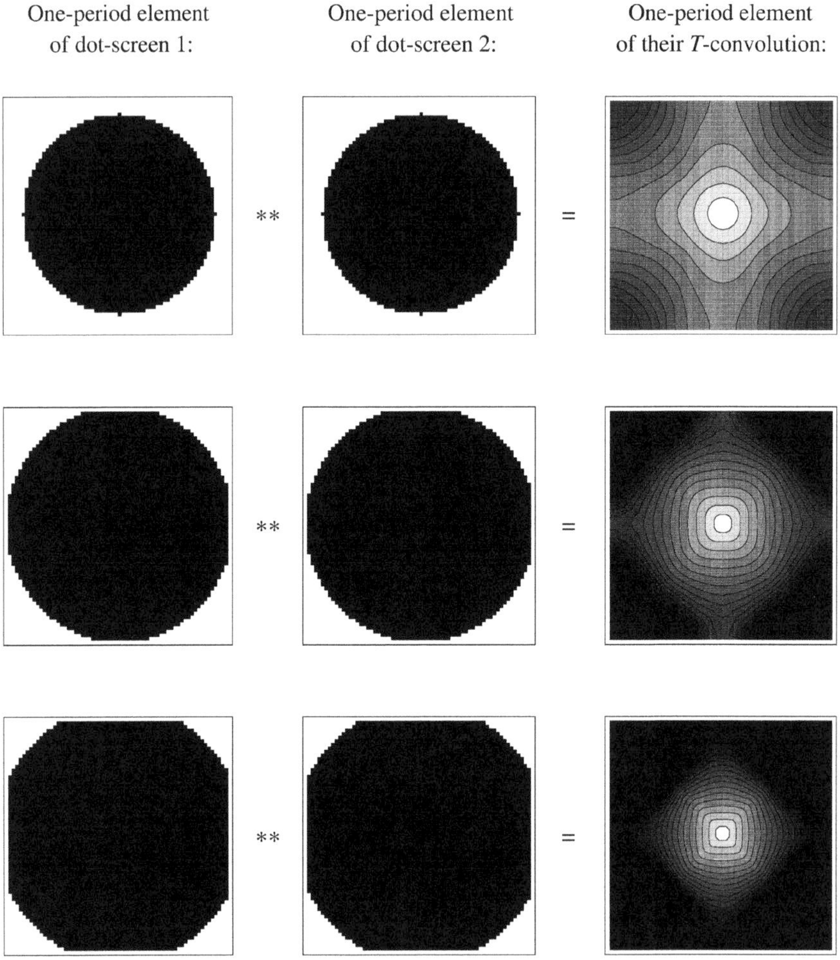


**Figure 4.7:** $T$-convolution of two identical, circular black dot-screens: each row shows the $T$-convolution at a different gray level (screen dot size). The $T$-convolution in each of the rows is calculated digitally by multiplying the FFTs of the two screen elements and taking the inverse FFT of their product. It clearly appears that at darker gray levels the forms obtained by the $T$-convolution are rather squarish; this agrees perfectly with the moiré profile forms actually obtained in the screen superposition (Fig. 4.6) at the respective gray levels.

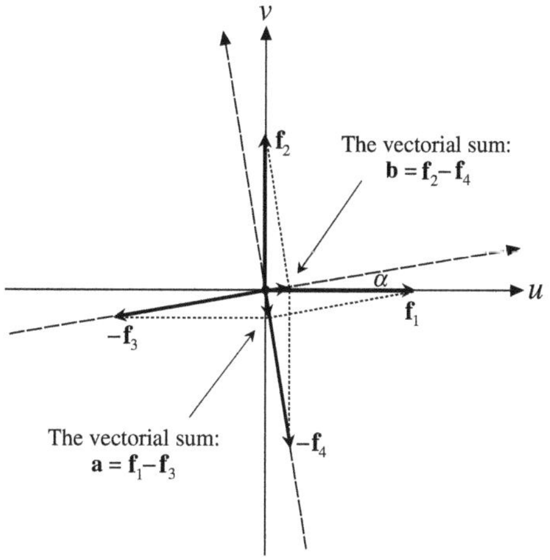


**Figure 4.8:** A detail from Fig. 4.3(c) showing the spectral interpretation (vector diagram) of the (1,0,-1,0)-moiré between two dot-screens with identical frequencies and a small angle difference $\alpha$ (for the sake of clarity the angle $\alpha$ is shown here slightly larger than in Fig. 4.3(c)). It is clearly seen that the low frequency vectorial sums **a** and **b** (which are the geometric locations of the two fundamental impulses of the (1,0,-1,0)-moiré cluster) are closely perpendicular to the directions of the two original screens: **a** is perpendicular to the bisecting direction between $\mathbf{f}_1$ and $\mathbf{f}_3$, and **b** is perpendicular to the bisecting direction between $\mathbf{f}_2$ and $\mathbf{f}_4$.

### 4.4.2 Orientation and size of the moiré cells

As we can see in Fig. 4.4, although the (1,0,-1,0)-moiré cells inherit the forms of the original screen cells, they do not necessarily inherit their orientations. Rather than having the same direction as the cells of the original screens (or an intermediate orientation), the moiré cells in this figure appear in a perpendicular direction. This fact may seem surprising at first, but in fact it can easily be understood using the theory developed in Sec. 4.3: As we already know, the orientation and the size of the moiré are determined by the location of the fundamental impulses of the moiré-cluster in the spectrum, i.e., by the location of the basis vectors **a** and **b** (Eq. (4.9)). We have seen in Example 4.1 that in the case of the (1,0,-1,0)-moiré these vectors are reduced to:

$$\begin{aligned} \mathbf{a} &= \mathbf{f}_1 - \mathbf{f}_3 \\ \mathbf{b} &= \mathbf{f}_2 - \mathbf{f}_4 \end{aligned} \qquad (4.17)$$

And indeed, as we can see in Figs. 4.3(c) and 4.8, when the two original screens have the same frequency, these basis vectors are perpendicular to the bisectors of the angles formed between the frequency vectors $\mathbf{f}_1$, $\mathbf{f}_3$ and $\mathbf{f}_2$, $\mathbf{f}_4$. This means that the (1,0,-1,0)-moiré cluster (and the corresponding moiré profile in the image domain) are closely perpendicular to the original screens $r_1(x,y)$ and $r_2(x,y)$. Note that the precise period and angle of this moiré can be found by formulas (2.10) which were derived for the (1,-1)-moiré between two line-gratings with identical periods $T$ and angle difference of $\alpha$.[10] This perpendicularity of the (1,0,-1,0)-moiré with respect to the original screens (see Fig. 4.12) is the 2D analog of the (1,-1)-moiré in the superposition of two gratings with $T_1 = T_2$ and $\theta_1 \approx \theta_2$, where the moiré bands are approximately perpendicular to both original gratings (see Fig. 2.5).

Obviously, the fact that the direction of the moiré profile is almost perpendicular to the direction of the original screens is a particular property of the (1,0,-1,0)-moiré between two regular screens having identical frequencies; in other cases the angle of the moiré may be different. For example, if $\theta_1 = \theta_2$ and $T_1 \neq T_2$ the resulting moiré is parallel to the original screens (see Fig. C.24 in Appendix C). In the most general case the moiré angle can be found by Eqs. (2.28) and (2.8). A further discussion on the general case is provided in Sec. C.15 of Appendix C; it also includes an example with hexagonal screens.

## 4.5 The case of more complex and higher order moirés

As we have seen in Sec. 4.3 above, the general moiré case differs from the elementary (1,0,-1,0)-moiré in that in Proposition 4.5 the $(k_1,k_2)$-subnailbed of $R_1(u,v)$ and the $(k_3,k_4)$-subnailbed of $R_2(u,v)$ no longer coincide with the nailbeds $R_1(u,v)$ and $R_2(-u,-v)$ themselves. Equivalently, from the image domain point of view, the moiré-profile is no longer a (normalized) $T$-convolution of the original images $r_1(x,y)$ and $r_2(-x,-y)$ themselves, but rather a $T$-convolution of their derived images $\hat{r}_1(x,y)$ and $\hat{r}_2(x,y)$, whose spectra are the $(k_1,k_2)$-subnailbed of $R_1(u,v)$ and the $(k_3,k_4)$-subnailbed of $R_2(u,v)$. This means that in the general case the intensity profile form of the $(k_1,k_2,k_3,k_4)$-moiré cannot be expected to reflect the original forms of the screen elements, but rather a more complex relationship between them.

We will illustrate this using the case of the (1,0,-1,1)-moiré, which occurs between two dot-screens (of circular black dots) with a frequency ratio of $\sqrt{2}\approx 1.4142$ and an angle difference $\alpha$ close to 45° (see Figs. 4.9(a),(b)). In this case the moiré-cluster which surrounds the spectrum origin has the basis vectors:

$$\begin{aligned} \mathbf{a} &= \mathbf{f}_1 - \mathbf{f}_3 + \mathbf{f}_4 \\ \mathbf{b} &= \mathbf{f}_2 - \mathbf{f}_3 - \mathbf{f}_4 \end{aligned} \tag{4.18}$$

and according to Eq. (4.11) it contains all the impulses of the full nailbed-convolution whose index-vectors are of the type: $(m, n, -m-n, m-n)$. The amplitude of the $(m,n)$-th impulse in this cluster is $d_{m,n} = a^{(1)}{}_{m,n}\, a^{(2)}{}_{-m-n,m-n}$.

[10] Remember that the 2D (1,0,-1,0)-moiré between two screens is geometrically equivalent to the moiré between two pairs of gratings; referring to Fig. 2.10(a), the gratings A and C generate a (1,-1) moiré, and the gratings B and D generate a second, perpendicular (1,-1)-moiré.

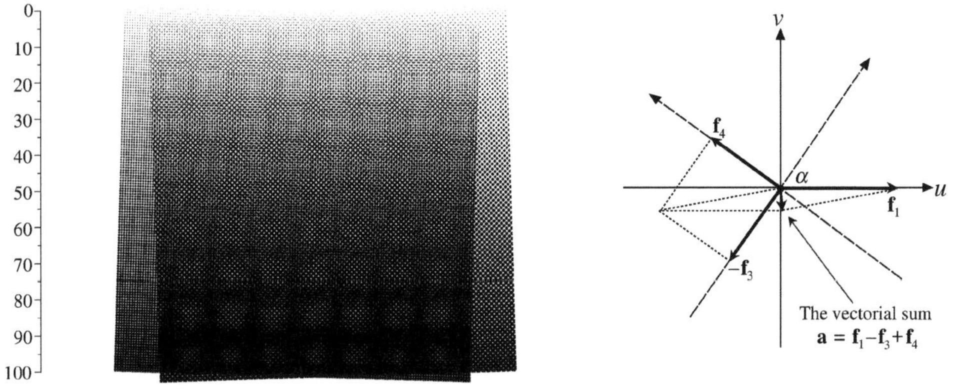


(a) (b)

**Figure 4.9:** (a) The (1,0,-1,1)-moiré between two dot-screens of gradually increasing black circular dots, whose frequency ratio is $|\mathbf{f}_1|/|\mathbf{f}_3| = \sqrt{2} \approx 1.4142$ and whose angle difference $\alpha$ is close to 45°. (b) The spectral interpretation (vector diagram) of the moiré in question (for the sake of clarity, only the frequency vectors in one of the two perpendicular directions are shown). The low frequency vectorial sum **a** is the impulse location of one of the two perpendicular fundamental impulses of the (1,0,-1,1)-moiré impulse cluster.

Fig. 4.10 shows this moiré-cluster for two different dot-size combinations of the original dot-screens, and the moiré intensity profiles obtained by taking the inverse Fourier transform of each of these spectra. As we can see, these results accurately predict the moiré intensity profile forms that are actually obtained in the screen superposition (Fig. 4.9) at the corresponding gray levels.

Note that, in general, the more complex the moiré (that is, the more superposed layers it involves, or the higher its $k_i$-indices or harmonics are), the more blurred, low-contrast and washed-out its profile form looks. The most visually impressive moiré profile forms are normally obtained in low-order moirés between few superposed layers.

Finally, it should be noted that although we have only presented in this chapter our analysis method for the case of two superposed layers (line-gratings or dot-screens), this approach is completely general and it can be used for deriving the intensity profile form of any order moirés between any number of superposed layers.

## PROBLEMS

4-1. *T-convolution.* Explain the moiré profiles in Fig. 2.9 in terms of *T*-convolution of the two original gratings (see Proposition 4.2).

4-2. *T-convolution.* Why is a normalization by $1/T$ (or by $1/T_xT_y$, in the 2D case) required in the definition of $T$-convolution? (see Eqs. (4.5) and (4.16)). What would be, otherwise, the $T$-convolution of two periodic images with periods of, say, $T = 10$, which consist of a small black point on a white background?

4-3. *2D T-convolution.* Suppose that the "1"-shaped cell in the center of Figs. 4.5(a),(b) is replaced by its inverse video, namely: by a white "1"-shaped dot on a black background. How will the $T$-convolution at the right-hand side of each figure be modified?

4-4. *2D T-convolution vs. 2D convolution.* It is sometimes carelessly said that the form of a single 2D period of the (1,0,-1,0)-moiré in the superposition of two dot-screens is given by a (normalized) convolution of a single dot of the first screen with a single dot of the second screen. Explain why this formulation is imprecise. *Hint*: This statement is only correct when the convolution of the two single dots does not exceed the size of a single 2D period $(T_x,T_y)$; otherwise the outer ends of the neighbouring convolution periods inevitably penetrate (additively) into the area of the current period. This cyclical wrap-around effect is automatically taken care of by $T$-convolution, but not by the simple convolution of single screen dots.

4-5. *The use of DFT as a Fourier transform approximation* (see [Brigham88 Sec. 6.4]). Many numeric software packages include routines for performing 1D or 2D DFT (Discrete Fourier Transform), usually using the FFT (Fast Fourier Transform) algorithm. Generate two 256×256 matrices consisting of discrete line gratings having angles and periods similar to the gratings shown in Fig. 2.5, and multiply them element-by-element to obtain the matrix of their superposition. Now, apply DFT to each of the three matrices, and compare your results with the spectra shown in Fig. 2.5. Do you notice any differences? Can you identify in your DFT results any DFT artifacts such as folding-over or leakage [Brigham88 pp. 101–103, 172–173]? Can you extract from the DFT of the superposition the comb of the (1,-1)-moiré and obtain, by applying on it an inverse DFT, a faithful image-domain representation of the moiré, like in Fig. 4.2? Do you have any suggestion how to improve your results? (*Hint*: If you can restrict yourself to discrete gratings with rational angles, whose 256×256 matrices perfectly wrap around and tile the plane without discontinuities on the matrix boundaries, most of the DFT artifacts will be eliminated and you will obtain clean and clear spectra. Alternatively, various filtering methods may be also used to reduce the DFT artifacts [Brigham88 Sec. 9.2]).

4-6. *The use of DFT as a Fourier transform approximation.* Suppose that you use 2D DFT to find the spectrum of a periodic grating or dot-screen. How can you distinguish in the spectrum obtained by DFT between a "true impulse" that belongs to a periodicity within the image, and a "false impulse" which is owed to folding-over (aliasing)? *Hint*: What happens to "true impulses" and to "false impulses" when you slightly rotate the original image before applying to it the DFT?

4-7. *Moiré magnification.* Suppose that periodic dot-screen A consisting of tiny pinholes is superposed on top of periodic dot-screen B that consists of black "1"-shaped dots (see Fig. 4.4(a)). Both screens have identical frequencies. What do you expect to see in the superposition:

(a) When both layers are oriented exactly to the same direction?

(b) When layer A is slightly rotated counterclockwise, like in Fig. 4.4(a)?

(c) When layer A is slightly rotated clockwise? (*Hint*: The "1"-shaped moiré will be inversed by 180°.)

(a)

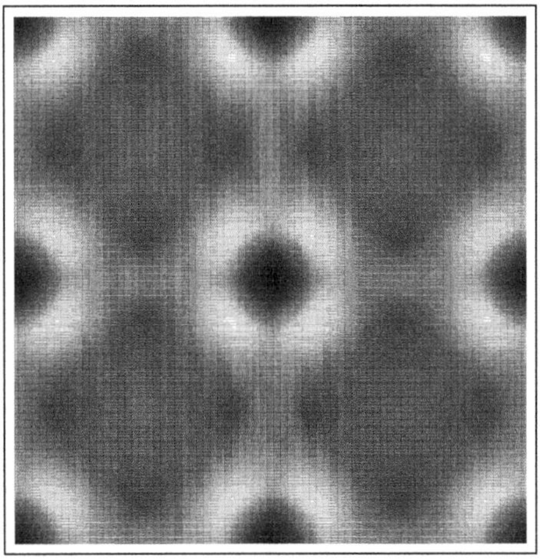

(b)

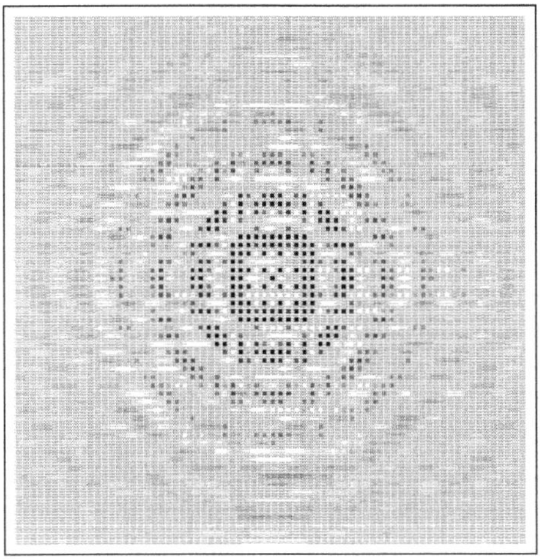

(c)

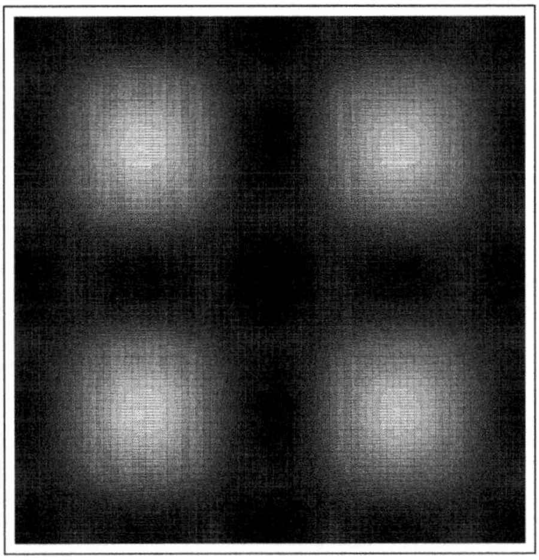

(d)

**Figure 4.10:** Left images (spectral domain): the impulse-cluster of the (1,0,-1,1)-moiré between two identical dot-screens with circular black dots, analytically calculated (up to 32 harmonics) for two different dot sizes (screen gray levels). Spectra (a) and (c) only differ in their impulse amplitudes (note that the impulse amplitudes in these spectra are indicated by their relative darkness). Right images (image domain): reconstruction of the corresponding moiré-profiles, obtained by taking the inverse FFT of each of these spectra. Note that these results agree perfectly with the moirés actually obtained in the screen superposition at the respective gray levels (levels 55 and 70 on the scale at Fig. 4.9).

The best way to visualize these effects is by printing dot screens A and B on two transparencies, and slowly rotating transparency A on top of transparency B on a light-table or against any diffuse light source. How do you explain cases (a)–(c), using Eq. (4.17) and the spectral domain interpretation given in Fig. 4.8?

4-8. What do you expect to see in cases (a)–(c) of the previous problem when the frequency of screen A is slightly smaller than that of screen B? And when the frequency of screen A is slightly larger than that of screen B? And when screens A and B are irregular (see Sec. 2.12 and Fig. 2.13(b))?

4-9. Suppose that periodic dot-screen B consists of black "1"-shaped dots, as shown under a large magnification in Fig. 4.11(a). The screen frequency to both directions is $f$. What do you expect to see when you superpose on top of this screen a periodic dot-screen A of tiny pinholes: (a) If dot-screen A has the frequency $2f$? (b) If dot-screen A has the frequency $f/2$?

4-10. Given the dot-screen shown in Fig. 4.11(b), how would you design the corresponding pinholes layer A in order to obtain a visible moiré having the profile form of Fig. 4.11(b)? (*Hint*: The periodicity of screen B consists in this case of 2×2 letters.)

4-11. *Polychromatic moirés.* What would you expect to see when a periodic dot-screen consisting of a polychromatic period is superposed by a periodic dot-screen of the same frequency consisting of (a) tiny pinholes, or (b) tiny black dots? (See Chapter 9).

4-12. *Microlens arrays.* A microlens array is an optical device made of a sheet of tiny lenslets that are geometrically arranged on a given lattice like the dots of a dot-screen [Hutley91]. Microlens arrays have the particularity that each of their lenslets focuses on a very small region of the underlying image, and therefore they behave much like screens of small pinholes (see Plate 1). What are the advantages and the shortcomings of using a microlens array rather than a pinhole screen for a typical moiré application? (*Hint*: Since the lenslets and the substrate between them are translucent, microlens arrays have the advantage of letting most of the incident light pass through the array. They can therefore be used for producing moiré effects either by reflection or by transmission. On the other hand, microlens arrays are more sensitive than pinhole screens to optical parameters such as focal distance and parallax (the viewing angle), and their manufacturing is less flexible and more expensive.)

4-13. *Document security.* Can you think of an application of the moiré effect between dot-screens for document authentication and anti-counterfeiting? *Hint*: The document can be protected by the high-quality printing of a halftoned image that consists of a specially designed halftone screen. This special screen is made of tiny halftone dots having a predefined shape that remains unchanged throughout a wide range of gray levels; for example, the halftone dots may have the shape of the letter pair "US" in varying sizes and linewidths, to allow for the various gray levels of the image. When the appropriate revealer (pinhole screen or microlens array) is superposed on this halftoned image, a highly visible moiré (in our example a repetitive "US" pattern) will appear within the superposed area. Since the detail of the tiny halftone dots will not resist photocopying, scanning, or any other digital or analog copying method, any such counterfeited document will be immediately recognized by the absence or by the corrupted shape of the moiré pattern when the revealer is superposed on the document. For more details see [Amidror02; Amidror01; Amidror99].

4-14. *Document security (continued).* Suppose that some random noise is added to the dot-screen of Fig. 4.11(a), so that its "1"-shaped dots can be hardly identified through this noise. What do you expect to see when you superpose on top of this corrupted dot-screen the corresponding uncorrupted pinholes layer (or microlens array)? Can you

| | | | | | |
|---|---|---|---|---|---|
| 1 | 1 | 1 | 1 | 1 | 1 |
| 1 | 1 | 1 | 1 | 1 | 1 |
| 1 | 1 | 1 | 1 | 1 | 1 |
| 1 | 1 | 1 | 1 | 1 | 1 |
| 1 | 1 | 1 | 1 | 1 | 1 |
| 1 | 1 | 1 | 1 | 1 | 1 |

(a)

| | | | | | |
|---|---|---|---|---|---|
| 1 | 2 | 1 | 2 | 1 | 2 |
| 2 | 1 | 2 | 1 | 2 | 1 |
| 1 | 2 | 1 | 2 | 1 | 2 |
| 2 | 1 | 2 | 1 | 2 | 1 |
| 1 | 2 | 1 | 2 | 1 | 2 |
| 2 | 1 | 2 | 1 | 2 | 1 |

(b)

**Figure 4.11:** (a) A periodic dot-screen consisting of tiny "1"-shaped dots. (b) A variant of (a) consisting of both "1"-shaped dots and "2"-shaped dots. Both dot-screens are shown here very considerably magnified.

think of an application for a covert document anti-counterfeiting method? For more details see [Amidror99].

4-15. *Document security (continued).* Suppose that you superpose a periodic dot-screen A consisting of tiny "USA"-shaped dots over a similar screen B consisting of "$50"-shaped dots at the same frequency. What do you expect to see when a pinhole layer C of the same frequency is overlaid on top of the superposed layers A and B and slowly rotated between 0° and 90°, if the angle difference between layers A and B is: (a) 45°; (b) 10°; (c) 0°? Can you think of an application for a covert document anti-counterfeiting method? For more details see [Amidror99].

4-16. *Deconvolution.* According to Proposition 4.5, the intensity profile of the (1,0,-1,0)-moiré between two periodic dot-screens is, up to a certain normalization, a $T$-convolution of the two original dot-screens: $m(x,y) = r_1(x,y) ** r_2(x,y)$ (a more rigorous formulation of this result is given later in Proposition 10.5). Consequently, the following interesting question may be naturally posed: Suppose that dot-screen $r_1(x,y)$ consists of "1"-shaped dots; what should layer $r_2(x,y)$ be in order that the resulting moiré $m(x,y)$ consist of "2"-shaped periods? This is, in fact, a classical deconvolution problem, which can be formally solved by considering the respective spectra. According to the convolution theorem we have:

$$M(u,v) = R_1(u,v)\, R_2(u,v)$$

and therefore:

$$R_2(u,v) = \frac{M(u,v)}{R_1(u,v)}$$

which gives, by applying an inverse Fourier transform:

$$r_2(x,y) = \mathcal{F}^{-1}\left[\frac{M(u,v)}{R_1(u,v)}\right]$$

Unfortunately, however, this formal solution is unrealizable in most non-trivial cases, and the task of finding such a layer $r_2(x,y)$ remains practically impossible. Can you explain why? *Hint*: Apart from the inherent instability of the solution, due to the fact that the spectrum $R_1(u,v)$ normally contains many zero-valued points [Kunt86 Sec. 8.7.4; Russ95 pp. 336–339], there is also no guarantee that the hypothetical solution $r_2(x,y)$ be a physically realizable reflectance (or transmittance) function, i.e., a purely real-valued function whose values vary between 0 and 1. (Note that "1"-shaped and "2"-shaped periods are not symmetric, so that the spectra $M(u,v)$ and $R_1(u,v)$ are both complex-valued). A detailed discussion about deconvolution can be found in [Jansson97].

4-17. *Higher order moirés.* Can you design two dot screens that generate in their superposition a $(k_1,k_2,k_3,k_4)$-moiré other than the (1,0,-1,0)-moiré (for example, a moiré of order 2), which has a "1"-shaped profile as in Fig. 4.4? Explain.

4-18. *A geometric image-domain explanation for the magnification and rotation properties of the (1,0,-1,0)-moire between two dot screens.* Although the Fourier theory provides a full qualitative and quantitative explanation of the moiré phenomenon and its various properties, some simple cases can be explained qualitatively directly in the image domain, by considering geometric relations between the superposed layers. For example, Fig. 1.1(b) provides an image-domain explanation of the (1,-1)-moiré between two superposed gratings. Based on its geometry, the period and the angle of the moiré bands (Eqs. (2.9) and (2.10)) can be directly derived [Nishijima64]. Can you find, using Fig. 4.12, a geometric image-domain explanation for the magnification and the rotation properties of the (1,0,-1,0)-moiré between two dot-screens as shown in Fig. 4.4(a)? (*Hint*: note how the moiré image is formed by sampling the "1"-shaped dots of layer B through the tiny pinholes of layer A.) Can you also explain in a similar way the inverse-video case of Fig. 4.4(d)? Can such geometric considerations replace the Fourier-based explanation in more complex cases, such as the (1,0,-1,1)-moiré shown in Fig. 4.9, or higher order moirés between two (or more) dot-screens?

**Figure 4.12:** A schematic magnification of the dot-screen superposition of Fig. 4.4(a). The black background between the tiny pinholes of layer A has been removed to make the details of layer B clearly visible. Gray areas show the dark parts of the moiré.

# Chapter 5

## The algebraic foundation of the spectrum properties[1]

### 5.1 Introduction

We have seen in the previous chapters that the spectrum-convolution (i.e., the spectrum of the layer superposition) consists of a "forest" of impulses with real- or complex-valued amplitudes, depending on the symmetry properties in the image domain. We have also seen that the occurrence of a moiré phenomenon in the image superposition is associated with the appearance of 1D or 2D impulse clusters in the spectrum (see Figs. 2.5 and 4.3). By now, we have already explained the role of the main cluster, the one which appears around the spectrum origin; but we did not yet characterize the other clusters which are simultaneously generated in the spectrum of the superposition.

The aim of the present chapter is to help us acquire a full understanding of the spectrum of the superposition and its structural properties. We will pursue this goal by formalizing the structure of the spectrum (the impulse "forest" and "clusters") using an algebraic approach which is based on the theory of *geometry of numbers* (see the Glossary in Appendix D). One of the distinctive characteristics of geometry of numbers is that it combines concepts from both continuous and discrete mathematics. This theory will help us to fully understand the clusterization phenomenon in the spectrum of the superposition, which is a discrete phenomenon, using our knowledge from linear algebra of vector spaces, that belongs to the realm of continuous mathematics. In particular, this approach will provide a complete identification of all impulses which participate in each of the clusters in the spectrum. An extensive set of illustrative examples is given in Sec. 5.7.

Note that throughout this algebraic discussion we will completely ignore the amplitudes of the impulses in the spectrum, and we will only concentrate on their indices, their geometric locations, and the relations between them. Only then, based on the algebraic results obtained in the present chapter, we will reintroduce in Chapter 6 the impulse amplitudes, and relate the algebraic structure of the spectrum, via the Fourier theory, to properties of the layer superposition and its moirés back in the image domain.

### 5.2 The support of a spectrum; lattices and modules

From the algebraic point of view, the spectrum-plane $u,v$ is considered as a 2D Euclidean vector space $\mathbb{R}^2$. The geometric location of each impulse is therefore a point (or a vector; we will not distinguish between points and their corresponding vectors) with coordinates $(u,v)$ in this plane (see Sec. 2.2 and Fig. 2.1).

---

[1] This chapter can be skipped or browsed rapidly upon first reading, and revisited later when required.

The set of the geometric locations on the $u,v$ plane of all the impulses in a given spectrum (either the spectrum of a single layer or the spectrum of a layer superposition) is called the *support* of that spectrum. It is important to note that the support of a spectrum contains the geometric locations of *all* the impulses in the spectrum, including those whose amplitudes happen to be zero; this ensures that there are no "gaps" or "holes" in the algebraic structure of the support.[2] As we have already seen in Chapter 2 the support of the spectrum is only determined by the frequencies and angles of the superposed layers, but it is invariant under changes in the profile shape within the period of each layer; such changes do not influence the impulse *locations* in the spectrum, but only their *amplitudes* (and the fact that an impulse amplitude may become 0 or depart from 0 does not influence the spectrum support, either).

### 5.2.1 Lattices and modules in $\mathbb{R}^n$

Let us define here two algebraic structures that will frequently occur in the following discussions concerning the support of a spectrum.

**Definition 5.1:** Let $\mathbf{v}_1,...,\mathbf{v}_m$ be $m$ vectors in $\mathbb{R}^n$ that are linearly independent[3] over $\mathbb{R}$ (obviously, $m \le n$). The set of all the points (vectors) in $\mathbb{R}^n$ given by:

$$L = \{n_1\mathbf{v}_1 + ... + n_m\mathbf{v}_m \mid n_i \in \mathbb{Z}\}$$

(i.e., all the linear combinations of the vectors $\mathbf{v}_1,...,\mathbf{v}_m$ with *integer* coefficients) is called a *lattice* (or a *dot-lattice*) in $\mathbb{R}^n$ [Cassels71 p. 9]. ■

The vectors $\mathbf{v}_1,...,\mathbf{v}_m$ are called a *basis* or an *integral basis* (over $\mathbb{Z}$) of the lattice $L$; like in a vector space the basis of a lattice is not unique, but the expression of any vector of the lattice according to a given basis is unique. The number $m$ of vectors in the basis is called the *rank* of the lattice. It is interesting to note that unlike in vector spaces, $r$ linearly independent vectors in a lattice of rank $r$ are not always a basis of the lattice; for example, (1,0) and (0,2) are linearly independent vectors in $\mathbb{Z}^2$, but they do not span the whole of $\mathbb{Z}^2$.

$\mathbb{Z}^n$ is a trivial example of a lattice of rank $n$ in $\mathbb{R}^n$; in fact, any other lattice of rank $n$ in $\mathbb{R}^n$ is obtained from $\mathbb{Z}^n$ by a non-singular linear transformation $\Phi$: $\mathbb{R}^n \rightarrow \mathbb{R}^n$ [Siegel89 p. 18; Gruber93 p. 741].

**Definition 5.2:** Let $\mathbf{v}_1,...,\mathbf{v}_m$ be $m$ arbitrary vectors in $\mathbb{R}^n$. The set of all the points (vectors) in $\mathbb{R}^n$ given by:

$$M = \{n_1\mathbf{v}_1 + ... + n_m\mathbf{v}_m \mid n_i \in \mathbb{Z}\}$$

is called a $\mathbb{Z}$*-module* (or in short: a *module*)[4] in $\mathbb{R}^n$. ■

---

[2] Note that this definition is different from the usual one, according to which the support of a function $f(x)$ is the closure of the set of $x$ values for which $f(x) \neq 0$.

[3] Non-zero vectors $\mathbf{v}_1,...,\mathbf{v}_m$ in $\mathbb{R}^n$ are called *linearly independent* over $\mathbb{R}$ (or over $\mathbb{Z}$, etc.) if $t_1\mathbf{v}_1 + ... + t_m\mathbf{v}_m = \mathbf{0}$ with $t_i \in \mathbb{R}$ (respectively, $t_i \in \mathbb{Z}$) implies that $t_1 = ... = t_m = 0$.

[4] Note that this definition of a $\mathbb{Z}$-module is more restrictive than the classical definitions of a module (like in [Siegel89 p. 43] or [Artin91 p. 450]), according to which a module is a generalization of a vector space so that any vector space (including $\mathbb{R}^n$ itself) is also considered as a module.

The vectors $\mathbf{v}_1,\ldots,\mathbf{v}_m$ are called *generating vectors* of the module $M$, but they are not generally a basis, since they are not necessarily linearly independent in $\mathbb{R}^n$ (and in fact, their number $m$ may be even larger than $n$). The maximum number $r$ of linearly independent (over $\mathbb{R}$) vectors in a module $M$ is called the *rank* of $M$ (denoted: $\text{rank}_{\mathbb{R}}M = r$, or simply: $\text{rank}M = r$); it is clear that $r \le m$ and $r \le n$ [Siegel89 p. 44]. We will call the maximum number $z$ of linearly independent vectors over $\mathbb{Z}$ in a module $M$ the *integral rank* of $M$ (denoted: $\text{rank}_{\mathbb{Z}}M = z$); it is clear that $r \le z \le m$.[5]

Clearly, a lattice is a special case of a module, in which the $m$ generating vectors are linearly independent; in the case of a lattice we have, therefore, $r = z = m$. This means that every lattice is also a module; but not every module is a lattice. While a lattice is always a *discrete* subset of $\mathbb{R}^n$ (meaning that it does not contain arbitrarily close points), a $\mathbb{Z}$-module may be *dense* in $\mathbb{R}^n$ (even though it is not continuous).[6] Consider, for example, the following module in $\mathbb{R}^2$: $M_1 = \{k(1,0) + l(\sqrt{2},0) \mid k,l \in \mathbb{Z}\}$. $M_1$ is generated by the vectors $(1,0)$ and $(\sqrt{2},0)$, and its integral rank is 2; however, its rank is only 1, since all its members are located within $\mathbb{R}^2$ on a straight line (the $x$ axis). Moreover, the module $M_1$ is dense on this line, although it does not fully cover the entire continuous line: for example, $(\frac{1}{2},0) \notin M_1$. As a second example, consider the following module in $\mathbb{R}^2$: $M_2 = \{k(1,0) + l(\frac{1}{2},\frac{1}{2}) + m(0,1) \mid k,l,m \in \mathbb{Z}\}$. Although $M_2$ is generated by three vectors in $\mathbb{R}^2$, the third vector is actually redundant in this case (since it can be obtained as an integral linear combination of the two others), and the module $M_2$ coincides with an oblique lattice in $\mathbb{R}^2$ having the basis: $(1,0)$, $(\frac{1}{2},\frac{1}{2})$.

These two examples can be summarized as follows:

| | $\text{rank}_{\mathbb{R}}$ | | $\text{rank}_{\mathbb{Z}}$ | Description: |
|---|---|---|---|---|
| $M_1$ | 1 | < | 2 | 1D dense module |
| $M_2$ | 2 | = | 2 | 2D discrete lattice |

In fact, the following general property holds:

**Proposition 5.1:** A module in $\mathbb{R}^n$ is a lattice *iff* it is discrete [Siegel89 p. 44]; and a module in $\mathbb{R}^n$ is not a lattice *iff* it is dense in a subgroup of $\mathbb{R}^n$. Moreover, using the notation $r = \text{rank}_{\mathbb{R}}M$ and $z = \text{rank}_{\mathbb{Z}}M$, a module $M$ is a lattice (and therefore discrete) *iff* $z = r$; the module is not a lattice (and is dense in a subgroup of $\mathbb{R}^n$) *iff* $z > r$.[5] ■

It is interesting to note that a module does not necessarily have a basis (over $\mathbb{R}$). For example, we have seen that the module $M_1$ in the example above is of $\text{rank}_{\mathbb{R}} = 1$; but still, it

[5] Note that $z < r$ is impossible, since linear independence over $\mathbb{R}$ implies linear independence over $\mathbb{Z}$ (and linear dependence over $\mathbb{Z}$ implies linear dependence over $\mathbb{R}$).

[6] Formally, a subset $D$ of $\mathbb{R}^n$ is called *discrete* if there exists a number $d > 0$ such that for any points $a,b \in D$ the distance between $a$ and $b$ is larger than $d$. A subset $S$ of $\mathbb{R}^n$ is called *dense* or *everywhere dense* in $\mathbb{R}^n$ if $[S] = \mathbb{R}^n$, where $[S]$ denotes the closure of $S$, i.e., the set containing $S$ and all its limit points [EncMath88, Vol. 3, p. 434]. Examples: (1) The set of all integer numbers is discrete. (2) Both the set of all rational numbers and the set of all irrational numbers are dense in $\mathbb{R}$, although none of them is continuous in $\mathbb{R}$.

cannot be generated by a single vector (since a single vector generates a superset of $M_1$, the entire line). This means that there exists no basis to $M_1$. But although a module $M$ does not necessarily have a basis (over $\mathbb{R}$), it does always have an *integral basis* (over $\mathbb{Z}$) which spans it: If the $m$ generating vectors $\mathbf{v}_1,...,\mathbf{v}_m$ of the module $M$ are linearly independent (over $\mathbb{Z}$), they are themselves an integral basis of $M$, and $\text{rank}_{\mathbb{Z}}M = m$. Otherwise, we take the minimal subset $\mathbf{v}_1,...,\mathbf{v}_z$ from the $m$ generating vectors which still spans the module $M$; $\mathbf{v}_1,...,\mathbf{v}_z$ are linearly independent over $\mathbb{Z}$ (since otherwise one of them is a linear combination over $\mathbb{Z}$ of the others, and $\mathbf{v}_1,...,\mathbf{v}_z$ is not minimal). Therefore $\mathbf{v}_1,...,\mathbf{v}_z$ are an integral basis (over $\mathbb{Z}$) of $M$, and their number $z$ is the integral rank of $M$.

**Notation:** Let $\mathbf{v}_1,...,\mathbf{v}_m$ be $m$ arbitrary vectors in $\mathbb{R}^n$. We denote the *vector space* and the *module* which are spanned (generated) by these vectors by:

$$\text{Sp}(\mathbf{v}_1,...,\mathbf{v}_m) = \{n_1\mathbf{v}_1 + ... + n_m\mathbf{v}_m \mid n_i \in \mathbb{R}\}$$

$$\text{Md}(\mathbf{v}_1,...,\mathbf{v}_m) = \{n_1\mathbf{v}_1 + ... + n_m\mathbf{v}_m \mid n_i \in \mathbb{Z}\}$$

$\text{Sp}(\mathbf{v}_1,...,\mathbf{v}_m)$ is the set of all the linear combinations over $\mathbb{R}$ of the vectors $\mathbf{v}_1,...,\mathbf{v}_m \in \mathbb{R}^n$, and $\text{Md}(\mathbf{v}_1,...,\mathbf{v}_m)$ is the set of all their linear combinations over $\mathbb{Z}$. The notations Sp( ) and Md( ) can be also used in the case of an infinite set of vectors $\mathbf{v}_1,\mathbf{v}_2,... \in \mathbb{R}^n$. ■

Clearly, $\text{Sp}(\mathbf{v}_1,...,\mathbf{v}_m)$ is a subspace of the vector space $\mathbb{R}^n$, whereas $\text{Md}(\mathbf{v}_1,...,\mathbf{v}_m)$ is a module within this subspace: $\text{Md}(\mathbf{v}_1,...,\mathbf{v}_m) \subset \text{Sp}(\mathbf{v}_1,...,\mathbf{v}_m) \subseteq \mathbb{R}^n$. While $\text{Sp}(\mathbf{v}_1,...,\mathbf{v}_m)$ is continuous and has the cardinality of the continuum, the module $\text{Md}(\mathbf{v}_1,...,\mathbf{v}_m)$ is only a denumerable infinite set which is imbedded within $\text{Sp}(\mathbf{v}_1,...,\mathbf{v}_m)$, and it is either discrete or dense in it. Moreover, we have:

$$\text{Sp}(\text{Md}(\mathbf{v}_1,...,\mathbf{v}_m)) = \text{Sp}(\mathbf{v}_1,...,\mathbf{v}_m)$$

This means that $\text{Sp}(\mathbf{v}_1,...,\mathbf{v}_m)$ is the smallest subspace of $\mathbb{R}^n$ which includes the module $\text{Md}(\mathbf{v}_1,...,\mathbf{v}_m)$; we will call it the *continuous extension* of the module. It is clear that "filling the gaps" inside the module $\text{Md}(\mathbf{v}_1,...,\mathbf{v}_m)$ by admitting $n_i \in \mathbb{R}$ rather than $n_i \in \mathbb{Z}$ does not change the number of independent vectors over $\mathbb{R}$, so that we have:

$$\text{rank}_{\mathbb{R}}\, \text{Md}(\mathbf{v}_1,...,\mathbf{v}_m) = \dim \text{Sp}(\mathbf{v}_1,...,\mathbf{v}_m) \tag{5.1}$$

Using these new terms we can now reformulate some results which were obtained earlier in this section:

Since linear independence over $\mathbb{R}$ implies linear independence over $\mathbb{Z}$, it is clear that for any set of vectors $\mathbf{v}_1,...,\mathbf{v}_m$ the maximum number of linearly independent vectors over $\mathbb{Z}$ is greater than or equal to the maximum number of linearly independent vectors over $\mathbb{R}$:

$$\text{rank}_{\mathbb{Z}}\, \text{Md}(\mathbf{v}_1,...,\mathbf{v}_m) \geq \text{rank}_{\mathbb{R}}\, \text{Md}(\mathbf{v}_1,...,\mathbf{v}_m)$$

and by Eq. (5.1):
$$\text{rank}_{\mathbb{Z}}\, \text{Md}(\mathbf{v}_1,...,\mathbf{v}_m) \geq \dim \text{Sp}(\mathbf{v}_1,...,\mathbf{v}_m) \tag{5.2}$$

And furthermore, reformulating Proposition 5.1 we obtain:

The module $M = \text{Md}(\mathbf{v}_1,...,\mathbf{v}_m)$ is a lattice (and therefore discrete) *iff* the equality in Eq. (5.2) holds, i.e.: $\text{rank}_{\mathbb{Z}}\, \text{Md}(\mathbf{v}_1,...,\mathbf{v}_m) = \dim \text{Sp}(\mathbf{v}_1,...,\mathbf{v}_m)$
and conversely, $M$ is not a lattice (and is dense on a subgroup of $\mathbb{R}^n$) *iff* the inequality in Eq. (5.2) holds, i.e.: $\text{rank}_{\mathbb{Z}}\, \text{Md}(\mathbf{v}_1,...,\mathbf{v}_m) > \dim \text{Sp}(\mathbf{v}_1,...,\mathbf{v}_m)$.

### 5.2.2 Application to the frequency spectrum

Let us now proceed from the general case (with vectors $\mathbf{v}_1,...,\mathbf{v}_m \in \mathbb{R}^n$) to our particular case of interest, in which $\mathbf{f}_1,...,\mathbf{f}_m \in \mathbb{R}^2$ are frequency vectors in the spectrum plane $u,v$. Let us start with some examples:

**Example 5.1:** The support of the spectrum of any *periodic* function of two variables $p(x,y)$ is a lattice in $\mathbb{R}^2$, i.e., in the $u,v$ plane; this follows from the decomposition of the periodic function into a Fourier series (see Appendix A). If $p(x,y)$ is 2-fold periodic, the support of its spectrum is a 2D lattice (see, for example, Fig. A.2 in Appendix A). If $p(x,y)$ is 1-fold periodic, like a line-grating, the support of its spectrum is a 1D lattice on a straight line through the origin of the $u,v$ plane (Fig. A.1). This 1D lattice consists of all the points $k\mathbf{f}$ where $\mathbf{f}$ is the fundamental frequency of $p(x,y)$ and $k$ runs through all integers. Note that all functions with the same period have an identical spectrum support, even when some (or even most) of the impulses in their spectra happen to have a zero amplitude, as in the case of $p(x) = \cos(2\pi x/T)$. ■

**Example 5.2:** Let $r_1(x,y)$ and $r_2(x,y)$ be line gratings, with fundamental frequency vectors $\mathbf{f}_1$ and $\mathbf{f}_2$, respectively, as in Fig. 2.5. As we have seen, the spectrum of each of them is an impulse comb; and if we superpose (i.e., multiply) $r_1(x,y)$ and $r_2(x,y)$, the spectrum of their superposition is the convolution of these two combs. The support of this spectrum convolution (see Fig. 2.5(f)) is given by: $\text{Md}(\mathbf{f}_1,\mathbf{f}_2) = \{n_1\mathbf{f}_1 + n_2\mathbf{f}_2 \mid n_i \in \mathbb{Z}\}$, which is a module in the spectrum plane $u,v$. If the vectors $\mathbf{f}_1$ and $\mathbf{f}_2$ are linearly independent (over $\mathbb{R}$) in $\mathbb{R}^2$, they are also linearly independent over $\mathbb{Z}$, so that $z = r = 2$, and therefore this module is in fact a lattice of rank 2, as in Fig. 2.5(f). Otherwise, i.e., if $\mathbf{f}_1$ and $\mathbf{f}_2$ are collinear (=linearly dependent over $\mathbb{R}$), there exist two possible cases:

(1) If $\mathbf{f}_1$ and $\mathbf{f}_2$ are also linearly dependent over $\mathbb{Z}$ (so that $z = r = 1$), or in other words if $\mathbf{f}_1$ and $\mathbf{f}_2$ are *commensurable* (i.e., the ratio of their lengths is rational),[7] then $\text{Md}(\mathbf{f}_1,\mathbf{f}_2)$ is a lattice of rank 1 which is located on the line spanned by $\mathbf{f}_1$ and $\mathbf{f}_2$.

(2) If $\mathbf{f}_1$ and $\mathbf{f}_2$ are linearly independent over $\mathbb{Z}$ (so that $z > r$), or in other words if $\mathbf{f}_1$ and $\mathbf{f}_2$ are *incommensurable*, then $\text{Md}(\mathbf{f}_1,\mathbf{f}_2)$ becomes a dense set of points on the line spanned by $\mathbf{f}_1$ and $\mathbf{f}_2$, namely: a module of rank 1 and integral rank of 2. ■

---

[7] Two vectors $\mathbf{v}_1,\mathbf{v}_2 \in \mathbb{R}^2$ (or real numbers in $\mathbb{R}$) are called *commensurable* if there exist non-zero integers $m,n$ such that $\mathbf{v}_2=(m/n)\mathbf{v}_1$. This means that both $\mathbf{v}_1$ and $\mathbf{v}_2$ can be measured as integer multiples of the same length unit, say $(1/n)\mathbf{v}_1$. More generally, $k$ vectors $\mathbf{v}_1,...,\mathbf{v}_k$ in $\mathbb{R}^n$ (or real numbers in $\mathbb{R}$) are called *commensurable* if they are linearly dependent over $\mathbb{Q}$ (the set of all rational numbers); note that this is identical to linear dependence over $\mathbb{Z}$. And conversely, vectors (or real numbers) that are linearly independent over $\mathbb{Q}$ (or over $\mathbb{Z}$) are called *incommensurable* [EncMath88 "Quasi-periodic motion", Vol. 7 p. 436]. Note that $\mathbf{v}_1,...,\mathbf{v}_k$ are commensurable *iff* $\text{rank}_{\mathbb{Z}}\, \text{Md}(\mathbf{v}_1,...,\mathbf{v}_k) < k$; they are incommensurable *iff* $\text{rank}_{\mathbb{Z}}\, \text{Md}(\mathbf{v}_1,...,\mathbf{v}_k) = k$.

**Example 5.3:** In the general case, if we superpose $m$ line gratings whose frequency vectors are $\mathbf{f}_i$, then the support of their spectrum convolution is given by the module:

$$\text{Md}(\mathbf{f}_1,...,\mathbf{f}_m) = \{n_1\mathbf{f}_1 + ... + n_m\mathbf{f}_m \mid n_i \in \mathbb{Z}\} \quad \text{where } \mathbf{f}_i \in \mathbb{R}^2. \tag{5.3}$$

The rank of this module is obviously $r \leq 2$, since it is imbedded in the 2D spectral plane $u,v$, but as for its integral rank $z$ we only know that $r \leq z \leq m$. Therefore, here again, there exist two possible cases: If $z > r$ then the spectrum support $\text{Md}(\mathbf{f}_1,...,\mathbf{f}_m)$ is not a lattice but rather a dense module. But as we will see below, in some cases it may happen that $z = r$, so that the spectrum support $\text{Md}(\mathbf{f}_1,...,\mathbf{f}_m)$ does coincide in the $u,v$ plane with a 2D or 1D lattice, and is discrete. ■

In the discussion below we will also need the continuous counterpart of $\text{Md}(\mathbf{f}_1,...,\mathbf{f}_m)$, namely:

$$\text{Sp}(\mathbf{f}_1,...,\mathbf{f}_m) = \{n_1\mathbf{f}_1 + ... + n_m\mathbf{f}_m \mid n_i \in \mathbb{R}\} \quad \text{where } \mathbf{f}_i \in \mathbb{R}^2. \tag{5.4}$$

It is clear that $\text{Sp}(\mathbf{f}_1,...,\mathbf{f}_m)$ is a subspace of $\mathbb{R}^2$ (it may either coincide with $\mathbb{R}^2$, if $\dim \text{Sp}(\mathbf{f}_1,...,\mathbf{f}_m)=2$, or be a line through its origin, if $\dim \text{Sp}(\mathbf{f}_1,...,\mathbf{f}_m)=1$; $\dim \text{Sp}(\mathbf{f}_1,...,\mathbf{f}_m)=0$ is a degenerate case which occurs when the spectrum only contains the DC impulse and represents a constant image). We therefore have: $\text{Md}(\mathbf{f}_1,...,\mathbf{f}_m) \subset \text{Sp}(\mathbf{f}_1,...,\mathbf{f}_m) \subseteq \mathbb{R}^2$.

## 5.3 The mapping between the impulse indices and their geometric locations

We return now to the fundamental Eq. (2.26) which specifies for every $(n_1,...,n_m)$-impulse in the spectrum-convolution its impulse location in the $u,v$ plane:

$$\mathbf{f}_{n_1,...,n_m} = n_1\mathbf{f}_1 + ... + n_m\mathbf{f}_m$$

Note that throughout the discussion which follows the index $m$ counts 1-fold periodic layers (gratings) in the superposition, and each 2-fold periodic layer is counted as two 1-fold periodic layers. Let now $\mathbf{f}_{n_1,...,n_m} = n_1\mathbf{f}_1 + ... + n_m\mathbf{f}_m$ be a point (vector) in $\text{Md}(\mathbf{f}_1,...,\mathbf{f}_m)$; this point is the geometric location in the $u,v$ plane of the $(n_1,...,n_m)$-impulse of the spectrum convolution. As we can see, the index-vector $(n_1,...,n_m)$ of this impulse defines a point in $\mathbb{Z}^m$, the lattice of all the points in $\mathbb{R}^m$ having integer coordinates: $\mathbb{Z}^m = \{(n_1,...,n_m) \mid n_i \in \mathbb{Z}\}$. This lattice will henceforth be called the *indices-lattice*. The $(n_1,...,n_m)$-impulse can therefore be referred to in two different ways: either by its index-vector $(n_1,...,n_m) \in \mathbb{Z}^m$, or by its geometric location in the $u,v$ spectrum plane, $\sum n_i\mathbf{f}_i \in \text{Md}(\mathbf{f}_1,...,\mathbf{f}_m)$ (see for example Eqs. (4.10) vs. (4.11)).

Moreover, for any given set of frequency vectors $\mathbf{f}_1,...,\mathbf{f}_m \in \mathbb{R}^2$ (i.e., for any given superposition of $m$ gratings) there exists a natural mapping between the indices of the impulses and their geometric locations. This mapping from the indices-lattice $\mathbb{Z}^m$ to the corresponding module (spectrum support) $\text{Md}(\mathbf{f}_1,...,\mathbf{f}_m)$ in the $u,v$ plane is given by the linear transformation (homomorphism) $\Psi_{\mathbf{f}_1,...,\mathbf{f}_m} : \mathbb{Z}^m \rightarrow \text{Md}(\mathbf{f}_1,...,\mathbf{f}_m)$ which is defined by:

$$\Psi_{\mathbf{f}_1,...,\mathbf{f}_m}(n_1,...,n_m) = n_1\mathbf{f}_1 + ... + n_m\mathbf{f}_m \quad (5.5)$$

We will see below that this transformation is closely related to the moirés generated in the superposition of the $m$ gratings defined by the frequency vectors $\mathbf{f}_1,...,\mathbf{f}_m$. For example, we will see that the transformation $\Psi_{\mathbf{f}_1,...,\mathbf{f}_m}$ is singular *iff* the vectors $\mathbf{f}_1,...,\mathbf{f}_m$ represent a singular moiré (see Sec. 5.4.1).

Note that although this linear transformation is only defined here for integer coordinates $n_i$, i.e., between $\mathbb{Z}^m$ and Md$(\mathbf{f}_1,...,\mathbf{f}_m)$, it has a natural continuous extension to their full enclosing vector spaces $\mathbb{R}^m$ and Sp$(\mathbf{f}_1,...,\mathbf{f}_m)$: By admitting that $n_i \in \mathbb{R}$ rather than $n_i \in \mathbb{Z}$, $\Psi_{\mathbf{f}_1,...,\mathbf{f}_m}$ becomes a continuous linear transformation $\Phi_{\mathbf{f}_1,...,\mathbf{f}_m} : \mathbb{R}^m \rightarrow$ Sp$(\mathbf{f}_1,...,\mathbf{f}_m)$ (where Sp$(\mathbf{f}_1,...,\mathbf{f}_m) \subseteq \mathbb{R}^2$), which is defined the same way as $\Psi_{\mathbf{f}_1,...,\mathbf{f}_m}$ above.

Obviously, each choice of the vectors $\mathbf{f}_1,...,\mathbf{f}_m \in \mathbb{R}^2$ (the fundamental frequency vectors of the $m$ superposed gratings) defines a different linear transformation $\Psi_{\mathbf{f}_1,...,\mathbf{f}_m}$, which maps the $(n_1,...,n_m)$-impulse to the point (geometric location) $n_1\mathbf{f}_1 + ... + n_m\mathbf{f}_m$ in the spectrum plane $u,v$. We will first consider $\Psi_{\mathbf{f}_1,...,\mathbf{f}_m}$ as a function of $(n_1,...,n_m)$ alone, with an arbitrary fixed set of frequency vectors $\mathbf{f}_1,...,\mathbf{f}_m$. Only then, in Sec. 5.6.3 below, we will consider $\Psi_{\mathbf{f}_1,...,\mathbf{f}_m}$ as a function of the frequency vectors $\mathbf{f}_1,...,\mathbf{f}_m$ as well, and we will see what happens in the spectrum when $\mathbf{f}_1,...,\mathbf{f}_m$ are being varied.

## 5.4 A short reminder from linear algebra

In order to better understand the properties of the discrete linear transformation (5.5), we will first study its continuous extension $\Phi_{\mathbf{f}_1,...,\mathbf{f}_m}$, whose properties can easily be determined using some basic notions from linear algebra. For this end, we will briefly review in the present section the needed algebraic notions concerning vector spaces and linear transformations between them (see, for example, [Lipschutz91]). Then, in the following section we will return to the original discrete transformation (5.5) and study its properties by considering it as the *restriction* of the continuous transformation $\Phi_{\mathbf{f}_1,...,\mathbf{f}_m}$, i.e., with $n_i \in \mathbb{Z}$ rather than $n_i \in \mathbb{R}$.

### 5.4.1 The image and the kernel of a linear transformation

Let $\Phi$ be a linear transformation from a vector space $V$ to a vector space $W$, i.e., $\Phi: V \rightarrow W$. The *image* of $\Phi$ and the *kernel* of $\Phi$ are defined as:

$$\text{Im}\Phi = \{\mathbf{w} \in W \mid \mathbf{w} = \Phi(\mathbf{v}), \mathbf{v} \in V\}$$

$$\text{Ker}\Phi = \{\mathbf{v} \in V \mid \Phi(\mathbf{v}) = \mathbf{0}\}$$

Both Im$\Phi$ and Ker$\Phi$ are vector spaces (subspaces of $W$ and $V$, respectively), and moreover, there exists between their dimensions the following relationship (see, for example, [Lipschutz91 p. 318]):

$$\dim \mathrm{Ker}\Phi + \dim \mathrm{Im}\Phi = \dim V \tag{5.6}$$

In the case of our particular transformation $\Phi_{\mathbf{f}_1,...,\mathbf{f}_m}$, $V$ and $W$ are respectively $\mathbb{R}^m$ and $\mathbb{R}^2$ and $\mathrm{Im}\Phi_{\mathbf{f}_1,...,\mathbf{f}_m}$ is the subspace $\mathrm{Sp}(\mathbf{f}_1,...,\mathbf{f}_m) \subseteq \mathbb{R}^2$, where $m$ is the number of superposed gratings. Therefore $\dim V = m$ and $\dim \mathrm{Im}\Phi_{\mathbf{f}_1,...,\mathbf{f}_m} = 2$ (or: $\dim \mathrm{Im}\Phi_{\mathbf{f}_1,...,\mathbf{f}_m} = 1$; the degenerate case of $\dim \mathrm{Im}\Phi_{\mathbf{f}_1,...,\mathbf{f}_m} = 0$, in which the spectrum only contains the DC impulse, can be ignored). Hence we obtain:

$$\begin{aligned} \dim \mathrm{Ker}\Phi_{\mathbf{f}_1,...,\mathbf{f}_m} &= \dim V - \dim \mathrm{Im}\Phi_{\mathbf{f}_1,...,\mathbf{f}_m} \\ &= m - 2 \qquad \text{(or: } m - 1\text{)} \end{aligned} \tag{5.7}$$

Therefore, when $m \geq 3$ (or respectively: $m \geq 2$) we obtain $\dim \mathrm{Ker}\Phi_{\mathbf{f}_1,...,\mathbf{f}_m} \geq 1$. This means that a non-trivial subspace of $\mathbb{R}^m$ is mapped, under the transformation $\Phi_{\mathbf{f}_1,...,\mathbf{f}_m}$, to the location (0,0) in the $u,v$ plane, i.e., to the spectrum origin. We will see shortly (in Sec. 5.5) the significance of this fact.

A linear transformation $\Phi : V \to W$ is called *singular* if there exists a non-zero $\mathbf{v} \in V$ that is mapped by $\Phi$ to $\mathbf{0} \in W$ (i.e., if $\dim \mathrm{Ker}\Phi \geq 1$). The linear transformation $\Phi$ is called *regular* or *non-singular* if the only vector in $V$ that it maps to $\mathbf{0} \in W$ is $\mathbf{0} \in V$, i.e., if $\mathrm{Ker}\Phi = \{\mathbf{0}\}$ (or still in other words, if $\dim \mathrm{Ker}\Phi = 0$).

In our case, therefore, the linear transformation $\Phi_{\mathbf{f}_1,...,\mathbf{f}_m}$ as well as its discrete counterpart $\Psi_{\mathbf{f}_1,...,\mathbf{f}_m}$ of Eq. (5.5) are singular *iff* there exist $k_i \in \mathbb{Z}$ not all of them 0 such that $\sum k_i\mathbf{f}_i = \mathbf{0}$, which means that the vectors $\mathbf{f}_1,...,\mathbf{f}_m$ are linearly dependent over $\mathbb{Z}$. But according to Sec. 2.9 this also means that $\Phi_{\mathbf{f}_1,...,\mathbf{f}_m}$ and $\Psi_{\mathbf{f}_1,...,\mathbf{f}_m}$ are singular *iff* the superposition is singular (i.e., it contains a singular moiré).

### 5.4.2 Partition of a vector space into equivalence classes

Let $V$ be a vector space and let $U$ be a subspace in $V$. For any vector $\mathbf{v} \in V$ we define:

$$\mathbf{v} + U = \{\mathbf{v} + \mathbf{u} \mid \mathbf{u} \in U\}$$

$\mathbf{v} + U$ is a copy of the subspace $U$ inside $V$, parallel to $U$, which is shifted (translated) from the origin by the vector $\mathbf{v}$. The set $\mathbf{v} + U$ is called the *equivalence class* (or *coset*) of the vector $\mathbf{v}$ in $V$ modulo $U$ (i.e., with respect to $U$). The set of all the equivalence classes in $V$ gives a disjoint and exhaustive partition of $V$ (that is: for any $\mathbf{v} \in V$, $\mathbf{v}$ is a member of exactly one equivalence class modulo $U$). All the vectors within the same equivalence class are called *equivalent modulo U.*[8] An element chosen from an equivalence class is called a *representative* of its equivalence class.

As an illustration, let $V$ be the 3D Euclidean space $\mathbb{R}^3$, and let $U$ be the 2D subspace of $\mathbb{R}^3$ defined by: $U = \{(x,y,z) \mid z = x + y\}$ (i.e., a plane through the origin). Then, each

[8] Formally, this means that if $\mathbf{a}$ and $\mathbf{b}$ belong to the same class $\mathbf{v} + U$, then there exists a vector $\mathbf{u} \in U$ such that $\mathbf{a} = \mathbf{b} + \mathbf{u}$ (or in other words, $\mathbf{a} - \mathbf{b} \in U$). This means that $\mathbf{a}$ and $\mathbf{b}$ belong to the same "translated copy of the subspace $U$" within $V$. This relation between $\mathbf{a}$ and $\mathbf{b}$ is indeed an equivalence relation, i.e., it is reflexive, symmetric and transitive.

equivalence class (modulo $U$) in $\mathbb{R}^3$ is a parallel translation of the $U$ plane within $\mathbb{R}^3$: $(x_0,y_0,z_0) + U = \{(x_0,y_0,z_0) + (x,y,z) \mid z = x+y\}$.

Note that the only equivalence class in $V$ which is itself a vector space is $\mathbf{0} + U$ (the equivalence class of the vector $\mathbf{0}$), i.e., the subspace $U$ itself; all the other classes are parallel translations of $U$ within the vector space $V$, and they do not contain the vector $\mathbf{0}$. Nevertheless, we will still say that each of the translated equivalence classes has the same dimension as the original, unshifted subspace $U$: $\dim(\mathbf{v} + U) = \dim U$.

The set of all the equivalence classes modulo $U$ in $V$ is itself a vector space, which is called the *quotient space* and denoted $V/U$. Between the dimensions of the vector spaces $V$, $U$ and $V/U$ there exists the following relationship [Lipschutz91 pp. 386–387]:

$$\dim U + \dim (V/U) = \dim V \tag{5.8}$$

Finally, it is important to note that for each subspace $U$ of $V$ we get a different partition of $V$ into equivalence classes (modulo $U$). We say that each subspace $U$ of $V$ *induces* a different partition of $V$ into equivalence classes. In the following we will concentrate on one particular partition of $V$, which has some special properties.

### 5.4.3 The partition of $V$ into equivalence classes induced by $\Phi$

Let $\Phi$ be a linear transformation from a vector space $V$ to a vector space $W$, i.e., $\Phi : V \to W$. Since $\mathrm{Ker}\Phi$ is a subspace of $V$, it follows that $\mathrm{Ker}\Phi$ induces a partition of $V$ into equivalence classes. This particular partition of $V$ has an important property: the whole equivalence class of $\mathbf{0}$ within $V$, i.e., $\mathrm{Ker}\Phi$, is mapped by $\Phi$ onto $\mathbf{0} \in W$; and moreover, each of the other equivalence classes within $V$, $\mathbf{v} + \mathrm{Ker}\Phi$, is mapped by $\Phi$ onto a single point within $\mathrm{Im}\Phi$: $\Phi(\mathbf{v} + \mathrm{Ker}\Phi) = \Phi(\mathbf{v}) + \mathbf{0} = \Phi(\mathbf{v})$. Furthermore, two vectors $\mathbf{a},\mathbf{b} \in V$ are mapped by $\Phi$ to the same point in $\mathrm{Im}\Phi$ *iff* they belong to the same equivalence class of $V$ (modulo $\mathrm{Ker}\Phi$). This means that to each equivalence class in this particular partition of $V$ there belongs one point in $\mathrm{Im}\Phi$, and vice versa; the quotient space $V/\mathrm{Ker}\Phi$ (i.e., the space of all the equivalence classes induced by $\Phi$ in $V$) and the subspace $\mathrm{Im}\Phi$ in $W$ are therefore isomorphic. This is indeed proven by the First Isomorphism theorem [Adkins92 pp. 113–114].

These results can be interpreted, loosely speaking, as a "dimension preservation law" under the linear transformation $\Phi$: Assume, for example, that $\Phi : \mathbb{R}^3 \to \mathbb{R}^2$ (where $\Phi$ is surjective). Here $\dim \mathrm{Im}\Phi = 2$, and therefore by Eq. (5.6) $\dim \mathrm{Ker}\Phi = 3-2 = 1$. Therefore $\mathrm{Ker}\Phi$ is a 1D subspace of $\mathbb{R}^3$, namely: a certain straight line $S$ through the origin. $\mathrm{Ker}\Phi$ (or simply $\Phi$) induces a disjoint and exhaustive partition of the 3D space $\mathbb{R}^3$ into a 2D set (quotient space) containing all the 1D shifted lines parallel to $S$ (equivalence classes). Each of these 1D lines is "collapsed" by $\Phi$ onto a single point in $\mathrm{Im}\Phi$; and hence the 2D set of lines within $\mathbb{R}^3$ is mapped (isomorphically) onto the entire 2D plane $\mathrm{Im}\Phi$. It can be said, loosely speaking, that if 2 dimensions out of the 3 dimensions of $\mathbb{R}^3$ are "used" by $\Phi$ to span $\mathrm{Im}\Phi$, then the $3-2 = 1$ remaining dimensions are "invested" in each point of Im$\Phi$ by mapping onto it an entire 1D portion (shifted line) of $\mathbb{R}^3$. Thus, although the image of $\Phi$ is only 2D, each point in it "absorbs" an entire 1D portion of $\mathbb{R}^3$, so that all the 3 dimensions of $\mathbb{R}^3$ have actually been "used" by $\Phi$. The aim of this "naive" description of the "collapsing" process will become clear later, in Sec. 5.6.

Note that this "collapsing" effect only occurs if $\Phi$ is a singular transformation. If $\Phi$ is non-singular then Ker$\Phi = \{\mathbf{0}\}$, and therefore each point of $V$ forms an equivalence class of its own, containing only a single point. Since dim Ker$\Phi = 0$ it follows from Eq. (5.6) that dim Im$\Phi$ = dim $V$, and therefore the transformation $\Phi$ in this case is an isomorphism which simply maps every single point (equivalence class) of $V$ into a single point in Im$\Phi$. Obviously, this is only possible when dim $V \leq$ dim $W$.

### 5.4.4 The application of these results to our continuous case

Let us return back to our continuous linear transformation $\Phi_{\mathbf{f}_1,...,\mathbf{f}_m} : \mathbb{R}^m \to \mathrm{Sp}(\mathbf{f}_1,...,\mathbf{f}_m) \subseteq \mathbb{R}^2$. Since Ker$\Phi_{\mathbf{f}_1,...,\mathbf{f}_m}$ is a subspace of $\mathbb{R}^m$ (with dimension $m-2$ or $m-1$), it follows from the above discussion that Ker$\Phi_{\mathbf{f}_1,...,\mathbf{f}_m}$ induces a partition of $\mathbb{R}^m$ into equivalence classes (of dimension $m-2$, or respectively, $m-1$). And furthermore, this partition of $\mathbb{R}^m$ has the following special property: the linear transformation $\Phi_{\mathbf{f}_1,...,\mathbf{f}_m}$ maps the whole $m-2$ (or $m-1$) dimensional equivalence class of $\mathbf{0}$ in $\mathbb{R}^m$, Ker$\Phi_{\mathbf{f}_1,...,\mathbf{f}_m}$, into the origin (0,0) of the $u,v$ plane; and similarly, for every $\mathbf{v} \in \mathbb{R}^m$ the transformation $\Phi_{\mathbf{f}_1,...,\mathbf{f}_m}$ maps the whole $m-2$ (or $m-1$) dimensional equivalence class of $\mathbf{v}$ in $\mathbb{R}^m$, $\mathbf{v}$ + Ker$\Phi_{\mathbf{f}_1,...,\mathbf{f}_m}$, into a single point in the $u,v$ plane, $\Phi_{\mathbf{f}_1,...,\mathbf{f}_m}(\mathbf{v})$.

## 5.5 The discrete mapping $\Psi$ vs. the continuous mapping $\Phi$

Let us return to our original discrete transformation $\Psi_{\mathbf{f}_1,...,\mathbf{f}_m} : \mathbb{Z}^m \to \mathrm{Md}(\mathbf{f}_1,...,\mathbf{f}_m)$ given in Eq. (5.5). Looking now at $\Psi_{\mathbf{f}_1,...,\mathbf{f}_m}$ as the *restriction* of $\Phi_{\mathbf{f}_1,...,\mathbf{f}_m}$ with $n_i \in \mathbb{Z}$ rather than $n_i \in \mathbb{R}$, we can get a better insight into the properties of the discrete mapping $\Psi_{\mathbf{f}_1,...,\mathbf{f}_m}$.

Many of the algebraic notions we defined above for vector spaces in the continuous case have a similar counterpart also in the discrete case with $n_i \in \mathbb{Z}$ (see Table 5.1):

As in the continuous case, we can define for the transformation $\Psi_{\mathbf{f}_1,...,\mathbf{f}_m} : \mathbb{Z}^m \to \mathrm{Md}(\mathbf{f}_1,...,\mathbf{f}_m)$ the *image* of $\Psi_{\mathbf{f}_1,...,\mathbf{f}_m}$ and the *kernel* of $\Psi_{\mathbf{f}_1,...,\mathbf{f}_m}$; Ker$\Psi_{\mathbf{f}_1,...,\mathbf{f}_m}$ is a sub-lattice of the indices-lattice $\mathbb{Z}^m$, and Im$\Psi_{\mathbf{f}_1,...,\mathbf{f}_m}$ is the module $\mathrm{Md}(\mathbf{f}_1,...,\mathbf{f}_m)$, i.e., the spectrum support in the $u,v$ plane. Furthermore, given a sub-lattice $L$ of $\mathbb{Z}^m$ we can also define the partition of the lattice $\mathbb{Z}^m$ into equivalence classes modulo $L$. The set of all the equivalence classes $(n_1,...,n_m) + L$ in $\mathbb{Z}^m$ gives a disjoint and exhaustive partition of $\mathbb{Z}^m$, where each equivalence class is a parallel translation of $L$ within $\mathbb{Z}^m$.

Now, if we take as sub-lattice $L$ the kernel of the transformation $\Psi_{\mathbf{f}_1,...,\mathbf{f}_m}$, we get a special partition of $\mathbb{Z}^m$ which has the following property, as in the continuous case: the transformation $\Psi_{\mathbf{f}_1,...,\mathbf{f}_m}$ maps the whole equivalence class of $\mathbf{0}$ in $\mathbb{Z}^m$, Ker$\Psi_{\mathbf{f}_1,...,\mathbf{f}_m}$, into the

| Continuous terms (with $n_i \in \mathbb{R}$) | Equivalent terms in the discrete case (with $n_i \in \mathbb{Z}$) |
|---|---|
| $\mathbb{R}^m$ | $\mathbb{Z}^m$ |
| Vector independence over $\mathbb{R}$ | Vector independence over $\mathbb{Z}$ |
| $\Phi_{\mathbf{f}_1,\dots,\mathbf{f}_m} : \mathbb{R}^m \to \mathrm{Sp}(\mathbf{f}_1,\dots,\mathbf{f}_m)$ $(=\mathbb{R}^2$ or $\mathbb{R}^1)$ | $\Psi_{\mathbf{f}_1,\dots,\mathbf{f}_m} : \mathbb{Z}^m \to \mathrm{Md}(\mathbf{f}_1,\dots,\mathbf{f}_m)$ |
| $\mathrm{Im}\Phi_{\mathbf{f}_1,\dots,\mathbf{f}_m} = \mathrm{Sp}(\mathbf{f}_1,\dots,\mathbf{f}_m)$ | $\mathrm{Im}\,\Psi_{\mathbf{f}_1,\dots,\mathbf{f}_m} = \mathrm{Md}(\mathbf{f}_1,\dots,\mathbf{f}_m)$ |
| $\mathrm{Ker}\Phi_{\mathbf{f}_1,\dots,\mathbf{f}_m}$ is a subspace of $\mathbb{R}^m$ | $\mathrm{Ker}\,\Psi_{\mathbf{f}_1,\dots,\mathbf{f}_m}$ is a subspace of $\mathbb{Z}^m$ |
| $\mathbf{v} + \mathrm{Ker}\Phi_{\mathbf{f}_1,\dots,\mathbf{f}_m}$ | $(n_1,\dots,n_m) + \mathrm{Ker}\,\Psi_{\mathbf{f}_1,\dots,\mathbf{f}_m}$ |

**Table 5.1:** Summary of the continuous terms with $n_i \in \mathbb{R}$ and their discrete restrictions with $n_i \in \mathbb{Z}$.

origin (0,0) of the $u,v$ plane; and similarly, for every $(n_1,\dots,n_m) \in \mathbb{Z}^m$ the transformation $\Psi_{\mathbf{f}_1,\dots,\mathbf{f}_m}$ maps the whole equivalence class of $(n_1,\dots,n_m)$ in $\mathbb{Z}^m$, $(n_1,\dots,n_m) + \mathrm{Ker}\,\Psi_{\mathbf{f}_1,\dots,\mathbf{f}_m}$, into a single point $\Psi_{\mathbf{f}_1,\dots,\mathbf{f}_m}(n_1,\dots,n_m)$ in the module $\mathrm{Md}(\mathbf{f}_1,\dots,\mathbf{f}_m)$ on the $u,v$ plane.

Furthermore, the equivalent of equality (5.6) for a discrete linear transformation $\Psi$ between two modules, $\Psi: M_1 \to M_2$, is given by:

$$\mathrm{rank}_{\mathbb{Z}}\,\mathrm{Ker}\,\Psi + \mathrm{rank}_{\mathbb{Z}}\,\mathrm{Im}\,\Psi = \mathrm{rank}_{\mathbb{Z}}\,M_1 \tag{5.9}$$

This can be proved in the same way as in the proof of Eq. (5.6) for the continuous case of vector spaces (see [Lipschutz91 p. 331 No. 9.23]), by replacing throughout the proof the term "linear independence over $\mathbb{R}$" by the term "linear independence over $\mathbb{Z}$".[9]

Now, in the case of our transformation $\Psi_{\mathbf{f}_1,\dots,\mathbf{f}_m} : \mathbb{Z}^m \to \mathrm{Md}(\mathbf{f}_1,\dots,\mathbf{f}_m)$, both $M_1$ and $\mathrm{Ker}\,\Psi_{\mathbf{f}_1,\dots,\mathbf{f}_m}$ are in fact lattices (and hence by Proposition 5.1 they have $\mathrm{rank}_{\mathbb{Z}} = \mathrm{rank}_{\mathbb{R}}$, which can be simply denoted by "rank"). We get, therefore, from Eq. (5.9):

$$\mathrm{rank}\,\mathrm{Ker}\,\Psi_{\mathbf{f}_1,\dots,\mathbf{f}_m} + \mathrm{rank}_{\mathbb{Z}}\,\mathrm{Md}(\mathbf{f}_1,\dots,\mathbf{f}_m) = \mathrm{rank}\,\mathbb{Z}^m = m \tag{5.10}$$

whereas for the continuous transformation $\Phi_{\mathbf{f}_1,\dots,\mathbf{f}_m}$ we have by Eq. (5.6):

$$\dim\,\mathrm{Ker}\Phi_{\mathbf{f}_1,\dots,\mathbf{f}_m} + \dim\,\mathrm{Sp}(\mathbf{f}_1,\dots,\mathbf{f}_m) = \dim\,\mathbb{R}^m = m \tag{5.11}$$

(where $\dim\,\mathrm{Sp}(\mathbf{f}_1,\dots,\mathbf{f}_m) = 2$ or 1).

[9] Note that a similar equality with $\mathrm{rank}_{\mathbb{R}}$ instead of $\mathrm{rank}_{\mathbb{Z}}$ is in general false. The proof fails in this case since a module does not generally have a basis, but only an integral basis (see Sec. 5.2), so that only linear independence over $\mathbb{Z}$ can be guaranteed.

It is important to note, however, that the original dimension of $\mathrm{Ker}\Phi_{\mathbf{f}_1,\ldots,\mathbf{f}_m}$ in $\mathbb{R}^m$ is not necessarily preserved in its restriction to $\mathbb{Z}^m$, $\mathrm{Ker}\Psi_{\mathbf{f}_1,\ldots,\mathbf{f}_m}$. In fact, since $\mathrm{Ker}\Psi_{\mathbf{f}_1,\ldots,\mathbf{f}_m} \subset \mathrm{Ker}\Phi_{\mathbf{f}_1,\ldots,\mathbf{f}_m}$, it is clear that:

$$\mathrm{rank}\ \mathrm{Ker}\Psi_{\mathbf{f}_1,\ldots,\mathbf{f}_m} \le \dim\ \mathrm{Ker}\Phi_{\mathbf{f}_1,\ldots,\mathbf{f}_m} \tag{5.12}$$

However, the equality in (5.12) does not always hold. This can be illustrated by an example in the 3D case: Let $\mathrm{Ker}\Phi_{\mathbf{f}_1,\ldots,\mathbf{f}_m}$ be a 2D subspace of $\mathbb{R}^3$ (i.e., a plane through its origin); its restriction to $\mathbb{Z}^3$, $\mathrm{Ker}\Psi_{\mathbf{f}_1,\ldots,\mathbf{f}_m}$, is the lattice of all points of $\mathbb{Z}^3$ included in $\mathrm{Ker}\Phi_{\mathbf{f}_1,\ldots,\mathbf{f}_m}$. It is clear that depending on the plane inclinations in the space $\mathbb{R}^3$ $\mathrm{Ker}\Psi_{\mathbf{f}_1,\ldots,\mathbf{f}_m}$ may have rank = 2 (e.g., if the plane $\mathrm{Ker}\Phi_{\mathbf{f}_1,\ldots,\mathbf{f}_m}$ contains both the $x$ and $y$ axes of $\mathbb{R}^3$), rank = 1 (e.g., if the plane only contains the $x$ axis but forms an *irrational angle* [see Sec. 8.6] with the $y$ axis), or rank = 0 (if the only integral point in the plane is (0,...,0)).

This is, indeed, an important difference between the continuous and the discrete cases. In the continuous case the dimension of $\mathrm{Ker}\Phi_{\mathbf{f}_1,\ldots,\mathbf{f}_m}$ is automatically determined by equality (5.6). In the discrete case, however, the rank of $\mathrm{Ker}\Psi_{\mathbf{f}_1,\ldots,\mathbf{f}_m}$ (the discrete counterpart of $\mathrm{Ker}\Phi_{\mathbf{f}_1,\ldots,\mathbf{f}_m}$) is only bounded by inequality (5.12), but its exact value depends also on other parameters, namely: the inclinations of the continuous subspace $\mathrm{Ker}\Phi_{\mathbf{f}_1,\ldots,\mathbf{f}_m}$ within $\mathbb{R}^m$, as shown in the example above. Consequently, the rank of the translated lattice (equivalence class) which "collapses" on each point of $\mathrm{Im}\Psi_{\mathbf{f}_1,\ldots,\mathbf{f}_m} = \mathrm{Md}(\mathbf{f}_1,\ldots,\mathbf{f}_m)$ in the discrete case may be smaller than the dimension of the full, continuous translated subspace (equivalence class) which "collapses" on each point of $\mathrm{Im}\Phi_{\mathbf{f}_1,\ldots,\mathbf{f}_m} = \mathrm{Sp}(\mathbf{f}_1,\ldots,\mathbf{f}_m)$ in the corresponding continuous case.

The question is, therefore, what happens to the "dimension preservation law" in the discrete case? In fact, the "lost" dimensions of $\mathrm{Ker}\Psi_{\mathbf{f}_1,\ldots,\mathbf{f}_m}$ are not really lost, and they are simply taken care of elsewhere, in $\mathrm{Im}\Psi_{\mathbf{f}_1,\ldots,\mathbf{f}_m}$. Since the right hand sides of Eqs. (5.10) and (5.11) are equal, inequality (5.12) implies that:

$$\mathrm{rank}_{\mathbb{Z}}\ \mathrm{Md}(\mathbf{f}_1,\ldots,\mathbf{f}_m) \ge \dim\ \mathrm{Sp}(\mathbf{f}_1,\ldots,\mathbf{f}_m) \tag{5.13}$$

where $\dim\ \mathrm{Sp}(\mathbf{f}_1,\ldots,\mathbf{f}_m) = 2$ or 1. More precisely, if we note the difference by $d$, we obtain:

$$\dim\ \mathrm{Ker}\Phi_{\mathbf{f}_1,\ldots,\mathbf{f}_m} - \mathrm{rank}\ \mathrm{Ker}\Psi_{\mathbf{f}_1,\ldots,\mathbf{f}_m} = \mathrm{rank}_{\mathbb{Z}}\ \mathrm{Md}(\mathbf{f}_1,\ldots,\mathbf{f}_m) - \dim\ \mathrm{Sp}(\mathbf{f}_1,\ldots,\mathbf{f}_m) = d$$

This means that if due to the inclinations of the subspace $\mathrm{Ker}\Phi_{\mathbf{f}_1,\ldots,\mathbf{f}_m}$ in $\mathbb{R}^m$ it happens, as in the example above, that $\mathrm{Ker}\Psi_{\mathbf{f}_1,\ldots,\mathbf{f}_m}$ cannot attain the full dimension of $\mathrm{Ker}\Phi_{\mathbf{f}_1,\ldots,\mathbf{f}_m}$, then this "loss" of $d$ units in $\mathrm{rank}\ \mathrm{Ker}\Psi_{\mathbf{f}_1,\ldots,\mathbf{f}_m}$, the first term of Eq. (5.10), is automatically "balanced" by an identical increase in the second term of Eq. (5.10): $\mathrm{rank}_{\mathbb{Z}}\ \mathrm{Md}(\mathbf{f}_1,\ldots,\mathbf{f}_m)$ is increased by $d$ units with respect to $\dim\ \mathrm{Sp}(\mathbf{f}_1,\ldots,\mathbf{f}_m)$. This means that the number of independent over $\mathbb{Z}$ vectors which span $\mathrm{Im}\Psi_{\mathbf{f}_1,\ldots,\mathbf{f}_m}$ is higher by $d$ than the number of independent over $\mathbb{R}$ vectors which span its enclosing continuous space, $\mathrm{Im}\Phi_{\mathbf{f}_1,\ldots,\mathbf{f}_m}$. This situation is illustrated in Examples 5.8 and 5.9 of Sec. 5.7.

Furthermore, if we only look at the right hand side of the above equation we have:

$$\text{rank}_{\mathbb{Z}}\,\text{Md}(\mathbf{f}_1,...,\mathbf{f}_m) - \dim\,\text{Sp}(\mathbf{f}_1,...,\mathbf{f}_m) = d \tag{5.14}$$

or equivalently, by Eq. (5.1):

$$\text{rank}_{\mathbb{Z}}\,\text{Md}(\mathbf{f}_1,...,\mathbf{f}_m) - \text{rank}_{\mathbb{R}}\,\text{Md}(\mathbf{f}_1,...,\mathbf{f}_m) = d \tag{5.15}$$

According to Proposition 5.1 we obtain, therefore, the following result:

**Proposition 5.2:** The module Md($\mathbf{f}_1,...,\mathbf{f}_m$), the support of the spectrum, is a lattice (and therefore discrete) *iff* $d = 0$ (i.e., *iff* the continuous and discrete dimensions are identical); and conversely, Md($\mathbf{f}_1,...,\mathbf{f}_m$) is a dense module in Im$\Phi_{\mathbf{f}_1,...,\mathbf{f}_m}$ *iff* $d > 0$. (We remember from Eq. (5.4) that Im$\Phi_{\mathbf{f}_1,...,\mathbf{f}_m}$, i.e., Sp($\mathbf{f}_1,...,\mathbf{f}_m$), can be either the entire $u,v$ plane, or a 1D line through its origin). ■

As we will see later (in Sec. 6.2), this important result provides a criterion for the periodicity of the superposition of periodic layers (functions).

Two interesting consequences follow immediately:

(a) The spectrum support of a non-singular superposition can be a discrete lattice (meaning that the superposition is periodic; see Sec. 6.2(b)) only in the case of $m = 2$ non-collinear gratings (as in Fig. 2.5). If $m \geq 3$ then $\dim \text{Ker}\Phi = m - \dim \text{Im}\Phi > 0$ (since $\dim \text{Im}\Phi = 2$ or 1), and therefore if $\text{rank Ker}\Psi = 0$ (=non-singular state) then $d = \dim \text{Ker}\Phi - \text{rank Ker}\Psi > 0$. Therefore for $m \geq 3$ gratings, any non-singular case has a dense spectrum support.

(b) The spectrum support of a *singular* superposition can be a discrete lattice even if $m \geq 3$. This occurs when $d = 0$. In other words, if the spectrum support is 2D this occurs when $\text{rank}_{\mathbb{Z}}\,\text{Im}\Psi_{\mathbf{f}_1,...,\mathbf{f}_m} = 2$ and $\text{rank Ker}\Psi_{\mathbf{f}_1,...,\mathbf{f}_m} = m-2$; and if the spectrum support is 1D ($\mathbf{f}_1,...,\mathbf{f}_m$ are collinear) this occurs when $\text{rank}_{\mathbb{Z}}\,\text{Im}\Psi_{\mathbf{f}_1,...,\mathbf{f}_m} = 1$ and $\text{rank Ker}\Psi_{\mathbf{f}_1,...,\mathbf{f}_m} = m-1$.

The various possible cases which may occur in the spectrum support in the superposition of $m = 2,...,6$ gratings are summarized in Table 5.2. Several illustrative examples are given in Sec. 5.7.

## 5.6 The algebraic interpretation of the impulse locations in the spectrum support

### 5.6.1 The global spectrum support

Using the terminology introduced in the previous sections it now becomes clear that the set of all the impulse locations in the spectrum convolution (the support of the impulse "forest") is in fact the module Md($\mathbf{f}_1,...,\mathbf{f}_m$), i.e., the image of the indices-lattice $\mathbb{Z}^m$ under the transformation $\Psi_{\mathbf{f}_1,...,\mathbf{f}_m}$. We have seen that this spectrum support can be either a dense module or a discrete lattice, and we found necessary and sufficient conditions for either case. Table 5.2 gives a systematic summary of the different possible cases in the

| $m$ | The frequency vectors: | dim Im$\Phi$ | rank$_{\mathbb{Z}}$ Im$\Psi$ | spectrum support | dim Ker$\Phi$ | rank $L$ | Sing. / Not | Examples | Remarks |
|---|---|---|---|---|---|---|---|---|---|
| 1 | $\mathbf{f}_1$ | 1 | = 1 | 1D-L | 0 | 0 | **N** | Example 5.1 | (1) |
| 2 | $\mathbf{f}_1, \mathbf{f}_2$ coplanar: | 2 | = 2 | 2D-L | 0 | 0 | **N** | Example 5.2 | (2) |
| | $\mathbf{f}_1, \mathbf{f}_2$ collinear: | 1 | = 1 | 1D-L | 1 | 1 | S | Example 5.2 | (3) |
| | | 1 | < 2 | 1D-M | 1 | 0 | **N** | Example 5.2 | |
| 3 | $\mathbf{f}_1, \mathbf{f}_2, \mathbf{f}_3$ coplanar: | 2 | = 2 | 2D-L | 1 | 1 | S | Example 5.5 | |
| | | 2 | < 3 | 2D-M | 1 | 0 | **N** | | (4) |
| | $\mathbf{f}_1, \mathbf{f}_2, \mathbf{f}_3$ collinear: | 1 | = 1 | 1D-L | 2 | 2 | S | Example 5.8 | |
| | | 1 | < 2 | 1D-M | 2 | 1 | S | Example 5.9 | |
| | | 1 | < 3 | 1D-M | 2 | 0 | **N** | | |
| 4 | $\mathbf{f}_1, \mathbf{f}_2, \mathbf{f}_3, \mathbf{f}_4$ coplanar: | 2 | = 2 | 2D-L | 2 | 2 | S | Example 5.10 | |
| | | 2 | < 3 | 2D-M | 2 | 1 | S | | |
| | | 2 | < 4 | 2D-M | 2 | 0 | **N** | | |
| | $\mathbf{f}_1, \mathbf{f}_2, \mathbf{f}_3, \mathbf{f}_4$ collinear: | 1 | = 1 | 1D-L | 3 | 3 | S | | |
| | | 1 | < 2 | 1D-M | 3 | 2 | S | | |
| | | 1 | < 3 | 1D-M | 3 | 1 | S | | |
| | | 1 | < 4 | 1D-M | 3 | 0 | **N** | | |
| 5 | $\mathbf{f}_1, \mathbf{f}_2, \mathbf{f}_3, \mathbf{f}_4, \mathbf{f}_5$ coplanar: | 2 | = 2 | 2D-L | 3 | 3 | S | | |
| | | 2 | < 3 | 2D-M | 3 | 2 | S | | |
| | | 2 | < 4 | 2D-M | 3 | 1 | S | Example 5.11 | (5) |
| | | 2 | < 5 | 2D-M | 3 | 0 | **N** | | |
| | $\mathbf{f}_1, \mathbf{f}_2, \mathbf{f}_3, \mathbf{f}_4, \mathbf{f}_5$ collinear: | 1 | = 1 | 1D-L | 4 | 4 | S | | |
| | | 1 | < 2 | 1D-M | 4 | 3 | S | | |
| | | 1 | < 3 | 1D-M | 4 | 2 | S | | |
| | | 1 | < 4 | 1D-M | 4 | 1 | S | | |
| | | 1 | < 5 | 1D-M | 4 | 0 | **N** | | |
| 6 | $\mathbf{f}_1, \mathbf{f}_2, \mathbf{f}_3, \mathbf{f}_4, \mathbf{f}_5, \mathbf{f}_6$ coplanar: | 2 | = 2 | 2D-L | 4 | 4 | S | Example 5.12 | |
| | | 2 | < 3 | 2D-M | 4 | 3 | S | | |
| | | 2 | < 4 | 2D-M | 4 | 2 | S | Example 5.13 | (6) |
| | | 2 | < 5 | 2D-M | 4 | 1 | S | | |
| | | 2 | < 6 | 2D-M | 4 | 0 | **N** | | |
| | $\mathbf{f}_1, \mathbf{f}_2, \mathbf{f}_3, \mathbf{f}_4, \mathbf{f}_5, \mathbf{f}_6$ collinear: | 1 | = 1 | 1D-L | 5 | 5 | S | | |
| | | 1 | < 2 | 1D-M | 5 | 4 | S | | |
| | | 1 | < 3 | 1D-M | 5 | 3 | S | | |
| | | 1 | < 4 | 1D-M | 5 | 2 | S | | |
| | | 1 | < 5 | 1D-M | 5 | 1 | S | | |
| | | 1 | < 6 | 1D-M | 5 | 0 | **N** | | |

**Table 5.2:** (*continued on the opposite page*)

**Legend:**

1D = one dimensional; 2D = two dimensional; L = discrete lattice; M = dense module; S = singular; N = non-singular. By "coplanar" is meant: coplanar but non-collinear.

**Remarks:**

(1) A single grating; no superposition (and no moiré).

(2) This is the only non-singular superposition with a discrete spectrum support.

(3) A singular moiré between two gratings occurs *iff* $\mathbf{f}_1,\mathbf{f}_2$ are collinear (i.e., $\alpha = 0°$ or $180°$) and commensurable.

(4) Note that 2D-M includes also the hybrid case in which the 2D spectrum support is dense in one direction and discrete in the other. For instance, in the case of three coplanar frequency vectors this may occur when two of the vectors are collinear but incommensurable, while the third vector is oriented in a different direction (see the module $M_4$ in Problem 5-2).

(5) To this category belongs the singular superposition of 5 identical gratings with equal angle differences of 72°.

(6) To this category belongs the singular superposition of three identical screens with angle differences of 30° (or 60°), which is the traditional screen combination used in colour printing.

Note that each pair of non-collinear gratings may be counted also as one 2D screen. For example, $m = 4$ corresponds either to 4 superposed gratings or to 2 superposed screens, etc.

**Table 5.2:** (*continued.*) Summary of the algebraic structural properties of the various possible cases for $m = 1,...,6$ superposed gratings (the legend and the remarks for the table are given above). The interpretation of these properties in terms of the image domain is discussed in Chapter 6.

superposition of $m = 2,...,6$ gratings (or equivalently, up to three 2-fold periodic layers such as dot-screens). The interpretation of the algebraic structure of the spectrum support in terms of the superposition in the image domain will be discussed in Sec. 6.2.

### 5.6.2 The individual impulse-clusters

We now proceed from the global spectrum support to the support of each of the individual impulse clusters. The cluster of impulse-locations which fall on the spectrum origin when the $(k_1,...,k_m)$-moiré reaches a singular state is simply the image under $\Psi_{\mathbf{f}_1,...,\mathbf{f}_m}$ of the lattice $L = \text{Ker}\,\Psi_{\mathbf{f}_1,...,\mathbf{f}_m}$, i.e., Im$L$. Similarly, the other clusters of impulse-locations which are simultaneously formed in the spectrum plane are the images of the other equivalence classes $(r_1,...,r_m) + L$ in the indices-lattice $\mathbb{Z}^m$ (where $(r_1,...,r_m)$ denotes a representative point of the equivalence class). Let us now explain this in more detail.

In Sec. 2.9 we defined a singular moiré as a configuration of the superposed layers in which the moiré period is infinitely large (i.e., its frequency is zero). More formally, a $(k_1,...,k_m)$-moiré reaches a singular state whenever the location of its fundamental impulse, the $(k_1,...,k_m)$-impulse in the spectrum convolution, coincides with the spectrum origin (0,0) (i.e., whenever the frequency vectors $\mathbf{f}_1,...,\mathbf{f}_m$ of the superposed layers are such that

$\Sigma k_i \mathbf{f}_i = \mathbf{0}$). We have seen, however, in Sec. 4.3 that when a $(k_1,...,k_m)$-moiré reaches a singular state, not only the $(k_1,...,k_m)$-impulse itself falls on the spectrum origin, but rather, a whole infinite impulse-cluster. This cluster clearly contains the 1D comb formed by the $(nk_1,...,nk_m)$-impulses with all integer values of $n$, but in the general case this cluster may contain other impulses, too, and it may be 2D or even of a higher rank. How can we characterize all the impulses which belong to this cluster (i.e., fall on the spectrum origin)? Using our new terminology, when the frequency vectors $\mathbf{f}_1,...,\mathbf{f}_m$ are such that a $(k_1,...,k_m)$-singular moiré occurs, the linear transformation $\Psi_{\mathbf{f}_1,...,\mathbf{f}_m}$ maps to the spectrum origin not only the point $(k_1,...,k_m)$ but the entire sub-lattice $L \subset \mathbb{Z}^m$ induced by $\Psi_{\mathbf{f}_1,...,\mathbf{f}_m}$, namely: $L = \mathrm{Ker}\,\Psi_{\mathbf{f}_1,...,\mathbf{f}_m}$. The sub-lattice $L$ corresponds, therefore, to the impulse-cluster which collapses onto the spectrum origin at the $(k_1,...,k_m)$-singular state, and its points (integer $m$-tuples) are the indices of the impulses of this cluster (see for example Eq. (4.10)). This is illustrated by several examples in Sec. 5.7.

However, whenever a $(k_1,...,k_m)$-singular moiré state occurs, apart from the main cluster of the $(k_1,...,k_m)$-impulse which is formed at the spectrum origin, other impulse clusters are also simultaneously formed elsewhere in the spectrum. Let us see now in detail what is the nature of these impulse clusters, and how we can characterize the impulses which belong to each of the clusters. We have seen above that the transformation $\Psi_{\mathbf{f}_1,...,\mathbf{f}_m}$ induces a partition of the indices-lattice $\mathbb{Z}^m$ into disjoint and exhaustive equivalence classes $(r_1,...,r_m) + L$, which are translations of the sub-lattice $L$ in $\mathbb{Z}^m$ (the sub-lattice $L$ itself is the equivalence class $\mathbf{0} + L$ which contains all the points of $\mathbb{Z}^m$ that are mapped by $\Psi_{\mathbf{f}_1,...,\mathbf{f}_m}$ to the spectrum origin). We have also seen that the transformation $\Psi_{\mathbf{f}_1,...,\mathbf{f}_m}$ has the special property that it maps every equivalence class $(r_1,...,r_m) + L$ of the indices-lattice $\mathbb{Z}^m$ into a different *single point* within the spectrum plane. This explains why an infinite (but still denumerable) number of clusters are formed in the spectrum simultaneously with the main cluster of the $(k_1,...,k_m)$-moiré: each of these clusters is simply the image under $\Psi_{\mathbf{f}_1,...,\mathbf{f}_m}$ of a different equivalence class $(r_1,...,r_m) + L$ of the indices-lattice $\mathbb{Z}^m$. The indices of the impulses in each of these clusters are therefore a translated replica of the indices of the impulses of $L$, each of which being incremented by a "cluster representative" $(r_1,...,r_m)$ (see figures in the examples below). The location of each cluster $(r_1,...,r_m) + L$ in the spectrum is given by $\Psi_{\mathbf{f}_1,...,\mathbf{f}_m}(r_1,...,r_m) + \mathbf{0} = \Psi_{\mathbf{f}_1,...,\mathbf{f}_m}(r_1,...,r_m)$, i.e., it is shifted from the spectrum origin by $\Psi_{\mathbf{f}_1,...,\mathbf{f}_m}(r_1,...,r_m)$.

As for the relationship in the singular state between the rank of a single cluster and the rank of the entire spectrum support, $\mathrm{Md}(\mathbf{f}_1,...,\mathbf{f}_m)$, we have from Eq. (5.10):

$$\mathrm{rank}\,L + \mathrm{rank}_{\mathbb{Z}}\,\mathrm{Md}(\mathbf{f}_1,...,\mathbf{f}_m) = m \tag{5.16}$$

These ranks depend, of course, on the specific choice of the frequency vectors $\mathbf{f}_1,...,\mathbf{f}_m$ of the superposed layers in the singular state: since the module $\mathrm{Md}(\mathbf{f}_1,...,\mathbf{f}_m)$ is generated by the frequency vectors $\mathbf{f}_1,...,\mathbf{f}_m \in \mathbb{R}^2$, $\mathrm{rank}_{\mathbb{Z}}\,\mathrm{Md}(\mathbf{f}_1,...,\mathbf{f}_m)$ is simply the maximum number of vectors among $\mathbf{f}_1,...,\mathbf{f}_m \in \mathbb{R}^2$ which are still linearly independent over $\mathbb{Z}$. Rank $L$ complements this number to $m$, the number of superposed gratings, so that it indicates the "redundancy level" of the superposition, i.e., the number of *dependent* vectors (layers), which do not further enrich the spectrum support, but are rather "invested" in each of its existing points (and hence enrich each of the clusters).

At this point we may introduce a new criterion for the singularity of a given superposition. As we have seen in Sec. 5.4.1, a superposition of $m$ gratings is singular (or in other words: at least one of its moirés is singular) *iff* the frequency vectors $\mathbf{f}_1,...,\mathbf{f}_m$ of the superposed layers are linearly dependent over $\mathbb{Z}$, i.e., there exist $k_i \in \mathbb{Z}$ not all of them 0 such that $\sum_{i=1}^{m} k_i \mathbf{f}_i = \mathbf{0}$. This can be reformulated now as follows:

**Proposition 5.3:** A superposition of $m$ gratings (or $m/2$ dot-screens, etc.) is singular *iff* $\text{rank}_{\mathbb{Z}}\,\text{Md}(\mathbf{f}_1,...,\mathbf{f}_m) < m$. ■

And according to Eq. (5.16) we obtain also:

**Proposition 5.4:** An equivalent criterion for the singularity of a superposition of periodic layers is that $L = \text{Ker}\,\Psi_{\mathbf{f}_1,...,\mathbf{f}_m}$ contains points of $\mathbb{Z}^m$ other than (0,...,0), or in other words: $\text{rank}\,L > 0$. ■

**Remark 5.1:** Note that $\text{rank}\,L$, which is also the rank of the moiré cluster, is precisely the *order of the singularity* of the moiré (see Remark 2.7 in Sec. 2.9), i.e., the "redundancy level" of the superposition. In other words: this is the degree of redundancy in $\sum k_i \mathbf{f}_i = \mathbf{0}$, i.e., the number of vectors among $\mathbf{f}_1,...,\mathbf{f}_m$ which are linearly dependent on the others (over $\mathbb{Z}$) and do not further enrich the spectrum support $\text{Md}(\mathbf{f}_1,...,\mathbf{f}_m)$, but are rather "invested" in each of the collapsed clusters. ■

It is interesting to note that for different singular moirés different configurations of clusters are formed in the spectrum (in general, either the assignment of impulses to each cluster or the cluster locations in the spectrum or both may differ). This is because for different sub-lattices $\text{Ker}\,\Psi_{\mathbf{f}_1,...,\mathbf{f}_m}$ the indices-lattice $\mathbb{Z}^m$ is partitioned into different sets of equivalence classes. This is illustrated by Examples 5.5–5.7 in Sec. 5.7.

### 5.6.3 The spread-out clusters slightly off the singular state

Finally, let us see what happens in the spectrum when we start moving away from the $(k_1,...,k_m)$-singular state. When we slightly modify one or more of the frequency vectors $\mathbf{f}_1,...,\mathbf{f}_m$ of the superposed layers, each of the clusters in the spectrum starts "spreading out", revealing thus the infinity of points from which it is composed (Figs. 5.2–5.10).[10] In particular, the main cluster which spreads out around the spectrum origin enables us to visualize the impulses which correspond to the moiré (which originate from $\text{Ker}\,\Psi_{\mathbf{f}_1,...,\mathbf{f}_m}$ in

[10] Formally speaking, when the superposition moves away from the singular state, $\text{Ker}\,\Psi$ becomes $\{\mathbf{0}\}$, so that each point in $\mathbb{Z}^m$ becomes its own one-member equivalence class. Therefore the spread-out clusters in the spectrum no longer correspond to the current equivalence classes. However, we will still consider the "spread-out clusters" in the spectrum to be traces of the clusters of the singular state that we have just left, and we will continue to call them "clusters" in this sense. Note that $L$ is a sub-lattice of $\mathbb{Z}^m$, which is defined by the original $\text{Ker}\,\Psi$ *at the moment of singularity*, and it remains unchanged even when the vectors $\mathbf{f}_1,...,\mathbf{f}_m$ and $\text{Ker}\,\Psi$ are later modified to $\mathbf{f}'_1,...,\mathbf{f}'_m$ and $\text{Ker}\,\Psi'$, and $L \neq L' = \text{Ker}\,\Psi'$.

the singular state of the moiré). Depending on which of the vectors $\mathbf{f}_1,...,\mathbf{f}_m$ have been changed and how, the clusters in the spectrum may be *partially spread-out* (for example, when only one dimension of the cluster has been spread out, and each point still represents an infinity of impulses); or *fully spread-out* (when each point of the cluster represents exactly one single impulse, so that no two impulses in the cluster fall on the same point in the spectrum). It should be noted that although in the examples we have seen until now the spread-out moiré clusters in the $u,v$ plane were always 1D or 2D discrete lattices (see, for example, Figs. 2.5 and 4.3), in the general case each spread-out cluster in the spectrum may be also a dense module. This only happens, however, when $\text{rank}\,L > 2$. In fact, $\text{rank}\,L$ of the singular state is precisely the rank of the fully spread-out cluster in the spectrum, i.e., the number of dimensions which have been "invested" in each of the individual clusters (see Problem 5-12). Therefore, when the rank of $L$ is $r > 2$, each fully spread-out cluster becomes a dense module in the 2D spectrum plane (since $\text{rank}_{\mathbb{Z}}\,\text{Im}L = \text{rank}\,L$ is larger than the dimension of the $u,v$ plane in which it is imbedded). See for example the clusters in Fig. 5.9, where $\text{rank}\,L = 6-2 = 4$. The interpretation of this situation in terms of the image domain will be discussed in Sec. 6.3.

Note that even when each of the clusters in itself is a discrete lattice, their intermingled impulses throughout the spectrum are not necessarily located on a common lattice, and the global spectrum support may be an everywhere dense module.

## 5.7 Examples

In this section we present a number of examples to illustrate the above discussion, and to demonstrate the contribution of the algebraic approach developed in this chapter to the understanding of the structure of the spectrum support. In particular, these examples illustrate the clusterization phenomenon, and the identification of the impulses which participate in each of the clusters in the spectrum. We start in Example 5.4 with the simplest possible case, the superposition-moiré between two gratings; in this case the algebraic situation is straightforward, and it is presented rather informally, by way of introduction. Then in Examples 5.5–5.9 we present various moiré configurations between three gratings, since in the case of three gratings all the algebraic structures occur in the 3D space and are therefore easy to understand. Examples 5.10–5.14 illustrate some more interesting cases which occur in higher dimensions. Note that each of the examples comes to illustrate some different features or properties, as indicated in each case. It may be instructive to try to find out for each of the examples to which entry in Table 5.2 it corresponds.

**Example 5.4:** The simplest possible example consists of the superposition of two gratings. Let us illustrate this situation with the case of the (3,-2)-moiré, a 3-rd order moiré which becomes visible when the (3,-2)-impulse in the spectrum convolution is located inside the visibility circle, i.e. when the frequency vector $\mathbf{f}_2$ of the second grating is close to $\frac{3}{2}\mathbf{f}_1$ (see the vector diagram in Fig. 5.1(b)). This (3,-2)-impulse is the fundamental impulse

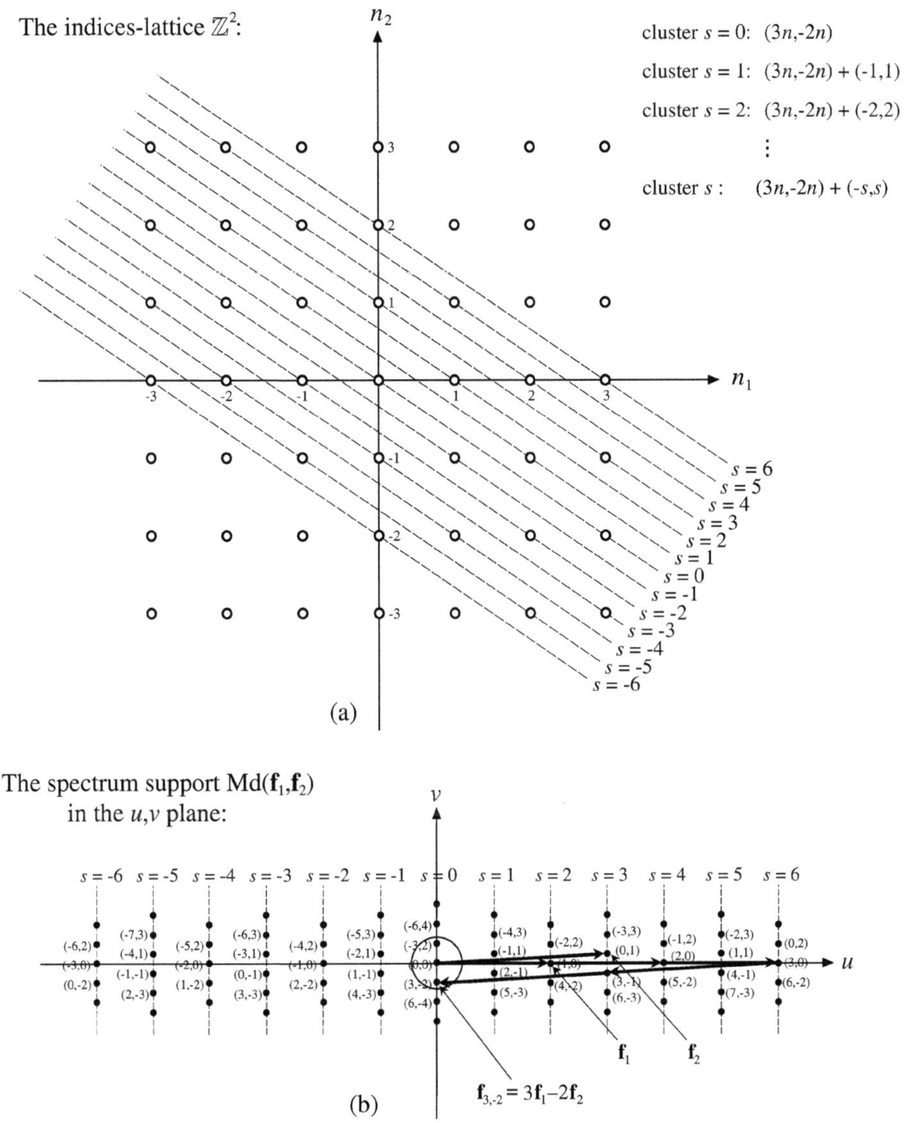


**Figure 5.1:** A schematic illustration of the transformation $\Psi_{\mathbf{f}_1,\mathbf{f}_2}(n_1,n_2) = n_1\mathbf{f}_1 + n_2\mathbf{f}_2$ which maps the indices-lattice $\mathbb{Z}^2$ (top) into the $u$,$v$ spectrum-plane (bottom), in the case of a two grating superposition with $f_2 = \frac{3}{2}f_1$ and $\alpha \approx 0°$. (a) Schematic view of the indices-lattice, $\mathbb{Z}^2$. The dashed lines illustrate the $2n_1+3n_2 = s$ diagonals (=equivalence classes). (b) The image of the mapping $\Psi_{\mathbf{f}_1,\mathbf{f}_2}$ in the $u$,$v$ plane, showing the corresponding impulse clusters in the spectrum support, slightly before $\alpha$ reaches 0°; black dots indicate the impulse locations. The $s$-th diagonal in (a) is mapped into the $s$-th comb in the $u$,$v$ spectrum (b). Note that the 0-th comb represents here the (3,-2)-moiré.

of a 1D-cluster through the spectrum origin, which represents the moiré in question; but in the same time other 1D clusters are also formed in the spectrum, in parallel to the main 1D cluster. Note that when $\alpha = 0°$ and the frequency vector $\mathbf{f}_2$ attains exactly the point $\mathbf{f}_2 = \frac{3}{2}\mathbf{f}_1$ each of the 1D clusters collapses into a single point on the $u$ axis, and in particular, the main cluster collapses into the spectrum origin, so that the (3,-2)-moiré becomes singular (and hence invisible in the layer superposition).

Let us analyze this example to illustrate the algebraic discussion of the preceding sections. In this case the indices-lattice (the lattice of all the indices of the impulses obtained in the spectrum convolution) is $\mathbb{Z}^2$, and the linear transformation $\Psi_{\mathbf{f}_1,\mathbf{f}_2}$ which maps each index pair $(n_1,n_2) \in \mathbb{Z}^2$ into the geometric location of the $(n_1,n_2)$-impulse in the $u,v$ plane is given according to Eq. (5.5) by:

$$\Psi_{\mathbf{f}_1,\mathbf{f}_2}(n_1,n_2) = n_1\mathbf{f}_1 + n_2\mathbf{f}_2$$

Fig. 5.1(a) illustrates the indices-lattice $\mathbb{Z}^2$ and its partition into equivalence classes induced by the sub-lattice $(3n,-2n)$. This sub-lattice itself becomes the cluster $s = 0$ of the partition, containing the indices of the fundamental impulse of the (3,-2)-moiré and all its harmonics. The indices of this 0-th cluster are given by: $L = \{(3n,-2n) \mid n \in \mathbb{Z}\}$, and the indices of the $s$-th cluster are given in this case by: $(-s,s) + L$. Fig. 5.1(b) shows the image of the transformation $\Psi_{\mathbf{f}_1,\mathbf{f}_2}$ in the $u,v$ plane, i.e., the spectrum support, when the vectors $\mathbf{f}_1,\mathbf{f}_2$ are *almost* in the singular position ($\alpha$ is almost 0°). When $\mathbf{f}_1,\mathbf{f}_2$ are *exactly* in the singular position, $\Psi_{\mathbf{f}_1,\mathbf{f}_2}$ maps each equivalence class of $\mathbb{Z}^2$ into a single point on the $u$ axis (the point into which the respective 1D cluster in the spectrum collapses). But as soon as $\mathbf{f}_1,\mathbf{f}_2$ start moving away from the singular state, each of these 1D clusters starts spreading out and gives a comb of impulses in the spectrum (as in Fig. 5.1(b)). ■

**Example 5.5:** (1D clusters on a 2D support in the $u,v$ plane): Consider the (1,1,1)-singular moiré which occurs between three gratings when their frequency vectors are given, in polar coordinates, by: $\mathbf{f}_1 = (0°,32)$, $\mathbf{f}_2 = (120°,32)$, $\mathbf{f}_3 = (240°,32)$, i.e., in Cartesian coordinates: $\mathbf{f}_1 = (32,0)$, $\mathbf{f}_2 = -16,16\sqrt{3})$, $\mathbf{f}_3 = (-16,-16\sqrt{3})$ (see Fig. 5.2(a)).[11] Since in this case $\mathbf{f}_3$ is a linear combination, both over $\mathbb{Z}$ and over $\mathbb{R}$, of $\mathbf{f}_1$ and $\mathbf{f}_2$ (i.e: $\mathbf{f}_3 = -\mathbf{f}_1-\mathbf{f}_2$), we have here: $\text{rank}_{\mathbb{Z}}\, \text{Md}(\mathbf{f}_1,\mathbf{f}_2,\mathbf{f}_3) = \dim \text{Sp}(\mathbf{f}_1,\mathbf{f}_2,\mathbf{f}_3) = 2$. This means by Eq. (5.14) that $d = 0$, and therefore the spectrum support, $\text{Md}(\mathbf{f}_1,\mathbf{f}_2,\mathbf{f}_3)$, is in this singular case a discrete lattice of rank 2 (see Fig. 5.2(a)). Furthermore, from Eq. (5.16) we learn that each point of this lattice represents a collapsed lattice (cluster) whose rank is: $\text{rank}\, L = 3-2 = 1$. And indeed, when the three superposed gratings slightly move away from the singular moiré state (i.e., when their frequency vectors $\mathbf{f}_i$ are slightly modified), each of the 1D clusters in the spectrum starts spreading out, and in the image domain a 1D moiré becomes visible in the superposition, as indicated by the low frequencies of the 1D spread-out cluster around the spectrum origin (see Fig. 5.2(b)).

[11] Note that the choice of $\mathbf{f}_1 = (32,0)$ for the first layer is arbitrary, and for any other choice, $\mathbf{f}_2$ and $\mathbf{f}_3$ could be adapted accordingly. However, for the sake of consistency and to facilitate comparisons between the spectra we will use the same convention in most of the following examples, too.

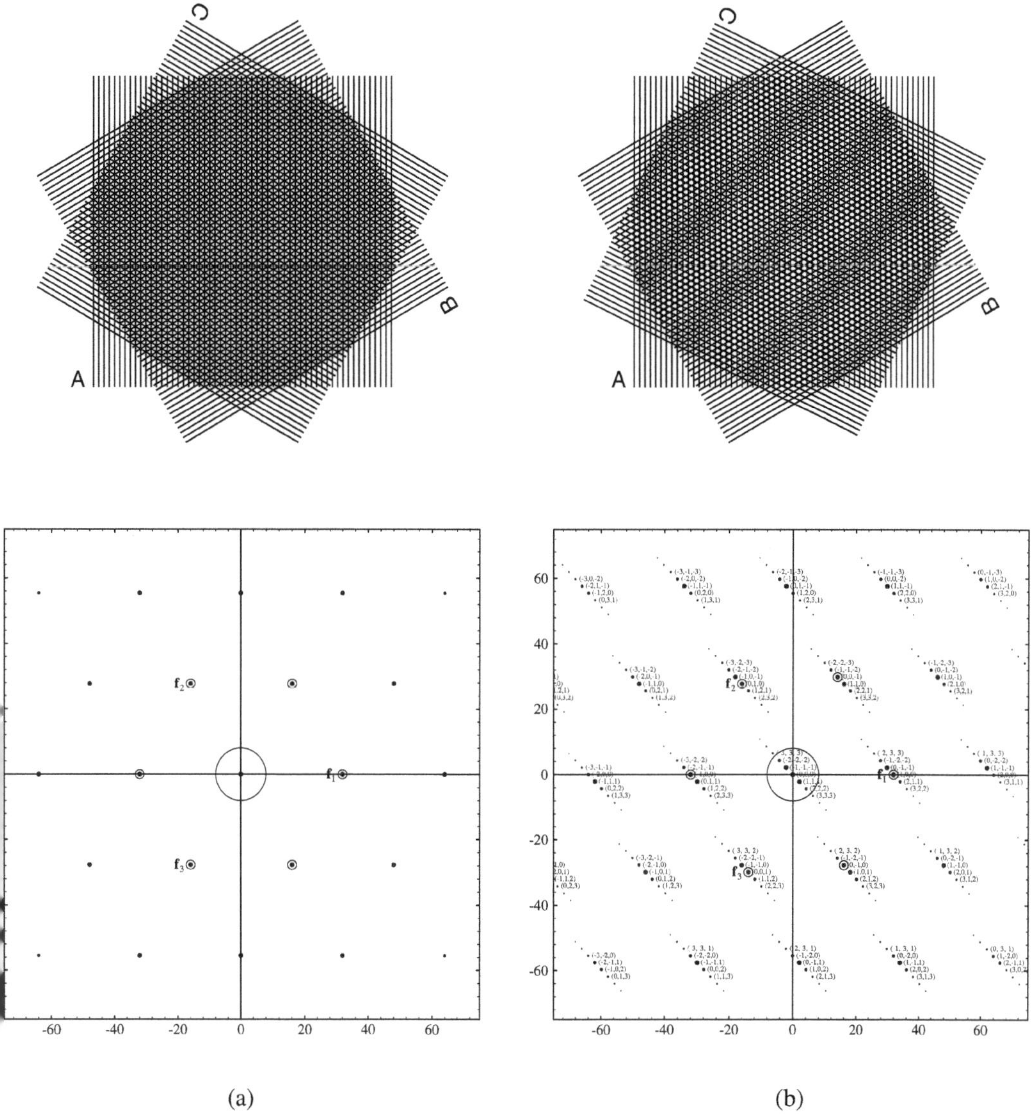


**Figure 5.2:** The singular 3-grating superposition of Example 5.5 (top) and its spectrum support (bottom). (a) Exactly at the singular state: the spectrum support forms here a 2D lattice, each point of which represents a collapsed cluster. (b) Slightly off the singular state: each of the clusters in the spectrum is spread out, clearly demonstrating its 1D nature. Encircled points denote the locations of the fundamental impulses of the three original combs. Large points represent convolution impulses of the first order, and smaller points represent convolution impulses of higher orders. Only impulses up to the 5-th order are shown. (A magnified version of this figure can be found in the Internet site of the book.)

This explanation already shows how the structural properties of the spectrum support can be determined using Eqs. (5.14) and (5.16). However, in order to illustrate the algebraic discussion of the preceding sections, and particularly, to illustrate the assignment of impulses to each cluster, we will analyze this example in full detail. The linear transformation $\Psi_{\mathbf{f}_1,\mathbf{f}_2,\mathbf{f}_3}$ is given in this singular case by:

$$\Psi_{\mathbf{f}_1,\mathbf{f}_2,\mathbf{f}_3}(n_1,n_2,n_3) = n_1\mathbf{f}_1 + n_2\mathbf{f}_2 + n_3\mathbf{f}_3$$
$$= n_1(32,0) + n_2(-16,16\sqrt{3}) + n_3(-16,-16\sqrt{3}) \quad (5.17)$$

Let us compare the transformation $\Psi_{\mathbf{f}_1,\mathbf{f}_2,\mathbf{f}_3}$ itself with its continuous counterpart, $\Phi_{\mathbf{f}_1,\mathbf{f}_2,\mathbf{f}_3} : \mathbb{R}^3 \to \mathbb{R}^2$. $\mathrm{Ker}\Phi_{\mathbf{f}_1,\mathbf{f}_2,\mathbf{f}_3}$, i.e., the subspace of $\mathbb{R}^3$ which is mapped by $\Phi_{\mathbf{f}_1,\mathbf{f}_2,\mathbf{f}_3}$ into the origin (0,0) of the $u,v$ plane, contains all the points $(n_1,n_2,n_3) \in \mathbb{R}^3$ which solve the following set of two linear equations, obtained from Eq. (5.17):

$$\begin{cases} 32n_1 - 16n_2 - 16n_3 = 0 \\ 16\sqrt{3}n_2 - 16\sqrt{3}n_3 = 0 \end{cases}$$

The solution of this set of equations is: $\mathrm{Ker}\Phi_{\mathbf{f}_1,\mathbf{f}_2,\mathbf{f}_3} = \{(n_1,n_2,n_3) \mid n_1 = n_2 = n_3,\ n_i \in \mathbb{R}\}$ which means that $\mathrm{Ker}\Phi_{\mathbf{f}_1,\mathbf{f}_2,\mathbf{f}_3}$ is the diagonal line $z = y = x$ of $\mathbb{R}^3$. Therefore, $\mathbb{R}^3$ is partitioned by $\Phi_{\mathbf{f}_1,\mathbf{f}_2,\mathbf{f}_3}$ into an infinite 2D set of translated lines (1D equivalence classes) parallel to the line $z = y = x$. Since $\dim \mathrm{Ker}\Phi_{\mathbf{f}_1,\mathbf{f}_2,\mathbf{f}_3} = 1$ we have here $\dim \mathrm{Im}\Phi_{\mathbf{f}_1,\mathbf{f}_2,\mathbf{f}_3} = 3-1 = 2$, and indeed the continuous transformation $\Phi_{\mathbf{f}_1,\mathbf{f}_2,\mathbf{f}_3}$ maps into each point of the 2D $u,v$ plane an entire 1D line from this decomposition of $\mathbb{R}^3$.

Returning now to the discrete case where $n_i \in \mathbb{Z}$, it is clear that in this example the lattice $L = \mathrm{Ker}\Psi_{\mathbf{f}_1,\mathbf{f}_2,\mathbf{f}_3} = \mathrm{Ker}\Phi_{\mathbf{f}_1,\mathbf{f}_2,\mathbf{f}_3} \cap \mathbb{Z}^3$ is indeed a lattice of rank 1 on the diagonal line $z = y = x$, given by: $L = \{(n_1,n_2,n_3) \mid n_1 = n_2 = n_3,\ n_i \in \mathbb{Z}\}$, so there is no loss of dimensions in this case. The lattice $L$ consists of the indices of all the impulses of the 1D cluster which collapses, precisely at the singular state, on the origin of the spectrum: {..., (-1,-1,-1), (0,0,0), (1,1,1), ...}. This cluster can be seen spread-out about the spectrum origin in Fig. 5.2(b), which shows the spectrum slightly off the singular state.

Each of the other clusters in this spectrum consists of the impulses of one parallel translation of $L$ within $\mathbb{Z}^3$: $(r_1,r_2,r_3) + L = \{(r_1,r_2,r_3) + (n_1,n_2,n_3) \mid n_1 = n_2 = n_3,\ n_i \in \mathbb{Z}\}$; each of these translated lattices of rank 1 is mapped by $\Psi_{\mathbf{f}_1,\mathbf{f}_2,\mathbf{f}_3}$ into a single point $\Psi_{\mathbf{f}_1,\mathbf{f}_2,\mathbf{f}_3}(r_1,r_2,r_3)$ within the $u,v$ spectrum-plane. For example (see Fig. 5.2): on top of the fundamental impulse of the first grating, which is the (1,0,0)-impulse in the spectrum convolution (located in the $u,v$ plane at $\mathbf{f}_1 = (32,0)$), collapses the entire 1D cluster $(1,0,0) + L$, i.e: {..., (0,-1,-1), (1,0,0), (2,1,1), ...}. This cluster can be seen spread-out about the impulse $\mathbf{f}_1$ in Fig. 5.2(b).

Let us consider now the image of the discrete and the continuous transformations. The image of the transformation $\Psi_{\mathbf{f}_1,\mathbf{f}_2,\mathbf{f}_3}$, i.e., the support of all the collapsed clusters in the spectrum precisely at the singular state, is the module given by Eq. (5.3):

$$\mathrm{Md}(\mathbf{f}_1,\mathbf{f}_2,\mathbf{f}_3) = \{n_1(32,0) + n_2(-16,16\sqrt{3}) + n_3(-16,-16\sqrt{3}) \mid n_i \in \mathbb{Z}\}$$

The image of the *continuous* transformation $\Phi_{\mathbf{f}_1,\mathbf{f}_2,\mathbf{f}_3}$ is the entire $u,v$ plane:

$$\mathrm{Sp}(\mathbf{f}_1,\mathbf{f}_2,\mathbf{f}_3) = \mathbb{R}^2$$

Therefore, as we have already seen above, we have in this example: $\mathrm{rank}_{\mathbb{Z}}\, \mathrm{Md}(\mathbf{f}_1,\mathbf{f}_2,\mathbf{f}_3) = \dim \mathrm{Sp}(\mathbf{f}_1,\mathbf{f}_2,\mathbf{f}_3) = 2$. This means, as we can see in Fig. 5.2(a), that the support of the spectrum forms in this case a lattice of rank 2; and each point of this lattice is in fact a collapsed 1D cluster of impulses, representing one equivalence class from the partition induced by $\Psi_{\mathbf{f}_1,\mathbf{f}_2,\mathbf{f}_3}$ in the indices-lattice $\mathbb{Z}^3$. And indeed, when the superposed gratings slightly move away from the singular moiré state (i.e., when their frequency vectors $\mathbf{f}_i$ are slightly modified), each of the 1D clusters in the spectrum starts spreading out, and in the image domain a 1D moiré becomes visible in the superposition, as indicated by the low frequencies of the 1D spread-out cluster around the spectrum origin (see Fig. 5.2(b)). ■

**Example 5.6:** (The same clusters as in Example 5.5, but at different locations): Consider the (1,1,1)-singular moiré which occurs between three gratings when their frequency vectors have the values: $\mathbf{f}_1 = (32,0)$, $\mathbf{f}_2 = (0,32)$, $\mathbf{f}_3 = (-32,-32)$ (see Fig. 5.3(a)). In this case the linear transformation $\Psi_{\mathbf{f}_1,\mathbf{f}_2,\mathbf{f}_3}$ is given by:

$$\Psi_{\mathbf{f}_1,\mathbf{f}_2,\mathbf{f}_3}(n_1,n_2,n_3) = n_1(32,0) + n_2(0,32) + n_3(-32,-32)$$

Here, $L = \mathrm{Ker}\,\Psi_{\mathbf{f}_1,\mathbf{f}_2,\mathbf{f}_3}$ is again the same 1D lattice as in Example 5.5: $\{(n_1,n_2,n_3) \mid n_1 = n_2 = n_3,\ n_i \in \mathbb{Z}\}$, so that the partition of $\mathbb{Z}^3$ (and the assignment of impulses to each cluster) remain the same as in Example 5.5. Only the locations of the clusters in the spectrum are modified, as shown in Fig. 5.3. ■

**Example 5.7:** (The same support, i.e., cluster locations, as in Example 5.5, but a different assignment of impulses to each cluster): Consider the (1,-1,1)-singular moiré which occurs between three gratings when their frequency vectors have the polar coordinates: $\mathbf{f}_1 = (0°,32)$, $\mathbf{f}_2 = (60°,32)$, $\mathbf{f}_3 = (120°,32)$, i.e., in Cartesian coordinates: $\mathbf{f}_1 = (32,0)$, $\mathbf{f}_2 = (16,16\sqrt{3})$, $\mathbf{f}_3 = (-16,16\sqrt{3})$ (see Fig. 5.4(a)). In this case the transformation $\Psi_{\mathbf{f}_1,\mathbf{f}_2,\mathbf{f}_3}$ is:

$$\Psi_{\mathbf{f}_1,\mathbf{f}_2,\mathbf{f}_3}(n_1,n_2,n_3) = n_1(32,0) + n_2(16,16\sqrt{3}) + n_3(-16,16\sqrt{3})$$

Here the kernel $L = \mathrm{Ker}\,\Psi_{\mathbf{f}_1,\mathbf{f}_2,\mathbf{f}_3}$ is again a 1D lattice, but not the same 1D lattice as in Example 5.5: $L = \{(n_1,n_2,n_3) \mid n_1 = -n_2 = n_3,\ n_i \in \mathbb{Z}\}$. Therefore the partition of the indices-lattice $\mathbb{Z}^3$ into equivalence classes (and hence the assignment of impulses to each cluster) are different than in Example 5.5. It so happens, however, that $\Psi_{\mathbf{f}_1,\mathbf{f}_2,\mathbf{f}_3}$ in this case maps the equivalence classes into the same support in the $u,v$ plane as in Example 5.5 (compare Fig. 5.4(a) with Fig. 5.2(a)). ■

**Example 5.8:** (2D clusters on a 1D support in the $u,v$ plane): Consider the ((1,1,1),(1,-1,0))-singular moiré[12] which occurs between three gratings whose frequency

[12] We use here the full-length notation of this combined 2D moiré (see Sec. 2.8) in order to avoid any possible confusion.

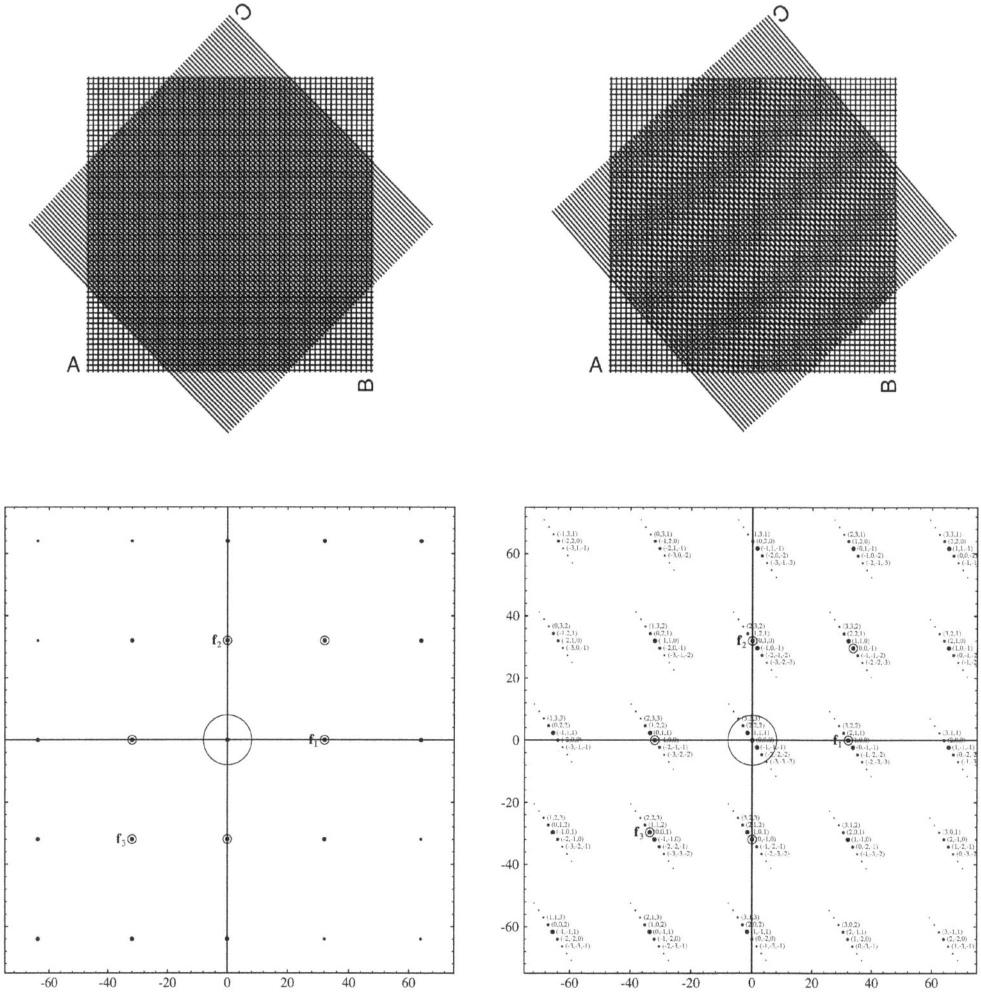


(a) (b)

**Figure 5.3:** The singular 3-grating superposition of Example 5.6 (top) and its spectrum support (bottom). (a) Exactly at the singular state: the spectrum support forms here a 2D lattice, each point of which represents a collapsed cluster. (b) Slightly off the singular state: each of the clusters in the spectrum is spread out, clearly demonstrating its 1D nature. With respect to Fig. 5.2(a), only the cluster locations have been modified, but not the assignment of impulses to the clusters. Only impulses up to the 5-th order are shown. (A magnified version of this figure can be found in the Internet site of the book.)

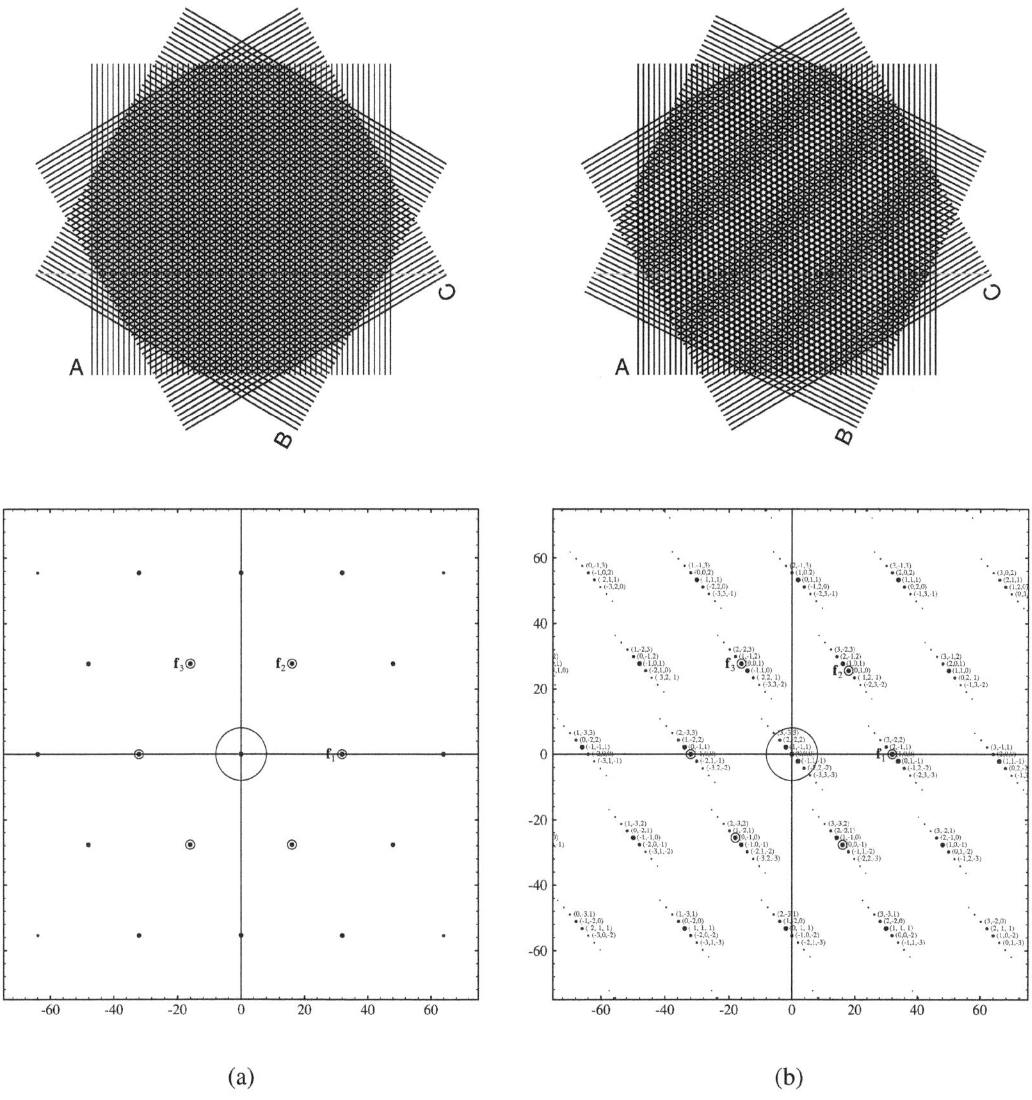


**Figure 5.4:** The singular 3-grating superposition of Example 5.7 (top) and its spectrum support (bottom). (a) Exactly at the singular state: the spectrum support forms here a 2D lattice, each point of which represents a collapsed cluster. (b) Slightly off the singular state: each of the clusters in the spectrum is spread out, clearly demonstrating its 1D nature. In this case the cluster locations are the same as in Fig. 5.2(a), but the assignment of impulses to the clusters is different. Only impulses up to the 5-th order are shown. (A magnified version of this figure can be found in the Internet site of the book.)

vectors are: $\mathbf{f}_1 = \mathbf{f}_2 = (32,32)$, $\mathbf{f}_3 = (-64,-64)$ (see Fig. 5.5(a)). In this case the vectors $\mathbf{f}_1$, $\mathbf{f}_2$ and $\mathbf{f}_3$ in the $u,v$ plane are collinear (linearly dependent over $\mathbb{R}$), and also commensurable (linearly dependent over $\mathbb{Z}$). We have, therefore: $\dim \text{Sp}(\mathbf{f}_1,\mathbf{f}_2,\mathbf{f}_3) = \text{rank}_\mathbb{Z} \text{Md}(\mathbf{f}_1,\mathbf{f}_2,\mathbf{f}_3) = 1$, so that the spectrum support, $\text{Md}(\mathbf{f}_1,\mathbf{f}_2,\mathbf{f}_3)$, is in this singular case a discrete lattice of rank 1. Moreover, from Eq. (5.16) we find that each point of this lattice represents, in fact, a collapsed lattice (cluster) whose rank is: $\text{rank}\, L = 3-1 = 2$. And indeed, when the three superposed gratings move a little from the singular moiré state, each of the 2D clusters in the spectrum spreads out, and in the image domain a 2D moiré becomes visible in the superposition, as indicated by the low frequencies of the 2D spread-out cluster around the spectrum origin (see Fig. 5.5(b)).

Let us analyze this example in more detail. In this case the linear transformation $\Psi_{\mathbf{f}_1,\mathbf{f}_2,\mathbf{f}_3}$ is given by:

$$\Psi_{\mathbf{f}_1,\mathbf{f}_2,\mathbf{f}_3}(n_1,n_2,n_3) = n_1(32,32) + n_2(32,32) + n_3(-64,-64)$$

We will consider, again, both this transformation and its continuous counterpart, $\Phi_{\mathbf{f}_1,\mathbf{f}_2,\mathbf{f}_3}$. The kernel of $\Phi_{\mathbf{f}_1,\mathbf{f}_2,\mathbf{f}_3}$ is the solution of the equation $32n_1 + 32n_2 - 64n_3 = 0$ with $n_i \in \mathbb{R}$, which is the 2D plane $2z = x + y$ in $\mathbb{R}^3$. Therefore $\mathbb{R}^3$ is partitioned by $\Phi_{\mathbf{f}_1,\mathbf{f}_2,\mathbf{f}_3}$ into an infinite 1D set of translated planes (2D equivalence classes) which are parallel to the plane $2z = x + y$. Since $\dim \text{Ker}\Phi_{\mathbf{f}_1,\mathbf{f}_2,\mathbf{f}_3} = 2$, we have here $\dim \text{Im}\Phi_{\mathbf{f}_1,\mathbf{f}_2,\mathbf{f}_3} = 3 - 2 = 1$, and indeed the image of $\Phi_{\mathbf{f}_1,\mathbf{f}_2,\mathbf{f}_3}$ is the 1D line which is spanned in the $u,v$ plane by $\mathbf{f}_1$, $\mathbf{f}_2$, $\mathbf{f}_3$, namely: $v = u$. Therefore, the continuous transformation $\Phi_{\mathbf{f}_1,\mathbf{f}_2,\mathbf{f}_3}$ maps into each point of this line in the $u,v$ plane an entire 2D plane from the decomposition of $\mathbb{R}^3$.

Returning now to the discrete case, it is clear that the kernel $L$ of the mapping $\Psi_{\mathbf{f}_1,\mathbf{f}_2,\mathbf{f}_3}$ is the 2D lattice: $L = \{(n_1,n_2,n_3) \mid 2n_3 = n_1+n_2,\ n_i \in \mathbb{Z}\}$ which is imbedded in the plane $2z = x + y$. Therefore in this case there is no loss of dimensions, and the cluster which falls on the spectrum origin (as well as all the other clusters) is of a 2D nature; see Fig. 5.5(b), which shows the spread-out clusters in the spectrum slightly off the singular state.[13]

The support of all the collapsed clusters in the spectrum precisely at the singular state is the image of the transformation $\Psi_{\mathbf{f}_1,\mathbf{f}_2,\mathbf{f}_3}$, i.e., the module given by Eq. (5.3):

$$\text{Md}(\mathbf{f}_1,\mathbf{f}_2,\mathbf{f}_3) = \{n_1(32,32) + n_2(32,32) + n_3(64,64) \mid n_i \in \mathbb{Z}\}$$

The image of the *continuous* transformation $\Phi_{\mathbf{f}_1,\mathbf{f}_2,\mathbf{f}_3}$ in the $u,v$ plane is the line $v = u$:

$$\text{Sp}(\mathbf{f}_1,\mathbf{f}_2,\mathbf{f}_3) = \{(u,v) \in \mathbb{R}^2 \mid v = u\}$$

In this example we have, therefore: $\text{rank}_\mathbb{Z} \text{Md}(\mathbf{f}_1,\mathbf{f}_2,\mathbf{f}_3) = \dim \text{Sp}(\mathbf{f}_1,\mathbf{f}_2,\mathbf{f}_3) = 1$. This means, as we can see in Fig. 5.5(a), that the support of the spectrum forms in this case a lattice of

[13] Note that (1,1,1) and (1,-1,0) form a basis of the 2D lattice $L$ (see Fig. 5.5(b)). This means that in the superposition of this example not only the 3-layer (1,1,1)-moiré is singular, but also the 2-layer (1,-1,0)-moiré. And indeed, slightly away from the singular state both of them are simultaneously visible in the superposition, and they form together a combined 2D moiré, as shown in Fig. 5.5(b).

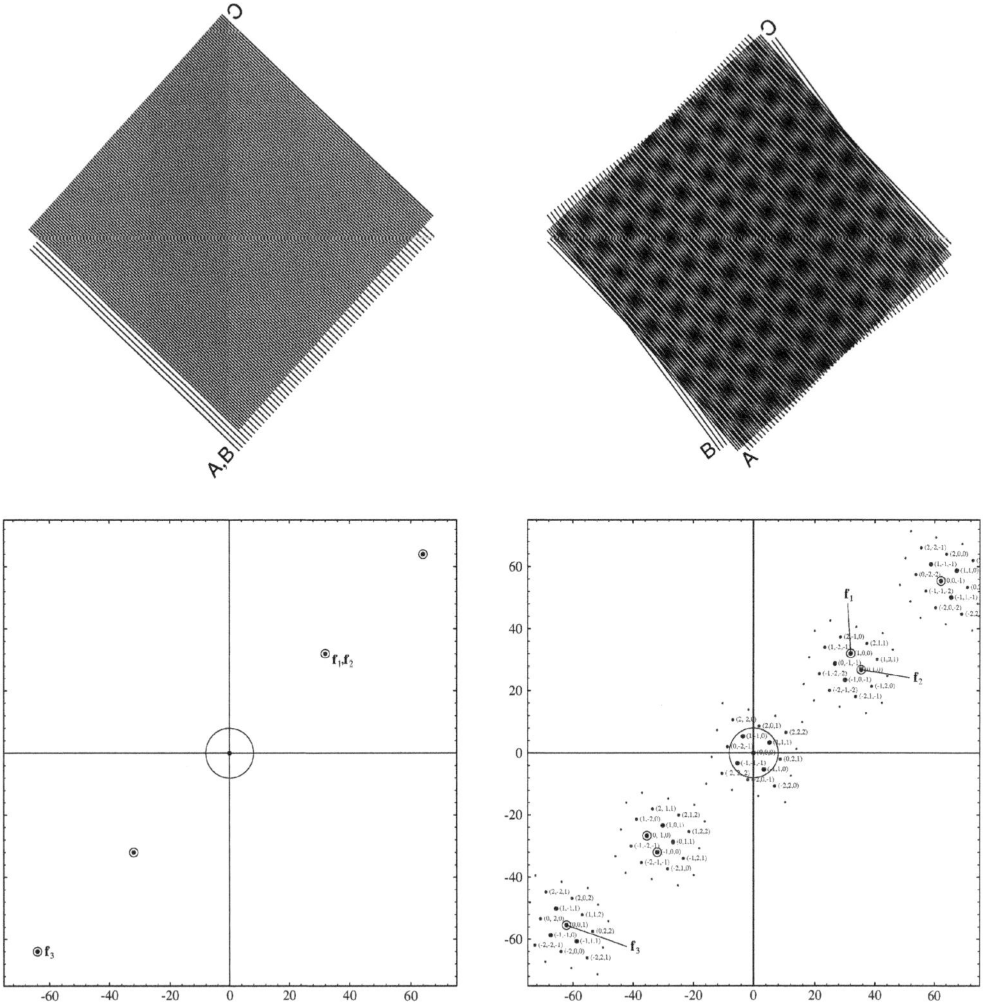


(a) (b)

**Figure 5.5:** The singular 3-grating superposition of Example 5.8 (top) and its spectrum support (bottom). (a) Exactly at the singular state: the spectrum support forms here a 1D lattice, each point of which represents a collapsed cluster. (b) Slightly off the singular state: each of the clusters in the spectrum is spread out, clearly demonstrating its 2D nature. Only impulses up to the 3-rd order are shown. (A magnified version of this figure can be found in the Internet site of the book.)

rank 1; and each point of this lattice is in fact a collapsed 2D cluster of impulses, representing one equivalence class (translation of the 2D lattice $L$) from the partition induced by $\Psi_{\mathbf{f}_1,\mathbf{f}_2,\mathbf{f}_3}$ in the indices-lattice $\mathbb{Z}^3$. And indeed, when the superposed gratings in the image domain move a little from the singular moiré state (i.e., when their frequency vectors $\mathbf{f}_i$ in the spectrum are slightly modified), each of the 2D clusters in the spectrum starts spreading out. This means that in the image domain a 2D moiré becomes visible in the superposition, as indicated by the low frequencies of the 2D spread-out cluster around the spectrum origin (see Fig. 5.5(b)). ■

**Example 5.9:** (1D clusters on a 1D support in the $u,v$ plane): This example illustrates what happens in a case similar to Example 5.8 when the lattice $L = \text{Ker}\Psi_{\mathbf{f}_1,\mathbf{f}_2,\mathbf{f}_3}$ which collapses on the spectrum origin has a lower rank than its continuous counterpart $\text{Ker}\Phi_{\mathbf{f}_1,\mathbf{f}_2,\mathbf{f}_3}$, due to an irrational inclination of the plane $\text{Ker}\Phi_{\mathbf{f}_1,\mathbf{f}_2,\mathbf{f}_3}$ within $\mathbb{R}^3$. Consider the singular moiré which occurs between three gratings whose frequency vectors are: $\mathbf{f}_1 = \mathbf{f}_2 = (32,32)$, $\mathbf{f}_3 = q\mathbf{f}_1$ where $q$, unlike in Example 5.8, is an irrational number, say $-\sqrt{2}$ (see Fig. 5.6(a)). In this case the vectors $\mathbf{f}_1$, $\mathbf{f}_2$, $\mathbf{f}_3$ in the $u,v$ plane are linearly dependent over $\mathbb{R}$, but only two of them are linearly dependent over $\mathbb{Z}$. We therefore have: $\dim \text{Sp}(\mathbf{f}_1,\mathbf{f}_2,\mathbf{f}_3) = 1 < \text{rank}_{\mathbb{Z}} \text{Md}(\mathbf{f}_1,\mathbf{f}_2,\mathbf{f}_3) = 2$, so that the spectrum support, $\text{Md}(\mathbf{f}_1,\mathbf{f}_2,\mathbf{f}_3)$, is in this case a dense module of $\text{rank}_{\mathbb{Z}} = 2$ imbedded on the 1D line $\text{Sp}(\mathbf{f}_1,\mathbf{f}_2,\mathbf{f}_3)$. Moreover, from Eq. (5.16) we see that each point of this module is, in fact, a collapsed cluster whose rank is: $\text{rank}\,L = 3-2 = 1$. And indeed, if the three superposed gratings move a little from the singular moiré state, each of the 1D clusters in the spectrum spreads out, and in the image domain a 1D moiré becomes visible in the superposition, as indicated by the low frequencies of the 1D spread-out cluster around the spectrum origin (see Fig. 5.6(b)).

Let us analyze this example in more detail, comparing it to Example 5.8. In the present case the linear transformation $\Psi_{\mathbf{f}_1,\mathbf{f}_2,\mathbf{f}_3}$ is given by:

$$\Psi_{\mathbf{f}_1,\mathbf{f}_2,\mathbf{f}_3}(n_1,n_2,n_3) = n_1(32,32) + n_2(32,32) + n_3(-32\sqrt{2},-32\sqrt{2})$$

The kernel of the continuous counterpart of this transformation, $\Phi_{\mathbf{f}_1,\mathbf{f}_2,\mathbf{f}_3}$, is the solution of the equation $32n_1 + 32n_2 - 32\sqrt{2}n_3 = 0$ with $n_i \in \mathbb{R}$, which is the 2D plane $\sqrt{2}z = x + y$ in $\mathbb{R}^3$. Therefore $\mathbb{R}^3$ is partitioned by $\Phi_{\mathbf{f}_1,\mathbf{f}_2,\mathbf{f}_3}$ into an infinite 1D set of translated planes (2D equivalence classes) which are parallel to the plane $\sqrt{2}z = x + y$. The image of $\Phi_{\mathbf{f}_1,\mathbf{f}_2,\mathbf{f}_3}$ is the 1D line which is spanned in the $u,v$ plane by $\mathbf{f}_1$, $\mathbf{f}_2$, $\mathbf{f}_3$, namely: $v = u$. Therefore, like in Example 5.8, the continuous transformation $\Phi_{\mathbf{f}_1,\mathbf{f}_2,\mathbf{f}_3}$ maps into each point of this line in the $u,v$ plane an entire 2D plane from the decomposition of $\mathbb{R}^3$.

Let us now return to the discrete case. Although the kernel of the continuous $\Phi_{\mathbf{f}_1,\mathbf{f}_2,\mathbf{f}_3}$ is still a 2D plane in $\mathbb{R}^3$, as in Example 5.8, we see that in the present case, owing to the irrational inclination of this plane, its discrete restriction to $n_i \in \mathbb{Z}$ (i.e., the lattice $L = \text{Ker}\Psi_{\mathbf{f}_1,\mathbf{f}_2,\mathbf{f}_3}$ which collapses to the spectrum origin) has a lower rank than 2. In fact, the only points of the indices-lattice $\mathbb{Z}^3$ which fall on the 2D plane $\sqrt{2}z = x + y$ are those for which $z = 0$, so that we get: $L = \text{Ker}\Psi_{\mathbf{f}_1,\mathbf{f}_2,\mathbf{f}_3} = \{(n_1,n_2,n_3) \mid n_2 = -n_1,\ n_3 = 0,\ n_i \in \mathbb{Z}\}$.

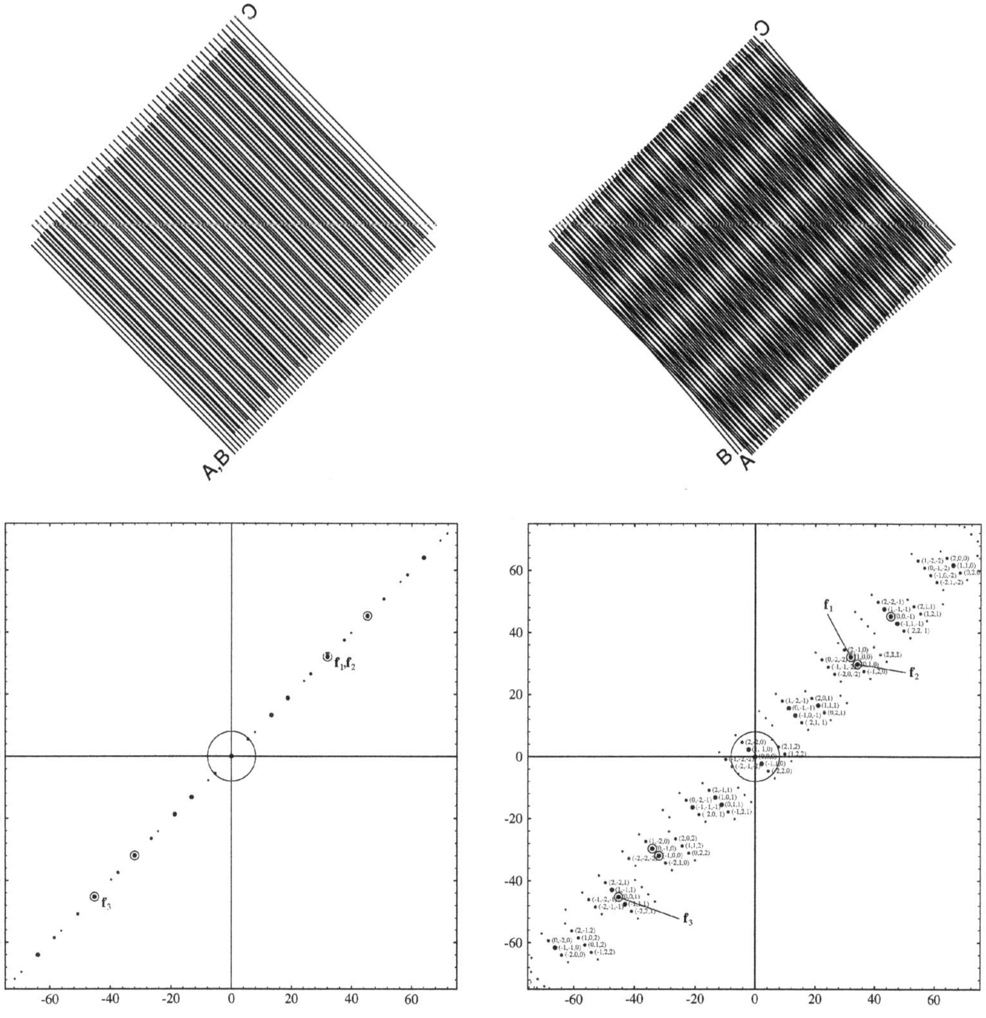


(a) (b)

**Figure 5.6:** The singular 3-grating superposition of Example 5.9 (top) and its spectrum support (bottom). (a) Exactly at the singular state: the spectrum support forms here a 1D module (of integral rank 2), each point of which represents a collapsed cluster. (b) Slightly off the singular state: each of the clusters in the spectrum is spread out, clearly demonstrating its 1D nature. Only impulses up to the 3-rd order are shown. (A magnified version of this figure can be found in the Internet site of the book.)

Therefore, in this case the cluster which falls on the spectrum origin is of rank 1: $L = \{..., (-1,1,0), (0,0,0), (1,-1,0), ...\}$ (see Fig. 5.6(b), which shows the spread-out clusters slightly off the singular state). Similarly, a 1D cluster which consists of one parallel translation of $L$ within $\mathbb{Z}^3$ collapses on each other point in the spectrum support. For example (see Fig. 5.6(b)), on the fundamental impulse of the first grating, which is the (1,0,0)-impulse in the spectrum convolution, collapses the whole 1D cluster $(1,0,0) + L$, i.e.: $\{..., (0,1,0), (1,0,0), (2,-1,0), ...\}$.

The support of all these collapsed clusters in the spectrum precisely at the singular state is the image of the transformation $\Psi_{\mathbf{f}_1,\mathbf{f}_2,\mathbf{f}_3}$, i.e. the module of $\text{rank}_{\mathbb{Z}} = 2$ given by Eq. (5.3):

$$\text{Md}(\mathbf{f}_1,\mathbf{f}_2,\mathbf{f}_3) = \{n_1(32,32) + n_2(32,32) + n_3(-32\sqrt{2},-32\sqrt{2}) \mid n_i \in \mathbb{Z}\}$$

This module is imbedded in the image of the continuous transformation $\Phi_{\mathbf{f}_1,\mathbf{f}_2,\mathbf{f}_3}$ in the $u,v$ plane, which is the same 1D line as in Example 5.8:

$$\text{Sp}(\mathbf{f}_1,\mathbf{f}_2,\mathbf{f}_3) = \{(u,v) \in \mathbb{R}^2 \mid v = u\}$$

In this example we therefore have: $\dim \text{Sp}(\mathbf{f}_1,\mathbf{f}_2,\mathbf{f}_3) = 1 < \text{rank}_{\mathbb{Z}} \text{Md}(\mathbf{f}_1,\mathbf{f}_2,\mathbf{f}_3) = 2$. This means, as we can see in Fig. 5.6(a), that in this case the support of the spectrum forms a dense module of $\text{rank}_{\mathbb{Z}} = 2$ which is imbedded on the 1D line $v = u$; and each point of this module consists of a whole 1D cluster, representing one equivalence class (translation of the 1D lattice $L$) from the partition induced by $\Psi_{\mathbf{f}_1,\mathbf{f}_2,\mathbf{f}_3}$ in the indices-lattice $\mathbb{Z}^3$. And indeed, if the superposed gratings move a little from the singular moiré state, each of the 1D clusters in the spectrum spreads out, and in the image domain a 1D moiré becomes visible in the superposition, as indicated by the low frequencies of the 1D spread-out cluster around the spectrum origin (see Fig. 5.6(b)).

As we can see in this example, the "loss" of one dimension in the discrete $\text{Ker}\Psi_{\mathbf{f}_1,\mathbf{f}_2,\mathbf{f}_3}$ due to an irrational inclination of the 2D plane $\text{Ker}\Phi_{\mathbf{f}_1,\mathbf{f}_2,\mathbf{f}_3}$ in $\mathbb{R}^3$ (i.e., the loss of one dimension in each cluster) is "compensated" in the image of $\Psi_{\mathbf{f}_1,\mathbf{f}_2,\mathbf{f}_3}$ in the $u,v$ plane by an increment of 1 in the integral rank of this module: whereas in Example 5.8 $\text{Im}\Psi_{\mathbf{f}_1,\mathbf{f}_2,\mathbf{f}_3}$ was a module of $\text{rank}_{\mathbb{Z}} = 1$ imbedded on the 1D line $\text{Im}\Phi_{\mathbf{f}_1,\mathbf{f}_2,\mathbf{f}_3}$ (namely: a lattice of rank 1), in the present case $\text{Im}\Psi_{\mathbf{f}_1,\mathbf{f}_2,\mathbf{f}_3}$ is a dense module of $\text{rank}_{\mathbb{Z}} = 2$ which is imbedded on the same line $\text{Im}\Phi_{\mathbf{f}_1,\mathbf{f}_2,\mathbf{f}_3}$. (Note that the continuous $\text{Ker}\Phi_{\mathbf{f}_1,\mathbf{f}_2,\mathbf{f}_3}$ and $\text{Im}\Phi_{\mathbf{f}_1,\mathbf{f}_2,\mathbf{f}_3}$ have both the same dimensions as in Example 5.8; only the dimensions of their discrete counterparts have changed.) ■

**Example 5.10:** (2D clusters on a 2D support in the $u,v$ plane): Consider the (1,1,-1,0)-singular moiré which occurs between two screens (or four gratings) when their frequency vectors are given by: $\mathbf{f}_1 = (32,0)$, $\mathbf{f}_2 = (0,32)$, $\mathbf{f}_3 = (32,32)$ and $\mathbf{f}_4 = (-32,32)$ (see Fig. 5.7(a)). It is easy to see that in this case $\mathbf{f}_3$, $\mathbf{f}_4$ are linear combinations, both over $\mathbb{Z}$ and $\mathbb{R}$, of $\mathbf{f}_1$, $\mathbf{f}_2$ (namely: $\mathbf{f}_3 = \mathbf{f}_1+\mathbf{f}_2$, $\mathbf{f}_4 = \mathbf{f}_2-\mathbf{f}_1$), while $\mathbf{f}_1$ and $\mathbf{f}_2$ are independent. Therefore we have here: $\text{rank}_{\mathbb{Z}} \text{Md}(\mathbf{f}_1,\mathbf{f}_2,\mathbf{f}_3,\mathbf{f}_4) = \dim \text{Sp}(\mathbf{f}_1,\mathbf{f}_2,\mathbf{f}_3,\mathbf{f}_4) = 2$. This means that the spectrum support, $\text{Md}(\mathbf{f}_1,\mathbf{f}_2,\mathbf{f}_3,\mathbf{f}_4)$, is in this singular case a discrete lattice of rank 2. And furthermore, from

Eq. (5.16) we see that each point of this lattice represents, in fact, a collapsed cluster whose rank is: $\text{rank}\,L = 4-2 = 2$. ■

It should be noted that in general it is not always practical to find arithmetically the Cartesian coordinates of the frequency vectors $\mathbf{f}_i$ and to determine the lattice $L$. In such cases, a computer program which calculates the comb convolutions in the spectral domain can be helpful. Given the polar coordinates of the frequency vectors $\mathbf{f}_i$ (i.e., the frequencies and the directions of each superposed layer) this program calculates the spectral convolution (up to a specified number of harmonics), using the rules of comb or nailbed convolution (Proposition 2.3). The resulting impulse configuration (=spectrum support) is graphically displayed in the $u,v$ spectrum, showing the location (and optionally also the index) of each impulse. This is how the figures illustrating the examples of this section have been prepared. This method is useful both for getting a general overview of the spectrum support, and for finding out the indices of any particular impulses in the spectrum. This is demonstrated in the following example:

**Example 5.11:** (1D clusters on a dense 2D support in the $u,v$ plane): Consider the (1,1,1,1,1)-singular moiré which occurs in the superposition of 5 gratings with identical frequencies, and angle differences of 360°/5 = 72°. In this case the arithmetic calculation of the Cartesian coordinates is more tricky (the values $\sin 72° = \frac{1}{4}\sqrt{10+2\sqrt{5}}$ and $\cos 72° = \frac{1}{4}(\sqrt{5}-1)$ can be obtained from the radiuses of the circumscribed and the inscribed circles in a regular polygon [EncMath88 Vol. 7 p. 221, "Polygon"]). However, the spectrum support obtained by computer immediately gives us an insight into the nature of this case. Fig. 5.8(a) shows the spectrum support exactly at the specified singular configuration (using our usual convention that the frequency of each layer is 32). Visibly, this spectrum support is not a discrete lattice, but rather an everywhere dense module on the $u,v$ plane. In order to visually identify the individual impulses belonging to each of the collapsed clusters in the singular state, we move slightly off the singular state (by modifying the values of one or more of the frequency vectors) so that the impulse clusters in the spectrum become fully spread out (see Fig. 5.8(b)). As we can see here, each cluster is only of rank 1; this implies according to Eq. (5.16) that $\text{rank}_{\mathbb{Z}}\, \text{Im}\, \Psi_{\mathbf{f}_1,\ldots,\mathbf{f}_5} = 5-1 = 4$, and since $\dim \text{Im}\, \Phi_{\mathbf{f}_1,\ldots,\mathbf{f}_5}$ is obviously only 2, it follows, indeed, that the spectrum support of this singular state is everywhere dense on the $u,v$ plane (see Table 5.2). ■

**Example 5.12:** (2D dense clusters on a discrete 2D support in the $u,v$ plane): Consider the singular rational superposition of three dot-screens,[14] whose frequency vectors are given by (based on our usual convention that $\mathbf{f}_1$ is located on the horizontal axis and its frequency is normalized to 32): $\mathbf{f}_1 = (32,0)$, $\mathbf{f}_2 = (0,32)$, $\mathbf{f}_3 = (\frac{4}{5}\cdot 32, \frac{3}{5}\cdot 32)$, $\mathbf{f}_4 = (-\frac{3}{5}\cdot 32, \frac{4}{5}\cdot 32)$, $\mathbf{f}_5 = (\frac{4}{5}\cdot 32, -\frac{2}{5}\cdot 32)$ and $\mathbf{f}_6 = (\frac{2}{5}\cdot 32, \frac{4}{5}\cdot 32)$; see Fig. 5.9(a). Therefore the linear transformation $\Psi_{\mathbf{f}_1,\ldots,\mathbf{f}_6}$ is given here by:

---

[14] For rational superpositions see Sec. 8.6. This example is based on a rational screen combination which has been proposed by Adobe Systems Inc. for colour printing at device resolution of 1270 dpi with screen frequencies of about 90 lpi: $(\Delta x_1,\Delta y_1) = (12,4)$ pixels, $(\Delta x_2,\Delta y_2) = (4,12)$ pixels, $(\Delta x_3,\Delta y_3) = (10,10)$ pixels. After normalizing the three screens to fit our conventions for 3-screen superpositions (see Sec. 3.4.2 and Fig. 3.4 there), we obtain the spectrum support shown in Fig. 5.9(a).

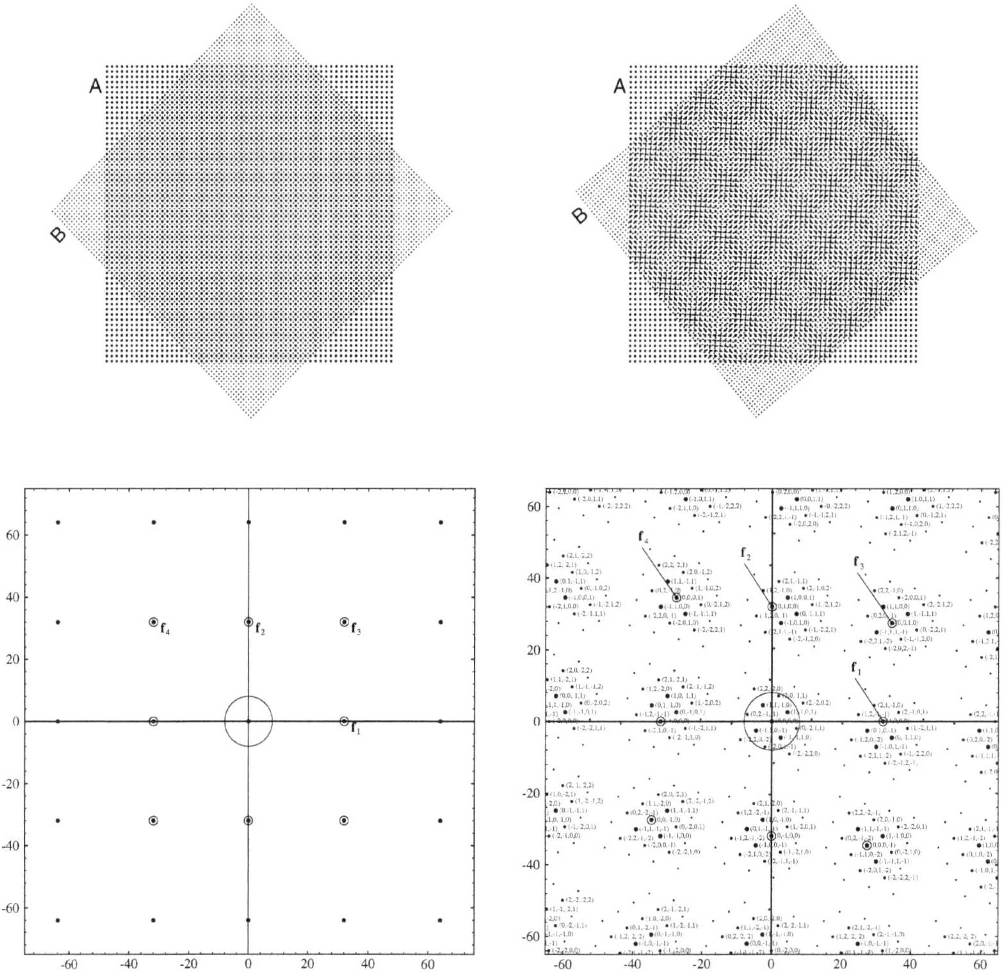


(a) (b)

**Figure 5.7:** The singular 2-screen superposition of Example 5.10 (top) and its spectrum support (bottom). (a) Exactly at the singular state: the spectrum support forms here a 2D lattice, each point of which represents a collapsed cluster. (b) Slightly off the singular state: each of the clusters is spread out, clearly demonstrating its 2D nature. Only impulses up to the 3-rd order are shown. (A magnified version of this figure can be found in the Internet site of the book.)

 

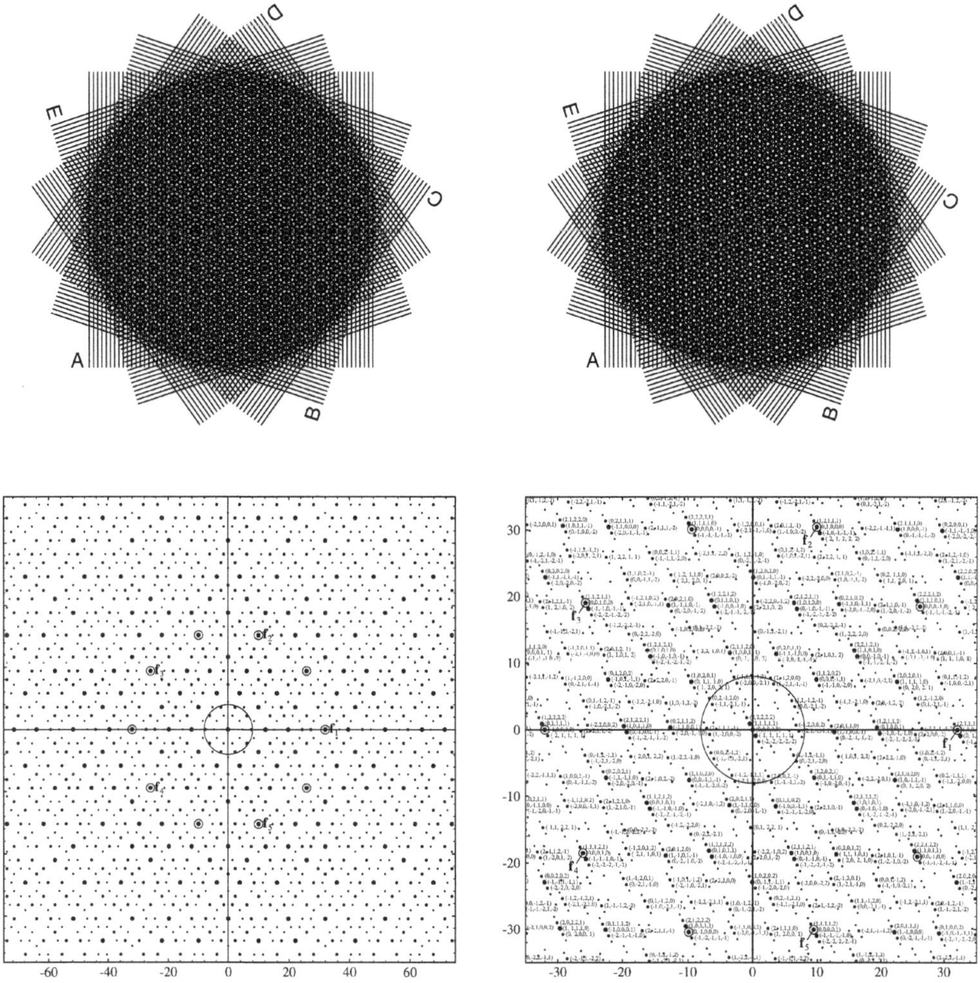


(a) (b)

**Figure 5.8:** The singular 5-grating superposition of Example 5.11 (top) and its spectrum support (bottom). (a) Exactly at the singular state: the spectrum support forms here an everywhere dense 2D module, each point of which represents a collapsed cluster. The spectrum in (b) shows an enlarged view of the central part of spectrum (a), slightly off the singular state: each of the clusters in the spectrum is spread out, clearly demonstrating its 1D nature. Only impulses up to the 3-rd order are shown. (A magnified version of this figure can be found in the Internet site of the book.)

$\Psi_{\mathbf{f}_1,\ldots,\mathbf{f}_6}(n_1,n_2,n_3,n_4,n_5,n_6)$

$$= n_1(32,0) + n_2(0,32) + n_3(\tfrac{4}{5}\cdot 32, \tfrac{3}{5}\cdot 32) + n_4(-\tfrac{3}{5}\cdot 32, \tfrac{4}{5}\cdot 32) + n_5(\tfrac{4}{5}\cdot 32, -\tfrac{2}{5}\cdot 32) + n_6(\tfrac{2}{5}\cdot 32, \tfrac{4}{5}\cdot 32)$$

In order to find $\text{Ker}\Phi_{\mathbf{f}_1,\ldots,\mathbf{f}_6}$ of the continuous transformation we have to solve the following set of two linear equations for $n_1,\ldots,n_6 \in \mathbb{R}$:

$$\begin{cases} 32n_1 + \tfrac{4}{5}\cdot 32n_3 - \tfrac{3}{5}\cdot 32n_4 + \tfrac{4}{5}\cdot 32n_5 + \tfrac{2}{5}\cdot 32n_6 = 0 \\ 32n_2 + \tfrac{3}{5}\cdot 32n_3 + \tfrac{4}{5}\cdot 32n_4 - \tfrac{2}{5}\cdot 32n_5 + \tfrac{4}{5}\cdot 32n_6 = 0 \end{cases}$$

The solution of this set of equations is: $\{(n_1,n_2,n_3,n_4,n_5,n_6) \mid 2n_5 = -2n_1+n_2-n_3+2n_4,$ $2n_6 = -n_1-2n_2-2n_3-n_4,\ n_i \in \mathbb{R}\}$. This is clearly a 4D volume (having four free variables) in the 6D space $\mathbb{R}^6$. Furthermore, the lattice $L = \text{Ker}\Psi_{\mathbf{f}_1,\ldots,\mathbf{f}_6}$, which is the discrete solution for $\Psi_{\mathbf{f}_1,\ldots,\mathbf{f}_6}$ (i.e., with $n_i \in \mathbb{Z}$), is also a 4D lattice which is imbedded in this volume. This means that there is no loss of dimensions, so that the spectrum support is indeed a 2D discrete lattice; and since $\text{rank}\,L = 4$ each cluster in the 2D spectrum is a dense module with $\text{rank}_{\mathbb{Z}} = 4$ (see Sec. 5.6.3).

If we closely look at the points (impulses) of the main cluster around the origin (for example, all its impulses up to order 2; see Fig. 5.9(b)), we can see that in this case there occur simultaneously several different moirés all of which having the same singular state: First, we have a 3-screen moiré, spanned by the (0,1,-1,0,1,0)-impulse and its orthogonal counterpart, (1,0,0,1,0,-1); and then we have a 2-screen moiré between each of the three screen pairs: a moiré spanned by the (1,2,-2,-1,0,0)-impulse and its orthogonal counterpart; a moiré spanned by the (2,0,0,0,-2,-1)-impulse and its orthogonal counterpart; and a moiré spanned by the (0,0,2,0,-1,-2)-impulse and its orthogonal counterpart. To each of these moirés belongs a 2D sub-cluster (sub-lattice of the 4D lattice $L$); clearly, all of them collapse together onto the spectrum origin at the singular state. ■

**Example 5.13:** (2D discrete clusters on a dense 2D support in the $u,v$ plane): Consider the singular superposition of three screens with identical frequencies and equal angle differences of 30° (this is the conventional screen combination traditionally used in colour printing; see Fig. 5.10(a)). It is interesting to note that this superposition manifests a 12-fold rotational symmetry, which is clearly seen both in the image domain and in the spectrum. And yet, whenever the three superposed screens move slightly off the singular state, the generated moiré is two-dimensional and it only presents a 4-fold rotational symmetry. The algebraic analysis of this example provides the explanation of this phenomenon: the cluster which is collapsed on the spectrum origin in this singular superposition is indeed a 2D lattice (see the spread-out cluster in Fig. 5.10(b)).

In this example we have: $\mathbf{f}_1 = (32,0)$, $\mathbf{f}_2 = (0,32)$, $\mathbf{f}_3 = (16\sqrt{3},16)$, $\mathbf{f}_4 = (-16,16\sqrt{3})$, $\mathbf{f}_5 = (16\sqrt{3},-16)$ and $\mathbf{f}_6 = (16,16\sqrt{3})$. Note that $\mathbf{f}_1$, $\mathbf{f}_2$, $\mathbf{f}_3$ and $\mathbf{f}_4$ are linearly independent over $\mathbb{Z}$, but $\mathbf{f}_5 = \mathbf{f}_3-\mathbf{f}_2$ and $\mathbf{f}_6 = \mathbf{f}_1+\mathbf{f}_4$. This means that $\text{rank}_{\mathbb{Z}}\text{Md}(\mathbf{f}_1,\ldots,\mathbf{f}_6) = 4$, while $\dim \text{Sp}(\mathbf{f}_1,\ldots,\mathbf{f}_6) = 2$, so that the number of "lost" dimensions is $d = 2$, and the rank of a single cluster is according to Eq. (5.16) rank $L = 6-4 = 2$. The same result can be also obtained by analyzing Ker $\Psi_{\mathbf{f}_1,...,\mathbf{f}_6}$; the linear transformation $\Psi_{\mathbf{f}_1,...,\mathbf{f}_6}$ is, in this case:

$$\Psi_{\mathbf{f}_1,...,\mathbf{f}_6}(n_1,n_2,n_3,n_4,n_5,n_6)$$
$$= n_1(32,0) + n_2(0,32) + n_3(16\sqrt{3},16) + n_4(-16,16\sqrt{3}) + n_5(16\sqrt{3},-16) + n_6(16,16\sqrt{3})$$

In order to find Ker $\Phi_{\mathbf{f}_1,...,\mathbf{f}_6}$ of the continuous transformation we have to solve the following set of two linear equations for $n_1,...,n_6 \in \mathbb{R}$:

$$\begin{cases} 32n_1 + 16\sqrt{3}n_3 - 16n_4 + 16\sqrt{3}n_5 + 16n_6 = 0 \\ 32n_2 + 16n_3 + 16\sqrt{3}n_4 - 16n_5 + 16\sqrt{3}n_6 = 0 \end{cases}$$

The solution of this set of equations is: $\{(n_1,n_2,n_3,n_4,n_5,n_6) \mid 2n_5 = -\sqrt{3}n_1+n_2-n_3+\sqrt{3}n_4,\ 2n_6 = -n_1-\sqrt{3}n_2-\sqrt{3}n_3-n_4,\ n_i \in \mathbb{R}\}$. This is clearly a 4D volume (having four free variables) in the 6D space $\mathbb{R}^6$. However, the lattice $L$, which is the discrete solution for $\Psi_{\mathbf{f}_1,...,\mathbf{f}_6}$ (i.e., with $n_i \in \mathbb{Z}$), is *not* a 4D lattice but rather only a 2D lattice (since for $n_5$ and $n_6$ to be integers it is required that $n_3 = -n_5$ and $n_4 = -n_6$ in order that all the roots be cancelled out). This means that in this case there is a loss of two dimensions in $L = \text{Ker}\,\Psi_{\mathbf{f}_1,...,\mathbf{f}_6}$ with respect to Ker $\Phi_{\mathbf{f}_1,...,\mathbf{f}_6}$, and therefore the spectrum support of this singular case is an everywhere dense module (see Table 5.2). And again, according to Eq. (5.16) we see that each point of this module (at the singular state) represents, indeed, a collapsed 2D cluster: rank $L = 6-4 = 2$. ■

## 5.8 Concluding remarks

We have seen in this chapter how an algebraic approach, based on the theory of geometry of numbers, can deepen our understanding of the spectrum of any superposition of periodic layers (such as gratings, screens, etc.). This algebraic approach explains the formation of impulse-clusters in the spectrum in any singular state or in its proximity, i.e., whenever a moiré occurs in the superposition. It clarifies the global structure of the spectrum support, as well as the internal structure and location of each individual cluster, and it clearly explains which impulse goes to which cluster and why. The key point of this approach is the algebraic formalization provided by the mapping $\Psi_{\mathbf{f}_1,...,\mathbf{f}_m}(n_1,...,n_m) = n_1\mathbf{f}_1 + ... + n_m\mathbf{f}_m$ from $\mathbb{Z}^m$ to $\mathbb{R}^2$, which maps the impulse indices $(n_1,...,n_m)$ into the impulse locations in the $u,v$ spectrum plane. The clusterization of the impulse locations in the spectrum-support simply reflects the partition of the lattice $\mathbb{Z}^m$ of the impulse indices into equivalence classes, which is induced by $\Psi_{\mathbf{f}_1,...,\mathbf{f}_m}$ (or simply, by the frequency vectors $\mathbf{f}_1,...,\mathbf{f}_m$ that define the layer superposition).

Based on these results we will reintroduce in Chapter 6 the impulse amplitudes, too, in order to relate the algebraic structure of the spectrum, via the Fourier theory, to properties of the layer superposition and its moirés back in the image domain.

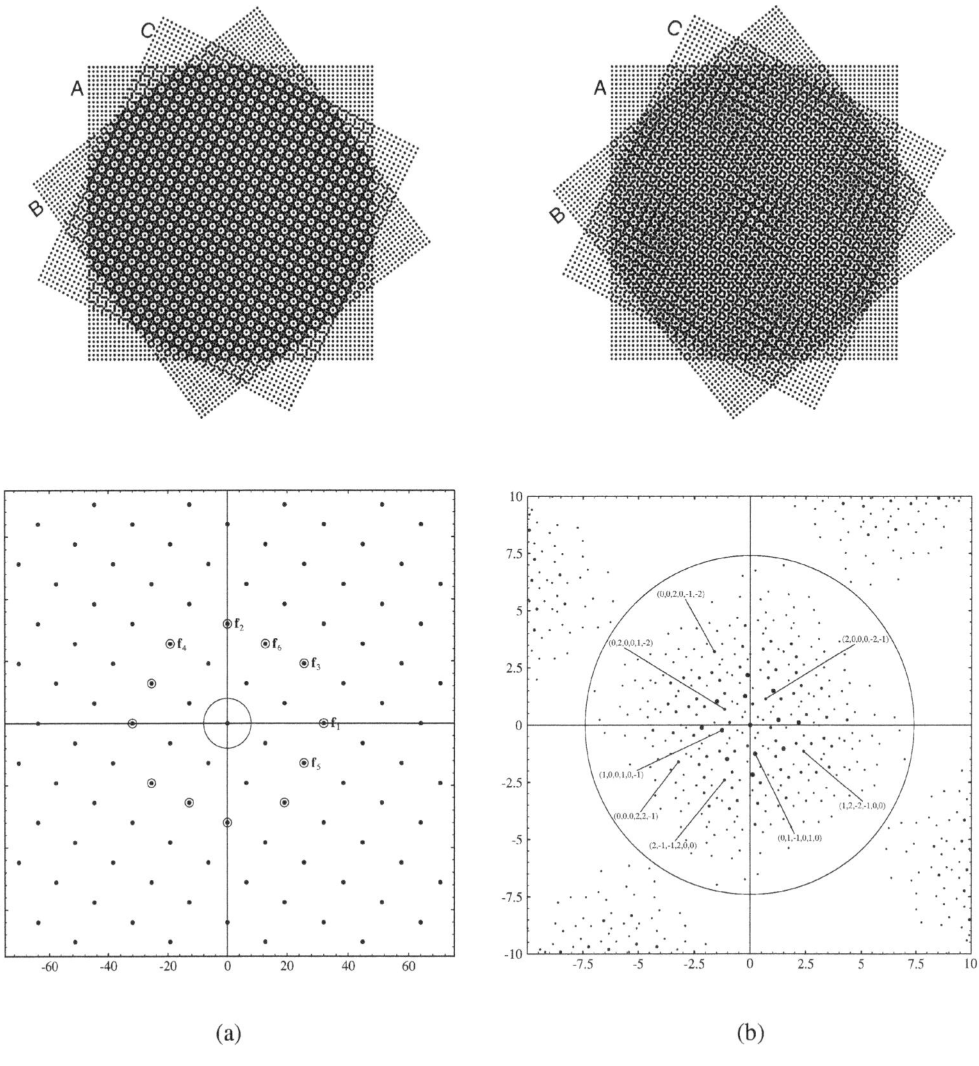


**Figure 5.9:** The singular 3-screen superposition of Example 5.12 (top) and its spectrum support (bottom). (a) Exactly at the singular state: the spectrum support forms here a 2D lattice, each point of which represents a collapsed cluster. The spectrum in (b) shows an enlarged view of the central part of spectrum (a), showing the spread-out main cluster slightly off the singular state: the cluster forms in the $u,v$ plane a dense 2D module. Only impulses up to the 3-rd order are shown. (A magnified version of this figure can be found in the Internet site of the book.)

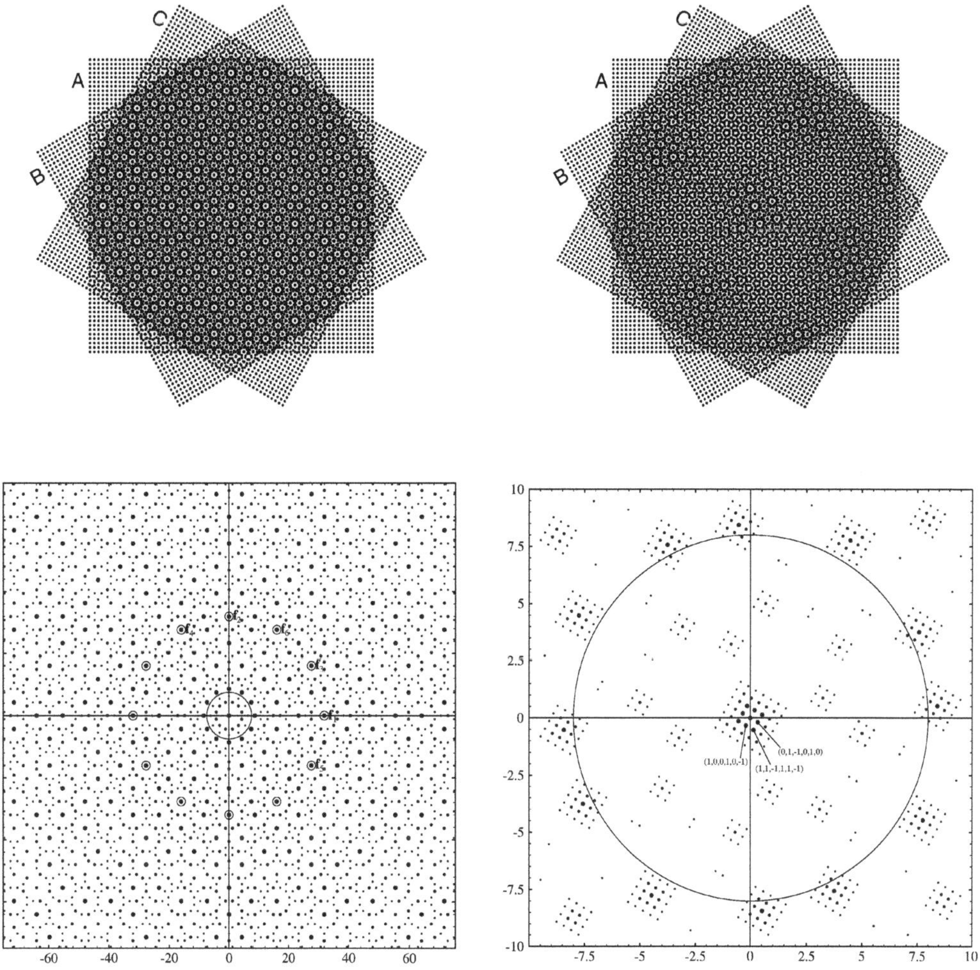


(a) (b)

**Figure 5.10:** The singular 3-screen superposition (top) and the spectrum support (bottom) of Example 5.13: the traditional 3-screen combination used for colour printing. (a) Exactly at the singular state: the spectrum support forms here an everywhere dense 2D module, each point of which represents a collapsed cluster. The spectrum in (b) shows an enlarged view of the central part of spectrum (a), slightly off the singular state: each of the clusters is spread out, clearly demonstrating its 2D lattice structure. Only impulses up to the 3-rd order are shown. (A magnified version of this figure can be found in the Internet site of the book.)

## PROBLEMS

5-1. *2D dense module.* In Sec. 5.2.1 two examples of modules in $\mathbb{R}^2$ have been given, $M_1$ and $M_2$. As a third example, consider the following module in $\mathbb{R}^2$: $M_3 = \{k(1,0) + l(\sqrt{2},0) + m(0,1) + n(0,\sqrt{2}) \mid k,l,m,n \in \mathbb{Z}\}$. This module consists of an infinite number of parallel copies of $M_1$, whose spacing in the vertical direction is dense. The module $M_3$ is, therefore, dense throughout the whole plane $\mathbb{R}^2$. What are its rank and its integral rank?

5-2. *2D hybrid module.* As a fourth example, consider the following module in $\mathbb{R}^2$: $M_4 = \{k(1,0) + l(\sqrt{2},0) + m(0,1) \mid k,l,m \in \mathbb{Z}\}$. This module consists of an infinite number of parallel copies of $M_1$, whose vertical coordinates are integer numbers. The module $M_4$ is, therefore, dense on any horizontal line $y=n$, $n \in \mathbb{Z}$, and null everywhere else; this is an example of a hybrid module in $\mathbb{R}^2$ which is dense in one direction but discrete in the other. What are its rank and its integral rank?

5-3. Fill in the table of Sec. 5.2.1 the modules $M_3$ and $M_4$ from the previous problems. Can you think of other types of modules in $\mathbb{R}^2$ to complete this table?

5-4. What happens to the module $M_2 = \{k(1,0) + l(\frac{1}{2}, \frac{1}{2}) + m(0,1) \mid k,l,m \in \mathbb{Z}\}$ when we replace its generating vector $(\frac{1}{2},\frac{1}{2})$ by $(1,1)$?

5-5. Table 5.2 provides examples for many of the superposition configurations that it contains. Complete the table by providing similar superposition examples for all the other cases in the table.

5-6. Is it true that whenever two impulses or more have a common geometric location $(u,v)$ in the spectrum-convolution the layer superposition is singular? And conversely, is it true that in any singular superposition there exist at least two impulses in the spectrum-convolution that have a common geometric location $(u,v)$?

5-7. Give at least three different necessary and sufficient conditions for the singularity of a given superposition, and show that they are all equivalent.

5-8. Let $d$ denote the difference between the discrete and the continuous dimensions of the spectrum support. Show that the following claims are all equivalent:

(a) The spectrum support is a discrete lattice *iff* $d = 0$; it is a dense module *iff* $d > 0$ (Proposition 5.2).

(b) The spectrum support is a discrete lattice *iff* $\text{rank}_{\mathbb{Z}}\,\text{Md}(\mathbf{f}_1,...,\mathbf{f}_m) = \dim \text{Sp}(\mathbf{f}_1,...,\mathbf{f}_m)$; it is a dense module *iff* $\text{rank}_{\mathbb{Z}}\,\text{Md}(\mathbf{f}_1,...,\mathbf{f}_m) > \dim \text{Sp}(\mathbf{f}_1,...,\mathbf{f}_m)$.

(c) The spectrum support is a discrete lattice *iff* $\text{rank Ker}\,\Psi_{\mathbf{f}_1,...,\mathbf{f}_m} = \dim \text{Ker}\Phi_{\mathbf{f}_1,...,\mathbf{f}_m}$; it is a dense module *iff* $\text{rank Ker}\,\Psi_{\mathbf{f}_1,...,\mathbf{f}_m} < \dim \text{Ker}\Phi_{\mathbf{f}_1,...,\mathbf{f}_m}$.

(d) The spectrum support is a discrete lattice *iff* $\text{rank}_{\mathbb{Z}}\,\text{Im}\,\Psi_{\mathbf{f}_1,...,\mathbf{f}_m} = \dim \text{Im}\Phi_{\mathbf{f}_1,...,\mathbf{f}_m}$; it is a dense module *iff* $\text{rank}_{\mathbb{Z}}\,\text{Im}\,\Psi_{\mathbf{f}_1,...,\mathbf{f}_m} > \dim \text{Im}\Phi_{\mathbf{f}_1,...,\mathbf{f}_m}$.

5-9. Normally, the spectrum support $\text{Md}(\mathbf{f}_1,...,\mathbf{f}_m)$ can be a discrete lattice only in singular superpositions. Show that the only non-singular cases having a discrete support are (see Table 5.2): (1) the case of $m = 1$ (i.e., a single grating, and hence no superposition); (2) the case of $m = 2$ superposed gratings with non-collinear frequency vectors $\mathbf{f}_1$, $\mathbf{f}_2$ (or equivalently: a single 2D screen).

5-10. It is clear, however, that not in all singular cases the spectrum support is a discrete lattice: when $\text{rank}_{\mathbb{Z}}\,\text{Md}(\mathbf{f}_1,...,\mathbf{f}_m) > \dim \text{Sp}(\mathbf{f}_1,...,\mathbf{f}_m)$, the spectrum support is dense even if the superposition is singular. Give at least two different examples of this type.

5-11. *The rank of the clusters in a singular state.* Show that the rank of each of the clusters in the spectrum of a singular superposition is given by $m - z$, where $m$ is the number of superposed gratings and $z$ is the maximum number of vectors among the frequency vectors $\mathbf{f}_1,...,\mathbf{f}_m \in \mathbb{R}^2$ which are linearly independent over $\mathbb{Z}$.

5-12. *The rank of the fully spread-out moiré cluster.* Suppose that we are given a singular layer superposition having frequency vectors $\mathbf{f}_1,...,\mathbf{f}_m$ and $L = \mathrm{Ker}\Psi_{\mathbf{f}_1,...,\mathbf{f}_m}$. When the superposed layers are slightly moved away from their singular state, they have new frequency vectors $\mathbf{f}'_1,...,\mathbf{f}'_m$ and a new $L' = \mathrm{Ker}\Psi'_{\mathbf{f}'_1,...,\mathbf{f}'_m}$. Show that when the moiré cluster is fully spread-out in the $u,v$ plane (see Sec. 5.6.3) its integral rank equals the rank of the original $L = \mathrm{Ker}\Psi_{\mathbf{f}_1,...,\mathbf{f}_m}$ in $\mathbb{Z}^m$ at the singular state, namely: $\mathrm{rank}_{\mathbb{Z}}\ \mathrm{Im}L = \mathrm{rank}\,L$. *Hint*: If we denote by $\Psi'_{\mathbf{f}'_1,...,\mathbf{f}'_m}|_L$ the restriction of transformation $\Psi'_{\mathbf{f}'_1,...,\mathbf{f}'_m}$ which is only defined between $L \subset \mathbb{Z}^m$ and the image of $L$ under $\Psi'_{\mathbf{f}'_1,...,\mathbf{f}'_m}$, $\mathrm{Im}L \subset \mathrm{Md}(\mathbf{f}'_1,...,\mathbf{f}'_m)$, then when the moiré cluster is fully spread-out we have $\Psi'_{\mathbf{f}'_1,...,\mathbf{f}'_m}|_L = \{\mathbf{0}\}$ and therefore from Eq. (5.9): $\mathrm{rank}_{\mathbb{Z}}\ \mathrm{Im}\,L = \mathrm{rank}\,L$.

5-13. *The rank of the fully spread-out moiré cluster (continued).* Show that when the moiré cluster is fully spread-out its integral rank is: $\mathrm{rank}_{\mathbb{Z}}\ \mathrm{Im}\,L = m - \mathrm{rank}_{\mathbb{Z}}\,\mathrm{Md}(\mathbf{f}_1,...,\mathbf{f}_m)$. *Hint*: This follows from the previous problem by Eq. (5.16).

5-14. The subspace $\mathrm{Ker}\Phi_{\mathbf{f}_1,...,\mathbf{f}_m}$ of $\mathbb{R}^m$ obviously contains the point $(0,...,0) \in \mathbb{Z}^m$; but if $\dim \mathrm{Ker}\Phi_{\mathbf{f}_1,...,\mathbf{f}_m} \geq 1$ it may contain also other points of $\mathbb{Z}^m$. Let us denote by $L$ the set of all points of $\mathbb{Z}^m$ which are included in $\mathrm{Ker}\Phi_{\mathbf{f}_1,...,\mathbf{f}_m}$, i.e., $L = \mathrm{Ker}\Phi_{\mathbf{f}_1,...,\mathbf{f}_m} \cap \mathbb{Z}^m$. Show that $L$ is indeed a sub-lattice of $\mathbb{Z}^m$. (The sub-lattice $L$ is the restriction of $\mathrm{Ker}\Phi_{\mathbf{f}_1,...,\mathbf{f}_m}$ to $\mathbb{Z}^m$, i.e., $L = \mathrm{Ker}\Psi_{\mathbf{f}_1,...,\mathbf{f}_m}$.)

5-15. Suppose that the moiré cluster is a lattice. What is the algebraic structure of the other clusters which are simultaneously generated with it? *Hint*: Note that the only equivalence class in the lattice $\mathbb{Z}^m$ which is itself a lattice is $\mathbf{0} + L$ (the equivalence class of the vector $(0,...,0)$), i.e., the sub-lattice $L$ itself. All the other classes are parallel translations of $L$ within the lattice $\mathbb{Z}^m$, and they do not contain the vector $(0,...,0)$ and therefore have no origin. Nevertheless, we still say that each of the translated equivalence classes has the same rank as the unshifted sub-lattice $L$: $\mathrm{rank}\ ((r_1,...,r_m) + L) = \mathrm{rank}\,L$. Compare with the continuous counterpart of Sec. 5.4.2.

5-16. Give the explicit expression for the transformations $\Phi_{\mathbf{f}_1,\mathbf{f}_2,\mathbf{f}_3,\mathbf{f}_4}$ and $\Psi_{\mathbf{f}_1,\mathbf{f}_2,\mathbf{f}_3,\mathbf{f}_4}$ in the case of Example 5.10. Find $\mathrm{Ker}\Phi_{\mathbf{f}_1,\mathbf{f}_2,\mathbf{f}_3,\mathbf{f}_4}$ and $\mathrm{Ker}\Psi_{\mathbf{f}_1,\mathbf{f}_2,\mathbf{f}_3,\mathbf{f}_4}$; what are their algebraic structures and their dimensions?

5-17. Find in Example 5.10 the indices of the impulses which participate: (a) in the main cluster; (b) in the cluster around the point $\mathbf{f}_1$; and (c) in the cluster around $\mathbf{f}_3$ (see Fig. 5.7(b)). The cluster which surrounds the point $\mathbf{f}_1+\mathbf{f}_3$ is not fully shown in this figure (it is partially beyond its border); what are the indices of its impulses?

5-18. Fig. 4.9(b) shows the vector diagram of the (1,0,-1,1)-moiré between two dot-screens. How is this moiré related to the (1,1,-1,0)-moiré which is discussed in Example 5.10 and illustrated in Fig. 5.7?

5-19. Give a detailed analysis of the (1,2,-2,-1)-moiré which appears between two dot-screens with identical frequencies ($|\mathbf{f}_1| = |\mathbf{f}_2| = |\mathbf{f}_3| = |\mathbf{f}_4|$) when the angle difference between them is around $\alpha = \arctan\frac{3}{4} \approx 36.87°$. (The spectrum of the singular superposition is shown in Fig. 8.3(a) in Chapter 8, and the spectrum of the superposition slightly off the singular state is shown in Fig. 4.3(b) in Chapter 4.)

5-20. Show that a superposition of two regular dot-screens (or line-grids) is periodic *iff* it is singular (i.e., *iff* it generates a singular moiré). Is this also true for a superposition of three or more dot-screens?

5-21. Show that in any *periodic* superposition of three or more regular dot-screens, more than one 2D moiré impulse-clusters collapse onto the spectrum origin, generating together a cluster of rank > 2. (*Hint*: Since all the screens in the superposition have a common period, in particular, each subset of two screens in the superposition is periodic. Therefore, according to the previous problem, each pair of screens constitutes a singular superposition, and generates a 2D cluster that collapses onto the spectrum origin.) We see, therefore, as a result, that each singular superposition of 3 or more regular screens in which only one 2D cluster collapses onto the spectrum origin is necessarily not periodic (and hence it is almost-periodic). As an example, compare Figs. 5.9 and 5.10.

5-22. *Computer program for drawing the spectrum convolution.* Write a computer program that, given the polar coordinates of the frequency vectors $\mathbf{f}_i$ (i.e., the frequencies and the directions of each superposed layer), calculates the spectral convolution (up to a specified number of harmonics), using the rules of comb or nailbed convolution (Proposition 2.3). The resulting impulse configuration (=spectrum support) should be graphically displayed, showing the location (and optionally also the index) of each impulse in the $u,v$ spectrum, like in Figs. 5.2–5.10. This program may prove very useful for elucidating the nature of moiré effects which occur in various layer superpositions, especially in rather complex cases.

5-23. In many important cases, like the screen superpositions of Chapter 4 or the traditional 3-screen combination used for colour printing (see Fig. 5.10), each of the spread-out clusters in the spectrum convolution is in itself a discrete lattice. However, if $\operatorname{rank} L > 2$ then each of the clusters, when fully spread out, becomes in itself a dense module in the $u,v$ plane. What can you say in such cases about the moiré effect, which is represented by the spread-out cluster around the spectrum origin? (We will return to this question in Sec. 6.3).

5-24. Note that even when each spread-out cluster is in itself a discrete lattice in the $u,v$ plane, the overall intermingled cluster combination which forms the spectrum support is usually dense (see for example Figs. 4.3(b),(c)). What does this say about the superposition in the image domain? (We will see in Chapter 6 that even when the moiré itself, i.e., its extracted profile-form, is periodic, the overall layer superposition still may be non-periodic.)

# Chapter 6

# Fourier-based interpretation of the algebraic spectrum properties

## 6.1 Introduction

In the previous chapter we analyzed the properties of the spectrum convolution (i.e., the spectrum of the layer superposition) from a pure algebraic point of view, concentrating only on the spectrum support, and ignoring the impulse amplitudes. In the present chapter we will "augment" these algebraic foundations by reintroducing the impulse amplitudes on top of their geometric locations in the spectrum. We will investigate in Secs. 6.2–6.3 the properties of the impulse amplitudes that are associated with the algebraic structures, and through the Fourier theory, we will see how both the structural and the amplitude properties of the spectrum are related to properties of the layer superposition and its moiré effects back in the image domain. Finally, in Secs. 6.4–6.8 we will analyze the layer superpositions when their moirés become singular, and we will see what happens in the Fourier expressions when each of the impulse clusters collapses down (i.e. coalesces) into a single compound impulse. This chapter is, in fact, a generalization of the basic ideas developed in Chapter 4, based on the new algebraic notions of Chapter 5.

Note that starting from this chapter we will usually use exponential Fourier expansions instead of the trigonometric cosine and sine form that we preferred until now for didactic reasons (see Secs. 2.5–2.6 and Sec. A.2 in Appendix A). As we will see below, the "marriage" of this exponential form with the vector notations of Chapter 5 gives an extremely useful mathematical formulation for the description of periodic layers and their superpositions — a Fourier vectorial formulation which contains all the available information in a concise, clear and elegant way. Furthermore, this exponential form also allows for layers with non-symmetric periods and for shifted layers; we will return to this point in detail in Chapter 7.

## 6.2 Image domain interpretation of the algebraic structure of the spectrum support

As we have seen in Eq. (2.27), the amplitude of the $(k_1,...,k_m)$-impulse in the spectrum convolution is a product of the amplitudes of the individual impulses contributed by the spectrum of each of the superposed layers: $a_{k_1,...,k_m} = a^{(1)}{}_{k_1} \cdot \ ... \ \cdot a^{(m)}{}_{k_m}$. By reintroducing the amplitude values of the spectrum impulses on top of their geometric locations, we get again a full description of the spectrum. This permits us to use the Fourier theory to transform the structural results that we have algebraically obtained in the spectral domain back into the image domain as well. We start, in this section, by considering the structure of the global spectrum support, and interpreting its influence on the image domain. The structure of the individual impulse-clusters and its image domain interpretation will be discussed in Sec. 6.3.

As we have seen in Table 5.2, the spectrum convolution (i.e., the spectrum of the layer superposition) can have four different types of spectrum support, which are denoted in the table by 2D-L, 2D-M, 1D-L and 1D-M. These four types are the four possible combinations of two basic and independent properties of the spectrum support: (a) it can be either 2D or 1D; (b) it can be either a discrete lattice or a dense module. Let us see now what is the image domain interpretation of each of these two basic, independent properties:

(a) A 2D spectrum support indicates that the structure in the image domain is indeed of a 2D nature. On the other hand, a 1D spectrum support in the $u,v$ plane means that all the "action" in the image domain takes place only in one direction, while in the perpendicular direction the image is constant. This happens in a grating superposition when all the original gratings are parallel (their frequency vectors are collinear); although they are 2D functions in the $x,y$ image plane, they vary only along one direction, while in the perpendicular direction all of them, and hence also their superposition and their moirés, remain constant. This is in fact a case of 1D nature which is artificially extended to the 2D $x,y$ image plane.

(b) The support of the spectrum convolution is a discrete lattice *iff* the layer superposition in the image domain is a periodic function (either 1D or 2D). This follows from the Fourier series decomposition of periodic functions (see Appendix A). What is, however, the image domain interpretation of cases in which the spectrum support is a dense module? On the one hand it is clear that the layer superposition in these cases is not periodic; but on the other hand their spectrum is still impulsive and not continuous, meaning that the superposition is not aperiodic, either. In fact, such cases belong to an intermediate class of functions, which is known as *almost-periodic* functions. In Appendix B we shortly review the subject of almost-periodicity, and we present some of the main properties of almost-periodic functions. According to these results, a spectrum formed by a dense module of impulses represents a *generalized Fourier series* expansion that belongs to an almost-periodic function. This means that in such cases the layer superposition back in the image domain is an almost-periodic function.

The four possible types of spectrum support and their interpretations in the image domain are summarized in Table 6.1.

Based on these facts we can now reformulate Proposition 5.2 of Sec. 5.5 as a criterion for the periodicity of the superposition of periodic layers (functions):

**Proposition 6.1:** The superposition of $m$ gratings (or $m/2$ grids, dot-screens, etc.) is periodic *iff* $\text{rank}_{\mathbb{Z}}\,\text{Md}(\mathbf{f}_1,...,\mathbf{f}_m) = \dim \text{Sp}(\mathbf{f}_1,...,\mathbf{f}_m)$, or equivalently: *iff* $\text{rank}_{\mathbb{Z}}\,\text{Md}(\mathbf{f}_1,...,\mathbf{f}_m) = \text{rank}_{\mathbb{R}}\,\text{Md}(\mathbf{f}_1,...,\mathbf{f}_m)$. The superposition is almost-periodic *iff* $\text{rank}_{\mathbb{Z}}\,\text{Md}(\mathbf{f}_1,...,\mathbf{f}_m) > \dim \text{Sp}(\mathbf{f}_1,...,\mathbf{f}_m)$, or equivalently: *iff* $\text{rank}_{\mathbb{Z}}\,\text{Md}(\mathbf{f}_1,...,\mathbf{f}_m) > \text{rank}_{\mathbb{R}}\,\text{Md}(\mathbf{f}_1,...,\mathbf{f}_m)$. (Note that the case of '<' is impossible; see Footnote 5 in Sec. 5.2. This means that the two conditions above are exhaustive; and indeed, the superposition of periodic functions is either periodic or almost-periodic.) ■

| Spectrum support | Superposition in the image domain | Examples |
|---|---|---|
| 2D-L | 2-fold periodic | Example 5.5 |
| 1D-L | 1-fold periodic | Example 5.8 |
| 2D-M | 2-fold almost-periodic* | Example 5.13 |
| 1D-M | 1-fold almost-periodic | Example 5.9 |

* Note that the case of 2D-M also includes the hybrid case in which the 2D spectrum is dense in one direction and discrete in the other; this case corresponds in the image domain to a 2D function which is almost-periodic in one direction and periodic in the other. This hybrid case may occur, for instance, in the superposition of three gratings, two of which have the same direction, but with incommensurable frequencies. See, for example, the module $M_4$ in Problem 5-2.

**Table 6.1:** The four possible types of spectrum support in the $u,v$ plane and their interpretation in the image domain.

## 6.3 Image domain interpretation of the impulse-clusters in the spectrum

We have already seen in Chapter 4 that the main cluster in the spectrum (the impulse-cluster which is centered on the spectrum origin) is the Fourier transform of the isolated (extracted) moiré in question. When this cluster is slightly spread-out and its fundamental impulses are located within the visibility circle, the corresponding moiré may be clearly visible in the image domain (if the amplitudes are not too weak); but when the singular state is reached and the cluster impulses collapse onto the DC, the moiré in the image domain gets an infinite period and disappears.

However, the algebraic discussion of Chapter 5 provides further insight into the image-domain interpretation of the main cluster. As we have seen, the support of this cluster in the $u,v$ plane is not necessarily a discrete lattice (as was the case in Chapter 4), and in the more general case it can even be a dense module (like in Example 5.12 of Sec. 5.7). In fact, here, too, there exist four different cases, whose interpretation back in the image domain is summarized in Table 6.2.

It is interesting to note that the algebraic structure of the moiré cluster is not necessarily of the same type as the algebraic structure of the overall spectrum of the superposition. The moiré cluster may have a lower dimension (a 1D cluster imbedded in an overall 2D spectrum, as in Example 5.5 in Sec. 5.7) or a simpler structure (a 2D-L cluster within a 2D-M spectrum, such as in Figs. 4.3(b),(c) or in example 5.13 of Sec. 5.7), or even both

| Main cluster | Extracted moiré in the image domain* | Examples |
|---|---|---|
| 2D-L | 2-fold periodic | Example 5.10 |
| 1D-L | 1-fold periodic | Example 5.5 |
| 2D-M | 2-fold almost-periodic | Example 5.12 |
| 1D-M | 1-fold almost-periodic | ** |

* Note that the moiré is visible in the image domain only slightly off the singular state, when the cluster is spread-out. At the singular state itself the main cluster collapses on the spectrum origin, and the moiré is not visible.

** This case occurs, for instance, when we superpose on top of a 1-fold periodic moiré in the image domain a new grating in the same direction as the 1D moiré, but with an incommensurable period.

**Table 6.2:** The four possible types of cluster support in the *u*,*v* plane and their interpretation in the image domain.

(a 1D-L cluster within a 2D-M spectrum, as in Example 5.11 of Sec. 5.7).[1] This means, back in the image domain, that even when the overall superposition of the periodic layers is not periodic but rather almost-periodic, a moiré generated in this superposition may still be periodic. This is, in fact, a very common situation, which is clearly illustrated, for instance, in Figs. 2.10 or 3.1(a): although the overall superposition is not periodic (notice the microstructure!), the intensity profile of the isolated moiré is indeed periodic.

Our algebraic approach of Chapter 5 provides also information regarding the other clusters which are formed in the spectrum simultaneously with the main cluster: These clusters are simply translated replicas of the main cluster, both in terms of impulse indices and in terms of impulse locations in the *u*,*v* plane. Note, however, that the impulse amplitudes within each of the clusters are calculated according to Eq. (2.27), meaning that the impulse amplitudes in the shifted clusters are *not* identical to the impulse amplitudes in the main cluster. The shifted clusters contribute higher frequencies to the global structure of the spectrum support, and in terms of the image domain, they take part in the generation of the microstructure details in the superposition (see Chapter 8).

## 6.4 The amplitude of the collapsed impulse-clusters in a singular state

In order to better understand the spectrum of a singular state, let us adopt once again the dynamic approach. As the superposed layers gradually approach a singular state, the

[1] Obviously, since the moiré cluster is a subset of the overall spectrum, its structure can never be *more* complex than that of the overall spectrum.

spectrum undergoes an "inverse playback" of the cluster spreading-out process. The singular state itself is the limit case where each of the spread-out impulse clusters collapses down into a single compound impulse. A *compound impulse* is an impulse which results in the spectrum when several impulses fall together on the same impulse location. A comb or a nailbed whose individual impulses are compound will be called a *compound comb* or a *compound nailbed.* The question is, what happens to the amplitudes of the impulses of each cluster at the limit point when the singular state is attained, and each of the impulse clusters in the spectrum fuses down into a single compound impulse?

Since the spectrum of the superposed layers is the convolution of the individual spectra, the answer to this question follows from the properties of convolution. The convolution of a function $f(x,y)$ with an impulsive function such as a nailbed or a comb is simply a sum of replicas of $f(x,y)$, which are copied on top of each impulse of the nailbed or the comb [Gaskill78 pp. 295–296]:

$$f(x,y) ** \sum_{m=-\infty}^{\infty} \sum_{n=-\infty}^{\infty} c_{m,n}\, \delta(x-mT_x\, , y-nT_y) = \sum_{m=-\infty}^{\infty} \sum_{n=-\infty}^{\infty} c_{m,n} f(x-mT_x\, , y-nT_y)$$

Hence, if an overlapping occurs between several replicas of $f(x,y)$, the overlap behaves *additively*. Since in our case $f(x,y)$, too, is impulsive, it follows that when several impulses in the convolution fall on the same point, their amplitudes are summed. We obtain, therefore, the following result:

**Proposition 6.2:** When an impulse cluster collapses down into a single impulse, the amplitude of the resulting compound impulse is the sum of the individual amplitudes of all the collapsed impulses. (More loosely speaking, this can be considered as an "amplitude preservation law" which states that when impulses coincide on the same geometric location in the spectrum, the sum of their amplitudes is preserved.)[2] ■

In the following sections we will illustrate this result by analyzing a number of singular cases. For the sake of simplicity we will first consider the case of two superposed gratings (see Fig. 2.5). Then, in the subsequent sections, we will illustrate Proposition 6.2 by analyzing more general singular cases. The purpose of the discussion in the sections which follow is not only to illustrate Proposition 6.2, but also to shed more light on the relationship between the algebraic structure of the spectrum support and the image domain properties, via the Fourier theory.

## 6.5 The exponential Fourier expression for two-grating superpositions

Let $p_1(x,y)$ and $p_2(x,y)$ be two periodic gratings on the $x,y$ plane. We have already seen in Secs. 2.6 and 4.2 the expansion of such gratings into trigonometric Fourier series.

---

[2] This *additive* behaviour of the impulse amplitudes should not be confused with the *multiplicative* behaviour of the individual impulse amplitudes in the convolution process: Each individual impulse amplitude in the convolution is obtained by Eq. (2.27) as a *product*; but if several impulses thus obtained happen to fall on the same geometric location, their individual amplitudes are then *summed.*

However, as explained in Sec. A.2 of Appendix A, Fourier series expansions can be also expressed in an equivalent exponential form, which is much more compact and easier to manipulate. As shown in Sec. A.3.3 the Fourier series expansion of gratings $p_1(x,y)$ and $p_2(x,y)$ is given in the exponential form by:

$$p_1(x,y) = \sum_{n_1=-\infty}^{\infty} c^{(1)}{}_{n_1}\, e^{i2\pi n_1(u_1x+v_1y)}$$

$$p_2(x,y) = \sum_{n_2=-\infty}^{\infty} c^{(2)}{}_{n_2}\, e^{i2\pi n_2(u_2x+v_2y)}$$

This can be written even more compactly in vector notation (see Eq. (A.20)):

$$\begin{aligned} p_1(\mathbf{x}) &= \sum_{n_1=-\infty}^{\infty} c^{(1)}{}_{n_1}\, e^{i2\pi n_1\mathbf{f}_1\cdot\mathbf{x}} \\ p_2(\mathbf{x}) &= \sum_{n_2=-\infty}^{\infty} c^{(2)}{}_{n_2}\, e^{i2\pi n_2\mathbf{f}_2\cdot\mathbf{x}} \end{aligned} \tag{6.1}$$

where $\mathbf{f}_1 = (u_1,v_1)$ and $\mathbf{f}_2 = (u_2,v_2)$ are the frequency vectors of the two gratings, and $\mathbf{x} = (x,y)$. Note that while in the trigonometric Fourier series the $n$-th term corresponds (in the Fourier transform sense) to the $n$-th impulse pair in the spectrum of the periodic function, in the exponential Fourier series the $n$-th exponential term corresponds to a single impulse in the spectrum, the $n$-th impulse. The $n$-th exponential term fully determines this impulse by specifying both its geometric location, $n\mathbf{f}$, and its amplitude, $c_n$.

Using this vector notation the Fourier expansion of the superposition of two gratings is given, therefore, by:

$$\begin{aligned} p_1(\mathbf{x})\cdot p_2(\mathbf{x}) &= \Big(\sum_{n_1=-\infty}^{\infty} c^{(1)}{}_{n_1}\, e^{i2\pi n_1\mathbf{f}_1\cdot\mathbf{x}}\Big)\Big(\sum_{n_2=-\infty}^{\infty} c^{(2)}{}_{n_2}\, e^{i2\pi n_2\mathbf{f}_2\cdot\mathbf{x}}\Big) \\ &= \sum_{n_1=-\infty}^{\infty}\sum_{n_2=-\infty}^{\infty} c^{(1)}{}_{n_1}\, c^{(2)}{}_{n_2}\, e^{i2\pi(n_1\mathbf{f}_1+n_2\mathbf{f}_2)\cdot\mathbf{x}} \end{aligned} \tag{6.2}$$

It is interesting to note that the $(n_1,n_2)$-th term in this double sum, i.e., the term whose indices are $n_1$ and $n_2$, fully corresponds to the $(n_1,n_2)$-th impulse in the spectrum of the superposition (the spectrum convolution): it explicitly defines both the geometric location of this impulse, $\mathbf{f}_{n_1,n_2} = n_1\mathbf{f}_1 + n_2\mathbf{f}_2$, and its amplitude, $c_{n_1,n_2} = c^{(1)}{}_{n_1}c^{(2)}{}_{n_2}$ (see Proposition 2.3).

Now, the $(k_1,k_2)$-moiré can be isolated from superposition (6.2) by extracting from this double sum only those terms which correspond to the spectrum-convolution impulses that are spanned by the $(k_1,k_2)$-impulse, namely: the terms which belong to the impulses of the $(k_1,k_2)$-comb. The $(k_1,k_2)$-moiré is expressed, therefore, by:

$$m_{k_1,k_2}(\mathbf{x}) = \sum_{n=-\infty}^{\infty} c^{(1)}{}_{nk_1}\, c^{(2)}{}_{nk_2}\, e^{i2\pi n(k_1\mathbf{f}_1+k_2\mathbf{f}_2)\cdot\mathbf{x}} \tag{6.3}$$

(compare this concise and elegant expression with its cumbersome trigonometric counterpart given in Eq. (4.3)). For example, the (1,-1)-moiré, which contains only the terms that correspond to the impulses of the (1,-1)-comb, is given by:

$$m_{1,-1}(\mathbf{x}) = \sum_{n=-\infty}^{\infty} c^{(1)}{}_{n}\, c^{(2)}{}_{-n}\, e^{i2\pi n(\mathbf{f}_1-\mathbf{f}_2)\cdot\mathbf{x}} \tag{6.4}$$

Its frequency is, indeed, the difference $\mathbf{f}_1-\mathbf{f}_2$, and its impulse amplitudes are $c_n = c^{(1)}{}_n c^{(2)}{}_{-n}$.

## 6.6 Two-grating superpositions and their singular states

As we already know, the spectrum of each periodic grating is a comb of impulses, whose step and direction represent the frequency and the orientation of the grating, and whose impulse amplitudes are the coefficients of the Fourier series representation of the grating. Suppose that we are given two gratings $p_1(\mathbf{x})$ and $p_2(\mathbf{x})$. For the sake of simplicity we may assume, without loss of generality, that the first grating is fixed at the angle of $\theta_1 = 0$, so that its comb is located on the horizontal $u$ axis, while the second grating and its comb are oriented to $\theta_2 = \alpha$ (see Fig. 2.5). We may also assume that each of the gratings is centered about the origin (the more general case which also involves layer shifts will be discussed in Chapter 7). According to the convolution theorem, the spectrum of the superposition of the two gratings is the convolution of their individual spectra, i.e., the convolution of the two combs. This convolution operation can be thought of as a process of placing a parallel copy of the rotated comb on top of each impulse of the horizontal comb (after an appropriate amplitude scaling). This gives a nailbed whose impulses are located on an oblique 2D lattice with an angular opening of $\alpha$ (see Fig. 2.5(f)). This spectrum corresponds, indeed, term by term to the Fourier series (6.2) of the superposition. What happens, now, when the angle difference $\alpha$ between the two gratings tends to 0? As we can see in the spectrum convolution (Fig. 2.5(f)), each copy of the rotated comb pivots about its central impulse, and approaches the horizontal axis. When the limit of $\alpha = 0$ is attained, all the impulses of the oblique nailbed collapse into a single line, the $u$ axis. As we will see below, the results obtained at this precise moment depend on the frequencies of the two gratings in question (see also Example 5.2 in Sec. 5.2.2).

### 6.6.1 Two gratings with identical frequencies

Let us start with the simplest case in which the frequencies of the two gratings, $f_1 = |\mathbf{f}_1|$ and $f_2 = |\mathbf{f}_2|$, are identical: $f_1 = f_2 = f$. In the *image domain*, as $\alpha$ decreases and approaches 0° a moiré pattern becomes visible between the two gratings (see Figs. 3.3(a)–(c), and the parameter space $(\alpha,q)$ of Fig. 3.2 with $q = f_2/f_1 = 1$). The period of the moiré becomes larger and larger as $\alpha$ gets smaller, until it becomes infinitely large and hence invisible when the singular state is reached, i.e., when $\alpha = 0°$. At this moment the two gratings are superposed precisely in match on top of one another, at the same angle 0° and with the same period $T$, so that the superposition in the image domain becomes in fact a single 1-fold periodic grating, that we will call the *product-grating*. The product-grating has the same period and angle as the two superposed gratings, and its intensity profile is the product of their intensity profiles.

In the *spectral domain*, as $\alpha$ tends to 0° each of the impulses in the spectrum approaches the $u$ axis, so that the entire oblique nailbed (see Fig. 2.5(f)) becomes squeezed closer and closer to the $u$ axis. Let us first consider the (1,-1)-impulse of the spectrum convolution. As $\alpha$ tends to 0° the (1,-1)-impulse gradually approaches the spectrum origin, and at a certain moment it enters into the visibility circle, meaning that the (1,-1)-moiré becomes visible. As shown in Fig. 2.5(f) the (1,-1)-impulse spans in the spectrum a whole comb (or cluster) of impulses whose indices are: $\{(n,-n) \mid n \in \mathbb{Z}\}$; this cluster is in fact the 1D cluster of the (1,-1)-moiré. However, as we have seen in Sec. 5.6.2, the other impulses of the spectrum convolution are also simultaneously grouped into clusters.[3] Until the singular state is attained, these 1D clusters are fully spread-out, and have the form of parallel impulse-combs (see Fig. 2.5(f)). As $\alpha$ tends to 0° all of these impulse-combs get squeezed closer and closer to the horizontal $u$ axis; and in particular, the main (0-th) cluster gets squeezed towards the DC impulse. When the singular state is attained, at $\alpha = 0°$, each of these parallel 1D clusters collapses into a single compound impulse on the $u$ axis of the spectrum, and the previously 2D spectrum support suddenly becomes 1D. All of the compound impulses thus obtained form together a single compound comb on the $u$ axis, whose step equals $f$. This compound comb is the spectrum of the product-grating that is obtained in the image domain at the singular point $\alpha = 0°$.

Let us see now what happens to the impulse amplitudes during this collapsing process in the spectrum. While the impulses get squeezed towards the $u$ axis, as $\alpha \to 0°$, their amplitudes are not affected (see Proposition 2.2 in Sec. 2.6). But according to Proposition 6.2, at the limit point when $\alpha$ reaches 0° and each of the impulse-clusters collapses or fuses down into a single impulse on the $u$ axis, the amplitude of the resulting compound impulse is the sum of the amplitudes of all the collapsed impulses.

Let us illustrate this by explicitly analyzing the superposition (6.2). Using Cauchy's rule of summing a double series "diagonally" (see Sec. C.12.2 in Appendix C), the terms of this double series can be also summed along the $n_1+n_2 = s$ diagonals ($s \in \mathbb{Z}$):

$$p_1(\mathbf{x})\cdot p_2(\mathbf{x}) = \sum_{s=-\infty}^{\infty} \sum_{n_1+n_2=s} c^{(1)}_{n_1} c^{(2)}_{n_2} e^{i2\pi(n_1\mathbf{f}_1+n_2\mathbf{f}_2)\cdot\mathbf{x}}$$

$$= \sum_{s=-\infty}^{\infty} \sum_{n_1=-\infty}^{\infty} c^{(1)}_{n_1} c^{(2)}_{s-n_1} e^{i2\pi(n_1\mathbf{f}_1+(s-n_1)\mathbf{f}_2)\cdot\mathbf{x}}$$

As we can see in Fig. 2.5(f), this last expression corresponds to a summation of the Fourier terms along the diagonal clusters in the spectrum convolution. Note in particular that the cluster $s = 0$ (whose impulses are: ..., (-1,1), (0,0), (1,-1), ...) corresponds to the extracted (1,-1)-moiré, which is given, indeed, by the partial sum along the diagonal $s = 0$ (see Eq. (6.4)).

Now, when $\alpha = 0°$ and the two superposed gratings $p_1(\mathbf{x})$ and $p_2(\mathbf{x})$ are brought to a perfect match with $\mathbf{f}_1 = \mathbf{f}_2 = \mathbf{f}$, we obtain:

$$p_1(\mathbf{x})\cdot p_2(\mathbf{x}) = \sum_{s=-\infty}^{\infty} \Big( \sum_{n_1=-\infty}^{\infty} c^{(1)}_{n_1} c^{(2)}_{s-n_1} \Big) e^{i2\pi s\mathbf{f}\cdot\mathbf{x}}$$

or according to Eq. (2.27):

$$= \sum_{s=-\infty}^{\infty} \Big( \sum_{n_1=-\infty}^{\infty} c_{n_1,s-n_1} \Big) e^{i2\pi s\mathbf{f}\cdot\mathbf{x}}$$

[3] The $s$-th 1D cluster in the spectrum consists of the impulses: $(n,-n) + (s,0) = (n-s, s-n) + (s,0) = (n, s-n)$ for all $n \in \mathbb{Z}$; as we have seen in Sec. 5.6.2, these impulse-indices are a translated version of the impulse-indices $(n,-n)$ of the main, 0-th cluster.

Hence, we succeeded in this case to bring the superposition $p_1(\mathbf{x})\cdot p_2(\mathbf{x})$ to the form (A.20) of a Fourier series expansion, where the $s$-th Fourier coefficient is $\sum_{n_1=-\infty}^{\infty} c_{n_1,s-n_1}$. As we can see, this Fourier series represents, indeed, a compound comb in the spectrum; its $s$-th impulse has the frequency $s\mathbf{f}$ and the amplitude $\sum_{n_1=-\infty}^{\infty} c_{n_1,s-n_1}$, which is precisely the sum of the impulse amplitudes of the $s$-th collapsed cluster in the spectrum.

Note in particular that the amplitude of the compound DC impulse after the 0-th cluster has collapsed onto it is: $\sum_{n_1=-\infty}^{\infty} c_{n_1,-n_1}$, which is indeed the sum of the impulse amplitudes of the 0-th cluster, the comb of the (1,-1)-moiré (see Eq. (6.4)).

### 6.6.2 Two gratings with different frequencies

Assume now that the two gratings $p_1(\mathbf{x})$ and $p_2(\mathbf{x})$ have different frequencies $f_1 \neq f_2$ (where $f_1 = |\mathbf{f}_1|$ and $f_2 = |\mathbf{f}_2|$). Once again, when $\alpha = 0°$ the frequency vectors $\mathbf{f}_1$ and $\mathbf{f}_2$ are collinear, both being located on the $u$ axis. As we have seen in Example 5.2 of Sec. 5.2.2, we can distinguish here between two different cases:

(1) If the frequencies $\mathbf{f}_1$ and $\mathbf{f}_2$ of the two gratings are *commensurable* (i.e., there exist $p,q \in \mathbb{Z}$ so that $\mathbf{f}_2 = \frac{p}{q}\mathbf{f}_1$), then the spectrum support $\mathrm{Md}(\mathbf{f}_1,\mathbf{f}_2) = \{n_1\mathbf{f}_1+n_2\mathbf{f}_2 \mid n_1,n_2 \in \mathbb{Z}\}$ is a lattice of rank 1 spanned by $\mathbf{f}_1$ and $\mathbf{f}_2$ on the $u$ axis. This means that when $\alpha = 0°$ the impulses of the oblique nailbed collapse onto a single compound comb on the $u$ axis. In the image domain this means that the grating superposition is 1-fold periodic in the $x$ direction. Since both $\mathbf{f}_1$ and $\mathbf{f}_2$ are integer multiples of a common frequency $\mathbf{f} = \mathbf{f}_1/q = \mathbf{f}_2/p$ (namely: $\mathbf{f}_1 = q\mathbf{f}$, $\mathbf{f}_2 = p\mathbf{f}$), expression (6.2) can be written here as:

$$p_1(\mathbf{x})\cdot p_2(\mathbf{x}) = \sum_{n_1=-\infty}^{\infty} \sum_{n_2=-\infty}^{\infty} c_{n_1,n_2}\, e^{i2\pi(qn_1+pn_1)\mathbf{f}\cdot\mathbf{x}}$$

and using a variant of Cauchy's diagonal summation, where the terms are summed along the $qn_1+pn_2 = s$ diagonals (note that $q$ and $p$ are constant integers, so that the summation is over all $(n_1,n_2)$ with $qn_1+pn_2 = s$, i.e., all $(n_1,n_2)$ on the $s$-th 1D cluster):

$$= \sum_{s=-\infty}^{\infty} \Big( \sum_{qn_1+pn_2=s} c_{n_1,n_2} \Big)\, e^{i2\pi s\mathbf{f}\cdot\mathbf{x}}$$

$$= \sum_{s=-\infty}^{\infty} \Big( \sum_{\substack{(n_1,n_2)\,\in \\ s\text{-th cluster}}} c_{n_1,n_2} \Big)\, e^{i2\pi s\mathbf{f}\cdot\mathbf{x}}$$

This is illustrated in Fig. 5.1 for the case of $\mathbf{f}_2 = \frac{3}{2}\mathbf{f}_1$. Thus, we succeeded here, too, to bring the superposition $p_1(\mathbf{x})\cdot p_2(\mathbf{x})$ when $\alpha = 0°$ into the form (A.20) of a Fourier series expansion, where the $s$-th Fourier coefficient is given by: $c_s = \sum_{\substack{(n_1,n_2)\,\in \\ s\text{-th cluster}}} c_{n_1,n_2}$. As we can see, this Fourier series represents in the spectral domain a compound comb on the $u$ axis; this compound comb is the spectrum of the product-grating (the grating superposition) when $\alpha = 0°$. The step of this compound comb is $f = f_1/q = f_2/p$, i.e., it is narrower than the combs of the two original gratings. In the image domain this means that the product-grating in this case is indeed a periodic function, whose period $T = 1/f = qT_1 = pT_2$ is larger than the original grating periods, $T_1$ and $T_2$. The amplitude of the $s$-th impulse of the compound comb is given by $c_s$ above, which is the sum of the impulse amplitudes of the $s$-th cluster in the spectrum. We see, therefore, that as $\alpha$ tends to 0° each of the 1D clusters in the spectrum convolution collapses onto the $u$ axis, so that the $s$-th cluster generates the $s$-th compound impulse of the compound comb (Fig. 5.1(b) shows an example with $q = 2$, $p = 3$). In particular, the 0-th cluster in the spectrum convolution, which contains the $(pn,-qn)$-impulses, collapses onto the DC, and its amplitude is indeed $\sum_{n=-\infty}^{\infty} c_{pn,-qn}$, the sum of the impulse amplitudes of the 0-th cluster. Note that this 0-th cluster is precisely the comb of the $(p,-q)$-moiré, which is spanned by the $(p,-q)$-impulse in the spectrum convolution. The collapsing of this cluster onto the DC impulse indicates that $\alpha = 0°$ is, indeed, a singular state of the $(p,-q)$-moiré. Fig. 5.1 shows the example of the (3,-2)-moiré. Note that the case of two gratings with identical frequencies that we have seen in Sec. 6.6.1 is simply the special case in which $p = 1$, $q = 1$ (so that $f_1 = f_2 = f$).

(2) If the frequencies $f_1$ and $f_2$ of the two gratings are *incommensurable*, then as we have seen in Example 5.2 of Sec. 5.2.2, the spectrum support $\mathrm{Md}(\mathbf{f}_1,\mathbf{f}_2)$ becomes when $\alpha = 0°$ a 1D module which is everywhere dense on the $u$ axis. In this case, all the impulses of the spectrum convolution fall, again, on the $u$ axis, but this time no two impulses fall on the same point (since otherwise there would exist $p,q \in \mathbb{Z}$ such that $f_2 = \frac{p}{q}f_1$, and $f_1$ and $f_2$ would be commensurable). In particular, no impulse falls exactly on top of the DC impulse, meaning that the resulting superposition is not singular. In this case expression (6.2) cannot be simplified any further, and it remains a generalized Fourier series expansion. This means that in this case when $\alpha = 0°$ the superposition $p_1(\mathbf{x}) \cdot p_2(\mathbf{x})$ in the image domain is not periodic, but rather a 1-fold almost-periodic function in the $x$ direction (see Table 6.1 above, and Fig. B.2 in Appendix B).

## 6.7 Two-screen superpositions and their singular states

After having analyzed the case of two superposed gratings, let us consider now a more interesting case: the superposition of two screens (see Sec. 4.3), or equivalently, the superposition of four gratings.

Let $p_1(\mathbf{x})$ and $p_2(\mathbf{x})$ be two 2-fold periodic functions such as dot-screens or line-grids. As shown in Sec. A.3.4 of Appendix A, their Fourier series expansions are given by:

$$p_1(\mathbf{x}) = \sum_{n_1=-\infty}^{\infty} \sum_{n_2=-\infty}^{\infty} c^{(1)}_{n_1,n_2} \, e^{i2\pi(n_1\mathbf{f}_1+n_2\mathbf{f}_2)\cdot\mathbf{x}}$$

$$p_2(\mathbf{x}) = \sum_{n_3=-\infty}^{\infty} \sum_{n_4=-\infty}^{\infty} c^{(2)}_{n_3,n_4} \, e^{i2\pi(n_3\mathbf{f}_3+n_4\mathbf{f}_4)\cdot\mathbf{x}}$$

Now, when $p_1(\mathbf{x})$ and $p_2(\mathbf{x})$ are superposed (= multiplied), the resulting function over the $x,y$ plane is given by:

$$p_1(\mathbf{x})\cdot p_2(\mathbf{x}) = \Big(\sum_{n_1=-\infty}^{\infty}\sum_{n_2=-\infty}^{\infty} c^{(1)}_{n_1,n_2}\, e^{i2\pi(n_1\mathbf{f}_1+n_2\mathbf{f}_2)\cdot\mathbf{x}}\Big)\Big(\sum_{n_3=-\infty}^{\infty}\sum_{n_4=-\infty}^{\infty} c^{(2)}_{n_3,n_4}\, e^{i2\pi(n_3\mathbf{f}_3+n_4\mathbf{f}_4)\cdot\mathbf{x}}\Big)$$

$$= \sum_{n_1=-\infty}^{\infty}\sum_{n_2=-\infty}^{\infty}\sum_{n_3=-\infty}^{\infty}\sum_{n_4=-\infty}^{\infty} c^{(1)}_{n_1,n_2} c^{(2)}_{n_3,n_4}\, e^{i2\pi(n_1\mathbf{f}_1+n_2\mathbf{f}_2+n_3\mathbf{f}_3+n_4\mathbf{f}_4)\cdot\mathbf{x}}$$

The same result holds also for the superposition of four 1-fold periodic functions such as line-gratings, with the only difference that each of the inseparable Fourier coefficients $c^{(1)}_{n_1,n_2}$ or $c^{(2)}_{n_3,n_4}$ is replaced in this case by a product $c^{(1)}_{n_1}c^{(2)}_{n_2}$ or $c^{(3)}_{n_3}c^{(4)}_{n_4}$. In both cases we have, therefore, according to Eq. (2.27):

$$= \sum_{n_1=-\infty}^{\infty}\sum_{n_2=-\infty}^{\infty}\sum_{n_3=-\infty}^{\infty}\sum_{n_4=-\infty}^{\infty} c_{n_1,n_2,n_3,n_4}\, e^{i2\pi(n_1\mathbf{f}_1+n_2\mathbf{f}_2+n_3\mathbf{f}_3+n_4\mathbf{f}_4)\cdot\mathbf{x}} \qquad (6.5)$$

**Remark 6.1:** It is interesting to note that when the indices $n_i$ run through all integers, the frequency component in the exponent of Eq. (6.5) precisely represents the spectrum-support of the superposition in the $u,v$ plane, i.e., the module $\mathrm{Md}(\mathbf{f}_1,\mathbf{f}_2,\mathbf{f}_3,\mathbf{f}_4) = \{n_1\mathbf{f}_1+n_2\mathbf{f}_2+n_3\mathbf{f}_3+n_4\mathbf{f}_4 \mid n_i \in \mathbb{Z}\}$ which is spanned by the four frequency vectors $\mathbf{f}_1,...,\mathbf{f}_4$. As we can see, the generalized Fourier expansion (6.5) is tightly related to the algebraic structure of the spectrum: the exponential part of the Fourier expansion specifies the frequencies of the spectrum support, i.e., the geometric location of each $(n_1,n_2,n_3,n_4)$-impulse in the spectrum, while the amplitude of the impulse is given by the coefficient $c_{n_1,n_2,n_3,n_4}$. This remarkable correspondence between the exponential vector Fourier notation of Eq. (6.5) and the explicit form of the spectrum support, $\mathrm{Md}(\mathbf{f}_1,\mathbf{f}_2,\mathbf{f}_3,\mathbf{f}_4)$, clearly shows the fundamental relationship between the Fourier series and the algebraic structure of the spectrum support. In Chapter 7 we will see how this exponential vector Fourier notation can allow for layer shifts as well. ■

At this point we can clearly see the significance of the results obtained in the algebraic analysis of Chapter 5. If the frequency vectors $\mathbf{f}_1,...,\mathbf{f}_4$ (the generating vectors of the module) are incommensurable, i.e., linearly independent over $\mathbb{Z}$, meaning that $\mathrm{rank}_{\mathbb{Z}}\, \mathrm{Md}(\mathbf{f}_1,\mathbf{f}_2,\mathbf{f}_3,\mathbf{f}_4) = 4$, then from Eq. (5.16) we have $\mathrm{rank}L = 0$, which means in turn that no two impulses in the spectrum fall on the same point. In particular, no impulse falls exactly on top of the DC impulse, so that the resulting spectrum is not singular. In this case expression (6.5) cannot be simplified any further, and it remains a generalized Fourier series, meaning that the superposition is not periodic, but rather an almost-periodic function. But if, on the contrary, the frequency vectors $\mathbf{f}_1,...,\mathbf{f}_4$ are commensurable (i.e., linearly dependent over $\mathbb{Z}$), then $\mathrm{rank}L > 0$, and the spectrum impulses are partitioned into clusters: the main cluster falls on the spectrum origin, and its translated counterparts fall in different locations throughout the $u,v$ plane. This is, of course, a singular state, since all the impulses of the main cluster fall on the spectrum origin. Now, according to Sec. C.12.2 of Appendix C (or rather a generalization thereof), we can sum the terms of Eq. (6.5) by clusters, i.e., by grouping them according to the partition of the impulses into clusters.

Let us illustrate this for the most familiar case of a 2-screen superposition, where each cluster in the spectrum convolution, including the moiré cluster, is a 2D lattice; this means that $\mathrm{rank}L = 2$, and hence, by Eq. (5.16), we have in the singular state: $\mathrm{rank}_{\mathbb{Z}}\, \mathrm{Md}(\mathbf{f}_1,\mathbf{f}_2,\mathbf{f}_3,\mathbf{f}_4) = 4-2 = 2$ (see Sec. 4.3, and also Example 5.10 in Sec. 5.7). Note that in this singular case the set of all collapsed-down clusters in the $u,v$ plane forms a 2D lattice; we will henceforth call this lattice, which is the support of a compound nailbed, the *lattice-of-clusters*.[4] Let the vectors $\mathbf{g}$ and $\mathbf{h}$ be an integral basis (over $\mathbb{Z}$) of the lattice-of-clusters in the $u,v$ plane. Therefore, any cluster in the lattice-of-clusters has the geometric location:

$$r\mathbf{g} + s\mathbf{h} \qquad (r,s \in \mathbb{Z})$$

in the $u,v$ plane. The pair $(r,s)$ is the index (or the label) of each cluster within the lattice-of-clusters; for example, the pair (0,0) belongs to the main cluster, situated on the spectrum origin. Therefore, using the same technique as in Sec. 6.6, we can rewrite here Eq. (6.5) as follows:

$$= \sum_{r=-\infty}^{\infty} \sum_{s=-\infty}^{\infty} \Big( \sum_{\substack{(n_1,n_2,n_3,n_4) \in \\ (r,s)\text{-cluster}}} c_{n_1,n_2,n_3,n_4} \Big) \, e^{i2\pi(r\mathbf{g}+s\mathbf{h})\cdot\mathbf{x}} \tag{6.6}$$

Hence, we succeeded in this singular case to bring the screen superposition $p_1(\mathbf{x}) \cdot p_2(\mathbf{x})$ to the form (A.26) of a proper 2D Fourier series expansion, whose $(r,s)$-coefficient is:

$$c_{r,s} = \sum_{\substack{(n_1,n_2,n_3,n_4) \in \\ (r,s)\text{-cluster}}} c_{n_1,n_2,n_3,n_4} \tag{6.7}$$

This means that our singular superposition $p_1(\mathbf{x}) \cdot p_2(\mathbf{x})$ is, indeed, a 2-fold periodic function. And as we can see, Fourier series (6.6) represents, indeed, a compound 2D nailbed in the spectrum: Its $(r,s)$-th compound impulse has the geometric location $r\mathbf{g}+s\mathbf{h}$ and the amplitude $c_{r,s}$, which is the sum of the impulse amplitudes of the $(r,s)$-th cluster in the spectrum. This singular superposition corresponds to the singular state of the 2-fold $(k_1,k_2,k_3,k_4)$-moiré, where $(k_1,k_2,k_3,k_4)$ and its orthogonal counterpart are a basis of the (0,0)-th cluster, the 2D cluster which has collapsed on the spectrum origin.

Let us see now, as a generalization of Sec. 6.5, how we can find here the explicit expression of the 2-fold $(k_1,k_2,k_3,k_4)$-moiré, whose impulses belong to the 2D (0,0)-th cluster in the spectrum convolution. Consider the 2D partial sum of Eq. (6.5) which consists of all the terms of this quadruple sum that are spanned by the $(k_1,k_2,k_3,k_4)$-term and its orthogonal counterpart, the $(-k_2,k_1,-k_4,k_3)$-term. As we have seen in Sec. 4.3, this 2D partial sum corresponds to the (0,0)-th cluster of impulses in the spectrum convolution, which represents the $(k_1,k_2,k_3,k_4)$-moiré. This partial sum consists of all the terms whose indices are $mk_1-nk_2$, $mk_2+nk_1$, $mk_3-nk_4$ and $mk_4+nk_3$, since (see Eq. (4.10)):

$$m(k_1,k_2,k_3,k_4) + n(-k_2,k_1,-k_4,k_3) = (mk_1-nk_2,\ mk_2+nk_1,\ mk_3-nk_4,\ mk_4+nk_3)$$

The 2-fold $(k_1,k_2,k_3,k_4)$-moiré between dot-screens $p_1(\mathbf{x})$ and $p_2(\mathbf{x})$ is therefore given by:

$$m_{k_1,k_2,k_3,k_4}(x,y) = \sum_{m=-\infty}^{\infty} \sum_{n=-\infty}^{\infty} c^{(1)}_{mk_1-nk_2,\, mk_2+nk_1}\, c^{(2)}_{mk_3-nk_4,\, mk_4+nk_3}$$
$$\times\ e^{i2\pi[(mk_1-nk_2)\mathbf{f}_1 + (mk_2+nk_1)\mathbf{f}_2 + (mk_3-nk_4)\mathbf{f}_3 + (mk_4+nk_3)\mathbf{f}_4]\cdot\mathbf{x}}$$

---

[4] Cases in which the set of all collapsed-down clusters is not a lattice but rather a dense module will be discussed in the next section.

$$= \sum_{m=-\infty}^{\infty} \sum_{n=-\infty}^{\infty} c^{(1)}{}_{mk_1-nk_2,\, mk_2+nk_1}\, c^{(2)}{}_{mk_3-nk_4,\, mk_4+nk_3}$$
$$\times\ e^{i2\pi[m(k_1\mathbf{f}_1+k_2\mathbf{f}_2+k_3\mathbf{f}_3+k_4\mathbf{f}_4) + n(-k_2\mathbf{f}_1+k_1\mathbf{f}_2-k_4\mathbf{f}_3+k_3\mathbf{f}_4)]\cdot\mathbf{x}} \qquad (6.8)$$

Note that this expression remains true even when the moiré has moved away from its singular state. As an example, the (1,0,-1,0)-moiré is defined by the partial sum consisting of all the terms of the quadruple sum (6.5) whose indices are $m$, $n$, $-m$, $-n$, namely:

$$m_{1,0,-1,0}(x,y) = \sum_{m=-\infty}^{\infty} \sum_{n=-\infty}^{\infty} c^{(1)}{}_{m,n}\, c^{(2)}{}_{-m,-n}\ e^{i2\pi(m\mathbf{f}_1+n\mathbf{f}_2-m\mathbf{f}_3-n\mathbf{f}_4)\cdot\mathbf{x}}$$
$$= \sum_{m=-\infty}^{\infty} \sum_{n=-\infty}^{\infty} c^{(1)}{}_{m,n}\, c^{(2)}{}_{-m,-n}\ e^{i2\pi[m(\mathbf{f}_1-\mathbf{f}_3) + n(\mathbf{f}_2-\mathbf{f}_4)]\cdot\mathbf{x}} \qquad (6.9)$$

This is, indeed, the 2D equivalent of the (1,-1)-moiré between two gratings (see Eq. (6.4)); its fundamental frequency vectors are $\mathbf{f}_1-\mathbf{f}_3$ and $\mathbf{f}_2-\mathbf{f}_4$ (see Eq. (4.17)), and its impulse amplitudes are $c_{m,n} = c^{(1)}{}_{m,n}\, c^{(2)}{}_{-m,-n}$ (see Example 4.2 in Sec. 4.3).

## 6.8 The general superposition of $m$ layers and its singular states

The extension of Secs. 6.6 and 6.7 to the general case, i.e., to the singular superposition of $m$ 1-fold periodic layers, is rather straightforward. The general idea remains the same as in the transition from expression (6.5) to (6.6), namely: to separate the summation of all the impulses of the spectrum convolution into an *external* summation over the clusters, and an *internal* summation within a cluster.

Let $p_1(\mathbf{x})$ ,...., $p_m(\mathbf{x})$ be $m$ 1-fold periodic layers. Their superposition is given by:

$$p_1(\mathbf{x}) \cdot \ldots \cdot p_m(\mathbf{x}) =$$
$$= \Big(\sum_{n_1=-\infty}^{\infty} c^{(1)}{}_{n_1}\, e^{i2\pi n_1\mathbf{f}_1\cdot\mathbf{x}}\Big) \cdot \ldots \cdot \Big(\sum_{n_m=-\infty}^{\infty} c^{(m)}{}_{n_m}\, e^{i2\pi n_m\mathbf{f}_m\cdot\mathbf{x}}\Big)$$
$$= \sum_{n_1=-\infty}^{\infty} \cdots \sum_{n_m=-\infty}^{\infty} c^{(1)}{}_{n_1}\cdot\ldots\cdot c^{(m)}{}_{n_m}\ e^{i2\pi(n_1\mathbf{f}_1+\ \ldots\ +n_m\mathbf{f}_m)\cdot\mathbf{x}} \qquad (6.10)$$

(Note that if an inseparable 2-fold periodic function, such as a screen with circular dots, takes part in the superposition instead of two of the 1-fold periodic functions, then the only difference in Eq. (6.10) is that two of the coefficients, $c^{(i)}{}_{n_i}$ and $c^{(j)}{}_{n_j}$, will be replaced by a single inseparable coefficient, $c^{(i,j)}{}_{n_i,n_j}$.) Now, according to Eq. (2.27) we obtain:

$$= \sum_{n_1=-\infty}^{\infty} \cdots \sum_{n_m=-\infty}^{\infty} c_{n_1,\ldots,n_m}\ e^{i2\pi(n_1\mathbf{f}_1+\ \ldots\ +n_m\mathbf{f}_m)\cdot\mathbf{x}} \qquad (6.11)$$

Note, once again, the tight relationship that exists between the exponential vector Fourier series expansion of the superposition and the algebraic structure of the spectrum convolution: the Fourier expansion of the superposition explicitly contains in its exponential part the vectorial expression of the spectrum support, including all the impulse locations, and the corresponding impulse amplitudes are given by the coefficients $c_{n_1,\ldots,n_m}$.

Now, if this superposition is singular the impulses of the spectrum convolution are partitioned into collapsed clusters. Assume that the rank of each collapsed cluster is $r$ (obviously, $r > 0$). Then, according to Eq. (5.16) the rank of the entire spectrum support is $z = m - r$. This means that the spectrum support $\mathrm{Md}(\mathbf{f}_1,...,\mathbf{f}_m)$ is spanned by $z < m$ generating vectors, say, $\mathbf{g}_1,...,\mathbf{g}_z$. Therefore, by separating the summation of all the impulses of the spectrum convolution (Eq. (6.11)) into an external summation over the clusters and an internal summation within a cluster, we obtain:

$$= \sum_{s_1=-\infty}^{\infty} \dots \sum_{s_z=-\infty}^{\infty} \Big( \sum_{\substack{(n_1,...,n_m) \in \\ (s_1,...,s_z)\text{-cluster}}} c_{n_1,...,n_m} \Big)\; e^{i2\pi(s_1\mathbf{g}_1 + \dots + s_z\mathbf{g}_z)\cdot\mathbf{x}} \tag{6.12}$$

where the index-vectors $(s_1,...,s_z)$ enumerate the clusters. For example, in the singular superposition of three identical screens with angle differences of 30° (see Example 5.13 in Sec. 5.7) we have $m = 6$, $z = 4$ and $r = m - z = 2$, and Eq. (6.12) represents a 2D almost-periodic singular state with an everywhere dense spectrum support.

In cases where $z = 2$ the spectrum support is a 2D lattice; if the vectors $\mathbf{g}$ and $\mathbf{h}$ are an integral basis of this lattice of clusters, then Eq. (6.12) is simplified into a form similar to Eq. (6.6):

$$\sum_{s_1=-\infty}^{\infty} \sum_{s_2=-\infty}^{\infty} \Big( \sum_{\substack{(n_1,...,n_m) \in \\ (s_1,s_2)\text{-cluster}}} c_{n_1,...,n_m} \Big)\; e^{i2\pi(s_1\mathbf{g} + s_2\mathbf{h})\cdot\mathbf{x}} \tag{6.13}$$

In this case the singular layer superposition is, indeed, 2-fold periodic. A case of this type has been given in Example 5.12 in Sec. 5.7.

Finally, in cases where $z = 1$ the spectrum support is a 1D lattice; if the vector $\mathbf{g}$ is an integral basis of this lattice of clusters, then Eq. (6.12) is simplified into:

$$\sum_{s=-\infty}^{\infty} \Big( \sum_{\substack{(n_1,...,n_m) \in \\ s\text{-th cluster}}} c_{n_1,...,n_m} \Big)\; e^{i2\pi s\mathbf{g}\cdot\mathbf{x}} \tag{6.14}$$

In this case the singular layer superposition is 1-fold periodic. A case of this type has been given in Example 5.8 in Sec. 5.7.

Let us see now how we can explicitly express the $(k_1,...,k_m)$-moiré, whose impulses belong to the $\mathbf{0}$-th cluster in the spectrum convolution, the cluster that collapses at the moment of singularity on the DC.

Suppose, first, that $r = 1$, so that the moiré cluster is 1D. In this case the $(k_1,...,k_m)$-moiré extracted from the multiple sum (6.11) contains only the impulses of the $(k_1,...,k_m)$-comb, and is therefore given by:

$$m_{k_1,...,k_m}(\mathbf{x}) = \sum_{n=-\infty}^{\infty} c^{(1)}{}_{nk_1} \cdot \ldots \cdot c^{(m)}{}_{nk_m}\; e^{i2\pi n(k_1\mathbf{f}_1 + \dots + k_m\mathbf{f}_m)\cdot\mathbf{x}} \tag{6.15}$$

If $r = 2$, so that the moiré cluster is 2D, let $(k_1^{(1)},...,k_m^{(1)})$ and $(k_1^{(2)},...,k_m^{(2)})$ be an integral basis of this cluster. Following the same considerations as in the case of a 2-fold moiré between two dot-screens (see at the end of the previous section), we see that the partial sum of Eq. (6.11) which corresponds to our moiré consists of the terms whose indices are given by:

$$n_1(k_1^{(1)},...,k_m^{(1)}) + n_2(k_1^{(2)},...,k_m^{(2)}) = (n_1k_1^{(1)}+n_2k_1^{(2)}, \ldots, n_1k_m^{(1)}+n_2k_m^{(2)})$$

The general 2-fold periodic moiré between the $m$ layers, the $((k_1^{(1)},...,k_m^{(1)}), (k_1^{(2)},...,k_m^{(2)}))$-moiré (see Sec. 2.8), is given, therefore, by:

$$\begin{aligned} m_{((k_1^{(1)},...,k_m^{(1)}),\,(k_1^{(2)},...,k_m^{(2)}))}(x,y) &= \sum_{n_1=-\infty}^{\infty}\sum_{n_2=-\infty}^{\infty} c^{(1)}_{n_1k_1^{(1)}+n_2k_1^{(2)}} \cdot \ldots \cdot c^{(m)}_{n_1k_m^{(1)}+n_2k_m^{(2)}} \\ &\quad \times e^{i2\pi[(n_1k_1^{(1)}+n_2k_1^{(2)})\mathbf{f}_1 + \ldots + (n_1k_m^{(1)}+n_2k_m^{(2)})\mathbf{f}_m]\cdot\mathbf{x}} \\ &= \sum_{n_1=-\infty}^{\infty}\sum_{n_2=-\infty}^{\infty} c^{(1)}_{n_1k_1^{(1)}+n_2k_1^{(2)}} \cdot \ldots \cdot c^{(m)}_{n_1k_m^{(1)}+n_2k_m^{(2)}} \\ &\quad \times e^{i2\pi[n_1(k_1^{(1)}\mathbf{f}_1+\ldots+k_m^{(1)}\mathbf{f}_m) + n_2(k_1^{(2)}\mathbf{f}_1+\ldots+k_m^{(2)}\mathbf{f}_m)]\cdot\mathbf{x}} \end{aligned} \tag{6.16}$$

This can be also extended to cases with $r > 2$, where the moiré cluster has a higher rank than 2 and the extracted moiré is therefore no longer periodic but rather almost-periodic.

## PROBLEMS

6-1. Is it possible for the spectrum of a superposition of periodic layers to contain both compound and simple impulses simultaneously? Explain. *Hint*: If in Fig. 2.3(d) $\mathbf{f}_3$ is chosen so that it equals $2(\mathbf{f}_2-\mathbf{f}_1)$ then $\mathbf{f}_2-\mathbf{f}_1+\mathbf{f}_3$ falls on top of $\mathbf{f}_1-\mathbf{f}_2$, and yet no new impulses fall on top of the DC impulse; what would happen if the same gratings had a square profile rather than a cosinusoidal profile?

6-2. May the spectrum of a non-singular superposition contain a compound impulse? May the spectrum of a singular superposition contain a simple impulse?

6-3. Normally, superposing a new layer on top of a given superposition causes the appearance of new impulses in the spectrum due to the spectrum convolution process (see, for example, Figs. 2.2 and 2.3). In some particular cases, however, superposing an additional layer does not influence the spectrum support. (For example: superposing a 60°-oriented grating on top of a 0°-oriented grating will enrich the spectrum support; but superposing a third grating at 120° on top of them will not further enrich the spectrum support.) How can you characterize such cases? Give a necessary and sufficient condition.

6-4. Let $p(x,y)$ be a binary square-wave grating that is centered on the origin and oriented to angle $\theta$. The trigonometric form of its Fourier series development has been given in Eq. (2.20), using the coefficients (2.16) (see Secs. 2.5 and 2.6).

(a) Reformulate this Fourier series using the exponential vector Fourier series notation (see also Secs. A.2 and A.3 in Appendix A).

(b) What happens in each of these two Fourier series notations when the grating is not symmetric with respect to the origin? (Consider, for example, a sawtooth grating having the asymmetric profile form: /|/|/|/|/| ).

(c) Sketch the spectrum of the $n$-th terms in the trigonometric Fourier series and in the exponential Fourier series. How do they relate to each other? (*Hint*: Consider Eqs. (A.7) in Sec. A.2.) Show that the spectra of both Fourier series representations of the grating $p(x,y)$ are, indeed, identical.

6-5. Compare the exponential Fourier series representation of the $(k_1,k_2)$-moiré (Eq. (6.3)) with its trigonometric counterpart given in Eq. (4.3). What are the main advantages of the exponential notation?

6-6. According to the convolution theorem, the spectrum of the superposition of two gratings $p_1(\mathbf{x})$ and $p_2(\mathbf{x})$ is the convolution of their individual spectra, i.e., the convolution of the two combs $P_1(\mathbf{f})$ and $P_2(\mathbf{f})$ (see Appendix A.3.3). Show that the Fourier series of the grating superposition, given by Eq. (6.2), corresponds, indeed, to this spectrum convolution.

6-7. *The product-grating.* Let $p_1(\mathbf{x})$ and $p_2(\mathbf{x})$ be two centered gratings with the same frequency $f$, that are superposed with an angle difference $\alpha = 0$. What is the resulting product-grating and what is its spectrum:

(a) If $p_1(\mathbf{x})$ and $p_2(\mathbf{x})$ are "raised" cosinusoidal gratings, like in Fig. 2.2?

(b) If $p_1(\mathbf{x})$ and $p_2(\mathbf{x})$ are binary gratings, like in Fig. 2.5? What happens if the grating openings $\tau_1$ and $\tau_2$ are not equal?

6-8. If $p_1(\mathbf{x})$ and $p_2(\mathbf{x})$ are identical binary gratings, their superposition when $\alpha = 0$ is identical to $p_1(\mathbf{x})$ and $p_2(\mathbf{x})$. Using Eq. (2.16), show that the sum of the $n$-th comb in the spectrum convolution is, indeed, equal to the amplitude of the $n$-th impulse of $p_1(\mathbf{x})$.

6-9. Give the explicit expression $m_{1,0,-1,1}(\mathbf{x})$ of the 2-fold periodic (1,0,-1,1)-moiré between two dot-screens (see Sec. 4.5).

6-10. Give the explicit expression $m_{1,1,-1,0}(\mathbf{x})$ of the 2-fold periodic (1,1,-1,0)-moiré between two dot-screens (see Example 5.10 in Sec. 5.7). How does it differ from the case of the previous problem? Give also the explicit expression of the superposition at the moment when the (1,1,-1,0)-moiré is singular, using Eq. (6.6).

6-11. Give the explicit expression of the 2-fold periodic moiré in the conventional 3-screen superposition with identical frequencies and equal angle differences of 30° (see Example 5.15 in Sec. 5.7). Give also the explicit expression of the superposition at the moment of singularity, using Eq. (6.12).

6-12. Find the vectors **g** and **h** which span the lattice-of-clusters in the spectrum of the singular (1,2,-2,-1)-superposition of two identical dot-screens (see Fig. 8.3 in Chapter 8). Using Eq. (6.6) Give the explicit expression of this singular 2-screen superposition. Find also the explicit expression $m_{1,2,-2,-1}(\mathbf{x})$ of the (1,2,-2,-1)-moiré that becomes visible slightly away from the singular superposition.

6-13. Consider the spectrum of a singular superposition of two dot-screens $p_1(x,y)$ and $p_1(x,y)$. See, for example, Figs. 4.3(b) or 8.6(a) which show the spectrum of the superposition slightly off the singular (1,2,-2,-1)-moiré; when the singular point is attained, each of the impulse-clusters collapses down into a single compound impulse which is located in the center of the cluster, as shown in Fig. 8.3(a). Suppose that the frequency vectors of the two superposed screens at the singular point are given by $\mathbf{f}_1,...,\mathbf{f}_4$. How can you find the vectors **g** and **h** which span the lattice-of-clusters in the spectrum of the $(k_1,k_2,k_3,k_4)$-singular superposition (see Eq. (6.6))? Clearly, since the frequency vectors $\mathbf{f}_1,...,\mathbf{f}_4$ span the support of the spectrum convolution, **g** and **h** are integral linear combinations of these vectors. Of course, **g** and **h** can be found graphically by sketching the spectrum convolution using the computer program of Problem 5-18. However, is there a general way to find explicitly the vectors **g**, **h** as a function of $\mathbf{f}_1,...,\mathbf{f}_4$ at the singular point of the $(k_1,k_2,k_3,k_4)$-moiré, i.e., when (see Fig. 4.3(b)):

$$\mathbf{a} = k_1\mathbf{f}_1 + k_2\mathbf{f}_2 + k_3\mathbf{f}_3 + k_4\mathbf{f}_4 = \mathbf{0}$$
$$\mathbf{b} = -k_2\mathbf{f}_1 + k_1\mathbf{f}_2 - k_4\mathbf{f}_3 + k_3\mathbf{f}_4 = \mathbf{0}$$

6-14. Derive Eq. (6.8) as a particular case of Eq. (6.16).

# Chapter 7

## The superposition phase

### 7.1 Introduction

In all the previous chapters up to now we intentionally avoided the question of the phase of the superposed periodic layers (or functions). Note, in particular, that until now we have only discussed the behaviour of moiré effects when the superposed layers were centered about the origin; and moreover, we only considered *rotations* and *scalings* of the superposed layers (i.e., transformations which preserve the origin), but *shift* operations were excluded.

It is therefore our aim in the present chapter to analyze what happens to the superposition (and in particular to each of its moiré effects) when we mutually displace the superposed layers. And indeed, we will see that if we slide the superposed layers on top of one another, without changing their angles or their periods, the moiré patterns simply undergo a lateral shift across the superposition, without changing their form, their period or their angle. But because the structures involved here (the original layers and the resulting moirés) are periodic, a shift of an integer number of periods has no visible effect, and we only need to consider its residue, or the *phase*. Another question that we will touch in this chapter (and revisit in more detail in Chapter 8) is how layer shifts influence the microstructure of the superposition, even when no moiré effects are visible.

In order to investigate these questions we will need to further extend our Fourier-based approach by incorporating in the image-domain the notion of shifts, and admitting non-centered and non-symmetric layers. The dual, spectral-domain counterpart of this image-domain generalization is that impulse-pairs in the spectrum will no longer have equal, purely real-valued amplitudes, like in Fig. 2.1, but rather complex-conjugate amplitudes (see Sec. 2.2). This generalization will allow the superposed layers to be freely located (or shifted) on top of each other, with any desired phases or layer positions. This generalization will be done by adding to each periodic function a new parameter which specifies its displacement from the initial position. We will see how this new parameter fits into the exponential vector Fourier notation that we have adopted in Chapter 6, and how it allows us, in a simple and elegant way, to cope with the quite complex question of the phases in the superposition of any $m$ periodic layers.

But before proceeding to questions concerning the phases in the superposition and in its eventual moirés (Secs. 7.6–7.7), we will first review the phase of a single periodic function (Sec. 7.2), and define our phase terminology: first for the 1D case (Sec. 7.3), and then for the 2D case (Secs. 7.4–7.5). Complementary information on the connection between the phase in the sense of complex number theory and the phase in periodic functions is provided in Appendix C.4.

## 7.2 The phase of a periodic function

Let $p(x,y)$ (or in short, $p(\mathbf{x})$) be a 2-fold periodic function with fundamental frequency vectors $\mathbf{f}_1$, $\mathbf{f}_2$. As explained in Sec. A.3.4 of Appendix A, if $p(x,y)$ satisfies some convergence conditions, which are normally met by all our cases of interest, it can be represented in the form of a 2-fold Fourier series:

$$p(\mathbf{x}) = \sum_{m=-\infty}^{\infty} \sum_{n=-\infty}^{\infty} c_{m,n}\, e^{i2\pi(m\mathbf{f}_1+n\mathbf{f}_2)\cdot\mathbf{x}} \tag{7.1}$$

where the Fourier series coefficients $c_{m,n}$ are determined by:

$$c_{m,n} = \frac{1}{A}\iint_A p(\mathbf{x})\, e^{-i2\pi(m\mathbf{f}_1+n\mathbf{f}_2)\cdot\mathbf{x}}\, d\mathbf{x}$$

The spectrum of $p(\mathbf{x})$ is an oblique impulse-nailbed, whose $(m,n)$-th impulse has the frequency $m\mathbf{f}_1+n\mathbf{f}_2$ and the amplitude $c_{m,n}$, where $\mathbf{f}_1 = (u_1,v_1)$ and $\mathbf{f}_2 = (u_2,v_2)$ are the fundamental frequency-vectors of the nailbed. This spectrum is given in vector form by Eq. (A.30) in Appendix A, where $\mathbf{f} = (u,v)$:

$$P(\mathbf{f}) = \sum_{m=-\infty}^{\infty} \sum_{n=-\infty}^{\infty} c_{m,n}\, \delta(\mathbf{f}-(m\mathbf{f}_1+n\mathbf{f}_2)) \tag{7.2}$$

As mentioned in Sec. 2.2, if $p(\mathbf{x})$ is symmetric about the origin then the coefficients $c_{m,n}$ are all real numbers and the spectrum $P(\mathbf{f})$ is purely real. If, however, $p(\mathbf{x})$ is non-symmetric, then the coefficients $c_{m,n}$ (or some of them) have non-zero imaginary parts and the spectrum $P(\mathbf{f})$ is therefore complex-valued. It may be in order to remind here that any complex-valued function can be represented either by its real part and its imaginary part, or by its magnitude (modulus) and its phase (argument); see Appendix C.4 for a more detailed review. The magnitude and the argument of the complex spectrum are called, respectively, the *magnitude-spectrum* and the *phase-spectrum.*

Assume now that we shift $p(x,y)$ in the image domain by the vector $\mathbf{a} = (a,b)$, namely, by $a$ units in the $x$ direction and by $b$ units in the $y$ direction. We obtain for the shifted periodic function $p(x-a,\ y-b)$ the following expression:

$$\begin{aligned} p(\mathbf{x}-\mathbf{a}) &= \sum_{m=-\infty}^{\infty} \sum_{n=-\infty}^{\infty} c_{m,n}\, e^{i2\pi(m\mathbf{f}_1+n\mathbf{f}_2)\cdot(\mathbf{x}-\mathbf{a})} \\ &= \sum_{m=-\infty}^{\infty} \sum_{n=-\infty}^{\infty} e^{-i2\pi(m\mathbf{f}_1+n\mathbf{f}_2)\cdot\mathbf{a}}\, c_{m,n}\, e^{i2\pi(m\mathbf{f}_1+n\mathbf{f}_2)\cdot\mathbf{x}} \end{aligned} \tag{7.3}$$

As we can see, each term of Eq. (7.1) has been simply multiplied here by a complex number $e^{-i2\pi(m\mathbf{f}_1+n\mathbf{f}_2)\cdot\mathbf{a}}$, which is independent of the variable $\mathbf{x}$. The spectrum of the shifted function is given, therefore, by:

$$P_{\mathbf{a}}(\mathbf{f}) = \sum_{m=-\infty}^{\infty} \sum_{n=-\infty}^{\infty} e^{-i2\pi(m\mathbf{f}_1+n\mathbf{f}_2)\cdot\mathbf{a}}\, c_{m,n}\, \delta(\mathbf{f}-(m\mathbf{f}_1+n\mathbf{f}_2)) \tag{7.4}$$

This is, in fact, a particular case of the 2D shift theorem [Bracewell95 p. 156], which is applied here to the case of 2D periodic functions having nailbed spectra. The 2D shift theorem says that if the spectrum of a function $f(x,y)$ is $F(u,v)$, then the spectrum of the shifted function $f(x-a,\ y-b)$ is $e^{-i2\pi(ua+vb)}\cdot F(u,v)$. This means that a shift of $\mathbf{a} = (a,b)$ in the image domain multiplies the spectrum at each frequency $\mathbf{f} = (u,v)$ by the complex factor $e^{-i2\pi\mathbf{f}\cdot\mathbf{a}}$. And indeed, a comparison of Eqs. (7.2) and (7.4) shows that in our case, in which the spectrum is an impulse-nailbed, each $(m,n)$-impulse in the spectrum remains in its original location, at the frequency $\mathbf{f} = m\mathbf{f}_1+n\mathbf{f}_2$, and only its amplitude is multiplied by the complex factor $e^{-i2\pi(m\mathbf{f}_1+n\mathbf{f}_2)\cdot\mathbf{a}}$, as predicted by the shift theorem. (Note that the magnitude of this complex factor is 1, so that it only influences the *phase* of the impulse.)

It follows from the shift theorem as a corollary that the increment generated in the phase-spectrum as a result of a shift of **a** in the image domain is a linear function of the frequency, meaning that the increment in the phase-spectrum due to the shift has the form of a continuous linear plane through the origin, whose slopes are determined by $\mathbf{a} = (a,b)$:[1]

$$\varphi(\mathbf{f}) = -2\pi\mathbf{f}\cdot\mathbf{a}$$

namely:
$$\varphi(u,v) = -2\pi(ua + vb) \tag{7.5}$$

In our case, however, the spectrum of $p(\mathbf{x})$ can only be non-zero at the points of the nail-bed, i.e. at the frequencies of the $(m,n)$-th impulses: $\mathbf{f} = m\mathbf{f}_1+n\mathbf{f}_2$, or in other words: $(u,v) = m(u_1,v_1) + n(u_2,v_2) = (mu_1+nu_2\,,\, mv_1+nv_2)$. The phase increment generated at the $(m,n)$-th impulse in the spectrum as a result of the shift of **a** in the image domain is, therefore:

$$\varphi(m\mathbf{f}_1+n\mathbf{f}_2) = -2\pi(m\mathbf{f}_1+n\mathbf{f}_2)\cdot\mathbf{a}$$

namely:
$$\varphi(mu_1+nu_2\,,\, mv_1+nv_2) = -2\pi[(mu_1+nu_2)a + (mv_1+nv_2)b] \tag{7.6}$$

which is simply the restriction of the linear plane (7.5) to the points of our nailbed. In other words, Eq. (7.6) samples the continuous plane (7.5) of the phase-spectrum increment, which is due to the shift theorem, at all the impulse locations $m\mathbf{f}_1+n\mathbf{f}_2$. This is clearly seen in the spectrum of $p(\mathbf{x}-\mathbf{a})$ above (Eq. (7.4)); see also Sec. C.4 in Appendix C.

As we can see, an image-domain shift of the periodic function $p(\mathbf{x})$ only influences in the spectral domain the *phases* of the impulses; the *impulse locations* in the spectrum, as well as the impulse magnitudes (the *absolute values* of their amplitudes), are not influenced by the shift. This could be, indeed, expected, since a shift does not modify the periods, the angles or the intensity profile of the function $p(\mathbf{x})$.

The shifted periodic function (7.3) can be rewritten yet in another form, as follows:

$$p(\mathbf{x}-\mathbf{a}) = \sum_{m=-\infty}^{\infty}\sum_{n=-\infty}^{\infty} c_{m,n}\, e^{i2\pi(m\mathbf{f}_1+n\mathbf{f}_2)\cdot\mathbf{x} - i2\pi(m\mathbf{f}_1+n\mathbf{f}_2)\cdot\mathbf{a}}$$

or:
$$p(\mathbf{x}-\mathbf{a}) = \sum_{m=-\infty}^{\infty}\sum_{n=-\infty}^{\infty} c_{m,n}\, e^{i2\pi(m\mathbf{f}_1+n\mathbf{f}_2)\cdot\mathbf{x} - i2\pi\phi_{m,n}} \tag{7.7}$$

where $-2\pi\phi_{m,n} = -2\pi(m\mathbf{f}_1+n\mathbf{f}_2)\cdot\mathbf{a}$ is the phase increment of the $(m,n)$-th impulse in the spectrum due to the shift. We can see, therefore, that the phase of the periodic function $p(x,y)$ may be expressed in the image domain, in the Fourier series representation of $p(x,y)$, in two equivalent ways:

---

[1] Note that the converse is also true: it follows from the shift theorem that a linear increment occurs in the phase-spectrum *iff* the original function has undergone a shift in the image domain.

(a) Either explicitly in the exponential part, as in Eq. (7.7) above;

(b) Or implicitly, lumped together with the complex-valued Fourier coefficients $c_{m,n}$ (the impulse amplitudes).

The connection between these two representations of the phase can be clearly seen from Eq. (7.3), where the exponential constant representing the phase can be incorporated either in the main exponential part of the expression, as in (a), or in the coefficient $c_{m,n}$, as in (b). The significance of this point will become clear soon.

If the periodic function $p(x,y)$ is symmetric with respect to a certain point $(x,y)$, it is natural to choose its "initial phase" (or its "in-phase" position) as the position in which its center of symmetry coincides with the origin. This natural choice is also advantageous in terms of the spectral domain: In this position the impulse amplitudes in the spectrum are purely real (or, in terms of magnitude and phase, their phase components are constantly zero). But when the symmetric function is shifted from this position by a non-integral number of periods, its impulse amplitudes in the spectrum get an imaginary component (or, in terms of magnitude and phase, their phase component becomes a linear function of the frequency, according to the shift theorem).

However, if the periodic function $p(x,y)$ has no point of symmetry (like the screen with "1"-shaped dots in Fig. 4.4), there is no longer a "privileged" position which can be considered in a natural way as the "initial phase" or the "in-phase" position of the function. Even in terms of the spectral domain there is no longer any "privileged" position, since in every position of the function its impulse amplitudes will have an imaginary component (i.e., a non-zero phase component, which is not even a linear function of the frequency). Therefore, in the case of an asymmetric function we will arbitrarily choose a certain shift position of the function as its initial phase, and we will fix this position by inserting its phase components inside the coefficients $c_{m,n}$, as in point (b) above. Once this initial phase and its corresponding complex-valued coefficients $c_{m,n}$ have been fixed, they will be kept frozen, and from that moment on any shifts of $p(x,y)$ relative to this initial position will be represented only in the exponential part, as in point (a), without modifying the fixed coefficients $c_{m,n}$.

It should be remembered, however, that since we are only dealing with real-valued images, their spectrum is always Hermitian [Bracewell86 p. 15], which means that the amplitudes of the impulse-twins (which are symmetrically located to both sides of the origin, as in Fig. 2.1) are always complex-conjugates.

## 7.3 The phase terminology for periodic functions in the 1D case

Let us introduce now some notations and terms in connection with the phase of periodic functions. We will start, for the sake of simplicity, with the 1D case. Assume that $p(x)$ is a periodic function (symmetric or not) of period $T$. As a simple example of such a function one may consider $p(x) = \cos(2\pi x/T)$. Clearly, since $p(x)$ is periodic, $p(x-nT)$ is identical to $p(x)$ for any integer $n$, so that a shift of $p(x)$ by an integer multiple of the period $T$ is indistinguishable from the unshifted function. This fact suggests that any shift of $p(x)$ by $a$ should be considered as being composed of an integer number of periods $T$ plus a residue $t$:

$$a = nT + t, \qquad \text{with:} \qquad 0 \le t < T$$

where only the residue $t$ affects $p(x-a)$:

$$p(x-a) = p(x-nT-t) = p(x-t)$$

Consequently, we may introduce here the following notational conventions:

(1) The total distance by which $p(x)$ has been displaced will be called the *shift* or the *displacement* of $p(x)$, and denoted by $a$ (as in: $p(x-a)$).

(2) The residue $t = a \bmod T$, where: $0 \le t < T$, will be called the *effective shift* (or *effective displacement*) of $p(x)$.

The shift and the effective shift of the periodic function $p(x)$ may be also expressed in terms of periods rather than in terms of distance units:

(3) The shift of $p(x)$ expressed as a number (integer or not) of periods $T$ will be called the *period-shift* of $p(x)$, and denoted by: $\phi = \frac{a}{T} = fa$. We have, therefore, $a = \phi T$.

(4) The residue $r = \frac{t}{T} = \phi \bmod 1$, a number between 0 and 1 denoting the fraction of $T$ by which $p(x)$ has been effectively shifted, will be called the *effective period-shift* of $p(x)$.

As for the term *phase* itself, we will continue using it in the most general sense, including in the context of complex numbers (e.g., the magnitude and phase representation of the spectrum, the phase component of an impulse, etc.).

As we can see, the influence of a shift of $a$ on the phase of the periodic function $p(x)$ can be expressed in terms of the period-shift $\phi$ as follows:

$$\begin{aligned} p(x-a) &= \sum_{n=-\infty}^{\infty} c_n \, e^{i2\pi nf(x-a)} = \sum_{n=-\infty}^{\infty} c_n \, e^{i2\pi nfx - i2\pi nfa} \\ &= \sum_{n=-\infty}^{\infty} c_n \, e^{i2\pi nfx - i2\pi n\phi} \end{aligned} \tag{7.8}$$

The phase increment of the $n$-th impulse in the spectrum of $p(x-a)$ due to the shift of $a$ is, therefore, $\varphi(nf) = -2\pi n\phi = -2\pi\phi_n$. Note that shifting $p(x)$ does not influence the impulse locations in its spectrum, which are always given by $\{nf \mid n \in \mathbb{Z}\}$.

## 7.4 The phase terminology for 1-fold periodic functions in the 2D case

We proceed now to the 2D case, and start with the simplest 2D case — that of a 1-fold periodic function in an arbitrary direction in the $x,y$ plane (such as a rotated grating; see

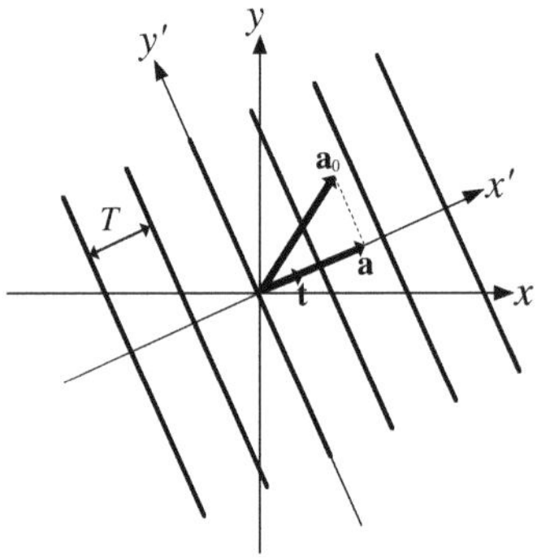


**Figure 7.1:** A schematic plot of a 1-fold periodic function $p(\mathbf{x})$ in the image domain. A shift of $\mathbf{a}_0$ from the origin is equivalent to a shift of $\mathbf{a}$ or to an effective shift of $\mathbf{t}$.

Fig. 7.1, and Sec. A.3.3 in Appendix A). We will see now how the 1D phase-related terms defined above can be generalized into this 2D case by considering them as vector quantities in the $x,y$ plane.

Assume that $p(\mathbf{x})$ is a 1-fold periodic function, symmetric or not, whose fundamental period-vector is $\mathbf{T}$. Note that the fundamental period-vector $\mathbf{T}$ of $p(\mathbf{x})$ is the shortest non-zero vector attached to the origin in the $x,y$ plane which satisfies $p(\mathbf{x}+\mathbf{T}) = p(\mathbf{x})$. This shortest period-vector is located on the $x'$ axis, i.e., in the main direction of $p(\mathbf{x})$. Clearly, shifting $p(\mathbf{x})$ in the $y'$ direction (i.e., perpendicularly to its main direction, $x'$) has no effect, since $p(\mathbf{x})$ is constant in that direction. Therefore, the effect of any arbitrary shift $\mathbf{a}_0$ of $p(\mathbf{x})$ (that is, of moving the origin of $p(\mathbf{x})$ to the point $\mathbf{a}_0$ in the $x,y$ plane) is equivalent to a shift of $p(\mathbf{x})$ by $\mathbf{a} = \text{proj}(\mathbf{a}_0)_\mathbf{T}$, the orthogonal projection of $\mathbf{a}_0$ on the period-vector $\mathbf{T}$ or on the $x'$ axis; note that $\mathbf{a}$ is the vector component of $\mathbf{a}_0$ in the main direction of $p(\mathbf{x})$. Therefore we will henceforth consider only shifts of $p(\mathbf{x})$ along its main direction, and any other shift $\mathbf{a}_0$ will be represented by its component $\mathbf{a}$ in that direction (Fig. 7.1).

Since $p(\mathbf{x})$ is periodic with the fundamental period-vector $\mathbf{T}$, $p(\mathbf{x}-n\mathbf{T})$ is identical to $p(\mathbf{x})$ for any integer $n$, so that a shift of $p(\mathbf{x})$ by an integer multiple of the vector-period $\mathbf{T}$ is indistinguishable from the unshifted function. This fact suggests that any shift of $p(\mathbf{x})$ by $\mathbf{a}$ should be considered as being composed of an integer number of period-vectors $\mathbf{T}$ plus a vectorial residue $\mathbf{t}$:[2]

$$\mathbf{a} = n\mathbf{T} + \mathbf{t}, \qquad \text{with:} \qquad \mathbf{t} = r\mathbf{T}, \quad 0 \le r < 1$$

where only the residue $\mathbf{t}$ affects $p(\mathbf{x}-\mathbf{a})$:

$$p(\mathbf{x}-\mathbf{a}) = p(\mathbf{x}-n\mathbf{T}-\mathbf{t}) = p(\mathbf{x}-\mathbf{t})$$

[2] Note that the vectors $\mathbf{T}$ and $\mathbf{a}$ are collinear, since both of them are located on the $x'$ axis; the vector $\mathbf{t}$ is also collinear with them.

Consequently, we may introduce here the following notational conventions as an extension of the 1D case of Sec. 7.3:

(1) The total distance by which $p(\mathbf{x})$ has been displaced will be called the *shift* or the *displacement* of $p(\mathbf{x})$, and denoted by $\mathbf{a}$ (as in: $p(\mathbf{x}-\mathbf{a})$).

(2) The residue $\mathbf{t} = \mathbf{a} \bmod \mathbf{T}$, where: $\mathbf{0} \le \mathbf{t} < \mathbf{T}$ (meaning: $\mathbf{t} = r\mathbf{T}$, $0 \le r < 1$), will be called the *effective shift* (or *effective displacement*) of $p(\mathbf{x})$.

The shift and the effective shift of the periodic function $p(\mathbf{x})$ may be also expressed in terms of periods rather than in terms of distance units:

(3) The shift of $p(\mathbf{x})$ expressed as a number (integer or not) of periods $\mathbf{T}$ will be called the *period-shift* of $p(\mathbf{x})$, and denoted by: $\phi = \frac{|\mathbf{a}|}{|\mathbf{T}|}$. We have, therefore: $\mathbf{a} = \phi\mathbf{T}$. Remembering that $\mathbf{T}\cdot\mathbf{T}^{-1} = 1$ (see Eq. (A.42) in Appendix A), we multiply both sides (in the sense of scalar product) by $\mathbf{T}^{-1}$, and hence we obtain $\mathbf{a}\cdot\mathbf{T}^{-1} = \phi$, which can be expressed, as an extension of the 1D case, as:

$$\phi = \mathbf{f}\cdot\mathbf{a} \tag{7.9}$$

where $\mathbf{f} = \mathbf{T}^{-1}$ is the fundamental frequency-vector of the 1-fold periodic function $p(\mathbf{x})$; see Eq. (A.43) in Appendix A.

(4) The residue $r = \frac{|\mathbf{t}|}{|\mathbf{T}|} = \phi \bmod 1$, a number between 0 and 1 denoting the fraction of $\mathbf{T}$ by which $p(\mathbf{x})$ has been effectively shifted, will be called the *effective period-shift* of $p(\mathbf{x})$.

As in the 1D case, we will continue to use the term *phase* in the most general sense, including in the context of complex numbers.

As we can see, the influence of a shift of $\mathbf{a}$ on the phase of the 1-fold periodic function $p(\mathbf{x})$ can be expressed in terms of the period-shift $\phi$ as follows:

$$p(\mathbf{x}-\mathbf{a}) = \sum_{n=-\infty}^{\infty} c_n\, e^{i2\pi n\mathbf{f}\cdot(\mathbf{x}-\mathbf{a})} = \sum_{n=-\infty}^{\infty} c_n\, e^{i2\pi n\mathbf{f}\cdot\mathbf{x}\, -\, i2\pi n\mathbf{f}\cdot\mathbf{a}}$$
$$= \sum_{n=-\infty}^{\infty} c_n\, e^{i2\pi n\mathbf{f}\cdot\mathbf{x}\, -\, i2\pi n\phi} \tag{7.10}$$

The phase increment of the $n$-th impulse in the spectrum of $p(\mathbf{x}-\mathbf{a})$ due to the shift of $\mathbf{a}$ is, therefore, $\varphi(n\mathbf{f}) = -2\pi n\phi = -2\pi\phi_n$. Note, again, that shifting $p(\mathbf{x})$ does not influence the impulse locations in its spectrum, which remain as before $\{n\mathbf{f} \mid n \in \mathbb{Z}\}$.

## 7.5 The phase terminology for the general 2D case: 2-fold periodic functions

We proceed now to the generalization of the above phase terminology to the general 2D case. Assume that $p(\mathbf{x})$ is a 2-fold periodic function (symmetric or not) whose fundamental frequency vectors in the spectrum are $\mathbf{f}_1$ and $\mathbf{f}_2$. As shown in Appendix A (Sec. A.6), the periodic properties of $p(\mathbf{x})$ can be expressed in the image domain in two alternative ways: either in terms of the classical period-vectors $\mathbf{P}_1$ and $\mathbf{P}_2$, or in terms of the step-vectors $\mathbf{T}_1$ and $\mathbf{T}_2$. Accordingly, the phase in the 2D case can be also presented in two ways; we will see, however, that for the sake of studying layer superpositions the second approach is by far more advantageous.

### 7.5.1 Using the period-vector notation

Let $\mathbf{P}_1$ and $\mathbf{P}_2$ be the fundamental period-vectors of $p(\mathbf{x})$ which correspond to $\mathbf{f}_1$, $\mathbf{f}_2$ (see Fig. 7.2(a)). Therefore, as explained in Sec. A.3.4 in Appendix A, $p(\mathbf{x}-m\mathbf{P}_1-n\mathbf{P}_2)$ is identical to $p(\mathbf{x})$ for any integers $m$ and $n$, so that a shift of $p(\mathbf{x})$ by an integer linear combination of $\mathbf{P}_1$ and $\mathbf{P}_2$ is indistinguishable from the unshifted function. This fact suggests that any shift of $p(\mathbf{x})$ by $\mathbf{a}$ should be considered as being composed of an integer linear combination of the periods $\mathbf{P}_1$ and $\mathbf{P}_2$, plus a residue $\mathbf{t}$, as follows:

$$\begin{aligned}\mathbf{a} &= s_1\mathbf{P}_1+s_2\mathbf{P}_2 = (m+r_1)\mathbf{P}_1 + (n+r_2)\mathbf{P}_2\\ &= (m\mathbf{P}_1+n\mathbf{P}_2) + (r_1\mathbf{P}_1+r_2\mathbf{P}_2), \qquad \text{with:} \quad m,n \in \mathbb{Z},\ \ 0 \le r_1,r_2 < 1\end{aligned}$$

where only the residue $\mathbf{t} = r_1\mathbf{P}_1+r_2\mathbf{P}_2$ affects $p(\mathbf{x})$:

$$p(\mathbf{x}-\mathbf{a}) = p(\mathbf{x}-m\mathbf{P}_1-n\mathbf{P}_2-\mathbf{t}) = p(\mathbf{x}-\mathbf{t})$$

Consequently, we may introduce here the following notational conventions, as an extension of the 1-fold periodic case:

(1) The total distance by which $p(\mathbf{x})$ has been displaced will be called the *shift* or the *displacement* of $p(\mathbf{x})$, and denoted by $\mathbf{a}$ (as in: $p(\mathbf{x}-\mathbf{a})$).

(2) The residue $\mathbf{t} = r_1\mathbf{P}_1+r_2\mathbf{P}_2$ (see above) will be called the *effective shift* (or *effective displacement*) of $p(\mathbf{x})$.

The shift and the effective shift of the periodic function $p(\mathbf{x})$ may be also expressed in terms of periods rather than in terms of distance units:

(3) The shift of $p(\mathbf{x})$ expressed as a pair of numbers (integer or not) representing the number of periods $\mathbf{P}_1$, $\mathbf{P}_2$ will be called the *period-shift* of $p(\mathbf{x})$, and denoted by: $\boldsymbol{\phi} = (m+r_1,n+r_2) = (s_1,s_2)$. We have, therefore: $\mathbf{a} = \boldsymbol{\phi}\begin{pmatrix}\mathbf{P}_1\\ \mathbf{P}_2\end{pmatrix}$. Note that $\boldsymbol{\phi}$ is simply the coordinate pair $(s_1,s_2)$ of the point $\mathbf{a}$ in terms of the oblique coordinate system defined by $\mathbf{P}_1$ and $\mathbf{P}_2$ (see Fig. 7.2(a)).

(4) The residue $\mathbf{r} = (r_1,r_2)$, a pair of numbers between 0 and 1 denoting the fractions of $\mathbf{P}_1$, $\mathbf{P}_2$ by which $p(\mathbf{x})$ has been effectively shifted, will be called the *effective period-shift* of $p(\mathbf{x})$.

For example, if the 2-fold periodic function of Fig. 7.2(a) is shifted by $\mathbf{a}$ (i.e., its origin is displaced to $\mathbf{a}$) where $\mathbf{a} = 2.5\mathbf{P}_1 + 3.5\mathbf{P}_2$, then according to the above terminology its effective shift is $\mathbf{t} = 0.5\mathbf{P}_1 + 0.5\mathbf{P}_2$, its period-shift is $\boldsymbol{\phi} = (2.5,3.5)$ and its effective period-shift is $\mathbf{r} = (0.5,0.5)$.

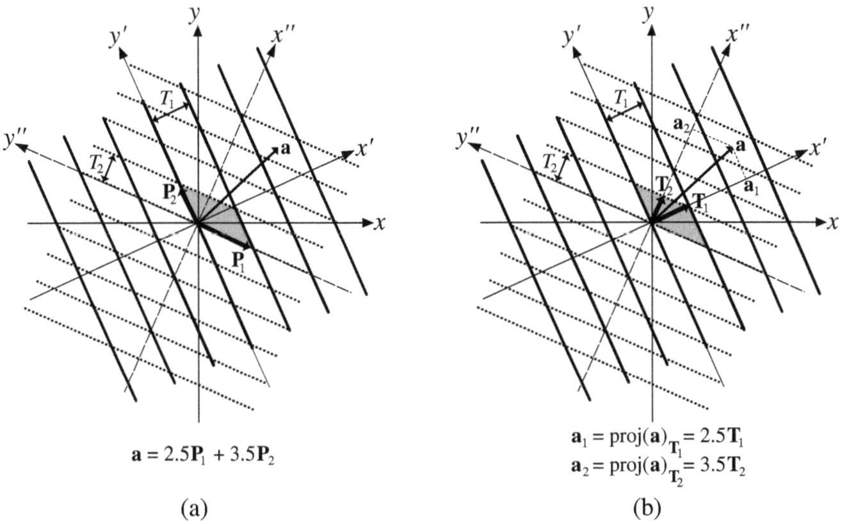


**Figure 7.2:** A schematic plot of the unshifted 2-fold periodic function $p(\mathbf{x})$ in the image domain (see Fig. A.2 in Appendix A). (a) A vector **a**, expressed in terms of the period-vectors $\mathbf{P}_1$, $\mathbf{P}_2$. (b) The same vector **a**, expressed in terms of the step-vectors $\mathbf{T}_1$, $\mathbf{T}_2$. Using a more technical lingo, the vector **a** is expressed in (a) in terms of its *contravariant* coordinates in the $y',y''$ coordinate system, while in (b) it is expressed in terms of its *covariant* coordinates in the $x',x''$ coordinate system [Bronshtein97 p. 192].

### 7.5.2 Using the step-vector notation

Let $p(\mathbf{x})$ be the same 2-fold periodic function as above (see Fig. 7.2(b)). Consider the fundamental period-parallelogram of $p(\mathbf{x})$ defined by the vectors $\mathbf{P}_1$, $\mathbf{P}_2$. If the given 2-fold periodic function $p(\mathbf{x})$ is separable, i.e., if it can be presented as a product of two 1D functions, then let $p_1(\mathbf{x})$ and $p_2(\mathbf{x})$ be the two 1-fold periodic components of $p(\mathbf{x})$: $p(\mathbf{x}) = p_1(\mathbf{x}) \cdot p_2(\mathbf{x})$. If, however, $p(\mathbf{x})$ is inseparable (like a screen with circular dots), we choose $p_1(\mathbf{x})$ and $p_2(\mathbf{x})$ to be the two 1-fold periodic gratings defined by the parallelogram borders (see the solid and the dotted gratings in Fig. 7.2). Now, instead of the original function $p(\mathbf{x})$ we consider the separable superposition $p'(\mathbf{x}) = p_1(\mathbf{x}) \cdot p_2(\mathbf{x})$; we will call $p_1(\mathbf{x})$ and $p_2(\mathbf{x})$ the *virtual gratings* of $p(\mathbf{x})$. Note that the period-vectors (or rather step-vectors[3]) $\mathbf{T}_1$ and $\mathbf{T}_2$ of the 1-fold periodic functions $p_1(\mathbf{x})$ and $p_2(\mathbf{x})$ are precisely the step-vectors of the 2-fold periodic function $p(\mathbf{x})$ (see Sec. A.6 in Appendix A).

Clearly, $p'(\mathbf{x})$ and $p(\mathbf{x})$ have the same periods and the same phases, and in fact, they differ in their Fourier series representations by only their coefficients $c_{m,n}$ (see Eq. (7.1)). Therefore, as far as periodicity and phases are concerned, $p(\mathbf{x})$ and $p'(\mathbf{x})$ are equivalent,

[3] As mentioned in Sec. A.6 of Appendix A, the period-vector and the step-vector of a 1-fold periodic function are identical.

and we may represent the inseparable function $p(\mathbf{x})$ by the separable function $p'(\mathbf{x})$. For the sake of simplicity we will proceed with the discussion below using $p'(\mathbf{x})$; in the case of a separable function $p(\mathbf{x})$, we may simply replace $p'(\mathbf{x})$ by $p(\mathbf{x})$ itself.

Now, the idea is to look at each of the two superposed 1-fold periodic virtual gratings separately: Obviously, any shift of $p'(\mathbf{x})$ by $\mathbf{a}$ can be considered as a simultaneous shift of both $p_1(\mathbf{x})$ and $p_2(\mathbf{x})$ by $\mathbf{a}$. However, as we have seen in Sec. 7.4, a shift of $p_1(\mathbf{x})$ by $\mathbf{a}$ is equivalent to a shift of $p_1(\mathbf{x})$ by $\mathbf{a}_1 = \text{proj}(\mathbf{a})_{\mathbf{T}_1}$; and a shift of $p_2(\mathbf{x})$ by $\mathbf{a}$ is equivalent to a shift of $p_2(\mathbf{x})$ by $\mathbf{a}_2 = \text{proj}(\mathbf{a})_{\mathbf{T}_2}$. For example, in the case shown in Fig. 7.2 we have: $\mathbf{a} = 2.5\mathbf{P}_1 + 3.5\mathbf{P}_2$, and hence: $\mathbf{a}_1 = 2.5\mathbf{T}_1$, $\mathbf{a}_2 = 3.5\mathbf{T}_2$.

We see that the shift of $p'(\mathbf{x})$ by $\mathbf{a}$ is identical to a shift of the grating $p_1(\mathbf{x})$ by $\mathbf{a}_1 = \phi_1\mathbf{T}_1$ and a shift of the grating $p_2(\mathbf{x})$ by $\mathbf{a}_2 = \phi_2\mathbf{T}_2$, where $\phi_1$ and $\phi_2$ are the individual period-shifts of the shifted gratings $p_1(\mathbf{x}-\mathbf{a}_1)$ and $p_2(\mathbf{x}-\mathbf{a}_2)$ (see Sec. 7.4, (3)). This is equivalent, in turn, to an effective shift of the grating $p_1(\mathbf{x})$ by $\mathbf{t}_1 = r_1\mathbf{T}_1$ and an effective shift of the grating $p_2(\mathbf{x})$ by $\mathbf{t}_2 = r_2\mathbf{T}_2$, where $r_1$ and $r_2$ represent the fractional parts of $\phi_1$ and $\phi_2$.

As we can see, we have reduced here the phase of the 2-fold periodic function into the phases of its two 1-fold periodic virtual gratings, and hence we can simply use here the phase terminology of Sec. 7.4 above, applied to each of the virtual gratings separately. Consequently, we may introduce here the following notational conventions as an extension of the 1-fold periodic case of Sec. 7.4, using the step-vector approach:

(1) The total distance by which $p(\mathbf{x})$ has been displaced is called the *shift* or the *displacement* of $p(\mathbf{x})$, and denoted by $\mathbf{a}$ (as in: $p(\mathbf{x}-\mathbf{a})$). As we have seen, this is equivalent to shifts of $\mathbf{a}_1$ and $\mathbf{a}_2$ in the virtual gratings $p_1(\mathbf{x})$ and $p_2(\mathbf{x})$, where $\mathbf{a}_1$ and $\mathbf{a}_2$ are the projections of the vector $\mathbf{a}$ in the directions of $\mathbf{T}_1$ and $\mathbf{T}_2$.

(2) The residue $\mathbf{t}$, called the *effective shift* (or *effective displacement*) of $p(\mathbf{x})$, can be expressed in a similar way by $\mathbf{t}_1$ and $\mathbf{t}_2$, where $\mathbf{t}_1$ and $\mathbf{t}_2$ are the effective shifts of the virtual gratings $p_1(\mathbf{x})$ and $p_2(\mathbf{x})$ (see Sec. 7.4, (2)): $\mathbf{t}_1 = \mathbf{a}_1 \bmod \mathbf{T}_1$, $\mathbf{t}_2 = \mathbf{a}_2 \bmod \mathbf{T}_2$. In other words, $\mathbf{t}_1$ and $\mathbf{t}_2$ are the projections of the vector $\mathbf{t}$ in the directions of $\mathbf{T}_1$ and $\mathbf{T}_2$.

The shift and the effective shift of $p(\mathbf{x})$ may be also expressed in terms of periods rather than in terms of distance units:

(3) The shift of $p(\mathbf{x})$ expressed as a pair of numbers (integer or not) representing the number of periods $\mathbf{T}_1$, $\mathbf{T}_2$ of $p_1(\mathbf{x})$ and $p_2(\mathbf{x})$ is called the *period-shift* of $p(\mathbf{x})$, and denoted by: $\boldsymbol{\phi} = (\phi_1,\phi_2)$, where $\phi_1$ and $\phi_2$ are the period-shifts of each of the two virtual gratings $p_1(\mathbf{x})$ and $p_2(\mathbf{x})$. By Eq. (7.9) we have, therefore: $\phi_1 = \mathbf{f}_1 \cdot \mathbf{a}_1$, $\phi_2 = \mathbf{f}_2 \cdot \mathbf{a}_2$, where $\mathbf{f}_1 = \mathbf{T}_1^{-1}$ and $\mathbf{f}_2 = \mathbf{T}_2^{-1}$ are the fundamental frequency-vectors of the 2-fold periodic function $p(\mathbf{x})$. (As we will see in Sec. 7.7, it is also true here that $\phi_1 = \mathbf{f}_1 \cdot \mathbf{a}$, $\phi_2 = \mathbf{f}_2 \cdot \mathbf{a}$ since $\mathbf{a}_1$ and $\mathbf{a}_2$ are projections of the same vector $\mathbf{a}$ on the directions of $\mathbf{f}_1$ and $\mathbf{f}_2$.)

(4) The residue $\mathbf{r} = (r_1,r_2)$, a pair of numbers between 0 and 1 denoting the fractions of $\mathbf{T}_1$, $\mathbf{T}_2$ by which $p_1(\mathbf{x})$ and $p_2(\mathbf{x})$ have been effectively shifted, is called the *effective period-shift* of $p(\mathbf{x})$.

Note that these definitions are identical to those given in Sec. 7.5.1, and they are only presented from a different point of view.

Now, the influence of a shift **a** on the phase of the 2-fold periodic function $p'(\mathbf{x})$ can be expressed in terms of the period-shifts $\phi_1$ and $\phi_2$ of $p_1(\mathbf{x})$ and $p_2(\mathbf{x})$ as follows:

$$
\begin{aligned}
p'(\mathbf{x}-\mathbf{a}) &= p_1(\mathbf{x}-\mathbf{a}_1)\cdot p_2(\mathbf{x}-\mathbf{a}_2) \\
&= \Big(\sum_{m=-\infty}^{\infty} c^{(1)}_m\, e^{i2\pi m\mathbf{f}_1\cdot(\mathbf{x}-\mathbf{a}_1)}\Big)\Big(\sum_{n=-\infty}^{\infty} c^{(2)}_n\, e^{i2\pi n\mathbf{f}_2\cdot(\mathbf{x}-\mathbf{a}_2)}\Big) \\
&= \Big(\sum_{m=-\infty}^{\infty} c^{(1)}_m\, e^{i2\pi m\mathbf{f}_1\cdot\mathbf{x}\,-\,i2\pi m\mathbf{f}_1\cdot\mathbf{a}_1}\Big)\Big(\sum_{n=-\infty}^{\infty} c^{(2)}_n\, e^{i2\pi n\mathbf{f}_2\cdot\mathbf{x}\,-\,i2\pi n\mathbf{f}_2\cdot\mathbf{a}_2}\Big) \\
&= \sum_{m=-\infty}^{\infty}\sum_{n=-\infty}^{\infty} c^{(1)}_m c^{(2)}_n\, e^{i2\pi(m\mathbf{f}_1+n\mathbf{f}_2)\cdot\mathbf{x}\,-\,i2\pi(m\mathbf{f}_1\cdot\mathbf{a}_1+n\mathbf{f}_2\cdot\mathbf{a}_2)} \\
&= \sum_{m=-\infty}^{\infty}\sum_{n=-\infty}^{\infty} c^{(1)}_m c^{(2)}_n\, e^{i2\pi(m\mathbf{f}_1+n\mathbf{f}_2)\cdot\mathbf{x}\,-\,i2\pi(m\phi_1+n\phi_2)}
\end{aligned}
\tag{7.11}
$$

where $\phi_1$ and $\phi_2$ are the period-shifts of the two shifted 1-fold periodic functions $p_1(\mathbf{x}-\mathbf{a}_1)$ and $p_2(\mathbf{x}-\mathbf{a}_2)$.

Once we obtained the phase of the separable 2-fold periodic function $p'(\mathbf{x}) = p_1(\mathbf{x})\cdot p_2(\mathbf{x})$ after the shift, all that we need to do in order to obtain the equivalent expression for the original, inseparable function $p(\mathbf{x})$ itself is to replace in Eq. (7.11) the Fourier coefficients $c^{(1)}_m c^{(2)}_n$ which belong to $p'(\mathbf{x})$ by the inseparable Fourier coefficients $c_{m,n}$ of the original function $p(\mathbf{x})$. We obtain:

$$p(\mathbf{x}-\mathbf{a}) = \sum_{m=-\infty}^{\infty}\sum_{n=-\infty}^{\infty} c_{m,n}\, e^{i2\pi(m\mathbf{f}_1+n\mathbf{f}_2)\cdot\mathbf{x}\,-\,i2\pi(m\phi_1+n\phi_2)} \tag{7.12}$$

Comparing this expression with Eq. (7.7), which was directly obtained by the shift theorem, we see that the phase increment of the $(m,n)$-th impulse in the spectrum of $p(\mathbf{x}-\mathbf{a})$ due to the shift of **a** is $\varphi(m\mathbf{f}_1+n\mathbf{f}_2) = -2\pi(m\phi_1+n\phi_2) = -2\pi\phi_{m,n}$, where $\phi_1$, $\phi_2$ are the period-shifts of the two shifted virtual gratings $p_1(\mathbf{x}-\mathbf{a}_1)$ and $p_2(\mathbf{x}-\mathbf{a}_2)$. This is the 2D equivalent of the phase increment $\varphi(nf) = -2\pi n\phi = -2\pi\phi_n$ in the 1D case and in the 1-fold periodic case (see at the end of Secs. 7.3, 7.4).

We see, again, that shifting $p(\mathbf{x})$ does not influence the impulse locations in its spectrum, which are determined by $\{m\mathbf{f}_1 + n\mathbf{f}_2 \mid m,n \in \mathbb{Z}\}$.

Note that the above generalization of the phase concepts from scalar quantities in the 1D case (Sec. 7.3) to the equivalent vectorial quantities in the 2D case, with scalar product replacing the simple number multiplication (e.g., $\phi = \mathbf{f}\cdot\mathbf{a}$ replacing $\phi = fa$), was made possible thanks to the step-vector approach. This further illustrates the fact that the **T**-based notations are more flexible and better suited for our needs than the **P**-based notations (see also Sec. A.6 in Appendix A). We will see below that the **T**-based notation can be also generalized to the superposition of *any* number of periodic layers.

## 7.6 Moiré phases in the superposition of periodic layers

As we have seen in the previous sections, each periodic function has also a shift parameter, which indicates its lateral displacement relative to a given "initial" location; in symmetric functions this initial location is taken as the centered position about the origin. This shift parameter belongs, in fact, to the family of *geometric* parameters, like the period and the angle of the periodic function. Although it does influence the impulse amplitudes in the spectrum (their phase components), it does not affect in the image-domain the intensity levels or the profile form of the function, but only its relative position (or phase-shift) with respect to its initial location.

Suppose, now, that we superpose two or more periodic functions in their initial phase; this gives us the initial phase of the superposition and the initial phase of its moirés. Note that the initial phase is characterized by the absence of a phase component in the exponent of the Fourier series, the initial phase being incorporated in the coefficients $c_{m,n}$. What happens now if we shift one or more of the superposed layers? It should be noted, first, that since shifts in any individual layer do not influence the impulse locations in its own spectrum, they do not influence the impulse locations in the spectrum convolution, either. Therefore we obtain the following result:

**Proposition 7.1:** Shifts in any of the superposed layers have no influence on the support of the spectrum of the superposition and on its algebraic structure as presented in Chapter 5. In particular, such shifts have no influence on the spectrum support of any moiré generated in the superposition, meaning that the period and angle of the moiré remain unchanged. ■

In fact, we will see below that even the intensity levels and the profile form of each moiré remain unchanged under shifts in the original layers, and the moiré only undergoes a lateral shift. We will see, however, that this is not generally true for the superposition as a whole.

Let us now study in detail the influence of shifts in the individual superposed layers on their superposition.

Let $p_1(\mathbf{x}), \ldots, p_m(\mathbf{x})$ be $m$ 1-fold periodic functions given in their initial phases, and let $p(\mathbf{x}) = p_1(\mathbf{x}) \cdot \ldots \cdot p_m(\mathbf{x})$ be their superposition. Note that each pair of non-collinear 1-fold periodic functions may represent in the superposition one 2-fold periodic function, such as a dot-screen. As we have seen in Sec. 6.8, we have, therefore:

$$\begin{aligned} p(\mathbf{x}) &= p_1(\mathbf{x}) \cdot \ldots \cdot p_m(\mathbf{x}) \\ &= \Big(\sum_{n_1=-\infty}^{\infty} c^{(1)}_{n_1}\, e^{i2\pi n_1 \mathbf{f}_1 \cdot \mathbf{x}}\Big) \cdot \ldots \cdot \Big(\sum_{n_m=-\infty}^{\infty} c^{(m)}_{n_m}\, e^{i2\pi n_m \mathbf{f}_m \cdot \mathbf{x}}\Big) \\ &= \sum_{n_1=-\infty}^{\infty} \ldots \sum_{n_m=-\infty}^{\infty} c^{(1)}_{n_1} \cdot \ldots \cdot c^{(m)}_{n_m}\, e^{i2\pi(n_1\mathbf{f}_1 + \ldots + n_m\mathbf{f}_m)\cdot\mathbf{x}} \end{aligned} \qquad (7.13)$$

The $(k_1,\ldots,k_m)$-moiré extracted from this superposition contains only the impulses of the $(k_1,\ldots,k_m)$-comb, and is therefore given by:

$$m_{k_1,\dots,k_m}(\mathbf{x}) = \sum_{n=-\infty}^{\infty} c^{(1)}_{nk_1}\cdot\dots\cdot c^{(m)}_{nk_m}\ e^{i2\pi n(k_1\mathbf{f}_1+\dots+k_m\mathbf{f}_m)\cdot\mathbf{x}} \tag{7.14}$$

Assume now that each of the layers $p_i(\mathbf{x})$ is shifted from its initial position by $\mathbf{a}_i$ (for any layer $i$ which remains unshifted we simply take $\mathbf{a}_i = \mathbf{0}$). We have, therefore:

$$p_1(\mathbf{x}-\mathbf{a}_1)\cdot\dots\cdot p_m(\mathbf{x}-\mathbf{a}_m)$$

$$= \Big(\sum_{n_1=-\infty}^{\infty} c^{(1)}_{n_1}\ e^{i2\pi n_1\mathbf{f}_1\cdot(\mathbf{x}-\mathbf{a}_1)}\Big)\cdot\dots\cdot\Big(\sum_{n_m=-\infty}^{\infty} c^{(m)}_{n_m}\ e^{i2\pi n_m\mathbf{f}_m\cdot(\mathbf{x}-\mathbf{a}_m)}\Big)$$

$$= \Big(\sum_{n_1=-\infty}^{\infty} c^{(1)}_{n_1}\ e^{i2\pi n_1\mathbf{f}_1\cdot\mathbf{x} - i2\pi n_1\mathbf{f}_1\cdot\mathbf{a}_1}\Big)\cdot\dots\cdot\Big(\sum_{n_m=-\infty}^{\infty} c^{(m)}_{n_m}\ e^{i2\pi n_m\mathbf{f}_m\cdot\mathbf{x} - i2\pi n_m\mathbf{f}_m\cdot\mathbf{a}_m}\Big)$$

$$= \sum_{n_1=-\infty}^{\infty}\dots\sum_{n_m=-\infty}^{\infty} c^{(1)}_{n_1}\cdot\dots\cdot c^{(m)}_{n_m}\ e^{i2\pi(n_1\mathbf{f}_1+\dots+n_m\mathbf{f}_m)\cdot\mathbf{x} - i2\pi(n_1\mathbf{f}_1\cdot\mathbf{a}_1+\dots+n_m\mathbf{f}_m\cdot\mathbf{a}_m)}$$

$$= \sum_{n_1=-\infty}^{\infty}\dots\sum_{n_m=-\infty}^{\infty} c^{(1)}_{n_1}\cdot\dots\cdot c^{(m)}_{n_m}\ e^{i2\pi(n_1\mathbf{f}_1+\dots+n_m\mathbf{f}_m)\cdot\mathbf{x} - i2\pi(n_1\phi_1+\dots+n_m\phi_m)} \tag{7.15}$$

We see, again, that shifting the superposed layers $p_i(\mathbf{x})$ does not influence the impulse locations in the spectrum convolution, which are given by the module $\mathrm{Md}(\mathbf{f}_1,\dots,\mathbf{f}_m) = \{n_1\mathbf{f}_1+\dots+n_m\mathbf{f}_m \mid n_1,\dots,n_m \in \mathbb{Z}\}$. It is instructive to note here how the generalized Fourier expansion (7.15) is tightly related to the structure of the spectrum, and compactly expresses all the spectral information: For any $(n_1,\dots,n_m)$-impulse in the spectrum, $c^{(1)}_{n_1}\cdot\dots\cdot c^{(m)}_{n_m}$ gives its original amplitude, $e^{i2\pi(n_1\mathbf{f}_1+\dots+n_m\mathbf{f}_m)\cdot\mathbf{x}}$ determines its location in the spectrum, and $e^{-i2\pi(n_1\phi_1+\dots+n_m\phi_m)}$ determines its phase increment due to the layer shifts. This, is, indeed, a further extension of Remark 6.1 in Sec. 6.7.

Note that if an inseparable 2-fold periodic function (such as a screen with circular dots) takes part in the superposition instead of two of the 1-fold periodic functions, then the only difference in Eq. (7.15) is that two of the coefficients, $c^{(i)}_{n_i}$ and $c^{(j)}_{n_j}$, will be replaced by a single inseparable coefficient, $c^{(i,j)}_{n_i,n_j}$.

It is also interesting to mention here that if we denote the frequency, the period-shift and the amplitude of the $(n_1,\dots,n_m)$-impulse in the spectrum convolution by:

$$\mathbf{f}_{n_1,\dots,n_m} = n_1\mathbf{f}_1+\dots+n_m\mathbf{f}_m \tag{7.16}$$

$$\phi_{n_1,\dots,n_m} = n_1\phi_1+\dots+n_m\phi_m \tag{7.17}$$

and:

$$c_{n_1,\dots,n_m} = c^{(1)}_{n_1}\cdot\dots\cdot c^{(m)}_{n_m} \tag{7.18}$$

then we can rewrite Eq. (7.15) in the form:

$$= \sum_{n_1=-\infty}^{\infty}\dots\sum_{n_m=-\infty}^{\infty} c_{n_1,\dots,n_m}\ e^{i2\pi\mathbf{f}_{n_1,\dots,n_m}\cdot\mathbf{x} - i2\pi\phi_{n_1,\dots,n_m}} \tag{7.19}$$

or even more compactly, using the vector notation $\mathbf{n} = (n_1,\dots,n_m)$:

$$= \sum_{\mathbf{n}} c_{\mathbf{n}}\ e^{i2\pi\mathbf{f}_{\mathbf{n}}\cdot\mathbf{x} - i2\pi\phi_{\mathbf{n}}} \tag{7.20}$$

which is clearly an elegant generalization of Eqs. (7.8) and (7.10).

Now, the $(k_1,...,k_m)$-moiré extracted from this superposition contains only the impulses of the $(k_1,...,k_m)$-comb, and is therefore given by:

$$\sum_{n=-\infty}^{\infty} c^{(1)}{}_{nk_1}\cdot...\cdot c^{(m)}{}_{nk_m}\; e^{i2\pi n(k_1\mathbf{f}_1+\;...\;+k_m\mathbf{f}_m)\cdot\mathbf{x}\;-\;i2\pi n(k_1\phi_1+\;...\;+k_m\phi_m)} \tag{7.21}$$

Note that this $(k_1,...,k_m)$-moiré is a 1-fold periodic function, whose period is: $\mathbf{f}_{k_1,...,k_m} = k_1\mathbf{f}_1+\;...\;+k_m\mathbf{f}_m$, and whose period-shift is $\phi_{k_1,...,k_m} = k_1\phi_1+\;...\;+k_m\phi_m$ (see Eq. (7.10)).

Comparing this expression with Eq. (7.14) we see that the exponent here is simply decremented by $i2\pi n(k_1\phi_1+\;...\;+k_m\phi_m)$, where the expression in parentheses is a constant number, $\phi_{k_1,...,k_m}$. This suggests, by the shift theorem, that expression (7.21) simply represents a shifted version of Eq. (7.14); therefore, if we denote by $\mathbf{b}_{k_1,...,k_m}$ the unknown shift that the $(k_1,...,k_m)$-moiré $m_{k_1,...,k_m}(\mathbf{x})$ itself has undergone, we can rewrite expression (7.21) as follows:

$$m_{k_1,...,k_m}(\mathbf{x}-\mathbf{b}_{k_1,...,k_m}) = \sum_{n=-\infty}^{\infty} c^{(1)}{}_{nk_1}\cdot...\cdot c^{(m)}{}_{nk_m}\; e^{i2\pi n(k_1\mathbf{f}_1+\;...\;+k_m\mathbf{f}_m)\cdot\mathbf{x}\;-\;i2\pi n(k_1\phi_1+\;...\;+k_m\phi_m)} \tag{7.22}$$

But according to Eq. (7.3), the effect of shifting $m_{k_1,...,k_m}(\mathbf{x})$ (Eq. (7.14)) by $\mathbf{b}_{k_1,...,k_m}$ is:

$$m_{k_1,...,k_m}(\mathbf{x}-\mathbf{b}_{k_1,...,k_m}) = \sum_{n=-\infty}^{\infty} c^{(1)}{}_{nk_1}\cdot...\cdot c^{(m)}{}_{nk_m}\; e^{i2\pi n(k_1\mathbf{f}_1+\;...\;+k_m\mathbf{f}_m)\cdot(\mathbf{x}-\mathbf{b}_{k_1,...,k_m})}$$

$$= \sum_{n=-\infty}^{\infty} c^{(1)}{}_{nk_1}\cdot...\cdot c^{(m)}{}_{nk_m}\; e^{i2\pi n(k_1\mathbf{f}_1+\;...\;+k_m\mathbf{f}_m)\cdot\mathbf{x}\;-\;i2\pi n(k_1\mathbf{f}_1+\;...\;+k_m\mathbf{f}_m)\cdot\mathbf{b}_{k_1,...,k_m}} \tag{7.23}$$

By comparing the exponential parts of Eqs. (7.23) and (7.22) we find that for every $n$, i.e., for every term in the summation, we have:

$$(k_1\mathbf{f}_1+\;...\;+k_m\mathbf{f}_m)\cdot\mathbf{b}_{k_1,...,k_m} = k_1\phi_1+\;...\;+k_m\phi_m$$

or, using Eqs. (7.15) and (7.16):

$$\mathbf{f}_{k_1,...,k_m}\cdot\mathbf{b}_{k_1,...,k_m} = \phi_{k_1,...,k_m} \tag{7.24}$$

Remembering now that $\mathbf{f}\cdot\mathbf{f}^{-1} = 1$ (see Sec. A.6 in Appendix A), and that $\mathbf{f}_{k_1,...,k_m}$ and $\mathbf{b}_{k_1,...,k_m}$ are collinear, we multiply both sides (in the sense of scalar product) by $(\mathbf{f}_{k_1,...,k_m})^{-1}$:

$$\mathbf{b}_{k_1,...,k_m} = (\mathbf{f}_{k_1,...,k_m})^{-1}\phi_{k_1,...,k_m}$$

and by Eq. (A.42):

$$= \mathbf{T}_{k_1,...,k_m}\;\phi_{k_1,...,k_m} \tag{7.25}$$

where $\mathbf{T}_{k_1,...,k_m}$ is the period-vector (or rather the step-vector) of the $(k_1,...,k_m)$-moiré in the image domain, and $\phi_i = \frac{|\mathbf{a}_i|}{|\mathbf{T}_i|}$. The shift that the $(k_1,...,k_m)$-moiré undergoes is given, therefore, by the period of the moiré multiplied by the scalar coefficient $\phi_{k_1,...,k_m} = k_1\phi_1+\;...\;+k_m\phi_m$, which is a linear combination of the period-shifts $\phi_1,...,\phi_m$ of the individual superposed gratings. Note the similarity with $\mathbf{a} = \phi\mathbf{T}$ in the case of a single grating (Sec. 7.4, (3)). We obtain, therefore, the following important result:

**Proposition 7.2:** When the original $m$ superposed layers $p_1(\mathbf{x}),\;...\;,p_m(\mathbf{x})$ are shifted by $\mathbf{a}_1,...,\mathbf{a}_m$, respectively, each of the $(k_1,...,k_m)$-moirés generated in the superposition merely undergoes a shift of $\mathbf{b}_{k_1,...,k_m}$, which is a multiple of its step-vector by a scalar, as stated by Eq. (7.25). The size and the direction of the shift are therefore different for each of the $(k_1,...,k_m)$-moirés in the superposition. The periods, the angles, and the profile-shapes of the moirés, on their part, remain unchanged. ■

**Example 7.1:** In the particular case of the (1,-1)-moiré between $m = 2$ gratings we have:

$$\mathbf{b}_{1,-1} = \mathbf{T}_{1,-1}\,(\phi_1 - \phi_2) \tag{7.26}$$

and:

$$\phi_{1,-1} = \phi_1 - \phi_2 \tag{7.27}$$

Fig. 7.3 illustrates what happens to the (1,-1)-moiré between two superposed non-parallel gratings when we shift the first grating A along its main direction (=horizontally) while keeping the second grating B at its initial phase (so that $\mathbf{a}_2 = \mathbf{0}$): When $\mathbf{a}_1 = \mathbf{0}$ we obtain $\mathbf{b}_{1,-1} = \mathbf{0}$ and the moiré is in-phase; and as we gradually increase the shift $\mathbf{a}_1$ from $\mathbf{0}$ to $\mathbf{T}_1 = \mathbf{f}_1^{-1}$ (one full period of the first grating), the moiré is gradually shifted along its main direction from $\mathbf{b}_{1,-1} = \mathbf{0}$ to $\mathbf{b}_{1,-1} = \mathbf{T}_{1,-1}$ (namely, one full period of the moiré). Note that the shift of the moiré is larger than the shift of the grating (in the same proportion as their periods), and is in a different direction, that of the moiré period (or frequency). ■

**Remark 7.1:** It follows therefore from Eqs. (2.10)–(2.11) that the (1,-1)-moiré between two gratings with $T_1 = T_2$ and $\alpha \neq 0$ moves *almost perpendicularly* to the shift of the original grating (see Fig. 7.3), whereas when $T_1 \neq T_2$ and $\alpha = 0$ it moves *in parallel* to the shift of the original grating (see Fig. 7.9); in all other cases the (1,-1)-moiré moves in an intermediate direction, as determined by Eq. (2.9). The same goes also for the (1,0,-1,0)-moiré between two dot screens, since it can be considered as two (1,-1)-moirés between two pairs of virtual line gratings (see Fig. 2.10(a), Sec. 2.12, and Sec. 4.4.2). ■

As we can see, we once again encounter here, like in Sec. 4.4.1, the magnification property of the moiré effect: a microscopic shift in the original layers may give rise to a significantly larger shift of the moiré. This property is indeed used in various applications for tasks such as precision alignment, detection and measuring of microscopic displacements, etc. (see, for example, [Patorski93 pp. 99–139]).

**Example 7.2:** Consider the (1,-1)- and (1,-2)-moirés of Fig. 7.4(a), between $m = 2$ gratings. If we shift grating B along its main direction by 1/4 period ($\mathbf{a}_2 = \mathbf{T}_2/4$) while keeping grating A at its initial phase ($\mathbf{a}_1 = \mathbf{0}$), we find from Eq. (7.25) that the (1,-1)-moiré is shifted by 1/4 moiré-period ($\mathbf{b}_{1,-1} = -\mathbf{T}_{1,-1}/4$), while the (1,-2)-moiré is shifted by 1/2 moiré-period ($\mathbf{b}_{1,-2} = -\mathbf{T}_{1,-2}/2$, since $k_2 = 2$). This is indeed confirmed in Fig. 7.4(b). ■

## 7.7 The influence of layer shifts on the overall superposition

It is important to note, however, that unlike each of the individual moirés, the overall superposition in the image domain after the original layers have been shifted by $\mathbf{a}_1,...,\mathbf{a}_m$ is *not*, in general, a shifted version (a rigid motion) of the original superposition. On the contrary, the microstructure of the superposition may be modified, too. In the present section we will only have a first glance at this question, and we will return to it in more detail in Chapter 8; see, for example, Fig. 8.2(a),(b).

It is clear that every shift $\mathbf{a}$ of the superposition $p_1(\mathbf{x}) \cdot ... \cdot p_m(\mathbf{x})$ can be seen as a combination of individual shifts $\mathbf{a}_1,...,\mathbf{a}_m$ of the gratings $p_1(\mathbf{x}), ... ,p_m(\mathbf{x})$ along their main directions. However, the converse is not necessarily true: Let, for example, $\mathbf{a}_1$, $\mathbf{a}_2$, $\mathbf{a}_3$ be the individual shifts of the gratings $p_1(\mathbf{x})$, $p_2(\mathbf{x})$, $p_3(\mathbf{x})$ which correspond to the shift $\mathbf{a}$ of the entire superposition $p_1(\mathbf{x}){\cdot}p_2(\mathbf{x}){\cdot}p_3(\mathbf{x})$. Obviously, $\mathbf{a}_1$, $\mathbf{a}_2$, and $\mathbf{a}_3$ are linearly dependent, since once $\mathbf{a}_1$ and $\mathbf{a}_2$ (the components of $\mathbf{a}$ in the main directions of $p_1(\mathbf{x})$ and $p_2(\mathbf{x})$) have been fixed, $\mathbf{a}_3$ must join them in order that the superposition be shifted as a whole. Therefore $\mathbf{a}_3$ is a linear combination of $\mathbf{a}_1$ and $\mathbf{a}_2$: $\mathbf{a}_3 = s_1\mathbf{a}_1 + s_2\mathbf{a}_2$ with certain constants $s_1,s_2 \in \mathbb{R}$. However, if we choose $\mathbf{a}_3$ differently, the superposition $p_1(\mathbf{x}){\cdot}p_2(\mathbf{x}){\cdot}p_3(\mathbf{x})$ will not be shifted as a whole, and its internal structure will be modified. We have, therefore, the following result:

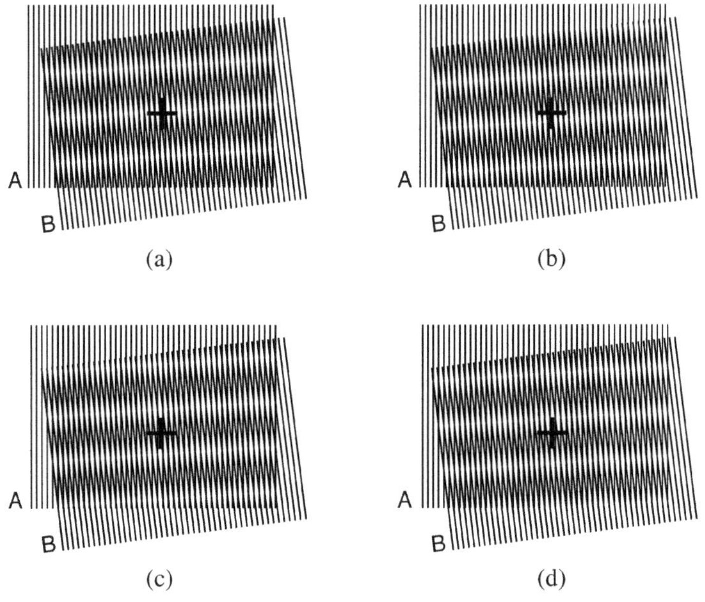


**Figure 7.3:** A (1,-1)-moiré between two gratings and its phase shifts due to a shift in the grating A. The origin of each image is indicated by a cross. (a) Both gratings and the moiré are in their initial phase. (b) Grating A is shifted by 1/4 period to the right; the moiré is consequently shifted 1/4 moiré-period downward. (c) Grating A is shifted by 1/2 period to the right; the moiré is consequently shifted 1/2 moiré-period downward. (d) Grating A is shifted by 3/4 period to the right; the moiré is consequently shifted 3/4 moiré-period downward. See Fig. 7.9 and Remark 7.1 for the case of parallel gratings (i.e. where $\alpha = 0$).

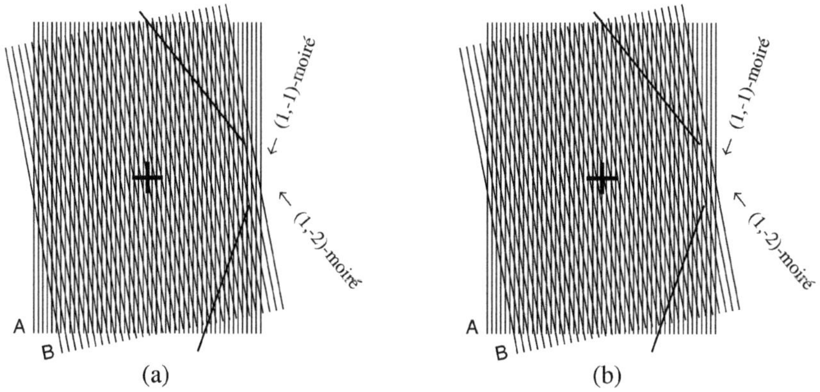


**Figure 7.4:** The (1,-1)- and (1,-2)-moirés between the two gratings of Fig. 2.7, and their phase shifts due to a shift in grating B. (a) Both gratings and the moirés are in their initial phase. (b) Grating B is shifted by 1/4 period to the right; consequently, the (1,-1)-moiré is shifted by 1/4 of its period, while the (1,-2)-moiré is shifted by 1/2 of its period (note that the cross and the lines indicating the moiré directions have not been moved). As in Fig. 2.7, the different moirés are best seen holding the figure at grazing angle, turning it and looking along the respective lines.

**Proposition 7.3:** Non-trivial shifts of $\mathbf{a}_1,...,\mathbf{a}_m$ in the gratings $p_1(\mathbf{x})$, ... ,$p_m(\mathbf{x})$ in their respective main directions (i.e., shifts $\mathbf{a}_1,...,\mathbf{a}_m$ which are not integer multiples of the respective periods $\mathbf{T}_1,...,\mathbf{T}_m$) do not, in general, correspond to a shift $\mathbf{a}$ of the entire superposition $p_1(\mathbf{x}) \cdot ... \cdot p_m(\mathbf{x})$. They do, however, *iff* the shifts $\mathbf{a}_i$ are projections of the same vector $\mathbf{a}$ on the main directions of $p_i(\mathbf{x})$; or, in other words, *iff* the normal lines defined by each of the shifts $\mathbf{a}_i$ perpendicularly to the main axis of $p_i(\mathbf{x})$ (see Fig. 7.2(b)) meet in a single point $\mathbf{a}$. ■

The proof of this proposition is based on the following observation [Vygodski73 p. 142]: The scalar product $\mathbf{v}\cdot\mathbf{w}$ of vectors $\mathbf{v}$ and $\mathbf{w}$ can be understood as a number which equals the product of the length of vector $\mathbf{v}$ by the length of the projection of vector $\mathbf{w}$ on the direction of $\mathbf{v}$: $\mathbf{v}\cdot\mathbf{w} = |\mathbf{v}|\,|\text{proj}(\mathbf{w})_\mathbf{v}|$. It follows from this observation that for any point $\mathbf{a}$ in the $x,y$ plane we have:

$$\mathbf{f}_i\cdot\mathbf{a} = \mathbf{f}_i\cdot\mathbf{a}_i \tag{7.28}$$

where $\mathbf{a}_i$ is the projection of $\mathbf{a}$ on the direction of $\mathbf{f}_i$.

Therefore, recalling Eq. (7.15) we have:

$$p_1(\mathbf{x}-\mathbf{a}_1) \cdot ... \cdot p_m(\mathbf{x}-\mathbf{a}_m)$$

$$= \sum_{n_1=-\infty}^{\infty} ... \sum_{n_m=-\infty}^{\infty} c^{(1)}_{n_1}\cdot...\cdot c^{(m)}_{n_m}\; e^{i2\pi(n_1\mathbf{f}_1+\,...\,+n_m\mathbf{f}_m)\cdot\mathbf{x}\,-\,i2\pi(n_1\mathbf{f}_1\cdot\mathbf{a}_1+\,...\,+n_m\mathbf{f}_m\cdot\mathbf{a}_m)} \tag{7.29}$$

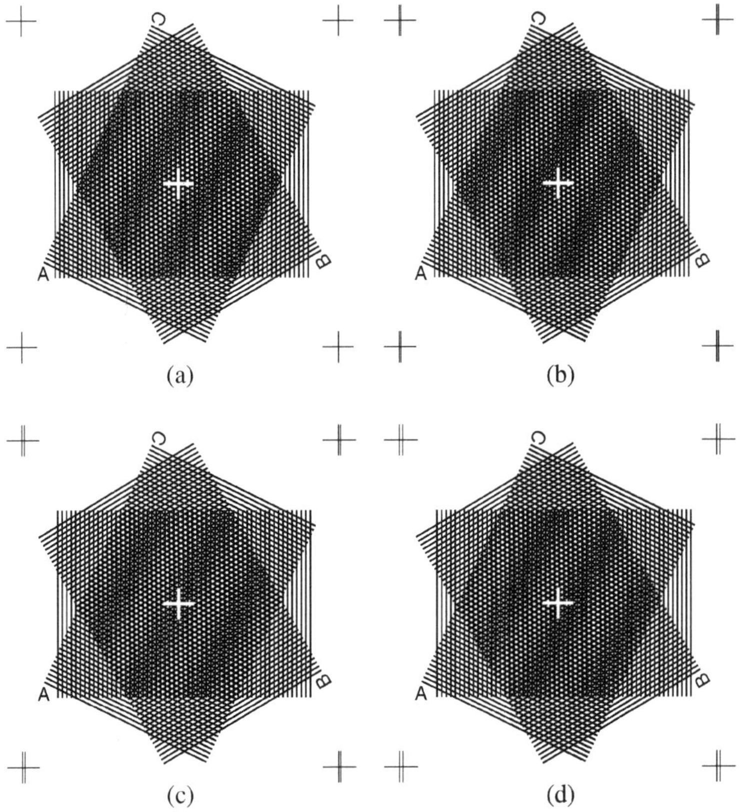


**Figure 7.5:** The 3-grating (1,1,1)-moiré of Fig. 2.8(h),(i) and its phase shifts due to a shift in grating A. The origin is indicated in each image by a white cross. (a) The three gratings and the moiré are in their initial phase. (b) Grating A is shifted by 1/4 period to the right; consequently, the moiré is shifted 1/4 moiré-period in the main direction of the moiré. (c) Grating A is shifted by 1/2 period to the right; consequently, the moiré is shifted 1/2 moiré-period in the main direction of the moiré. (d) Grating A is shifted by 3/4 period to the right; consequently, the moiré is shifted 3/4 moiré-period in the main direction of the moiré. The tiny shifts of grating A can be best perceived in the hairline crosses that surround each case.

But thanks to Eq. (7.28), which is satisfied for $\mathbf{a}_1,...,\mathbf{a}_m$ *iff* the shifts $\mathbf{a}_i$ are projections of the same vector $\mathbf{a}$ on the directions of $\mathbf{f}_i$, the phase component of the last expression:

$$-2\pi(k_1\mathbf{f}_1\cdot\mathbf{a}_1 + \ ... \ + k_m\mathbf{f}_m\cdot\mathbf{a}_m) \tag{7.30}$$

can be rewritten, under the same conditions, in the form:

$$-2\pi(k_1\mathbf{f}_1 + \ ... \ + k_m\mathbf{f}_m)\cdot\mathbf{a} \tag{7.31}$$

Hence, under these conditions — the conditions of Proposition 7.3 — Eq. (7.29) becomes:

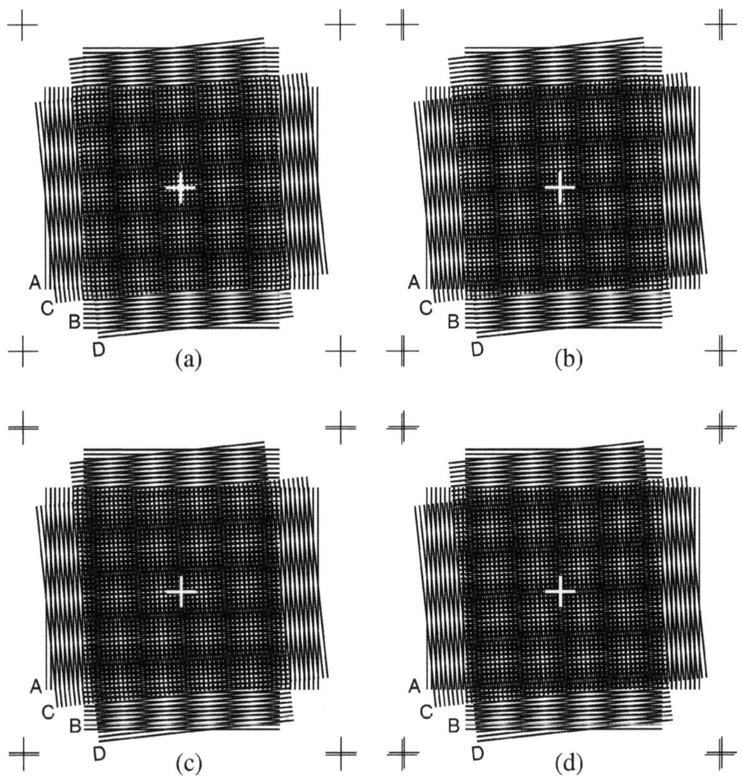


**Figure 7.6:** The combined ((1,0,-1,0),(0,1,0,-1))-moiré between two line grids (=four gratings), and its phase shifts due to a shift in the first grid (gratings A,B). (a) Both grids and the two perpendicular moirés are in their initial phase. (b) The first grid (gratings A,B) is shifted by 1/2 period to the right; consequently, the moiré is shifted 1/2 moiré-period downward. (c) The first grid (gratings A,B) is shifted by 1/2 period upward; consequently, the moiré is shifted 1/2 moiré-period to the right. (d) The first grid (gratings A,B) is shifted by 1/2 period to the right and 1/2 period upward; consequently, the moiré is shifted 1/2 moiré-period downward and 1/2 period to the right. The tiny shifts of the first grid can be best perceived in the hairline crosses that surround each case. See Remark 7.1 and Problem 7-19 for the case of parallel grids (i.e. $\alpha = 0$).

$$= \sum_{n_1=-\infty}^{\infty} \ldots \sum_{n_m=-\infty}^{\infty} c^{(1)}_{n_1} \cdot \ldots \cdot c^{(m)}_{n_m} \; e^{i2\pi(n_1\mathbf{f}_1 + \ldots + n_m\mathbf{f}_m)\cdot\mathbf{x} \, - \, i2\pi(n_1\mathbf{f}_1 + \ldots + n_m\mathbf{f}_m)\cdot\mathbf{a}}$$

$$= \sum_{n_1=-\infty}^{\infty} \ldots \sum_{n_m=-\infty}^{\infty} c^{(1)}_{n_1} \cdot \ldots \cdot c^{(m)}_{n_m} \; e^{i2\pi(n_1\mathbf{f}_1 + \ldots + n_m\mathbf{f}_m)\cdot(\mathbf{x} - \mathbf{a})}$$

$$= p_1(\mathbf{x}-\mathbf{a}) \cdot \ldots \cdot p_m(\mathbf{x}-\mathbf{a})$$

where **a** is the global shift of the entire superposition.

Note that in this case (and only in this case!) the period-shift of each individual layer can be expressed either by $\phi_i = \mathbf{f}_i \cdot \mathbf{a}_i$ (using Eq. (7.9)) or by $\phi_i = \mathbf{f}_i \cdot \mathbf{a}$.

As an example to illustrate Proposition 7.3, note that in the case of $m > 2$ gratings shifting just one layer (by a non-trivial shift) does not correspond to a global shift of the entire superposition, and it rather modifies the microstructure of the superposition. It is also interesting to note that in the case of $m = 2$ gratings Proposition 7.3 can be simplified as follows: Non-trivial shifts of $\mathbf{a}_1$ in $p_1(\mathbf{x})$ and of $\mathbf{a}_2$ in $p_2(\mathbf{x})$, in their respective main directions, correspond to a shift $\mathbf{a}$ of the entire superposition $p_1(\mathbf{x}) \cdot p_2(\mathbf{x})$ *iff* $p_1(\mathbf{x})$ and $p_2(\mathbf{x})$ (and hence, their periods $\mathbf{T}_1$ and $\mathbf{T}_2$) are not collinear. Therefore, if the two gratings are not collinear, any individual shifts $\mathbf{a}_1$, $\mathbf{a}_2$ will always correspond to a shift $\mathbf{a}$ of the entire superposition.

The simplest example of layer shifts which do not give a rigid motion of the superposition but rather a microstructure modification is the (1,-1)-singular superposition of two identical gratings with an angle difference of $\alpha = 0$ (see Fig. 7.7): When the two gratings are superposed in-phase the lines of the two original gratings fall on top of each other, and the superposition is identical to the original gratings. But when one of the gratings is shifted by half a period the lines of the two original gratings fall between each other, thus filling the white spaces better and giving a darker superposition.

We will return to this subject again in detail (but from a different angle) in Chapter 8, where we will also obtain a further refinement of Proposition 7.3.

Note that Propositions 7.2 and 7.3 may seem at first sight contradictory: Suppose that $p_1(\mathbf{x})$ and $p_2(\mathbf{x})$ are two superposed gratings (with non-collinear periods $\mathbf{T}_1$, $\mathbf{T}_2$). How can it be that shifts of $\mathbf{a}_1$ and $\mathbf{a}_2$ in $p_1(\mathbf{x})$ and $p_2(\mathbf{x})$ cause a shift of the entire superposition by $\mathbf{a}$, while, on the other hand, each of the different $(k_1,k_2)$-moirés which are present in that superposition undergoes a different shift and in a different direction? In fact, as we can also see in Figs. 7.3 and 7.4, there exists no contradiction here: The entire superposition is indeed shifted by the same vector $\mathbf{a}$, but as we have seen in Sec. 7.4 (see Fig. 7.1), this common shift is then "translated" into the proper language of each individual moiré by considering its projection on the moiré's own main direction.

The strength of Proposition 7.2 is that it remains true in *all* cases: each of the $(k_1,...,k_m)$-moirés in the superposition is individually shifted (without changing its period, angle, or profile form), even when the shifts of the individual gratings do not correspond to a rigid motion of the superposition as a whole but rather modify its microstructure. This is illustrated, for instance, in Fig. 7.6.

Proposition 7.3 can be also formulated, thanks to the shift theorem, in terms of the phase increment in the impulses of the spectrum-convolution: Shifts of the individual gratings cause a rigid motion of the entire superposition *iff* the phase increments in the impulses of the spectrum-convolution due to the shifts are *linear* with respect to the frequency (see also Appendix C.4).

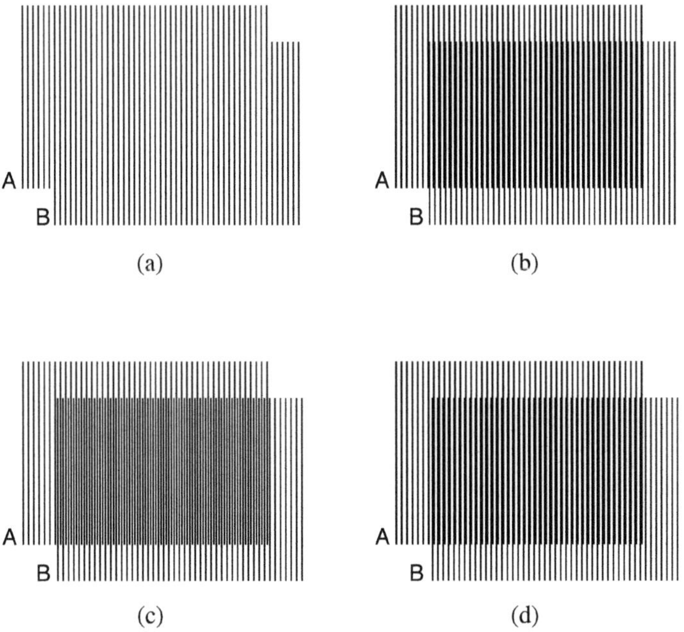


**Figure 7.7:** The superposition of two identical gratings with angle difference of $\alpha = 0$: (a) Both gratings are in their initial phase. (b) The first grating is shifted by 1/4 period to the right. (c) The first grating is shifted by 1/2 period to the right. (d) The first grating is shifted by 3/4 period to the right. As we can clearly see, these layer shifts do not cause a rigid motion of the superposition, but rather change its microstructure.

Finally, if we consider each of the $(k_1,...,k_m)$-moirés in itself as a grating (or a 1-fold periodic function), and their superposition as a combined moiré (see Sec. 2.8), then the same reasoning of Proposition 7.3 applies also to combined moirés. As we have seen, non-trivial shifts in the superposed gratings cause each $(k_1,...,k_m)$-moiré in itself to be rigidly shifted in the superposition. A combined moiré composed of two $(k_1,...,k_m)$-moirés will undergo a rigid motion *iff* its two moirés are not collinear (as in Fig. 7.5). In the general case of more than two $(k_1,...,k_m)$-moirés, the combined moiré will undergo a rigid motion *iff* the shifts $\mathbf{b}_i$ of its individual moirés are all projections of the same vector $\mathbf{b}$ on the main directions of the moirés. Otherwise, the individual rigid motions of each of the $(k_1,...,k_m)$-moirés can no longer be combined into a rigid motion of the full combined moiré.

Note that in our discussion in the present chapter we have extended the scope of the term "phase" from its classical context, the phase of 1D or 2D periodic functions, to the *superposition* of such periodic functions, which may already belong to the realm of almost-periodic functions. However, in this larger context the phase loses its original meaning that was tightly related to the concept of the period; it no longer automatically means a rigid motion of the function, and it may, rather, involve a change in its internal structure (which we will call in Chapter 8 the *microstructure*).

## PROBLEMS

7-1. *Measurement of microscopic displacements.* Assume that the two gratings in Example 7.1 are parallel (i.e., the angle between them is $\alpha = 0$) and that their periods are $T_1 = 100$ µm, $T_2 = 110$ µm.

(a) Use Eq. (2.11) to find the period $T_{1,-1}$ of the (1,-1)-moiré which is generated in the superposition. What is the angle of the moiré bands?

(b) Assume that the first grating is shifted to the right by half a period, i.e., by $a = 50$ µm. What happens to the moiré bands?

(c) Show that in the case of $\alpha = 0$ (i.e., two parallel gratings) Eq. (7.26) is reduced into:

$$b_{1,-1} = a_1\frac{T_{1,-1}}{T_1} \qquad \text{or:} \qquad b_{1,-1} = -a_2\frac{T_{1,-1}}{T_2}$$

depending on which of the two gratings has been shifted. These expressions show that the shift of the moiré is a magnification of the shift of the original grating, the magnification rate being the ratio of their periods. What is the meaning of the + or – signs?

(d) Since very small displacements of the gratings cause large shifts of the moiré bands, this can be utilized to measure extremely small movements. Based on this idea, can you design a method for high precision measuring of microscopic displacements? What is the displacement magnification power in the present example? How would you further improve the precision of this method? Can the magnification be increased to any arbitrarily large value? Explain.

7-2. *Precision alignment.* Based on Example 7.1, propose a precision alignment method using two line gratings. Would you choose for this application two gratings with identical periods and angles? Explain.

7-3. Can you upgrade the precision alignment method of the previous problem into a 2D aligning method by using two line-grids instead of two line gratings? Would you prefer to use two dot-screens rather than two line-grids? Explain.

7-4. *High precision measurement of distances of translation.* Suppose that we want to make accurate measurements of the translation of a carriage (in a plotter, a milling machine, etc.) along its longitudinal way. We may use for this end two gratings with a slightly different period, one which is fixed along the way, and the other fixed to the carriage. A photocell on the carriage detects the moiré bands as the carriage moves along its way, as well as their intensity level (this is required in order to find out the precise position within one moiré period). How can you calculate the position of the carriage along its way, and the distance it travelled? How would you distinguish between movements in the two opposite directions? (See, for example, [Post94 pp. 104–105].)

7-5. *Measurements of refractive index (refractometry).* A simple refractometer could consist of a thin rectangular cell whose opposite walls are two identical binary gratings with a small angle difference $\alpha$. Light passing through the two gratings generates a moiré effect whose period and angle are determined by Eq. (2.10). When the material to be effect whose period and angle are determined by Eq. (2.10). When the material to be examined is introduced between the two gratings, the change in the refractive index slightly deviates the light passing through the cell, and the image of the first grating is slightly displaced with respect to the second grating. This slight displacement causes a large shift of the moiré bands with respect to their original position, in accordance with Eq. (7.26) (see Example 7.1). How can you use this technique to quantitatively measure the refractive index of a given material? (More on this subject can be found, for example, in [Nishijima64].)

7-6. *Moiré kinematics.* Suppose that two transmission gratings (i.e., gratings of opaque black lines on a transparent support) having slightly different periods $T_2 > T_1$ are superposed with angle difference of $\alpha = 0$, like in problem 7.1. When the first grating is translated on top of the other, what one sees is moiré bands that move by much faster than the moving grating itself (in what direction?). Using Eqs. (2.11) and (7.26) deduce the velocity of the moiré bands as a function of the velocity of the moving grating. Which other parameters will influence the moiré velocity? And what happens if $T_2 < T_1$? Find an application to this velocity magnification property of the moiré.

7-7. *Moiré kinematics (continued).* Suppose that two identical transmission gratings (with $T_1 = T_2$) are superposed with angle difference of $\alpha = 0$. What happens to the moiré bands when one of the gratings is rotated on top of the other? Using Eqs. (2.10) and (7.26) deduce the radial velocity and the angular velocity of the moiré bands as a function of the angular velocity of the rotating grating. Which other parameters will influence the moiré velocity? Do we obtain in this case a magnification of the angular velocity of the moiré bands with respect to the angular velocity of the rotating grating? (See, for example, [Theocaris73 p. 988]; do you obtain the same results?)

7-8. *Latent images.* This is an application of the simplest singular case, the singular (1,-1)-moiré between two identical gratings. Let A and B be two identical gratings with period *T*. Assume that within the borders of a given shape, say, a triangle, the parallel lines of grating A are laterally shifted by half a period (see Fig. 7.8(a)). Since the gray level outside the triangle is identical to the gray level inside it, the presence of the triangular shape within grating A is not easily detected by the eye (look at Fig. 7.8(a) from a distance of two meters or more; this shows how Fig. 7.8(a) would be perceived from a normal reading distance if the grating were printed in a much higher frequency, which is beyond the visibility limits of the eye). Now, assume that grating B having the same period *T* but no latent images (see Fig. 7.8(b)) is printed on a transparency and superposed on top of grating A at an angle difference of $\alpha = 0$. Fig. 7.8(c) shows the superposition when the two gratings are superposed in-phase, and Fig. 7.8(d) shows the superposition when grating B is shifted by half a period. As we can see, thanks to grating B the shape of the latent image hidden in grating A becomes clearly visible.

(a) Explain this phenomenon in terms of the microstructure modifications which occur in the grating superposition resulting from layer shifts in the (1,-1)-singular moiré state between two gratings (see Fig. 7.7).

(b) How would the latent image be perceived when grating B is shifted by a quarter of a period? More generally, how would the latent image evolve when grating B is slowly shifted to the right across a few periods?

(c) What would happen in Figs. 7.8(c),(d) if the latent image within grating A were shifted by $T/4$ rather than by $T/2$?

(d) What would happen in Figs. 7.8(c),(d) if gratings A and B had different openings $\tau_1 \neq \tau_2$, while still having the same equal period $T$?

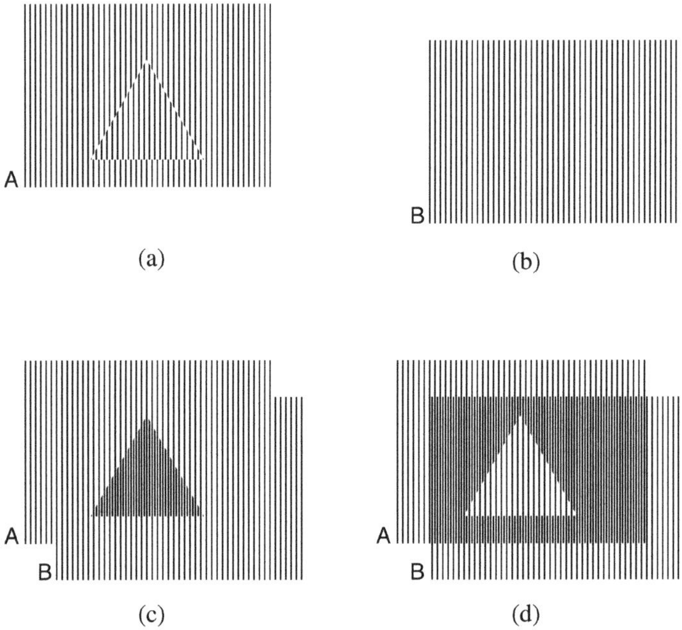


**Figure 7.8:** A latent image and its detection in the superposition of two identical gratings with angle difference of $\alpha = 0$ (see Problem 7-8): (a) Grating A and its latent image. (b) Grating B, the detecting layer. (c) In-phase superposition of grating B on top of grating A. (d) The superposition when grating B is shifted by half a period to the right. Compare with Fig. 7.7.

7-9. *Document security.* How can this phenomenon be used for document authentication? (Hint: assume that layer A figures with its latent image on the document itself, and that layer B serves as a detecting device.) If the grating frequency used, $f = 1/T$, is sufficiently high, the gratings themselves will not be visible by the naked eye and they will appear as a constant gray area, but the latent image will be clearly visible once the detecting device is laid on top of the document. Owing to its high frequency, layer A and its latent image will be reproduced by any photocopier as a constant gray area; what will be seen when the detecting layer B is laid on a photocopy of the original document?

7-10. *Multiple latent images in colour printing.* The method described in the previous problems can be extended also to the 2-fold periodic case of dot-screens, i.e., to the singular (1,0,-1,0)-moiré between two dot-screens with identical periods and angles. Let A and B be two identical regular screens with period $T$ to both directions, and assume that within the borders of a given shape inside dot-screen A (say, a triangle, like in the previous problems) the screen dots are shifted by half a period to both directions. Here, again, the presence of the latent image within screen A is not easily detected by the eye, but when a detecting transparency consisting of screen B (of the same period $T$) is superposed on top of screen A at the same orientation, the latent image becomes clearly visible.

(a) How can this method be used in black and white images printed with a standard halftone screen at 45° (see Sec. 3.2)? Does the size of the halftone dots, which varies according to the gray level in the different regions of the image, influence the perception of the latent image through the detecting screen? Can the latent image be perceived in areas where the image is completely black, or completely white? Can the perception of the latent image be improved by compressing the dynamic range of the printed image, for example into the range [0.2...0.8] rather than [0...1]?

(b) How can this method be used in colour printing using CMYK halftone screens for hiding several different latent images within the same area of the document? Assuming the conventional screen combination described in Sec. 3.2, and a detecting screen which is printed in black ink on a transparency, what will be seen as the detecting screen is rotated on top of the colour image (the document), at what angles and in what colours?

7-11. *Scrambled Indicia®*. In 1997 the U.S. Postal service started issuing stamps incorporating hidden text or images intended to deter counterfeiters and to offer an interesting design element for collectors. The first stamp incorporating this feature was the 32¢ stamp issued in September 1997 to commemorate the 50th anniversary of the U.S. Air Force, which contains the hidden text "USAF". This text is not visible to the naked eye, but can be viewed by using a special decoder lens, available through the U.S. Postal Service, consisting of a transparent microlens grating. When the decoder lens is positioned on the stamp at the appropriate angle, the latent image becomes visible.

(a) How does the use of a 1-fold periodic grating microlens rather than a 2-fold periodic dot-screen as a decoder influence the method described in the previous problem? Assuming the conventional screen combination of Sec. 3.2, how many different latent images can be encoded into the same area of the stamp?

(b) Can you think of a hierarchical security system with a public decoder that detects one message hidden on the stamp and a private decoder for the use of the authorities only, that detects another, secret message using a different grating frequency? Is such a security system compatible with the conventional screen combination traditionally used in colour printing? Is it compatible with a stable moiré-free screen combination (see Sec. 3.5 and Fig. 3.8)? Explain.

**Note:** *Scrambled Indicia®* is a registered trademark of Graphic Security Systems Corporation, and it is protected by U.S. Patent no. 5,708,717 [Alasia98]. More information on this method can be found in GSSC's internet site at the address `http://www.graphicsecurity.com`, and in the patent description.

7-12. Is it possible, using one of the methods described in the previous problems, to encode a full-colour latent image in a document? Is the colour of the observed latent image stable? Can you design, for example, a latent image with the Red Cross flag, consisting of a red cross on a white background? What happens to the foreground and background colours of the latent image as the detecting device is slightly shifted on top of the document?

7-13. All the latent image methods described in the preceding problems are based on a singular state of the superposition. What is the angular tolerance in such cases, namely, the maximum angular error permitted in positioning the detecting transparency on top of the document? What happens to the latent image when the detecting transparency is positioned with an angle deviation slightly exceeding this limit?

7-14. Can similar latent image methods be devised for non-singular superpositions, too?

7-15. Is it possible, using the method described in Chapters 4 and 9, to encode a full-colour hidden image in a document? Compare with the method described in the previous problems according to the following guidelines:

(a) The possibility to generate any hidden message, periodic or not (e.g., a single occurrence of the hidden image).

(b) The size and the orientation of the viewed hidden image.

(c) The stability and fidelity of the colour of the viewed hidden image.

(d) The tolerance of the method to slight misalignments: is the hidden image still visible when the detecting device is positioned on the document with a slight angle error?

(e) The behaviour of the viewed hidden image when the detecting device is laterally shifted on the document.

7-16. Give an expression analog to Eq. (7.28) for the case where the singular superposition of the $m$ 1-fold periodic layers is 1-fold periodic (see Eq. (6.12)).

7-17. Suppose that each of the original layers in the superposition undergoes a combination of rotation, scaling and shift transformations. What happens to the moiré effects in the superposition? Can you devise a general rule for the behaviour of the $(k_1,...,k_m)$-moiré when each of the $m$ superposed gratings undergoes an affine transformation? (This is, in fact, a particular case of a more general question that will be treated in Chapter 10.)

7-18. Fig. 7.9 shows the superposition of two parallel periodic line gratings, the second grating (B) having a slightly bigger period (i.e. $T_2 > T_1$), as the first grating (A) is slightly being shifted to the right. Explain the behaviour of the (1,-1)-moiré in the superposition, and its shift magnification property. What happens if $T_2 < T_1$?

7-19. What happens in Fig. 7.6 when $\alpha = 0$ and $T_2 > T_1$? And when $\alpha = 0$ and $T_2 < T_1$?

7-20. What happens in Figs. C.23 and C.24 as the pinhole screen is being shifted to the left?

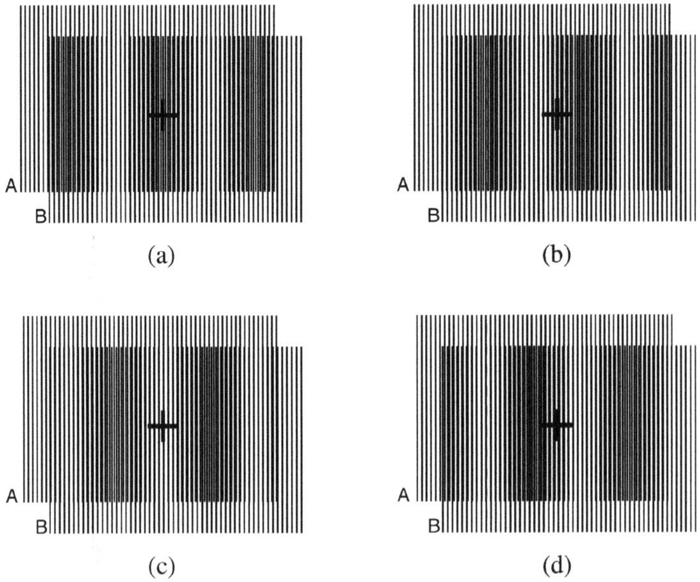


**Figure 7.9:** A (1,-1)-moiré between two parallel line gratings. See Problem 7-18.

# Chapter 8

## Macro- and microstructures in the superposition

### 8.1 Introduction

As we have already seen earlier in Sec. 2.12 and in Chapter 7, when periodic layers (gratings, grids, dot-screens, etc.) are superposed, new structures of two distinct levels may appear in the superposition, which do not exist in any of the original layers: the *macrostructures* and the *microstructures*.

The macrostructures, i.e., the moiré patterns proper, are, of course, the most prominent; being much coarser than the detail of the original layers they are clearly visible even when observed from a distance. The microstructures, on the contrary, are almost as small as the periods of the original layers (typically, just 2–5 times larger), and therefore they are only visible when examining the superposition from a close distance or through a magnifying glass. These tiny structures are also called *rosettes* owing to the various flower-like shapes they often form in the superposition of dot-screens [Yule67 p. 339].

However, in spite of their tiny size, the microstrucrures which occur in the superposition are very rich in detail, and their study appears to be no less fascinating than the study of the macrostructures. As we can see in Figs. 8.1 and 8.2, quite attractive rosette-forms often appear in the superposition, and a look through a magnifying glass may reveal an amazing, subtle and delicate micro-world, full of surprising geometrical forms.

As we will see in this chapter, macrostructures and microstructures may coexist in the same superposition. However, while microstructures exist practically in any superposition, except for the most trivial cases, macro moiré effects are not always present (*cf.* stable and unstable moiré-free states). In fact, we will see in Sec. 8.4 that the macrostructures, whenever they exist, are constructed from the microstructures of the superposition.

In the present chapter we will investigate the microstructures generated in the superposition of periodic layers and their properties both in the image domain and in the spectral domain. Note that our approach in this chapter is completely general, and not only limited to the rosette morphology in the case of the classical 3-screen superposition used for colour printing, which has already been studied in [Yule67], [Delabastita92 pp. 57–59], and [Daels94]. We start by describing the behaviour of the microstructure in different superposition cases: In Secs. 8.2–8.3 we discuss the rosettes in singular states; in Sec. 8.4 we explain what happens to the rosettes slightly off the singular state, and in Sec. 8.5 we describe the microstructure in stable moiré-free superpositions. Then, in the following sections, we proceed to the formal explanation of these phenomena.

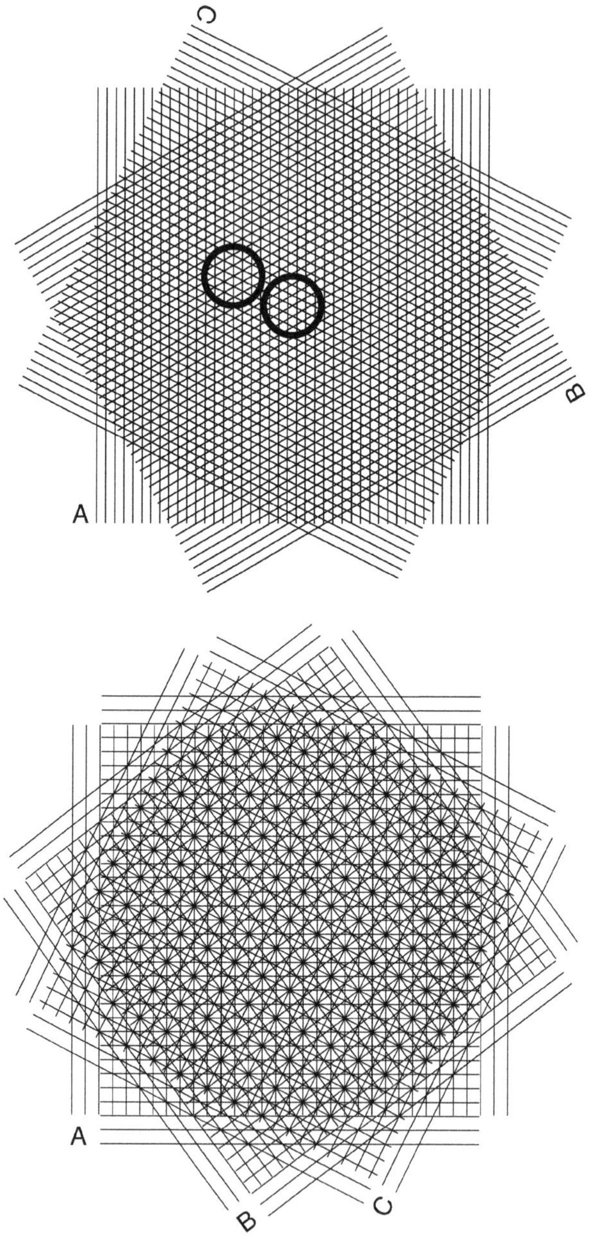


(a)

(b)

**Figure 8.1:** The superposition of periodic layers may yield very spectacular microstructures (rosettes). (a) A magnification of the 3-grating superposition of Fig. 2.8(h). Note the star-like rosettes which form the bright areas of the macro-moiré and the triangular microstructure which forms the darker areas. (b) A magnification of a singular superposition of 3 grids (=6 gratings) with $\theta_1 = 0°$, $\theta_2 = 36.8699°$, $\theta_3 = 63.4349°$, $T_1 = T_2$, $T_3 = 1.118\,T_1$. This is an example of a periodic, singular superposition.

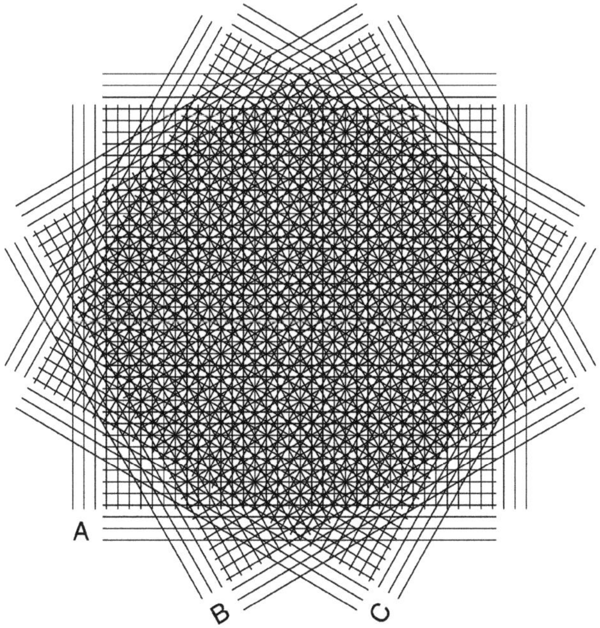


(a)

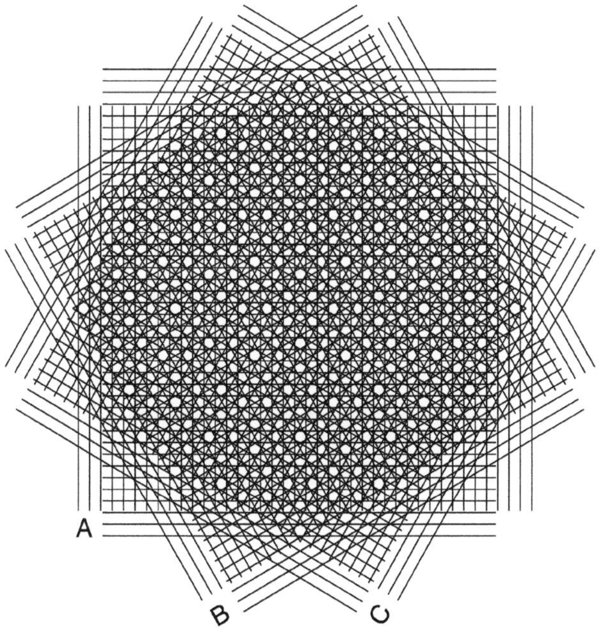


(b)

**Figure 8.2:** A magnification of the superposition of 3 grids (=6 gratings) with identical periods and equal angle differences of 30°. This is an example of an almost-periodic, singular superposition. In (a) all the grids are superposed in their initial phase, while in (b) the grid A has been shifted by half a period in both directions; note the substantial change in the form of the microstructure due to this shift.

**Remark 8.1:** Since the microstructures (rosettes) are most neatly perceived in the superposition of dot-screens (see Fig. 2.10), and also because of our special interest in screens in the context of colour printing, we will prefer, wherever possible, to illustrate our discussion using examples of dot-screen superpositions.[1] Note also that for the sake of convenience we will usually use in this chapter screens with circular dots, and we will adopt the convention that in the initial phase of each individual layer (see Sec. 7.2) a 0-valued element (i.e., a black dot or line) is centered on top of the origin. This will facilitate the visual illustration of matching points between the different layers. ■

## 8.2 Rosettes in singular states

Let us start with the microstructure in moiré-free singular cases, where the superposition looks uniform and no macro-moirés are visible. Since in these cases the only structure which appears in the image domain is the microstructure, it is clear that their spectra only represent the microstructure. Such cases will serve us as a starting point for studying the spectral representation of the microstructure. The microstructure in the case of stable moiré-free superpositions will be discussed later, in Sec. 8.5.

We remember that each impulse in the spectrum of a singular state is in fact a compound impulse representing a full cluster of impulses which has collapsed into a single location. According to the algebraic structure of the compound spectrum we can distinguish here between two types of singular cases: singular cases in which the spectrum support is a discrete lattice and the layer superposition is periodic; and singular cases in which the spectrum support is a dense module and the layer superposition yields an almost-periodic image. It has been shown in Chapter 5 and in Proposition 6.1 that the first case occurs when: $m > \text{rank}_{\mathbb{Z}}\, \text{Md}(\mathbf{f}_1,...,\mathbf{f}_m) = \dim \text{Sp}(\mathbf{f}_1,...,\mathbf{f}_m)$, while the second case occurs when: $m > \text{rank}_{\mathbb{Z}}\, \text{Md}(\mathbf{f}_1,...,\mathbf{f}_m) > \dim \text{Sp}(\mathbf{f}_1,...,\mathbf{f}_m)$; in both cases the inequality on the left is the condition for a superposition to be singular. We will illustrate the first case by the singular (1,2,-2,-1)-moiré between two identical screens with angle difference of $\alpha = \arctan \frac{3}{4} \approx 36.87°$; and the second case by the singular {1,1,1}-moiré between three identical screens with equal angle differences of 30° (i.e., the conventional singular screen combination traditionally used in colour printing).

### 8.2.1 Rosettes in periodic singular states

Let us consider the microstructure which occurs in a periodic singular case, such as the (1,2,-2,-1)-singular superposition of two screens (Fig. 8.3). As we can see, the superposition in this case is periodic, and the rosettes are ordered in a perfectly repetitive pattern. And indeed, the spectrum of this superposition is a compound nailbed (Fig. 8.3(a)), where each impulse represents a collapsed cluster. Since the only structure which

---

[1] Of course, no dot-screen equivalent exists for superpositions with an odd number $m$ of gratings, like the case of $m = 3$ in Fig. 8.1(a).

appears here in the superposition is the microstructure, it is clear that this nailbed represents the periodic microstructure of the superposition. And indeed, the two fundamental (compound) impulses of this nailbed (whose frequency vectors **g** and **h** are a basis of the lattice-of-clusters in the spectrum support; see Sec. 6.7) determine the frequency and the direction of the microstructure in the image domain. In our example of the (1,2,-2,-1)-singular state the frequency of the microstructure is, by the Pythagoras theorem, $g = f_1/\sqrt{5}$ (see Fig. 8.3(a)), and hence its period is $\sqrt{5} \approx 2.236$ times larger than the screen period; its orientation is $\varphi = \arctan(2) \approx 63.435°$ with respect to $\mathbf{f}_1$.[2]

It is interesting to note that the same lattice (spectrum support) as in Fig. 8.3(a) can be also obtained in other singular two-screen superpositions having other sets of frequency vectors. Consider, for example, the (2,0,-1,2)-singular screen superposition whose frequency vectors are $\mathbf{f}_1,\mathbf{f}_2,\mathbf{f}'_3,\mathbf{f}'_4$ as shown in Fig. 8.4. The spectrum obtained in this case is a compound nailbed with the same spectrum support as in Fig. 8.3(a), but the impulse clusters which collapse onto each of the compound impulses in these two cases are not the same. Therefore, in spite of their common spectrum support these spectra are different — since their respective compound impulses have different amplitudes (being the amplitude sums of different impulse families). And indeed, although these two singular 2-screen superpositions have in the image-domain the same microstructure periodicity, the rosette shapes within their periods are different (compare Figs. 8.3 and 8.4).

It should be noted that there exist, in fact, infinitely many different "equi-support" singular superpositions; each of them is obtained by a different choice of the basis vectors $\mathbf{f}_1,\mathbf{f}_2,\mathbf{f}_3,\mathbf{f}_4$ from among the points of the same lattice (nailbed support) in the spectrum.

### 8.2.2 Rosettes in almost-periodic singular states

Let us now consider the microstructure obtained in an almost-periodic singular case, such as the classical singular 3-screen superposition used for colour printing (see Fig. 8.5). Obviously, in this case there is no rosette periodicity in the superposed image. Rather, like in Fig. B.1 of Appendix B, we can detect here in the image domain "almost" periodicities, and the rosette forms are only almost-repetitive. This explains the fuzzy and elusive look of the microstructure in this case: looking at any location in the superposition, the eye is tempted at first to believe that the rosette structures are repetitive; but after a deeper examination it realizes that this repetition is just an illusion.

For example, let us look carefully at the almost-periodic rosette pattern of Fig. 8.5(b), in which the three screens are superposed in-phase (i.e., they have a common dot at the origin). Clearly, apart from the origin, nowhere else in the superposition does there occur again a precise 3-screens dot match (otherwise the superposition would be periodic). But

[2] Obviously, the period of the microstructure is always greater than or equal to the original screen periods: Since the impulses of the original screen frequencies are included in the compound nailbed, it is clear that the fundamental impulses of the compound nailbed can either coincide with the original screen frequencies (as in the singular (1,0,-1,0)-moiré or in the singular moiré shown in Fig. 5.7(a)), or fall even closer to the DC (as in the singular (1,2,-2,-1)-moiré; see Fig. 8.3(a)).

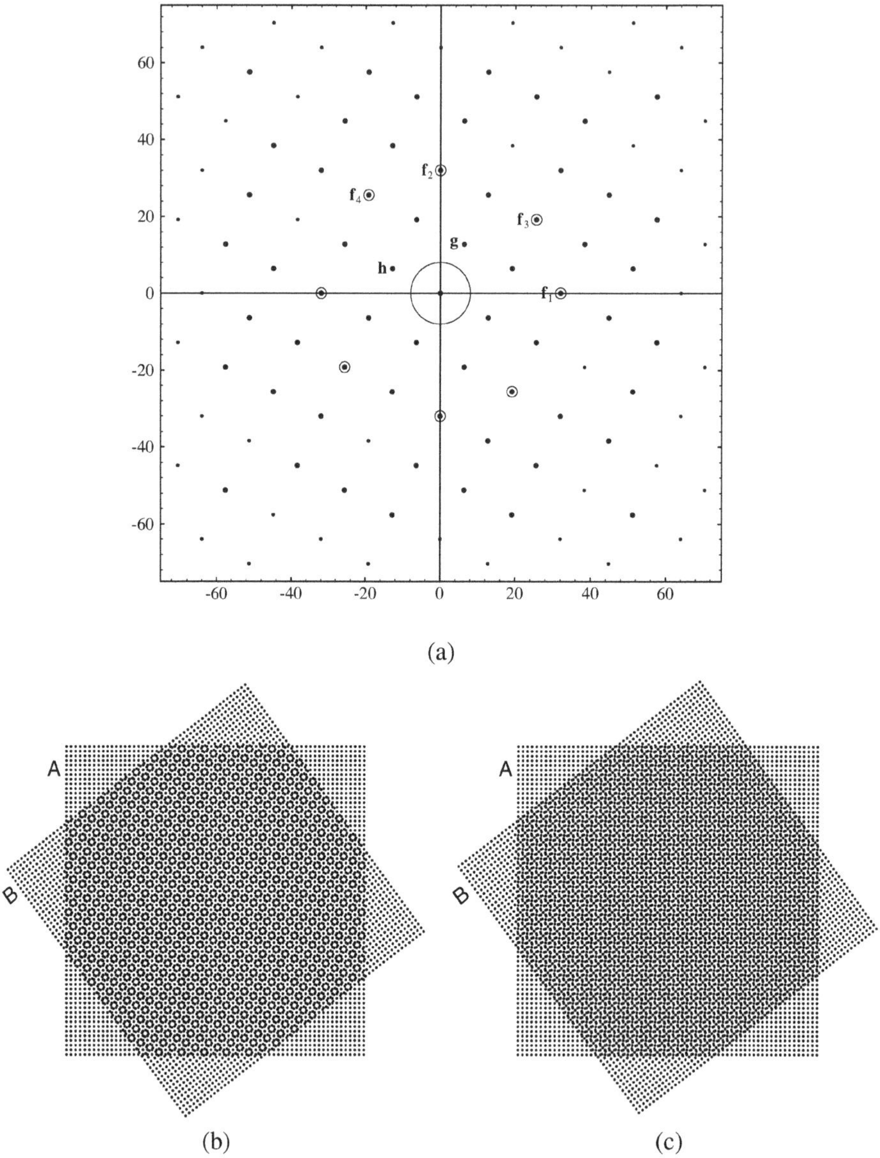


**Figure 8.3:** The singular (1,2,-2,-1)-state between two identical screens with angle difference of $\alpha = \arctan\frac{3}{4} \approx 36.87°$. (a) The spectrum support; **g** and **h** are the fundamental compound impulses of the microstructure. (b), (c) The screen superposition in the image domain: in-phase superposition in (b), and counter-phase superposition in (c). Note the uniformity of the microstructure, and the difference between the rosette shapes in (b) and (c).

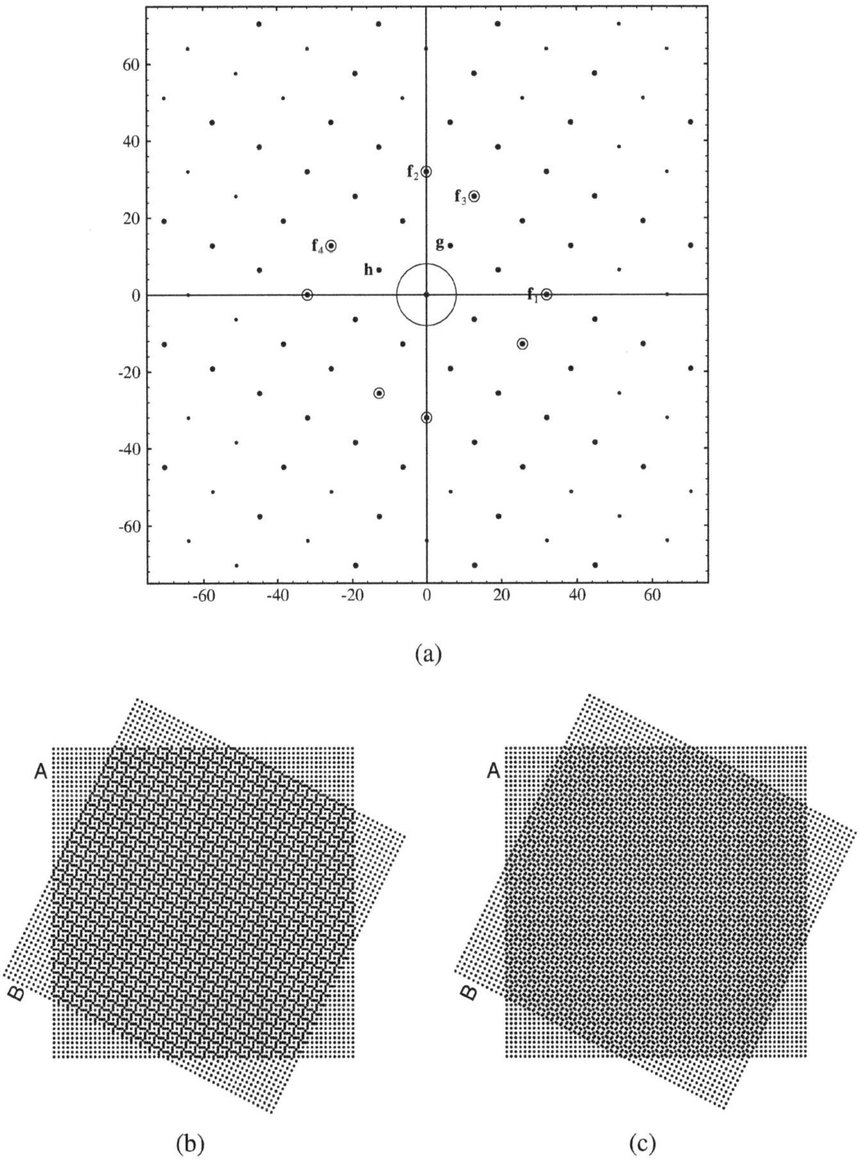


**Figure 8.4:** The singular (2,0,-1,2)-state between two screens with frequency ratio of 0.894 and angle difference of $\alpha = \arctan(2) \approx 63.435°$. (a) The spectrum support; **g** and **h** are the fundamental compound impulses of the microstructure. (b), (c) The screen superposition in the image domain: in-phase superposition in (b), and counter-phase superposition in (c). Note the uniformity of the microstructure, and the difference between the rosette shapes in (b) and (c).

at an infinite number of locations in the superposition there occurs an "almost" 3-screens dot match. The farther we go from the origin, the better the "almost" matches that we can find. This is, indeed, a characteristic property of almost-periodic functions.

In the spectral domain, the spectrum of an almost-periodic singular case is no longer a compound nailbed whose support is a discrete lattice, but rather a "forest" of compound impulses (each of which representing a collapsed cluster), whose support is a dense module (see Fig. 8.5(a)). And again, since the only structure which appears here in the superposition is the microstructure, it is clear that this "compound module" represents the almost-periodic microstructure of the superposition.

## 8.3 The influence of layer shifts on the rosettes in singular states

As we have seen in Chapter 7, shifts in the individual superposed layers may cause, depending on the case, either a global shift (a rigid motion) of the superposition as a whole, or a real modification in the microstructure of the superposition. Figs. 8.3(b),(c), 8.4(b),(c) and 8.5(b),(c) illustrate the microstructure modifications which occur due to such layer shifts in three different singular screen superpositions; Figs. 8.3 and 8.4 show cases in which the screen superposition is periodic, and Fig. 8.5 shows a case in which the superposition is almost-periodic. It is important to note that when the superposition is periodic, as in Fig. 8.3, the microstructure modifications that are caused by the layer shifts do not influence this periodicity or its orientation, but only the internal structure within each period (namely, the rosette shapes).

If we examine the forms of the rosettes which are generated as the phase of the original layers is being modified, we find two extreme types of rosettes, as well as all the possible intermediate types which occur between them. One extreme type occurs when the original layers are superposed "in-phase", i.e., when each layer has a black element (dot or line) centered on the origin; and the other extreme type occurs when the original layers are superposed in counter-phase. A gradual transition between these extreme rosette shapes occurs in the intermediate phase positions.

These two extreme rosette shapes are illustrated in Fig. 8.3(b),(c) for the case of the periodic (1,2,-2,-1)-singular moiré, in Fig. 8.4(b),(c) for the periodic (2,0,-1,2)-singular moiré, and in Fig. 8.5(b),(c) for the almost-periodic case of the classical 3-screen superposition with identical frequencies and angle differences.

The precise rosette shapes and their variations due to lateral shifts in the superposed layers are characteristic properties (like "fingerprints") of each particular singular state. Note that even "equi-support" singular cases have different rosette shapes, although their rosette periodicities are identical (see Figs. 8.3 and 8.4). Most famous are the rosette forms obtained in the classical superposition of three identical screens with equal angle differences; these rosette forms are well known in the printing industry and they have been

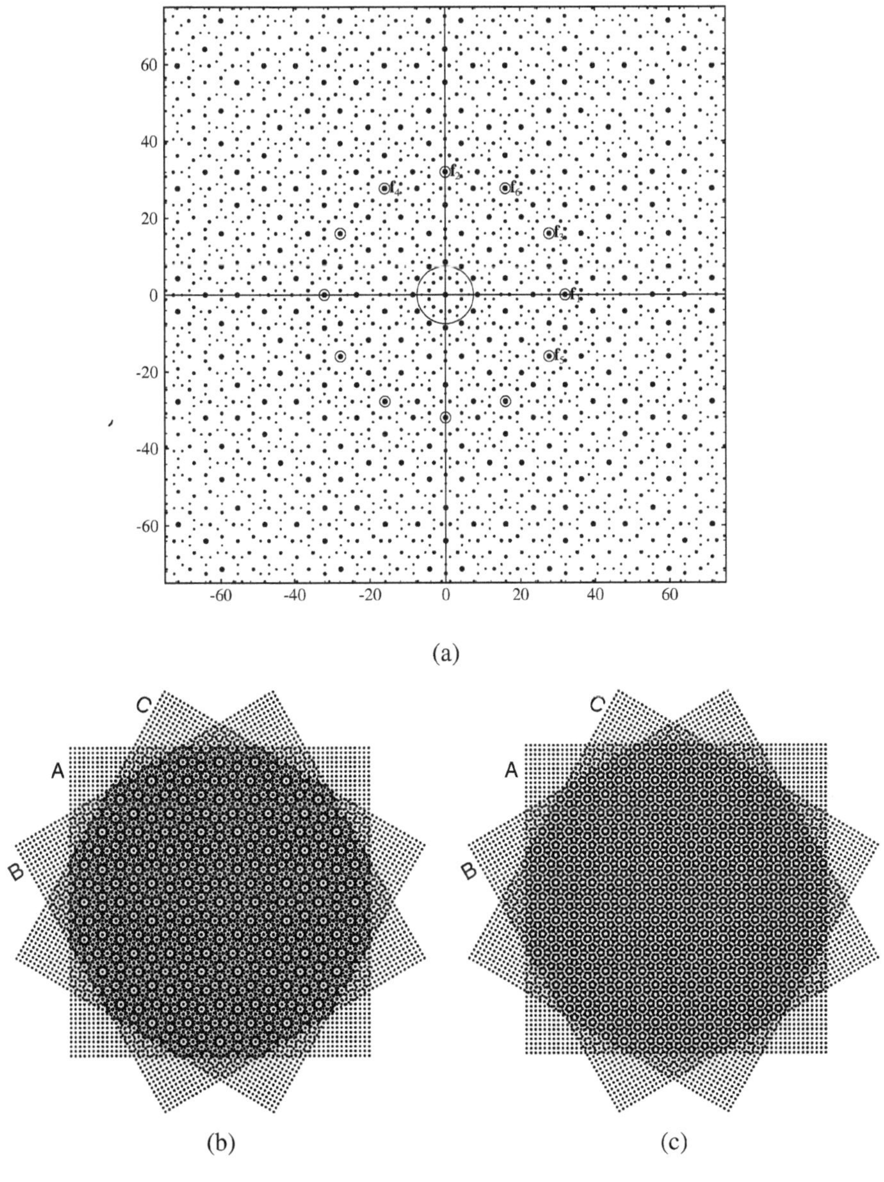


**Figure 8.5:** The singular {1,1,1}-state between three identical screens at equal angle differences of 30°. (a) The spectrum support (including only impulses up to order 3). (b), (c) The screen superposition in the image domain: in-phase superposition in (b), and counter-phase superposition in (c). Note the uniformity of the microstructure, and the difference between the rosette shapes in (b) and (c).

widely described in literature [Yule67 pp. 339–341; Delabastita92 pp. 57–59; Daels94]. As illustrated in Fig. 8.5(b), when the three screens are superposed in-phase, i.e., with a black dot centered on the origin, a perfect match of one screen-dot from each layer occurs at the origin. This generates at the origin a "dot-centered" rosette. Due to the almost-periodicity, "almost-perfect" copies of this dot-centered rosette can be found at any distance from the origin, thus generating a uniform microstructure with almost-dot-centered rosettes throughout. However, when the screens are superposed in counter-phase, a "clear-centered" rosette pattern is generated (see Fig. 8.5(c)). It is interesting to note also the similar behaviour of the microstructure in the equivalent case which consists of a superposition of line-grids rather then dot-screens; the "dot-centered" and the "clear-centered" rosettes generated in this case are clearly seen in Fig. 8.2(a),(b). It should be emphasized, however, that the rosette shapes obtained in other singular states may be completely different; and as we can see in the case of the singular (1,2,-2,-1)-moiré (Fig. 8.3(b),(c)), even the terms "dot-centered" and "clear-centered" may no longer be appropriate for the in-phase and counter-phase rosettes.

It is interesting to ask now how are such variations in the rosette shapes, due to layer shifts in the superposition, reflected in the spectrum of the singular case? And furthermore, why in some singular cases the difference between the two extreme rosette types is very significant, while in other singular cases the difference is hardly distinguishable? As we have seen earlier (Proposition 6.2), the amplitude of each impulse in the spectrum of a singular superposition (i.e., the amplitude of each compound impulse) is the sum of the amplitudes of all the individual impulses which collapsed onto the same location; see, for example, Eq. (6.7). On the other hand, we also remember that a shift in any of the superposed layers modifies the complex amplitudes of the impulses in its own spectrum. The answer to the above questions is found, therefore, in the way in which variations in the complex amplitudes of the individual impulses within each collapsed cluster influence the summed-up complex amplitude of the resulting compound impulse: in some cases the variations in the summed-up amplitudes may be significant, while in other cases they may be cancelled out. The variations in the complex amplitudes of the compound impulses due to the shifts in the superposed layers reflect, therefore, the variations in the rosette shapes as a function of the shifts $\mathbf{a}_1,...,\mathbf{a}_m$ in the individual layers.

## 8.4 The microstructure slightly off the singular state; the relationship between macro- and microstructures

At this point we come to one of the most interesting subjects in the behaviour of the micro- and the macrostructures in the superposition. As we already know, when we slightly move away from the singular state of a given moiré, this moiré becomes visible in the superposition in the form of a macro-moiré, with a large, visible period. Looking now at this superposition through a magnifying glass, we discover that in fact, the visible macro-structures are constructed from the microstructures of the superposition. The key point in the relationship between macro- and microstructures in the superposition can be stated as follows:

**Proposition 8.1:** When the microstructures of the superposition are similar and uniformly distributed throughout the superposed image, the resulting superposition looks from a distance uniform and smooth, and no moiré is visible (see, for instance, Figs. 8.3–8.5). However, if different types of rosettes are generated in alternate areas of the superposed image, the eye observes a different gray level in each of these areas (due to the different surface-covering rates of the dots in the different rosette types), and a macro-moiré becomes visible (see Fig. 2.8(h) and its magnification in Fig. 8.1(a), and Fig. 2.10). This is, indeed, the microscopic interpretation of the macroscopic moiré patterns. ■

However, this is not yet all. Looking carefully at the microstructure of any given macro-moiré, we discover that the relationship between the micro- and the macrostructures is even deeper than what is stated in Proposition 8.1. In fact, we have:

**Proposition 8.2:** The microstructure alternations which make up a macro-moiré are, to a very close approximation, nothing else but the microstructure forms which are obtained at the singular state of that macro-moiré by all possible phase shifts. The two extreme "in-phase" and "counter-phase" microstructures (e.g., the "dot-centered" and the "clear-centered" rosettes in the case of the classical 3-screen superposition) generate the two extreme intensity levels of the visible macro-moiré (its brightest and darkest areas), and the intermediate forms between them generate all the in-between intensity levels of the macro-moiré.[3] ■

This can be clearly illustrated for the (1,2,-2,-1)-moiré by comparing Fig. 8.6(b) with Figs. 8.3(b),(c), and for the classical 3-screen superposition by comparing Fig. 8.7(b) with Figs. 8.5(b),(c).

It should be emphasized, however, that Proposition 8.2 is only a close approximation. The reason is that as the angles or the frequencies are slightly modified in order to move our macro-moiré slightly away from its singular state, the microstructures are also slightly modified. However, the closer the macro-moiré is to its singular state, the better the approximation provided by Proposition 8.2.

## 8.5 The microstructure in stable moiré-free superpositions

Let us now consider in more detail the microstructures which occur in stable moiré-free superpositions, such as the superposition of two identical screens with an angle difference of 30° (see Fig. 8.8). Just like singular moiré-free states (Sec. 8.2), stable moiré-free superpositions have no visible macro-moirés, and they show a uniform-looking micro-

---

[3] Singular states in which there is no clear visual distinction between "in-phase" and "counter-phase" microstructures do not produce off the singular state a visible macro-moiré in the superposition. This often happens in moirés of high orders, or in moirés involving many superposed layers.

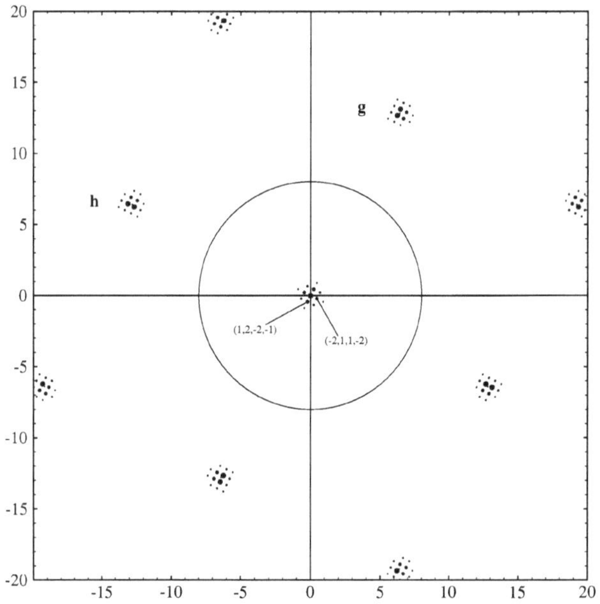


(a)

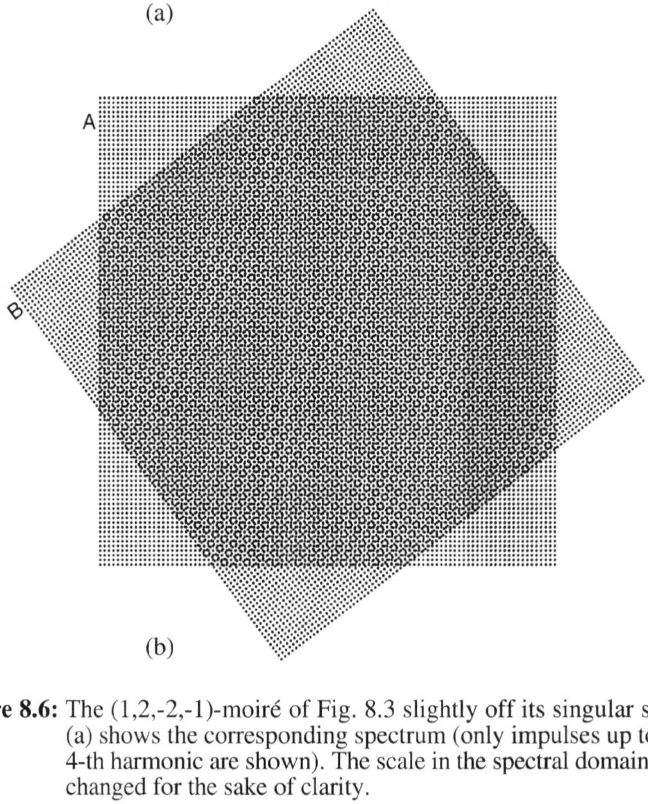


(b)

**Figure 8.6:** The (1,2,-2,-1)-moiré of Fig. 8.3 slightly off its singular state; (a) shows the corresponding spectrum (only impulses up to the 4-th harmonic are shown). The scale in the spectral domain was changed for the sake of clarity.

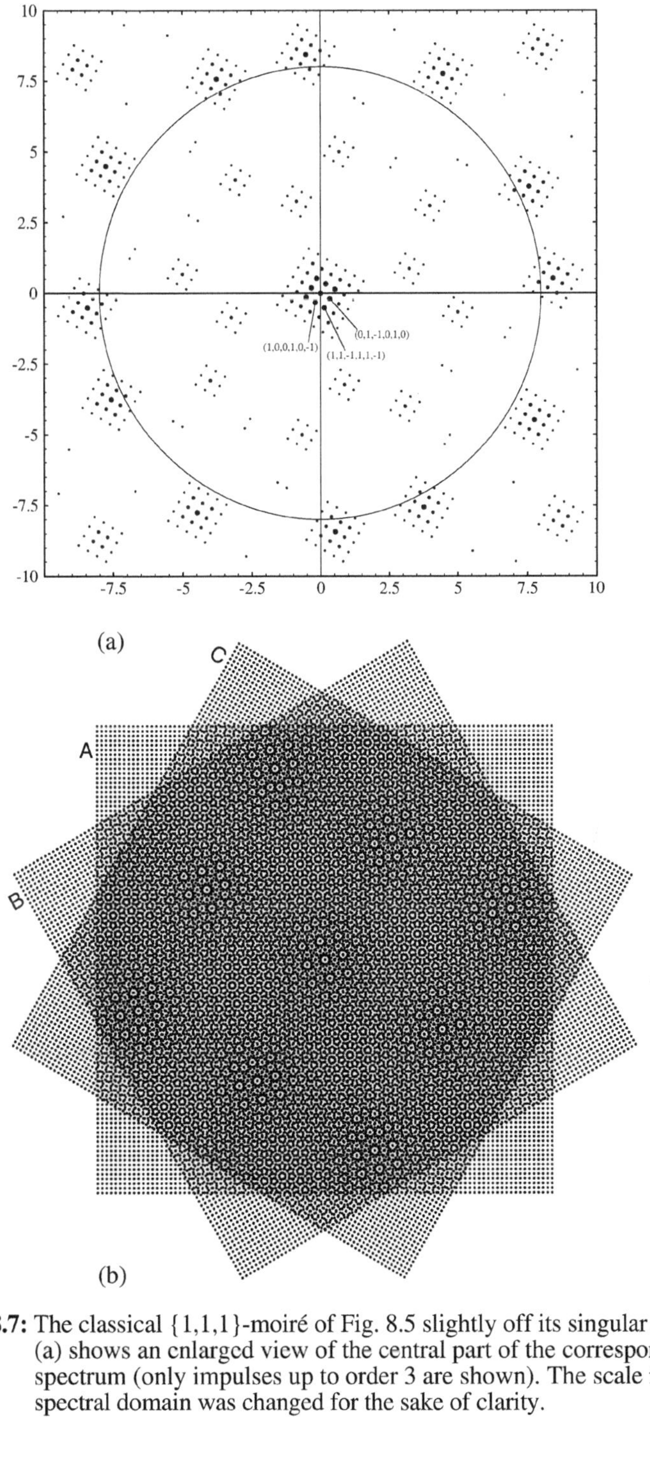


**Figure 8.7:** The classical {1,1,1}-moiré of Fig. 8.5 slightly off its singular state; (a) shows an enlarged view of the central part of the corresponding spectrum (only impulses up to order 3 are shown). The scale in the spectral domain was changed for the sake of clarity.

structure. However, this is also where the similarity between these two types of moiré-free superpositions ends. Stable moiré-free cases are not singular superpositions, and therefore their tolerance to layer rotations, scalings and shifts is significantly higher (see Chapter 3). This means that in all the neighbouring layer combinations which are still included within the tolerance limits no macro-moirés are visible, and hence, in terms of Proposition 8.1, no significant microstructure variations occur in the superposition. The microstructure of such cases seems to be "uniformly disordered", meaning that it consists of a uniform but non-periodic blend of various types of rosettes. Moreover, although this microstructure varies when the superposed layers are modified within the tolerance limits, its overall look remains unchanged. In particular, no visible rosette-type changes occur in such cases owing to layer shifts; this can be clearly seen in Fig. 8.8(b),(c), in contrast to Figs. 8.3–8.5 where rosette-type changes owing to layer shifts are clearly visible. This curious difference in the microstructure behaviour between singular and non-singular moiré-free cases will be fully elucidated in the sections which follow.

The difference between singular and stable moiré-free superpositions is also remarkable in the spectral domain: while in singular states each impulse in the spectrum is in fact a compound-impulse representing a full cluster of impulses which have collapsed into a single location, in stable moiré-free cases each impulse in the spectrum has its own distinct location, and different impulses never fall together on the same point. This fact provides, indeed, the spectral domain interpretation of the microstructure invariance under layer shifts in stable moiré-free superpositions (see the last paragraph in Sec. 8.3).

Fig. 3.8 of Chapter 3 shows an example of a 3-screen stable moiré-free superposition. As we can see, its microstructure has, again, the same basic properties: it looks "uniformly disordered", and it does not present substantial changes under layer shifts (as well as under layer rotations and scalings within the specified tolerance limits).

## 8.6 Rational vs. irrational screen superpositions; rational approximants

Before we proceed to the explanation of the various interesting microstructure phenomena that we have encountered in the preceding sections, let us introduce here the concept of rational approximations that will be needed in the discussions which follow.

Consider the lattice $\mathbb{Z}^2$ of the integer points $(k,l)$ in the $x,y$ plane. Suppose that we draw a straight line passing through the origin (0,0) and through a certain integer point $(m,n) \in \mathbb{Z}^2$. Clearly, the slope of this line, $\tan\alpha = \frac{n}{m}$, will be a rational number (i.e., a ratio between two integers). It is often said, by abuse of language, that *angle* $\alpha$ is rational or that the *line* is rational; but what is really meant in both cases is that the slope $\tan\alpha$ is rational.[4]

It is clear that if the slope of a line passing through the origin is rational, say $\tan\alpha = \frac{n}{m}$, then this line passes through the integer point $(m,n)$, and more generally, through all the

[4] Note that this abuse of language may be quite misleading, since an angle $\alpha$ (in degrees) may be a rational number while $\tan\alpha$ is irrational, and vice versa.

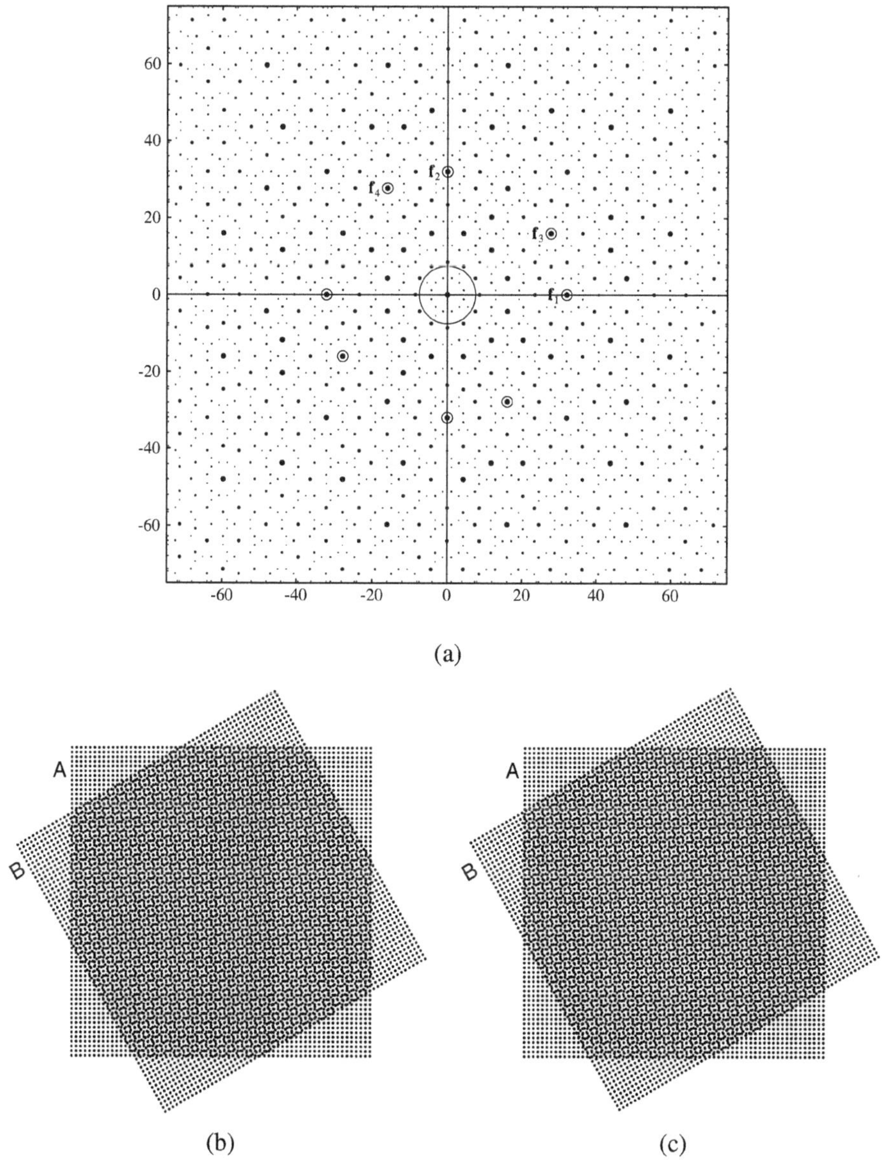


**Figure 8.8:** The stable moiré-free superposition of two identical screens with angle difference of $\alpha = 30°$. (a): The spectrum support (showing only impulses up to order 5). (b), (c) The screen superposition in the image domain: in-phase superposition in (b), and counter-phase superposition in (c). Note the uniformity of the microstructure; however, unlike in Fig. 8.5, no visible differences exist between the rosette shapes in (b) and (c). The spectrum support (a) is the same as in the singular 3-screen superposition of Fig. 8.5(a), but this time it consists of simple impulses and not of compound impulses (collapsed clusters).

integer points of the form $k(m,n)$ with $k \in \mathbb{Z}$ (see Fig. 8.9(a)). And to the contrary, if $\tan\alpha$ is irrational (for example: if $\alpha = 30°$, so that $\tan\alpha = \frac{\sqrt{3}}{3}$; see Fig. 8.9(b)), then the line will not pass through any point of $\mathbb{Z}^2$ except for the origin (0,0). In this case, too, it is often said, by abuse of language, that *angle* $\alpha$ is irrational or that the *line* is irrational, meaning in reality that $\tan\alpha$ is irrational. It should be noted, however, as we can see in the example of Fig. 8.9(b), that although an irrational line through the origin does not pass through any other integer points of $\mathbb{Z}^2$, still it does pass very close to points of $\mathbb{Z}^2$ (more precisely: for any positive $\varepsilon$, be it as small as we may wish, we can find points of $\mathbb{Z}^2$ whose distance from the line is smaller than $\varepsilon$, provided that we go far enough from the origin). This means that although lines with irrational slopes cannot be drawn using points of $\mathbb{Z}^2$, they can be still approximated in $\mathbb{Z}^2$ by lines with rational slopes which are very close to the desired irrational slope. In fact, for any given error $\varepsilon$, be it as small as we may want, we can find an integer point $(m,n) \in \mathbb{Z}^2$ such that the line defined by this point and by the origin has a slope differing by less than $\varepsilon$ from the desired irrational slope.[5] In general, however, the smaller the required error, the farther from the origin such points will be found. We can conclude, therefore, that to any irrational line there exist infinitely many approximating rational lines (or in short: *rational approximations* or *rational approximants*), up to any desired degree of proximity. Note that these concepts are easily extended to higher dimensions (for example: rational or irrational lines and planes in $\mathbb{Z}^3$, or even in $\mathbb{Z}^m$).

Let us proceed now to the 2D case of *rational or irrational dot-screens* (or *grids*).[6] This case arises, for instance, in the field of modern printing, which is based on digital techniques. While in the traditional, analog halftoning process it was possible, using photographic techniques, to generate halftone screens of virtually any desired periods and angles, in digital systems we are limited to an imposed underlying pixel-grid at the device resolution. For example, in a 300 dpi laser printer only dots on a pixel-grid whose period is 1/300 of an inch can be printed, and no in-between points or pixel-fractions can be addressed. This device-resolution grid can be seen as a lattice $\mathbb{Z}^2$ whose integer points $(k,l)$ correspond to the device pixel locations. The question is, therefore, how can we generate halftone screens of any given periods and angles on this underlying lattice? The answer is, of course, that not all period and angle combinations can be obtained using points of $\mathbb{Z}^2$. If the required angle is rational, meaning that $\tan\alpha = \frac{n}{m}$, then any period which is an integer multiple of the triangle hypotenuse $p = \sqrt{m^2 + n^2}$ can be realized (see Fig. 8.10(a));[7] all other periods, however, are excluded. But if the required angle is irrational then we are completely out of luck, since no period can bring us back to a point of $\mathbb{Z}^2$. As we can see, the set of "permitted" angle and period combinations in $\mathbb{Z}^2$, or in other words, the set of the realizable dot-screens in $\mathbb{Z}^2$, is extremely sparse with respect to the continuum of all possible cases. Even the classic halftone screens with angles such as 30°, 60° or 15° are not realizable on digital devices. However, here, too, any irrational dot-screen (i.e., any

---

[5] This follows directly from the fact that any irrational number, including $\tan\alpha$, can be approximated by a series of closer and closer rational fractions $\frac{n}{m}$, for example by taking more and more digits from its infinite decimal representation.

[6] We mention this case just for the sake of completeness, although we will not need it here.

[7] Note that such periods are often themselves irrational numbers, just like the angle $\alpha$ itself (in degrees).

angle and period combination which cannot be realized in $\mathbb{Z}^2$) can be still approximated by a rational screen, simply by truncation to the nearest pixel (Fig. 8.10). Note however that unlike in the case of irrational lines, in the case of screens the number of available rational approximations on a given underlying lattice is limited, and better approximations of the required screen up to any desired degree of proximity can only be obtained by using more sophisticated methods (for example by sacrificing the strict periodicity of the cells and allowing them to have a certain number of slightly different forms). Useful, illustrated discussions on this subject can be found, for example, in [Fink92, Chapters 6–7] or in [Kang97 Sec. 9.3.3].

We arrive now to the case that will interest us in our discussion on superposed layers, namely: *rational or irrational screen* (or *grid*) *superpositions*. Note that in our case the existence or inexistence of an underlying device pixel grid is irrelevant, since our moiré theory works in the continuous world and treats the superposition of any given periodic screens, be they digital or analog. What interests us here, instead, is screen (or grid) superpositions. Assume that we are given two superposed regular dot-screens $p_1(x,y)$ and $p_2(x,y)$ whose angles and periods are defined, respectively, by the period-vectors (or step-vectors) $\mathbf{T}_1$, $\mathbf{T}_2$ and $\mathbf{T}_3$, $\mathbf{T}_4$.[8] We assume for the sake of convenience that these dot-screens are superposed with a common black dot centered on the origin. Now, if the angle and period combinations of the screens are such that throughout the superposition plane there exists no perfect dot superposition other than at the origin, we say that the two dot-screens are mutually incommensurable, or equivalently, that *the screen superposition is irrational.*

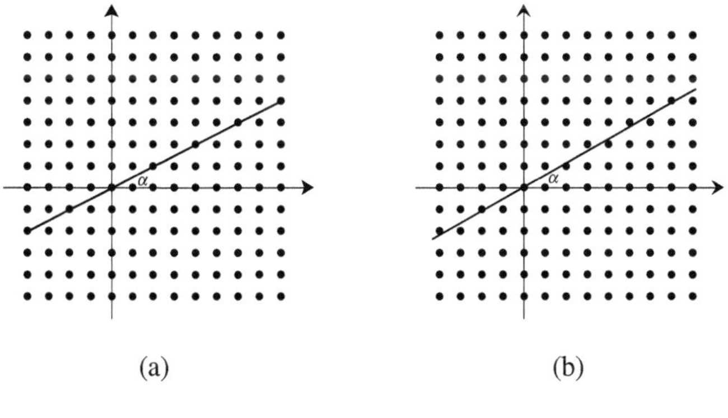

**Figure 8.9:** A rational line (a) and an irrational line (b) through the origin of the lattice $\mathbb{Z}^2$. The slope of rational line (a) is $\tan\alpha = \frac{1}{2}$ so that its angle is $\alpha = \arctan\frac{1}{2} \approx 26.565°$, while the slope of irrational line (b) is $\tan\alpha = \frac{\sqrt{3}}{3}$ so that its angle is $\alpha = \arctan\frac{\sqrt{3}}{3} = 30°$. Rational line (a) is a rational approximation of irrational line (b).

[8] Note that the above discussion on rational vs. irrational screens with respect to an underlying device-pixel grid is, in fact, a particular case in which $\mathbf{T}_1$, $\mathbf{T}_2$ correspond to the device-pixel grid.

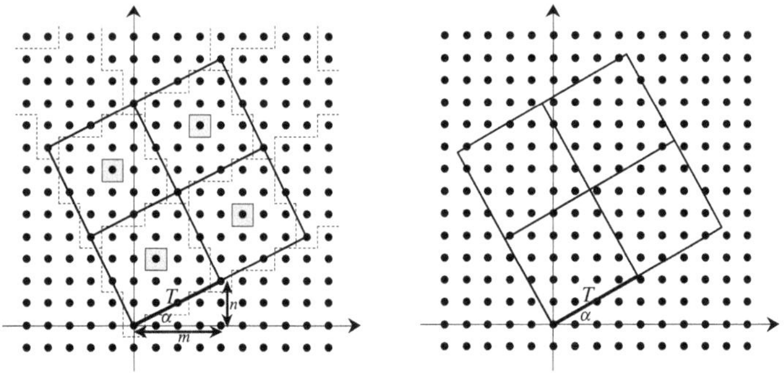


(a) (b)

**Figure 8.10:** A rational grid (a) and an irrational grid (b) on the lattice $\mathbb{Z}^2$. The slope of rational grid (a) is $\tan\alpha = \frac{1}{2}$ (so that its angle is $\alpha \approx 26.565°$), and its period is $T = \sqrt{4^2 + 2^2} = 2\sqrt{5}$; irrational grid (b) has the slope $\tan\alpha = \frac{\sqrt{3}}{3}$ (so that its angle is $\alpha = 30°$) and its period is the same as in (a). Grid (a) is a *rational approximation* of irrational grid (b). The dashed cells in (a) show how a rational halftone screen can be obtained from the rational grid.

If, on the contrary, there exist perfect dot superpositions other than at the origin we say that the dot-screens are mutually commensurable, or that *their superposition is rational.* Note that this definition can be immediately extended to any number of superposed layers. Examples of irrational screen superpositions are given in Figs. 8.8(b) and 8.5(b), which show, respectively, a 2-screen and a 3-screen irrational superpositions. An example of a rational 2-screen superposition is given in Fig. 8.3(b), and an example of a rational 3-line-grid superposition is shown in Fig. 8.1(b).

The following result is immediately obtained from the above definition:

**Proposition 8.3:** The superposition of any number of periodic screens (or grids) is rational *iff* it is periodic. The superposition is irrational *iff* it is almost-periodic. ■

Note that the periodicity of a superposition implies its singularity whenever the equivalent grating number $m$ is bigger than 2 (see consequence (a) of Proposition 5.2 and Remark 2 in Table 5.2, Chapter 5); but this condition is always satisfied in any screen (or grid) superposition, since even in a 2-screen superposition we already have $m = 4$. However, the inverse claim is not necessarily true; for example, the conventional 3-screen superposition is singular but not periodic. We can schematically summarize the situation as follows:

**Proposition 8.4:** For any screen- (or grid-) superpositions we have:

$$\text{Rationality} \Leftrightarrow \text{Periodicity} \Rightarrow \text{Singularity.} \quad ■$$

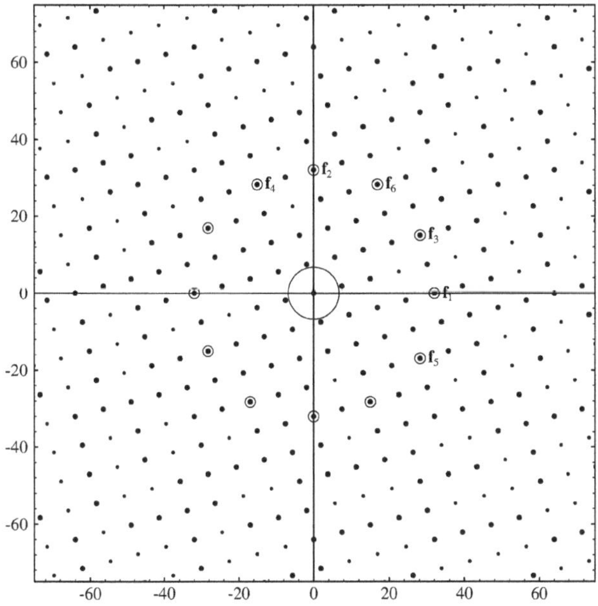


(a)

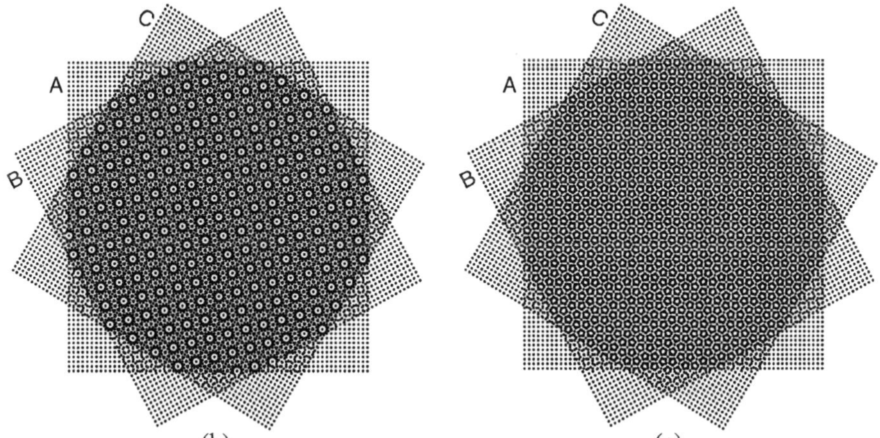


(b) (c)

**Figure 8.11:** A rational approximation of the singular {1,1,1}-state of Fig. 8.5, which is obtained by slightly rotating or scaling layers B and C: $\theta_2 = 28.0725°$, $T_2 = T_1$; $\theta_3 = -30°$, $T_3 = 0.9718T_1$, so that dots $(4T_2,-T_2)$ and $(3T_3,3T_3)$ of layers B and C fall exactly on dot $(4T_1,T_1)$ of layer A. This rational approximant is still a singular {1,1,1}-state, although the collapsed impulse clusters are not the same as before. (a) The spectrum support (including only impulses up to order 3). (b), (c) The screen superposition in the image domain: in-phase superposition in (b), and counter-phase superposition in (c). Note that the spectrum impulses in (a) are arranged here in a perfect lattice of compound impulses, and the screen superposition is fully periodic. Note also the similarity between the rosette shapes in Fig. 8.5(b),(c) and (b),(c) in the present figure.

Now, although an irrational screen superposition contains no perfect dot matches other than at the origin, it still does contain an infinite number of almost-perfect dot matches. This can be clearly seen, for example, in the irrational 3-screen superpositon of Fig. 8.5(b): the central dot in each of the almost-dot-centered rosettes indicates an almost-perfect superposition of all layers, with different degrees of mismatch. The fact that at infinitely many locations in the irrational superposition there occurs an almost-perfect dot coincidence between all layers suggests that the irrational (almost-periodic) dot-screen superposition can be approximated by rational (periodic) screen superpositions in an infinite number of ways, up to any desired degree of proximity. All that we have to do is to find in the superposition an almost-perfect dot coincidence, other than at the origin, where the mismatch is sufficiently small. In general, the smaller the permitted mismatch, the farther from the origin such almost-perfect dot superpositions will be found. Once we have chosen an almost-perfect dot superposition, we keep one of the screens fixed and slightly rotate or scale each of the other superposed screens such that our chosen dot becomes a point of perfect match between all of the layers. In other words, we use one of the superposed screens as a reference screen (much like the device-pixel grid in the case of digital systems), and we slightly adjust the angle and the period of each of the remaining screens so that it perfectly coincides with the reference screen at the selected point. The result is a rational, periodic superposition whose periods and angles, as well as rosette shapes, approximate those of the original irrational superposition. Fig. 8.11 shows a rational approximation (or *rational approximant*) for the irrational 3-screen superposition of Fig. 8.5(b), which has been obtained by aligning the second and the third screens to a perfect match in the center of the first rosette to the east-north-east of the origin, i.e., at the point $(x,y) = (4T,T)$. Clearly, if we choose another rosette having a smaller mismatch, the required adjustments in the second and in the third superposed screens will be smaller, and the rational approximant will become even more faithful to the original irrational superposition.

## 8.7 Algebraic formalization

Having described the various interesting phenomena related to the microstructure of the superposition, we are ready now to introduce the mathematics that will help us to elucidate these phenomena.

We have seen in Chapter 5 how an algebraic formalization of the structure of the spectrum could give us new, profound insights into the various phenomena in question. We will see now that here, too, an algebraic formalization may prove to be very useful, and open the way to a better understanding of the relationship between the microstructure and the phase in the layer superposition.

Let us start with a simple example to illustrate our line of thought and to motivate our algebraic approach.

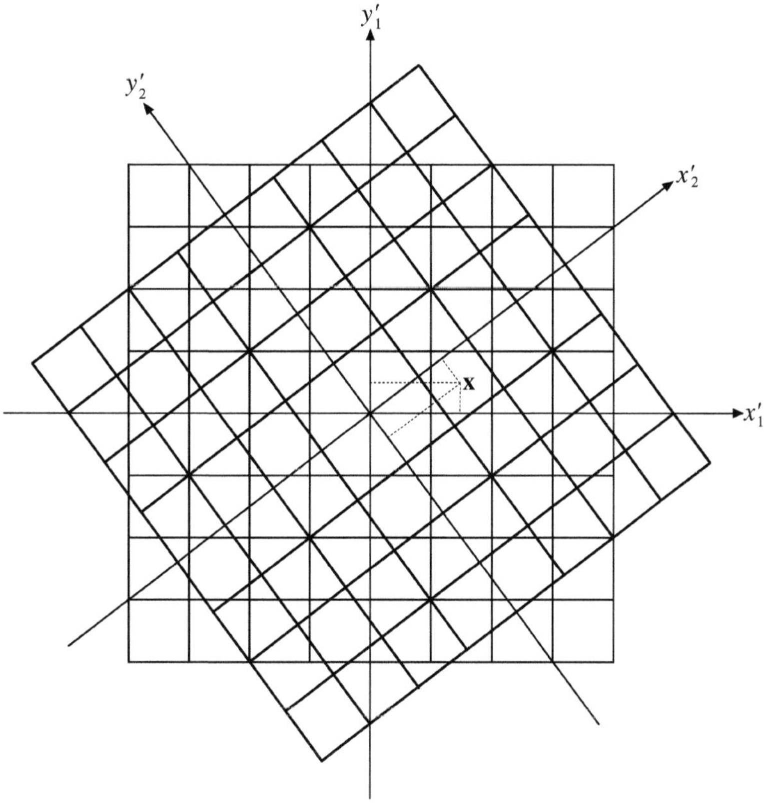


**Figure 8.12:** A magnified view of the superposition of two identical square grids with an angle difference of $\alpha = \arctan\frac{3}{4} \approx 36.87°$ (compare with Fig. 8.3). The period-coordinates of point **x** in the superposition are $\xi_1 = \frac{3}{2}$, $\xi_2 = \frac{1}{2}$, $\xi_3 = \frac{3}{2}$, and $\xi_4 = -\frac{1}{2}$. For the sake of simplicity we chose the $x'_1,y'_1$ coordinates to coincide with the $x$ and $y$ axes of the $x,y$ plane.

**Example 8.1:** Consider the superposition of two identical square grids (or dot-screens) with an angle difference of $\alpha = \arctan\frac{3}{4} \approx 36.87°$, like in Fig. 8.3 (see the magnified view in Fig. 8.12). Clearly, each point **x** in the $x,y$ plane (i.e., in the superposition) can be expressed in terms of the coordinate system $x'_1,y'_1$ of the first grid, as well as in terms of the coordinate system $x'_2,y'_2$ of the second grid. However, by analogy with the phase terminology of Chapter 7, we will find it advantageous to express point **x** in the coordinate system of each of these square grids in terms of the grid's own period $T_i$. Hence, for each square grid $i$ in the superposition ($i = 1,2$) we define the *period-coordinates* $\xi_{2i-1}$ and $\xi_{2i}$ at the point **x** as the coordinates of point **x** in the coordinate system $x'_i,y'_i$ of that grid, expressed in terms of periods $T_i$. Table 8.1 gives the period-coordinates of the two grids of Fig. 8.12 at various points $\mathbf{x} = (x,y)$ in the superposition, along with a verbal description of the microstructure of the superposition at these points.

Note that the period-coordinates $\xi_i$ should not be confused with the period-shifts $\phi_i$ of Chapter 7. The period-shifts $\phi_i$ have been introduced in Secs. 7.4–7.5 for expressing *shifts* of periodic layers in terms of number of periods. The period-coordinates $\xi_i$, for their part, express in terms of number of periods the coordinates of any point $\mathbf{x}$ within a static superposition. Note that when a layer shift occurs the origin and the coordinate system of the shifted layers are displaced within the $x,y$ plane, so that the period-coordinate $\xi_i$ of any point $\mathbf{x}$ in the superposition is decremented by the period-shift $\phi_i$ which corresponds to that layer shift. For instance, assume that the second grid of our example is shifted by half a period in each of its two main directions; this layer shift is expressed by the period-shifts $(\phi_1,\phi_2,\phi_3,\phi_4) = (0,0,\frac{1}{2},\frac{1}{2})$. Therefore, at any point $\mathbf{x}$ in the superposition the new period-coordinates after the shift are given by:

$$(\xi_1,\xi_2,\xi_3,\xi_4)_{\text{new}} = (\xi_1,\xi_2,\xi_3,\xi_4)_{\text{old}} - (\phi_1,\phi_2,\phi_3,\phi_4)$$

By analogy with the phase terminology of Sec. 7.5.2 we call the vector $(\xi_1,\xi_2,\xi_3,\xi_4)$ the *period-coordinate vector of the superposition* at the point $(x,y)$. As we can see, the period-coordinate vector $(\xi_1,\xi_2,\xi_3,\xi_4)$ at any point $(x,y)$ is strongly related to the local microstructure of the superposition at that point. For example, whenever $(\xi_1,\xi_2,\xi_3,\xi_4)$ is purely integer, i.e., $(\xi_1,\xi_2,\xi_3,\xi_4) \in \mathbb{Z}^4$, the superposition at $(x,y)$ contains a meeting point of full periods in all layers, which means, according to our convention from Remark 8.1, that point $(x,y)$ is the center of a dot-centered rosette. Similarly, whenever the $\xi_i$ values are all half-integers (i.e., $\xi_i = k_i+\frac{1}{2}$, $k_i \in \mathbb{Z}$), the point $(x,y)$ in the superposition is the center of a clear-centered rosette.

Now, if we run throughout all the points $(x,y) \in \mathbb{R}^2$ (i.e., throughout the whole given superposition), which parts of $\mathbb{R}^4$ will be occupied by the corresponding points $(\xi_1,\xi_2,\xi_3,\xi_4)$? It is clear that $\mathbb{R}^4$ will not be completely filled; for instance, in the present

| $(x,y)$ | $(\xi_1,\xi_2,\xi_3,\xi_4)$ | Microstructure at $(x,y)$: |
|---|---|---|
| (0,0) | (0,0,0,0) | Center of a dot-centered rosette |
| $(\frac{3}{2}T,\frac{1}{2}T)$ | $(\frac{3}{2},\frac{1}{2},\frac{3}{2},-\frac{1}{2})$ | Center of a clear-centered rosette |
| $(\frac{1}{2}T,T)$ | $(\frac{1}{2},1,1,\frac{1}{2})$ | — |
| $(T,2T)$ | (1,2,2,1) | Center of a dot-centered rosette |

**Table 8.1:** The period-coordinates of the two grids in Fig. 8.12 at various points $\mathbf{x} = (x,y)$.

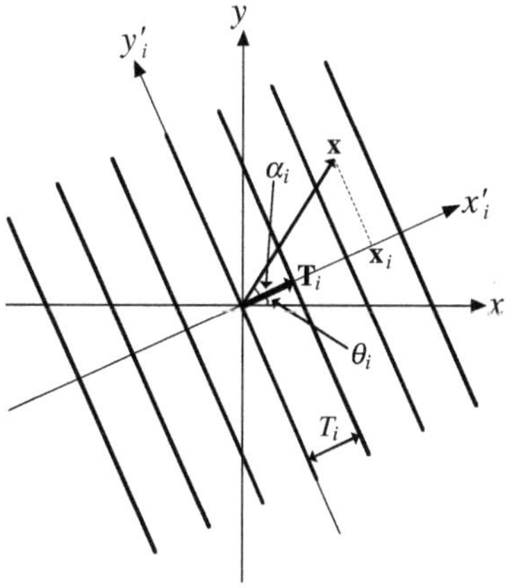


**Figure 8.13:** A schematic view of layer $i$ in the superposition, showing $\mathbf{x}_i$, the projection of point $\mathbf{x} = (x,y)$ on axis $x'_i$. $\theta_i$ is the orientation of axis $x'_i$, and $\alpha_i$ is the angle formed between the direction of point $\mathbf{x}$ and axis $x'_i$. The coordinates of point $\mathbf{x}_i$ are $(x_i,y_i)$ in terms of the $x,y$ plane, and $(x'_i,0)$ in terms of the $x'_i,y'_i$ coordinates of the $i$-th layer. The period-coordinate of point $\mathbf{x}$ with respect to axis $x'_i$ is $\xi_i = \frac{5}{2}$.

superposition (see Fig. 8.12) the point $(0,0,-\frac{1}{2},-\frac{1}{2})$ cannot be obtained — it can only be obtained when the second layer is shifted by half a period in each of its two main directions. In order to investigate this (and other) questions, we find it useful to define a transformation $\Xi : \mathbb{R}^2 \rightarrow \mathbb{R}^4$, which gives for any point $(x,y)$ in the superposition plane its corresponding point $(\xi_1,\xi_2,\xi_3,\xi_4)$ in $\mathbb{R}^4$. And indeed, we will see below that the investigation of this transformation and of its properties will shed a new light onto the microstructure and the phase relationships of the layer superposition. ■

Having explained the motivation for the proposed algebraic formalization, we are ready now to go back to the general case and to introduce our new formal approach.

Let $p_1(\mathbf{x}), ... ,p_m(\mathbf{x})$ be $m$ 1-fold periodic functions (gratings) given in their initial phase, so that their origins coincide with the origin of the $x,y$ plane, and let $p(\mathbf{x}) = p_1(\mathbf{x})\cdot ... \cdot p_m(\mathbf{x})$ be their superposition. (As usual, a pair of 1-fold periodic functions may represent in the superposition one 2-fold periodic function, such as a dot-screen.) We remember that the main periodicity direction of the $i$-th grating is the direction $\theta_i$ along which the grating has the smallest period > 0. Now, let $\mathbf{x}$ be a point in the $x,y$ superposition plane. For each grating $i$ of the superposition we define the period-coordinate $\xi_i$ at point $\mathbf{x}$ as the number (integer or not) of periods $\mathbf{T}_i$ between the grating origin and $\mathbf{x}_i$, the projection of $\mathbf{x}$ on the axis defining the main periodicity direction of grating $i$. In other words, $\xi_i$ is the 1D coordinate of the point $\mathbf{x}$ on this axis, expressed in period units (see Fig. 8.13). If $\alpha_i$ is the angle formed between the direction of point $\mathbf{x}$ and the main periodicity direction of grating $i$ we have, therefore:

$$\xi_i = \frac{|\mathbf{x}|\cos\alpha_i}{|\mathbf{T}_i|} = \frac{|\mathbf{x}_i|}{|\mathbf{T}_i|}$$

and hence: $\mathbf{x}_i = \xi_i\mathbf{T}_i$. Remembering that $\mathbf{T}_i\!\cdot\!\mathbf{T}_i^{-1} = 1$ (see Eq. (A.42) in Appendix A) we multiply both sides (in the sense of scalar product) by $\mathbf{T}_i^{-1}$, and hence we obtain $\mathbf{x}_i\!\cdot\!\mathbf{T}_i^{-1} = \xi_i$. Using $\mathbf{f}_i = \mathbf{T}_i^{-1}$ (see Eq. (A.43) in Appendix A), where $\mathbf{f}_i$ is the frequency vector of the 1-fold periodic function $p_i(\mathbf{x})$, we obtain:

$$\xi_i = \mathbf{f}_i\!\cdot\!\mathbf{x}_i \tag{8.1}$$

(notice the analogy with $\phi_i = \mathbf{f}_i\!\cdot\!\mathbf{a}_i$ in Sec. 7.4!).

Now, we remember that the scalar product $\mathbf{v}\!\cdot\!\mathbf{w}$ can be understood as a number which gives the product of the length of vector $\mathbf{v}$ by the length of the projection of vector $\mathbf{w}$ on the direction of $\mathbf{v}$ (or vice versa) [Vygodski73 p. 142]:

$$\mathbf{v}\!\cdot\!\mathbf{w} = |\mathbf{v}|\;|\mathrm{proj}(\mathbf{w})_{\mathbf{v}}|$$

This means that for any point $\mathbf{x}$ in the $x,y$ plane we have:

$$\mathbf{f}_i\!\cdot\!\mathbf{x} = \mathbf{f}_i\!\cdot\!\mathbf{x}_i \tag{8.2}$$

where $\mathbf{x}_i$ is the projection of $\mathbf{x}$ on the direction of $\mathbf{f}_i$. Therefore Eq. (8.1) can be reformulated as:

$$\xi_i = \mathbf{f}_i\!\cdot\!\mathbf{x} \tag{8.3}$$

The period-coordinate $\xi_i$ can be also expressed in the form $\xi_i = g_i(x,y)$ as a function of the plane coordinates $x,y$: Let $\mathbf{x} = (x,y)$ be a point in the plane, and let the $x'_i$ axis through the origin represent the main periodicity direction $\theta_i$ of the $i$-th grating. We also denote by $y'_i$ the axis perpendicular to $x'_i$ through the origin (see Fig. 8.13). The coordinates of point $\mathbf{x}$ in terms of the rotated axes $x'_i,y'_i$ are:

$$x'_i = x\cos\theta_i + y\sin\theta_i$$

$$y'_i = -x\sin\theta_i + y\cos\theta_i$$

and therefore the projection of the point $\mathbf{x}$ on the $x'_i$ axis is given in terms of these rotated coordinates by: $\mathbf{x}_i = (x'_i,0)$. This means that $\xi_i$ is explicitly given in the form $\xi_i = g_i(x,y)$ by:

$$\xi_i = g_i(x,y) = \frac{x'_i}{T_i} = x\,\frac{\cos\theta_i}{T_i} + y\,\frac{\sin\theta_i}{T_i} \tag{8.4}$$

As we can see, for each layer $i$ of the superposition, the period-coordinate $\xi_i$ is uniquely defined at any point $\mathbf{x} = (x,y)$ of the plane. Therefore we may define a transformation $\Xi : \mathbb{R}^2 \to \mathbb{R}^m$, called the *period-coordinate function*, which gives for any point $\mathbf{x} = (x,y)$ in the plane the period-coordinate $\xi_i$ of this point in each of the $m$ superposed 1-fold periodic layers:

$$\Xi(x,y) = (\xi_1, \dots, \xi_m) \tag{8.5}$$

In other words, this transformation gives for any point $\mathbf{x} = (x,y)$ in the superposition plane its coordinates in the main direction of each of the $m$ layers, in terms of each layer's period. Since each of the functions $\xi_i = g_i(x,y)$ is linear, i.e., $\xi_i = a_i x + b_i y$ (see Eq. (8.4)) it follows that $(\xi_1, \dots, \xi_m)$ too is linear in $x$ and $y$, so that $\Xi$ is a linear transformation. Therefore, the image of $\Xi$ is a linear subspace within $\mathbb{R}^m$ whose dimension is 2, namely: $\Xi$ maps the $x,y$ superposition plane into a plane $\mathrm{Im}(\Xi)$ within $\mathbb{R}^m$ which passes through the origin.[9] Note that the subspace $\mathrm{Im}(\Xi)$ may have a lower dimension than 2 if the transformation $\Xi$ is degenerate; for example, if all the $m$ superposed gratings have the same orientation, so that $\xi_2, \dots, \xi_m$ are constant multiples of $\xi_1$, then all the vectors $(\xi_1, \dots, \xi_m) \in \mathbb{R}^m$ are collinear and $\dim \mathrm{Im}(\Xi) = 1$. Such degenerate cases will generally be ignored in the discussions which follow.[10]

Let us now consider the plane $\mathrm{Im}(\Xi)$ which is defined by the transformation $\Xi$ within $\mathbb{R}^m$. Points $(\xi_1, \dots, \xi_m)$ in $\mathrm{Im}(\Xi)$ which are only composed of integer values have a special significance, since they indicate that the corresponding point $(x,y)$ in the superposition is located on a junction of full periods from the origin in all of the superposed layers. Since we have assumed that the 1-fold periodic functions $p_1(\mathbf{x}), \dots, p_m(\mathbf{x})$ are given in their initial phase, we know that the plane $\mathrm{Im}(\Xi)$ contains at least the point $(0,\dots,0)$; but does it contain any other integer point $(k_1,\dots,k_m)$? Clearly, if $\mathrm{Im}(\Xi)$ contains an integer point $(k_1,\dots,k_m) \neq (0,\dots,0)$, then it contains also the whole 1D lattice $L$ defined by the integer multiples $n(k_1,\dots,k_m)$, and the superposition is 1-fold periodic; and if $\mathrm{Im}(\Xi)$ contains two integer points $(k_1^{(1)},\dots,k_m^{(1)}) \neq (0,\dots,0)$ and $(k_1^{(2)},\dots,k_m^{(2)}) \neq (0,\dots,0)$ which are not on the same line through the origin, then it contains the whole 2D lattice $L$ defined by their integer linear combinations: $i(k_1^{(1)},\dots,k_m^{(1)}) + j(k_1^{(2)},\dots,k_m^{(2)})$, and the superposition is 2-fold periodic. Depending on the plane inclinations within the space $\mathbb{R}^m$ the lattice $L \subset \mathrm{Im}(\Xi)$ may have rank = 2 (in the case of $m = 3$ this happens, for instance, if the plane $\mathrm{Im}(\Xi)$ contains both the $x$ and $y$ axes of $\mathbb{R}^3$); rank = 1 (e.g., if the plane only contains the $x$ axis of $\mathbb{R}^3$ but forms an irrational angle with the $y$ and $z$ axes); or rank = 0 (if the only integral point in the plane is the origin $(0,\dots,0)$).

**Example 8.2:** Consider the superposition of two identical periodic square grids (or dot-screens) which are rotated by angles 0 and $\alpha$, respectively. The transformation $\Xi$ is given in this case by:

---

[9] The explicit equation of this plane is most conveniently expressed in the parametric form by $\xi_1 = a_1 x + b_1 y, \dots, \xi_m = a_m x + b_m y$; by elimination of $x$ and $y$ it can be also expressed as a system of two linear equations $c_{1,1}\xi_1 + \dots + c_{m,1}\xi_m = 0,\ c_{1,2}\xi_1 + \dots + c_{m,2}\xi_m = 0$, i.e., as an intersection of two linear subspaces of dimension $m-1$.

[10] It is interesting to note that if the superposition in the $(x,y)$ plane consists of non-linearly curved layers (i.e., non-linear transformations of periodic functions; see Chapter 10), then the image of $\Xi$ is a curved 2D surface within $\mathbb{R}^m$.

$$\Xi\begin{pmatrix} x \\ y \end{pmatrix} = \begin{pmatrix} \xi_1 \\ \xi_2 \\ \xi_3 \\ \xi_4 \end{pmatrix} = \frac{1}{T}\begin{pmatrix} x \\ y \\ x\cos\alpha + y\sin\alpha \\ -x\sin\alpha + y\cos\alpha \end{pmatrix} \tag{8.6}$$

$$= \frac{1}{T}\begin{pmatrix} \begin{pmatrix} x \\ y \end{pmatrix} \\ \mathrm{M}\begin{pmatrix} x \\ y \end{pmatrix} \end{pmatrix}$$

where M is the 2×2 matrix which represents a rotation by angle $\alpha$:

$$\mathrm{M} = \begin{pmatrix} \cos\alpha & \sin\alpha \\ -\sin\alpha & \cos\alpha \end{pmatrix}$$

If the superposition is rational (as in the case of $\alpha = \arctan\frac{3}{4} \approx 36.87°$; see Example 8.1 and Fig. 8.3) then rank$L = 2$ and infinitely many points $(x,y)$ in the superposition possess an integer vector $(\xi_1,\xi_2,\xi_3,\xi_4)$; the superposition in this case is 2-fold periodic. Note that since an integer vector $(\xi_1,\xi_2,\xi_3,\xi_4)$ represents in the superposition a meeting point of full periods, its corresponding point $(x,y)$ in the superposition is a center of a dot-centered rosette.

If, however, the superposition is irrational, as in the case of $\alpha = 30°$ (see Fig. 8.8), then rank$L = 0$ and the plane Im($\Xi$) does not contain any integer $(\xi_1,\xi_2,\xi_3,\xi_4)$ except for the origin (0,...,0). This means that nowhere in the superposition except for the origin can a precise dot-centered rosette be formed. For similar reasons Im($\Xi$) contains no points $(\xi_1,\xi_2,\xi_3,\xi_4)$ with half-integers in all coordinates, meaning that at no point $(x,y)$ in the superposition does there exist a meeting point of half-periods, i.e., a clear-centered rosette. However, as we have seen in Sec. 8.6, the irrational superposition contains infinitely many approximations of such rosettes (of either type);[11] this can be clearly seen in Fig. 8.8. In such cases the screen superposition is not periodic but rather almost-periodic; and indeed, as we have already seen, this type of microstructure is a characteristic property of almost-periodic functions. ■

Let us see now a few properties of the transformation $\Xi$ that are related to lateral shifts of the superposed layers.

**Proposition 8.5:** Assume that the grating $p_i(\mathbf{x})$ in the superposition is laterally shifted by a vector $\mathbf{a}$; this shift can be expressed, as we have seen in Sec. 7.4, by the period-shift $\phi_i = \frac{|\mathbf{a}_i|}{|\mathbf{T}_i|}$, where $\mathbf{a}_i$ is the projection of $\mathbf{a}$ on the axis defining the direction of periodicity of $p_i(\mathbf{x})$, and $\mathbf{T}_i$ is the period of $p_i(\mathbf{x})$. Therefore, as a result of this shift, the period-coordinate $\xi_i$ of any point $(x,y)$ in the superposition is decremented by the period-shift $\phi_i$. This means that the plane Im($\Xi$) is shifted within $\mathbb{R}^m$ by $\phi_i$ along the axis of the $i$-th dimension. ■

[11] More precisely: for any positive $\varepsilon$, be it as small as we may desire, we can find in the superposition rosettes (of either type) with a mismatch smaller than $\varepsilon$, provided that we go far enough from the origin.

This result may be restated more formally as follows:

Let $\Xi(x,y)$ be the period-coordinate function which corresponds to the grating superposition $p_1(\mathbf{x}) \cdot \ldots \cdot p_m(\mathbf{x})$. Suppose now that the gratings $p_1(\mathbf{x}), \ldots, p_m(\mathbf{x})$ undergo shifts of $\mathbf{a}_1, \ldots, \mathbf{a}_m$, respectively, along their main periodicity directions. Then, the period-coordinate function which corresponds to the superposition after the shift is given by:

$$\begin{aligned}\Xi_A(x,y) &= \Xi(x,y) - S_A(x,y) \\ &= (\xi_1, \ldots, \xi_m) - (\phi_1, \ldots, \phi_m)\end{aligned} \tag{8.7}$$

where A denotes the multi-vector $(\mathbf{a}_1, \ldots, \mathbf{a}_m)$. Note that $\phi_i = \mathbf{f}_i \cdot \mathbf{a}_i$ (see Sec. 7.4) and $\xi_i = \mathbf{f}_i \cdot \mathbf{x}_i$, where $\mathbf{x}_i$ is the projection of the point $\mathbf{x}$ on the direction of $\mathbf{f}_i$, the periodicity direction of the grating $p_i(\mathbf{x})$. The function $S_A : \mathbb{R}^2 \to \mathbb{R}^m$ which defines the period-shifts of the $m$ gratings, $S_A(x,y) = (\phi_1, \ldots, \phi_m)$, is called the *period-shift function*; note that it returns the same constant vector for every point $\mathbf{x}$ in the superposition.

For example, if the second square grid (or dot-screen) of Example 8.2 above is shifted by half a period in each of its two main directions, the transformation $\Xi$ becomes:

$$\Xi_A\begin{pmatrix} x \\ y \end{pmatrix} = \begin{pmatrix} \xi_1 \\ \xi_2 \\ \xi_3 \\ \xi_4 \end{pmatrix} = \frac{1}{T}\begin{pmatrix} \begin{pmatrix} x \\ y \end{pmatrix} \\ \mathrm{M}\begin{pmatrix} x \\ y \end{pmatrix} \end{pmatrix} - \begin{pmatrix} 0 \\ 0 \\ \frac{1}{2} \\ \frac{1}{2} \end{pmatrix}$$

The period-coordinate of the superposition at the origin (0,0) will be, in this case, $(0, 0, -\frac{1}{2}, -\frac{1}{2})$. Clearly, if before the shift the plane Im($\Xi$) contained integer points of $\mathbb{Z}^4$, then after this shift Im($\Xi$) will contain none: the superposition will have no dot-centered rosettes.

**Proposition 8.6:** If grating $p_i(\mathbf{x})$ is shifted by an integer number of its periods, the superposition $p(\mathbf{x})$ and its microstructure remain, of course, unchanged. This is expressed in $\mathbb{R}^m$ by the fact that the plane Im($\Xi$) is shifted along the $i$-th axis of $\mathbb{R}^m$ by an integer number, so that the relative location of the plane with respect to points of $\mathbb{Z}^m$ remains unchanged. ■

**Proposition 8.7:** Assume that each of the individual gratings $p_i(\mathbf{x})$ is shifted by a non-integer number of periods. The combination of their shifts gives a rigid motion of the superposition as a whole (and hence only a lateral shift of the microstructure) *iff* these shifts cause the plane Im($\Xi$) to be shifted into itself in $\mathbb{R}^m$ (or in other words: *iff* the plane Im($\Xi$) is shifted within $\mathbb{R}^m$ by a vector which is included in this plane). ■

This result is easy to understand, since a rigid motion of the superposition by $(x_0,y_0)$ implies that every point $(\xi_1, \ldots, \xi_m)$ which used to be in Im($\Xi$) before the rigid transformation will still remain in Im($\Xi$), but now it will correspond in the superposition to the point $(x,y) - (x_0,y_0)$ rather than to the point $(x,y)$.

## 8.8 The microstructure of the conventional 3-screen superposition

As we have seen in Sec. 8.2.2, the in-phase superposition of three identical dot-screens (or square grids) with equal angle differences between them (for example, at orientations of $\theta_1 = 30°$, $\theta_2 = -30°$ and $\theta_3 = 0°$) generates an almost-periodic pattern of dot-centered rosettes (see Figs. 8.2(a) and 8.5(b)). But when one of the superposed layers is shifted by half a period in each of its two main directions, the microstructure of the superposition changes into a pattern of clear-centered rosettes (see Figs. 8.2(b) and 8.5(c)). How can we explain this interesting phenomenon mathematically, using our new algebraic formulation? And why, as we have seen in Fig. 8.8, does this phenomenon not occur when only two of the three layers are superposed?

The transformation $\Xi$ is defined for this 3-layer in-phase superposition by:

$$\Xi\begin{pmatrix} x \\ y \end{pmatrix} = \begin{pmatrix} \xi_1 \\ \xi_2 \\ \xi_3 \\ \xi_4 \\ \xi_5 \\ \xi_6 \end{pmatrix} = \frac{1}{T}\begin{pmatrix} \mathrm{M}_{30}\begin{pmatrix} x \\ y \end{pmatrix} \\ \mathrm{M}_{-30}\begin{pmatrix} x \\ y \end{pmatrix} \\ \mathrm{I}\begin{pmatrix} x \\ y \end{pmatrix} \end{pmatrix} = \frac{1}{T}\begin{pmatrix} \frac{\sqrt{3}}{2}x + \frac{1}{2}y \\ -\frac{1}{2}x + \frac{\sqrt{3}}{2}y \\ \frac{\sqrt{3}}{2}x - \frac{1}{2}y \\ \frac{1}{2}x + \frac{\sqrt{3}}{2}y \\ x \\ y \end{pmatrix} \tag{8.8}$$

where $\mathrm{M}_{30}$ and $\mathrm{M}_{-30}$ are the matrices which represent rotations by 30° and –30°, respectively, and I is the identity matrix:

$$\mathrm{M}_\theta = \begin{pmatrix} \cos\theta & \sin\theta \\ -\sin\theta & \cos\theta \end{pmatrix} \qquad \mathrm{I} = \mathrm{M}_0 = \begin{pmatrix} 1 & 0 \\ 0 & 1 \end{pmatrix}$$

Transformation $\Xi$ maps therefore the $x,y$ superposition plane into a plane $\mathrm{Im}(\Xi)$ within $\mathbb{R}^6$. Note that except for the point (0,0,0,0,0,0) the plane $\mathrm{Im}(\Xi)$ contains no integer point of $\mathbb{Z}^6$, since according to Eq. (8.8) whenever $\xi_5$ and $\xi_6$ are integers, $\xi_1,\xi_2,\xi_3,\xi_4$ are irrational numbers. This is not surprising, since we already know that our 3-screen superposition is not periodic, but rather almost-periodic (in the language of Chapter 5: the six frequency-vectors $\mathbf{f}_1 = (\frac{\sqrt{3}}{2}, \frac{1}{2})$, $\mathbf{f}_2 = (-\frac{1}{2}, \frac{\sqrt{3}}{2})$, $\mathbf{f}_3 = (\frac{\sqrt{3}}{2}, -\frac{1}{2})$, $\mathbf{f}_4 = (\frac{1}{2}, \frac{\sqrt{3}}{2})$, $\mathbf{f}_5 = (1,0)$, and $\mathbf{f}_6 = (0,1)$ span within the $u,v$ plane a module with rank = 4, since $\mathbf{f}_1$, $\mathbf{f}_2$, $\mathbf{f}_3$ and $\mathbf{f}_4$ are linearly independent over $\mathbb{Z}$, but $\mathbf{f}_5 = \mathbf{f}_4 - \mathbf{f}_2$ and $\mathbf{f}_6 = \mathbf{f}_1 - \mathbf{f}_3$).[12]

In order to better understand the microstructure of the conventional 3-screen superposition with equal angle differences of 30° (or 60°), let us return for a moment to the superposition of two identical screens with an angle difference of 30° (or 60°). As we have seen in Example 8.2 above and in Fig. 8.8, this superposition is characterized by the

[12] It is interesting to note that the superposition of the third screen on top of the initial 2-screen superposition does not add new impulse locations in the spectrum support (compare the 2-screen spectrum support in Fig. 8.8(a) with the 3-screen spectrum support in Fig. 8.5(a)). The reason is that the new frequency vectors $\mathbf{f}_5$ and $\mathbf{f}_6$ are linear combinations of the original frequency vectors $\mathbf{f}_1,\mathbf{f}_2,\mathbf{f}_3,\mathbf{f}_4$, and therefore all the new convolution impulses which are generated in the spectrum owing to the superposition of the third screen are located on top of already existing impulses. Thus, each impulse in the spectrum of the 2-screen superposition turns into a *compound impulse* in the spectrum of the 3-screen superposition, and the non-singular 2-screen superposition turns into a singular 3-screen superposition.

presence of approximate rosettes of all types (dot-centered, clear-centered, and all intermediate variants), which are uniformly distributed throughout the superposition plane, giving to the eye the impression of a uniform, regular microstructure. This situation is shown again in Fig. 8.14(a) and in its magnified version in Fig. 8.15(a). Figs. 8.14 and 8.15 also show in detail what happens when we superpose the third dot-screen on top of this 2-screen superposition (keeping our convention that in the initial phase of each layer a black dot is centered on the origin):

(1) If the third dot screen is superposed in-phase with the first two screens, so that all screens have a black dot centered on the origin, then wherever there used to be in the 2-screen superposition an almost-dot-centered rosette or an almost-clear-centered rosette, the third screen contributes a new dot of its own. This strengthens all the already existing dot-centered rosettes, but destroys all the 2-screen clear-centered rosettes. As a result, the 3-layer superposition no longer contains almost-clear-centered rosettes, and the microstructure becomes dominated by almost-dot-centered rosettes (compare the two layers in Figs. 8.14(a) and 8.15(a) with the three layers in Figs. 8.14(c) or 8.15(c), respectively).

(2) If the third dot-screen is superposed in counter-phase with respect to the first two screens, i.e., with a white space centered on the origin, then wherever there used to be in the 2-screen superposition an almost-clear-centered rosette or an almost dot-centered rosette, the third screen contributes a white space (which is obviously surrounded by four black dots). This strengthens all the already existing clear-centered rosettes, but destroys all the dot-centered rosettes. As a result, the 3-layer superposition no longer contains almost-dot-centered rosettes, and the microstructure becomes dominated by almost-clear-centered rosettes (compare the two layers in Figs. 8.14(a) and 8.15(a) with the three layers in Figs. 8.14(d) or 8.15(d), respectively).

(3) If all of the three screens are centered on the origin in counter-phase (i.e., with a white space centered on the origin), the addition of the third layer on top of the 2-screen superposition has the same effect as in case (2) (compare the 2-layers in Figs. 8.14(b) and 8.15(b) with the 3-layers in Figs. 8.14(e) or 8.15(e), respectively).

As we can see, the addition of the third layer significantly modifies the microstructure behaviour of the superposition: While in the 2-screen superposition almost-rosettes of all types are uniformly distributed throughout the plane, when the third layer is added on top, one type of almost-rosettes becomes dominant. Furthermore, in contrast to the 2-screen superposition, where shifts of the individual screens do not modify the nature of the microstructure (see Figs. 8.14(a),(b)), in the 3-screen superposition a shift of any of the layers may alter the dominant type of rosettes in the superposition and thus visibly modify the texture of the microstructure (see Figs. 8.14(c)–(f)).

Although this behaviour may seem surprising at first sight, in fact, there is nothing mysterious about it. As we already know, the plane Im($\Xi$) within $\mathbb{R}^4$ that contains all the period-coordinates $(\xi_1,\xi_2,\xi_3,\xi_4)$ of the two dot-screens at 30° and –30° is irrational, and therefore it contains no integer points of $\mathbb{Z}^4$ except for (0,0,0,0) — but it passes within $\mathbb{R}^4$

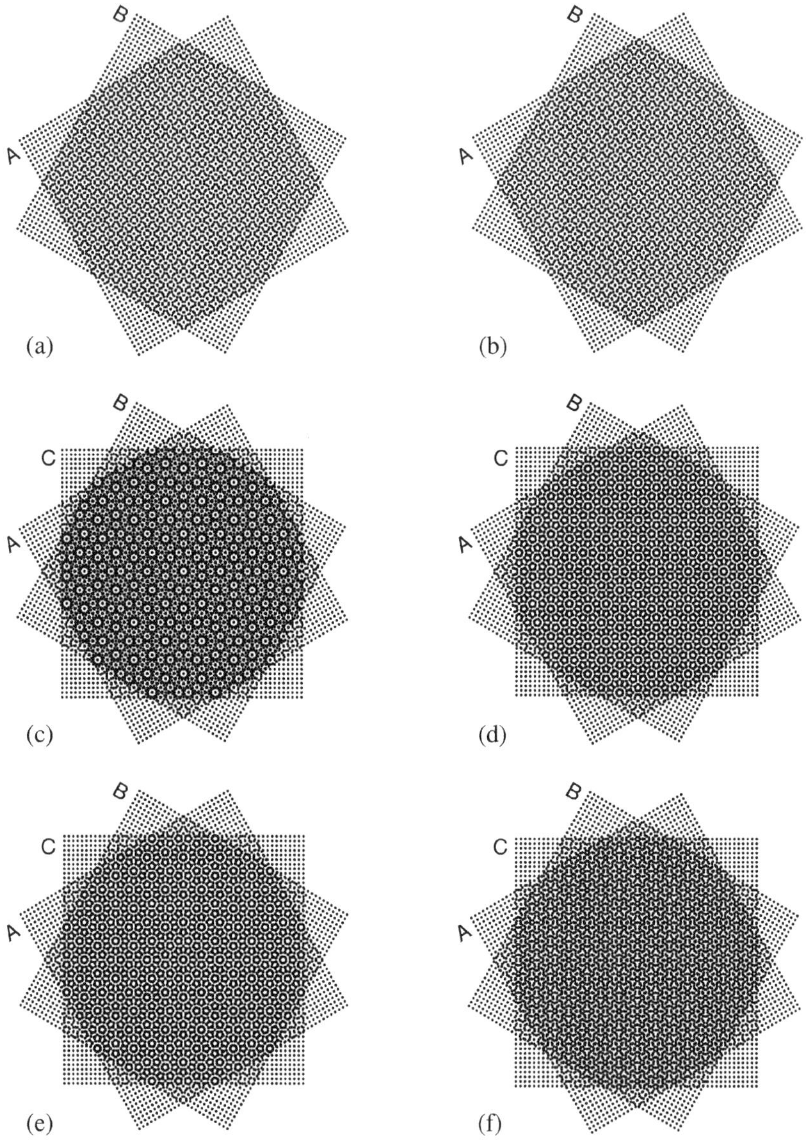


**Figure 8.14:** (a) In-phase superposition of two identical dot-screens at angles $\theta_1 = 30°$ and $\theta_2 = -30°$. (b) Counter-phase superposition of the same screens. (c) In-phase superposition of a third identical screen with angle $\theta_3 = 0°$ on top of (a); the period-shifts of the screens are (0,0,0,0,0,0). (d) Counter-phase superposition of a third identical screen with angle $\theta_3 = 0°$ on top of (a); the period-shifts of the screens are $(0,0,0,0,\frac{1}{2},\frac{1}{2})$. (e) Counter-phase superposition of a third identical screen with angle $\theta_3 = 0°$ on top of (b); the period-shifts of the screens are $(\frac{1}{2},\frac{1}{2},\frac{1}{2},\frac{1}{2},\frac{1}{2},\frac{1}{2})$. (f) Half-period shifted superposition of a third identical screen with angle $\theta_3 = 0°$ on top of (a); the period-shifts of the screens are $(0,0,0,0,\frac{1}{2},0)$.

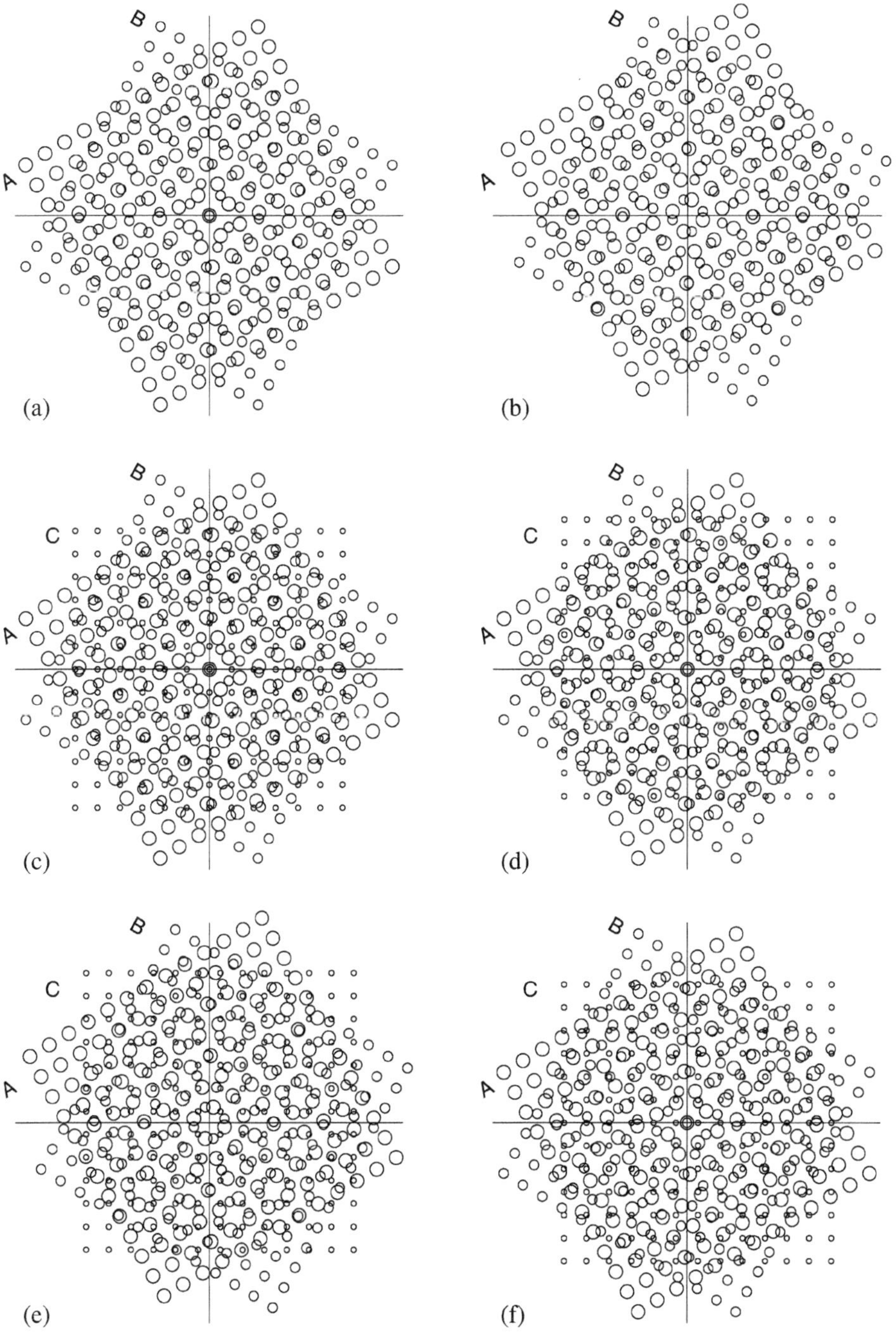


**Figure 8.15:** A magnified view of the screen superpositions of Fig. 8.14, permitting to distinguish between the different layers and their precise dot locations.

as close as we wish to integer points of $\mathbb{Z}^4$ and to half-integer points of $\mathbb{Z}^4+(\frac{1}{2},\frac{1}{2},\frac{1}{2},\frac{1}{2})$ (which correspond, respectively, to dot-centered or to clear-centered rosettes in the 2-screen superposition). Now, when we superpose a new dot-screen at 0° on top of the first two screens we increase the dimension of the period-coordinate vectors by 2, from $(\xi_1,\xi_2,\xi_3,\xi_4) \in \mathbb{R}^4$ to $(\xi_1,\xi_2,\xi_3,\xi_4,\xi_5,\xi_6) \in \mathbb{R}^6$. Denoting by $\Xi'$ the extension of the transformation $\Xi$ to $\mathbb{R}^6$, it is clear that Im($\Xi'$) remains a 2D plane within the extended period-coordinate space $\mathbb{R}^6$, where the first 4 coordinates $\xi_1,\xi_2,\xi_3,\xi_4$ of each point are the same as in Im($\Xi$) before. Let us see now what happens, for example, in case (a) above: In this case, wherever in $\mathbb{R}^4$ our plane Im($\Xi$) was close (say, up to $\varepsilon$) to a half-integer point, there come the two new coordinates $\xi_5$ and $\xi_6$ and destroy the candidacy of that point as an almost-half-integer within $\mathbb{R}^6$. As we will show below, this happens since the two new coordinates $\xi_5$ and $\xi_6$ are not independent of their predecessors $\xi_1,\xi_2,\xi_3,\xi_4$: as it can be seen from Eq. (8.8) above, $\xi_5 = \xi_4-\xi_2$ and $\xi_6 = \xi_1-\xi_3$. Note that if $\xi_1,\ldots,\xi_6$ were all independent of each other, then some of the almost-half-integer points in $\mathbb{R}^4$ would be, indeed, destroyed by $\xi_5$ and $\xi_6$, but infinitely many other almost-half-integer points would still remain almost-half-integer points in $\mathbb{R}^6$, too.

Let us see now how we can explain cases (1)–(3) above mathematically, using our new algebraic formulation.

Let us first consider the two superposed screens which are oriented to angles 30° and –30°. Assume at first that both screens have a black dot centered on the origin (see Fig. 8.14(a)). Since the superposition of these two screens is irrational (almost-periodic), at no point $(x,y)$ in the superposition except for the origin a precise dot superposition may occur; but at infinitely many points $(x,y)$ we have an almost-perfect dot superposition, where $\xi_1,\xi_2,\xi_3,\xi_4$ are almost-integers, or an almost-perfect white space superposition, where $\xi_1,\xi_2,\xi_3,\xi_4$ are almost-half-integers. Let $(x,y)$ be such a point (of either type); this means, therefore, that at this point:

$$\begin{pmatrix}\xi_1\\ \xi_2\end{pmatrix}-\begin{pmatrix}\xi_3\\ \xi_4\end{pmatrix}\approx\begin{pmatrix}m\\ n\end{pmatrix} \quad \text{where } m,n\in\mathbb{Z} \tag{8.9}$$

namely:

$$\tfrac{1}{T}\left[\mathrm{M}_{30}\begin{pmatrix}x\\ y\end{pmatrix}-\mathrm{M}_{-30}\begin{pmatrix}x\\ y\end{pmatrix}\right]\approx\begin{pmatrix}m\\ n\end{pmatrix}$$

$$\tfrac{1}{T}\left[\begin{pmatrix}\frac{\sqrt{3}}{2} & \frac{1}{2}\\ -\frac{1}{2} & \frac{\sqrt{3}}{2}\end{pmatrix}\begin{pmatrix}x\\ y\end{pmatrix}-\begin{pmatrix}\frac{\sqrt{3}}{2} & -\frac{1}{2}\\ \frac{1}{2} & \frac{\sqrt{3}}{2}\end{pmatrix}\begin{pmatrix}x\\ y\end{pmatrix}\right]\approx\begin{pmatrix}m\\ n\end{pmatrix}$$

$$\frac{1}{T}\begin{pmatrix}0 & 1\\ -1 & 0\end{pmatrix}\begin{pmatrix}x\\ y\end{pmatrix}\approx\begin{pmatrix}m\\ n\end{pmatrix}$$

$$\frac{1}{T}\begin{pmatrix}y\\ -x\end{pmatrix}\approx\begin{pmatrix}m\\ n\end{pmatrix}$$

and therefore:

$$\frac{1}{T}\begin{pmatrix}x\\ y\end{pmatrix}\approx\begin{pmatrix}k\\ l\end{pmatrix} \quad \text{with } k,l\in\mathbb{Z} \tag{8.10}$$

Now, if the third, 0°-screen is superposed in-phase (i.e., with a black dot centered on the origin, like in Fig. 8.14(c)), then (8.10) means:[13]

$$\begin{pmatrix} \xi_5 \\ \xi_6 \end{pmatrix} \approx \begin{pmatrix} k \\ l \end{pmatrix} \qquad \text{with } k,l \in \mathbb{Z} \tag{8.11}$$

This shows, therefore, that at any point $(x,y)$ in the superposition which satisfies condition (8.9), and in particular, at any point $(x,y)$ where $\xi_1,...,\xi_4$ are almost integers (giving a 2-layer almost-dot-centered rosette) or almost-half-integers (giving a 2-layer almost-clear-centered rosette), the period-coordinates $\xi_5,\xi_6$ of the third, 0°-screen are necessarily almost-integer. This means that the third screen contributes to the superposition a black dot of its own very close to $(x,y)$. This explains, indeed, case (1) above.

If, however, the third, 0°-screen is superposed in counter-phase with respect to the 30°- and the –30°-screens (i.e., with a white space centered on the origin, like in Fig. 8.14(d)), then Eq. (8.10) gives:

$$\begin{pmatrix} \xi_5 \\ \xi_6 \end{pmatrix} - \begin{pmatrix} \frac{1}{2} \\ \frac{1}{2} \end{pmatrix} \approx \begin{pmatrix} k \\ l \end{pmatrix} \qquad \text{with } k,l \in \mathbb{Z}$$

This means that at any point $(x,y)$ in the superposition where $\xi_1,\xi_2,\xi_3,\xi_4$ are almost-integers (giving a 2-layer almost-dot-centered rosette) or almost-half-integers (giving a 2-layer almost-clear-centered rosette), the period-coordinates $\xi_5,\xi_6$ of the third, 0°-screen are necessarily almost-half-integer, so that the third screen contributes a white-space centered very close to $(x,y)$. This is, indeed, the explanation of case (2) above.

Finally, in case (3), where the three screens are superposed with a white space centered on the origin, the demonstration remains the same as in case (2).

Note, however, that if the third screen were independent of the first two superposed screens, then the microstructure in the 3-screen superposition would remain "uniformly disordered" and invariant under layer shifts, like in the original 2-screen superposition.

## 8.9 Variance or invariance of the microstructure under layer shifts

As we can see, the microstructure of the conventional 3-screen superposition is not invariant under shifts of the individual layers because in this case the superposed layers are not independent of each other. When the superposed layers are independent — as in the case of the 2-screen superposition at 0° and 30° — all types of almost-rosettes are simultaneously present in the superposition, and no substantial microstructure changes occur when individual layers are shifted.

This illustrates, indeed, the following general result:

[13] Note that if one already observed from Eq. (8.8) that $\xi_5 = \xi_4 - \xi_2$ and $\xi_6 = \xi_1 - \xi_3$, then Eq. (8.11) can be directly deduced from Eq. (8.9).

**Proposition 8.8:** A non-trivial shift of individual layers in the superposition causes a substantial change in the microstructure of the superposition *iff* their frequency vectors $\mathbf{f}_i$ are linearly dependent over $\mathbb{Z}$, i.e., *iff* there exist $k_i \in \mathbb{Z}$ not all of them 0 such that $\Sigma k_i \mathbf{f}_i = \mathbf{0}$. But this precisely means that the superposition is *singular*.[14] ■

**Example 8.3:** Let us illustrate this result with a few singular or non-singular cases:

(i) *A 2-screen periodic, singular case*: The periodic 2-screen superposition of Example 8.1 above (Sec. 8.7) is singular, and therefore layer shifts may cause substantial changes in its microstructure (see also Sec. 8.2.1 and Figs. 8.3(b),(c)).

(ii) *A 3-screen almost-periodic, singular case*: The conventional 3-screen superposition is singular, and indeed, as we have seen above, layer shifts may cause substantial changes in its microstructure (see Fig. 8.5).

(iii) *A 3-screen periodic, singular case*: Any rational approximation of (ii) is singular (see Proposition 8.4), and therefore layer shifts may cause substantial changes in its microstructure (see Fig. 8.11).

(iv) *A 2-screen almost-periodic, non-singular case*: A stable moiré-free 2-screen superposition, like the superposition of two identical screens with angle difference of 30°, is non-singular; therefore its microstructure consists of a uniform blend of rosettes of all types, and it is not substantially influenced by layer shifts (see Fig. 8.8).

(v) *A 3-screen almost-periodic, non-singular case*: A stable moiré-free 3-screen superposition, like the screen combination discussed in Secs. 3.5–3.6 and shown in Fig. 3.8, is non-singular; therefore its microstructure consists of a uniform blend of rosettes of all types, and it is not substantially influenced by layer shifts.

(vi) *A 4-screen almost-periodic, non-singular case*: The superposition of four identical screens with equal angle differences of 22.5° (see Fig. 8.16 and also Remark 3.1 in Chapter 3) is non-singular. Therefore its microstructure consists of a blend of rosettes of all types, and it is not substantially influenced by layer shifts: Although each layer shift is distinct, all rosettes types are equally represented in all the different layer shifts, and there is no predominance of one particular rosette type in each layer shift.

It should be noted, however, that even when substantial microstructure modifications do occur, i.e., when the superposition is singular, they still may be more visible or less visible; but Proposition 8.8 does not say which cases give more significant or less significant modifications and why. In general, a moiré which is clearly visible with a strong amplitude has, by Propositions 8.1–8.2, significantly different in-phase and counter-phase microstructure. Therefore, when it becomes singular the microstructure variation (between

[14] Note that during the previous discussion we considered the linear dependence (or independence) over $\mathbb{Z}$ of the scalars $\xi_i$. However, this is equivalent to the linear dependence (or independence) over $\mathbb{Z}$ of the frequency vectors $\mathbf{f}_i$, since: $\Sigma k_i \mathbf{f}_i = \mathbf{0} \Leftrightarrow \forall \mathbf{x}\ (\Sigma k_i \mathbf{f}_i) \cdot \mathbf{x} = 0 \Leftrightarrow \forall \mathbf{x}\ \Sigma k_i \mathbf{f}_i \cdot \mathbf{x} = 0 \Leftrightarrow \forall \mathbf{x}\ \Sigma k_i \xi_i = 0$ (by Eq. (8.3)). An interesting result of this equivalence is that just as the spectral interpretation of a $(k_1,...,k_m)$-singular superposition is $\Sigma k_i \mathbf{f}_i = \mathbf{0}$, its image-domain interpretation is that, for any point $\mathbf{x}$ in the $x,y$ plane, $\Sigma k_i \xi_i = 0$ (provided that all the superposed layers are given in their initial phase). For example, in Fig. 8.12, which illustrates a (1,2,-2,-1)-singular superposition, any point $\mathbf{x}$ in the $x,y$ plane satisfies: $\xi_1 + 2\xi_2 - 2\xi_3 - \xi_4 = 0$. (In the spectral domain we have, of course, $\mathbf{f}_1 + 2\mathbf{f}_2 - 2\mathbf{f}_3 - \mathbf{f}_4 = \mathbf{0}$.)

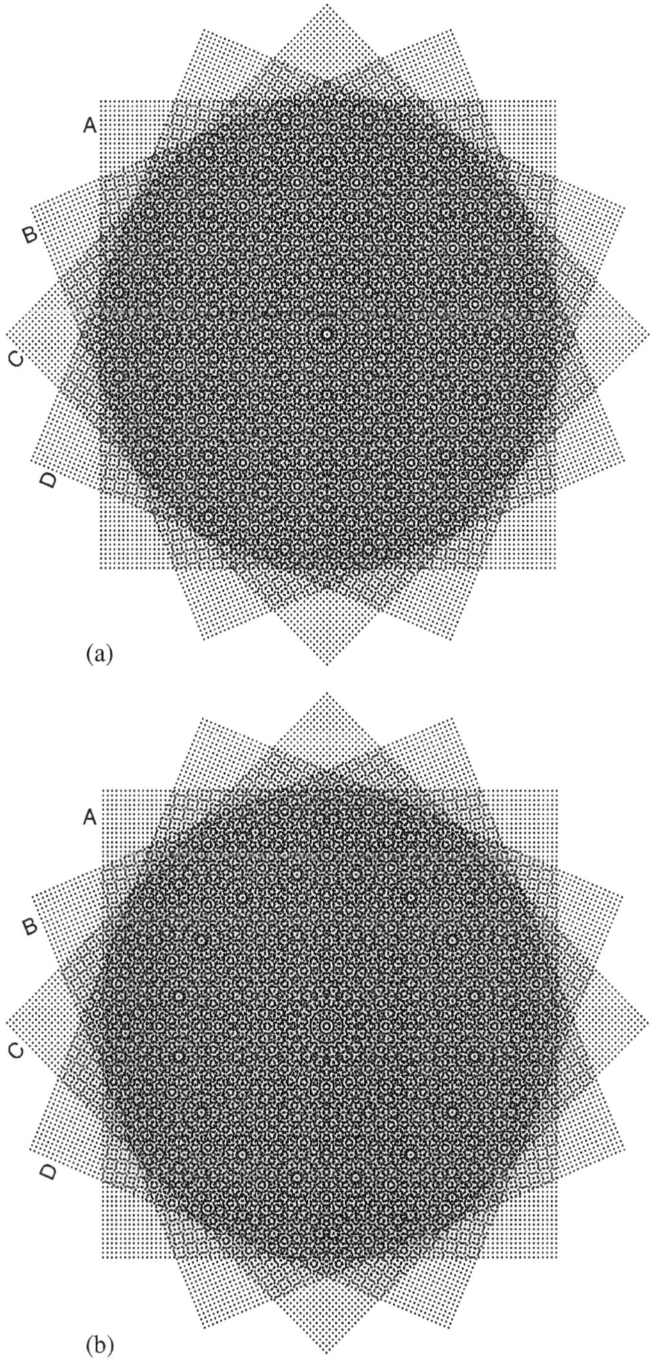


**Figure 8.16:** The superposition of four identical screens with equal angle differences of 22.5°: (a) in-phase superposition; (b) counter-phase superposition.

in-phase and counter-phase rosettes) due to layer shifts will be clearly visible. However, the higher the order of the singular state, the less visible are its microstructure changes in the superposition, and the less perceptible are the rosette-type changes which arise due to layer shifts in the singular state (see Proposition 8.2) — although they still exist according to Proposition 8.8. This can be illustrated by the following example:

(vii) *Periodic, stable moiré-free cases*: Strictly speaking, the only possible periodic, non-singular case is the superposition of two non-collinear gratings (see consequence (a) of Proposition 5.2 and Remark 2 in Table 5.2, Chapter 5). However, rational approximations of stable moiré-free cases such as (iv) or (v), which are, of course, periodic, still have a uniform looking microstructure which is very similar to the microstructure of the original stable moiré-free case and seems to be invariant under layer shifts. The reason is that although these rational approximants are, indeed, singular (see Proposition 8.4), the moirés which cause their singularities are of quite high orders. ■

Finally, it is interesting to compare Proposition 8.8 with Proposition 7.3 of Chapter 7. Proposition 7.3 tells us which layer shifts in the superposition will cause microstructure changes and which layer shifts will only cause a global shift (rigid motion) of the whole superposition. Proposition 8.8, for its part, tells us when microstructure changes that occur due to layer shifts will be *substantial*. Consider, for example, the superposition of two identical screens at 0° and 30°. According to Proposition 7.3, a non-trivial layer shift in this case will not cause a global rigid motion, but rather a microstructure change. And indeed, the microstructure resulting from this shift cannot be obtained by any global shift of the original superposition; for example, the perfect dot-centered rosette on the origin will be lost, and after the non-trivial shift there will exist no perfect dot-centered rosette in the superposition. And yet, according to Proposition 8.8, since the given superposition is not singular, this microstructure change is not substantial: the overall look of the microstructure is not altered by the shift, and all types of almost-rosettes still coexist in the superposition just as before the shift (see Fig. 8.8). However, a similar non-trivial layer shift in a singular superposition such as the conventional 3-screen superposition will indeed cause a substantial change in the microstructure (as shown in Fig. 8.5). Propositions 7.3 and 8.8 are schematically summarized in Fig. 8.17.

## 8.10 Period-coordinates and period-shifts in the Fourier decomposition

Having obtained our main results concerning the microstructure of the superposition, we return now to the roles of the period-coordinates $\xi_i$ and the function $\Xi(x,y)$. We have already noticed the remarkable analogy between the period-coordinates $\xi_i$ and the period-shifts $\phi_i$ defined in Chapter 7. In this section we will try to better understand this analogy, and to tie up all the remaining loose ends: We will show the precise roles of both $\xi_i$ and $\phi_i$ in the Fourier representations of the layer superposition and each of its moirés, and we will see how the functions $\Xi(x,y)$ and $S_A(x,y)$ are related to these Fourier representations.

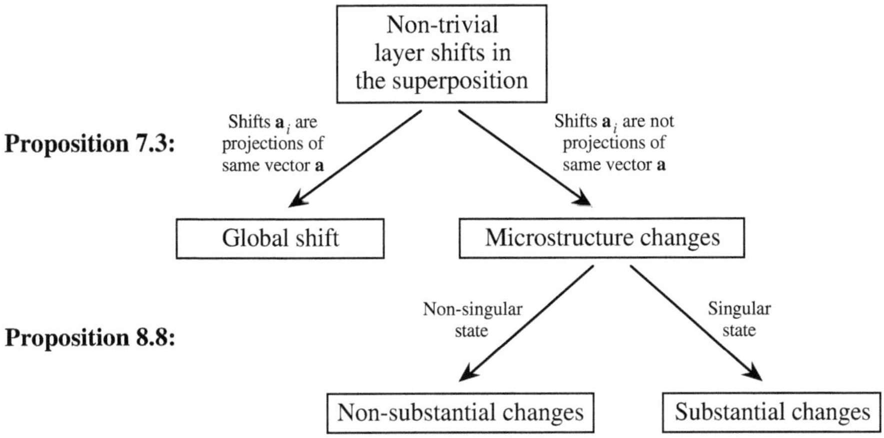


**Figure 8.17:** A schematic block diagram summarizing the effects of layer shifts on the microstructure of the superposition. (Note that *trivial* layer shifts, i.e., shifts of individual layers by multiples of their own periods, do not cause any distinguishable change in the superposition, not even a global shift.)

We have seen in Sec. 6.8 that the superposition of $m$ 1-fold periodic layers can be expressed in the form of a Fourier series as follows (Eq. (6.11)):

$$p_1(\mathbf{x}) \cdot \ldots \cdot p_m(\mathbf{x})$$

$$= \Big(\sum_{n_1=-\infty}^{\infty} c^{(1)}_{n_1}\, e^{i2\pi n_1 \mathbf{f}_1\cdot\mathbf{x}}\Big) \cdot \ldots \cdot \Big(\sum_{n_m=-\infty}^{\infty} c^{(m)}_{n_m}\, e^{i2\pi n_m \mathbf{f}_m\cdot\mathbf{x}}\Big)$$

$$= \sum_{n_1=-\infty}^{\infty} \ldots \sum_{n_m=-\infty}^{\infty} c^{(1)}_{n_1}\cdot\ldots\cdot c^{(m)}_{n_m}\; e^{i2\pi(n_1\mathbf{f}_1+\ \ldots\ +n_m\mathbf{f}_m)\cdot\mathbf{x}} \tag{8.12}$$

or in a more compact, vector notation:

$$= \sum_{\mathbf{n}} c_{\mathbf{n}}\, e^{i2\pi \mathbf{n}^{\mathrm{T}} \mathrm{F}\mathbf{x}} \tag{8.13}$$

where $\mathbf{n}$ is the index vector $(n_1,\ldots,n_m)$ (the superscript "T" indicating its role as a row vector in the matrix product), and F is the column multi-vector consisting of the frequency vectors $\mathbf{f}_1,\ldots,\mathbf{f}_m$, i.e., a $2{\times}m$ matrix.

As we have already seen in Sec. 7.6, Eqs. (8.12) or (8.13) can be interpreted in terms of frequency vectors as:

$$= \sum_{n_1=-\infty}^{\infty} \ldots \sum_{n_m=-\infty}^{\infty} c_{n_1,\ldots,n_m}\, e^{i2\pi \mathbf{f}_{n_1,\ldots,n_m}\cdot\mathbf{x}} \tag{8.14}$$

or in a more compact, vector form:

$$= \sum_{\mathbf{n}} c_{\mathbf{n}}\, e^{i2\pi \mathbf{f}_{\mathbf{n}}\cdot\mathbf{x}} \tag{8.15}$$

where:

$$c_{\mathbf{n}} = c_{n_1,\dots,n_m} = c^{(1)}_{n_1}\cdot\ldots\cdot c^{(m)}_{n_m}$$

$$\mathbf{f}_{\mathbf{n}} = \mathbf{f}_{n_1,\dots,n_m} = n_1\mathbf{f}_1+\ldots+n_m\mathbf{f}_m = \mathbf{n}\cdot(\mathbf{f}_1,\dots,\mathbf{f}_m) = \mathbf{n}^{\mathrm{T}}\,\mathrm{F} \tag{8.16}$$

However, using Eq. (8.3) we can also rewrite Eq. (8.12) as follows:

$$= \sum_{n_1=-\infty}^{\infty} \cdots \sum_{n_m=-\infty}^{\infty} c^{(1)}_{n_1}\cdot\ldots\cdot c^{(m)}_{n_m}\; e^{i2\pi(n_1\mathbf{f}_1\cdot\mathbf{x}_1+\,\dots\,+n_m\mathbf{f}_m\cdot\mathbf{x}_m)}$$

$$= \sum_{n_1=-\infty}^{\infty} \cdots \sum_{n_m=-\infty}^{\infty} c^{(1)}_{n_1}\cdot\ldots\cdot c^{(m)}_{n_m}\; e^{i2\pi(n_1\xi_1+\,\dots\,+n_m\xi_m)} \tag{8.17}$$

$$= \sum_{n_1=-\infty}^{\infty} \cdots \sum_{n_m=-\infty}^{\infty} c_{n_1,\dots,n_m}\; e^{i2\pi\xi_{n_1,\dots,n_m}} \tag{8.18}$$

or in a vector form which explicitly brings out the variable **x**:

$$= \sum_{\mathbf{n}} c_{\mathbf{n}}\, e^{i2\pi \mathbf{n}\cdot\Xi(\mathbf{x})} \tag{8.19}$$

where:

$$c_{\mathbf{n}} = c_{n_1,\dots,n_m} = c^{(1)}_{n_1}\cdot\ldots\cdot c^{(m)}_{n_m}$$

$$\xi_{\mathbf{n}} = \xi_{n_1,\dots,n_m} = n_1\xi_1+\ldots+n_m\xi_m = \mathbf{n}\cdot(\xi_1,\dots,\xi_m) = \mathbf{n}\cdot\Xi(\mathbf{x})$$

Here, $c_{n_1,\dots,n_m}$ and $\xi_{n_1,\dots,n_m}$ are, respectively, the Fourier coefficient and the period-coordinate of the $(n_1,\dots,n_m)$-term in the Fourier decomposition of the superposition. Note that this term corresponds in the spectrum of the superposition to the $(n_1,\dots,n_m)$-impulse which is located at the frequency given by Eq. (8.16).

We see, therefore, that the Fourier decomposition of the superposition, which is expressed by Eqs. (8.12) or (8.13), can be interpreted in two different ways:

(a) To highlight the role of the spectral-domain frequency vectors, the Fourier decomposition of the superposition can be written in the form:

$$= \sum_{\mathbf{n}} c_{\mathbf{n}}\; e^{i2\pi \mathbf{f}_{\mathbf{n}}\cdot\mathbf{x}} \tag{8.20}$$

(b) To highlight the role of the image-domain period-coordinates, the Fourier decomposition of the superposition can be written in the new form:

$$= \sum_{\mathbf{n}} c_{\mathbf{n}}\, e^{i2\pi \mathbf{n}\cdot\Xi(\mathbf{x})} \tag{8.21}$$

Comparing these two equations with Eq. (8.13) we see that in the first case the first two elements of $\mathbf{n}^{\mathrm{T}}\mathrm{F}\mathbf{x}$ in the exponent have been grouped together into: $\mathbf{n}^{\mathrm{T}}\mathrm{F} = n_1\mathbf{f}_1+\ldots+n_m\mathbf{f}_m = \mathbf{f}_{\mathbf{n}}$, whereas in the second case the two last elements of $\mathbf{n}^{\mathrm{T}}\mathrm{F}\mathbf{x}$ have been grouped together to give: $\mathrm{F}\mathbf{x} = (\mathbf{f}_1\cdot\mathbf{x},\dots,\mathbf{f}_m\cdot\mathbf{x}) = (\xi_1,\dots,\xi_m) = \Xi(\mathbf{x})$. The significance of this second interpretation will become more evident in the case of shifted layers, as we will see below.

Returning back to the superposition (8.12), the $(k_1,...,k_m)$-moiré extracted from this superposition contains only the impulses of the $(k_1,...,k_m)$-comb (see Eq. (6.15)), and it can be therefore rewritten now in a similar way either in the form (a):

$$\begin{aligned} m_{k_1,...,k_m}(\mathbf{x}) &= \sum_{n=-\infty}^{\infty} c^{(1)}{}_{nk_1}\cdot...\cdot c^{(m)}{}_{nk_m}\; e^{i2\pi n(k_1\mathbf{f}_1+\,...\,+k_m\mathbf{f}_m)\cdot\mathbf{x}} \\ &= \sum_{n=-\infty}^{\infty} c_{n(k_1,...,k_m)}\; e^{i2\pi n\mathbf{f}_{k_1,...,k_m}\cdot\mathbf{x}} \\ &= \sum_{n=-\infty}^{\infty} c_{n\mathbf{k}}\; e^{i2\pi n\mathbf{f}_\mathbf{k}\cdot\mathbf{x}} \end{aligned} \tag{8.22}$$

where $\mathbf{f}_{k_1,...,k_m}$ is the frequency vector of this moiré, or in the new form (b):

$$\begin{aligned} m_{k_1,...,k_m}(\mathbf{x}) &= \sum_{n=-\infty}^{\infty} c^{(1)}{}_{nk_1}\cdot...\cdot c^{(m)}{}_{nk_m}\; e^{i2\pi n(k_1\xi_1+\,...\,+k_m\xi_m)} \\ &= \sum_{n=-\infty}^{\infty} c_{n(k_1,...,k_m)}\; e^{i2\pi n\mathbf{k}\cdot\Xi(\mathbf{x})} \\ &= \sum_{n=-\infty}^{\infty} c_{n\mathbf{k}}\; e^{i2\pi n\mathbf{k}\cdot\Xi(\mathbf{x})} \end{aligned} \tag{8.23}$$

It is interesting to note that just as $\mathbf{f_k} = \mathbf{f}_{k_1,...,k_m} = k_1\mathbf{f}_1+...+k_m\mathbf{f}_m$ is the frequency vector of the $(k_1,...,k_m)$-moiré, $\xi_\mathbf{k} = \xi_{k_1,...,k_m} = k_1\xi_1+...+k_m\xi_m$ is the period-coordinate of the $(k_1,...,k_m)$-moiré at any point $\mathbf{x}$ in the $x,y$ plane.[15] For example, in the case of a (1,-1)-moiré between two gratings, the period-coordinate of the moiré at any point $\mathbf{x}$ is given by $\xi_{1,-1} = \xi_1-\xi_2$. Note the similarity with the period-shifts of the moiré in Eq. (7.21).

Suppose now that each of the layers $p_i(\mathbf{x})$ in the superposition has been shifted from its initial position by $\mathbf{a}_i$ (for any layer $i$ which remains unshifted, we simply take $\mathbf{a}_i = \mathbf{0}$). We have, therefore (see Eq. (7.15)):

$$\begin{aligned} &p_1(\mathbf{x}-\mathbf{a}_1)\cdot\,...\,\cdot p_m(\mathbf{x}-\mathbf{a}_m) \\ &= \Big(\sum_{n_1=-\infty}^{\infty} c^{(1)}{}_{n_1}\; e^{i2\pi n_1\mathbf{f}_1\cdot(\mathbf{x}-\mathbf{a}_1)}\Big)\cdot\,...\,\cdot\Big(\sum_{n_m=-\infty}^{\infty} c^{(m)}{}_{n_m}\; e^{i2\pi n_m\mathbf{f}_m\cdot(\mathbf{x}-\mathbf{a}_m)}\Big) \\ &= \sum_{n_1=-\infty}^{\infty}\cdots\sum_{n_m=-\infty}^{\infty} c^{(1)}{}_{n_1}\cdot...\cdot c^{(m)}{}_{n_m}\; e^{i2\pi(n_1\mathbf{f}_1+\,...\,+n_m\mathbf{f}_m)\cdot\mathbf{x}\,-\,i2\pi(n_1\mathbf{f}_1\cdot\mathbf{a}_1+\,...\,+n_m\mathbf{f}_m\cdot\mathbf{a}_m)} \end{aligned} \tag{8.24}$$

or, in a more compact vector notation:

$$= \sum_{\mathbf{n}} c_\mathbf{n}\; e^{i2\pi\mathbf{n}^\mathrm{T}(\mathrm{F}\mathbf{x}\,-\,\boldsymbol{\phi})} \tag{8.25}$$

where $\boldsymbol{\phi} = (\phi_1,...,\phi_m) = (\mathbf{f}_1\cdot\mathbf{a}_1,...,\mathbf{f}_m\cdot\mathbf{a}_m)$.

This can be written, of course, in a form similar to (a):

---

[15] Note that the period-coordinate of any 1-fold periodic function (such as a single layer $p_i(\mathbf{x})$ or the $(k_1,...,k_m)$-moiré $m_{k_1,...,k_m}(\mathbf{x})$) at point $\mathbf{x}$ is given by a single number, while the period-coordinate of a superposition is given by the vector $(\xi_1,...,\xi_m)$ of the period-coordinates of the individual layers at point $\mathbf{x}$.

$$= \sum_{\mathbf{n}} c_{\mathbf{n}}\, e^{i2\pi(\mathbf{f_n}\cdot\mathbf{x} - \mathbf{n}\cdot\boldsymbol{\phi})} \tag{8.26}$$

However, using Eq. (8.3) we may rewrite Eq. (8.24) in the more symmetric form:

$$= \sum_{n_1=-\infty}^{\infty} \dots \sum_{n_m=-\infty}^{\infty} c^{(1)}{}_{n_1}\cdot\ldots\cdot c^{(m)}{}_{n_m}\; e^{i2\pi(n_1\xi_1+\ \dots\ +n_m\xi_m) - i2\pi(n_1\phi_1+\ \dots\ +n_m\phi_m)} \tag{8.27}$$

$$= \sum_{n_1=-\infty}^{\infty} \dots \sum_{n_m=-\infty}^{\infty} c_{n_1,\dots,n_m}\; e^{i2\pi(\xi_{n_1,\dots,n_m} - \phi_{n_1,\dots,n_m})}$$

or in a vector notation which explicitly brings out the variable $\mathbf{x}$ and highlights the role of the transformations $\Xi(\mathbf{x}) = (\xi_1,\dots,\xi_m)$ and $S_{\mathbf{A}}(\mathbf{x}) = (\phi_1,\dots,\phi_m)$ in the image domain:

$$= \sum_{\mathbf{n}} c_{\mathbf{n}}\, e^{i2\pi\mathbf{n}\cdot(\Xi(\mathbf{x}) - S_{\mathbf{A}}(\mathbf{x}))} \tag{8.28}$$

and using Eq. (8.7):

$$= \sum_{\mathbf{n}} c_{\mathbf{n}}\, e^{i2\pi\mathbf{n}\cdot\Xi_{\mathbf{A}}(\mathbf{x})}$$

where:

$$c_{\mathbf{n}} = c_{n_1,\dots,n_m} = c^{(1)}{}_{n_1}\cdot\ldots\cdot c^{(m)}{}_{n_m}$$

$$\xi_{\mathbf{n}} = \xi_{n_1,\dots,n_m} = n_1\xi_1+\dots+n_m\xi_m = \mathbf{n}\cdot(\xi_1,\dots,\xi_m) = \mathbf{n}\cdot\Xi(\mathbf{x})$$

$$\phi_{\mathbf{n}} = \phi_{n_1,\dots,n_m} = n_1\phi_1+\dots+n_m\phi_m = \mathbf{n}\cdot(\phi_1,\dots,\phi_m) = \mathbf{n}\cdot S_{\mathbf{A}}(\mathbf{x})$$

$c_{n_1,\dots,n_m}$, $\xi_{n_1,\dots,n_m}$ and $\phi_{n_1,\dots,n_m}$ are, respectively, the Fourier coefficient, the period-coordinate and the period-shift of the $(n_1,\dots,n_m)$-term in the Fourier decomposition of the superposition.

It is instructive to compare Eq. (8.28) with its 1D equivalent, the Fourier decomposition of a periodic function $p(x)$ which has been shifted by the quantity $a$ (see Eq. (7.8)):

$$\begin{aligned} p(x-a) &= \sum_{n=-\infty}^{\infty} c_n\, e^{i2\pi nf(x-a)} \\ &= \sum_{n=-\infty}^{\infty} c_n\, e^{i2\pi nfx - i2\pi nfa} \\ &= \sum_{n=-\infty}^{\infty} c_n\, e^{i2\pi n(\xi-\phi)} \end{aligned} \tag{8.29}$$

As we can see, Eq. (8.28) is an extension of the 1D case, in which scalar values and indices are generalized into vector quantities.

Returning back to superposition (8.24), the $(k_1,\dots,k_m)$-moiré extracted from this superposition contains only the impulses of the $(k_1,\dots,k_m)$-comb, and is therefore given by:

$$\sum_{n=-\infty}^{\infty} c^{(1)}{}_{nk_1}\cdot\ldots\cdot c^{(m)}{}_{nk_m}\; e^{i2\pi n(k_1\mathbf{f}_1+\ \dots\ +k_m\mathbf{f}_m)\cdot\mathbf{x} - i2\pi n(k_1\mathbf{f}_1\cdot\mathbf{a}_1+\ \dots\ +k_m\mathbf{f}_m\cdot\mathbf{a}_m)} \tag{8.30}$$

This can be expressed, again, either in a form similar to (a):

$$= \sum_{n=-\infty}^{\infty} c_{n\mathbf{k}}\, e^{i2\pi n(\mathbf{f_k}\cdot\mathbf{x} - \mathbf{k}\cdot\boldsymbol{\phi})} \tag{8.31}$$

or in the new, more symmetric form (b):

$$= \sum_{n=-\infty}^{\infty} c_{n\mathbf{k}} \, e^{i2\pi n\mathbf{k}\cdot(\xi_{\mathbf{k}} - \phi_{\mathbf{k}})}$$

$$= \sum_{n=-\infty}^{\infty} c_{n\mathbf{k}} \, e^{i2\pi n\mathbf{k}\cdot(\Xi(\mathbf{x}) - S_{\mathbf{A}}(\mathbf{x}))} \qquad (8.32)$$

$$= \sum_{n=-\infty}^{\infty} c_{n\mathbf{k}} \, e^{i2\pi n\mathbf{k}\cdot\Xi_{\mathbf{A}}(\mathbf{x})}$$

which highlights the role of the transformations $\Xi(\mathbf{x})$ and $S_{\mathbf{A}}(\mathbf{x})$ in the image domain.

Eq. (8.32) is clearly a generalization of:

$$\sum_{n=-\infty}^{\infty} c_{nk} \, e^{i2\pi\, nk(\xi - \phi)} \qquad (8.33)$$

which is a partial sum of the 1D Fourier decomposition (8.29).

## PROBLEMS

8-1. How does the microstructure in the superposition of dot-screens depend on the size of the dots? (This can be checked, for example, by superposing screen gradations rather than constant dot-screens.)

8-2. How does the microstructure in the superposition of dot-screens depend on the shape of the dots (circular, triangular, "1"-shaped, etc.)? Suppose that the dot-screens of Fig. 8.5 are replaced by dot screens with triangular dots. How will this influence the shapes of the dot-centered rosettes and of the clear-centered rosettes in the superposition?

8-3. Is there any similarity between the rosette shapes in a dot-screen superposition and in the equivalent line-grid superposition? (See, for example, Fig. 2.10 in Sec. 2.11.)

8-4. Let $p_1(x,y)$ and $p_2(x,y)$ be two binary dot-screens (i.e., dot-screens having only the values of 0 and 1 for black and white, respectively), and suppose that they generate a visible moiré effect in their superposition. The spectrum of the layer superposition contains infinitely many new impulses that are generated in the spectrum convolution. These new impulses are responsible for all the new macro- and microstructures that appear in the superposition but not in the individual layers. As we already know from Chapter 4, if we extract from the spectrum of the layer superposition only the new, main impulse-cluster around the origin and take its inverse Fourier transform, we obtain back in the image domain the isolated contribution of the corresponding moiré effect.

(a) What do you expect to obtain by *eliminating* the impulses of the main cluster from the spectrum of the superposition, and taking the inverse Fourier transform of the remaining spectrum?

(b) What do you expect to obtain by extracting from the spectrum of the layer superposition all the new impulses which do not belong to the main cluster, and taking their inverse Fourier transform?

(c) Will the resulting images in (a) and (b) be binary, like the two original screens and their superposition?

8-5. Is it correct to say that whenever different types of rosettes are generated in alternate areas of the superposed image, the eye observes a different gray level in each of these areas and hence a macro-moiré becomes visible? What happens in high-order moirés that are hardly visible in the superposition? And what happens in the pseudo-moiré cases of Problems 2-10, 2-11 and 2-12?

8-6. According to Proposition 7.3, non-trivial layer shifts in the superposition of two line-gratings cause a rigid motion of the superposition as a whole *iff* their frequency vectors $\mathbf{f}_1$, $\mathbf{f}_2$ are not collinear. However, suppose that $\mathbf{f}_1$ and $\mathbf{f}_2$ *are* collinear. How does layer shifts influence this two-grating superposition:

(a) When the superposition is singular (for example, when $\mathbf{f}_2 = \mathbf{f}_1$)?

(b) When the superposition is not singular (for example, when $\mathbf{f}_2 = \sqrt{2}\mathbf{f}_1$)?

8-7. Verify Proposition 8.3.

8-8. A student suggests the following reasoning: "If a screen superposition is rational then the period-vectors $\mathbf{T}_i$ of the superposed screens are linearly dependent over $\mathbb{Z}$, i.e., there exist integers $n_i$ not all zeroes such that $\sum n_i\mathbf{T}_i = \mathbf{0}$. But this precisely means that the superposition is singular." Do you agree with this reasoning? In particular, we know that a superposition of periodic layers with frequency vectors $\mathbf{f}_i$ is singular *iff* there exist integers $k_i$ not all zeroes such that $\sum k_i\mathbf{f}_i = \mathbf{0}$; does this statement remain true if we replace $\mathbf{f}_i$ with $\mathbf{T}_i$? Explain.

8-9. Discuss the spectral domain interpretation of Proposition 8.8 (see also the last paragraphs in Secs. 8.3 and 8.5).

8-10. *Rational approximants for the 3-screen superposition with angle differences of 30°.* Fig. 8.11 gives a rational approximation of the singular, irrational 3-screen superposition with identical frequencies and angle differences of 30° that is shown in Fig. 8.10. Find and plot a few other rational approximants for the same irrational superposition. Observe that the better the approximation, the higher the similarity in the rosette shapes.

8-11. *Rational approximants for the 2-screen superposition with angle difference of 30°.* Fig. 8.8 shows the irrational 2-screen superposition with identical frequencies and an angle difference of 30°. Find and plot a few rational approximations for this irrational superposition. *Example*: The superposition of two identical screens with an angle difference of 28.0725° (in which the (1,-4,1,4)-moiré is singular). Plot this case in in-phase superposition and in counter-phase superposition, and observe the micro-structure difference between the two cases. Why does the rational approximant show such a microstructure variability, in contrast to the irrational case with angle difference of 30° (Fig. 8.8(b),(c))? Will closer approximants show a smaller difference between their in-phase and counter-phase microstructures?

8-12. *Microstructures in the superposition of non-regular screens.* Suppose that hexagonal screen B of Fig. 2.13(b) is rotated counterclockwise on top of hexagonal screen A. Analyze the microstructure of this superposition when the angle difference between the layers is 30°, 60° or 90°. How do these microstructures behave under layer shifts?

8-13. We have seen in Chapter 4 how one can design periodic layers that give in their superposition a moiré effect (i.e., a *macrostructure*) whose period has any desired shape, such as "1" (Fig. 4.4), "EPFL" (Plate 1), etc. Is it possible to design periodic layers that give in their superposition a *microstructure* of a given shape (say, rosettes having the form of "1")?

# Chapter 9

# Polychromatic moiré effects

## 9.1 Introduction

In the previous chapters we have outlined, based on the Fourier theory, the main principles of the moiré phenomenon in the superposition of any number of periodic structures. However, our attention was limited until now to the monochrome case, in which all the superposed structures (and hence also their moiré effects) consisted of black, white or intermediate gray levels. It is our aim now to proceed to the polychromatic case, too.

The first steps in the exploration of moiré effects in colour, based on the classical moiré approach, can be found in [Oster63] and in some applications to colour printing [Yule67; Wurzburg61]. More recent contributions include notably the work of Bryngdahl [Bryngdahl81] and some other applications [Patorski93 pp. 392–294; Hoy92]. However, as we have already seen in the monochrome case, the best approach for exploring phenomena in the superposition of periodic structures is the spectral approach, which is based on the Fourier theory. Our aim in the present chapter is, therefore, to extend our Fourier-based approach to the polychromatic case, and to provide thus a full quantitative analysis of the polychromatic moiré effects in the superposition of coloured periodic layers. This goal will be achieved by combining the Fourier theory that we have been using up till now with elements from colorimetry and colour vision.

In order to investigate polychromatic moirés in the superposition of any coloured periodic layers we have to consider the full colour spectrum of each point in any of the superposed layers. For this end we introduce both into the image domain and into the Fourier frequency domain a new dimension $\lambda$, representing the visible light wavelengths. In the image domain we represent each layer by the chromatic reflectance (or transmittance) function $r(x,y;\lambda)$, which is a generalization of the reflectance (or transmittance) function $r(x,y)$ in the monochrome case. Consequently, in the Fourier spectral domain each impulse amplitude becomes a function of $\lambda$. All the results obtained by our Fourier-based approach in the monochrome case remain valid in the polychromatic case, too, for every wavelength $\lambda$ separately. This enables us to find, for any given moiré and any point $(x,y)$, the full colour spectrum $\{r(x,y;\lambda) \mid 380 \le \lambda \le 750\}$ which expresses the visible colour of the moiré in question at the point $(x,y)$.

We start this chapter by reviewing in Sec. 9.2 some basic notions of colour theory. Then in Sec. 9.3 we show how the fundamental notions of our Fourier-based approach are extended from the monochrome case to the polychromatic case. In Sec. 9.4 we generalize the extraction of moiré profiles into the polychromatic case, and then, in Sections 9.5 and 9.6, we illustrate our theoretic results by several examples of polychromatic moirés: In Sec. 9.5 we discuss the case of the (1,-1)-moiré between colour line-gratings, and in Sec.

9.6 we discuss the two-dimensional (1,0,-1,0)-moiré between colour dot-screens. In Sec. 9.7 we briefly show how this discussion is generalized to higher order moirés as well.

## 9.2 Some basic notions from colour theory

The perception of colour is based on two main factors: the physical properties of light, and the physiological properties of the human visual system. Let us briefly review here some of the basic physical and physiological principles of colour perception.

### 9.2.1 Physical aspects of colour

Physically, visible light is an electromagnetic radiation, just like Gamma rays, X-rays, or radio waves. The various types of electromagnetic radiation differ from each other in their wavelengths. On the continuous range of electromagnetic wavelengths, Gamma rays occupy the range of extremely short wavelengths ($10^{-12}$ m and below), while radio waves are situated at much longer wavelengths (about 1 m to $10^4$ m). The range of visible light makes up a very small part of the electromagnetic spectrum, which extends between about 380 and 750 nanometers (where 1 nm = $1 \times 10^{-9}$ m). The wavelengths $\lambda$ within the range of visible light are perceived by the eye as different colours. The colours of purely monochromatic light (i.e., light consisting of precisely one wavelength $\lambda$) correspond to the colours which are observed in a rainbow. They gradually vary from violet (around 400 nm) through indigo, blue (around 440 nm), green (around 520 nm), yellow, orange, and red (around 700 nm). However, pure monochromatic colours only cover a small part of the totality of visible colours. In fact, by mixing together different quantities of light of different wavelengths, many new colours are obtained. Each such colour can be represented by its *colour spectrum* which specifies its precise light composition, namely, how much light it contains from each wavelength between $\lambda = 380$ nm and $\lambda = 750$ nm. A few schematic examples of such colour spectra (with values between 0 and 1, relative to the light source) and their corresponding colours are given in Fig. 9.1(a)–(f). Other important examples are the spectrum of white colour, which is constantly 1 (since it contains a full contribution from each light wavelength), and the spectrum of black, which is constantly 0 (since it contains no light at all). Note, however, that in reality colour spectra are rarely flat like in these examples, and they are more often represented by a continuous, curved function (whose values vary between 0 and 1). The spectrum of a pure spectral colour, however, consists of a single spike which is precisely located at the corresponding wavelength $\lambda$.

The colour of any physical object is obtained, therefore, by a mixture of light of many different wavelengths which is returned from the object and captured by our eye. If the object is a source of light (such as a lamp, the sun, etc.) the light coming from the object is, indeed, emitted by the object itself. However, in many other cases the object itself does not generate light, but just re-emits the light shed onto it by an external light source: either by reflection, if the object is opaque, or by transmission, if the object is transparent. Any object normally absorbs some of the light which illuminates it, and reflects (or transmits) to various degrees other parts of the light. The precise amount of light from each wavelength which is reflected (or transmitted) by an object is a characteristic property of the object, and it determines the spectrum of the light which is re-emitted from the object, and hence the visible colour of the object. The colour of an object depends, therefore, on the spectrum of the light source which illuminates it and on the proper spectral characteristics of the object itself.

The colour spectrum of the light which is reflected or transmitted by the given object can be measured by a spectrophotometer [McDonald97 pp. 59–61], and displayed graphically or numerically. This colour spectrum unambiguously determines the colour of the object in question. Note, however, that we did not yet specify how to deduce from a given spectrum the corresponding colour that is perceived by the eye; this question will be addressed in the following subsection. Other important notions that will be explained in later sections are the additive and the multiplicative composition rules of colour spectra, namely, the rules which determine the colours obtained by colour compositions.

### 9.2.2 Physiological aspects of colour

In the previous paragraphs we described some of the basic physical properties of light and their relationship to colour. However, it should be remembered that, after all, the light that we see is captured by our eyes, and its interpretation as a colour is done by our brain. This means that colour perception depends not only on the physical properties of light but also on physiological properties of the human visual system.

And, indeed, it turns out that the human eye has only three types of colour-sensitive light receptors (called *cones*), having peak sensitivities to blue, green and red light at about 440 nm, 545 nm and 580 nm [Foley90 pp. 576–577]. This is, indeed, the physiological basis of the tristimulus theory of colour perception, which tries to represent each visible colour as a point within a 3D space, whose coordinates correspond to the response of the three receptor types. Significant efforts have been done by the *Commission Internationale de l'Eclairage* (CIE) in the elaboration of such a 3D representation of the visible colours, in a way which would best correspond to human colour vision. This work was based on many experimental data which were collected in numerous experiments with human observers. These efforts resulted, over the years, in the definition of several such spaces, including the CIE tristimulus XYZ system. A precise mathematical formulation has been elaborated which permits to deduce from any given colour spectrum the XYZ coordinates of the colour perceived by the eye [Wyszecki82 pp. 156–168]. From these XYZ colorimetric values one can also find the colour coordinates in terms of the CIE L*a*b* colour space [*ibid.*], or in terms of the native device-dependent RGB colour space of any given colour display device [Foley90 pp. 585–589], thus permitting to visualize the colour in question on that device.

It is important to emphasize here the fundamental difference between the representation of a given colour in the form of a colour spectrum, and the trichromatic representation of the same colour as a point in a 3D space. While the first representation of the colour corresponds to its physical properties, the latter corresponds to its subjective perception by the human eye. Although the description of a given colour as a triplet of values is more concise, it does not preserve all the information which is provided by its full spectral representation. Moreover, it turns out that the mapping between the set of all colour spectra and the points of the 3D colour space is not injective, which means that many different colour spectra may produce the same point in the 3D volume: all of them are perceived by the human eye as the same colour. Such colours are called *metameric*. Although they are perceived by the eye as one and the same colour, they are physically different — and indeed (see, for example, [Wyszecki82 Sec. 3.3.10] or [Hunt87 pp. 72–73, 179–181]), two colour samples which are metameric under a certain illuminant may be perceived as having different colours under a different light source (which means that their colour spectra will be mapped, this time, into distinct points in the 3D colour space).

As we can see, the colour spectrum remains the most precise and unambiguous way to specify colours. And indeed, in our discussion below we will always represent colours by their full colour spectra, and not by triplets in a 3D space.

**Remark 9.1:** Note that in the present chapter we will simultaneously deal with the *colour spectrum* (the wavelengths spectrum of the visible light), and the *Fourier spectrum* (that we still use as in the monochrome case, but this time for each light wavelength $\lambda$ separately). Whenever this double use of the term "spectrum" may be ambiguous we will explicitly specify which of its two meanings is intended. ■

## 9.3 Extension of the spectral approach to the polychromatic case

As we have seen in the previous chapters, the spectral approach that we use for analyzing layer superpositions is based on the duality between 2D images in the $x,y$ plane and their 2D Fourier spectra in the $u,v$ frequency plane through the 2D Fourier transform. Let us see now how this duality can be extended to the polychromatic case, too.

### 9.3.1 The representation of images and image superpositions

We have seen in Sec. 2.2 that any image in the monochrome case is represented in the image domain by a *reflectance* function, which assigns to each point $(x,y)$ of the image a value between 0 and 1 representing its light reflectance: 0 for black (i.e., no reflected light), 1 for white (full light reflectance), and intermediate values for in-between shades. In the case of transparencies, the reflectance function is replaced by a *transmittance* function defined in a similar way. The fact that the superposition of black and any other shade always gives black suggested a *multiplicative* model for the superposition of monochrome images. Thus, when $m$ monochrome images are superposed, the reflectance of the resulting image is given by the product of the reflectance functions of the individual images (Eq. 2.1):

$$r(x,y) = r_1(x,y) \cdot \ldots \cdot r_m(x,y) \tag{9.1}$$

And hence, according to the convolution theorem, the spectrum of the superposition is given by the convolution (2.2):

$$R(u,v) = R_1(u,v) ** \ldots ** R_m(u,v) \tag{9.2}$$

In the polychromatic case, however, to each point $(x,y)$ of the image we can associate not only a single value between 0 and 1 representing its light reflectance (or transmittance), but rather a full colour spectrum. This colour spectrum gives for every wavelength $\lambda$ of the visible light (approximately between $\lambda = 380$ nm for violet and $\lambda = 750$ nm for red) a value between 0 and 1, which represents the reflectance (or the transmittance) of light of wavelength $\lambda$ at the point $(x,y)$ of the image. In other words, to each point $(x,y)$ of the image belongs a *reflectance* (or *transmittance*) *colour spectrum* $\{r(x,y;\lambda) \mid 380 \leq \lambda \leq 750\}$. The colour perceived by the eye at the point $(x,y)$ of the image can be deduced from this colour spectrum as already explained in Sec. 9.2.2. We will henceforth represent any colour image by its *chromatic* reflectance (or transmittance) function, $r(x,y;\lambda)$, which is a straightforward generalization of the reflectance (or transmittance) function $r(x,y)$ in the monochrome case.

When coloured layers are superposed (for example, by superposing transparencies or by overprinting), at any point $(x,y)$ of the superposition each of the individual layers ideally behaves as a colour filter, which blocks (or absorbs) light at some wavelengths, and reflects (or transmits) light, to various degrees, at other wavelengths.[1] This means that the multiplicative model for layer superpositions remains valid in the polychromatic case, too, *for every wavelength $\lambda$ separately*; this is schematically illustrated in Fig. 9.1. Therefore, the multiplicative superposition rule (9.1) becomes in the polychromatic case:[2]

$$r(x,y;\lambda) = r_1(x,y;\lambda) \cdot \ldots \cdot r_m(x,y;\lambda) \tag{9.3}$$

And according to the convolution theorem we get in the Fourier spectral domain, for every wavelength $\lambda$ separately:

$$R(u,v;\lambda) = R_1(u,v;\lambda) ** \ldots ** R_m(u,v;\lambda) \tag{9.4}$$

where $R_i(u,v;\lambda)$ is the 2D Fourier transform of the chromatic reflectance function $r_i(x,y;\lambda)$ for a given wavelength $\lambda$ (the Fourier transform being taken with respect to the first two variables).

This means that the theory that we have developed in previous chapters for the superposition of periodic layers in the monochrome case remains valid also in the

[1] We assume here an ideal situation in which each point $(x,y)$ in the superposition behaves independently of its neighbourhood. We do not consider effects such as light scattering inside the printed support, which may create some mutual influence between neighbouring points, or various interactions between the superposed inks which may cause them to deviate from the behaviour of ideal filters.

[2] This assumes, of course, that the overprinted inks are not opaque. In the case of opaque inks the multiplicative superposition rule is no longer valid, and it is replaced by another rule according to which the chromatic reflectance of the superposition at any point $(x,y)$ equals to the chromatic reflectance of the last printed ink at that point.

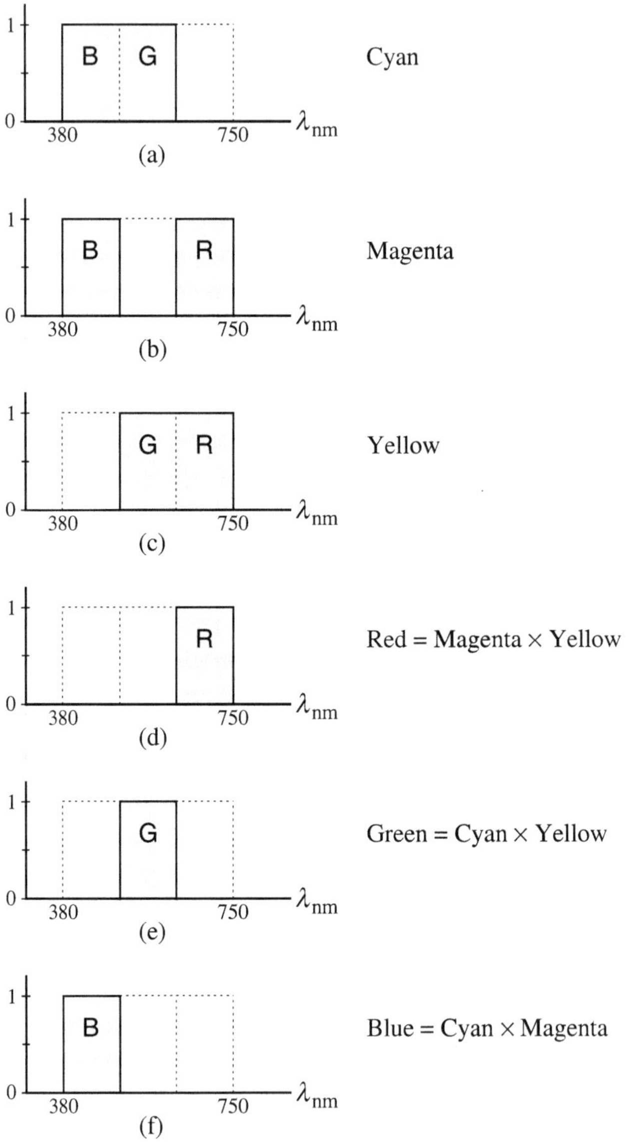


**Figure 9.1:** A schematic illustration of the multiplicative composition of colour spectra which occurs in the superposition of colour filters. (a)–(c): Schematic colour spectra of cyan, magenta and yellow filters; for instance, a cyan filter absorbs light in the red range of the spectrum and returns light in the green and blue ranges. (d)–(f): Superposition of pairs of these filters gives multiplicative combinations of their colour spectra, which correspond to the colours red, green and blue. (This phenomenon is called in colour theory a *subtractive colour combination*.)

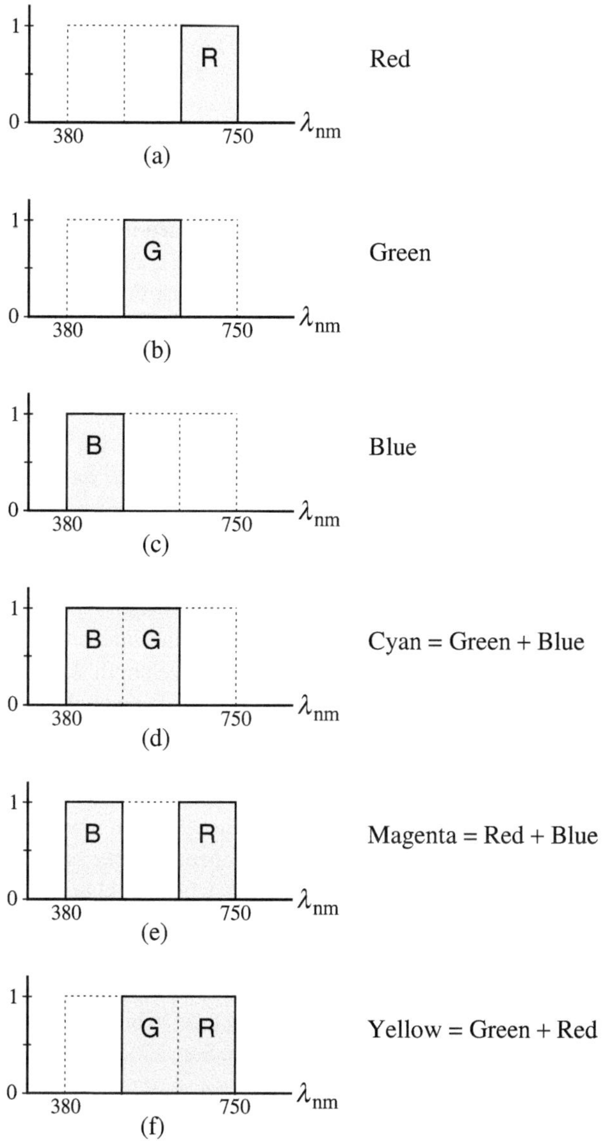


**Figure 9.2:** A schematic illustration of the additive composition of colour spectra, which is the basis for colour blending by averaging. (a)–(c): Schematic spectra representing the colours red, green and blue. (d)–(f): Blending of pairs of these spectra (for example, by juxtaposing small dots of the respective colours or by simultaneous projection) gives additive compositions of the original spectra, which correspond to the colours cyan, magenta and yellow. (This phenomenon is called in colour theory an *additive colour combination.*)

polychromatic case, but this time, for every wavelength $\lambda$ separately. In fact, $\lambda$ can be seen as a third dimension which is added to each 2D image description, both in the image domain and in the Fourier spectral domain.

As in the monochrome case, we are basically interested here in images $r(x,y;\lambda)$ which are *periodic* on the continuous $x,y$ plane, such as line-gratings or dot-screens, and their superpositions (obviously, in the polychromatic case a period of the image $r(x,y;\lambda)$ may consist of elements of various colours). As we have seen in Sec. 2.2 this implies that the Fourier spectrum of the image on the $u,v$ plane is not a continuous one but rather consists of impulses, which correspond to the frequencies in the Fourier series decomposition of the image. In the case of a 1-fold periodic image, such as a line-grating, the spectrum consists of a 1D comb of impulses through the origin; in the case of a 2-fold periodic image the spectrum is a 2D nailbed of impulses through the origin.

We have seen in Sec. 2.2 that in the monochrome case each of these impulses in the Fourier spectrum was characterized by three main properties: its *label* (which is its index in the Fourier series development); its *geometric location* (representing its frequency vector $\mathbf{f}$), and its *amplitude*. In the polychromatic case, the locations of the impulses remain independent of the wavelength $\lambda$, but the amplitude of each impulse is no longer a constant value, but rather a function of the light wavelength $\lambda$. We call this function the *chromatic amplitude* of the impulse; it represents the intensity of the corresponding periodic component in the image at each light wavelength $\lambda$. Note that the impulse amplitudes may be complex if the image is not symmetric about the origin.

### 9.3.2 The influence of the human visual system

The question of whether or not an impulse in the spectrum represents a *visible* periodic component in the image strongly depends on properties of the human visual system. We already know from the monochrome case that when looking at a small area from a sufficiently long viewing distance, our eyes no longer see the detail within the small area but rather *average* the fine detail and record only the overall intensity of the area. This property, known as the *spatial integration* of our visual system [Foley90 p. 568], is the basis for the halftoning technique which is widely used in bilevel printing devices to create a visual impression of continuous gray levels (see Sec. 3.2).[3]

In the polychromatic case, the spatial integration property of the eye is further extended and takes also a chromatic dimension: the averaging by the eye of fine details of different colours gives us a visual impression of a single uniform tint, whose colour spectrum is an average of the colour spectra of the original colours, weighted by the surface percentage of the different colours. This is based on the additive composition of colour spectra, which is schematically illustrated in Fig. 9.2; this principle is the basis for colour generation on display devices such as colour TV screens [Hunt87, Chapter 3].

[3] Note that in Sec. 2.2 we expressed the same property from the dual, Fourier point of view, in terms of cutoff frequencies in the spectral domain.

**Remark 9.2:** Note that the perception of colour involves two distinct principles of colour spectra composition: The *additive* composition of colour spectra occurs in the case of registered projection of light of different colours [Hunt87, pp. 10–11], or in its weighted-mean variant, in the juxtaposition of tiny coloured elements [*ibid.* pp. 20–24]. The *multiplicative* composition of colour spectra, represented by Eq. (9.3), occurs in the superposition of colour filters. Figs. 9.1 and 9.2 illustrate the two different cases. Note that in colour theory the multiplicative colour composition is usually called a *subtractive* colour combination [*ibid.* Chapter 4], since each of the superposed layers subtracts some portions from the spectrum of the original incident light. ■

### 9.3.3 The Fourier-spectrum convolution and the superposition moirés

We have seen in the monochrome case that when $m$ line-gratings are superposed in the image domain, the resulting Fourier spectrum is the convolution of their individual Fourier spectra; this is illustrated for the case of $m = 2$ in Fig. 2.5(a)–(f). This convolution of combs can be seen as an operation in which frequency vectors from the individual spectra are added vectorially, while the corresponding impulse amplitudes are multiplied. The geometric location of the general $(k_1,...,k_m)$-impulse in the spectrum-convolution is given by Eq. (2.26):

$$\mathbf{f}_{k_1,...,k_m} = k_1\mathbf{f}_1 + ... + k_m\mathbf{f}_m \tag{9.5}$$

and its amplitude is given by Eq. (2.27):

$$a_{k_1,...,k_m} = a^{(1)}{}_{k_1} \cdot ... \cdot a^{(m)}{}_{k_m} \tag{9.6}$$

In the polychromatic case these considerations still remain true, with the only difference that this time the impulse amplitudes should be understood as *chromatic amplitudes*, i.e., as functions of the light wavelength $\lambda$. Therefore, Eq. (9.6) will be rewritten here as follows:

$$a_{k_1,...,k_m}(\lambda) = a^{(1)}{}_{k_1}(\lambda) \cdot ... \cdot a^{(m)}{}_{k_m}(\lambda) \tag{9.7}$$

Eq. (9.5), for its part, remains unchanged; note that in particular, in the special case of $m = 2$ gratings, when a moiré effect occurs due to the (1,-1)-impulse in the convolution, Eqs. (2.9)–(2.11) are still valid.

## 9.4 Extraction of the moiré intensity profiles

Let us start by recalling the situation in the monochrome case. Assume that we are given two line-gratings (see Fig. 2.5(a),(b)). As we have seen in Chapter 2, the spectrum of each of these line-gratings (Fig. 2.5(d),(e)) consists of an infinite impulse-comb, in which the amplitude of the $n$-th impulse is given by the coefficient of the $n$-harmonic term in the Fourier series development of that line-grating. When we superpose (i.e., multiply) the two line-gratings (Fig. 2.5(c)) the spectrum of the superposition is the convolution of the two original combs, which gives an oblique nailbed of impulses (Fig. 2.5(f)). Each moiré which appears in the grating superposition is represented in the spectrum of the superposition by a comb of impulses passing through the origin which is included in this nailbed. If a moiré is visible in the superposition, it means that in the Fourier spectral domain the fundamental impulse-pair of the moiré-comb is located inside the visibility circle, close to the spectrum origin; this impulse-pair determines the period and the direction of the moiré. Now, as we have seen in Chapter 4, by extracting from the spectrum-convolution only this infinite moiré-comb and taking its inverse Fourier transform, we can reconstruct, back in the image domain, the isolated contribution of the moiré in question to the image superposition; this is the *intensity profile* of the moiré (see Fig. 4.2(a),(b)).

Moreover, we have also seen in Chapter 4 that thanks to the $T$-convolution theorem the extraction of the moiré intensity profiles can be done not only in the Fourier spectral domain, but also directly in terms of the image domain. To take the simplest examples, in the case of a (1,-1)-moiré between two line-gratings or a (1,0,-1,0)-moiré between two dot-screens the moiré intensity profile is simply a normalized $T$-convolution of the two original superposed layers, which is magnified and rotated by about 90° relative to the original layers. This is illustrated for the case of the (1,-1)-moiré in Figs. 2.5 and 4.2.

These considerations can be now extended also to the polychromatic case. The only difference is the introduction of the new dimension $\lambda$ into both domains: in the image domain the intensity at each point $(x,y)$ becomes a function of $\lambda$: $r(x,y;\lambda)$; and in the Fourier spectral domain, the impulse amplitudes are now considered as functions of $\lambda$.

In the following sections we illustrate the polychromatic extension of the moiré profile extraction by two simple examples: a (1,-1)-moiré between two colour line-gratings, and a (1,0,-1,0)-moiré between two colour dot-screens.

## 9.5 The (1,-1)-moiré between two colour line-gratings

When two colour gratings with similar periods are superposed with a small angle difference, a (1,-1)-moiré is generated which consists of chromatic bands. These moiré bands have, of course, the same period and angle as in the equivalent monochrome superposition, as predicted by Eqs. (2.9). The only difference between the monochrome and the polychromatic cases is in the colours of the macroscopic moiré bands and of the microstructure elements from which they are composed (see Plate 2).

The visible colours of these moiré bands could be found via the microstructure of the superposition (the small elements of constant colour in the superposition) by calculating or measuring the relative areas and the colours of the different micro-elements in the superposition, and computing at each location the additive composition of the contributing colours, weighted by their respective areas.

However, the Fourier-based theory we have developed permits us to find quantitatively the colours of the moiré bands in a much more general and effective way: As we already know, the chromatic intensity profile of the (1,-1)-moiré is simply a normalized and rotated *T*-convolution of the two original superposed layers. As shown by Figs. 9.3 and 9.4, the polychromatic *T*-convolution is similar to the monochrome *T*-convolution, except that it is performed for each wavelength $\lambda$ separately. This means that to each point of the period on the $x$ axis we add in the $y$ direction the colour-spectrum belonging to that point, ranging between $\lambda = 380...750$ nm. In both Figs. 9.3 and 9.4, (a) shows a sequence of two periods of the first original grating, (b) shows two periods of the second grating, and (c) shows two periods of the moiré intensity profile (= the resulting *T*-convolution). In each case, the $x$ axis shows periods in the image domain (in terms of length units), while the $y$ axis represents the wavelengths $\lambda$; a vertical section at any given point on the $x$ axis (i.e., at any point within the period) gives the full colour-spectrum which belongs to that given point. From this colour-spectrum one can deduce the colour perceived by the eye at that point, in terms of any colour space, as explained in Sec. 9.3.1 above.

Let us illustrate the polychromatic *T*-convolution by the two cases shown in Figs. 9.3 and 9.4:

Fig. 9.3 has been drawn for the colours of Plate 2(g); as we can see, the (1,-1)-moiré bands obtained between a grating of narrow white lines on a black background and a grating with a multicolour period are essentially a magnified and rotated version of the multicolour grating, where the intensity values of 1 are scaled down to a value $A$ which is determined in the *T*-convolution by the width of the narrow white lines. Note that the sharp step transitions of grating (b) have been replaced in the *T*-convolution (c) by softer ramps; the wider the white lines in grating (a), the wider become these ramps, causing some overlapping between the R,G,B stripes of the moiré in (c) and hence the appearance of intermediate colours between them. It is interesting to note that the polychromatic moiré shown in Plate 2(g) and Fig. 9.3 is, in fact, similar to the colourful moiré effect which often occurs on a colour TV screen when the image shown includes repetitive patterns (such as fine patterns of a jacket or bright stripes on a dark dress). The interference between these small periods in the image and the repetition period of the RGB phosphor dots on the TV screen may generate very colourful and spectacular moiré effects.[4]

Fig. 9.4 corresponds to the case of Plate 2(f), an example in which both of the superposed gratings are polychromatic. In this case, two identical gratings whose period consists of alternating red and green lines are superposed; the resulting moiré profiles (=the resulting *T*-convolution) consist of yellow and black bands.

It should be noted (see Remark 9.2) that while the colour of each microstructure element in the superposition is obtained as a *multiplicative composition* of the colour spectra of the

[4] In Trinitron tubes colour phosphors are indeed arranged as vertical stripes [Hunt87 p. 26]; in other types of TV screens the colour phosphors are arranged as repetitive dot-screens, but the resulting moiré effects are similar.

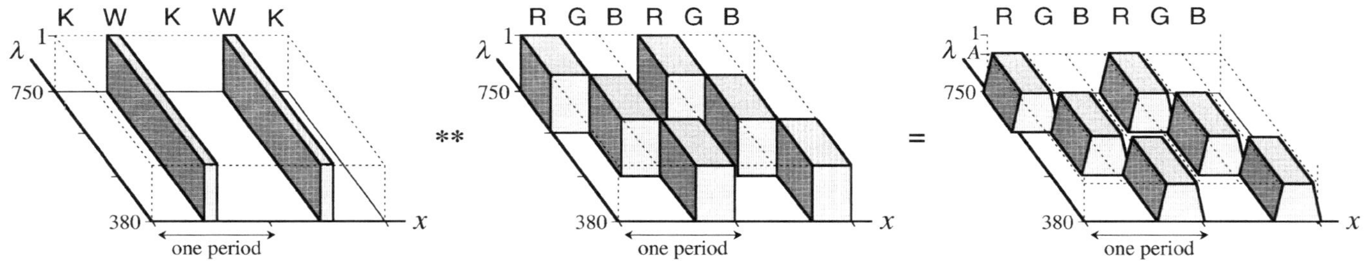


(a) (b) (c)

**Figure 9.3:** The chromatic intensity profile of a (1,-1) superposition moiré is the polychromatic $T$-convolution of the two superposed gratings. This figure corresponds to the colours of Plate 2(g), assuming ideal colour spectra as in Figs. 9.1 and 9.2. Parts (a) and (b) of the figure show the periods of the two superposed gratings, and (c) shows the periods of the resulting chromatic moiré intensity profile (the polychromatic $T$-convolution). In each case, the $x$ axis shows two periods, and the $\lambda$ axis represents the wavelengths. The $T$-convolution is performed for each wavelength $\lambda$ separately. A vertical section at any point on the $x$ axis (i.e. at any point within the period) gives the colour-spectrum at that point of the period.

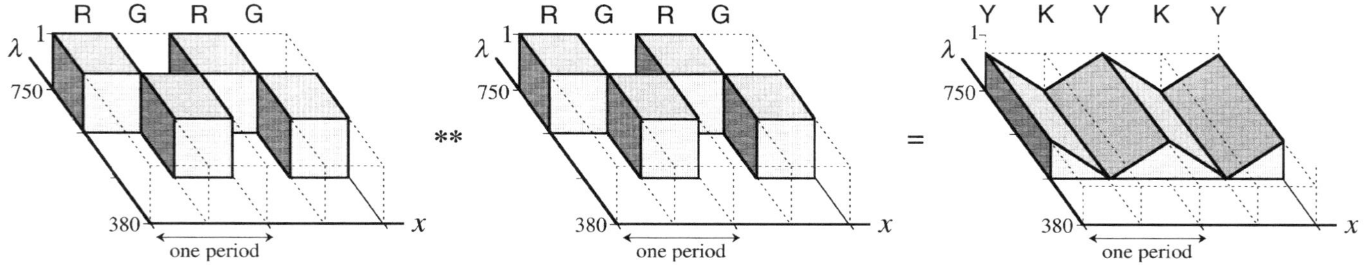


(a) (b) (c)

**Figure 9.4:** Same as Fig. 9.3, but this time showing the polychromatic $T$-convolution of the superposed gratings of Plate 2(f). The two superposed gratings consisting of alternating red and green stripes are represented by (a) and (b). As shown in (c), the resulting chromatic moiré intensity profile (the $T$-convolution) consists of yellow and black bands.

superposed layers, the overall colour of the moiré bands is obtained as an *additive composition* (or rather, a weighted mean) of the colour spectra of the individual microstructure elements in the superposition.[5] When the superposition is visualized from such a distance that the individual micro-elements are no longer distinguished, the chromatic spatial integration of the eye blends their colours additively, into a weighted mean, as explained above in Sec. 9.3.2. This is automatically taken care of by the polychromatic $T$-convolution, as is clearly illustrated in Fig. 9.4, which shows the polychromatic $T$-convolution in the case of Plate 2(f): As can be seen by a close look at Plate 2(f), the yellow moiré bands are composed of adjacent red and green microstructure elements, which are blended *additively* (as a weighted mean) to give the yellow moiré bands.[6] The black moiré bands consist of microstructure elements in which red and green stripes are superposed, resulting in black micro-elements (since all light wavelengths are blocked); this is a *multiplicative* combination of colour spectra. This case offers, indeed, a nice demonstration of both additive and multiplicative colour spectrum combinations simultaneously.

## 9.6 The (1,0,-1,0)-moiré between two colour dot-screens

A (1,0,-1,0)-moiré is generated when two dot-screens of similar periods are superposed with a small angle difference. As we have seen in Sec. 4.4, in this case, too, the intensity profile of the moiré is a normalized and rotated $T$-convolution of the two original superposed layers. The generalization to the polychromatic case is done, again, by introducing the new parameter $\lambda$; but since this time the periods involved are already two-dimensional, the wavelength $\lambda$ can no longer be presented graphically as in Fig. 9.3. The principle, however, remains the same, and the $T$-convolution is performed for each wavelength $\lambda$ separately.

A particularly interesting case of (1,0,-1,0)-moiré occurs when one of the superposed dot-screens consists of tiny "pinholes" on a black background: in this case, the moiré intensity profiles obtained are essentially a magnified and rotated version of the periods of the second screen, where the magnification rate is controlled by the superposition angle $\alpha$ (see Sec. 4.4). This phenomenon becomes even more spectacular in the chromatic case, as shown in Plate 3. This polychromatic moiré can be easily understood by performing the $T$-convolution for every wavelength $\lambda$ separately, like in the 1D case shown in Fig. 9.3. As we can see, in the upper part of the superposition in Plate 3 the black-and-white screen consists of tiny holes on a black background, so that the resulting moiré intensity profiles are indeed a magnified and rotated version of the coloured screen. In the lower part of the

---

[5] The additive and multiplicative composition rules for colour spectra (Remark 9.2) should not be confused with the additive and multiplicative layer superposition rules of Remark 2.1 (see Sec. 2.2).

[6] This effect of blending red and green microstructure elements into a yellow colour (as on a TV screen) can be best observed in the superposition of colour films on a light table; however, when printed on paper and observed by reflectance, the yellow colour appears darker, rather brownish, owing to the lower light intensity and the spectral impurities of the printing inks [Hunt87 pp. 222–225].

superposition, the coloured screen consists of tiny green triangles on a red background, and the observed moiré intensity profiles consist of red circles on a yellow background. This is explained by Fig. 9.5 which shows the $T$-convolution of the two screens of Plate 3 (with dot sizes corresponding to the bottom of the superposed area) for three representative wavelengths $\lambda$: (a) $\lambda$ in the red region of the colour-spectrum; (b) $\lambda$ in the green region; and (c) $\lambda$ in the blue region. Note that the white areas in the black-and-white screen (around the black circular dots) have the value of 1 in all wavelengths $\lambda$, while the black circular dots themselves have in all wavelengths the value 0. As we can see in the right hand side of Fig. 9.5, the contribution of the red wavelengths to the $T$-convolution consists of red circles on a lighter red background; the contribution of the green wavelengths to the $T$-convolution consists of a light green background around the circles; and the blue wavelengths of the colour-spectrum remain zero throughout the $T$-convolution. Therefore the resulting polychromatic $T$-convolution consists, indeed, of red circles on a yellow background.

Other interesting results can be obtained when both superposed layers consist of colour screens; all of these cases of (1,0,-1,0)-moiré are immediate consequences of the polychromatic $T$-convolution between the original layers in the superposition.

Note that like in the monochrome case, details in the second screen which are smaller than the pinhole size in the first screen will not be clearly visible in the resulting moiré, since they will be blurred and smoothed-out together with their background in the $T$-convolution.

## 9.7 The case of more complex and higher-order moirés

It should be finally mentioned that the polychromatic Fourier-based moiré theory presented in this chapter is completely general, and it remains valid for any number of superposed colour layers, and for moiré effects of any order. This generalization is done much like its black-and-white counterpart in Sec. 4.5, and it should not present any particular difficulties once the simple cases described in the last sections have been understood.

## PROBLEMS

9-1. *The composition rules of colour spectra.* Explain the precise roles of the additive and the multiplicative compositions of colour spectra in generating the colours of the moiré effect in the superposition of polychromatic layers.

9-2. *Different layer superposition rules* (see Remark 2.1 in Sec. 2.2). How would the polychromatic moiré effects of Plate 2 look like if the grating superposition were not based on the multiplicative superposition rule but rather on:

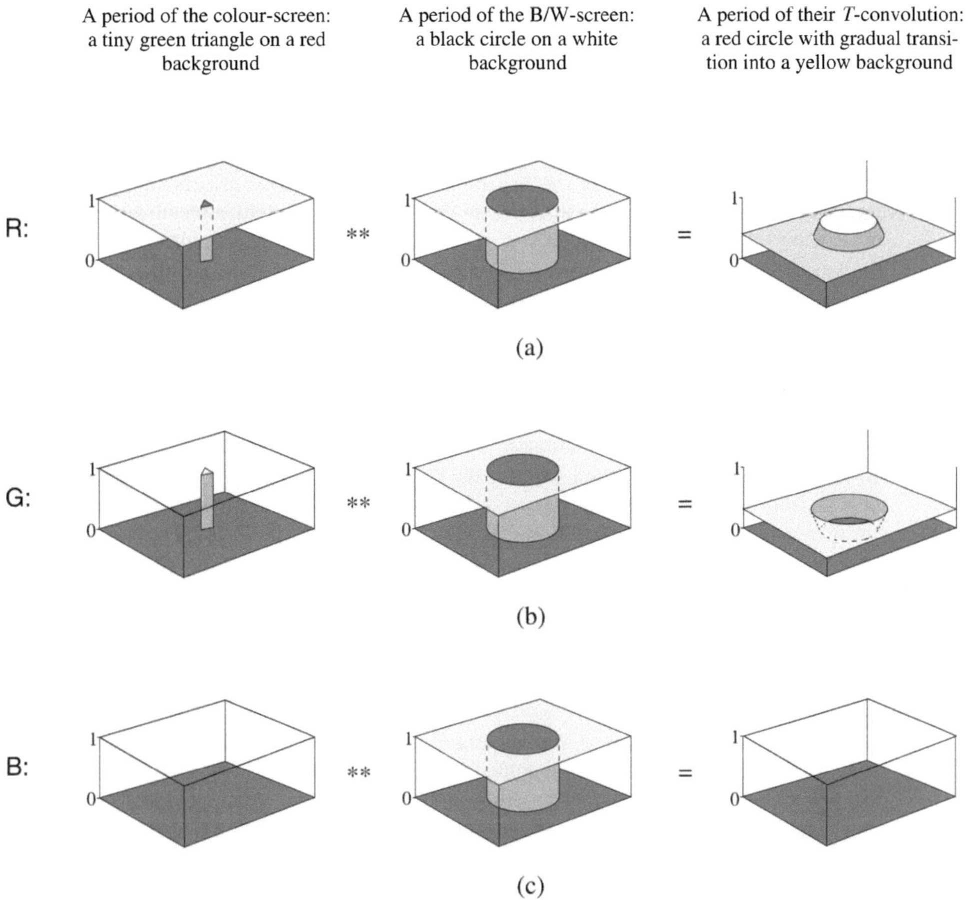


**Figure 9.5:** The $T$-convolution of the two screens of Plate 3 (at the bottom of the superposed area), at three representative wavelengths $\lambda$: (a) in the red region of the colour-spectrum; (b) in the green region; and (c) in the blue region. In each case the period of the first screen is shown in the left part of the figure, the period of the second (black-and-white) screen is shown in the center, and their $T$-convolution (i.e., the resulting moiré intensity profile) is shown to the right. As it can be seen in the right hand side of (a), (b) and (c), respectively, the contribution of the red region of the $T$-convolution consists of red circles on a lighter red background; the contribution of the green region of the $T$-convolution consists of a green background around the circles (and 0 inside them), and the blue region of the colour-spectrum remains zero throughout the $T$-convolution. Therefore, the resulting $T$-convolution consists of red circles on a yellow background (which is the additive composition of the green and the red contributions).

(a) The additive superposition rule?

(b) Any other superposition rule that you can think of?

9-3. What happens to the "1"-shaped moiré of Fig. 4.4 if the "1"-shaped dots of screen B appear in each period in a different colour (for example: if they appear intermittently in cyan, magenta and yellow, or in red, green and blue)?

9-4. *Document security.* How can the use of polychromatic moiré effects improve the application described in Problem 4-13 in the field of document authentication and anti-counterfeiting? For more details see [Amidror02; Amidror01; Amidror99].

9-5. The letters "EPFL" shown in Plate 4 are composed of a colour grating whose period $T$ consists of cyan, magenta and yellow vertical stripes of width $T/3$ each that are juxtaposed next to each other: C,M,Y,C,M,Y... When a transparency with a grating of the same period $T$ consisting of black lines is superposed on these letters with a small angle difference, a very colourful rainbow-like moiré effect appears within the letters.

(a) Explain the colours obtained within one period of this (1,-1)-moiré by drawing a schematic diagram of the polychromatic $T$-convolution, like in Fig. 9.3.

(b) If the period $T$ used is much smaller, the C,M,Y bands are no longer perceived by the eye from a normal viewing distance. But when the black grating transparency is superposed on the letters, the rainbow-like moiré effect is still produced. Explain.

(c) If the period $T$ used is below the resolution of standard colour photocopying machines, a colour photocopy of the image will preserve the overall colour of the letters, but their microstructure will be lost. What do you expect to see when the black grating transparency is superposed on the colour photocopy?

9-6. Can you think of other applications of polychromatic moiré effects in any scientific or technological field?

9-7. *Colorimetric stability of screen superpositions in colour printing.* It is well known in colour printing that using the same angle and frequency for the CMY halftone screens does not normally give satisfactory results. Even if the screen angles and frequencies are perfectly adjusted to keep the singular moiré invisible (i.e., with an infinite period), still the colour produced will depend on whether the dots fall on top of each other or side by side. The colour produced will therefore be extremely sensitive to minute variations in register between the CMY halftone screens. A detailed discussion on this subject, including a colour illustration, can be found in [Kang97 Sec. 9.5]; another good reference is [Yule67 pp. 335, 344]. In the case of the conventional screen combination traditionally used in colour printing, i.e., the superposition of three identical screens with angle differences of 30° (or 60°), colour variations due to registration shifts are much less significant, but they still exist due to the slight difference in dot coverage between dot-centered and clear-centered rosettes (see Sec. 8.3 and Fig. 8.5, and also [Yule67 pp. 339, 341–342] or [Dales94 pp. 13–15]). Stable moiré-free screen superpositions, in which no visible microstructure changes occur due to registration shifts (see Sec. 8.5 and Fig. 3.8), offer a still better colorimetric resistance to screen registration shifts (see for example [Schoppmeyer85 pp. 6–7]).

(a) How do you explain, using the additive and the multiplicative composition rules of colour spectra, the macroscopic colorimetric variations due to microscopic differences in dot registration in the superposed CMY screens?

(b) Do you expect the macroscopic colorimetric variations due to registration shifts to be zero in a superposition of CMY screens with random dot positions?

# Chapter 10

## Moirés between repetitive, non-periodic layers

### 10.1 Introduction

In the preceding chapters we basically restricted ourselves to the study of periodic layers and their superpositions. However, moiré effects are not only limited to the superposition of periodic layers, and, indeed, they often occur between other types of repetitive structures. Typical examples are the moiré effects which may be generated between families of curvilinear gratings (such as concentric circles), between straight gratings with variable spacings, or between any other repetitive geometric structures. Another example, that we have already implicitly used several times in previous chapters, is that of screen gradations, i.e., dot screens having a constant frequency but varying dot sizes and shapes (like in Figs. 4.1, 4.4, etc.).

It is but natural, therefore, to ask whether the scope of our Fourier-based approach can be extended to cover also superpositions of repetitive, non-periodic layers. And indeed, we will see in the present chapter that the basic concepts of our Fourier approach can be carried over to such cases, too. This includes the full duality between the image and the spectral domains, and in particular the possibility to extract the isolated contribution of any moiré either from the spectrum of the superposition or directly from the original layers in the image domain. However, as we will see, the spectrum of repetitive, non-periodic layers may be non-impulsive, or only partially impulsive, so that the spectrum of the superposition (the spectrum convolution) no longer consists of simple replicas of the spectra of the original layers. The analysis of the spectral domain may be therefore more complicated here than in the periodic case. Nevertheless, for most practical needs the *geometry* of moirés between repetitive layers can still be determined rather easily, purely in the image domain, using one of several well-known classical methods; these image-domain methods will be briefly reviewed in Chapter 11.

Our main aim in the present chapter is to understand the general rules which govern the layer superposition of repetitive, non-periodic layers and their resulting moirés, both in the image and in the Fourier spectral domains. We will show how these rules can be used in the *analysis* of moirés that appear in the superposition, as well as in the *synthesis* of moirés having any desired intensity profile and geometric layout. In particular, we will obtain the interesting result that the geometric layout and the periodic profile of the moiré are completely independent of each other: the geometric layout of the moiré is determined by the geometric layouts of the superposed layers, while the periodic profile of the moiré is determined by the periodic profiles of the superposed layers. Readers in a hurry will find the main results of this chapter, including the *fundamental moiré theorem*, in Sec. 10.9 (or in [Amidror98b]); but a leisurely stroll throughout the whole chapter and its many illustrated examples is still recommended for the interested readers.

## 10.2 Repetitive, non-periodic layers

Under the heading of "repetitive, yet non-periodic structures" one may lump together many different types of structures which, although not necessarily periodic, still show some kind of internal structural repetition that is governed by some given rules.

We will restrict ourselves here to three main types of repetitive, non-periodic structures: *coordinate-transformed structures*, *profile-transformed structures*, and *coordinate-and-profile transformed structures*. As their names indicate, such structures can be obtained from a certain initial periodic structure by the application of a *coordinate* transformation (as in the case of a curvilinear grating), a *profile* transformation (as in the case of a screen gradation), or a combination of both. (These three types of transformations are more formally explained in Remark 10.2 below.)

We start our discussion here with coordinate-transformed structures, that will occupy us in most of this chapter due to the special interest they present. Curvilinear gratings, straight gratings with varying frequency, curved line-grids and curved dot-screens all belong to this important category. The best way to deal with each of these structures is to consider it as a non-linear geometric transformation of a periodic, uncurved 2D structure. One may think of this transformation as an operation which "bends" or non-linearly stretches the original periodic structure according to a given mathematical rule.

We begin with the simplest case of coordinate-transformed structures, that of curvilinear gratings. Let $r(x,y)$ (denoting the reflectance or the transmittance at location $x,y$) be the curvilinear grating which is obtained by bending the 2D one-fold periodic grating $p(x')$, i.e., by replacing $x'$ with the function $g_1(x,y)$: $r(x,y) = p(g_1(x,y))$ (see various examples in Fig. 10.1). The intensity profile of the original, uncurved periodic grating $p(x')$ over the $x',y'$ plane, or sometimes its 1D section along the $x'$ axis, will be called the *periodic-profile* of the curvilinear grating $r(x,y)$. The periodic-profile of a curvilinear grating may be cosinusoidal, a square wave, a sawtooth wave, or any other periodic waveform. The function $x' = g_1(x,y)$ which bends $p(x')$ into the curvilinear grating $r(x,y)$ is called the *bending function.*[1, 2]

A curvilinear grating $r(x,y) = p(g_1(x,y))$ is therefore characterized by two basic and independent properties:

(a) Its *geometric layout* in the $x,y$ plane, i.e., the locus of the centers of its curvilinear corrugations in the $x,y$ plane, which is defined by the bending function $x' = g_1(x,y)$;

(b) The intensity behaviour across each of the curvilinear corrugations, which is determined by the periodic-profile $p(x')$.

---

[1] As we will see below, this is, in fact, the first component of a 2D transformation $(x',y') = \mathbf{g}(x,y) = (g_1(x,y),g_2(x,y))$ that is applied to the entire plane.

[2] Note that $x',y'$ are the coordinates of the original, periodic space, while $x,y$ are the coordinates of the target, transformed space; the bending functions are defined, therefore, in the inverse direction, from the transformed coordinates to the original coordinates. More details can be found in Appendix D of *Vol. II*.

Note that since in general the frequency of the curvilinear grating $r(x,y)$ varies throughout the plane, we may arbitrarily choose for its periodic-profile $p(x')$ any frequency. For the sake of convenience we will often choose this frequency to be $f = 1$ (or equivalently, we may include $f$ inside $g_1(x,y)$; for example, given $r(x,y) = \cos(2\pi ax^2)$ we may choose $g_1(x,y) = ax^2$ so that $p(x') = \cos(2\pi x')$ with $f = 1$ rather than $g_1(x,y) = x^2$ with $p(x') = \cos(2\pi ax')$, $f = a$. In such cases the term "*normalized* periodic-profile" will be used.

Before we proceed to a few examples, a general remark about our naming conventions for the different curvilinear gratings is in order:

**Remark 10.1:** In view of properties (a) and (b) above, to each grating there can be attributed two qualifiers: one describing the periodic profile of the grating, and the other describing the geometric layout of the grating. For instance: Fig. 10.1(d) shows a cosinusoidal circular grating, and Fig. 10.8(a) shows a square-wave parabolic grating. Note that the order of the two qualifiers in the grating name is immaterial; for example, we may also say that Fig. 10.1(d) shows a circular cosinusoidal grating. When only one qualifier is explicitly mentioned (for example: a square-wave grating, or an elliptic grating), there is normally no possible doubt as for which of the two qualifiers is intended. However, a remarkable exception is that of a cosinusoidal grating, in which case it may be unclear whether the term "cosinusoidal" refers to the periodic profile of the grating, like in Fig. 10.1(a), or to the geometric layout of the grating, like in Fig. 10.1(*l*). But since we will only rarely need the case of Fig. 10.1(*l*), we prefer to reserve the term "cosinusoidal grating" to the periodic profile of the grating, and to use for cases like Fig. 10.1(*l*) the more cumbersome but less ambiguous term "cosine-shaped grating". ■

**Example 10.1:** Assume that we are given a 2D cosinusoidal grating $p(x') = \cos(2\pi fx')$ over the $x',y'$ plane (Fig. 10.1(a)), and that we "bend" its parallel straight corrugations into parallel parabolas of the shape $y = ax^2+c$, without changing their cosinusoidal profile form (see Fig. 10.1(c)). This can be described mathematically as a non-linear transformation $x' = g_1(x,y) = y-ax^2$, where $a$ is a non-zero constant which defines the "bending rate" of the parabolas.[3] (Note that the level lines $x' = n$ of the surface $x' = y-ax^2$ over the $x,y$ plane are indeed the required parabolas $y = ax^2+n$.) The parabolic cosinusoidal grating obtained by applying this bending function to the original grating $p(x')$ is given, therefore, by: $r(x,y) = p(y-ax^2) = \cos(2\pi f(y-ax^2))$. Its geometric layout is given by the locus of its maxima in the $x,y$ plane, namely: $2\pi f(y-ax^2) = 2\pi k,\ k\in\mathbb{Z}$, and its normalized periodic-profile is $\cos(2\pi x')$. ■

**Example 10.2:** In the case of a circular grating with a cosinusoidal periodic-profile (Fig. 10.1(d)), the original, uncurved periodic wave (the periodic-profile of the circular grating) is again, this time using $r'$ rather than $x'$, $p(r') = \cos(2\pi fr')$. The circular grating $r(x,y)$ is obtained from $p(r')$ by replacing $r'$ with $\sqrt{x^2+y^2}$:

$$r(x,y) = p(\sqrt{x^2+y^2}) = \cos(2\pi f\sqrt{x^2+y^2})$$

[3] Notice that the term *curvature* is defined in mathematics in a different way (see [Courant88 p. 86]).

Note that if we take $r' = x^2 + y^2$ instead of $r' = \sqrt{x^2 + y^2}$ we still obtain a circular grating with a cosinusoidal periodic-profile form, but the concentric circles of this grating are no longer equidistant and they get closer to each other as their radius increases (see Fig. 10.1(g)). The reason is that the surface defined by $r' = x^2 + y^2$ is a top-opened paraboloid (rather than a top-opened cone, as is the surface $r' = \sqrt{x^2 + y^2}$) and therefore the level lines $r' = n$ of the surface $r' = x^2 + y^2$ are concentric circles $x^2 + y^2 = n$ whose radiuses are given by $\sqrt{n}$ (rather than concentric circles $x^2 + y^2 = n^2$ with radiuses $n$). ■

Let us proceed now to the case of curved dot-screens (or line-grids). Assume that the curved dot-screen $r(x,y)$ is obtained by bending a two-fold periodic dot-screen $p(x',y')$, i.e., by replacing $x'$ and $y'$ with the functions $x' = g_1(x,y)$ and $y' = g_2(x,y)$, respectively: $r(x,y) = p(g_1(x,y),g_2(x,y))$. An example of such a curved dot-screen $r(x,y)$ is shown in Fig. 10.2(b). The intensity profile of the original, uncurved two-fold periodic screen $p(x',y')$ is called the *periodic-profile* of the curved screen $r(x,y)$. The periodic-profile of a curved screen may be any two-fold periodic waveform; as in the case of curvilinear gratings, we will use the term "*normalized* periodic-profile" whenever we choose $p(x',y')$ to have a unit frequency (to both directions). The functions $x' = g_1(x,y)$ and $y' = g_2(x,y)$ which bend $p(x',y')$ into the curved screen $r(x,y)$ are called the *bending functions*.[2]

**Figure 10.1:** (*See following pages.*) Various curvilinear gratings $r(x,y)$ having a periodic-profile waveform of $\cos(2\pi fx)$, and their spectra $R(u,v)$ as obtained on computer by 2D DFT:
(a) Straight cosinusoidal grating: $\cos(2\pi fx)$.
(b) Rotated straight cosinusoidal grating: $\cos(2\pi f[x\cos\theta + y\sin\theta])$.
(c) Parabolic cosinusoidal grating: $\cos(2\pi f[y - 0.15x^2])$.
(d) Circular cosinusoidal grating: $\cos(2\pi f\sqrt{x^2 + y^2})$.
(e) Elliptic cosinusoidal grating: $\cos(2\pi f\sqrt{\tfrac{1}{4}x^2 + y^2})$.
(f) Hyperbolic cosinusoidal grating: $\cos(2\pi f\sqrt{|\tfrac{1}{2}x^2 - y^2|})$.
(g) Circular cosinusoidal zone grating: $\cos(2\pi f[x^2 + y^2]/8)$.
(h) Elliptic cosinusoidal zone grating: $\cos(2\pi f[\tfrac{1}{4}x^2 + y^2]/8)$.
(i) Hyperbolic cosinusoidal zone grating: $\cos(2\pi f[x^2 - y^2]/8)$.
(j) Linear cosinusoidal zone grating: $\cos(2\pi fx^2/8)$.
(k) Arg sinh($x$)-shaped cosinusoidal grating: $\cos(2\pi f[y - \operatorname{arg\,sinh}(x)])$.
(l) Cosine-shaped cosinusoidal grating: $\cos(2\pi f[y - \cos(2\pi fx/4)])$.
In all the centrosymmetric cases the imaginary part of the spectrum is identically zero. Note that the spectra of (g), (h) and (i) are purely continuous. In fact, these cases are almost *self-reciprocal transform pairs* [Bracewell95 p. 171], since the spectrum in these cases is, up to some scaling or phase details, identical to the image-domain function. The oscillations in the spectra of (c), (g), etc. fade out short of the spectrum border since the FFT cannot find higher frequencies in the corresponding finite-sized, sampled functions in the image-domain; in reality the spectra oscillate *ad infinitum* without fading out. Note also the FFT rippling artifacts in some of the spectra ((d), (e), etc.).

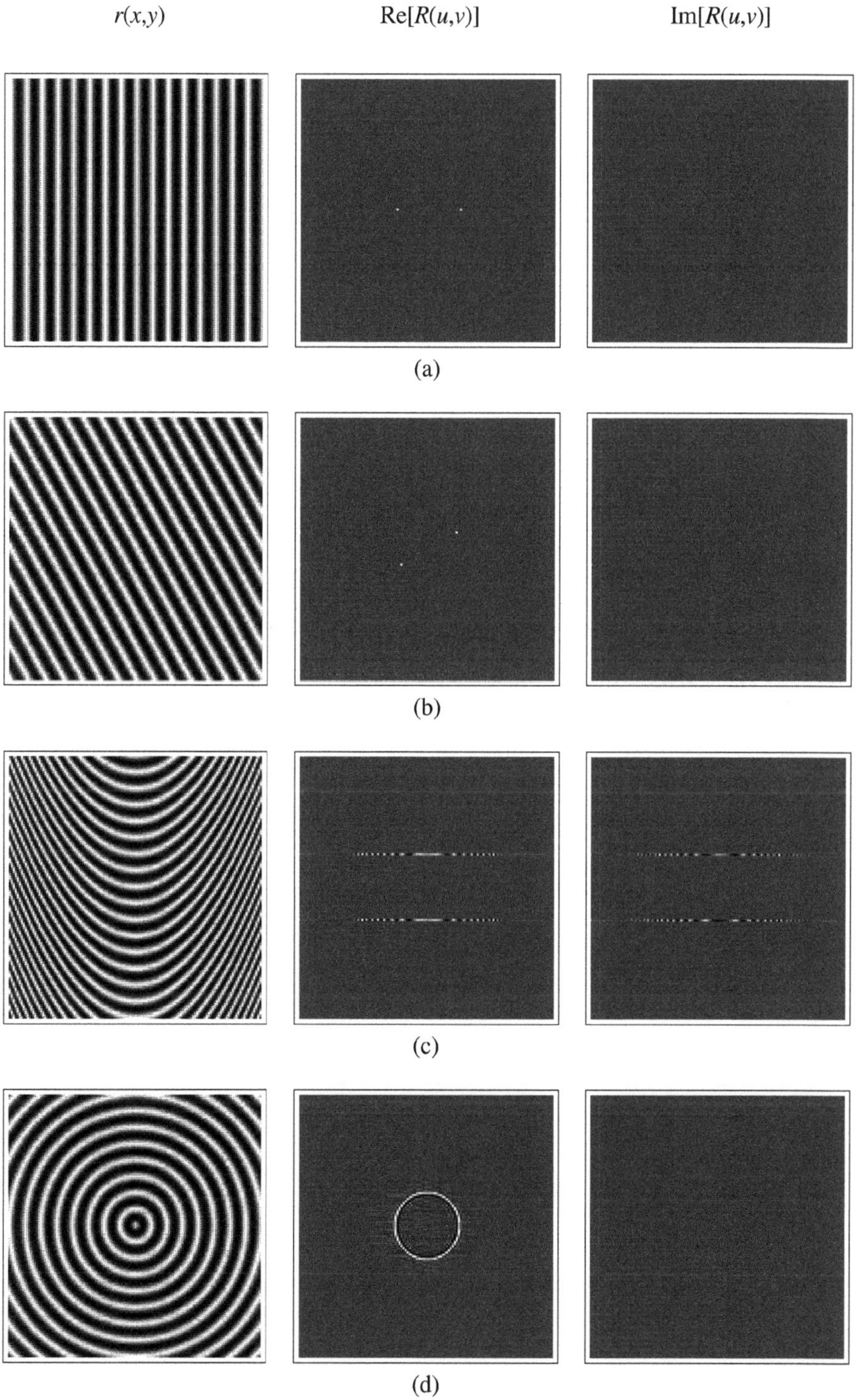
$r(x,y)$
$\mathrm{Re}[R(u,v)]$
$\mathrm{Im}[R(u,v)]$
(a)
(b)
(c)
(d)

$r(x,y)$ $\mathrm{Re}[R(u,v)]$ $\mathrm{Im}[R(u,v)]$

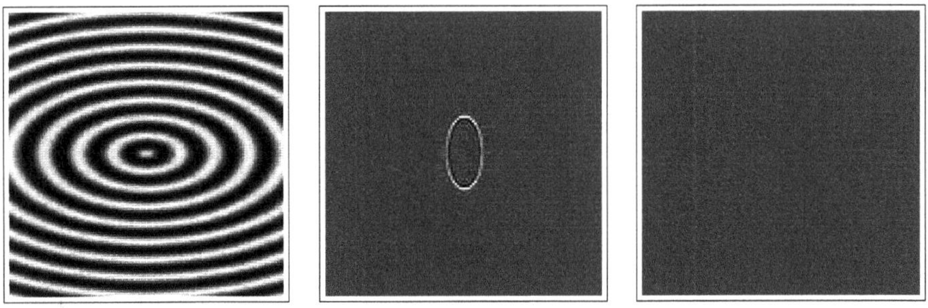

(e)

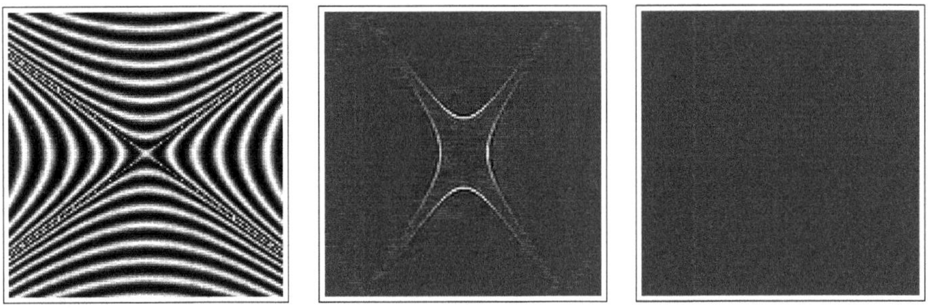

(f)

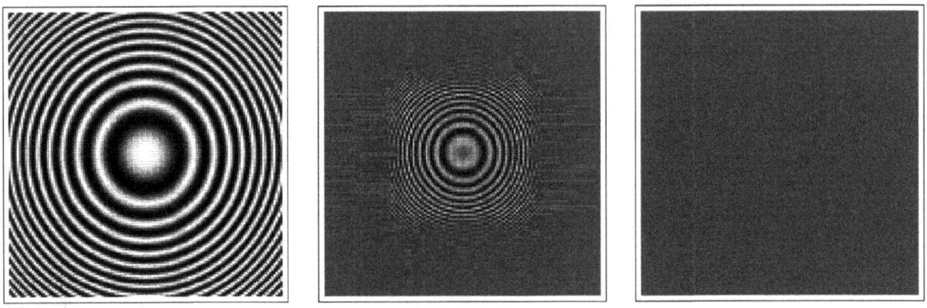

(g)

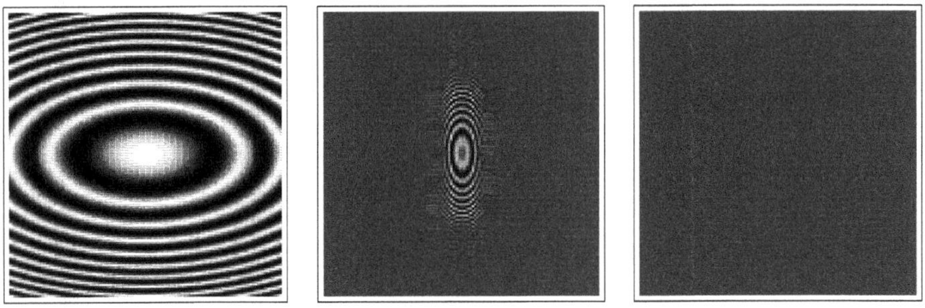

(h)

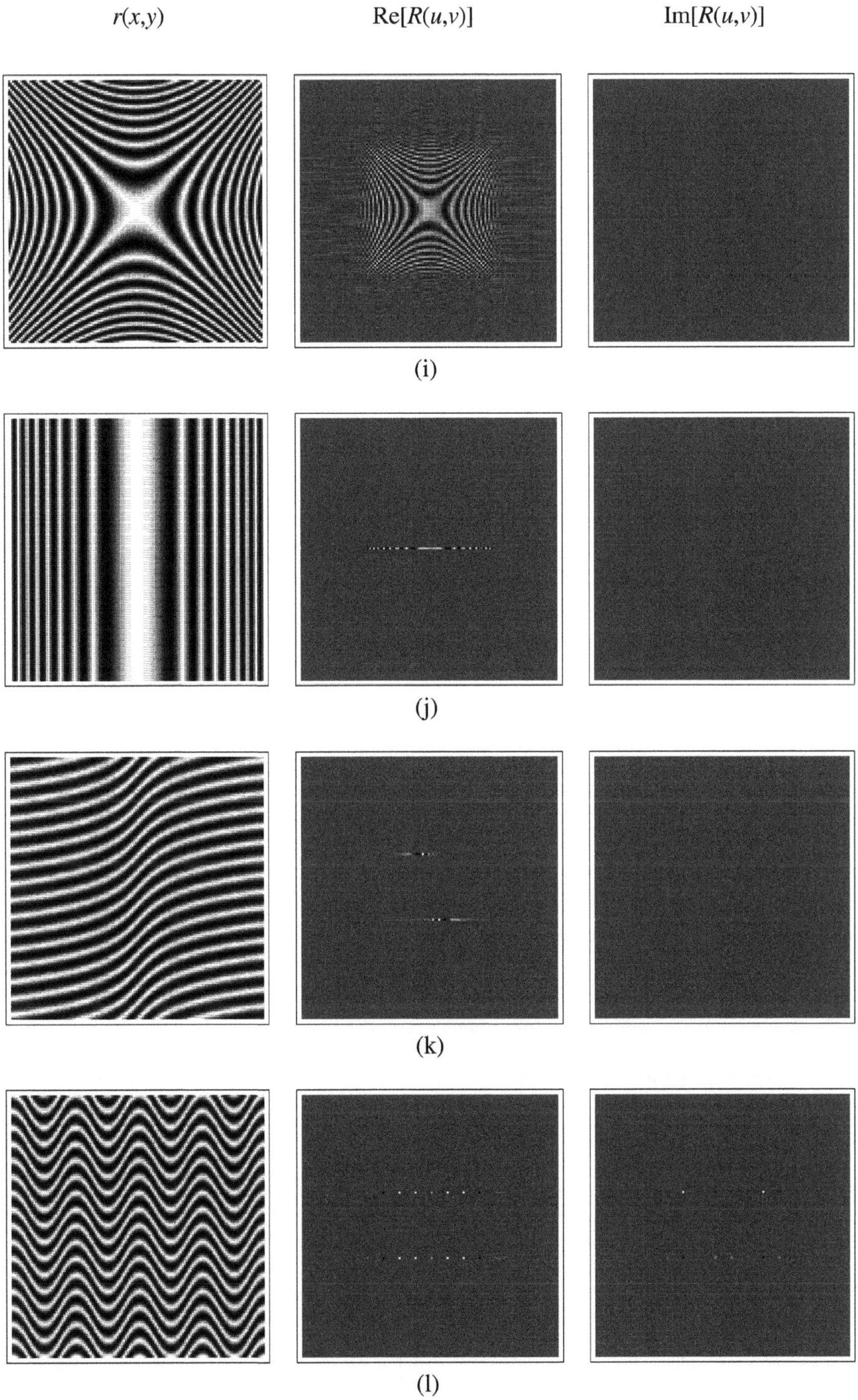
$r(x,y)$
$\mathrm{Re}[R(u,v)]$
$\mathrm{Im}[R(u,v)]$
(i)
(j)
(k)
(l)

A curved screen $r(x,y) = p(g_1(x,y),g_2(x,y))$ is therefore characterized by two basic and independent properties: its geometric layout, which is determined by the functions $g_1(x,y)$ and $g_2(x,y)$; and its intensity behaviour within each "curved period", which is determined by the two-fold periodic-profile $p(x',y')$.

In fact, this bending process can be interpreted as a mapping of $\mathbb{R}^2$ onto itself, or equivalently, as a *coordinate change* in $\mathbb{R}^2$ from the original $x',y'$ coordinate system into the $x,y$ system (see, for example, [Courant88 pp. 133–140]). This 2D coordinate transformation is specified for each of the two original $x',y'$ directions separately by the bending functions $x' = g_1(x,y)$ and $y' = g_2(x,y)$. The effect of this 2D coordinate transformation can be expressed, then, by:

$$\mathbf{g}: \begin{pmatrix} x \\ y \end{pmatrix} \rightarrow \begin{pmatrix} x' \\ y' \end{pmatrix} \qquad \text{where:} \qquad \begin{pmatrix} x' \\ y' \end{pmatrix} = \begin{pmatrix} g_1(x,y) \\ g_2(x,y) \end{pmatrix}$$

or in a more compact vector notation: $\mathbf{x}' = \mathbf{g}(\mathbf{x})$. Note that $\mathbf{g}$ is a mapping of $\mathbb{R}^2$ onto itself: $\mathbf{g}$: $\mathbb{R}^2 \rightarrow \mathbb{R}^2$; we denote it by a boldface letter since the value it returns, $\mathbf{g}(\mathbf{x})$, is a vector. Clearly, in order that the image of this mapping span the entire $x,y$ plane $\mathbb{R}^2$, the two individual bending functions $x' = g_1(x,y)$ and $y' = g_2(x,y)$ must be independent, i.e., there must exist no function $f(\ ,\ )$ other than $f(\ ,\ ) \equiv 0$ such that $f(g_1(x,y),g_2(x,y)) = 0$ is satisfied for all $(x,y)$.[4] Equivalently, this means that the Jacobian:

$$J(x,y) = \begin{vmatrix} \frac{\partial g_1}{\partial x} & \frac{\partial g_1}{\partial y} \\ \frac{\partial g_2}{\partial x} & \frac{\partial g_2}{\partial y} \end{vmatrix}$$

is not identically zero (see [Bronstein90 pp. 430–431], [Courant88 pp. 154–155]).[5] In order to avoid unnecessary mathematic complications we will generally assume that the coordinate transformation $\mathbf{g}$ is a diffeomorphism on $\mathbb{R}^2$, i.e., a one-to-one continuously-differentiable mapping of $\mathbb{R}^2$ onto itself whose inverse mapping is also continuously-differentiable. This ensures that our transformation has no abrupt jumps or other troublesome singularities.

**Example 10.3:** Assume that we are given a periodic binary line-grid $p(x',y')$ which is the superposition of a vertical square-wave grating $p_1(x',y') = \sum_{m=-\infty}^{\infty} \text{rect}(\frac{x' - mT}{\tau})$ and a horizontal square wave grating $p_2(x',y') = \sum_{n=-\infty}^{\infty} \text{rect}(\frac{y' - nT}{\tau})$, both having the same period $T$ and

---

[4] If $g_1(x,y)$ and $g_2(x,y)$ are dependent, for instance $g_2(x,y) = g_1(x,y)^2$, then the 2D transformation $\mathbf{g}(x,y)$ is degenerate, and it maps $\mathbb{R}^2$ into a 1D curve in $\mathbb{R}^2$ [Courant88 p. 155].

[5] If $g_1(x,y)$ and $g_2(x,y)$ satisfy also the Cauchy-Riemann conditions (a) $\frac{\partial g_1}{\partial x} = \frac{\partial g_2}{\partial y}$, $\frac{\partial g_1}{\partial y} = -\frac{\partial g_2}{\partial x}$ or (b) $\frac{\partial g_1}{\partial x} = -\frac{\partial g_2}{\partial y}$, $\frac{\partial g_1}{\partial y} = \frac{\partial g_2}{\partial x}$, then the transformation $\mathbf{g}(\mathbf{x})$ is *conformal* [Courant88 pp. 166–167], and it maps the straight lines $x = const.$, $y = const.$ into curve families $x' = const.$ and $y' = const.$ which intersect *at right angles*. This orthogonality is clearly stronger than the mere independence of $g_1(x,y)$ and $g_2(x,y)$; and indeed, condition (a) implies $J(x,y) > 0$, and condition (b) implies $J(x,y) < 0$. Such an orthogonality is not *required* for our needs (see for instance Fig. 10.2(b)), but it is advantageous; for example, it guarantees that the two curvilinear gratings which form together our curved grid $r(x,y)$ do not generate moirés between each other, within the curved grid itself (such moirés occur, for example, in Fig. 10.14).

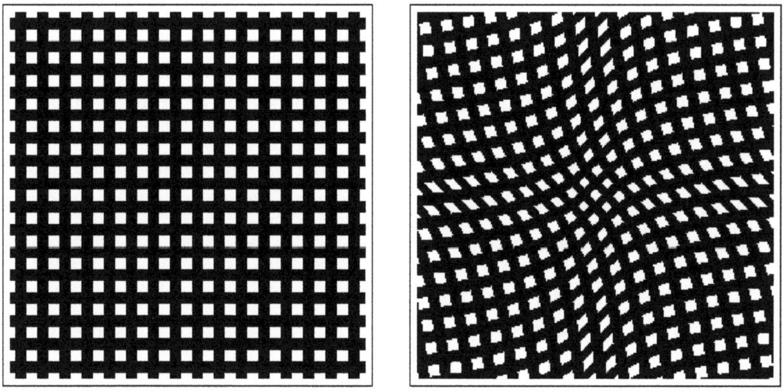

(a) (b)

**Figure 10.2:** (a) The periodic binary line-grid $p(x',y')$ of Example 10.3. (b) The curved binary grid $r(x,y)$ obtained by applying on $p(x',y')$ the 2D non-linear transformation $\mathbf{g}(x,y) = (x + \text{arg}\sinh(y),\ y - \text{arg}\sinh(x))$. Note that both line-grids can be seen also as dot-screens of white dots on a black background.

the same opening $\tau$; that is: $p(x',y') = p_1(x',y')p_2(x',y') = \sum_{m=-\infty}^{\infty}\sum_{n=-\infty}^{\infty} \text{rect}(\frac{x'-mT}{\tau})\text{rect}(\frac{y'-nT}{\tau})$.
We define the 2D non-linear transformation $\mathbf{g}(x,y)$ as follows:

$$\begin{pmatrix} x' \\ y' \end{pmatrix} = \mathbf{g}\begin{pmatrix} x \\ y \end{pmatrix} = \begin{pmatrix} g_1(x,y) \\ g_2(x,y) \end{pmatrix} = \begin{pmatrix} x + \text{arg}\sinh(y) \\ y - \text{arg}\sinh(x) \end{pmatrix}$$

By applying the non-linear transformation $\mathbf{g}(x,y)$ to the periodic binary line-grid $p(x',y')$ we obtain the curved binary grid $r(x,y)$, as shown in Fig. 10.2:

$$\begin{aligned} r(x,y) &= p(x + \text{arg}\sinh(y),\ y - \text{arg}\sinh(x)) \\ &= \sum_{m=-\infty}^{\infty}\sum_{n=-\infty}^{\infty} \text{rect}(\tfrac{x + \text{arg}\sinh(y) - mT}{\tau})\ \text{rect}(\tfrac{y - \text{arg}\sinh(x) - nT}{\tau}) \end{aligned}$$ ■

**Remark 10.2:** We assume here that each of our repetitive, non-periodic layers $r(\mathbf{x})$ belongs to the category of coordinate-transformed structures, i.e., that it has been obtained by gradually varying its geometric layout, while its periodic-profile remains unchanged throughout the $x,y$ plane; using our formal notation: $r(\mathbf{x}) = p(\mathbf{g}(\mathbf{x}))$. Repetitive, non-periodic layers of the other two categories, which are obtained by varying the periodic-profile of the layer, $r(\mathbf{x}) = t(p(\mathbf{x}))$, or by varying both the geometric layout and the periodic-profile, $r(\mathbf{x}) = t(p(\mathbf{g}(\mathbf{x})))$, will be discussed later, in Sec. 10.11. Note that $p(\mathbf{x})\rightarrow p(\mathbf{g}(\mathbf{x}))$ is a *domain transformation*, while $p(\mathbf{x})\rightarrow t(p(\mathbf{x}))$ is a *range transformation* (for more details on this subject see Sec. D.6 in Appendix D of *Vol. II*). ■

**Remark 10.3:** Note that the application of a nonlinear transformation (coordinate change) $\mathbf{g}(\mathbf{x})$ on a periodic function $p(\mathbf{x})$ such as a line-grating, line-grid, dot-screen etc., causes local variations in the size and in the orientation of its period, depending on the location of the point $\mathbf{x}$ in the $x,y$ plane. Thus, unlike in the periodic function $p(\mathbf{x})$, the *local period* (and the *local frequency*) of the transformed, curved function $r(\mathbf{x})$ are not constant and they vary according to the location in the $x,y$ plane. If $p(\mathbf{x})$ is a binary grating, its opening $\tau$ (the white line-width of the grating) is also affected by the transformation $\mathbf{g}(\mathbf{x})$ in the same way as the period $T$, so that the *local opening ratio* $\tau/T$ remains constant. This means that the line widths in the curvilinear grating $r(\mathbf{x})$ are no longer constant as in the original binary grating $p(\mathbf{x})$; this can be clearly seen, for example, in Fig. 10.2. Curvilinear gratings with constant line widths, like those drawn by a pen plotter, are mathematically obtained by also applying here a compensating non-linear *profile transformation* $t(x)$, which causes variations in the *local profile*, too, depending on the location of the point $\mathbf{x}$ in the $x,y$ plane: $r(\mathbf{x}) = t(p(\mathbf{g}(\mathbf{x})))$. ■

**Remark 10.4:** It may be tempting to suppose that the periodic function $\cos(2\pi fx)$ can be turned into a function of a varying frequency $f(x)$ by simply replacing the constant frequency $f$ with the function $f(x)$. According to this reasoning, any function $\cos(2\pi g(x))$ can be written in the form $\cos(2\pi f(x)x)$, whence one may deduce that its local frequency at any point $x$ is given by $f(x) = g(x)/x$. However, this reasoning is wrong: the local frequency of $\cos(2\pi g(x))$ is given by the *derivative* of $g(x)$, and in the 2D case by the *gradient* of $g(\mathbf{x})$. We will come back to this point in Sec. 11.4. ■

Since we are always interested in the spectral domain, too, the following natural question arises immediately: how does a coordinate change in the image domain affect the spectral domain? This is indeed an important question, and we will discuss it now in more detail.

## 10.3 The influence of a coordinate change on the spectrum

Assume that the function $f(x,y)$ has the spectrum (Fourier transform) $F(u,v)$. If $f(x,y)$ undergoes a coordinate change in the image domain:

$$\mathbf{g}: (x,y) \rightarrow (x',y')$$

what happens to its spectrum? Does it simply undergo a "reciprocal coordinate change" of some sort, owing to the duality between the image and the spectral domains? Or, at the least, is there a general rule or recipe which permits us to find the new spectrum directly from the original spectrum $F(u,v)$?

Unfortunately, the answer to this question is in general negative. The only remarkable exception is that of an affine coordinate transformation $\mathbf{g}(x,y)$, namely:

$$\mathbf{g}\begin{pmatrix} x \\ y \end{pmatrix} = \begin{pmatrix} a_1 x + b_1 y + c_1 \\ a_2 x + b_2 y + c_2 \end{pmatrix}$$

$$= \begin{pmatrix} a_1 & b_1 \\ a_2 & b_2 \end{pmatrix} \begin{pmatrix} x \\ y \end{pmatrix} + \begin{pmatrix} c_1 \\ c_2 \end{pmatrix}$$

or in a more compact notation: $\mathbf{g}(\mathbf{x}) = \mathrm{A}\mathbf{x}+\mathbf{c}$. In this case, if $f(\mathbf{x})$ has the spectrum $F(\mathbf{u})$, then $f(\mathbf{g}(\mathbf{x}))$, namely: $f(\mathrm{A}\mathbf{x}+\mathbf{c})$, has the spectrum:

$$\frac{1}{|\bar{A}|} e^{i2\pi(\mathbf{c}^{\mathrm{T}}\mathrm{A}^{-\mathrm{T}})\mathbf{u}} F(\mathrm{A}^{-\mathrm{T}}\mathbf{u}) \qquad (10.1)$$

where $\bar{A} = a_1b_2 - b_1a_2$ is the determinant of matrix A, $\mathbf{c}^{\mathrm{T}}$ indicates the role of $\mathbf{c}$ as a row vector in the matrix product, and $\mathrm{A}^{-\mathrm{T}}$ is the transposed inverse of matrix A:

$$\begin{pmatrix} a_1 & b_1 \\ a_2 & b_2 \end{pmatrix}^{-\mathrm{T}} = \frac{1}{\bar{A}} \begin{pmatrix} b_2 & -a_2 \\ -b_1 & a_1 \end{pmatrix}$$

This affine theorem for Fourier transforms is proved, using a somewhat different notation, in [Bracewell95 pp. 159–161]; it is interesting to mention the remarkable similarity between this theorem and its 1D counterpart which states that if $f(x) \leftrightarrow F(u)$ then:

$$f(ax+c) \leftrightarrow \frac{1}{|a|} e^{i2\pi cu/a} F(\tfrac{u}{a}) \qquad (10.2)$$

[Bracewell86 p. 126].[6]

Eq. (10.1) means, indeed, that any affine coordinate change $\mathbf{g}(\mathbf{x}) = \mathrm{A}\mathbf{x}+\mathbf{c}$ in the image domain causes a reciprocal coordinate change of $\mathrm{A}^{-\mathrm{T}}\mathbf{u}$ in the spectrum (plus a linear increment of $2\pi(\mathbf{c}^{\mathrm{T}}\mathrm{A}^{-\mathrm{T}})\mathbf{u}$ in the phase of the spectrum).[7]

However, if the coordinate change $\mathbf{g}(x,y)$ in the image domain is non-linear, no general rule exists which tells how the spectrum will be influenced. The following examples may illustrate the difficulty:

**Example 10.4:** A coordinate change of $g(x) = x^2$ in the 1D case:

We know that the spectrum (Fourier transform) of $p(x') = \cos(2\pi f x')$ is: $P(u) = \frac{1}{2}\delta(u-f) + \frac{1}{2}\delta(u+f)$; but after replacing $x'$ with $x^2$ in the image domain, the spectrum of $r(x) = p(g(x)) = \cos(2\pi f x^2)$ is: $R(u) = \frac{1}{2\sqrt{f}}(\cos(\frac{\pi}{2f}u^2) + \sin(\frac{\pi}{2f}u^2))$ (adapted from [Erdélyi54 p. 24]), or equivalently, using the known trigonometric identity $\cos\alpha + \sin\alpha = \sqrt{2}\cos(\alpha - \frac{\pi}{4})$: $R(u) = \frac{1}{\sqrt{2f}}\cos(\frac{\pi}{2f}u^2 - \frac{\pi}{4})$. Note that after applying the coordinate change in the image domain the grating $r(x)$ obtained is no longer periodic,[8] and therefore its spectrum $R(u)$ is no longer impulsive, but rather continuous and smooth.

---

[6] Note that in the special case of a purely linear transformation $\mathbf{g}$, i.e., when $\mathbf{c}=\mathbf{0}$, this rule simplifies into: $f(\mathrm{A}\mathbf{x}) \leftrightarrow \frac{1}{|\bar{A}|}F(\mathrm{A}^{-\mathrm{T}}\mathbf{u})$, which is a generalization of the 1D theorem: $f(ax) \leftrightarrow \frac{1}{|a|}F(\frac{u}{a})$ [Bracewell86 p. 122].

[7] On the reciprocity between the image and the spectral domains in the case of periodic functions see also Sec. A.4 and Fig. A.2 in Appendix A.

[8] In some references functions like $r(x) = \cos(2\pi f x^2)$ are said to be "periodic in $x^2$" (see, for example, [Lohman67 p. 62]) or "periodic in the $x^2$ space".

As another example, we know that the spectrum of $p(x') = e^{-|x'|}$ is: $P(u) = \frac{2}{1+(2\pi u)^2}$ [Bracewell86 p. 418]; but after replacing $x'$ with $x^2$ in the image domain, the spectrum of $r(x) = p(g(x)) = e^{-x^2}$ is: $R(u) = \sqrt{\pi}\, e^{-\pi^2 u^2}$ [Bracewell86 p. 123]. As we can see from these cases, there is no apparent general connection between the spectra $P(u)$ and $R(u)$. ■

**Example 10.5:** The effect of "bending" a 2D cosinusoidal grating into a parabolic cosinusoidal grating:

We know that the spectrum of the 2D cosinusoidal grating $p(x',y') = \cos(2\pi fx')$ is the impulse pair $P(u,v) = \frac{1}{2}\delta(u-f,v) + \frac{1}{2}\delta(u+f,v)$ (see Fig. 10.1(a)). What happens to the spectrum when we "bend" the grating into a parabolic shape by replacing $x'$ with $y-ax^2$ as in Fig. 10.1(c)? According to the well-known trigonometric identity $\cos(\alpha-\beta) = \cos\alpha\cos\beta + \sin\alpha\sin\beta$ we have:

$$r(x,y) = \cos(2\pi f(y-ax^2)) = \cos(2\pi fy)\cos(2\pi fax^2) + \sin(2\pi fy)\sin(2\pi fax^2)$$

Since each of these two products consists of one function of $x$ and one function of $y$, we can use here the separable-product theorem which says that if the 1D Fourier transforms of $f(x)$ and $g(x)$ are respectively $F(u)$ and $G(u)$, then the 2D Fourier transform of $f(x)g(y)$ is $F(u)G(v)$ [Bracewell95 p. 166]. We obtain, therefore, that:

$$\cos(2\pi fy)\cos(2\pi fax^2) \leftrightarrow [\tfrac{1}{2}\delta(v-f) + \tfrac{1}{2}\delta(v+f)]\, R_c(u)$$

where $R_c(u) = \frac{1}{2\sqrt{fa}}(\cos(\frac{\pi}{2fa}u^2) + \sin(\frac{\pi}{2fa}u^2)) = \frac{1}{\sqrt{2fa}}\cos(\frac{\pi}{2fa}u^2 - \frac{\pi}{4})$ is the continuous spectrum of $\cos(2\pi fax^2)$ from Example 10.4 above, and similarly:

$$\sin(2\pi fy)\sin(2\pi fax^2) \leftrightarrow i[\tfrac{1}{2}\delta(v-f) - \tfrac{1}{2}\delta(v+f)]\, R_s(u)$$

where $R_s(u) = \frac{1}{2\sqrt{fa}}(\cos(\frac{\pi}{2fa}u^2) - \sin(\frac{\pi}{2fa}u^2)) = \frac{1}{\sqrt{2fa}}\sin(\frac{\pi}{2fa}u^2 + \frac{\pi}{4})$ is the continuous spectrum of $\sin(2\pi fax^2)$ (adapted from [Erdélyi54 p. 23]). The functions $R_c(u)$ and $R_s(u)$ are shown in Fig. 10.3. Note that on the 2D $u,v$ plane $\delta(v-f)$ and $\delta(v+f)$ indicate horizontal *line-impulses* ("blades"); see [Bracewell95 pp. 199–122].

The spectrum of the bent grating $r(x,y)$ is, therefore:

$$R(u,v) = [\tfrac{1}{2}\delta(v-f) + \tfrac{1}{2}\delta(v+f)]\, R_c(u) + i[\tfrac{1}{2}\delta(v-f) - \tfrac{1}{2}\delta(v+f)]\, R_s(u)$$

$$= \tfrac{1}{2}[R_c(u) + iR_s(u)]\delta(v-f) + \tfrac{1}{2}[R_c(u) - iR_s(u)]\delta(v+f)$$

namely: a pair of continuous horizontal line-impulses, situated at a distance of $\pm f$ from the $u$ axis, whose continuous, modulated amplitudes are given by $\frac{1}{2}[R_c(u) + iR_s(u)]$ and $\frac{1}{2}[R_c(u) - iR_s(u)]$, respectively (see Fig. 10.1(c)).

Let us briefly mention here some interesting properties of this spectrum:

(a) Since the bent grating $r(x,y)$ is no longer symmetric with respect to the origin, its spectrum is no longer purely real.

(b) Unlike in spectra of periodic functions, in the present line-impulse spectrum there is no complete independence between the impulse amplitudes and the impulse locations in the spectrum. This is because the cosine frequency $f$ takes part not only in $\delta(v-f)$ and $\delta(v+f)$, which determine the locations of the two line-impulses in the spectrum, but also in $R_c(u)$ and $R_s(u)$, which determine their amplitudes (or amplitude modulations).

(c) The roles of $f$ and $a$ (the cosine frequency and the bending rate of the parabolas) in determining the amplitudes (or amplitude modulations) of the line-impulses are equivalent, since they always appear together in $R_c(u)$ and $R_s(u)$ as a product $fa$. This means that any increase in $f$ may be fully compensated by a proportional decrease in $a$ (or vice versa), without any influence on the amplitude of the line-impulses. However, the *locations* of the line-impulses in the spectrum are only determined by $f$, since $a$ does not take part in $\delta(v-f)$ and $\delta(v+f)$.

The significance of these properties will become clear later, in Sec. 10.7.3. ■

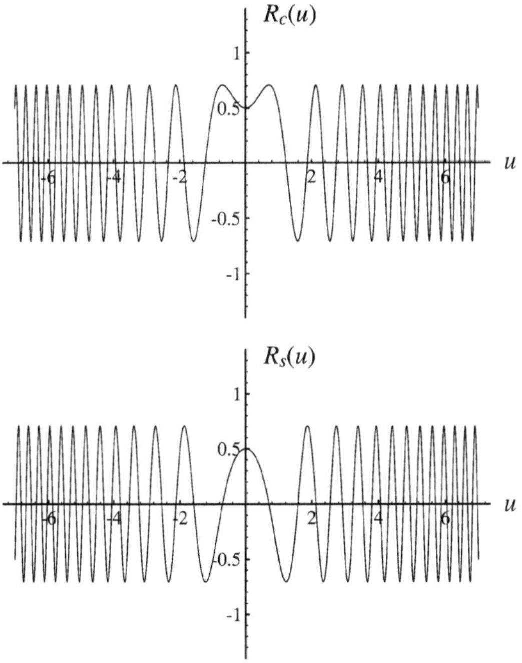


**Figure 10.3:** The functions $R_c(u)$ and $R_s(u)$, showing the real and the imaginary parts of the amplitude of the line-impulse situated at $v = f$ in the spectrum of the parabolic cosinusoidal grating (see Fig. 10.1(c)). The second line-impulse, situated in the spectrum at $v = -f$, has the same real part $R_c(u)$, but its imaginary part is $-R_s(u)$.

**Example 10.6:** A polar coordinate change in the 2D plane (see Fig. 10.4):

The spectrum of the 2D cosine function $p(r',\theta') = \cos(2\pi f r')$, before being bent, consists of the impulse pair $P(q,\varphi) = \frac{1}{2}\delta(q-f,\varphi) + \frac{1}{2}\delta(q+f,\varphi)$. By applying a polar coordinate change in the image domain $r'$ is replaced with $\sqrt{x^2+y^2}$, and hence $p(r',\theta')$ turns into the circular cosine function $r(x,y) = \cos(2\pi f\sqrt{x^2+y^2})$ which is, of course, no longer periodic.[9] Its spectrum $R(u,v)$ resembles the circular impulse ring $\delta(\sqrt{u^2+v^2} - f)$ which is obtained from $P(q,\varphi)$ by a polar coordinate change in the spectral domain, but in fact it has a more complicated structure which is not purely impulsive: It has a particular, dipole-like impulsive behaviour on the perimeter of a circle of radius $f$, but inside the circle it has a negative, continuous "wake" which gradually trails off toward the center (see [Amidror97]). Hence, the spectrum of the circular cosine $\cos(2\pi f\sqrt{x^2+y^2})$ does not represent a pure radial frequency of $f$ since it includes also lower radial frequencies (albeit of negligible amplitudes) due to its continuous wake. We see, therefore, that a polar coordinate transformation in the image domain does not result in a similar coordinate transformation in the spectral domain. (Note that the pure impulse ring $\delta(\sqrt{u^2+v^2} - f)$, for its part, is the spectrum of the Bessel function $2\pi f J_0(2\pi f\sqrt{x^2+y^2})$ [Bracewell86 p. 248].) ■

**Example 10.7:** The spectra of cosinusoidal zone gratings in the 2D plane:

If we take in the last example $r' = x^2+y^2$ instead of $r' = \sqrt{x^2+y^2}$ we obtain, as we have seen in Example 10.2, a circular cosine with a decreasing radial period (where the radius of the $n$-th circle is proportional to $\sqrt{n}$):[10]

$$r^+(x,y) = \cos(2\pi f(x^2+y^2))$$

We call this function (see Fig. 10.1(g)) a cosinusoidal zone grating.[11] As shown in Appendix C.6, the spectrum of this function is a continuous circular function with a decreasing radial period:

$$R^+(u,v) = \tfrac{1}{2f}\sin(\tfrac{\pi}{2f}(u^2+v^2)).$$

More generally, according to the 2D similarity theorem [Bracewell86 p. 244], the cosinusoidal *elliptic* zone grating (see Fig. 10.1(h)):

$$r^+(x,y) = \cos(2\pi f(ax^2+by^2))$$

has the spectrum:

$$R^+(u,v) = \tfrac{1}{2f\sqrt{ab}}\sin(\tfrac{\pi}{2f}(\tfrac{1}{a}u^2+\tfrac{1}{b}v^2)).$$

---

[9] We call such functions *radially periodic functions*; in some references they are called "periodic in the radius $r$" or "periodic in the $r$ space" (for example, in [Harburn75 p. 409]).

[10] Such functions are sometimes called in literature "periodic in $r^2$" or "periodic in the $r^2$ space" (for example, in [Harburn75 p. 409]).

[11] A zone grating (or zone plate) is a concentric circular grating where the radius of the $n$-th circle from the center is proportional to $\sqrt{n}$. In most applications the periodic-profile of the zone grating is a binary (black/white) square wave with opening ratio $\tau/T = 1/2$ [Patorski93 p. 18]. In the present example, however, we discuss zone gratings with a cosinusoidal periodic-profile.

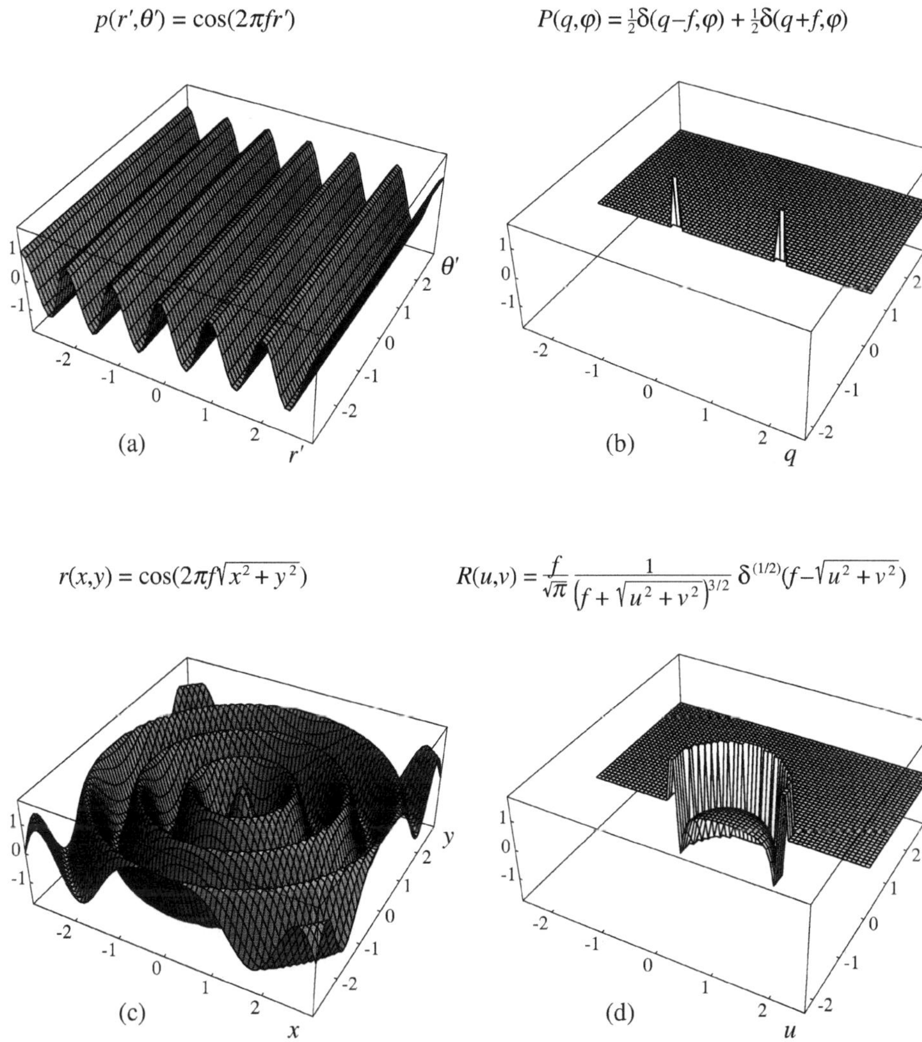


**Figure 10.4:** A circular cosinusoidal grating in the image domain, expressed (a) in terms of the polar coordinates $r',\theta'$ and (c) in terms of the Cartesian coordinates $x,y$; (b) and (d) are their respective spectra. (Note that both spectra have been cut along the horizontal axis in order to show their cross sections along this line.) Can the spectrum of the image after the coordinate change, $R(u,v)$, be expressed in terms of $P(q,\varphi)$, the spectrum of the original image?

Similarly, the hyperbolic counterpart of $r^{+}(x,y)$, which is called a cosinusoidal *hyperbolic* zone grating (see Fig. 10.1(i)):

$$r^{-}(x,y) = \cos(2\pi f(ax^2 - by^2))$$

has the spectrum:

$$R^{-}(u,v) = \frac{1}{2f\sqrt{ab}}\cos(\frac{\pi}{2f}(\frac{1}{a}u^2 - \frac{1}{b}v^2)).$$

The 2D functions $r(x,y) = p(x^2) = \cos(2\pi fx^2)$ and $r(x,y) = p(y^2) = \cos(2\pi fy^2)$ are sometimes called by analogy cosinusoidal *linear* zone gratings (see Fig. 10.1(j)). Their 2D spectra are given by the line-impulses $R_c(u)\delta(v)$ and $R_c(v)\delta(u)$, respectively, where $R_c(u) = \frac{1}{2\sqrt{f}}(\cos(\frac{\pi}{2f}u^2) + \sin(\frac{\pi}{2f}u^2)) = \frac{1}{\sqrt{2f}}\cos(\frac{\pi}{2f}u^2 - \frac{\pi}{4})$ (see Example 10.4 above).

Note that in order to complete the analogy for all conics, one may call the 2D parabolic grating $r(x,y) = p(y - ax^2) = \cos(2\pi f(y - ax^2))$ of Example 10.5 above (Fig. 10.1(c)) a cosinusoidal *parabolic* zone grating. ■

Although there exists no apparent rule which tells how the spectrum is influenced by a general non-linear transformation in the image domain, many cases of interest can be derived based on Fourier transform tables in literature, such as [Erdélyi54]. Furthermore, we will see below that at least in the case of curvilinear cosinusoidal gratings more can be often said to characterize their spectra. For this end we will adopt once again the intuitive approach of analyzing smooth and gradual transitions, just as we did in Chapter 5 to study the "collapsing" of impulse clusters in singular states.

## 10.4 Curvilinear cosinusoidal gratings and their different types of spectra

As we have seen in the examples above, applying a non-linear coordinate change to the 2D cosinusoidal grating $\cos(2\pi fx)$ may have a radical influence on its spectrum: its original, impulsive spectrum $\frac{1}{2}\delta(u-f, v) + \frac{1}{2}\delta(u+f, v)$ may turn into a semi-impulsive or even a completely non-impulsive spectrum. In fact, the main cases of interest can be qualitatively classified into the following classes:

(1) If after the coordinate change in the image domain the obtained curvilinear cosinusoidal grating remains periodic (like in Figs. 10.1(b),(*l*)), the resulting spectrum is still purely impulsive.

(1′) If after the coordinate change in the image domain the obtained curvilinear cosinusoidal grating becomes almost-periodic (see Appendix B), the resulting spectrum is still purely impulsive — but its impulses are no longer located on a common lattice, and they may even be everywhere dense.

(2) If the obtained curvilinear cosinusoidal grating is only periodic (with a constant period *T*) on every cross-section parallel to a given direction (like in Example 10.5 above and in Figs. 10.1(c),(k)), the resulting spectrum consists of a pair of amplitude-modulated line-impulses perpendicular to that direction, located at a distance of $\pm 1/T$ from the origin, each of which being continuous along its own support. The spectrum is

therefore impulsive in the direction of periodicity, but continuous in the perpendicular direction.

(2′) If the period $T$ of class (2) tends to infinity so that the cosinusoidal grating becomes constant in one direction (the original direction of periodicity), then the two line-impulses in the spectrum coincide and the spectrum becomes a single line-impulse through the origin (see Fig. 10.1(j)).

(3) If the obtained curvilinear cosinusoidal grating is radially periodic (a circular cosine; see Fig. 10.1(d) and Example 10.6 above), the resulting spectrum is no longer purely impulsive: it consists of a circular structure with a particular, dipole-like impulsive behaviour on its border, plus a negative continuous "wake" which gradually trails off toward the center.

(3′) More generally, if the obtained curvilinear cosinusoidal grating is radially periodic on any cross section through the origin, but the radial period varies according to the angle $\theta$ of the cross section (like in an elliptical grating or a hyperbolic grating; see Figs. 10.1(e),(f)), the resulting spectrum consists of a curvilinear dipole-like impulse which surrounds the origin along a certain curvilinear path, and gradually decays perpendicularly to this path into a continuous wake which trails off at the concave side of the path.

(4) If the obtained curvilinear cosinusoidal grating is no longer periodic on any cross section, as in Figs. 10.1(g)–(i), then its spectrum becomes purely continuous and completely loses the impulsive nature of the original spectrum prior to the coordinate change.

Clearly, classes (2) and (2′) present a higher degree of spectrum continuity than class (1), since they show a one-dimensional continuity in the spectrum; in the perpendicular direction, however, the spectrum is still impulsive. Similarly, classes (3) and (3′) present a still higher degree of spectrum continuity than classes (2) and (2′), since their spectra contain also purely continuous areas, i.e., 2D neighbourhoods in which the spectrum is continuous (but not identically zero!) and contains no impulsive behaviour at all. Class (4) has the highest spectrum continuity since its spectra are purely continuous and have no points with impulsive behaviour (=singularity points) at all.

Therefore, if we imagine a "continuity scale" for the different types of spectra, in which classes (1) and (1′) (the purely impulsive spectra) are ranked at the lowest grade and class (4) (purely continuous spectra) is situated at the highest grade, then we can say that:

$$(1),(1') \prec (2),(2') \prec (3),(3') \prec (4)$$

where $(i) \prec (j)$ means that class $(j)$ has *a higher degree of spectrum regularity* than class $(i)$, or equivalently, that class $(i)$ has *a higher degree of spectrum singularity* than class $(j)$.

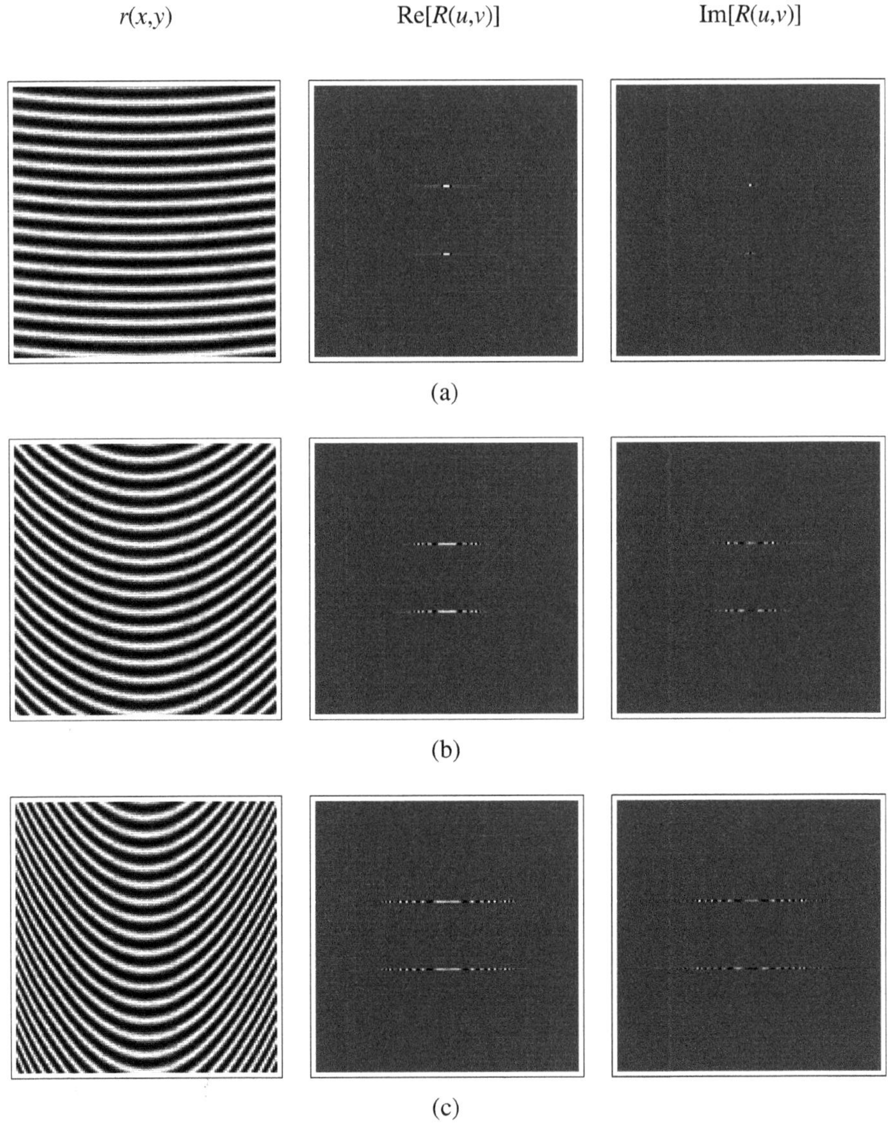


**Figure 10.5:** When the purely periodic function $\cos(2\pi fy)$ is gradually "bent" into a parabolic cosinusoidal grating $\cos(2\pi f(y-ax^2))$, its purely impulsive spectrum "leaks out" into a semi-impulsive spectrum of line-impulses. Conversely, the gradual transition in the spectrum from (c) to (a) can be seen as a sequence of line-impulses which give in the limit the impulses $\frac{1}{2}\delta(u,v-f)$ and $\frac{1}{2}\delta(u,v+f)$. Note that in reality the line-impulses in the spectra oscillate *ad infinitum* without fading out (see legend of Fig. 10.1). (a) $\cos(2\pi f(y-\frac{1}{128}x^2))$; (b) $\cos(2\pi f(y-\frac{1}{16}x^2))$; (c) $\cos(2\pi f(y-\frac{1}{8}x^2))$.

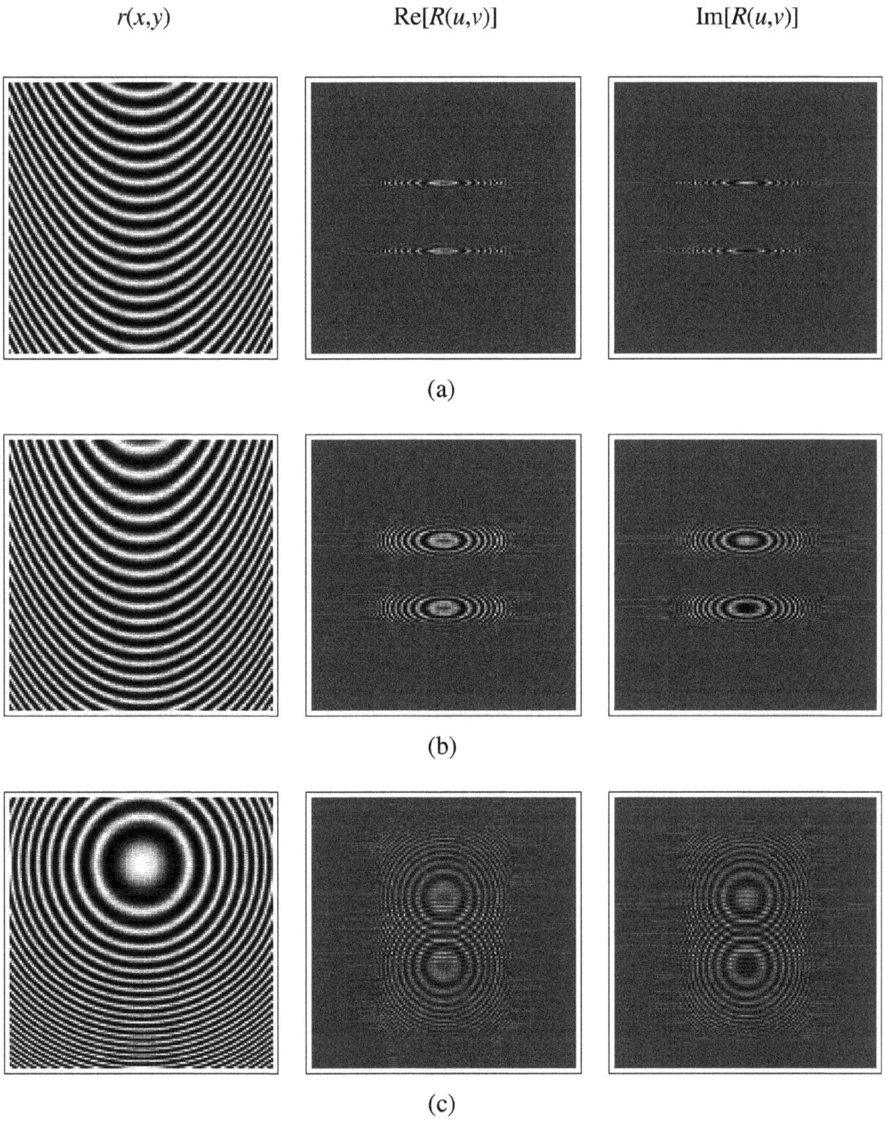


**Figure 10.6:** A gradual transition between a parabolic cosinusoidal grating and a shifted elliptic or circular cosinusoidal zone grating, which is obtained by varying $b$ in $\cos(2\pi f(y - by^2 - \frac{1}{8}x^2))$. In the spectral domain, the pair of line-impulses which forms the spectrum of the parabolic cosinusoidal grating "leaks out" into a pair of continuous humps. (a) $b = \frac{1}{200}$; (b) $b = \frac{1}{40}$; (c) $b = \frac{1}{8}$.

Note that cosinusoidal gratings whose frequency has a varying magnitude but a constant orientation are included in class (2′); cosinusoidal gratings whose frequency has a constant magnitude but a varying orientation are included in class (3).

### 10.4.1 Gradual transitions between cosinusoidal gratings of different types

In order to develop a better intuitive insight into the various degrees of spectrum regularity it may be instructive to analyze what happens in the spectrum when a given cosinusoidal grating in the image domain undergoes a gradual transition from one class into another.

Let us start with transitions between class (1) and class (2); for this end we return to Example 10.5 of Sec. 10.3, namely: the effect of "bending" a 2D cosinusoidal grating (see Fig. 10.5). It is interesting to observe in this example how a gradual change in the coefficient $a$, the "bending rate" of the parabolic grating $\cos(2\pi f(y-ax^2))$, influences the spectrum. Assume that we gradually modify the coefficient $a$: clearly, when $a > 0$ the grating corrugations have the form of upright, open top parabolas; when $a = 0$ the arms of the parabolas open out to become straight horizontal lines; and when $a < 0$ these lines bend down to form upside-down parabolas. As $a$ approaches zero (and the parabolas become more and more flattened), $\cos(2\pi fax^2)$ tends to 1 and $\sin(2\pi fax^2)$ tends to zero, and hence their spectra $R_c(u)$ and $R_s(u)$ tend in the limit to $\delta(u)$ and zero, respectively.[12] This means that as $a$ approaches zero, each of the two continuous horizontal line-impulses in the spectrum $R(u,v)$ becomes more concentrated around $u = 0$ while its "volume" elsewhere tends to zero (see Fig. C.2 in Appendix C). In other words: as $a \to 0$, each of the two continuous line-impulses in the spectrum "shrinks" into a point-impulse, and the spectrum tends in the limit to $\frac{1}{2}\delta(u,v-f) + \frac{1}{2}\delta(u,v+f)$, the impulsive spectrum of the straight cosinusoidal grating $\cos(2\pi fy)$. And conversely, when $a$ moves away from zero (and the parabolas become more and more curved), the original impulses $\frac{1}{2}\delta(u,v-f)$ and $\frac{1}{2}\delta(u,v+f)$ in the spectrum "leak out" in the horizontal direction to become continuous, amplitude-modulated line-impulses. This is illustrated in Fig. 10.5.

**Figure 10.7:** (*See opposite page.*) A gradual transition between circular, elliptic, linear and hyperbolic cosinusoidal gratings, which is obtained by varying $a$ in $\cos(2\pi f\sqrt{|ax^2 + y^2|})$. In the spectral domain, the pair of impulses which forms the spectrum of the straight cosinusoidal grating (when $a = 0$) "leaks out" into an elliptic impulsive shape (when $a$ gradually increases from zero) or into a hyperbolic impulsive shape (when $a$ gradually decreases from zero). (a) $a = 1$; (b) $a = \frac{1}{16}$; (c) $a = 0$; (d) $a = -\frac{1}{8}$.

---

[12] It is a remarkable fact that in spite of its undamped oscillatory nature, $R_c(u)$ tends as $a \to 0$ to $\delta(u)$. This follows from the fact that the total area under $R_c(u)$ is 1 independently of $a$, while its oscillatory area in the ranges $(u_0,\infty)$ and $(-\infty,-u_0)$ for any $u_0 > 0$ tends to zero when $a \to 0$. This is demonstrated in Appendix C.5. See also a similar case in [Saichev97 p. 13].

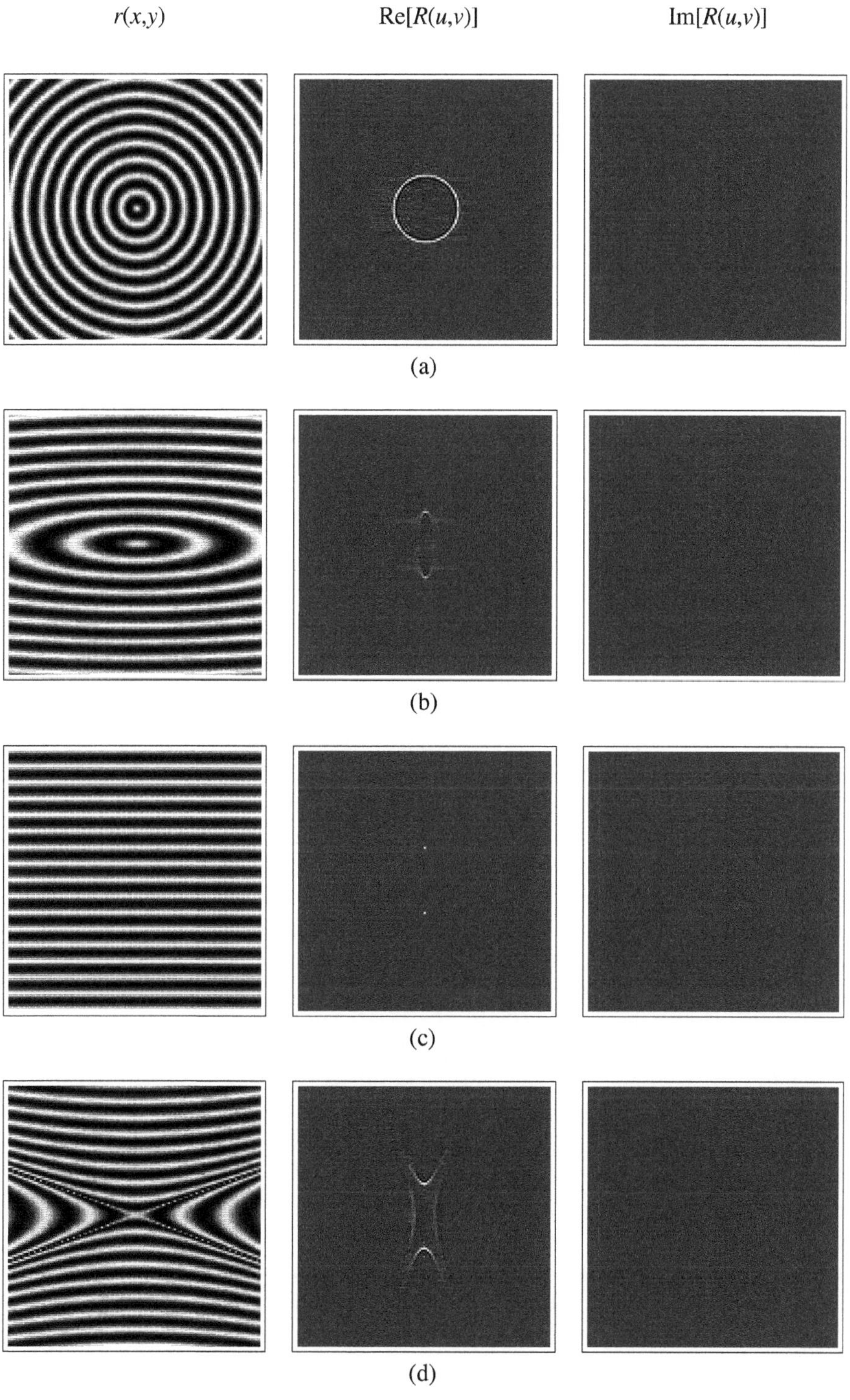
$r(x,y)$
$\mathrm{Re}[R(u,v)]$
$\mathrm{Im}[R(u,v)]$
(a)
(b)
(c)
(d)

This gradual bending process may be also used to give us an intuitive clue for understanding better the continuous spectra of class (4): Assume that $\cos(2\pi f g_1(x,y))$ is a curvilinear grating with a continuous spectrum. We establish a gradual transition between an original purely periodic grating $\cos(2\pi fy)$ and the curvilinear grating $\cos(2\pi f g_1(x,y))$, and observe in the spectral domain the transition from the impulsive spectrum of $\cos(2\pi fy)$ to the continuous spectrum of the curvilinear grating $\cos(2\pi f g_1(x,y))$. We may say, then, that as the grating $\cos(2\pi fy)$ in the image domain is gradually transformed into $\cos(2\pi f g_1(x,y))$, the two impulses $\frac{1}{2}\delta(u,v-f)$ and $\frac{1}{2}\delta(u,v+f)$ in the spectral domain gradually "melt down" and "leak out" in all directions, covering the whole $u,v$ plane with a thin, oscillating layer which forms the new continuous spectrum. One can also think of a gradual transition from a line-spectrum of class (2) into a continuous spectrum of class (4). For example, consider a gradual transition from $b = 0$ to $b = b_1 > 0$ in the curvilinear cosine $\cos(2\pi f(y-by^2-ax^2))$, as shown in Fig. 10.6. In the image domain this will smoothly change the curvilinear shape of the grating from a parabolic grating when $b = 0$ into a non-centered elliptic zone grating (or circular zone grating, if $b_1 = a$; see Fig. 10.6(c)). In the spectral domain, the two horizontally continuous line-impulses "leak out" in the vertical direction and generate a continuous spectrum. And conversely, when $b$ approaches zero the continuous spectrum gradually "shrinks" and tends in the limit into a pair of line-impulses.

As yet another example, let us see how gradual transitions can be also established between class (1) and classes (3) and (3'). Consider, for instance, the cosinusoidal grating $\cos(2\pi f\sqrt{|ax^2 + y^2|})$ and assume that we gradually vary its coefficient $a$ (see Fig. 10.7). Clearly, when $a = 1$ the grating corrugations have the form of concentric circles; when $a \to 0$ the circles gradually transform into horizontal ellipses which become flatter and flatter, until at the limit, when $a = 0$, they turn into straight horizontal lines: $\cos(2\pi fy)$. Finally, when $a$ becomes negative, the straight corrugations become curved again, but this time they get a concave, hyperbolic shape. In the spectral domain, when $a = 0$ (and the grating corrugations are horizontal straight lines) the spectrum consists of a pair of impulses $\frac{1}{2}\delta(u,v-f)$ and $\frac{1}{2}\delta(u,v+f)$. When $a$ moves away from zero in the positive direction, this impulse pair gradually "leaks out" to both sides into an elliptic, impulsive ring (with a continuous wake trailing off towards the origin), whose "mass" is still concentrated in its two extremes, around the "melting down" impulses. As $a$ approaches 1 (and the grating shape becomes circular) the elliptic impulse ring becomes more and more circular, and the amplitude peaks at its two extreme points gradually weaken until the ring amplitude becomes uniform all around. (The transition between an elliptic and a circular impulse is explained in [Bracewell95 pp. 130–131].) If $a$ moves away from zero in the negative direction (so that the grating shape gradually becomes hyperbolic) the original impulse pair "leaks out" to both sides into a hyperbolic, impulsive shape with a continuous wake which trails off *outwards* (see Fig. 10.7(d)).

Such gradual transitions give us, indeed, a useful qualitative insight into the influence of a coordinate change on the cosinusoidal grating and on its spectrum. They will also help us later on to extend our moiré theory from the cases of class (1) (i.e., periodic layers with

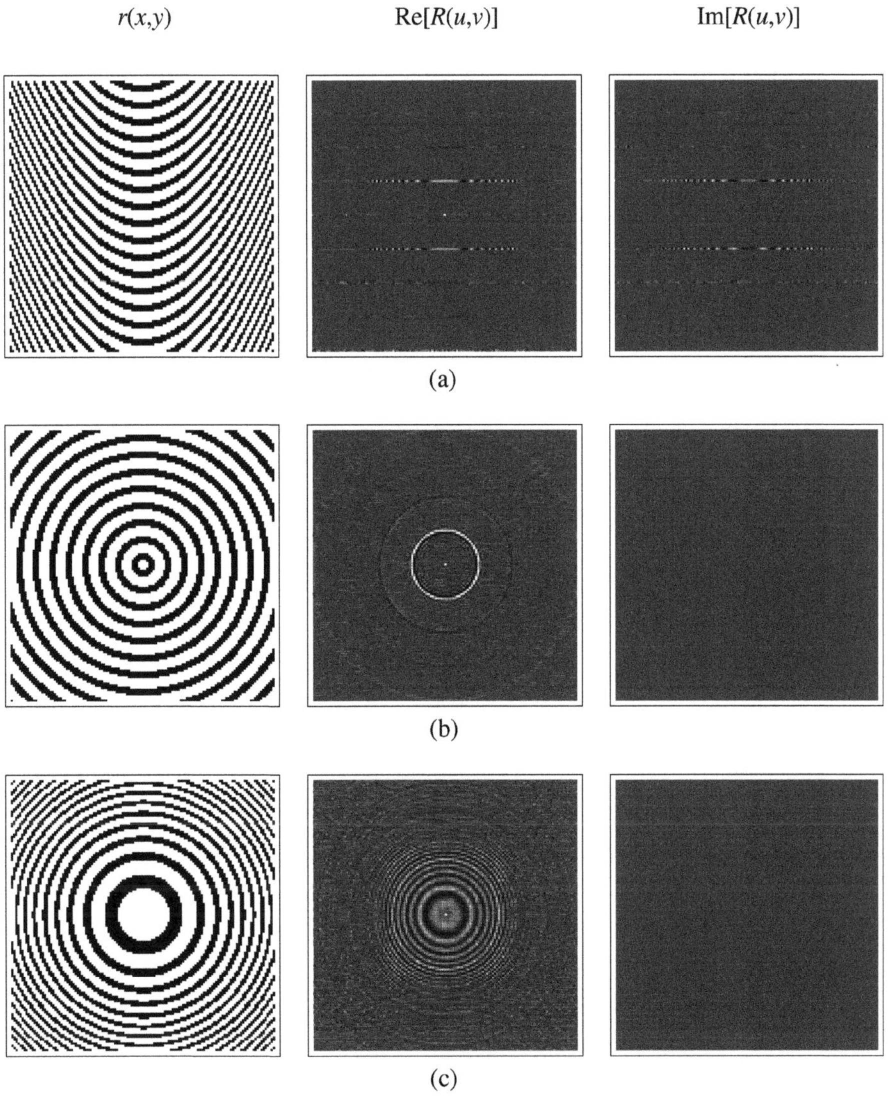


**Figure 10.8:** Some examples of curvilinear gratings $r(x,y)$ having a square wave periodic-profile (with opening ratio $\tau/T = 0.6$) and a bending function $g_1(x,y)$, and their respective spectra $R(u,v)$. (a) A parabolic grating: $g_1(x,y) = y - 0.15x^2$; (b) a circular grating: $g_1(x,y) = \sqrt{x^2+y^2}$; (c) a circular zone grating: $g_1(x,y) = (x^2+y^2)/8$. The amplitudes of the different harmonics in the spectra are weighted by the Fourier series coefficients $a_n$ of the square-wave (see Eq. (10.8)): $a_1 = 0.303$, $a_2 = -0.094$, $a_3 = -0.062$, etc; the sign inversions in the second and third harmonics are clearly visible in the spectra of cases (a) and (b). Notice the various DFT artifacts in the spectra (folding-over due to aliasing; rippling).

impulsive spectra) into cases of classes (2)–(4) as well (i.e., semi- or non-periodic layers with semi- or non-impulsive spectra). The approach based on this intuitive, gradual transition process will be called henceforth *the gradual transition approach.*

## 10.5 The Fourier decomposition of curved, repetitive structures

### 10.5.1 The Fourier decomposition of curvilinear gratings

Our next step in the analysis of repetitive structures with a given periodic-profile is based on the Fourier series development of their periodic-profiles. Let us start, again, with the case of curvilinear gratings. Assume that the curvilinear grating $r(x,y)$ is obtained by bending a periodic grating $p(x')$, i.e., by replacing $x'$ with a function $x' = g_1(x,y)$: $r(x,y) = p(g_1(x,y))$. A few examples of curvilinear gratings $r(x,y)$ with a square-wave periodic-profile form $p(x')$ are shown in Fig. 10.8. We first consider the Fourier series development of the original grating $p(x')$:

$$p(x') = \sum_{n=-\infty}^{\infty} c_n \, e^{i2\pi nfx'} \tag{10.3}$$

Then we replace $x'$ in this Fourier series with the function $g_1(x,y)$ which defines the curvilinear behaviour of the grating $r(x,y)$ throughout the plane (see Sec. 10.2), keeping the same coefficients $c_n$ as in the Fourier decomposition of $p(x')$:

$$r(x,y) = p(g_1(x,y)) = \sum_{n=-\infty}^{\infty} c_n \, e^{i2\pi nfg_1(x,y)} \tag{10.4}$$

This is, indeed, the generalized Fourier decomposition of our curvilinear grating $r(x,y)$. This approach has been introduced in [Lohmann67], where the periodic-profile $p(x')$ was a binary square wave.[13]

Note that if $p(x')$ is symmetric, the exponential series development (10.3) reduces into the corresponding cosine development (see Sec. A.2 in Appendix A):

$$\begin{aligned} p(x') &= \sum_{n=-\infty}^{\infty} a_n \cos(2\pi nx'/T) \\ &= a_0 + 2\sum_{n=1}^{\infty} a_n \cos(2\pi nx'/T) \end{aligned} \tag{10.5}$$

and therefore (10.4) becomes:

$$\begin{aligned} r(x,y) = p(g_1(x,y)) &= \sum_{n=-\infty}^{\infty} a_n \cos(2\pi ng_1(x,y)/T) \\ &= a_0 + 2\sum_{n=1}^{\infty} a_n \cos(2\pi ng_1(x,y)/T) \end{aligned} \tag{10.6}$$

[13] Such series decompositions are sometimes called in literature *quasi-Fourier series* (for example: in [Lohmann67 p. 1568]), and their curvilinear gratings are called *quasiperiodic structures* [Bryngdahl74 p. 1290]. But because this terminology may create confusion with quasi-periodic functions and their Fourier series (see Sec. B.5 in Appendix B), we will use instead the term *generalized Fourier series*, or simply, when no confusion may arise, *Fourier series*; see Appendix G in *Vol. II* for more details.

with the same coefficients $a_n$ as in (10.5).

**Remark 10.5:** For reasons of convenience we will sometimes prefer in the course of this chapter the cosine series development, which lends itself more easily to graphic interpretation. However, the simple cosine series development cannot be used for gratings with non-symmetric periodic-profile forms such as sawtooth waves etc., and in such cases the general exponential development (or equivalently, a development into a cosine and sine series) must be used. A further limitation of the cosine series development is discussed in Remark 10.8, Sec. 10.8. ■

**Example 10.8:** A parabolic grating with a square wave periodic-profile $p(x')$:

We recall from Sec. 2.5 that the square wave with period $T$ and opening $\tau$ is defined by:

$$p(x') = \begin{cases} 1 & |x'-nT| < \tau/2 \\ 0 & |x'-nT| > \tau/2 \end{cases} \tag{10.7}$$

and its Fourier series decomposition is:

$$p(x') = \sum_{n=-\infty}^{\infty} a_n \cos(2\pi n x'/T) \tag{10.8}$$

with: $\quad a_n = (\tau/T)\,\mathrm{sinc}(n\tau/T)$

Let now $r(x,y)$ be a parabolic grating having the square wave periodic-profile $p(x')$ with period $T$ and opening $\tau$ (see Fig. 10.8(a)). This curvilinear grating is obtained by bending $p(x')$, i.e., by applying to $p(x')$ the non-linear bending function $x' = y - ax^2$. The definition of $r(x,y)$ is obtained, therefore, by replacing $x'$ with $y - ax^2$ in Eq. (10.7):

$$r(x,y) = p(y-ax^2) = \begin{cases} 1 & |y-ax^2-nT| < \tau/2 \\ 0 & |y-ax^2-nT| > \tau/2 \end{cases}$$

The Fourier decomposition of $r(x,y)$ is obtained by replacing $x'$ with $y - ax^2$ in Eq. (10.8):

$$r(x,y) = p(y-ax^2) = \sum_{n=-\infty}^{\infty} a_n \cos(2\pi n(y-ax^2)/T)$$

keeping the same coefficients $a_n$ as in (10.8).

Alternatively, if we use the exponential notation we obtain the equivalent expression:

$$r(x,y) = p(y-ax^2) = \sum_{n=-\infty}^{\infty} c_n\, e^{i2\pi n f (y-ax^2)}$$

with the same coefficients $c_n = a_n$ (since $p(x')$ is symmetric; see Sec. A.2, Appendix A). ■

**Example 10.9:** A circular grating with a square wave periodic-profile $p(r')$ of period $T$ and opening $\tau$ (see Fig. 10.8(b)):

Once again, we first consider the Fourier development of the square wave periodic-profile (using its equivalent one-sided form; see Eq. (A.1) in Appendix A):

$$p(r') = a_0 + 2\sum_{n=1}^{\infty} a_n \cos(2\pi n r'/T)$$

with the same coefficients $a_n$ as above, and then we replace $r'$ with $\sqrt{x^2+y^2}$:

$$r(x,y) = p(\sqrt{x^2+y^2}) = a_0 + 2\sum_{n=1}^{\infty} a_n \cos(2\pi n\sqrt{x^2+y^2}/T)$$

This is, therefore, the decomposition of the circular grating $r(x,y)$ into a series of circular cosines with radial frequencies of $f_n = n/T$ (=radial periods of $T/n$). ■

### 10.5.2 The Fourier decomposition of curved line-grids and dot-screens

The situation in the 2D case of a curved dot-screen (or a curved line-grid) $r(x,y)$ is similar. Assume that the curved dot-screen $r(x,y)$ is obtained by bending a 2D periodic dot-screen $p(x',y')$, i.e., by replacing $x'$ and $y'$ with functions $x' = g_1(x,y)$ and $y' = g_2(x,y)$: $r(x,y) = p(g_1(x,y),g_2(x,y))$. An example of such a curved dot-screen $r(x,y)$ is shown in Fig. 10.2(b). According to the present approach we first consider the Fourier development of the original 2D periodic dot-screen $p(x',y')$:

$$p(x',y') = \sum_{m=-\infty}^{\infty} \sum_{n=-\infty}^{\infty} c_{m,n}\, e^{i2\pi(mx'/T_{x'} + ny'/T_{y'})} \qquad (10.9)$$

Then, we replace $x'$ and $y'$ in this Fourier series with the functions $g_1(x,y)$ and $g_2(x,y)$ which define the curved behaviour of the grating $r(x,y)$ throughout the plane (see Sec. 10.2), keeping the same coefficients $c_{m,n}$ as in the Fourier decomposition of $p(x',y')$:

$$r(x,y) = p(g_1(x,y),g_2(x,y)) = \sum_{m=-\infty}^{\infty} \sum_{n=-\infty}^{\infty} c_{m,n}\, e^{i2\pi(mg_1(x,y)/T_{x'} + ng_2(x,y)/T_{y'})} \qquad (10.10)$$

This is, therefore, the Fourier decomposition of our curved dot-screen $r(x,y)$.

Note that if $p(x',y')$ is symmetric, the exponential series development (10.9) reduces into the corresponding cosine development (see Sec. A.3.2 in Appendix A):

$$p(x',y') = \sum_{m=-\infty}^{\infty} \sum_{n=-\infty}^{\infty} a_{m,n} \cos 2\pi(mx'/T_{x'} + ny'/T_{y'}) \qquad (10.11)$$

and therefore Eq. (10.10) becomes:

$$r(x,y) = p(g_1(x,y),g_2(x,y))$$
$$= \sum_{m=-\infty}^{\infty} \sum_{n=-\infty}^{\infty} a_{m,n} \cos 2\pi(mg_1(x,y)/T_{x'} + ng_2(x,y)/T_{y'}) \qquad (10.12)$$

with the same coefficients $a_{m,n}$ as in (10.11).

**Remark 10.6:** We will usually prefer to choose $p(x',y')$ as a *normalized periodic-profile* with $T_{x'} = 1$, $T_{y'} = 1$; and if $T_{x'} \neq 1$ or $T_{y'} \neq 1$ we will consider them to be included within the functions $g_1(x,y)$, $g_2(x,y)$, leaving $p(x',y')$ itself normalized. Therefore $T_{x'}$ and $T_{y'}$ will usually be omitted from Eqs. (10.10), (10.12). ■

Eqs. (10.4), (10.6) or their 2D counterparts (10.10), (10.12) are simply a formal decomposition of the curved layer $r(x,y)$ into a sum of curvilinear exponentials (or a sum of curvilinear cosines) which were all subjected to the same coordinate transformation

**g**$(x,y)$ as $r(x,y)$ itself. This decomposition is simply an alternative representation of the curvilinear layer $r(x,y)$ in the image domain; an important advantage of this representation is that it allows one to approximate $r(x,y)$ by taking only its first term (or its first few terms), thus significantly simplifying its mathematical handling in the image domain. We will see in Chapter 11 how this can be used to find the geometric layout of moirés which appear in the superposition of curvilinear gratings. However, this series decomposition of the curved layer $r(x,y)$ has also another important role: it opens the way to the analysis of the spectrum of the curved layer $r(x,y)$, thus helping us to better understand things in the spectral domain, too. We will return to this point in the following section.

## 10.6 The spectrum of curved, repetitive structures

### 10.6.1 The spectrum of curvilinear gratings

We have seen in Sec. 10.5.1 that a curvilinear grating $r(x,y)$ can be represented in the image domain as a generalized Fourier series, i.e., as a sum of curvilinear cosines (or exponentials) which were all subjected to the same transformation $g_1(x,y)$ as the curvilinear grating $r(x,y)$ itself. This fact reduces the problem of finding $R(u,v)$, the spectrum of the curvilinear grating $r(x,y)$, into the question of finding the Fourier transform of a curvilinear cosine (or exponential). Just as in the classical case of periodic functions the Fourier transform pair:

$$\cos(2\pi fx) \leftrightarrow \tfrac{1}{2}\delta(u-f) + \tfrac{1}{2}\delta((u+f)$$

gives us: $$p(x) = \sum_{n=-\infty}^{\infty} a_n \cos(2\pi nx/T) \leftrightarrow P(u) = \sum_{n=-\infty}^{\infty} a_n\, \delta(u-n/T),$$

in the present case, if we know the Fourier spectrum $R_n(u,v)$ of the curvilinear cosine $r_n(x,y) = \cos(2\pi n g_1(x,y)/T)$, we obtain from Eq. (10.6):

$$r(x,y) = \sum_{n=-\infty}^{\infty} a_n \cos(2\pi n g_1(x,y)/T) \leftrightarrow R(u,v) = \sum_{n=-\infty}^{\infty} a_n\, R_n(u,v) \qquad (10.13)$$

Similarly, using the more general exponential notation we obtain from Eq. (10.4):

$$r(x,y) = \sum_{n=-\infty}^{\infty} c_n\, e^{i2\pi nf g_1(x,y)} \leftrightarrow R(u,v) = \sum_{n=-\infty}^{\infty} c_n\, R_n(u,v) \qquad (10.14)$$

where $R_n(u,v)$ is the spectrum of the curvilinear exponential function $r_n(x,y) = e^{i2\pi nf g_1(x,y)}$.

In other words, the spectrum of the curvilinear grating $r(x,y)$ is the sum of the spectra of the individual curvilinear cosines (or exponentials), where $a_n$ (or $c_n$) are the same coefficients as in the Fourier series decomposition of $r(x,y)$, and hence, according to Sec. 10.5.1, the same coefficients as in the decomposition of its periodic-profile, $p(x')$.[14]

[14] Although Eqs. (10.4) and (10.6) are only *generalized* Fourier series decompositions of $r(x,y)$, $R(u,v)$ is the Fourier spectrum of $r(x,y)$, since a sum of Fourier transforms is the Fourier transform of the sum. Note that we do not discuss here purely mathematical questions such as the precise conditions under which Eqs. (10.13) or (10.14) hold, or convergence issues. We simply note that in real-world cases of interest this result does correspond to the physical reality, and we limit ourselves here to such cases.

Therefore, in order to investigate the spectrum of the curvilinear grating $r(x,y)$ we first have to understand what happens to the spectrum of a 2D cosine function $\cos(2\pi fx)$ (or more generally, to the spectrum of an exponential function $e^{i2\pi fx}$) when the image domain undergoes a transformation or a coordinate change $g_1(x,y)$. As we have already seen in Sec. 10.3, there exists no such general rule when the transformation $g_1(x,y)$ is non-linear. However, for many cases of interest the Fourier transform of $\cos(2\pi g_1(x,y)/T)$ (or of $e^{i2\pi f g_1(x,y)}$) can be found based on Fourier transform tables in literature such as [Erdélyi54] which include Fourier transforms of functions of the form $\cos(g(x))$ or $e^{ig(x)}$. Some particular cases of interest have been discussed in Examples 10.5–10.7 of Sec. 10.3 and illustrated in Fig. 10.1; other cases are given in [Amidror98a]. As we have seen, in some situations the spectrum of the curvilinear cosine remains impulsive, while in other situations its spectrum may become semi-impulsive or even completely non-impulsive.

Eq. (10.13) is particularly useful when the individual spectra $R_n(u,v)$ of the curvilinear cosines are impulsive. In this case the spectrum $R(u,v)$ is composed of isolated, separately localized entities (impulses), and each of the terms $R_n(u,v)$ in the series represents indeed one of these isolated entities in the spectrum. If the individual spectra $R_n(u,v)$ are continuous, as in the case of Figs. 10.1(g)–(i), Eq. (10.13) is still valid — but it loses much of its practical usefulness: In this case the spectrum $R(u,v) = \sum a_n R_n(u,v)$ is a sum of continuous functions with overlapping supports, and its series representation no longer reflects a partition of the spectrum into spatially separated entities $R_n(u,v)$ with mutually exclusive supports on the $u,v$ plane, which can be individually localized, isolated and manipulated. Nevertheless, we will see in the following sections that one can still make use of such continuous spectra, for example in the extraction of a moiré effect from the spectrum of the superposition.

**Example 10.10:** The spectrum of a parabolic grating with a square wave periodic-profile:

We have seen in Example 10.8 that the Fourier development of the parabolic grating with a square wave periodic-profile is:

$$r(x,y) = \sum_{n=-\infty}^{\infty} a_n \cos(2\pi n(y-ax^2)/T)$$

with the same coefficients $a_n$ as in the square wave profile: $a_n = (\tau/T)\mathrm{sinc}(n\tau/T)$. Therefore the spectrum of this grating is:

$$R(u,v) = \sum_{n=-\infty}^{\infty} a_n R_n(u,v)$$

with the same coefficients $a_n$, where $R_0(u,v) = \delta(u,v)$ is the DC impulse and $R_n(u,v)$, the spectrum of $\cos(2\pi n(y-ax^2)/T)$, is a pair of continuous horizontal line-impulse which are vertically located at $v = \pm n/T$ (see Example 10.5 in Sec. 10.3):

$$R_n(u,v) = \tfrac{1}{2}[R_c(u) + iR_s(u)]\delta(v-nf) + \tfrac{1}{2}[R_c(u) - iR_s(u)]\delta(v+nf)$$

with:

$$R_c(u) = \frac{1}{2\sqrt{nfa}}\left(\cos\left(\frac{\pi}{2nfa}u^2\right) + \sin\left(\frac{\pi}{2nfa}u^2\right)\right)$$

$$R_s(u) = \frac{1}{2\sqrt{nfa}}\left(\cos\left(\frac{\pi}{2nfa}u^2\right) - \sin\left(\frac{\pi}{2nfa}u^2\right)\right)$$

where $a$ is the bending rate of the parabolic grating $r(x,y)$ and $f$ is its fundamental frequency $1/T$.

The spectrum of the parabolic grating with a square wave periodic-profile consists, therefore, of a DC impulse plus a series of such horizontal line-impulses which are vertically located at $v = n/T$, $n = \pm 1, \pm 2, \ldots$, and whose amplitudes are weighted by the coefficients $a_n$ (see Fig. 10.8(a)). It is interesting to note that the closer a line-impulse is to the spectrum origin, i.e., the smaller its index $n$, the denser are its oscillations; at $n = 0$ the oscillations are so dense that they completely collapse onto the DC impulse, which may be seen, therefore, as a degenerated line-impulse. ■

**Example 10.11:** The spectrum of a circular grating with a square wave periodic-profile:

According to Example 10.9 in Sec. 10.5.1 the Fourier development of the circular grating with square periodic-profile is (using for convenience the one-sided series form):

$$r(x,y) = a_0 + 2\sum_{n=1}^{\infty} a_n \cos(2\pi n\sqrt{x^2+y^2}/T)$$

with the same coefficients $a_n$ as in the square wave profile: $a_n = (\tau/T)\,\mathrm{sinc}(n\tau/T)$. Therefore, the spectrum of this circular grating is:

$$R(u,v) = a_0 R_0(u,v) + 2\sum_{n=1}^{\infty} a_n R_n(u,v)$$

where $R_0(u,v)$ is the DC impulse and $R_n(u,v)$, the spectrum of $\cos(2\pi n\sqrt{x^2+y^2}/T)$, is the peculiar dipole-like impulsive ring with a weak continuous "wake" trailing off toward the center that we have seen in Example 10.6 in Section 10.3 (see [Amidror97] for more details). The spectrum of a circular grating with a square wave periodic-profile is, therefore, a concentric series of such circular dipole-like rings with radiuses of $n/T$, whose amplitudes are weighted by the coefficients $a_n$ of the square wave profile (see Fig. 10.8(b)). Note that in this case the rings $R_n(u,v)$ are not completely spatially separable, since their weak, continuous "wakes" which trail off toward the spectrum center are overlapping. However, for many practical needs these continuous "wakes" can be considered as negligible, and we can say that the main frequency contribution of each of the rings $R_n(u,v)$ is concentrated on its impulsive (=singular) support, namely: on the perimeter of a circle with radius $n/T$ around the spectrum origin.[15] Note, however, that even on this singular support the impulsive behaviour of $R_n(u,v)$ is dipole-like, and hence more complex than that of a simple impulse ring $\delta(\sqrt{u^2+v^2} - f)$. ■

**Example 10.12:** The spectrum of a zone grating with square wave periodic-profile:

As we have seen in Example 10.7 of Sec. 10.3, a zone grating (zone plate) is a concentric circular grating where the radius of the $n$-th circle is proportional to $\sqrt{n}$. In many optical applications the periodic-profile of the zone grating has a binary (black/white)

[15] As stated for example in [Egorov93 p. 8], it is a known fact that singular points correspond to those phenomena which are most interesting from the point of view of each physical theory. The study of singularities is a most important problem, and in many mathematics disciplines one often examines functions modulo smooth ones, so that the points where a given function is smooth may be neglected.

square wave form. The Fourier development of this function is, therefore (using the one-sided series form):

$$r(x,y) = a_0 + 2\sum_{n=1}^{\infty} a_n \cos(2\pi n(x^2+y^2)/T)$$

with the same coefficients $a_n$ as in the square wave profile: $a_n = (\tau/T)\,\text{sinc}(n\tau/T)$. Therefore the spectrum of this circular grating is:

$$R(u,v) = a_0R_0(u,v) + 2\sum_{n=1}^{\infty} a_n R_n(u,v)$$

where $R_0(u,v)$ is the DC impulse $\delta(u,v)$ and $R_n(u,v)$, the spectrum of $\cos(2\pi n(x^2+y^2)/T)$, is according to Example 10.7:

$$R_n(u,v) = \tfrac{1}{2nf}\sin(\tfrac{\pi}{2nf}(u^2+v^2))$$

where $f$ is the fundamental frequency $1/T$. Note that the terms $R_n(u,v)$ of the spectrum $R(u,v)$ are not spatially separable, since each of them is a continuous sinusoidal zone grating which is centered on the origin and extends throughout the whole spectrum, and hence all of them are mutually overlapping at every point of the $u,v$ plane. ■

### 10.6.2 The spectrum of curved line-grids and dot-screens

We have seen in Sec. 10.5.2 that a curved dot-screen (or line-grid) can be represented in the image domain as a 2D generalized Fourier series, i.e., as a double sum of curved cosines (or exponentials) which were all subjected to the same transformation $\mathbf{g}(x,y)$ as the curved layer $r(x,y)$ itself.

Therefore, just as in the case of curvilinear gratings, the spectrum of the curved dot-screen $r(x,y)$ is the 2D sum of the spectra of the individual curved cosines (or exponentials), where $a_{m,n}$ (or $c_{m,n}$) are the same coefficients as in the 2D series decomposition of $r(x,y)$, and hence, according to Section 10.5.2, the same coefficients as in the decomposition of its periodic-profile, $p(x',y')$.

The 2D counterpart of Eq. (10.13) is, therefore, as follows: If $R_{m,n}(u,v)$ is the spectrum of the curved cosine $r_{m,n}(x,y) = \cos 2\pi(mg_1(x,y) + ng_2(x,y))$ then by Eq. (10.12):[16]

$$r(x,y) = \sum_{m=-\infty}^{\infty} \sum_{n=-\infty}^{\infty} a_{m,n} \cos 2\pi(mg_1(x,y) + ng_2(x,y))$$

$$\leftrightarrow R(u,v) = \sum_{m=-\infty}^{\infty} \sum_{n=-\infty}^{\infty} a_{m,n} R_{m,n}(u,v) \qquad (10.15)$$

Similarly, using the more general exponential notation we obtain from Eq. (10.10):

$$r(x,y) = \sum_{m=-\infty}^{\infty} \sum_{n=-\infty}^{\infty} c_{m,n}\, e^{i2\pi(mg_1(x,y) + ng_2(x,y))}$$

$$\leftrightarrow R(u,v) = \sum_{m=-\infty}^{\infty} \sum_{n=-\infty}^{\infty} c_{m,n} R_{m,n}(u,v) \qquad (10.16)$$

---

[16] Note that the constants $T_{x'}$, $T_{y'}$ (the original periods) have been incorporated here within the functions $g_1(x,y)$, $g_2(x,y)$; see Remark 10.6 above.

where $R_{m,n}(u,v)$ is the spectrum of the curvilinear exponential function $r_{m,n}(x,y) = e^{i2\pi(mg_1(x,y) + ng_2(x,y))}$.

## 10.7 The superposition of curved, repetitive layers

We arrive now to the main goal of the present chapter, namely, the investigation of superpositions of repetitive but non-periodic layers, such as curvilinear gratings. In the previous chapters we have seen in detail what happens, both in the image and in the spectral domains, when two or more periodic layers are superposed. When the superposed layers consist of repetitive but not necessarily periodic curvilinear structures, similar phenomena may occur in the superposition — but this time they are more complex and versatile (and hence often more interesting and visually attractive) than in the periodic case. Examples of moiré effects which occur in the superposition of curvilinear gratings can be found in the figures throughout this chapter, as well as in many references such as [Patorski93], [Oster69] etc. Two simple facts can be immediately observed:

(1) Since the original gratings are non-periodic, it is not surprising that the resulting moiré patterns are, in general, non-periodic;[17]

(2) Since a superposition of non-periodic curvilinear gratings contains throughout the $x,y$ plane many different angle/period combinations, various moirés may be visible simultaneously in different areas of the superposition.

In the present section we will analyze several superpositions of repetitive, non-periodic gratings with spectra of different types, and we will try to understand the connection between their spectral and their image domain properties. But first of all, let us see how our moiré definition is generalized to the case of repetitive, non-periodic layers.

### 10.7.1 Moirés in the superposition of curved, repetitive layers

The first step in the investigation of the superposition and its moirés can be done purely in the image domain, by analyzing the generalized Fourier series decompositions of the original layers and of the layer superposition.

Let us start once again with the simpler case of curvilinear gratings. Suppose that the original repetitive layers are given by the curvilinear gratings:

$$r_1(x,y) = p_1(g_1(x,y)) = \sum_{m=-\infty}^{\infty} c^{(1)}_m \, e^{i2\pi m g_1(x,y)}$$

$$r_2(x,y) = p_2(g_2(x,y)) = \sum_{n=-\infty}^{\infty} c^{(2)}_n \, e^{i2\pi n g_2(x,y)}$$

Their superposition is expressed therefore by the product:

[17] Note, however, that particular cases may be designed in which the resulting moiré is periodic although the original layers are not. We will return to this subject in Secs. 10.8 and 10.9 below.

$$r_1(x,y)\ r_2(x,y) = \Big(\sum_{m=-\infty}^{\infty} c^{(1)}_m\ e^{i2\pi m g_1(x,y)}\Big)\Big(\sum_{n=-\infty}^{\infty} c^{(2)}_n\ e^{i2\pi n g_2(x,y)}\Big)$$
$$= \sum_{m=-\infty}^{\infty}\sum_{n=-\infty}^{\infty} c^{(1)}_m c^{(2)}_n\ e^{i2\pi(m g_1(x,y) + n g_2(x,y))} \qquad (10.17)$$

We call the term in this double sum whose indices are $m$ and $n$ the $(m,n)$-term.

Consider now the partial sum which consists of all the terms of this double sum that are spanned by the $(k_1,k_2)$-term. This partial sum consists of all the $n(k_1,k_2)$-terms ($n \in \mathbb{Z}$), i.e., all the terms whose indices are $nk_1$ and $nk_2$:

$$m_{k_1,k_2}(x,y) = \sum_{n=-\infty}^{\infty} c^{(1)}_{nk_1}\ c^{(2)}_{nk_2}\ e^{i2\pi n(k_1 g_1(x,y) + k_2 g_2(x,y))} \qquad (10.18)$$

This partial sum corresponds to a repetitive structure which is present in the superposition $r_1(x,y)\ r_2(x,y)$, but is not present in either of the original layers $r_1(x,y)$ and $r_2(x,y)$ themselves. This structure is called the $(k_1,k_2)$-moiré. As an example, the (1,-1)-moiré is defined by the partial sum consisting of all the terms of the double sum (10.17) whose indices are $n$ and $-n$, namely:

$$m_{1,-1}(x,y) = \sum_{n=-\infty}^{\infty} c^{(1)}_n c^{(2)}_{-n}\ e^{i2\pi n(g_1(x,y) - g_2(x,y))} \qquad (10.19)$$

A schematic illustration of the (1,-1)-moiré between two straight gratings and between the two bent gratings obtained by $g_1(x,y)$, $g_2(x,y)$ is given in Fig. 11.1, Chapter 11.

Note that Eq. (10.18) is a generalization of its periodic counterpart, Eq. (6.3):

$$m_{k_1,k_2}(\mathbf{x}) = \sum_{n=-\infty}^{\infty} c^{(1)}_{nk_1}\ c^{(2)}_{nk_2}\ e^{i2\pi n(k_1\mathbf{f}_1 + k_2\mathbf{f}_2)\cdot\mathbf{x}}$$

And indeed, when $g_1(x,y)$ and $g_2(x,y)$ are linear functions of $x$ and $y$, i.e., $g_1(x,y) = u_1x+v_1y = \mathbf{f}_1\cdot\mathbf{x}$, $g_2(x,y) = u_2x+v_2y = \mathbf{f}_2\cdot\mathbf{x}$, the gratings $r_1(x,y)$ and $r_2(x,y)$ are periodic functions whose frequencies are given, respectively, by the vectors $\mathbf{f}_1 = (u_1,v_1)$ and $\mathbf{f}_2 = (u_2,v_2)$, and the $(k_1,k_2)$-moiré $m_{k_1,k_2}(x,y)$ becomes a periodic function whose frequency is $k_1\mathbf{f}_1+k_2\mathbf{f}_2$.

Now, all the substructures of the superposition that are defined by the different $(k_1,k_2)$-moirés coexist in the superposition simultaneously (they are simply different partial sums of the Fourier series decomposition of the superposition). However, not all of them are visible in the superposition simultaneously. As we have seen in Chapter 2, in the particular case where $r_1(x,y)$ and $r_2(x,y)$ are periodic the $(k_1,k_2)$-moiré is either visible or invisible throughout the entire $x,y$ plane depending on whether its frequency vector $k_1\mathbf{f}_1+k_2\mathbf{f}_2$ falls in the spectrum inside or outside the range of visible frequencies. For example, if the difference frequency $\mathbf{f}_1-\mathbf{f}_2$ falls inside the visible frequency range but the sum frequency $\mathbf{f}_1+\mathbf{f}_2$ falls beyond the visible frequency range, then the (1,-1)-moiré is visible throughout the superposition but the (1,1)-moiré is nowhere visible. This has been explained in detail in Sec. 2.3. In the case of curved gratings, where the frequency vector of each layer is no longer constant but rather varies throughout the $x,y$ plane, we consider at any point $(x,y)$ the local frequency vectors $\mathbf{f}_1(x,y)$, $\mathbf{f}_2(x,y)$ (see Sec. 10.10). The same visibility rule applies here as before — but this time it applies *locally*, i.e., at any point $(x,y)$ of the superposition individually. This means that the $(k_1,k_2)$-moiré becomes visible in the superposition (if its amplitude is sufficiently strong) wherever its local frequency is smaller than the maximal frequency that the eye can resolve at the corresponding viewing conditions. As a consequence, in the superposition of curved layers, unlike in the case of periodic layers, any moiré effect may be locally visible in some zones of the $x,y$ plane but invisible in other zones.

We proceed now to the superposition of curved dot-screens. Let $r_1(x,y)$ and $r_2(x,y)$ be two curved dot-screens defined by:

$$r_1(x,y) = p_1(g_1(x,y),g_2(x,y)) = \sum_{n_1=-\infty}^{\infty} \sum_{n_2=-\infty}^{\infty} c^{(1)}_{n_1,n_2}\, e^{i2\pi(n_1 g_1(x,y)+n_2 g_2(x,y))}$$

$$r_2(x,y) = p_2(g_3(x,y),g_4(x,y)) = \sum_{n_3=-\infty}^{\infty} \sum_{n_4=-\infty}^{\infty} c^{(2)}_{n_3,n_4}\, e^{i2\pi(n_3 g_3(x,y)+n_4 g_4(x,y))}$$

Their superposition is expressed therefore by the product:

$$\begin{aligned} r_1(x,y)\, r_2(x,y) &= \Big(\sum_{n_1=-\infty}^{\infty} \sum_{n_2=-\infty}^{\infty} c^{(1)}_{n_1,n_2}\, e^{i2\pi(n_1 g_1(x,y)+n_2 g_2(x,y))}\Big)\Big(\sum_{n_3=-\infty}^{\infty} \sum_{n_4=-\infty}^{\infty} c^{(2)}_{n_3,n_4}\, e^{i2\pi(n_3 g_3(x,y)+n_4 g_4(x,y))}\Big) \\ &= \sum_{n_1=-\infty}^{\infty} \sum_{n_2=-\infty}^{\infty} \sum_{n_3=-\infty}^{\infty} \sum_{n_4=-\infty}^{\infty} c^{(1)}_{n_1,n_2} c^{(2)}_{n_3,n_4}\, e^{i2\pi(n_1 g_1(x,y)+n_2 g_2(x,y)+n_3 g_3(x,y)+n_4 g_4(x,y))} \end{aligned} \tag{10.20}$$

We call the term in this sum whose indices are $n_1$, $n_2$, $n_3$ and $n_4$ the $(n_1,n_2,n_3,n_4)$-term.

Now, as we have done in the periodic case at the end of Sec. 6.7, we consider the 2D partial sum which consists of all the terms of this quadruple sum that are spanned by the $(k_1,k_2,k_3,k_4)$-term and its orthogonal counterpart, the $(-k_2,k_1,-k_4,k_3)$-term. This 2D partial sum consists of all the terms whose indices are $mk_1-nk_2$, $mk_2+nk_1$, $mk_3-nk_4$ and $mk_4+nk_3$, since (see Eq. (4.10)):

$$m(k_1,k_2,k_3,k_4) + n(-k_2,k_1,-k_4,k_3) = (mk_1-nk_2,\ mk_2+nk_1,\ mk_3-nk_4,\ mk_4+nk_3)$$

This partial sum is therefore given by:

$$\begin{aligned} m_{k_1,k_2,k_3,k_4}(x,y) &= \sum_{m=-\infty}^{\infty} \sum_{n=-\infty}^{\infty} c^{(1)}_{mk_1-nk_2,\ mk_2+nk_1}\, c^{(2)}_{mk_3-nk_4,\ mk_4+nk_3} \\ &\quad \times\ e^{i2\pi([mk_1-nk_2]g_1(x,y) + [mk_2+nk_1]g_2(x,y) + [mk_3-nk_4]g_3(x,y) + [mk_4-nk_3]g_4(x,y))} \\ &= \sum_{m=-\infty}^{\infty} \sum_{n=-\infty}^{\infty} c^{(1)}_{mk_1-nk_2,\ mk_2+nk_1}\, c^{(2)}_{mk_3-nk_4,\ mk_4+nk_3} \\ &\quad \times\ e^{i2\pi(m[k_1 g_1(x,y)+k_2 g_2(x,y)+k_3 g_3(x,y)+k_4 g_4(x,y)] + n[-k_2 g_1(x,y)+k_1 g_2(x,y)-k_4 g_3(x,y)+k_3 g_4(x,y)])} \end{aligned} \tag{10.21}$$

This partial sum corresponds to a 2D repetitive structure which is present in the superposition $r_1(x,y)\, r_2(x,y)$, but is not present in either of the original layers $r_1(x,y)$ and $r_2(x,y)$ themselves. This 2D structure is called the $(k_1,k_2,k_3,k_4)$-moiré. Clearly, this definition of the $(k_1,k_2,k_3,k_4)$-moiré between two curved dot-screens is a generalization of the $(k_1,k_2,k_3,k_4)$-moiré definition in the periodic case (see Eq. (6.8)). As an example, the

(1,0,-1,0)-moiré between two curved screens is defined by the partial sum consisting of all the terms of the quadruple sum (10.20) whose indices are $m$, $n$, $-m$, $-n$, namely:

$$m_{1,0,\text{-}1,0}(x,y) = \sum_{m=-\infty}^{\infty} \sum_{n=-\infty}^{\infty} c^{(1)}_{m,n} c^{(2)}_{-m,-n}\, e^{i2\pi(mg_1(x,y)+ng_2(x,y)-mg_3(x,y)-ng_4(x,y))}$$

$$= \sum_{m=-\infty}^{\infty} \sum_{n=-\infty}^{\infty} c^{(1)}_{m,n} c^{(2)}_{-m,-n}\, e^{i2\pi(m[g_1(x,y)-g_3(x,y)] + n[g_2(x,y)-g_4(x,y)])} \qquad (10.22)$$

This is, indeed, the 2D counterpart of the (1,-1)-moiré between two curvilinear gratings (Eq. (10.19)). We will return to moiré effects between curved screens in Sec. 10.9.2.

### 10.7.2 Image domain vs. spectral domain investigation of the superposition

The above considerations enable us to extract the Fourier series decomposition of any moiré from the product of the Fourier decompositions of the original layers, directly in the image domain. This approach has been introduced in the late 1960s (although not yet in the fully general form that we have presented here), and it has been used to investigate several simple cases of moirés between curvilinear gratings, notably between zone gratings (see, for example, [Lohmann67] and [Harburn75]).

Although this approach is based on concepts from the Fourier theory, it is completely free from any spectral domain considerations. And indeed, it appears that no real attempt has been made until now to investigate in depth the dual considerations in the spectral domain. A possible reason is that while the Fourier series decomposition of a curvilinear grating $r(x,y)$ in the image domain appears to be intuitive and straightforward, the dual situation in the spectral domain is far from being obvious, because of the difficulties in understanding the behaviour of the Fourier transform under non-linear coordinate transformations.

However, as we have already seen in our study of periodic layer superpositions in the preceding chapters, the spectral point of view may give us new insights into the nature of the problem, which cannot be acquired by pure image domain considerations. Since we have already studied in Sec. 10.6 the spectrum of curvilinear layers, we may try to use this information now in order to extend our spectral domain analysis into the case of non-periodic curvilinear layers. In the rest of this section we will illustrate the spectral domain analysis and the difficulties it presents by means of several simple examples. Readers who are only interested in the final, general results may skip at this point directly to Sec. 10.9.

Until now, our Fourier-based analysis of the various superposition phenomena between periodic layers benefited from the fact that the spectra of the original layers were always impulsive (either combs or nailbeds). Thus, the spectra of the layer superpositions (the spectrum convolutions) were also impulsive, and moreover, a straightforward indexing method permitted us to identify each impulse in the resulting spectrum individually. Each moiré phenomenon in the superposition was associated with an impulse cluster (or with its fundamental impulses) in the spectrum, and the different moirés could be identified using the index of their fundamental impulses in the spectrum.

However, as we have seen in Sec. 10.6, the spectra of non-periodic curvilinear layers may be non-impulsive or only partially impulsive. Therefore, apart from the difficulties in finding the analytic expressions of these spectra in a closed form, there exists here a more fundamental question: Can we identify in a continuous or a semi-continuous spectrum the contributions of each of the superposed layers to the generated moirés? And consequently, can we generalize the theory that we have developed for impulsive spectra in the previous chapters (including the moiré indexing and moiré extraction methods) into the more general cases involving continuous or semi-impulsive spectra?

Here, too, we will advance step by step, starting from the simplest cases. First, in Secs. 10.7.3 and 10.7.4 we discuss examples involving line-spectra; then, in Secs. 10.7.5 and 10.7.6 we discuss examples involving semi-impulsive spectra; and finally, in Secs. 10.7.7 and 10.7.8 we discuss examples involving continuous spectra. Note that since each layer in the superposition is represented by a reflectance (or transmittance) function taking values between 0 and 1, we will usually limit ourselves to functions whose profiles vary between 0 and 1, such as square waves, raised cosines of the form $\frac{1}{2}\cos(2\pi fx) + \frac{1}{2}$, etc. (see Sec. 2.3).

### 10.7.3 The superposition of a parabolic grating and a periodic straight grating

Assume that we superpose a parabolic grating $r_1(x,y)$ with a periodic, straight grating $r_2(x,y)$ (see Fig. 10.9). The spectrum $R_2(u,v)$ of the straight grating consists of an impulse-comb whose impulses are located in the $u,v$ plane at integer multiples of the grating frequency $\mathbf{f}_2$. As we have seen in Example 10.10 in Sec. 10.6.1, the spectrum $R_1(u,v)$ of the parabolic grating consists of a DC impulse plus a series of parallel continuous line-impulses which are located on non-zero integer multiples of the frequency $\mathbf{f}_1$ (the frequency of the parabolic grating along its periodicity axis).

When the two gratings are superposed (i.e., multiplied) in the image domain, their spectra in the $u,v$ plane are convolved. As we already know, the convolution of any object with a comb of impulses places a centered replica of that object on top of each impulse of the comb (after properly scaling its amplitude). Therefore, the spectrum of the superposition $r_1(x,y)r_2(x,y)$, namely: $R_1(u,v)**R_2(u,v)$, consists of an infinite number of replicas of the spectrum $R_1(u,v)$, each of which being centered on top of an impulse of the comb $R_2(u,v)$. The spectrum convolution $R_1(u,v)**R_2(u,v)$ is shown in Fig. 10.9(f).

Now, if we adopt the intuitive gradual transition approach presented in Sec. 10.4.1, we may think of $R_1(u,v)$ as a comb of impulses which "leaked out" perpendicularly to the comb direction to form parallel continuous line-impulses. Before the impulses "leak out", i.e., when the parabolic grating is still an uncurved periodic grating $p_1(x,y)$, its spectrum $P_1(u,v)$ is an impulse comb, just like $R_2(u,v)$, and therefore the convolution $P_1(u,v)**R_2(u,v)$ is an oblique lattice of impulses like in Fig. 2.5(f). We call this oblique lattice the *skeleton* of our line-impulse spectrum $R_1(u,v)**R_2(u,v)$. As the first grating starts bending into its parabolic form, all the impulses of this skeleton lattice (except for the impulses of the comb $R_2(u,v)$) start "leaking out" to both directions, forming the line-

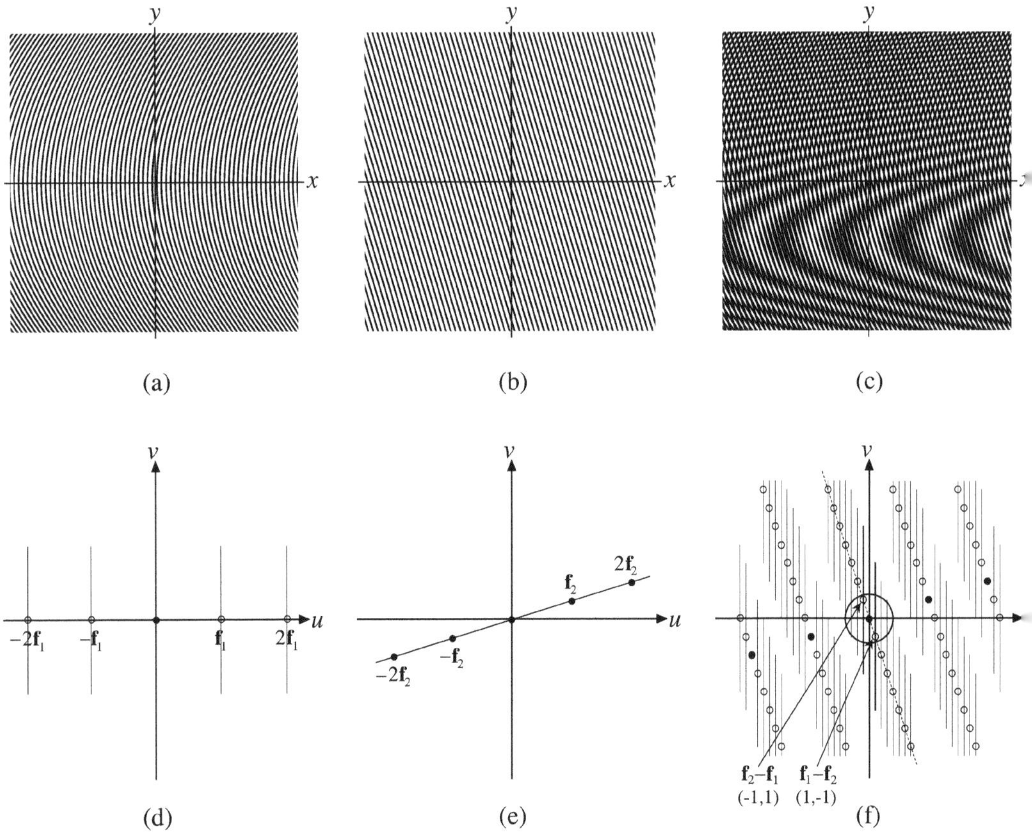


**Figure 10.9:** Parabolic grating (a) and periodic straight grating (b) and their superposition (c) in the image domain; their respective spectra are the infinite line-impulse comb (d), the infinite impulse comb (e) and their convolution (f). Black dots in the spectra represent impulses, while open dots indicate the skeleton locations of the line-impulses. Notice the (1,-1)-moiré which appears in the superposition. Compare with the periodic case shown in Fig. 2.5.

impulse spectrum of Fig. 10.9(f). Note that the skeleton of our line-impulse spectrum is well-defined, since each of the modulated line-impulses has a unique central point, as it is clearly seen in Fig. 10.3.

This point of view permits us to identify each line-impulse in the spectrum of the superposition with the corresponding impulse in the skeleton lattice. Each line-impulse may thus "inherit" the properties of its original skeleton-impulse; the *skeleton location* (or the *center*) of a line-impulse in the spectrum will be defined as the location of its skeleton-impulse, and the *index* of a line-impulse will be defined as the index of its skeleton impulse. This permits us to carry over the important notions of *impulse location*

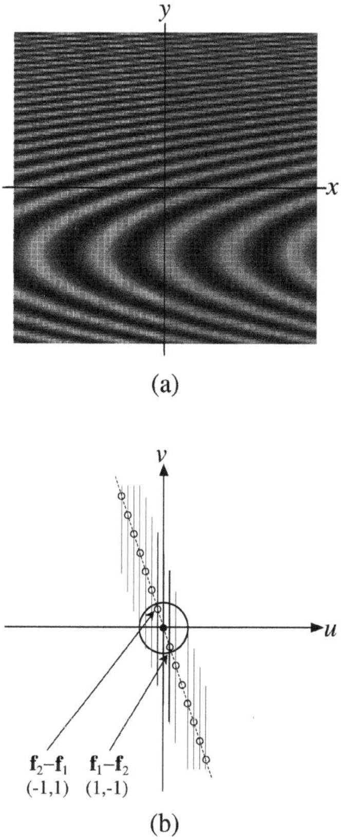


**Figure 10.10:** Extraction of the (1,-1)-moiré of Fig. 10.9: (b) shows the isolated line-impulse comb of the (1,-1)-moiré after its extraction from the full spectrum of Fig. 10.9(f). (a) shows the image domain function which corresponds to the spectrum (b). This is the intensity surface of the parabolic (1,-1)-moiré shown in Fig. 10.9(c); note that it only contains the isolated form of the extracted moiré (i.e., its isolated contribution to the superposition), but not the microstructure details of the original gratings and of the superposition. Compare with the periodic case shown in Fig. 4.2.

and *impulse index* to the case of continuous line-impulses, too. Hence, a $(k_1,k_2)$-moiré in the superposition is the moiré which is caused by the $(k_1,k_2)$-line-impulse in the spectrum convolution (or, in fact, by the $(k_1,k_2)$-cluster of line-impulses). This moiré becomes visible if the center of the $(k_1,k_2)$-line-impulse is located inside the visibility circle, close to the spectrum origin.

**Example 10.13:** As an illustration, let us consider the superposition shown in Fig. 10.9, where the parabolic grating $r_1(x,y)$ and the periodic straight grating $r_2(x,y)$ are given by:

$$r_1(x,y) = p_1(x-ay^2) = \sum_{m=-\infty}^{\infty} a^{(1)}{}_m \cos(2\pi m f_1 (x-ay^2))$$

$$r_2(x,y) = p_2(x\cos\theta + y\sin\theta) = \sum_{n=-\infty}^{\infty} a^{(2)}{}_n \cos(2\pi n f_2 (x\cos\theta + y\sin\theta))$$

In this case, the visible moiré effect in the superposition is represented in the spectrum convolution by the (1,-1)-line-impulse (whose center is located inside the visibility circle) and the whole (1,-1)-cluster of line-impulses which it spans. Each line-impulse of this cluster is centered on an impulse of the oblique (1,-1)-comb which is the skeleton of the (1,-1)-line-impulse cluster. However, as we can see in Figs. 10.9(f) and 10.10(b), this cluster of line-impulses differs from the spectrum of a parabolic grating of the type $p(x-ay^2)$ (see Fig. 10.9(d)) in that its parallel line-impulses are not orthogonal to the cluster direction (note that the cluster direction, i.e., the direction of the (1,-1)-skeleton comb, is given by the direction $\varphi_M$ of the vector $\mathbf{f}_1-\mathbf{f}_2$).

Let us try to find out what is the image-domain function whose spectrum is given by this non-orthogonal line-impulse cluster in the $u,v$ plane; the requested image-domain function will be, of course, the mathematical expression of the intensity surface of the isolated (1,-1)-moiré (see Fig. 10.10).[18] The non-orthogonal line-impulse cluster of Fig. 10.10(b) can be obtained from an orthogonal line-impulse cluster whose skeleton is located on the $u$ axis by a vertical shear transformation, namely: by replacing the vertical coordinate $v$ with $v+bu$ (where the coefficient $b$, in our case negative, is given by: $b = \tan\varphi_M$, $\varphi_M$ being the direction of the skeleton comb). Let us denote the parabolic grating which corresponds in the image domain to this orthogonal, unsheared line-impulse cluster by $p_M(x-a_My^2)$. Now, according to the shear theorem [Bracewell95 p. 158], a vertical shear in the spectral domain corresponds to a horizontal shear in the image domain, namely: if $f(x,y) \leftrightarrow F(u,v)$ then $f(x+by,y) \leftrightarrow F(u,v-bu)$. This means that our vertically sheared (1,-1)-line-impulse cluster corresponds in the image domain to the horizontally sheared parabolic moiré-grating $p_M((x+by)-a_My^2)$; but this is simply a vertically and horizontally shifted copy of the centered parabolic grating $p_M(x-a_My^2)$, namely: $p_M((x-x_0)-a_M(y-y_0)^2)$ with $x_0 = -b^2/4a_M$ and $y_0 = b/2a_M$ (see Fig. 10.11). And indeed, as it can be seen in Figs. 10.9(c) and 10.10(a), the (1,-1)-moiré generated in the present example forms a shifted parabolic grating pattern.

Note that the bending rate $a_M$ of this parabolic moiré is higher than the bending rate $a$ of the original parabolic grating $r_1(x,y)$. The value of $a_M$ can be found by the following spectral consideration: As we already know, the $n$-th line-impulse of the (1,-1)-cluster in the spectrum convolution is simply a replica of the $n$-th line-impulse in the spectrum of $r_1(x,y)$, which is scaled down by the amplitude of the $-n$-th impulse in the spectrum of the straight grating $r_2(x,y)$. Therefore, if we denote the skeleton location of the (1,-1)-line-impulse by $(u_0,v_0)$, then the $n$-th line-impulse of the (1,-1)-cluster, which is centered at $(u,v) = (nu_0,nv_0)$, is given by (see Example 10.10 in Sec. 10.6.1):[19]

---

[18] As we have seen in Sec. 10.7.1, the image-domain expression of this (1,-1)-moiré can be found directly by Eq. (10.19). But our declared aim here is to focus our attention on the spectral-domain analysis and its connection to the image domain.

[19] Note that this is only true for $n \neq 0$, since the 0-th element of the cluster is the DC impulse.

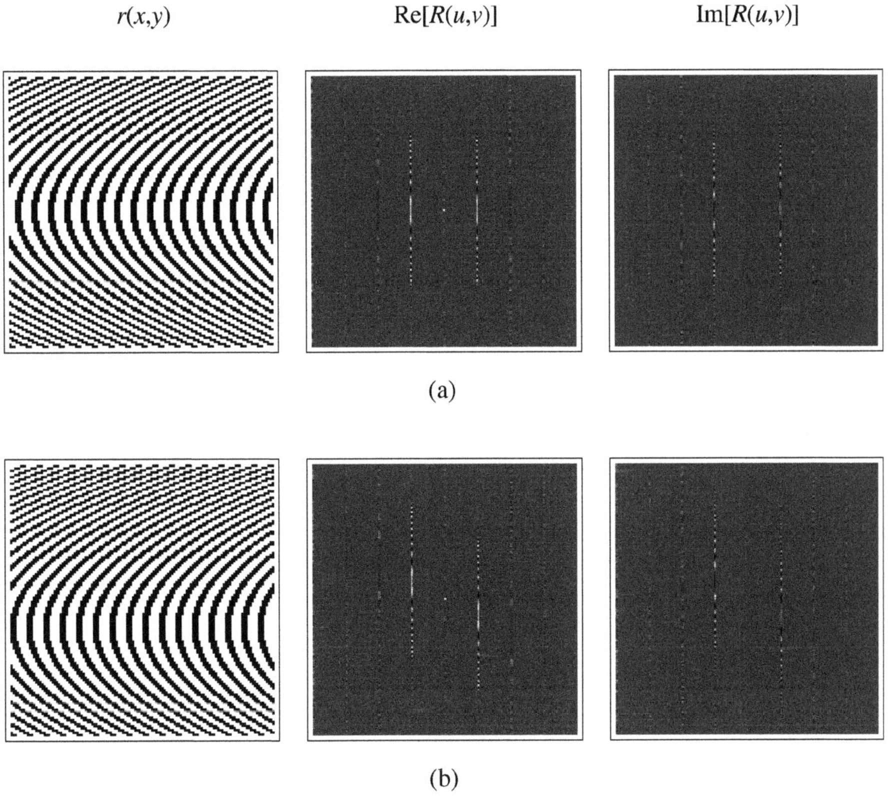


**Figure 10.11:** The effect of a vertical shear transformation on the spectrum is a horizontal shear transformation in the image domain; in the case of a parabolic grating this is equivalent to a horizontal and vertical shift. (a) $g_1(x,y) = x - 0.15y^2$; (b) $g_1(x,y) = (x+\frac{5}{12}) - 0.15(y+\frac{5}{3})^2$.

$$a_{n,-n}(u,v) = \begin{cases} \frac{1}{2}[R_c(v-nv_0) + iR_s(v-nv_0)]\ \delta(u-nu_0)\ a^{(1)}_n\ a^{(2)}_{-n} & n > 0 \\ \frac{1}{2}[R_c(v-nv_0) - iR_s(v-nv_0)]\ \delta(u-nu_0)\ a^{(1)}_n\ a^{(2)}_{-n} & n < 0 \end{cases} \tag{10.23}$$

with:

$$R_c(v) = \frac{1}{2\sqrt{nf_1a}}\left(\cos\left(\frac{\pi}{2nf_1a}v^2\right) + \sin\left(\frac{\pi}{2nf_1a}v^2\right)\right)$$

$$R_s(v) = \frac{1}{2\sqrt{nf_1a}}\left(\cos\left(\frac{\pi}{2nf_1a}v^2\right) - \sin\left(\frac{\pi}{2nf_1a}v^2\right)\right)$$

where $a$ is the bending rate of the original parabolic grating $r_1(x,y)$ and $f_1$ is its fundamental frequency. As we can see, the product $f_1a$ in all the line-impulses of the (1,-1)-moiré cluster is fully determined by $r_1(x,y)$. But since this line-impulse cluster is the spectrum of the parabolic (1,-1)-moiré, it is clear that the product $f_1a$ in its fundamental line-impulse, the (1,-1)-line-impulse, must be equal to the product $f_M a_M$ where $f_M$ and $a_M$ are the frequency and the bending rate of the parabolas of the (1,-1)-moiré. We have, therefore: $f_M a_M = f_1 a$, and hence: $a_M = a f_1 / f_M$. The frequency $f_M$ of the (1,-1)-moiré, as defined by the skeleton location of the (1,-1)-line-impulse, is smaller than $f_1$ (remember that $f_M$ is located inside the visibility circle), and therefore it follows, indeed, that $|a_M| > |a|$.

Note that as the rotation angle $\theta$ of $r_2(x,y)$ increases, a new (1,-2)-moiré gradually becomes visible: at first simultaneously with the (1,-1)-moiré, but then the (1,-1)-moiré fades out, and the (1,-2)-moiré takes its place as the predominant moiré (see Problem 10-1).

Finally, it is interesting to note that when the rotation angle $\theta$ of $r_2(x,y)$ approaches zero, the frequency $f_M$ of the (1,-1)-moiré decreases and consequently its bending rate $a_M$ increases. When $\theta = 0$ (assuming that $f_1 = f_2$) we reach a *singular state* in which the center of the (1,-1)-line-impulse collapses onto the spectrum origin together with its whole cluster. Consequently, in the image domain the bending rate $a_M$ of the (1,-1)-moiré parabolas reaches infinity, and the parabolas turn in the singular state into straight lines (or more precisely, into a horizontal linear zone grating). The situation in such singular cases will be examined below in detail. ■

It has been mentioned above that a $(k_1,k_2)$-moiré in the superposition becomes visible when the center of the $(k_1,k_2)$-line-impulse in the spectrum is located inside the visibility circle. However, it turns out that the moiré may still be visible even if the center of the line-impulse is slightly outside the visibility circle, but a segment of the line-impulse, which is relatively close to its skeleton location, falls inside the visibility circle. This explains, indeed, why our curvilinear $(k_1,k_2)$-moiré can still be visible in the superposition even when the skeleton location of the moiré is already outside the visibility circle. It also explains why in the present case more different moirés can be simultaneously observed than in a superposition of straight, periodic gratings. Similar results will be obtained in all cases of curvilinear grating superpositions, where the spectrum consists of 1D or 2D structures rather than simple impulses. This is, in fact, the spectral-domain interpretation of observation (2) in the beginning of Sec. 10.7.

Let us now examine the singular states of the moirés between a parabolic grating and a periodic straight grating. A significant difference between singular cases here and singular cases in the superposition of periodic layers is that in the present case the singularity in the image domain is *local* and no longer global: As we can see in Fig. 10.12(c), when a $(k_1,k_2)$-moiré becomes singular in the superposition of a parabolic grating and a straight grating, its singularity extends only along a straight line, and not throughout the whole $x,y$ plane, as was the case in the superposition of periodic layers. We will call the locus in the image domain in which a $(k_1,k_2)$-moiré is singular the *singular locus* of the $(k_1,k_2)$-moiré. The singular $(k_1,k_2)$-moiré locally disappears to the eye along its singular locus, its local period there being infinitely large. But at any other point of the $x,y$ plane outside the singular locus the singular $(k_1,k_2)$-moiré still has a finite local period, which gradually decreases as one moves away from the singular locus, until at a certain distance (where the

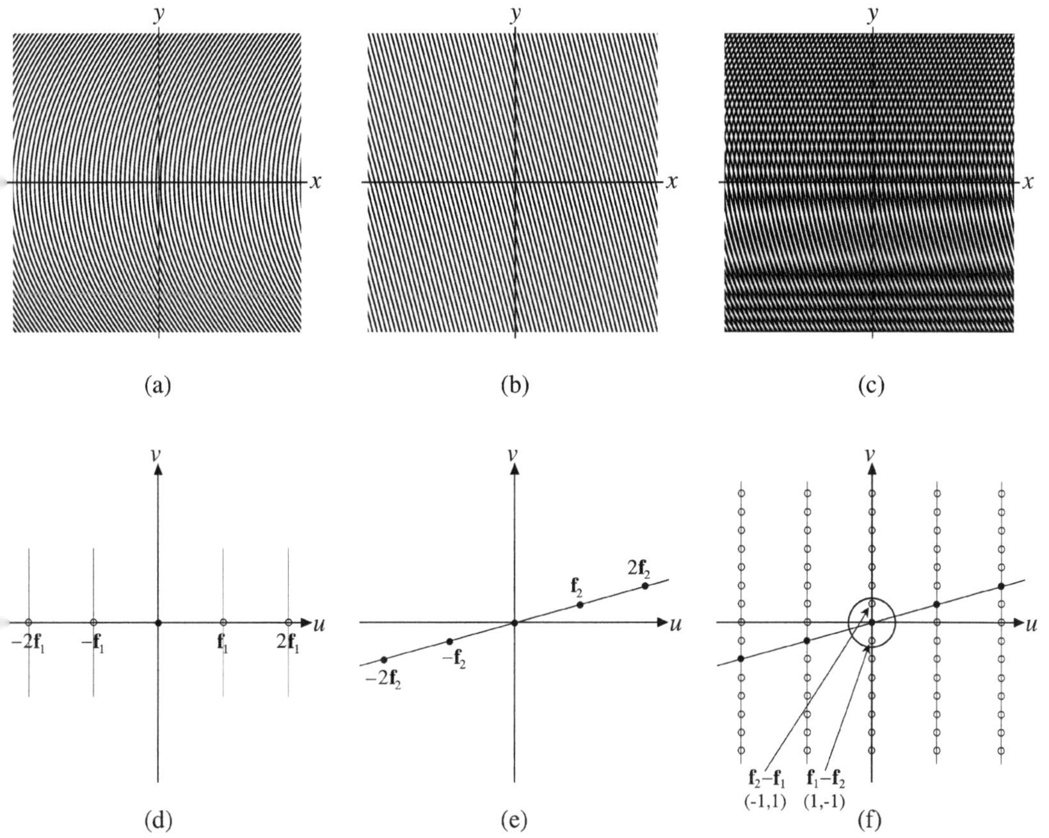


**Figure 10.12:** The singular state of the (1,-1)-moiré in the superposition of (a) parabolic grating $r_1(x,y)$ and (b) periodic straight grating $r_2(x,y)$ (with $f_2 > f_1$). Notice the singular locus of the (1,-1)-moiré in the image domain superposition (c). (d) and (e) are the spectra of (a) and (b), and (f) is the spectrum of (c), i.e., the convolution of (d) and (e). Note that all the line-impulses of the (1,-1)-cluster in (f) fall on the $v$ axis (compare with Fig. 10.9(f)).

local frequency of the moiré goes beyond the visibility circle) the eye can no longer resolve it and the moiré completely disappears.

Turning now to the spectral domain, in order that a $(k_1,k_2)$-moiré in the superposition be singular it is not necessarily required that the *center* of the $(k_1,k_2)$-line-impulse collapse on the spectrum origin, and it is enough that *any* point of this line-impulse fall on the origin. This is demonstrated in Fig. 10.12, which shows the singular (1,-1)-moiré between a parabolic grating $r_1(x,y)$ and a periodic straight grating $r_2(x,y)$ with frequencies $f_1 \neq f_2$. As we can see, when this singular state is reached (for example, by properly rotating the straight grating $r_2(x,y)$) all the line-impulses of the (1,-1)-cluster fall on the $v$ axis. More generally, the $(k_1,k_2)$-moiré in the superposition becomes singular when the line-impulses of the $(k_1,k_2)$-cluster in the spectrum convolution collapse on a single line through the origin (in the present case: the vertical $v$ axis).

Let us analyze the situation in the spectrum at this precise moment, as we did in Chapter 6 in the periodic case. When all the line-impulses of the $(k_1,k_2)$-cluster fall on a single line through the origin their amplitudes are summed up, and we obtain a new line-impulse whose amplitude is the sum of all the individual line-impulses. We call this line-impulse a *compound line-impulse*, by analogy with a compound impulse in the periodic case (see Sec. 6.4). As it is shown in Appendix C.8, this compound line-impulse through the spectrum origin corresponds in the image domain to a singular moiré in the form of a linear zone grating (a straight line grating whose local period decreases to both sides of its central line) which is shifted away from the image center. It is shown in Appendix C.8 that this shift is given by $y_0 = \frac{v_0}{2k_1 f_1 a}$, where $v_0$ is the *internal discrepancy* of the compound line-impulse, i.e., the distance of the center of the $(k_1,k_2)$-line-impulse (the fundamental impulse of the collapsed line-impulse cluster) from the spectrum origin, and $f_1$ and $a$ are the frequency and the bending rate of the parabolic grating $r_1(x,y)$.

### 10.7.4 The superposition of two parabolic gratings

Let us make now one step further, and assume that both of the superposed layers $r_1(x,y)$ and $r_2(x,y)$ consist of parabolic gratings. In this case, both of the spectra $R_1(u,v)$ and $R_2(u,v)$ consist of parallel continuous line-impulses, and hence the spectrum of the superposition is no longer a convolution of $R_1(u,v)$ with an impulse-comb, but rather a convolution of two series of parallel line-impulses (the 0-th element in each series being the DC impulse).

In order to understand what happens in such a spectrum convolution, let us consider first a simple, hypothetical case in which each of the two spectra to be convolved consists of a single continuous line-impulse. It may be instructive to see one of these line-impulses (say, the second one) as a continuum of individual impulses which make up together the continuous line-impulse. Using this convention, we may think of the convolution as an operation which places a centered replica of the first line-impulse (after properly scaling its amplitude) on top of each point of the second line-impulse. Therefore, if the two original line-spectra are not parallel, their convolution is a continuous surface extending over the whole $u,v$ plane, which is generated by all the parallel replicas of the first line-impulse. We will henceforth call the continuous surface obtained in the convolution of two non-parallel line-impulses a *hump*. If the two original line spectra are parallel, their convolution is a single line-impulse which is obtained by placing a centered replica of the first line-impulse on every point of the second line-impulse (where the amplitudes at overlapping points are summed up).

Having intuitively explained the convolution of two line-impulses, we can give now the precise, explicit expression for such a convolution: Suppose that we are given a horizontal

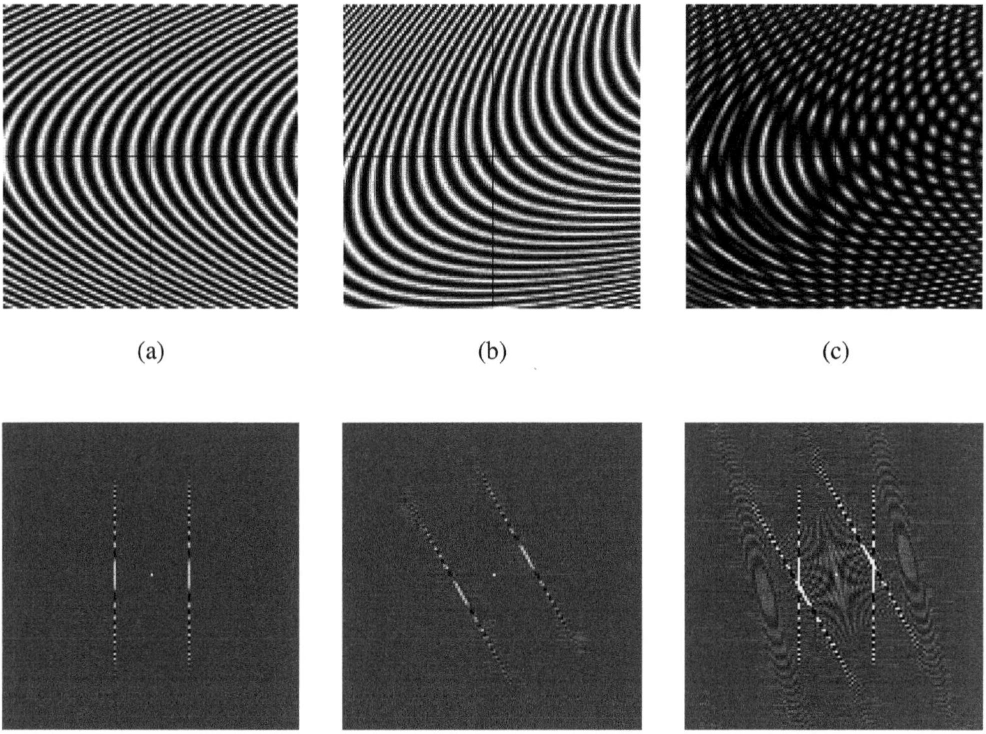

(a) (b) (c)

(d) (e) (f)

**Figure 10.13:** Two rotated parabolic, raised-cosinusoidal gratings (a) and (b) and their superposition (c) in the image domain; the respective spectra are shown in (d), (e) and their convolution (f). Notice the hyperbolic (1,-1)-moiré which appears in the superposition (c) and its hyperbolic humps in the spectrum (f). Compare with the periodic case shown in Fig. 2.2.

1D line-impulse $F(u)\delta(v)$, whose amplitude is defined by $F(u)$, and a vertical 1D line-impulse $G(v)\delta(u)$, whose amplitude is defined by $G(v)$. Both of these line-impulses are centered on the origin. It can be shown (see Appendix C.7) that their 2D convolution is given by the 2D function $F(u)G(v)$:

$$F(u)\delta(v) ** G(v)\delta(u) = F(u)G(v) \tag{10.24}$$

If the 1D line-impulse amplitudes $F(u)$ and $G(v)$ are continuous, it follows therefore that the line-impulse convolution is a continuous 2D function (a 2D hump centered on the origin).[20] If the original line-impulses are centered on the points $(u_1,v_1)$ and $(u_2,v_2)$, respectively, their convolution gives the same 2D hump, which is only shifted now to the point $(u_1+u_2, v_1+v_2)$:

$$F(u-u_1)\delta(v-v_1) ** G(v-v_2)\delta(u-u_2) = F(u-u_1-u_2)G(v-v_1-v_2) \qquad (10.25)$$

A similar result can be obtained also when the two line-impulses are not perpendicular, but rotated by arbitrary angles $\theta_1$ and $\theta_2$.

**Example 10.14:** The spectrum of the reflectance function defined by a raised parabolic cosinusoidal grating[21] $\frac{1}{2}\cos(2\pi f(y-ax^2)) + \frac{1}{2}$ consists of a pair of parallel line-impulses, plus a DC impulse on the spectrum origin. When two such reflection functions are superposed with different orientations, the resulting spectrum convolution consists of 4 line-impulse convolutions (i.e., continuous humps extending over the whole $u,v$ plane) which are centered on the (1,1), (1,-1), (-1,1) and (-1,-1) skeleton impulses, plus 4 line-impulses which are centered on the (1,0), (-1,0), (0,1) and (0,-1) skeleton impulses, and of course the DC impulse, (0,0) (see Fig. 10.13, and compare it with its uncurved counterpart, Fig. 2.2). This is a hybrid spectrum which consists of 4 continuous humps, 4 line-impulses, and a single impulse (the DC impulse at the origin). Note that the humps centered on the (1,-1) and (-1,1) skeleton impulses have intensity profiles of hyperbolic zone gratings, whereas the humps centered on the (1,1) and (-1,-1) skeleton impulses have intensity profiles of elliptic zone gratings. ■

Let us return now to the superposition of the parabolic gratings $r_1(x,y)$ and $r_2(x,y)$. If we adopt the intuitive gradual transition approach of Sec. 10.4.1, we may think of each of the spectra $R_1(u,v)$ and $R_2(u,v)$ as a comb of impulses which "leaked out" perpendicularly to the comb direction to form parallel continuous line-impulses. Before the impulses "leak out", i.e., when the two parabolic gratings are still uncurved periodic grating $p_1(x,y)$ and $p_2(x,y)$, each of the spectra $P_1(u,v)$ and $P_2(u,v)$ is an impulse comb, and therefore the convolution $P_1(u,v)**P_2(u,v)$ is an oblique lattice of impulses like in Fig. 2.5(f). This oblique lattice is the *skeleton* of our spectrum $R_1(u,v)**R_2(u,v)$. As the two gratings start bending into their parabolic forms, the impulses of $R_1(u,v)$ and $R_2(u,v)$ (except for their DC impulses) start "leaking out" to form line-impulses, and the spectrum convolution becomes a hybrid spectrum:

(a) The $(i,j)$ skeleton-impulses with both $i \neq 0$ and $j \neq 0$ turn into 2D line-impulse convolutions, i.e., continuous (elliptic or hyperbolic) humps over the whole $u,v$ plane;

(b) The $(i,j)$ skeleton-impulses with either $i = 0$ or $j = 0$ (i.e., the impulses of the original line-spectra $R_1(u,v)$ and $R_2(u,v)$) remain 1D line-impulses as in $R_1(u,v)$ and $R_2(u,v)$;

---

[20] A simple example of such a line-impulse convolution is illustrated in [Bracewell86 p. 243]. A more interesting case for contemplation consists of the line-impulses ("sinc blades") $G_1(u,v) = \mathrm{sinc}(u)\delta(v)$ and $G_2(u,v) = \mathrm{sinc}(v)\delta(u)$. Their 2D convolution is $G(u,v) = \mathrm{sinc}(u)\delta(v)**\mathrm{sinc}(v)\delta(u)$, which gives by virtue of Eq. (10.24) the 2D continuous surface $\mathrm{sinc}(u)\mathrm{sinc}(v)$.

[21] Remember that since reflectance functions always take values between 0 and 1, the cosinusoidal reflectance function is a raised cosine of the form: $\frac{1}{2}\cos(2\pi fx) + \frac{1}{2}$ (see Sec. 2.3).

(c) The DC impulse (0,0) remains a simple impulse.

Note that the skeleton lattice of our hybrid spectrum is well-defined, since each of the "leaked out" impulses (either 2D or 1D) has a unique central point.

This point of view permits us, again, to identify each line-impulse or hump in the spectrum of the superposition with the corresponding impulse in the skeleton cluster. Each line-impulse or hump may thus "inherit" the properties of its original skeleton-impulse: its *skeleton location* (=its *center*) and its *index*. This permits us to carry over the important notions of *impulse location* and *impulse index* to the continuous case, too. Hence, a $(k_1,k_2)$-moiré in the superposition is the moiré which is caused by the $(k_1,k_2)$-hump in the spectrum convolution (or, in fact, by the $(k_1,k_2)$-cluster of humps). The $(k_1,k_2)$-moiré becomes visible if the skeleton location of the $(k_1,k_2)$-hump is found within the visibility circle around the spectrum center. However, it may still be visible even if the skeleton location of the hump is slightly outside the visibility circle, but it still "leaks" into the visibility circle. This explains, indeed, why in the superposition of curvilinear gratings a curvilinear $(k_1,k_2)$-moiré can still be visible in the superposition even when the skeleton location of the $(k_1,k_2)$-hump is already outside the visibility circle (so that in the corresponding straight-grating superposition the $(k_1,k_2)$-moiré is no longer visible). But since this time the $(k_1,k_2)$-hump, the fundamental hump of the $(k_1,k_2)$-moiré, is two-dimensional and extends over the whole $u,v$ plane, we obtain the interesting result that theoretically in the present case several different $(k_1,k_2)$-moirés may be visible in the superposition simultaneously (see, for example, Fig. 10.14(a)). The only practical limitations are, of course, that in order to be clearly visible the amplitude of the $(k_1,k_2)$-hump (see below) must be sufficiently strong, and on the other hand, that the location of the moiré in the $x,y$ plane should not be too far away from the origin, in order not to fall beyond the borders of the layer superposition.

We have seen above that the *locations* of the humps and of the line-impulses in the spectrum convolution are given by the centers (=the skeleton locations) of these humps and line-impulses. These points form together the skeleton-lattice which is spanned by the skeleton-combs of the two original parabolic gratings; this lattice corresponds to the superposition of the *uncurved*, periodic gratings. This is, indeed, the generalization of Eq. (2.26) to the curvilinear case: Eq. (2.26) gives the *skeleton location* of all the humps and line-impulses in the spectrum convolution.

The question remains, therefore, how can we express the *amplitudes* of the entities in the spectrum convolution? Or, in other words, what is the extension of Eq. (2.27) to the curvilinear case? For the case of *line-impulses* (which are obtained by a convolution of a line-impulse with a point-impulse) the answer has been given in Sec. 10.7.3: the amplitude of the resulting line-impulse is the product of the amplitude of the original line-impulse and the amplitude of the point-impulse. For example, if the $m$-th vertical line-impulse in the first spectrum is given by: $a_m(v) = R_m(v)\delta(u-mf_1)$ and the $n$-th impulse in the comb of the second spectrum is $a_n\delta(u-nu_2, v-nv_2)$, where $\mathbf{f}_2 = (u_2,v_2)$ (see Fig. 10.9), then the $(m,n)$-th line-impulse resulting from the convolution is:

$$a_{m,n}(v) = R_m(v-nv_2)\, a_n\, \delta(u-mf_1-nu_2)$$

Its amplitude is $R_m(v-nv_2)a_n$, and its center is located at the point $(mf_1+nu_2, nv_2)$.

For the case of *humps* (which are obtained in the spectrum by a convolution of two orthogonal line-impulses) the amplitude is given by Eq. (10.25). For example, if the $m$-th horizontal line-impulse in the first spectrum is given by: $a^{(1)}{}_m(u) = R^{(1)}{}_m(u)\delta(v-mf_1)$ and the $n$-th vertical line-impulse in the second spectrum is given by: $a^{(2)}{}_n(v) = R^{(2)}{}_n(v)\delta(u-nf_2)$, then the $(m,n)$-th hump resulting from their convolution is:

$$a_{m,n}(u,v) = R^{(1)}{}_m(u-mf_1)\, R^{(2)}{}_n(v-nf_2)$$

whose center is located at the point $(mf_1,nf_2)$.

This is, indeed, the generalization of Eq. (2.27) that we were seeking. We can state now the generalization of Proposition 2.3 from Sec. 2.7 to cases of classes (1)–(2) (see Sec. 10.4), where the original spectra consist of simple impulses or line-impulses:

**Proposition 10.1:** The skeleton location (i.e., the geometric location of the center) of the $(k_1,...,k_m)$-line-impulse or the $(k_1,...,k_m)$-hump in the spectrum-convolution of the superposition of $m$ curvilinear gratings of classes (1)–(2) is given by the vectorial sum (2.26):

$$\mathbf{f}_{k_1,...,k_m} = k_1\mathbf{f}_1 + ... + k_m\mathbf{f}_m$$

and the amplitude of this line-impulse or hump is given by the product (2.27):

$$a_{k_1,...,k_m}(u,v) = a^{(1)}{}_{k_1}(u,v)\ ...\ a^{(m)}{}_{k_m}(u,v)$$

where $\mathbf{f}_i$ denotes the frequency vector of the fundamental skeleton location in the spectrum of the $i$-th curvilinear grating, and $k_i\mathbf{f}_i$ and $a^{(i)}{}_{k_i}(u,v)$ are respectively the frequency-vector and the amplitude of the $k_i$-th harmonic in the spectrum of the $i$-th curvilinear grating. If several line-impulses in the convolution happen to fall on the same line, their individual amplitudes are summed, and they form together a *compound line-impulse* (see Sec. 10.7.3). In the case of humps, however, such a summation of amplitudes always occurs, since each of the humps extends throughout the whole $u,v$ plane. ■

Returning now to the question of singular moirés, we find that in the present case, too, the singularity of a $(k_1,k_2)$-moiré in the image domain is *local*, and no longer global as in the periodic case. However, as we can see in Fig. 10.14(a), in the present case each $(k_1,k_2)$-moiré becomes singular only at a single point — and not along a straight line, as in the case of the previous subsection. In other words, the singular locus of a $(k_1,k_2)$-moiré in the present case is a 0D point in the $x,y$ plane, while in the previous case it was a 1D line in the plane, and in the superposition of periodic layers it consisted of the entire 2D $x,y$ plane.[22]

[22] It is interesting to note that in each of these cases the dimension of the singular locus of the moiré in the image domain (a 0D point, 1D line or the whole 2D plane) complements the dimension of the fundamental harmonic of the moiré in the spectral domain (2D hump, 1D line-impulse or 0D point-impulse) to $n = 2$, the space dimension.

Here, again, the singular $(k_1,k_2)$-moiré locally disappears at its singular locus point (its local period there being infinitely large); and at any other point of the $x,y$ plane surrounding its singular locus point the $(k_1,k_2)$-moiré has a finite local period, which gradually decreases as one moves away from the singular locus, until at a certain distance (where the local frequency of the moiré goes beyond the visibility circle) the eye can no longer resolve it and the moiré completely disappears. Such 2D locally visible moiré patterns will be nicknamed *eye-shaped moirés* or *moiré eyelets* due to their bull's-eye shape; each moiré eyelet is centered on its singular locus point, and it gradually disappears in all directions.

As we have seen above, in order that a $(k_1,k_2)$-moiré in the superposition be singular it is not necessarily required that the *center* of the $(k_1,k_2)$-hump fall on the spectrum origin, and it is enough that *any* point of this hump fall on the origin. But since this time the $(k_1,k_2)$-hump, the fundamental hump of the $(k_1,k_2)$-moiré, is two-dimensional and extends throughout the whole $u,v$ plane, we obtain the interesting result that in the present case every $(k_1,k_2)$-moiré is *always* singular (i.e., has a singular locus point in the $x,y$ plane), no matter how the original layers are superposed. This means also that *all* the different $(k_1,k_2)$-moirés in the superposition have a singular locus point somewhere in the $x,y$ plane simultaneously. When the superposed layers are manipulated (rotated, stretched, etc.), the centers of the different $(k_1,k_2)$-moiré eyelets (i.e., their singular locus points) are shifted to and fro within the $x,y$ plane, generating an interesting "ballet" of locally visible moiré eyelets.

**Example 10.15:** The superposition of two perpendicular parabolic gratings:

Let $r_1(x,y)$ be a horizontally oriented parabolic grating with frequency $f_1$ and bending rate $a_1$, and let $r_2(x,y)$ be a vertically oriented grating with frequency $f_2$ and bending rate $a_2$:

$$r_1(x,y) = p_1(x-a_1y^2) = \sum_{m=-\infty}^{\infty} a^{(1)}_m \cos(2\pi m f_1 (x-a_1y^2))$$

$$r_2(x,y) = p_2(y-a_2x^2) = \sum_{n=-\infty}^{\infty} a^{(2)}_n \cos(2\pi n f_2 (y-a_2x^2))$$

Fig. 10.14 shows the superposition of these two gratings and its spectrum (the spectrum convolution) for the simple case in which $f_1 = f_2$ and $a_1 = a_2$. As we can see, the spectrum origin is surrounded by an infinity of humps and of line-impulses, in accordance with points (a) and (b) above. Since all the line-impulses in this spectrum convolution are parallel to the main axes, none of them penetrates into the visibility circle. On the other hand, *all* of the $(k_1,k_2)$-humps in the spectrum extend throughout the whole $u,v$ plane, and therefore all of them penetrate into the visibility circle and pass through the spectrum origin. This means that in principle, all the humps which surround the spectrum origin (or rather the hump-clusters they span) generate visible moiré effects in the image domain, each of them being a moiré eyelet which is visible around its singular locus point. Clearly, the strongest moirés are generated by the (1,1)-cluster and by the (1,-1)-cluster (see Fig. 10.14(b)), and the next strongest are the (1,2)- and (2,1)-clusters to both sides of the (1,1)-cluster, and the (-1,2)- and (-2,1)-clusters to both sides of the (1,-1)-cluster.

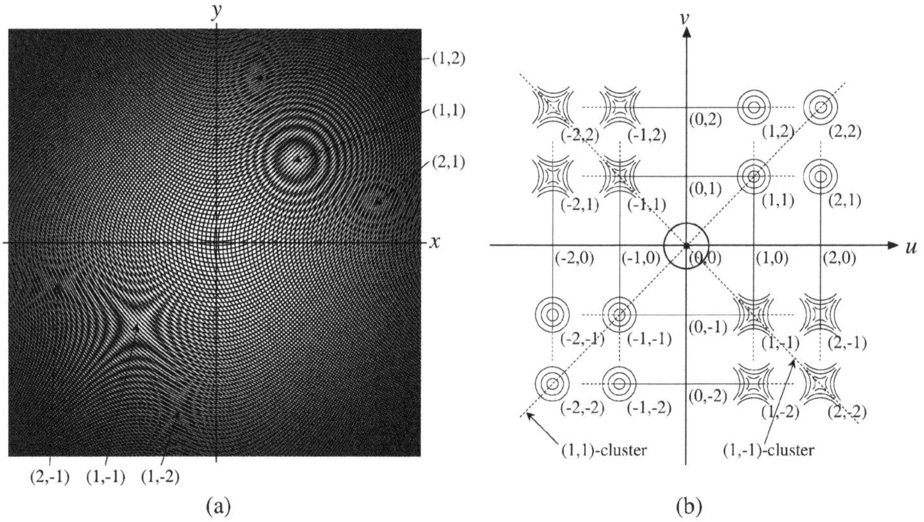


**Figure 10.14:** (a) The superposition of two perpendicular parabolic gratings (with identical frequencies $f_1 = f_2$ and bending rates $a_1 = a_2$). The visible moirés include the (1,1)-, (2,1)- and (1,2)-moiré eyelets, having the form of circular or elliptic zone gratings (top right), and the (1,-1)-, (2,-1)- and (1,-2)-moiré eyelets in the form of hyperbolic zone gratings (bottom left). (b) Schematic representation of the spectrum convolution which corresponds to this superposition; the straight lines represent line-impulses and the circular or hyperbolic concentric structures represent 2D humps. The two diagonal dotted lines indicate the humps which belong to the (1,1)-cluster or to the (1,-1)-cluster.

Let us find explicitly the amplitudes of the $(n,n)$-humps, the humps which make up the (1,1)-cluster in the spectrum convolution and correspond to the (1,1)-moiré eyelet in the superposition. From Eq. (10.25), Proposition 10.1 and Example 10.10 (Sec. 10.6.1) we obtain that the $(n,n)$-hump (for any $n \neq 0$),[23] which is centered at $(u,v) = (nu_0,nv_0)$, is given by:

$$a_{n,n}(u,v) = \qquad (10.26)$$

$$\begin{cases} \tfrac{1}{2}[R_{c,1}(v-nv_0) + iR_{s,1}(v-nv_0)]a^{(1)}{}_n \; \tfrac{1}{2}[R_{c,2}(u-nu_0) + iR_{s,2}(u-nu_0)]a^{(2)}{}_n & n > 0 \\ \tfrac{1}{2}[R_{c,1}(v-nv_0) - iR_{s,1}(v-nv_0)]a^{(1)}{}_n \; \tfrac{1}{2}[R_{c,2}(u-nu_0) - iR_{s,2}(u-nu_0)]a^{(2)}{}_n & n < 0 \end{cases}$$

where:

$$R_{c,1}(v) = \frac{1}{2\sqrt{nk_1 f_1 a_1}}\left(\cos\left(\frac{\pi}{2nk_1 f_1 a_1}v^2\right) + \sin\left(\frac{\pi}{2nk_1 f_1 a_1}v^2\right)\right)$$

$$R_{s,1}(v) = \frac{1}{2\sqrt{nk_1 f_1 a_1}}\left(\cos\left(\frac{\pi}{2nk_1 f_1 a_1}v^2\right) - \sin\left(\frac{\pi}{2nk_1 f_1 a_1}v^2\right)\right)$$

[23] Note that the 0-th element of the cluster is not a line-impulse but the DC impulse, so that Eq. (10.26) does not hold for $n = 0$.

$$R_{c,2}(v) = \frac{1}{2\sqrt{nk_2f_2a_2}}(\cos(\frac{\pi}{2nk_2f_2a_2}v^2) + \sin(\frac{\pi}{2nk_2f_2a_2}v^2))$$

$$R_{s,2}(v) = \frac{1}{2\sqrt{nk_2f_2a_2}}(\cos(\frac{\pi}{2nk_2f_2a_2}v^2) - \sin(\frac{\pi}{2nk_2f_2a_2}v^2))$$

By finding the expression of the (1,1)-hump cluster in the spectrum it can be shown that the (1,1)-moiré eyelet, in the image domain, has the form of a circular zone grating which is shifted by $(x_0,y_0) = (\frac{f_1}{2f_2a_2}, \frac{f_2}{2f_1a_1})$ from the origin (see Fig. 10.14(a)). Similarly, the (2,1)-hump cluster in the spectrum convolution corresponds in the image domain to the (2,1)-moiré eyelet, which is a weaker, elliptic zone grating located next to the (1,1)-moiré eyelet at a distance of $(\frac{2f_1}{2f_2a_2}, \frac{f_2}{4f_1a_1}) = (2x_0, y_0/2)$ from the origin. The (1,2)-moiré eyelet is symmetrically located in the other side of the (1,1)-moiré eyelet, at the point $(x_0/2, 2y_0)$.

In a similar way, the (1,-1)-hump cluster in the spectrum convolution corresponds to the (1,-1)-moiré eyelet in the superposition, which has the form of a hyperbolic zone grating shifted by $(-\frac{f_1}{2f_2a_2}, -\frac{f_2}{2f_1a_1}) = (-x_0, -y_0)$ from the origin. The (2,-1)- and (1,-2)-moiré eyelets can be seen in the superposition in the form of two weaker, elongated hyperbolic zone gratings placed to both sides of the (1,-1)-moiré eyelet, at the points $(-2x_0, -y_0/2)$ and $(-x_0/2, -2y_0)$ respectively. ■

### 10.7.5 The superposition of a circular grating and a periodic straight grating

Assume that we superpose a circular grating $r_2(x,y)$ with a periodic, straight grating $r_1(x,y)$; this is shown in Figs. 10.17, 10.18 and 10.19 for the cases of $f_2 < f_1, f_2 = f_1$ and $f_2 > f_1$, respectively. The spectrum $R_1(u,v)$ of the straight grating consists of an impulse-comb whose impulses are located in the $u,v$ plane at integer multiples of the grating frequency $\mathbf{f}_1$. As we have seen in Example 10.11 in Sec. 10.6.1, the spectrum $R_2(u,v)$ of the circular grating $r_2(x,y)$ consists of a DC impulse plus a concentric series of circular impulsive rings whose radiuses are non-zero integer multiples of the radial frequency $f_2$, where each of the rings has a weak, continuous "wake" trailing off towards the center of the spectrum.

When the two gratings are superposed (i.e., multiplied) in the image domain, their spectra in the $u,v$ plane are convolved. This means that a centered replica of the spectrum $R_2(u,v)$ is copied on top of each impulse of the comb $R_1(u,v)$, after being scaled down by the impulse amplitude. The spectrum convolution $R_1(u,v){*}{*}R_2(u,v)$ is schematically shown in Figs. 10.17(f), 10.18(f) and 10.19(f) for the cases of $f_2 < f_1, f_2 = f_1$ and $f_2 > f_1$, respectively.

If we adopt the intuitive gradual transition approach presented in Sec. 10.4.1, we may think of $R_2(u,v)$ as a comb of impulses where each impulse pair "leaked out" into a circular impulsive ring, as shown in the sequence of Figs. 10.7(c),(b) and (a). This is clearly demonstrated in Fig. 10.15. Before the impulses "leak out", i.e., when the circular

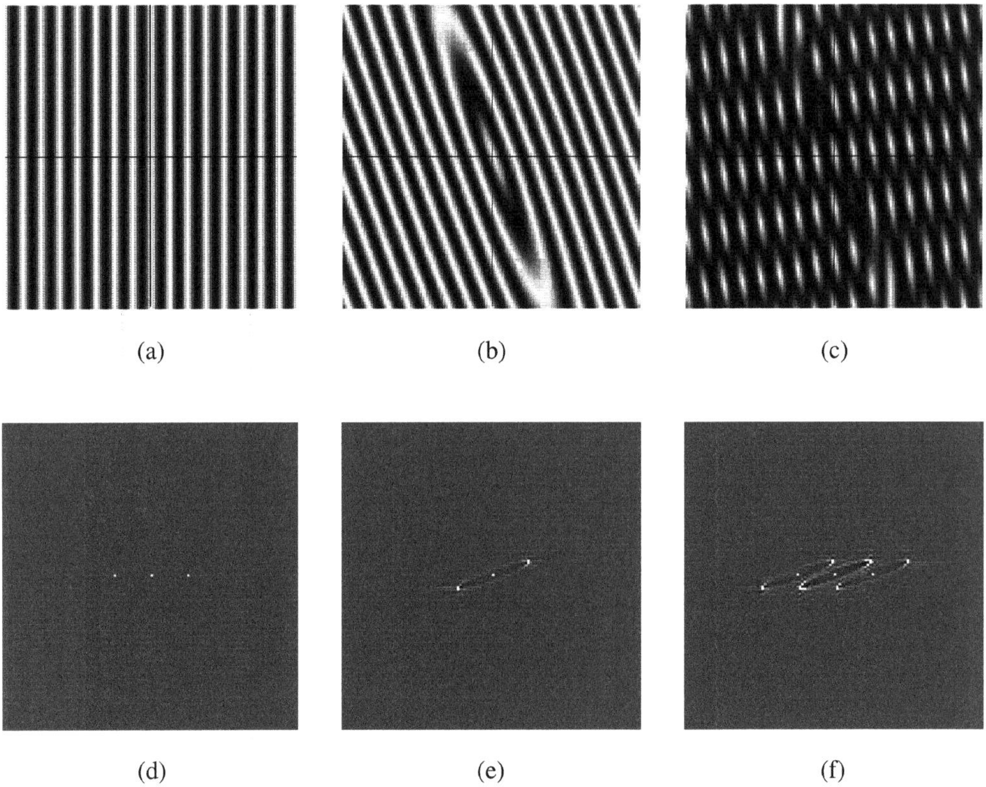

**Figure 10.15:** An intermediate stage in the gradual transition between the superposition of two raised cosinusoidal straight line gratings (Fig. 2.2) and the superposition of a raised cosinusoidal straight line grating with a raised cosinusoidal circular grating. The gratings (a) and (b) are given respectively by $\frac{1}{2}\cos(2\pi fx) + \frac{1}{2}$ and by a 20° rotation of $\frac{1}{2}\cos(2\pi f\sqrt{x^2 + \frac{1}{64}y^2}) + \frac{1}{2}$, and (c) is their superposition; (d), (e) and (f) show their respective spectra. The skeleton impulse pairs in spectrum (e) and hence also in the spectrum convolution (f) have just started to "leak out" into an elliptic impulsive ring, like in Fig. 10.7(b). Note that most of the "mass" of each impulsive ring is still concentrated in its two extremes, around the "melting down" skeleton impulses (see [Bracewell95 p. 131]).

grating is still an uncurved periodic grating $p_2(x,y)$ which is oriented to some direction $\theta$, its spectrum $P_2(u,v)$ is an impulse comb oriented in the same direction, and therefore the convolution $R_1(u,v)**P_2(u,v)$ is an oblique lattice of impulses like in Fig. 2.5(f). This oblique lattice is the *skeleton* of the spectrum $R_1(u,v)**R_2(u,v)$. As the second grating starts bending, first into an elliptic form (as in Fig. 10.15(b)) and then into a circular form, all the impulse pairs of this skeleton lattice (except for the impulses of the comb $R_1(u,v)$)

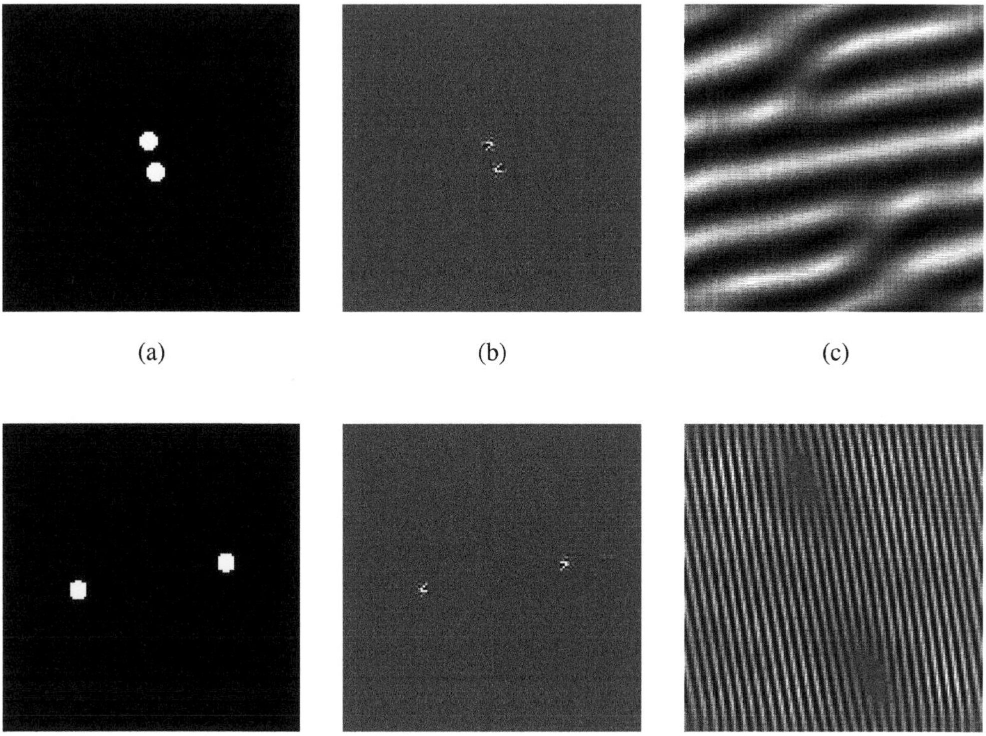

(a) (b) (c)

(d) (e) (f)

**Figure 10.16:** Top row: extraction of the "melting down" (1,-1) and (-1,1) impulse pair from the spectrum convolution of Fig. 10.15(f), obtained artificially by multiplying this convolution with a specially designed band-pass filter (a). The extracted "melted" impulses are shown in (b). Their inverse DFT, shown in (c), is the isolated contribution of the subtractive (1,-1)-moiré effect to the image domain superposition. Bottom row: extraction of the "melting down" (1,1) and (-1,-1) impulse pair from the spectrum convolution of Fig. 10.15(f), obtained artificially by multiplying this convolution with a specially designed band-pass filter (d). The extracted "melted" impulses are shown in (e). Their inverse DFT, shown in (f), is the isolated contribution of the additive (1,1)-moiré effect to the image domain superposition. Note that in fact the subtractive and the additive moirés are represented in the spectrum convolution by a common spectral ring, and their separation has been forced here by artificial means.

start "leaking out" to both directions, each pair forming first an elliptic impulsive ring oriented in the direction $\theta$ (like in Fig. 10.15(e),(f)), and finally a circular impulsive ring.

However, the skeleton of the resulting circular impulsive spectrum is no longer well-defined, since the two impulses $\pm n\mathbf{f}_2$ which "leak out" to form each circular ring are fused together into a single, non-separable ring. Consequently, we can no longer identify elements (rings) in the spectrum with single impulses in the skeleton lattice, but rather with *impulse pairs*. The *index* of an impulsive ring in the spectrum of the superposition will be defined as the index of its skeleton impulse pair, namely: $(m,\pm n)$, and there is no longer a distinction between the '+' and '–' signs in the second index.[24] Hence, a $(k_1,\pm k_2)$-moiré in the superposition is the moiré which is caused by the $(k_1,\pm k_2)$-ring in the spectrum convolution (or, in fact, by the $(k_1,\pm k_2)$-cluster of rings). This moiré becomes visible if the $(k_1,\pm k_2)$-ring, the fundamental ring of the $(k_1,\pm k_2)$-cluster, passes inside the visibility circle.

We may conclude, therefore, that in the present case there no longer exists a distinction between additive and subtractive moirés:[25] In the spectrum convolution both of the skeleton impulses $(k_1,k_2)$ and $(k_1,-k_2)$ are now parts of the same impulsive ring, so that the spectral representation of the moiré effect is no longer spatially separable into distinct additive and subtractive moirés.

It is interesting to note that this conclusion seems, at first, to be in conflict with the explicit moiré expressions that are obtained later in this chapter based on the fundamental moiré theorem (see Example 10.19 in Sec. 10.9.1). In fact, we will see there that in the superposition of a periodic straight grating and a circular grating the explicit expression of the isolated (1,1)-moiré (or rather, of its first-harmonic element) is $m_{1,1}(x,y) = \frac{1}{2}\cos(2\pi[f_2\sqrt{x^2+y^2}+f_1x])$, while the first-harmonic element of the isolated (1,-1)-moiré is $m_{1,-1}(x,y) = \frac{1}{2}\cos(2\pi[f_2\sqrt{x^2+y^2}-f_1x])$. These two functions and their spectra are shown in Fig. 10.20(a),(b) for the case of $f_2 < f_1$; a more detailed discussion on such functions and their spectra can be found in [Amidror98a pp. 911–912]. As we can see, the additive (1,1)-moiré $m_{1,1}(x,y)$ corresponds to the moiré curves which appear to the left of the origin, while the subtractive (1,-1)-moiré $m_{1,-1}(x,y)$ corresponds to the curves which appear to the right of the origin.[26] The clue for understanding this apparent contradiction can be found in the spectra of $m_{1,1}(x,y)$ and $m_{1,-1}(x,y)$: Since the functions $m_{1,1}(x,y)$ and $m_{1,-1}(x,y)$ are non-symmetric (but still purely real), their spectra are complex-valued but still Hermitian, meaning that their real part is symmetric, while their imaginary part is antisymmetric (see [Bracewell86 p. 15]). Now, since $m_{1,1}(x,y)$ and $m_{1,-1}(x,y)$ are mirror images of each other, so are their spectra, too. This means that the real parts of their spectra are identical, while the imaginary parts of their spectra are mutually sign-inversed (see Fig. 10.20(a),(b)). But since $m_{1,1}(x,y)$ and $m_{1,-1}(x,y)$ always appear in the superposition together (see Fig. 10.20(c)), the imaginary parts of their spectra cancel out each other and we are only left

---

[24] Note that the indices $(m,0)$ in the spectrum convolution belong to the impulses $m\mathbf{f}_1$ of the comb of the periodic grating $r_1(x,y)$, which are still present in the spectrum convolution.

[25] Note that in Chapters 10 and 11 we sometimes find it convenient to use the classical terms *subtractive moiré* and *additive moiré* (not to be confused with *additive superposition*!). These terms designate moirés which correspond, respectively, to frequency differences or frequency sums in the spectrum. For example, the (1,-1)-moiré is subtractive, while the (1,1)-moiré is an additive moiré.

[26] Note that this result has been also obtained, in a completely different way that is only based on image domain considerations, in [Firby84]; see in particular Fig. 5 there. See also [Post94 pp. 96–97].

with the real part. In other words: the spectrum of $m_{1,\pm1}(x,y) = m_{1,1}(x,y) + m_{1,-1}(x,y)$ is simply $2\text{Re}[M_{1,1}(u,v)]$, since $\text{Im}[M_{1,1}(u,v)] + \text{Im}[M_{1,-1}(u,v)] = 0$.

This resolves, indeed, the question concerning the inseparability of the $(k_1,\pm k_2)$-moiré into distinct additive and subtractive moirés: although both of the additive and subtractive moirés are present in the superposition, their individual spectra are exactly overlapping, so that the spectrum of their *sum* is not spatially separable, and the additive and subtractive

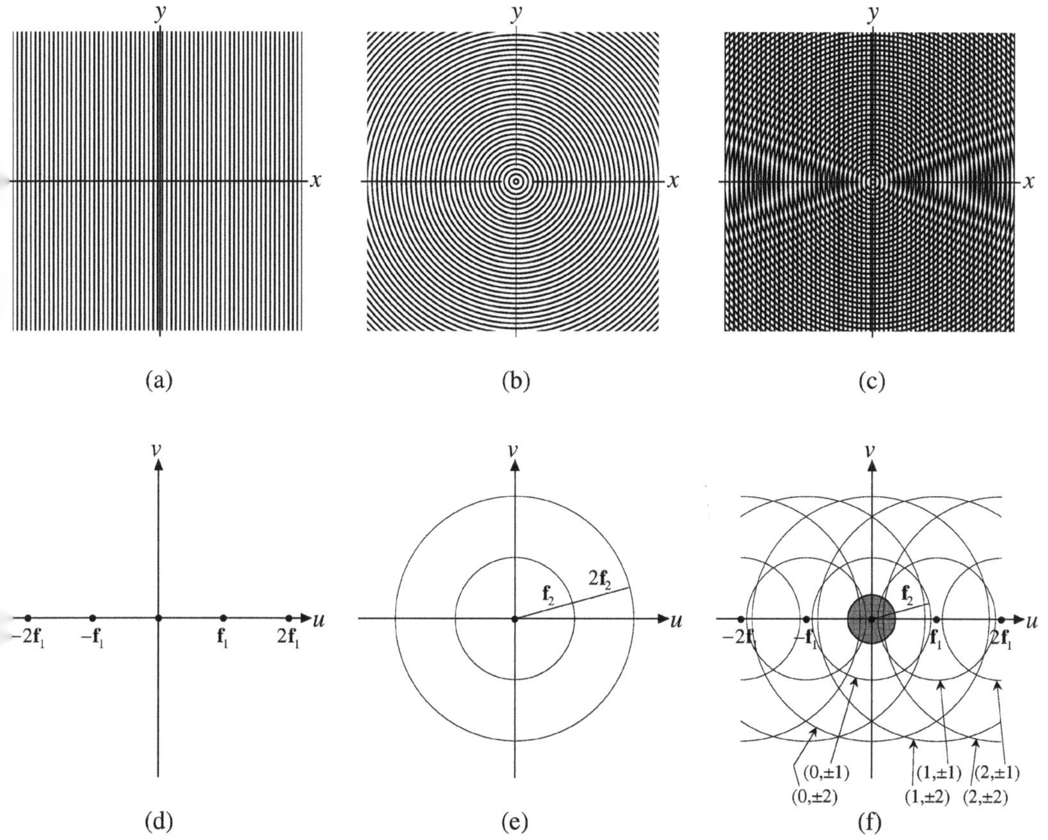


**Figure 10.17:** Periodic straight grating (a), circular grating (b) with $f_2 < f_1$, and their superposition (c) in the image domain. Their respective spectra (shown here schematically) are the infinite impulse comb (d), the concentric series of peculiar impulsive rings (e), and their convolution (f). Note that the ring wakes are not shown in the spectra. Some of the ring indices are indicated in the convolution (f); the gray circle in the center represents the visibility circle. Notice the hyperbolic (1,±1)-moiré which appears in the superposition (c).

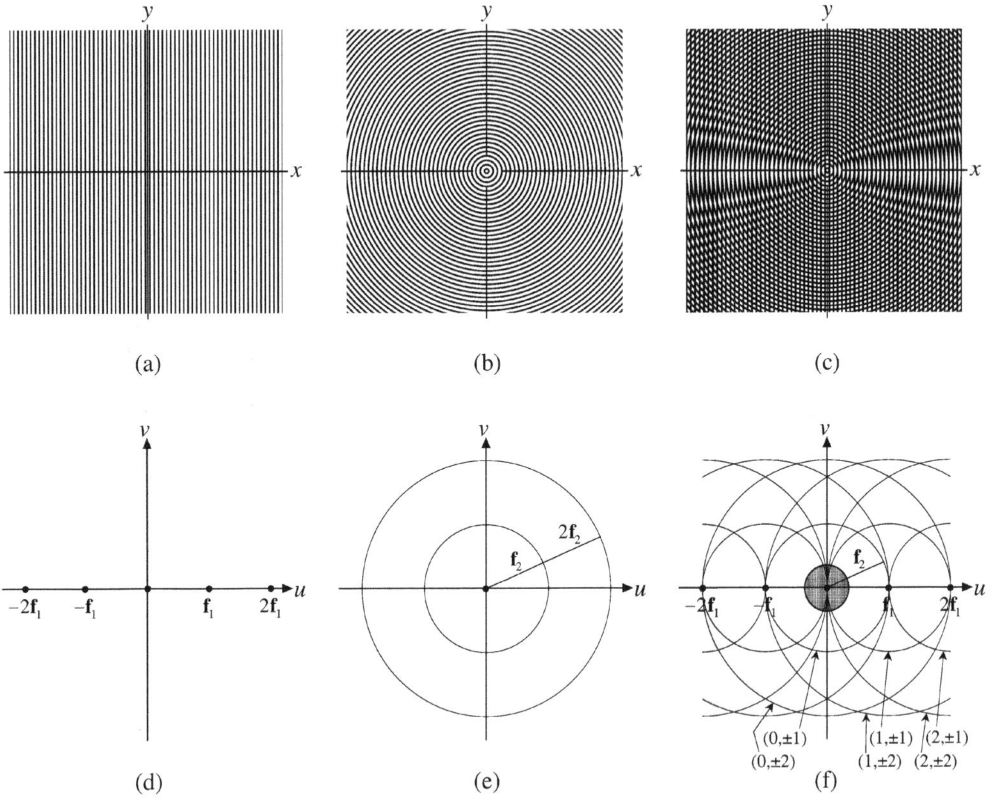


**Figure 10.18:** The singular state of the (1,-1)-moiré in the superposition of (a) periodic straight grating $r_1(x,y)$ and (b) circular grating $r_2(x,y)$ (with $f_2 = f_1$). Notice the singular locus of the (1,-1)-moiré along the $x$ axis of the superposition (c). (d) and (e) are schematic representations of the spectra of (a) and (b), and (f) is a schematic representation of the spectrum of (c), namely: the convolution of (d) and (e). Note that the ring wakes are not shown in the spectra.

moirés cannot be separately extracted from it by spectral filtering methods. (Note, however, that we can still *synthesize* them separately, as we have done, indeed, in Fig. 10.20(a),(b).)

**Example 10.16:** As an illustration, let us consider the superposition shown in Fig. 10.17, where the periodic straight grating $r_1(x,y)$ and the circular grating $r_2(x,y)$ are given by:

$$r_1(x,y) = p_1(x) = \sum_{m=-\infty}^{\infty} a^{(1)}_m \cos(2\pi m f_1 x)$$

$$r_2(x,y) = p_2(\sqrt{x^2+y^2}) = a_0 + 2\sum_{n=1}^{\infty} a^{(2)}_n \cos(2\pi n f_2 \sqrt{x^2+y^2})$$

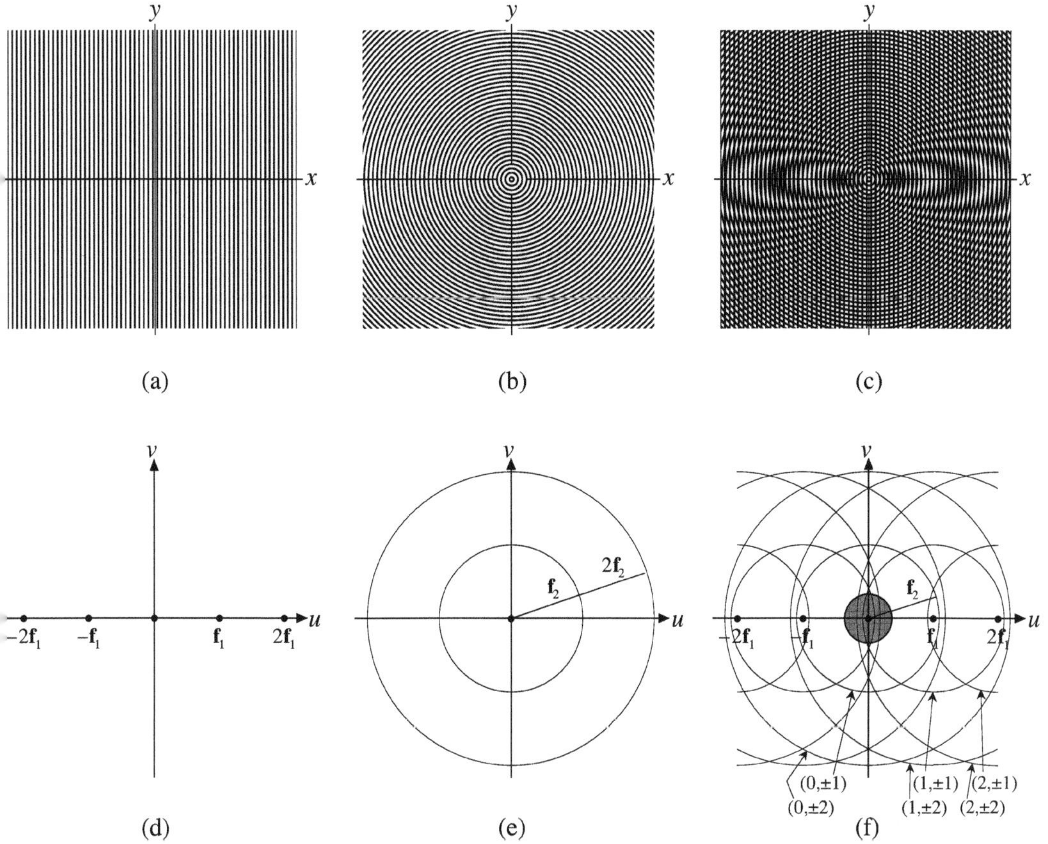


**Figure 10.19:** Periodic straight grating (a), circular grating (b) with $f_2 > f_1$, and their superposition (c) in the image domain. Their respective spectra (shown here schematically) are the infinite impulse comb (d), the concentric series of peculiar impulsive rings (e), and their convolution (f). The ring wakes are not shown in the spectra. Notice the elliptic (1,±1)-moiré which appears in the superposition (c); compare with Fig. 10.17 ($f_2 < f_1$) and Fig. 10.18 ($f_2 = f_1$).

with $f_2 < f_1$ (where $f_2 \approx f_1$). In this case, the visible moiré effect in the superposition is hyperbolic; it is represented in the spectrum convolution by the (1,±1)-impulsive ring, which passes inside the visibility circle, and the whole (1,±1)-cluster of rings that it spans. Note that this cluster contains the spectrum convolution elements whose indices are: ... (-2,±2), (-1,±1), (0,0), (1,±1), (2,±2), ..., i.e., the DC impulse (0,0) plus a series of non-concentric impulsive rings to both of its sides. But as we have just seen above, we may also say that this moiré effect is composed of the sum of the (1,1)-moiré, to which belong the left-hand hyperbolas, and the (1,-1)-moiré, to which belong the right-hand parabolas.

$m(x,y)$ $\qquad$ $\text{Re}[M(u,v)]$ $\qquad$ $\text{Im}[M(u,v)]$

(a)

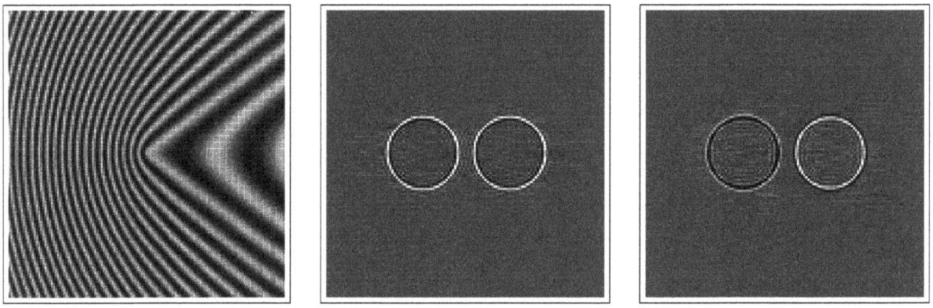

(b)

(c)

(d)

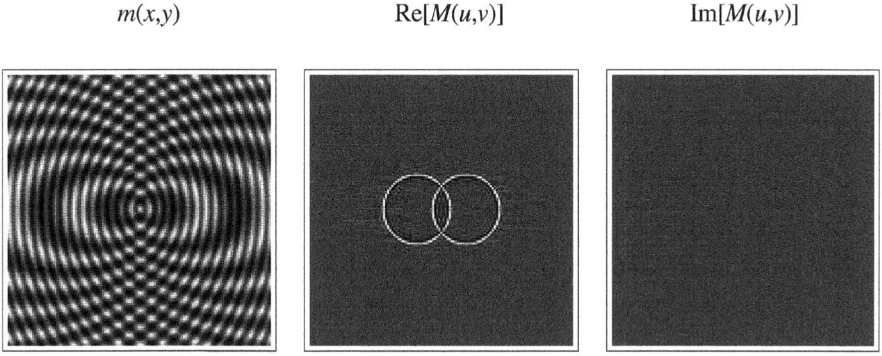


(e)

**Figure 10.20:** The isolated contribution of the (1,±1)-moirés $m(x,y)$ in the superposition of a straight cosinusoidal grating and a circular cosinusoidal grating with frequencies $f_1$ and $f_2$, and their spectra $M(u,v)$. (a) The (1,1)-moiré $m_{1,1}(x,y)$ and its spectrum, when $f_2 < f_1$. (b) The (1,-1)-moiré $m_{1,-1}(x,y)$ and its spectrum, when $f_2 < f_1$. (c) The (1,±1)-moiré $m_{1,\pm1}(x,y) = m_{1,1}(x,y) + m_{1,-1}(x,y)$ and its spectrum, when $f_2 < f_1$. (d) The (1,±1)-moiré $m_{1,\pm1}(x,y) = m_{1,1}(x,y) + m_{1,-1}(x,y)$ and its spectrum, when $f_2 = f_1$. (e) The (1,±1)-moiré $m_{1,\pm1}(x,y) = m_{1,1}(x,y) + m_{1,-1}(x,y)$ and its spectrum, when $f_2 > f_1$. Compare with Figs. 10.17 ($f_2 < f_1$), 10.18 ($f_2 = f_1$) and 10.19 ($f_2 > f_1$).

The spectrum of the (1,1)-moiré consists of the rings ... (-2,-2), (-1,-1), (0,0), (1,1), (2,2), ..., and the spectrum of the (1,-1)-moiré consists of the rings ... (-2,2), (-1,1), (0,0), (1,-1), (2,-2), ...; note, however, that these are not purely real rings as in the spectrum convolution, but rather Hermitian spectra whose imaginary parts only disappear in the spectrum of the superposition, where the two moirés are summed together. ■

Let us now examine the singular states of the moirés between a periodic straight grating and a circular grating. The moiré obtained in this superposition becomes singular when its fundamental impulse ring passes through the spectrum origin. For instance, in the example above the (1,±1)-moiré is singular when $f_2 = f_1$, so that the impulsive perimeter of the (1,±1)-ring passes through the origin. The geometric shape of this singular moiré effect is parabolic (see Fig. 10.18). Like in the superposition of a parabolic grating and a straight grating (Sec. 10.7.3), the singularity of the moiré in the present case is *local*, and the moiré is only singular along a straight line, the *singular locus* of the moiré. Along this line the singular moiré locally disappears to the eye, since its local period there is infinitely large. But at any other point of the $x,y$ plane outside the singular locus the singular moiré has a finite local period, which gradually decreases as one moves away from the singular locus, until at a certain distance the eye can no longer resolve it and the moiré completely disappears.

It is interesting to note that the (1,±1)-moiré in the superposition of a straight grating and a circular grating can leave its singular state (which occurs when $f_2 = f_1$) in two different ways: if the radius of the impulsive ring, $f_2$, becomes slightly smaller than $f_1$, so that the (1,±1)-ring moves slightly away from the spectrum origin, then the (1,±1)-moiré becomes hyperbolic (see Fig. 10.17); if, however, $f_2$ becomes slightly larger than $f_1$, so that the (1,±1)-ring encloses the spectrum origin, then the (1,±1)-moiré becomes elliptic (see Fig. 10.19). This phenomenon is explained in more detail in [Amidror98a pp. 911–912].

Figures 10.20(c)–(e) show the extracted (1,±1)-moiré in each of the three cases $f_2 < f_1$, $f_2 = f_1$ and $f_2 > f_1$. Similar hyperbolic, parabolic and elliptic moiré shapes occur also in any higher order $(k_1,\pm k_2)$-moiré between the straight and the circular gratings: when $k_2 f_2 = k_1 f_1$ the moiré effect is parabolic and singular, when $k_2 f_2 < k_1 f_1$ the moiré is hyperbolic, and when $k_2 f_2 > k_1 f_1$ the moiré is elliptic.

### 10.7.6 The superposition of two circular gratings

In the superposition of two circular gratings we may distinguish between two possible configurations: (i) when both of the circular gratings are superposed with a common center (which is located, for example, at the origin); (ii) when the centers of the two circular gratings are shifted with respect to each other. Although the first configuration is in fact a special case of the second, in which the distance between the grating centers is zero, it is still convenient to discuss it separately due to its particular interest.

For the sake of simplicity we will assume at first that both of the superposed circular gratings have raised cosinusoidal periodic-profiles of the form $\frac{1}{2}\cos(2\pi fx) + \frac{1}{2}$, with amplitudes varying between 0...1. The more general case with any periodic profile (square wave, etc.) will be then obtained through the Fourier series decomposition of the periodic profile.

Let us start with configuration (i), in which both of the superposed gratings share a common center (see Fig. 10.21). Since in this case the corrugations of both gratings remain constantly parallel, there is a strong analogy between this case and the superposition of two periodic gratings with an angle difference of $\alpha = 0$, and only the terms "period" and "frequency" are replaced here with "radial period" and "radial frequency", respectively. And indeed, the visible moiré in this case is a circularly periodic grating around the common center, whose radial frequency and radial period are given by: $f_M = |f_2 - f_1|$ and $T_M = 1/f_M = 1/|f_2 - f_1| = T_1 T_2/|T_2 - T_1|$, the same formulas as in the superposition of two parallel periodic gratings (cf. Eq. (2.11) in Sec. 2.4). As we can see in Fig. 10.21(f), the spectrum convolution which corresponds to the present superposition contains, in addition to the impulsive rings of the original layers (whose radial frequencies are $f_1$ and $f_2$) two new impulsive rings: the inner one, whose radial frequency is $|f_2 - f_1|$, corresponds to the subtractive (1,-1)-moiré which is visible in Fig. 10.21(c), and the outer one, whose radial frequency is $f_1 + f_2$, corresponds to the additive (1,1)-moiré which is simultaneously generated in the superposition (but which is not visible, since its radial frequency is far beyond the visibility circle). The extraction of each of these two moirés is clearly illustrated in Fig. 10.22.

Note that unlike in the case of Sec. 10.7.5 in which there was no possible distinction between subtractive and additive moirés, since both the $(k_1,k_2)$- and $(k_1,-k_2)$-skeleton impulses in the spectrum convolution were parts of the same spectral entity, in the present case each of the $(k_1,k_2)$- and $(k_1,-k_2)$-skeleton impulses generates a distinct ring in the spectrum convolution, and hence there is a clear distinction between the subtractive and the additive moirés. Note that the $(k_1,k_2)$-ring in the spectrum convolution contains the impulse locations of both the $(k_1,k_2)$ and $-(k_1,k_2)$ impulses of the skeleton spectrum.

Let us proceed now to configuration (ii), in which the centers of the two circular gratings are shifted with respect to each other. For the sake of simplicity we will assume that $f_1 = f_2$ and that the grating centers are symmetrically located to both sides of the origin, at the points $x = \pm x_0$ on the $x$ axis.

In order to understand the situation here, we start first with the case in which the shift $x_0$ is zero, namely: a superposition of two identical circular gratings both of which are centered on the origin (see Fig. 10.23). This superposition gives, of course, a circular grating of the same radial frequency. Note, however, that the periodic-profile of the superposition (=product) of the two raised cosinusoidal profiles is not precisely cosinusoidal, but a rather flat bottomed waveform which has a second harmonic component, too (as it is clearly seen in the spectrum of the superposition, (f)). This is explained using the trigonometric identity $\cos^2\alpha = \frac{1}{2} + \frac{1}{2}\cos 2\alpha$:

$$[\tfrac{1}{2}\cos(2\pi fr) + \tfrac{1}{2}]^2 = \tfrac{1}{4}\cos^2(2\pi fr) + \tfrac{1}{2}\cos(2\pi fr) + \tfrac{1}{4}$$

$$= \tfrac{3}{8} + \tfrac{1}{2}\cos(2\pi fr) + \tfrac{1}{8}\cos[2\pi(2f)r]$$

As we can see in Fig. 10.23(f), the spectrum convolution in this case contains no new elements within the visibility circle; this means that no visible moiré occurs in this case in the superposition, as shown, indeed, in Fig. 10.23(c).

Now, let us gradually shift the two circular gratings to both sides of the origin, one grating to the point $x = x_0$ and the other grating to the point $x = -x_0$. The situation in this case is shown in Fig. 10.24 for $x_0 = \pm 1$. According to the 2D shift theorem [Bracewell86 p. 244] the spectrum of a grating which has been shifted by $x_0$ is simply multiplied by $e^{-2\pi i x_0 u} = \cos(2\pi x_0 u) - i\sin(2\pi x_0 u)$; this can be clearly seen in the spectra shown in Fig. 10.24(d) and (e). The spectrum of the superposition, i.e., the convolution of the spectra (d) and (e), is shown in (f). As we can clearly see, this spectrum convolution contains two new elements which do not appear in the spectra (d) and (e): a second-harmonic ring which appears in the outer side of the spectrum; but also a new, continuous structure in the center of the spectrum, inside the visibility circle. This last structure is, of course, responsible for the moiré effect which is visible in the superposition (c); moreover, just like in the case of configuration (i) (see Fig. 10.21), we may say that the structure which appears inside the visibility circle corresponds to the subtractive (1,-1)-moiré, whereas the new outer ring in

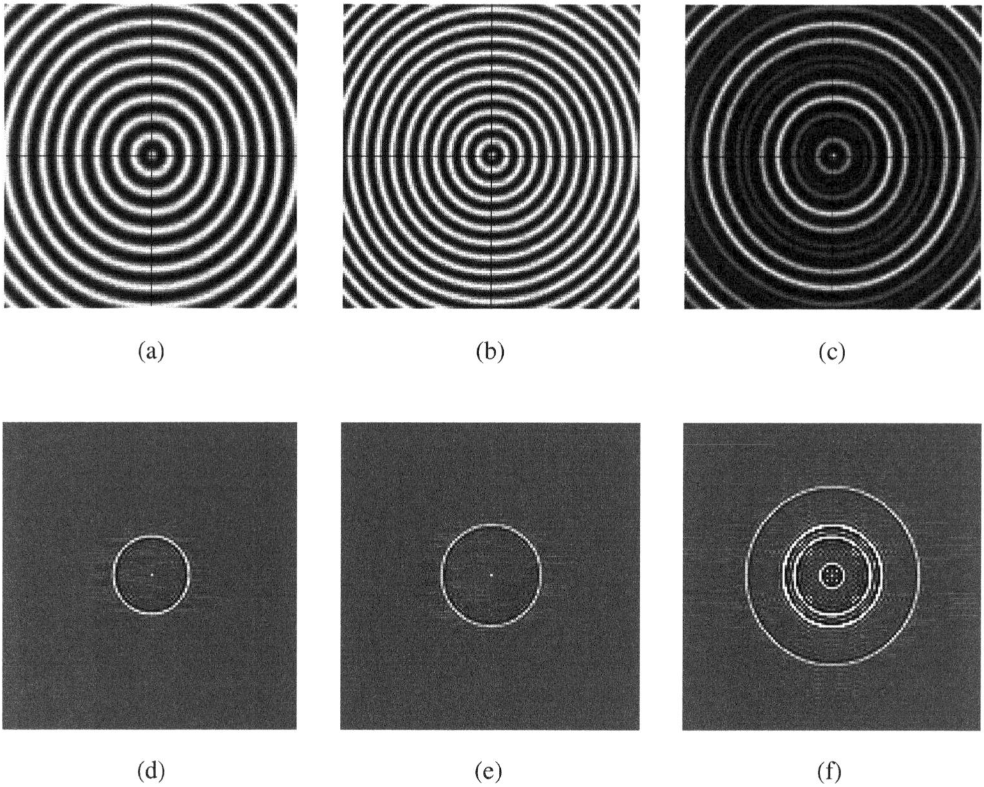


**Figure 10.21:** Two circular gratings (a) and (b) with raised cosinusoidal periodic-profiles $\frac{1}{2}\cos(2\pi fr) + \frac{1}{2}$, whose radial frequencies are $f_1$ and $f_2 = 1.3f_1$, and their superposition (c) with a common center. Their respective spectra are the peculiar impulsive rings (d) and (e), and their convolution (f). (Each of the spectra contains also a DC impulse owed to the constant $\frac{1}{2}$ in the original gratings.) Note that the convolution (f) contains the two original impulsive rings (d) and (e) (up to a certain amplitude scaledown), plus two new elements: an inner peculiar impulsive ring of radial frequency $|f_2-f_1|$, which is located inside the visibility circle and corresponds to the subtractive (1,-1)-moiré seen in the superposition (c), and an outer peculiar impulsive ring of radial frequency $f_1+f_2$, which corresponds to the additive (1,1)-moiré. The extracted moirés are shown in Fig. 10.22.

the spectrum corresponds to the additive (1,1)-moiré (which is not visible since its main radial frequency is far beyond the visibility circle). This can be confirmed by extracting the isolated elements in question from the spectrum convolution, as shown in Fig. 10.25, and taking their inverse Fourier transform: the structure obtained in each case is the

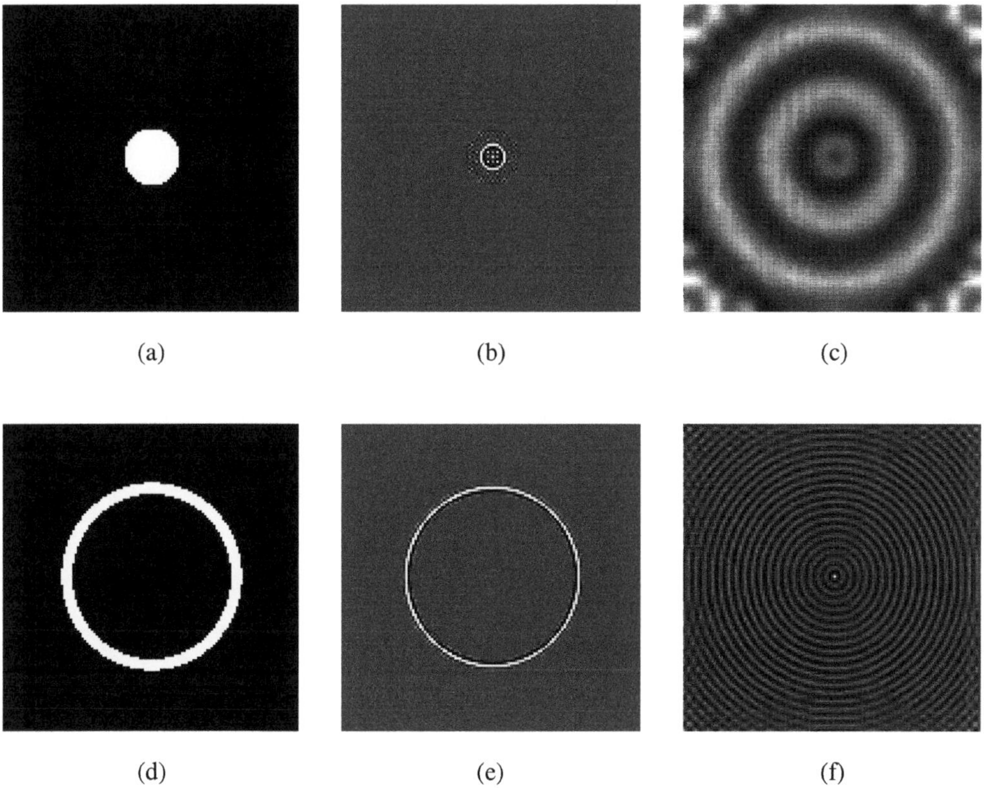


**Figure 10.22:** Top row: extraction of the inner impulsive ring from the spectrum convolution of Fig. 10.21(f), by multiplying this convolution with the circular low-pass filter (a). The extracted ring is shown in (b). Its inverse DFT, shown in (c), is the isolated contribution of the subtractive (1,-1)-moiré effect (whose radial frequency is $|f_2-f_1|$) to the image domain superposition. Bottom row: extraction of the outer impulsive ring from the spectrum convolution of Fig. 10.21(f), by multiplying this convolution with the circular band-pass filter (d). The extracted ring is shown in (e). Its inverse DFT, shown in (f), is the isolated contribution of the additive (1,1)-moiré effect (whose radial frequency is $f_1+f_2$) to the image domain superposition. Note that the filters (a), (d) are drawn using the image-domain conventions where white and black represent the values "1" and "0", while in the spectra (i.e. in (b) and (e)) white and black represent, as usual, positive and negative values.

isolated contribution of each of these spectral elements to the superposition in the image domain. As we can see, the (1,-1)-moiré generated by the internal spectral element consists of a hyperbolic grating, while the (1,1)-moiré generated by the external ring is a high-frequency elliptic grating.

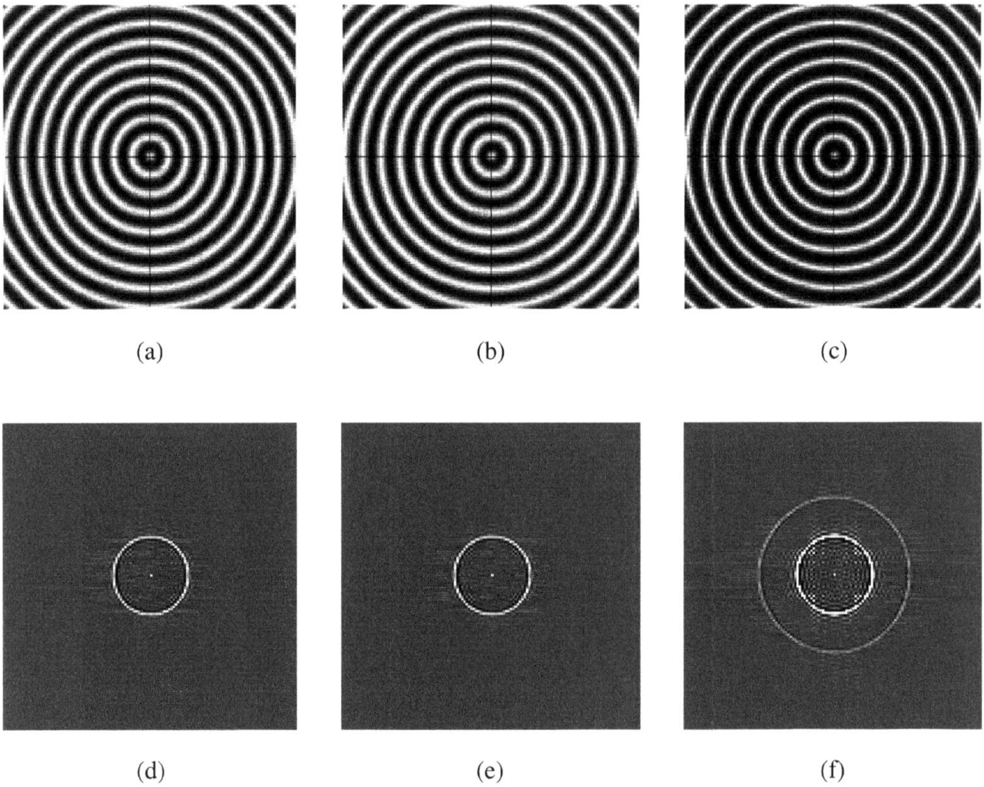

**Figure 10.23:** Two identical circular gratings (a) and (b) with raised cosinusoidal periodic-profiles $\frac{1}{2}\cos(2\pi fr) + \frac{1}{2}$, and their superposition (c) around the origin of the image domain. Their respective spectra are the peculiar impulsive rings (d) and (e), and their convolution (f). Note that the periodic-profile of the superposition (c) is not a simple raised cosine, and it contains also a second harmonic term, as clearly shown in the spectrum (f).

Now, if we continue shifting the two circular gratings away from each other, the elliptic (1,1)-moiré effect in the superposition becomes gradually more and more pronounced in the area between the centers of the superposed gratings, while the hyperbolic (1,-1)-moiré remains predominant in the external areas of the superposition. This can be seen in Fig. 10.26, in which the circular gratings have been shifted to $x_0 = \pm 4$. In this case we can clearly see that both of the new spectral elements in the spectrum convolution (Fig. 10.26(f)) are, in fact, continuous and overlap each other. This explains why these new spectral elements cannot be completely separated by filtering methods; any attempt to extract only one of these elements (i.e., the subtractive moiré or the additive moiré) from the spectrum convolution still remains "contaminated" by the other element, as is clearly shown in Fig. 10.27. Note that a similar contamination occurs in the case of Fig. 10.25, too, but it is less visible since for small shifts $x_0$ the outer spectral ring is still quite concentrated at the radial frequency of $2f$, and it does not really interfere with the other new spectral element inside the visibility circle.

Finally, it should be mentioned that if the periodic-profiles of the superposed circular gratings are not cosinusoidal but have a more complex waveform (such as a square waveform, etc.), each of the circular gratings can be decomposed into a Fourier series of circular cosinusoidal elements, and the respective spectra will simply contain additional harmonics of a similar shape, which result from the higher cosinusoidal terms of the Fourier series (see Example 10.11 in Sec. 10.6.1). Thus, the moiré in the superposition of two circular gratings with square wave periodic-profiles (see, for example, Fig. 11.4 in Chapter 11) have the same geometric layout as in the cosinusoidal case, and only the periodic-profile (waveform shape) of the moirés is affected. It is interesting to note that in the superposition of circular zone gratings (see Sec. 10.7.8) higher harmonics *do* generate new higher-order moirés which are simultaneously visible in the image domain, and which are not present if the periodic profiles are cosinusoidal and have no higher harmonics. This difference is explained at the end of Sec. 10.7.8.

**Remark 10.7:** It is interesting to note that the hyperbolic/elliptic geometric layouts of the subtractive/additive moirés in the present case can be derived using image domain methods, such as the fundamental moiré theorem (see Sec. 10.9.1), or the classical indicial equations method whose principles are reviewed below in Chapter 11 (see in particular Example 11.3). However, the mathematical derivation of these moirés directly from the spectral domain is not obvious. Even to the interesting question: "what is the convolution of two peculiar impulsive rings?" we do not give here a rigorous mathematical answer, and we just illustrate the situation by means of the results obtained by DFT (see the spectrum convolutions in Figs. 10.21, 10.23, 10.24 and 10.26). ■

#### 10.7.7 The superposition of a zone grating and a periodic straight grating

After having discussed superpositions of layers involving line-spectra (Secs. 10.7.3 and 10.7.4) or semi-impulsive spectra (Secs. 10.7.5 and 10.7.6), we proceed now to discuss superpositions of layers involving continuous spectra.

As a first example, assume that we superpose a circular zone grating $r_2(x,y)$ and a periodic grating $r_1(x,y)$ of vertical straight lines (this is illustrated in Fig. 10.28 for the case where both gratings have raised cosinusoidal periodic-profiles). The spectrum $R_1(u,v)$ of the straight grating consists of an impulse-comb whose impulses are located in the $u,v$ plane at integer multiples of the grating frequency $\mathbf{f}_1$. As we have seen in Example 10.12 in Sec. 10.6.1, the spectrum $R_2(u,v)$ of the circular zone grating $r_2(x,y)$ consists of a DC impulse plus a concentric series of continuous humps which are all centered on the spectrum origin, each having the form of a sinusoidal zone grating. Note that the humps in

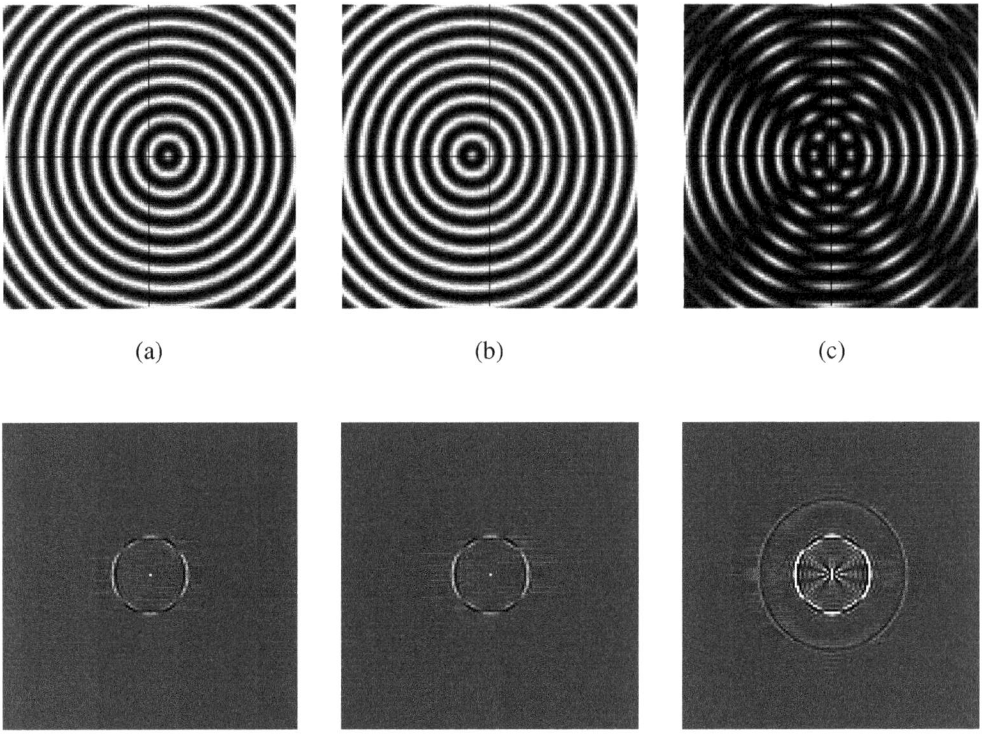

(a) (b) (c)

(d) (e) (f)

**Figure 10.24:** Two identical circular gratings with raised cosinusoidal periodic-profiles $\frac{1}{2}\cos(2\pi fr) + \frac{1}{2}$, which have been horizontally shifted from the origin to the points (a) $x = 1$ and (b) $x = -1$, and their superposition (c). Their respective spectra are the peculiar impulsive rings (d) and (e), and their convolution (f); only the real parts of the spectra are shown. (Each of the spectra contains also a DC impulse owed to the constant $\frac{1}{2}$ in the original gratings.) The new spectral element which appears inside the visibility circle in the spectrum convolution (f) corresponds to the subtractive (1,-1)-moiré which is clearly seen in the superposition (c), and the new outer ring corresponds to the additive (1,1)-moiré. The extracted moirés are shown in Fig. 10.25.

the spectrum $R_2(u,v)$ are not spatially separable, since all of them extend throughout the whole spectrum, mutually overlapping at every point of the $u,v$ plane.

Now, when the two gratings $r_1(x,y)$ and $r_2(x,y)$ are superposed (i.e., multiplied) in the image domain, their spectra in the $u,v$ plane are convolved. This means that a centered

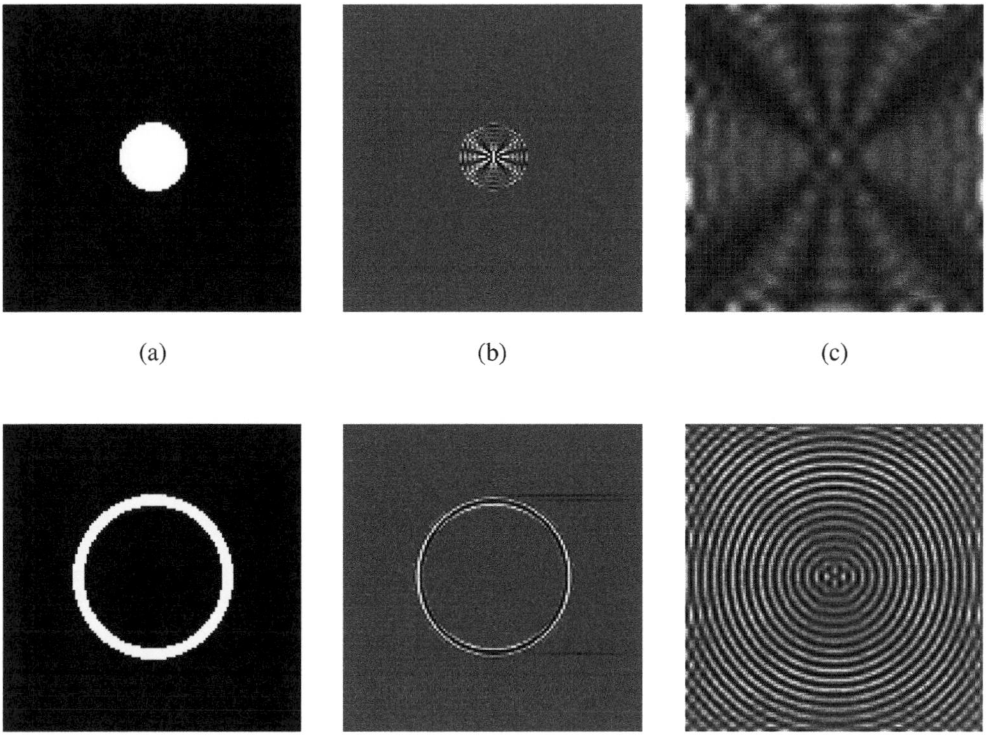

(a) (b) (c)

(d) (e) (f)

**Figure 10.25:** Top row: extraction of the internal spectral element from the spectrum convolution of Fig. 10.24(f), by multiplying this convolution with the circular low-pass filter (a). The extracted spectral element is shown in (b). Its inverse DFT, shown in (c), is the isolated contribution of the subtractive (1,-1)-moiré effect to the image domain superposition (namely: a hyperbolic grating). Bottom row: extraction of the outer ring from the spectrum convolution of Fig. 10.24(f), by multiplying this convolution with the circular band-pass filter (d). The extracted ring is shown in (e). Its inverse DFT, shown in (f), is the isolated contribution of the additive (1,1)-moiré effect to the image domain superposition (namely: a high-frequency elliptic grating).

replica of the spectrum $R_2(u,v)$ is copied on top of each impulse of the comb $R_1(u,v)$, after being scaled down by the impulse amplitude. The spectrum convolution $R_1(u,v)**R_2(u,v)$ is shown in Fig. 10.28(f).

If we adopt the intuitive gradual transition approach presented in Sec. 10.4.1, we may think of the first harmonic hump of $R_2(u,v)$ as a skeleton impulse pair located at $\pm\mathbf{f}_2$, which

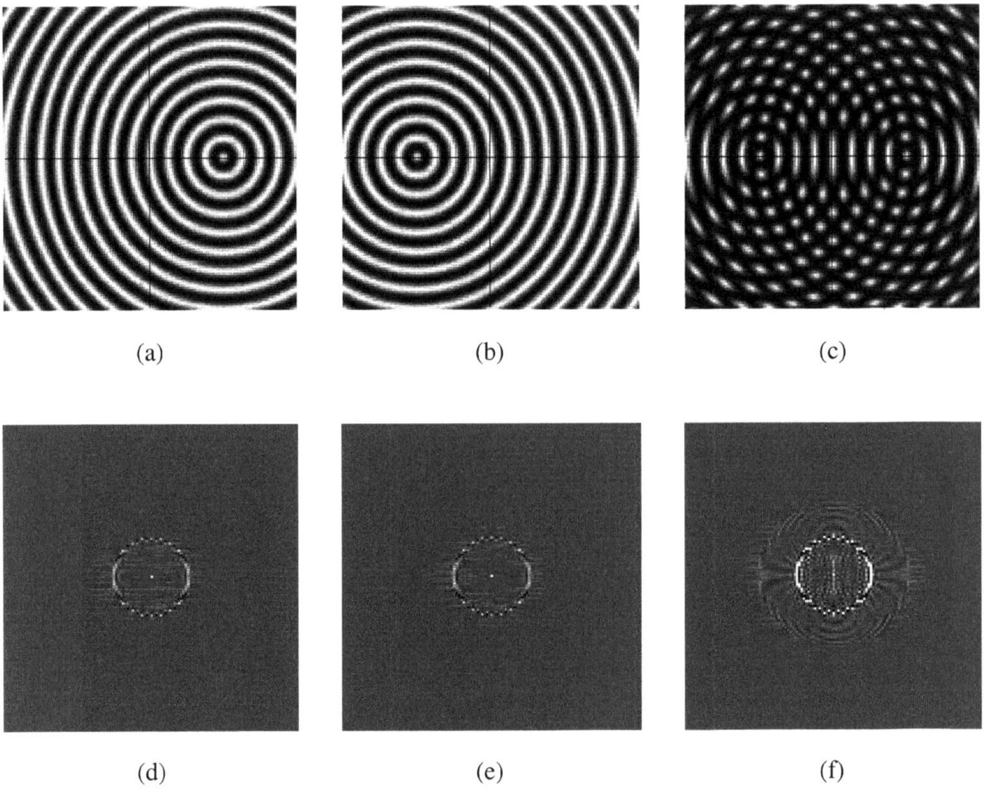


**Figure 10.26:** Same as Fig. 10.24, with a larger horizontal shift of $x = \pm 4$ instead of $x = \pm 1$. It is clearly seen that both of the new spectral elements which are generated in the spectrum convolution (f), the element inside the visibility circle which corresponds to the subtractive (1,-1)-moiré and the outer ring which corresponds to the additive (1,1)-moiré, are in fact *continuous* and overlap each other, so that they cannot be completely separated by applying filtering methods on the spectrum convolution (f). This is clearly shown in Fig. 10.27.

has "leaked out" into a continuous hump centered on the origin that covers the entire $u,v$ plane. Note that in this case, too, there is no distinction between the spectral entities that leaked out from the impulses $\mathbf{f}_2$ and $-\mathbf{f}_2$. Similarly, all the $n$ harmonics of $R_2(u,v)$ are also continuous humps centered on the origin. The spectrum convolution $R_1(u,v)**R_2(u,v)$ contains, therefore:

(a) The impulses of the comb of the first, periodic layer, which are located in the $u,v$ plane at the points $m\mathbf{f}_1$, and whose indices are $(m,0)$: ... (-2,0), (-1,0), (0,0), (1,0), (2,0), ...

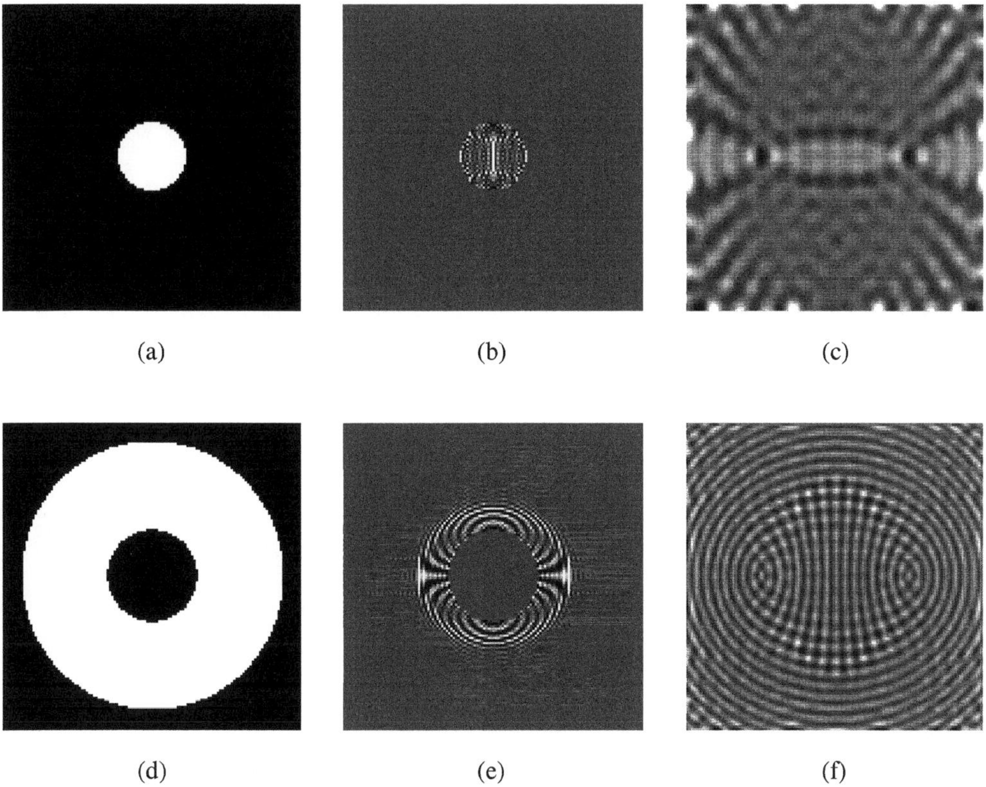

**Figure 10.27:** Top row: extraction of the internal spectral element from the spectrum convolution of Fig. 10.26(f), by multiplying this convolution with the circular low-pass filter (a). The extracted spectral element is shown in (b), and its inverse DFT is shown in (c). Bottom row: extraction of the outer element from the spectrum convolution of Fig. 10.26(f), by multiplying this convolution with the circular high-pass filter (d). The extracted element is shown in (e), and its inverse DFT is shown in (f). Since both the inner and the outer elements in the spectrum convolution of Fig. 10.26(f) are *continuous* and overlap each other, they cannot be completely separated from each other by applying simple filtering methods on the spectrum convolution.

(b) Centered on top of each of the impulses $(m,0)$ there is an infinite series of humps (sinusoidal zone gratings) whose indices are $(m,\pm n)$, namely: $(m,\pm 1)$, $(m,\pm 2)$, ...

As we can see, the spectrum convolution contains, in addition to the original spectral entities of the superposed images (the comb impulses $(m,0)$, and the humps $(0,\pm n)$ which are centered on the DC impulse) new continuous humps, whose indices are $(m,\pm n)$ with

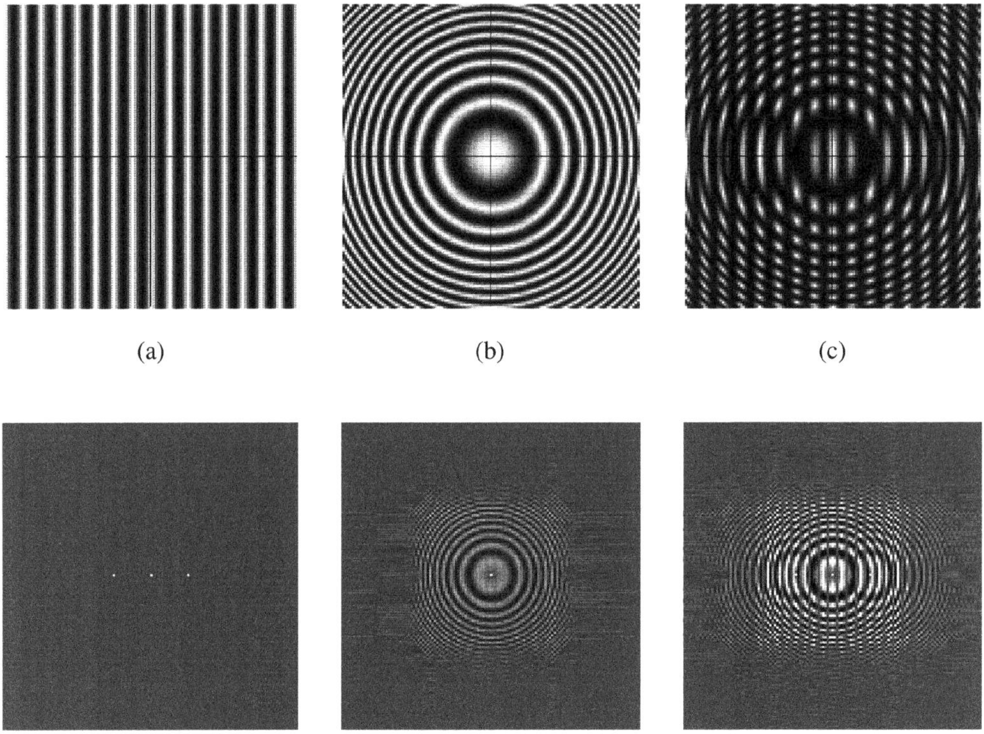


**Figure 10.28:** (a) Periodic raised cosinusoidal grating $\frac{1}{2}\cos(2\pi fx) + \frac{1}{2}$; (b) circular, raised cosinusoidal zone grating $\frac{1}{2}\cos(2\pi f[x^2+y^2]/8) + \frac{1}{2}$; and their superposition (c). The respective spectra are shown in (d), (e) and their convolution (f). The $(1,\pm1)$-moiré in the superposition (c) consists of two new cosinusoidal zone gratings which appear to both sides of the original one. The extracted moiré is shown in Fig. 10.29.

$m,n \neq 0$, which are centered on the other impulses of the comb. These new spectral entities are, indeed, responsible for the moiré effects which are generated in the superposition. This is illustrated in Fig. 10.29 for the case of gratings with raised cosinusoidal profiles shown in Fig. 10.28, where only the first harmonics ($m$ = 1,-1 and $n = \pm1$) are present in the spectra. Obviously, there is no way to extract by spectral filtering methods only the new elements from the spectrum convolution (Fig. 10.28(f)), since the spectral elements are continuous and overlapping. But if we artificially synthesize a spectrum which contains only these spectral elements (we may do it here analytically, since we know their explicit mathematical expressions), we can reconstruct, by applying an inverse Fourier

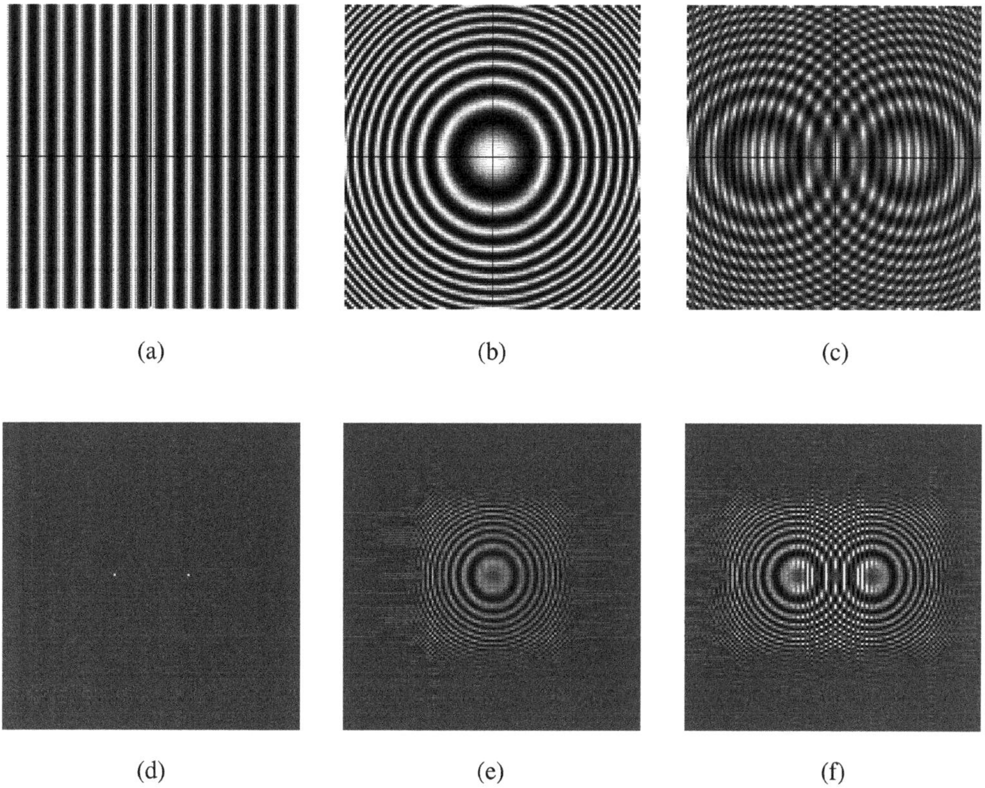


**Figure 10.29:** Extraction of the (1,±1)-moiré from Fig. 10.28: Since the two new humps in the convolution of Fig. 10.28(f) are overlapping and they cannot be isolated by spectral filtering methods, we reconstructed here the isolated contribution of these two humps alone by superposing the *unraised* gratings (a) $\frac{1}{2}\cos(2\pi fx)$ and (b) $\frac{1}{2}\cos(2\pi f[x^2+y^2]/8)$, as shown in (c); the respective spectra are (d), (e) and their convolution (f). As we can see, spectrum convolution (f) contains only the two new zone gratings from the spectrum convolution of Fig. 10.28(f), and hence its inverse Fourier transform (c) is the isolated contribution of the (1,±1)-moiré of Fig. 10.28(c) in the image domain.

transform, the isolated contribution of the moiré effect to the image domain superposition. This is clearly shown in Fig. 10.29, where only the first-order (1,±1)-moiré is present. This moiré has the form of a pair of zone gratings, which are centered at the points $\pm x_0$ to both sides of the origin (see the derivation below). Like in the case of Sec. 10.7.5, we may identify the left-side zone grating with the additive (1,1)-moiré, and the right-side zone grating with the subtractive (1,-1)-moiré; but here, too, their complex-valued spectra are exactly overlapping (see the spectrum of a shifted zone grating in Fig. 10.6(c)), so that the additive and subtractive moirés cannot be separately extracted from the spectrum by spectral filtering. We can therefore say that both zone gratings are considered as a single, non-separable (1,±1)-moiré effect. Since the two spectral humps of this moiré are continuous and extend throughout the whole $u,v$ plane, it is clear that they also pass through the spectrum origin, and hence the moiré has always two singular locus points in the image domain (i.e., two moiré-eyelets) — the centers of the two moiré zone gratings.

If the periodic-profiles of the original superposed gratings are not cosinusoidal but have a more complex waveform (for example: a square waveform, as in the case of the classical zone grating), further harmonic elements ($m = \pm2, \pm3, \ldots, n = \pm2, \pm3, \ldots$) are also present in the spectra, and they generate in the superposition higher-order moirés, all of which having the form of zone grating pairs which are centered along the $x$ axis to both sides of the origin. This is shown, for example, in Figs. 2 and 3 in [Walls75, p. 594]. The center points of the zone grating pair belonging to the $(k_1,\pm k_2)$-moiré are given by $x_0 = \pm\frac{k_1}{2k_2a}$, $y_0 = 0$.[27] This can be demonstrated as follows:

The superposition of a periodic grating of vertical straight lines with a circular zone grating is given, using the two-sided Fourier series decompositions of the gratings, by:

$$r_1(x,y)r_2(x,y) = \Big(\sum_{m=-\infty}^{\infty} a^{(1)}_m \cos(2\pi mfx)\Big)\Big(\sum_{n=-\infty}^{\infty} a^{(2)}_n \cos(2\pi nf[ax^2+by^2])\Big)$$

$$= \sum_{m=-\infty}^{\infty}\sum_{n=-\infty}^{\infty} a^{(1)}_m a^{(2)}_n \cos(2\pi mfx)\cos(2\pi nf[ax^2+by^2])$$

However, in order to simplify the derivation which follows, we prefer to use here for $r_1(x,y)$ and $r_2(x,y)$ their exponential Fourier series notation (which is, in fact, even more general than the cosine Fourier series notation, since it also allows non-symmetric periodic-profiles):

$$r_1(x,y)r_2(x,y) = \Big(\sum_{m=-\infty}^{\infty} c^{(1)}_m e^{i2\pi mfx}\Big)\Big(\sum_{n=-\infty}^{\infty} c^{(2)}_n e^{i2\pi nf(ax^2+by^2)}\Big)$$

$$= \sum_{m=-\infty}^{\infty}\sum_{n=-\infty}^{\infty} c^{(1)}_m c^{(2)}_n e^{i2\pi f[mx + n(ax^2+by^2)]}$$

The (1,1)-partial sum of this double sum (which corresponds to the new (1,1)-cluster in the spectrum convolution and to the (1,1)-moiré in the image domain) is:

$$m_{1,1}(x,y) = \sum_{n=-\infty}^{\infty} c^{(1)}_n c^{(2)}_n e^{i2\pi nf[x + (ax^2+by^2)]}$$

However, since $x + ax^2 + by^2$ equals $a(x-x_0)^2 + by^2 + c$ with $x_0 = -\frac{1}{2a}$ and $c = -\frac{1}{4a}$, we obtain:

$$= \sum_{n=-\infty}^{\infty} c^{(1)}_n c^{(2)}_n e^{i2\pi nf[a(x-x_0)^2 + by^2 + c]}$$

[27] Note that all the harmonics of the $(k_1,\pm k_2)$-moiré, the $(mk_1,\pm mk_2)$ components, will have the same center point, so they do not generate independent moiré eyelets. This is in accordance with point (2) in Sec. 2.8.

We see, therefore, that the (1,1)-moiré is simply a zone grating with a horizontal shift of $x_0 = -\frac{1}{2a}$ and an initial phase of $c = \frac{1}{4a}$. Note that this moiré is, indeed, a zone grating which is located to the left of the origin, as expected.

It can be shown in a similar way that the (1,-1)-moiré is a zone grating with a horizontal shift of $x_0 = \frac{1}{2a}$ and an initial phase of $c = -\frac{1}{4a}$. As expected, this zone grating is located to the right of the origin.

Similarly we obtain for the general $(k_1,k_2)$-moiré:

$$m_{k_1,k_2}(x,y) = \sum_{n=-\infty}^{\infty} c^{(1)}{}_{nk_1} c^{(2)}{}_{nk_2}\, e^{i2\pi nf[k_1x + k_2ax^2 + k_2by^2)]}$$

$$= \sum_{n=-\infty}^{\infty} c^{(1)}{}_{nk_1} c^{(2)}{}_{nk_2}\, e^{i2\pi nf[k_2a(x-x_0)^2 + k_2by^2 + k_1c]}$$

with $x_0 = -\frac{k_1}{2k_2a}$ and $c = \frac{k_1}{4k_2a}$. The $(k_1,k_2)$-moiré is therefore a zone grating which is shifted by $x_0 = -\frac{k_1}{2k_2a}$ from the origin, as we have expected, and whose initial phase is $c = \frac{k_1}{4k_2a}$.

It can be also shown that if $f$ is replaced in the layers $r_1(x,y)$ and $r_2(x,y)$ by different values $f_1$ and $f_2$, then the horizontal shift and the initial phase of the $(k_1,k_2)$-moiré become, respectively, $x_0 = -\frac{k_1f_1}{2k_2f_2a}$ and $c = \frac{k_1f_1}{4k_2f_2a}$. Note, in particular, that the shift $x_0$ is proportional to the frequency of the straight grating, $f_1$. An application of this property will be given in Problem 10-18.

It is interesting to note that the behaviour of the higher order moirés here is different than in the superposition of a circular grating and a periodic straight grating (Sec. 10.7.5): unlike in that case, several higher order moirés can be simultaneously visible here in the superposition. The reason is, of course, that each of the spectral components in the present case extends throughout the entire $u,v$ plane, including the spectrum origin and the visibility circle which surrounds it.

### 10.7.8 The superposition of two circular zone gratings

Like in the superposition of two circular gratings (Sec. 10.7.6), it may be convenient to distinguish here between two possible configurations: (i) when both of the circular zone gratings are superposed with a common center (which is located, for example, at the origin); (ii) when the centers of the two circular zone gratings are shifted apart.

For the sake of simplicity we will assume at first that both of the superposed circular zone gratings have raised cosinusoidal periodic-profiles of the form $\frac{1}{2}\cos(2\pi fx) + \frac{1}{2}$, with amplitudes varying between 0...1. The more general case with any periodic profile (square wave, etc.) will be then obtained through the Fourier series decomposition of the periodic profile.

We start with configuration (i), in which both of the superposed zone gratings share a common center (see Fig. 10.30). In this case the corrugations of both gratings remain

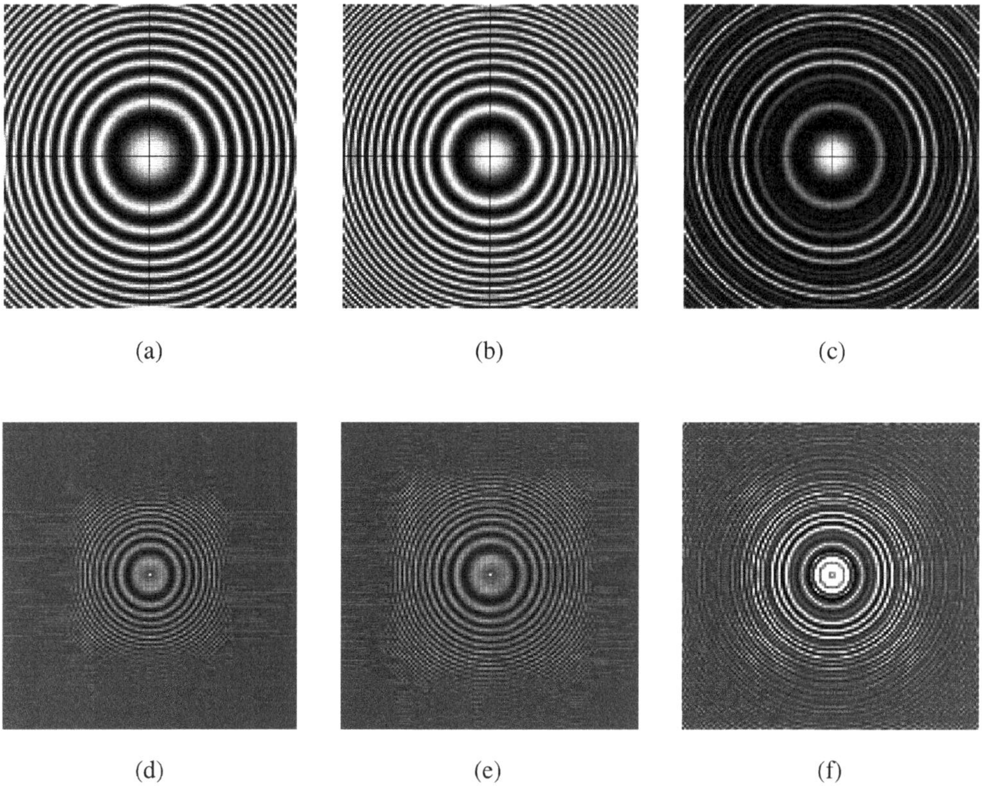


**Figure 10.30:** Two circular zone gratings (a) and (b) with raised cosinusoidal periodic-profiles $\frac{1}{2}\cos(2\pi fr) + \frac{1}{2}$, where $f_2 = 1.3f_1$, and their superposition (c); the zone grating shape of the moiré effect in (c) is best seen from a distance of about three meters. The respective spectra of (a), (b) and (c) are the zone gratings (d) and (e), and their convolution (f). (Each of the spectra contains also a DC impulse due to the constant $\frac{1}{2}$ in the original gratings.) Although the different components of the convolution (f) are overlapping and cannot be clearly seen here, we show in the text that (f) consists of the two zone gratings (d) and (e) (up to a certain amplitude scaledown), plus two new elements: an "inner" hump (zone grating) with $f = |f_2 - f_1|$, which is mainly concentrated inside the visibility circle and corresponds to the subtractive (1,-1)-moiré seen in the superposition (c), and an "outer" hump (zone grating) with $f = f_1 + f_2$, which corresponds to the additive (1,1)-moiré (not visible in (c)).

constantly parallel, and the visible moiré is a circular zone grating around the common center (see Fig. 10.30). And indeed, the spectrum convolution which corresponds to the present superposition contains, in addition to the two concentric continuous humps (zone gratings) of the original layers, whose basic radial frequencies are $f_1$ and $f_2$, two new concentric continuous humps: the "inner" one, whose basic radial frequency is $|f_2-f_1|$, corresponds to the subtractive (1,-1)-moiré which is visible in Fig. 10.30(c), and the "outer" one, whose basic radial frequency is $f_1+f_2$, corresponds to the additive (1,1)-moiré which is simultaneously generated in the superposition (but which is not visible, since its radial frequency is far beyond the visibility circle).

Since all of the humps in this spectrum convolution are continuous and centered around the spectrum origin, they are all overlapping and the two new humps which correspond to the subtractive and to the additive moirés cannot be clearly seen in Fig. 10.30(f). Instead, we give here the mathematical derivation which clearly shows their existence and their properties:

Consider the product of the two concentric, raised cosinusoidal zone gratings $r_1(x,y)$ and $r_2(x,y)$. Using the known trigonometric identity $\cos\alpha\cos\beta = \frac{1}{2}[\cos(\alpha-\beta) + \cos(\alpha+\beta)]$ we obtain:

$$r_1(x,y)\, r_2(x,y) = (\tfrac{1}{2}\cos(2\pi f_1 [x^2+y^2]) + \tfrac{1}{2})\,(\tfrac{1}{2}\cos(2\pi f_2 [x^2+y^2]) + \tfrac{1}{2})$$

$$= \tfrac{1}{8}\cos(2\pi(f_1-f_2)[x^2+y^2]) + \tfrac{1}{8}\cos(2\pi(f_1+f_2)[x^2+y^2])$$

$$+ \tfrac{1}{4}\cos(2\pi f_1 [x^2+y^2]) + \tfrac{1}{4}\cos(2\pi f_2 [x^2+y^2]) + \tfrac{1}{4}$$

Clearly, the first two terms in this sum correspond to the new subtractive and additive moirés which are generated in the superposition. Both of them are cosinusoidal zone gratings centered around the origin, and their basic radial frequencies are, indeed, $|f_1-f_2|$, and $f_1+f_2$.

We proceed now to configuration (ii), in which the centers of the two circular zone gratings are shifted with respect to each other. For the sake of simplicity we will assume that $f_1 = f_2$ and that the zone grating centers are symmetrically located to both sides of the origin, at the points $x = \pm x_0$ on the $x$ axis. As we can see in Figs. 10.31 and 10.32, the moiré effects obtained in this case consist of a subtractive (1,-1)-moiré in the form of a straight periodic grating, which is predominant when the shift $\pm x_0$ is small, and an additive (1,1)-moiré in the form of a zone grating centered around the origin, which is predominant when the shift $\pm x_0$ is larger. The detailed mathematical derivation of these moirés is given below in Example 10.18.

If the periodic-profiles of the original superposed gratings are not cosinusoidal but have a more complex waveform (for example: a square waveform, as in the case of the classical zone grating), further harmonic elements ($m = 2, 3, \ldots, n = 2, 3, \ldots$) are also present in the spectra, and they generate in the superposition higher-order moirés as well, all of which have the form of zone grating pairs which are centered to both sides of the origin along the $x$ axis. This is shown, for example, in Fig. 1 in [Leifer73, p. 34]. It is interesting to note that the behaviour of the higher order moirés here is different than in the superposition of circular gratings (Sec. 10.7.6): Unlike in that case, several higher order moirés can be simultaneously visible here in the superposition. The reason is, once again, that the

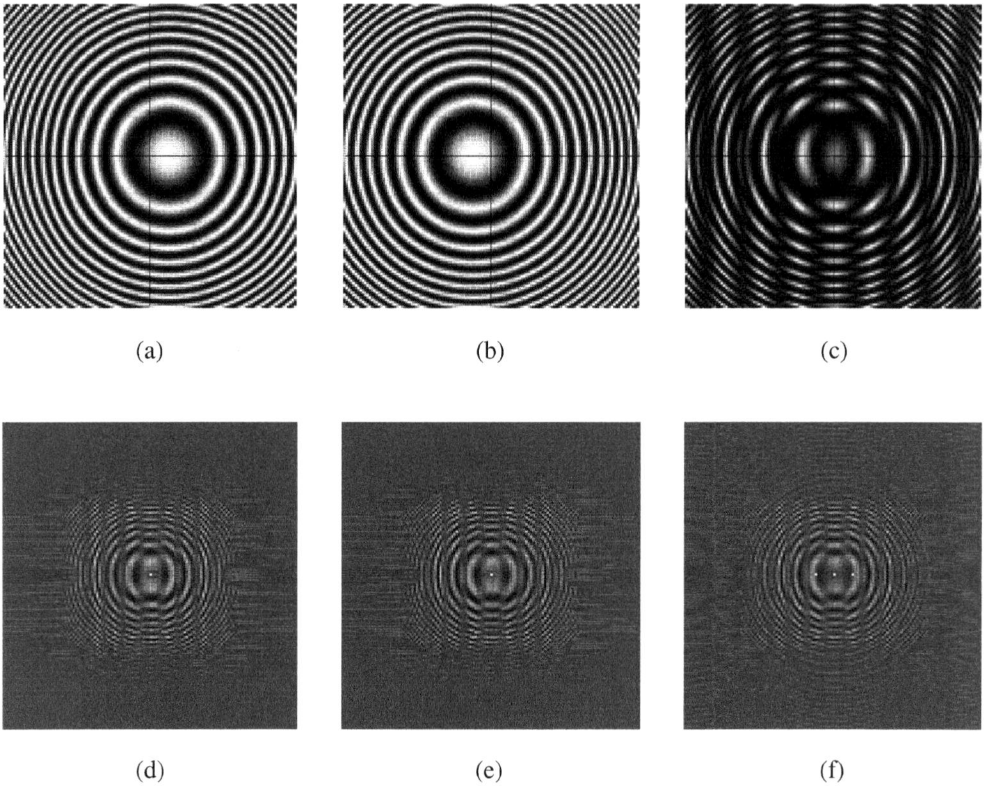


**Figure 10.31:** Two circular zone gratings with raised cosinusoidal periodic-profiles $\frac{1}{2}\cos(2\pi fr) + \frac{1}{2}$, which have been horizontally shifted from the origin to the points $x = 1$ (a) and $x = -1$ (b), and their superposition (c). Their respective spectra are shown (real parts only) in (d), (e) and their convolution (f). (Each of the spectra contains also a DC impulse due to the constant $\frac{1}{2}$ in the original gratings.) Note that the convolution (f) contains, in addition to the two zone gratings (d) and (e), two new elements: (1) A new pair of impulses, which is located inside the visibility circle and corresponds to the subtractive (1,-1)-moiré (which is clearly seen in the superposition (c) in the form of periodic, vertical bands); and (2), an outer zone grating which corresponds to the additive (1,1)-moiré (but which is not visible in the superposition (c)).

spectral components in the present case are 2D humps which extend throughout the entire $u,v$ plane, including the spectrum origin and the visibility circle which surrounds it. This means that several different moirés may be simultaneously visible in different locations of the superposition.

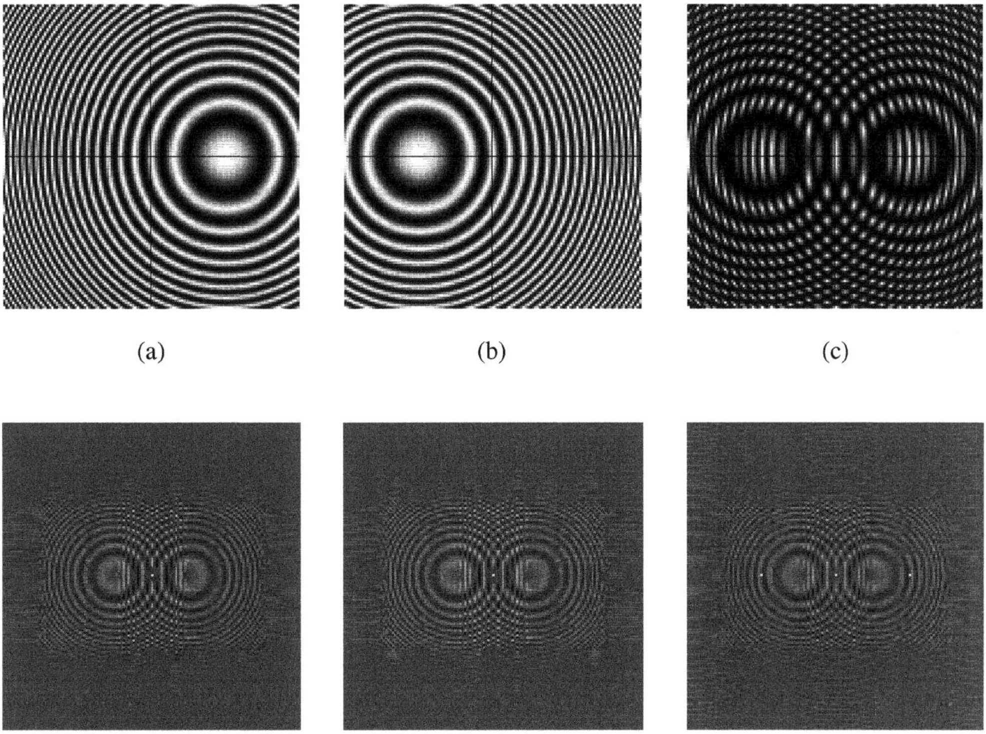

(a) (b) (c)

(d) (e) (f)

**Figure 10.32:** Same as Fig. 10.31, with a larger horizontal shift of $x = \pm 4$ instead of $x = \pm 1$. In this case the additive (1,1)-moiré can be seen in the superposition (c) in the form of a new zone grating around the origin; however, the subtractive (1,-1)-moiré, which consists of periodic, vertical bands, has a much higher frequency and is no longer visible in the superposition (note that its impulse pair in the spectrum convolution (f) is already beyond the visibility circle).

## 10.8 Periodic moirés in the superposition of non-periodic layers

As we have seen in Sec. 10.7, moiré patterns resulting from the superposition of non-periodic gratings are, in general, non-periodic. It is interesting, however, that particular cases can be designed in which the resulting moiré *is* periodic although the original layers are not. In terms of the spectral domain this gives the curious result that although the spectra of the original layers in such cases are non-impulsive or only partially impulsive, the moiré cluster obtained in their convolution is a purely impulsive comb. Two such examples are presented and analyzed in this section. The precise conditions under which such cases occur will be formally stated in Sec. 10.9.1.

**Example 10.17:** The superposition of two identical, laterally shifted parabolic gratings:

Consider the horizontally oriented parabolic grating given by:

$$r(x,y) = p(x-ay^2) = \sum_{n=-\infty}^{\infty} a_n \cos(2\pi nf(x-ay^2))$$

Let $r_1(x,y)$ and $r_2(x,y)$ be two identical instances of $r(x,y)$ which are vertically shifted by $+y_0$ and $-y_0$, respectively:

$$r_1(x,y) = r(x,y-y_0) = \sum_{m=-\infty}^{\infty} a^{(1)}_m \cos(2\pi mf[x-a(y-y_0)^2])$$

$$r_2(x,y) = r(x,y+y_0) = \sum_{n=-\infty}^{\infty} a^{(2)}_n \cos(2\pi nf[x-a(y+y_0)^2])$$

As shown in Fig. 10.33, the superposition $r_1(x,y)r_2(x,y)$ gives a periodic moiré effect in the form of horizontal straight bands. In order to understand this phenomenon, let us consider first the product of two shifted *cosinusoidal* parabolic gratings with the same frequencies $f$ and bending rate $a$. Using the known trigonometric identity $\cos\alpha\cos\beta = \frac{1}{2}[\cos(\alpha-\beta) + \cos(\alpha+\beta)]$ we have:

$$\cos(2\pi f[x-a(y-y_0)^2])\cos(2\pi f[x-a(y+y_0)^2])$$

$$= \tfrac{1}{2}\cos(2\pi f[x-a(y-y_0)^2 - x+a(y+y_0)^2]) + \tfrac{1}{2}\cos(2\pi f[x-a(y-y_0)^2 + x-a(y+y_0)^2])$$

$$= \tfrac{1}{2}\cos(2\pi f[-a(y^2-2y_0y+y_0^2) + a(y^2+2y_0y+y_0^2)])$$

$$+ \tfrac{1}{2}\cos(2\pi f[x-a(y^2-2y_0y+y_0^2) + x-a(y^2+2y_0y+y_0^2)])$$

$$= \tfrac{1}{2}\cos(2\pi fa\,4y_0y) + \tfrac{1}{2}\cos(2\pi f2[(x-ay_0^2)-ay^2]) \qquad (10.27)$$

This means that the product of the two shifted cosinusoidal parabolic gratings gives a sum of two cosinusoidal terms: the first term is a periodic cosinusoidal grating of horizontal lines (whose frequency $4fay_0$ depends on the shift $y_0$), and the second term is a horizontally oriented cosinusoidal parabolic grating (with frequency of $2f$). The spectrum of this product consists, therefore, of a pair of impulses located on the vertical $v$ axis at $v = \pm 4fay_0$, plus a pair of vertical line-impulses which are centered on the $u$ axis at $u = \pm 2f$. Now, if the vertical shifts $\pm y_0$ of the two original parabolic gratings are such that $|y_0| < \frac{1}{2a}$ it follows that $4fay_0 < 2f$, and the periodic cosine term has a smaller (and hence more visible) frequency. Otherwise, the parabolic cosinusoidal term of Eq. (10.27) has a smaller frequency and is more visible; it can be perceived in the superposition wherever its local period is larger than the local periods of the original superposed gratings.

Now, how is all this related to the spectrum convolution as depicted in Fig. 10.13?[28]

[28] Note that in the present case we are dealing with a product of pure cosines, and not with a product of raised cosines of the form $\frac{1}{2}\cos(...) + \frac{1}{2}$, as in Fig. 10.13; hence, the spectrum convolution here contains only the 4 terms (1,-1), (-1,1), (1,1) and (-1,-1), but not the terms (0,1), (0,-1), (1,0), (-1,0) and (0,0).

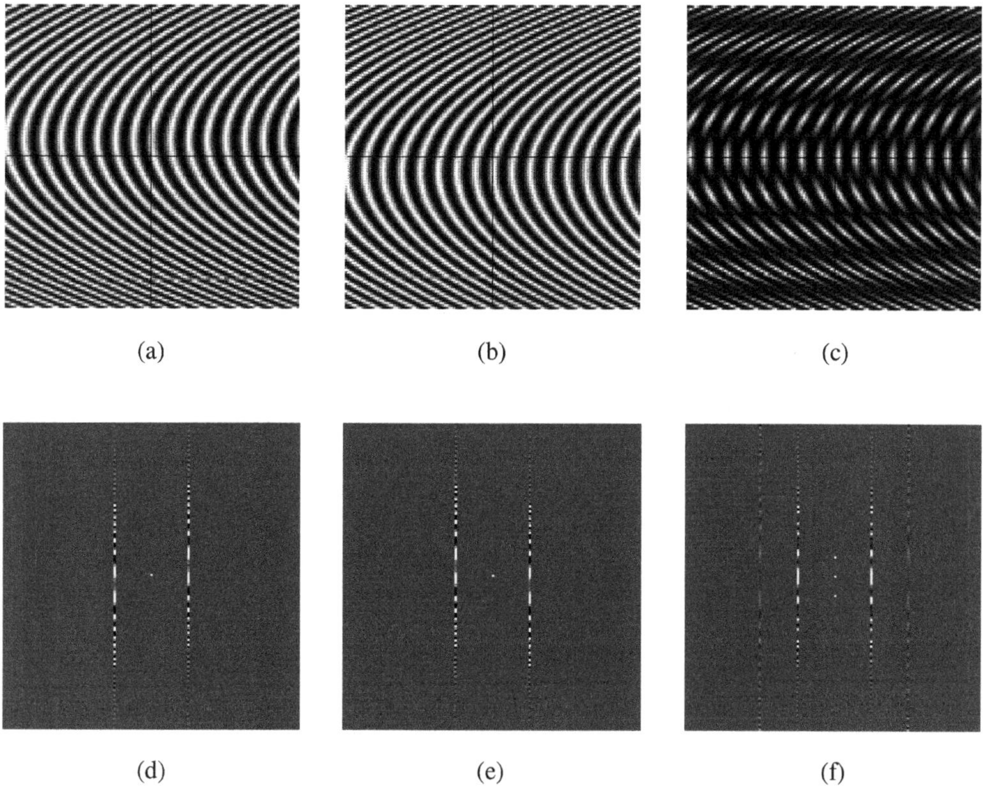


**Figure 10.33:** Two parabolic gratings with raised cosinusoidal periodic-profiles $\frac{1}{2}\cos(2\pi fr) + \frac{1}{2}$, which have been vertically shifted from the origin by $y = y_0$ in (a) and by $y = -y_0$ in (b), and their superposition (c). Their respective spectra are shown (real parts only) in (d), (e), and their convolution (f). (Each of the spectra contains also a DC impulse due to the constant $\frac{1}{2}$ in the original gratings.) Note that the convolution (f) contains two new elements: (1) A new pair of impulses, which is located inside the visibility circle and corresponds to the subtractive (1,-1)-moiré (which is clearly seen in the superposition (c) in the form of periodic, horizontal bands); and (2), an outer pair of line-impulses which corresponds to the additive (1,1)-moiré (but which is not visible in the superposition (c)).

We are dealing here with a limit case in which the superposition angle is $\theta = 0$. In this case the spectrum of each of the two original cosinusoidal parabolic gratings consists of a pair of vertical line-impulses centered on the $u$ axis at $u = \pm f$, but as a result of the shifts of $+y_0$ and $-y_0$ the amplitude of the line-impulses of the first spectrum is multiplied by $e^{-i2\pi y_0 v}$ (by virtue of the shift theorem) and the amplitude of the line-impulses of the second spectrum is multiplied by $e^{i2\pi y_0 v}$. Since all of the line-impulses here are vertical, the convolution operation does not give 2D humps, but rather 1D line-impulses. However, it turns out that the (1,-1)- and the (-1,1)-line-impulses of the convolution, which collapse together onto the vertical $v$ axis (both being centered on the spectrum origin) cancel each other almost everywhere and degenerate into a pair of point-impulses which is located on the $v$ axis at $v = \pm 4fay_0$; this impulse pair corresponds in the image domain to the first cosinusoidal term of Eq. (10.27). The (1,1)- and the (-1,-1)-line-impulses of the convolution, which are centered on the $u$ axis at $u = \pm 2f$, correspond in the image domain to the second term of Eq. (10.27), namely: the cosinusoidal parabolic moiré.

With this information at hand we may return now to the superposition of the shifted parabolic gratings $r_1(x,y)r_2(x,y)$. In order to simplify the derivation which follows, we prefer to use here for $r_1(x,y)$ and $r_2(x,y)$ their exponential Fourier series notation (which is more compact and easier to handle, and even more general than the cosine Fourier series notation since it also allows non-symmetric periodic-profiles):

$$r_1(x,y)r_2(x,y) = \Big(\sum_{m=-\infty}^{\infty} c^{(1)}_m\, e^{i2\pi mf(x - a(y-y_0)^2)}\Big)\Big(\sum_{n=-\infty}^{\infty} c^{(2)}_n\, e^{i2\pi nf(x - a(y+y_0)^2)}\Big)$$

$$= \sum_{m=-\infty}^{\infty}\sum_{n=-\infty}^{\infty} c^{(1)}_m c^{(2)}_n\, e^{i2\pi f(m[x - a(y-y_0)^2] + n[x - a(y+y_0)^2])}$$

The (1,-1)-partial sum of this double sum (which corresponds to the new (1,-1)-cluster in the spectrum convolution and to the (1,-1)-moiré in the image domain) is:

$$m_{1,-1}(x,y) = \sum_{n=-\infty}^{\infty} c^{(1)}_n c^{(2)}_{-n}\, e^{i2\pi nf([x - a(y-y_0)^2] - [x - a(y+y_0)^2])}$$

$$= \sum_{n=-\infty}^{\infty} c^{(1)}_n c^{(2)}_{-n}\, e^{i2\pi nf(-a(y^2 - 2y_0y + y_0^2) + a(y^2 + 2y_0y + y_0^2))}$$

$$= \sum_{n=-\infty}^{\infty} c^{(1)}_n c^{(2)}_{-n}\, e^{i2\pi nf(4ay_0y)}$$

Similarly, the (1,1)-partial sum of this double sum (which corresponds to the new (1,1)-cluster in the spectrum convolution and to the (1,1)-moiré in the image domain) is:

$$m_{1,1}(x,y) = \sum_{n=-\infty}^{\infty} c^{(1)}_n c^{(2)}_n\, e^{i2\pi nf([x - a(y-y_0)^2] + [x - a(y+y_0)^2])}$$

$$= \sum_{n=-\infty}^{\infty} c^{(1)}_n c^{(2)}_n\, e^{i2\pi nf(x - a(y^2 - 2y_0y + y_0^2) + x - a(y^2 + 2y_0y + y_0^2))}$$

$$= \sum_{n=-\infty}^{\infty} c^{(1)}_n c^{(2)}_n\, e^{i2\pi nf2((x - ay_0^2) - ay^2)}$$

These are, therefore, the Fourier decompositions of the (1,-1)- and (1,1)-moirés in the superposition. As we can see, the spectrum of the (1,-1)-moiré consists of a comb of impulses on the $v$ axis, which corresponds to the periodic horizontal bands of the (1,-1)-moiré; and the spectrum of the (1,1)-moiré consists of a comb of vertical line-impulses centered on the $u$ axis, which corresponds to the horizontally oriented parabolic grating of the (1,1)-moiré. As we have seen above, if $|y_0| < \frac{1}{2a}$ the periodic bands of the (1,-1)-moiré are more visible; otherwise the parabolic bands of the (1,1)-moiré are predominant in the superposition (wherever their local period is larger than the local periods of the superposed gratings). ■

**Remark 10.8:** Note that when we use the cosine Fourier series notation to express the original gratings, the subtractive (1,-1)-moiré and the additive (1,1)-moiré are obtained simultaneously. This can be seen, for example, in Eq. (10.27), or in the development based on the cosine notation at the end of Sec. 10.7.7 (note the "index acrobatics" that were required there to isolate the additive and the subtractive moirés separately). This happens since cosinusoidal terms in a product combine via the trigonometric identity $\cos\alpha\cos\beta = \frac{1}{2}[\cos(\alpha-\beta) + \cos(\alpha+\beta)]$. If we express the original gratings using the exponential Fourier series notation (which even allows for gratings with non-symmetric periodic-profiles), the exponential terms in the product combine via the identity $e^{ia}\, e^{ib} = e^{i(a+b)}$, and hence the expressions for the $(k_1,k_2)$-moiré and for the $(k_1,-k_2)$-moiré appear in the summation separately, and they can be extracted independently of each other. The advantage of the cosinusoidal notation is, however, that it lends itself more easily to graphic interpretation. ■

**Example 10.18:** The superposition of two identical, laterally shifted zone gratings:

It is a well known fact that the superposition of two laterally shifted circular zone gratings gives rise to a periodic moiré effect in the form of vertical straight bands, which are known as "Schuster fringes" (see for example Fig. 10.31 above, Fig. 1.11 in [Patorski93 p. 20] or Fig. 9a in [Harthong81 pp. 59-60]). In the present example we analyze and explain this fact and show that it is not only true for circular zone gratings but also for elliptic or hyperbolic zone gratings.

Consider the zone grating given, using the two-sided cosine Fourier series notation, by:

$$r(x,y) = p(ax^2 \pm by^2) = \sum_{n=-\infty}^{\infty} a_n \cos(2\pi nf(ax^2 \pm by^2))$$

Depending on the values of the coefficients $a$ and $b$ and on their signs, $r(x,y)$ represents a circular, an elliptic or a hyperbolic zone grating (see Example 10.7 in Sec. 10.3). Let $r_1(x,y)$ and $r_2(x,y)$ be two identical instances of $r(x,y)$ which are horizontally shifted by $+x_0$ and $-x_0$, respectively:

$$r_1(x,y) = r(x-x_0,y) = \sum_{m=-\infty}^{\infty} a^{(1)}_m \cos(2\pi mf[a(x-x_0)^2 \pm by^2])$$

$$r_2(x,y) = r(x+x_0,y) = \sum_{n=-\infty}^{\infty} a^{(2)}_n \cos(2\pi nf[a(x+x_0)^2 \pm by^2])$$

Again, as in the previous example, we consider first the product of two shifted *cosinusoidal* zone gratings with the same frequencies $f$ and coefficients $a$ and $b$. Using the known trigonometric identity $\cos\alpha\cos\beta = \frac{1}{2}[\cos(\alpha-\beta) + \cos(\alpha+\beta)]$ we have:

$$\cos(2\pi f[a(x-x_0)^2 \pm by^2])\cos(2\pi f[a(x+x_0)^2 \pm by^2]) =$$

$$= \tfrac{1}{2}\cos(2\pi f\,[a(x-x_0)^2 \pm by^2 - a(x+x_0)^2 \mp by^2])$$

$$+ \tfrac{1}{2}\cos(2\pi f\,[a(x-x_0)^2 \pm by^2 + a(x+x_0)^2 \pm by^2])$$

$$= \tfrac{1}{2}\cos(2\pi fa\,[x^2-2xx_0+x_0^2 - x^2-2xx_0-x_0^2])$$

$$+ \tfrac{1}{2}\cos(2\pi f\,[ax^2-2axx_0+ax_0^2 \pm by^2 + ax^2+2axx_0+ax_0^2 \pm by^2])$$

$$= \tfrac{1}{2}\cos(2\pi fa\,4x_0x) + \tfrac{1}{2}\cos(2\pi f\,2[ax^2 \pm by^2 + ax_0^2]) \tag{10.28}$$

This means that the product of the two shifted cosinusoidal zone gratings gives a sum of two cosinusoidal terms: the first term is a periodic cosinusoidal grating of vertical lines (whose frequency $4fax_0$ depends on the shift $x_0$), and the second term is a cosinusoidal zone grating (with basic frequency of $2f$) which is centered on the origin (see Appendix C.10). The spectrum of this product consists, therefore, of a pair of impulses located on the horizontal $u$ axis at $u = \pm 4fax_0$, which are added on top of a continuous background in the form of a sinusoidal zone grating (which is the spectrum of the cosinusoidal zone grating; see Appendix C.10). Now, if the horizontal shifts $\pm x_0$ of the original zone gratings are such that $|x_0| < \frac{1}{2a}$ it follows that $4fax_0 < 2f$, and the periodic cosine term has a smaller (and hence more visible) frequency. Otherwise, the centered cosinusoidal zone grating of Eq. (10.28) has a smaller basic frequency and is more visible; it can be perceived in the superposition wherever its local period is larger than the local periods of the superposed gratings.

We return now to the superposition of the shifted zone gratings $r_1(x,y)r_2(x,y)$. Like in the previous example, we prefer to use here for $r_1(x,y)$ and $r_2(x,y)$ the exponential Fourier series notation:

$$r_1(x,y)r_2(x,y) = \Big(\sum_{m=-\infty}^{\infty} c^{(1)}_m\, e^{i2\pi mf[a(x-x_0)^2 \pm by^2]}\Big)\Big(\sum_{n=-\infty}^{\infty} c^{(2)}_n\, e^{i2\pi nf[a(x+x_0)^2 \pm by^2]}\Big)$$

$$= \sum_{m=-\infty}^{\infty}\sum_{n=-\infty}^{\infty} c^{(1)}_m c^{(2)}_n\, e^{i2\pi f(m[a(x-x_0)^2 \pm by^2] + n[a(x+x_0)^2 \pm by^2])}$$

The (1,-1)-partial sum of this double sum (which corresponds to the new (1,-1)-cluster in the spectrum convolution and to the (1,-1)-moiré in the image domain) is:

$$m_{1,-1}(x,y) = \sum_{n=-\infty}^{\infty} c^{(1)}_n c^{(2)}_{-n}\, e^{i2\pi nf([a(x-x_0)^2 \pm by^2] - [a(x+x_0)^2 \pm by^2])}$$

$$= \sum_{n=-\infty}^{\infty} c^{(1)}_n c^{(2)}_{-n}\, e^{i2\pi nf(a(x^2 - 2x_0x + x_0^2) - a(x^2 + 2x_0x + x_0^2))}$$

$$= \sum_{n=-\infty}^{\infty} c^{(1)}_n c^{(2)}_{-n}\, e^{i2\pi nf(-4ax_0x)}$$

Similarly, the (1,1)-partial sum of this double sum (which corresponds to the new (1,1)-cluster in the spectrum convolution and to the (1,1)-moiré in the image domain) is:

$$m_{1,1}(x,y) = \sum_{n=-\infty}^{\infty} c^{(1)}_n c^{(2)}_n\, e^{i2\pi nf([a(x-x_0)^2 \pm by^2] + [a(x+x_0)^2 \pm by^2])}$$

$$= \sum_{n=-\infty}^{\infty} c^{(1)}_n c^{(2)}_n\, e^{i2\pi nf(a(x^2 - 2x_0x + x_0^2) \pm by^2 + a(x^2 + 2x_0x + x_0^2) \pm by^2)}$$

$$= \sum_{n=-\infty}^{\infty} c^{(1)}{}_n c^{(2)}{}_n \, e^{i2\pi n f 2(ax^2 \pm by^2 + ax_0{}^2)}$$

These are, therefore, the Fourier decompositions of the (1,-1)- and (1,1)-moirés in the superposition. As we can see, the spectrum of the (1,-1)-moiré consists of a comb of impulses on the $u$ axis, which corresponds to the periodic vertical bands of the (1,-1)-moiré; and the spectrum of the (1,1)-moiré consists of a comb of humps centered on the origin, which corresponds to the zone grating of the (1,1)-moiré. As we have seen above, if $|x_0| < \frac{1}{2a}$ the periodic bands of the (1,-1)-moiré are more visible, and otherwise it is the zone grating of the (1,1)-moiré which becomes predominant in the superposition. ■

## 10.9 Moiré analysis and synthesis in the superposition of curved, repetitive layers

In the present section we will see how the results which have been obtained in Chapter 4, namely: the analysis and the synthesis of moiré patterns between periodic layers, can be generalized to the case of repetitive, non-periodic layers, although the spectra in this case no longer consist of simple impulses but rather of more complex entities. In fact, we have already seen several such examples earlier in this chapter. For instance, Fig. 10.10 shows the (1,-1)-moiré which has been extracted from Fig. 10.9, the superposition of a parabolic grating and a straight grating. It may be instructive to compare Figs. 10.9 and 10.10 with their periodic counterparts, Figs. 2.5 and 4.2, whose spectra only contain the skeleton impulses of our line-impulses.

Note that a part of our generalization work was already done in Sec. 10.7.1, where we have generalized to the repetitive, non-periodic case the extraction of moirés from the layer superposition. It may be asked now whether this generalized moiré extraction in the image domain can be also interpreted in terms of the $T$-convolution theorem, as was the case in Chapter 4 for periodic layers. In other words: can Propositions 4.2, 4.3 and 4.5 of Chapter 4 be extended to the superposition of curvilinear, non-periodic layers, too?

The affirmative answer to this question has already been given by Harthong [Harthong81 pp. 30–33], using the theory of non-standard analysis. Here we will treat this subject from a different point of view, which is based on the Fourier approach.

### 10.9.1 The case of curvilinear gratings

We start by reformulating the $T$-convolution theorem (see Sec. 4.2) in terms of Fourier series [Zygmund68 p. 36]:

***T*-convolution theorem:** Let $p_1(x)$ and $p_2(x)$ be periodic functions of period $T$ integrable on a one-period interval $(0,T)$, and let their Fourier series representations be:

$$p_1(x) = \sum_{m=-\infty}^{\infty} c^{(1)}{}_m \, e^{i2\pi mx/T} \qquad\qquad p_2(x) = \sum_{n=-\infty}^{\infty} c^{(2)}{}_n \, e^{i2\pi nx/T}$$

Then their $T$-convolution:

$$h(x) = p_1(x) * p_2(x) = \tfrac{1}{T}\int_T p_1(x-x')\, p_2(x')\, dx'$$

is also periodic with the same period $T$, and its Fourier series representation is:

$$h(x) = \sum_{m=-\infty}^{\infty} c^{(1)}_m c^{(2)}_m\, e^{i2\pi mx/T} \qquad \blacksquare$$

Now, let $r_1(x,y)$ and $r_2(x,y)$ be two curvilinear gratings given by:

$$r_1(x,y) = p_1(g_1(x,y)) = \sum_{m=-\infty}^{\infty} c^{(1)}_m\, e^{i2\pi m g_1(x,y)}$$

$$r_2(x,y) = p_2(g_2(x,y)) = \sum_{n=-\infty}^{\infty} c^{(2)}_n\, e^{i2\pi n g_2(x,y)}$$

where $p_1(x') = \sum_{m=-\infty}^{\infty} c^{(1)}_m\, e^{i2\pi mx'}$ and $p_2(x') = \sum_{n=-\infty}^{\infty} c^{(2)}_n\, e^{i2\pi nx'}$ are the 1-fold periodic-profiles of the gratings $r_1(x,y)$ and $r_2(x,y)$, normalized to the period $\mathbf{T} = (1,0)$, and $g_1(x,y)$ and $g_2(x,y)$ are the bending functions which transform $p_1(x')$ and $p_2(x')$ into the gratings $r_1(x,y)$ and $r_2(x,y)$. (See also the note about layer normalizations in Sec. C.16 of Appendix C).

As we already know (see Sec. 10.7.1), the superposition of the two gratings is expressed by the product:

$$r_1(x,y)\; r_2(x,y) = \Big(\sum_{m=-\infty}^{\infty} c^{(1)}_m\, e^{i2\pi m g_1(x,y)}\Big)\Big(\sum_{n=-\infty}^{\infty} c^{(2)}_n\, e^{i2\pi n g_2(x,y)}\Big)$$

$$= \sum_{m=-\infty}^{\infty}\sum_{n=-\infty}^{\infty} c^{(1)}_m c^{(2)}_n\, e^{i2\pi(m g_1(x,y) + n g_2(x,y))} \qquad (10.29)$$

and the partial sum which corresponds to the (1,-1)-moiré consists of all the terms of this double sum whose indices are $n$ and $-n$, namely (see Eq. (10.19)):

$$m_{1,-1}(x,y) = \sum_{n=-\infty}^{\infty} c^{(1)}_n c^{(2)}_{-n}\, e^{i2\pi n(g_1(x,y) - g_2(x,y))}$$

Let us denote by $p_{1,-1}(x')$ the periodic profile of $m_{1,-1}(x,y)$ (normalized to the same period $\mathbf{T} = (1,0)$; see the note about layer normalizations in Sec. C.16 of Appendix C):

$$p_{1,-1}(x') = \sum_{n=-\infty}^{\infty} c^{(1)}_n c^{(2)}_{-n}\, e^{i2\pi nx'}$$

We see, therefore, by the $T$-convolution theorem that $p_{1,-1}(x')$ is simply the $T$-convolution:

$$p_{1,-1}(x') = p_1(x') * p_2(-x') \qquad (10.30)$$

where $x' = g_1(x,y) - g_2(x,y)$ is the bending function which brings $p_{1,-1}(x')$ into the geometric layout of the (1,-1)-moiré. In other words, we obtain the following generalization of Proposition 4.2:

**Proposition 10.2:** The (1,-1)-moiré $m_{1,-1}(\mathbf{x})$ in the superposition of two curvilinear gratings $r_1(\mathbf{x}) = p_1(g_1(\mathbf{x}))$ and $r_2(\mathbf{x}) = p_2(g_2(\mathbf{x}))$ is given by $m_{1,-1}(\mathbf{x}) = p_{1,-1}(g_{1,-1}(\mathbf{x}))$, where:

(1) $p_{1,-1}(x')$, the normalized periodic-profile of the (1,-1)-moiré, is the $T$-convolution of the normalized periodic-profiles of the original gratings:

$$p_{1,-1}(x') = p_1(x') * p_2(-x');$$ [29]

(2) $g_{1,-1}(\mathbf{x})$, the bending function which brings $p_{1,-1}(x')$ into the (1,-1)-moiré $m_{1,-1}(\mathbf{x})$, is given by:

$$g_{1,-1}(\mathbf{x}) = g_1(\mathbf{x}) - g_2(\mathbf{x}). \quad \blacksquare$$

Using less formal language, we may formulate this result as follows:

**Proposition 10.3:** The (1,-1)-moiré $m_{1,-1}(x,y)$ generated in the superposition of the two curvilinear gratings $r_1(x,y) = p_1(g_1(x,y))$ and $r_2(x,y) = p_2(g_2(x,y))$ can be seen from the image-domain point of view as the result of a 3-stage process:

(1) Normalization of the original curvilinear gratings by replacing in each of them $g_i(x,y)$ with $x'$ (i.e., by undoing in each of them the effect of the bending function), in order to straighten them into uncurved, normalized periodic gratings having identical periods $\mathbf{T} = (1,0)$. (See also Sec. C.16 in Appendix C.)

(2) $T$-convolution of these normalized periodic line-gratings. This gives the uncurved, normalized periodic-profile of the (1,-1)-moiré, with the same period $\mathbf{T} = (1,0)$.

(3) Bending this normalized periodic-profile of the moiré into the actual curvilinear geometric layout of the moiré, by replacing $x'$ with $g_1(x,y) - g_2(x,y)$ (i.e., by applying the non-linear bending function $x' = g_1(x,y) - g_2(x,y)$). $\blacksquare$

More generally, we have seen in Sec. 10.7.1 that the partial sum of (10.29) which corresponds to the $(k_1,k_2)$-moiré consists of all the terms of the double sum whose indices are $nk_1$ and $nk_2$, namely (see Eq. (10.18)):

$$m_{k_1,k_2}(x,y) = \sum_{n=-\infty}^{\infty} c^{(1)}_{nk_1} c^{(2)}_{nk_2} e^{i2\pi n(k_1 g_1(x,y) + k_2 g_2(x,y))}$$

If we denote by $p_{k_1,k_2}(x')$ the periodic-profile of $m_{k_1,k_2}(x,y)$ (normalized to the period $\mathbf{T} = (1,0)$):

$$p_{k_1,k_2}(x') = \sum_{n=-\infty}^{\infty} c^{(1)}_{nk_1} c^{(2)}_{nk_2} e^{i2\pi nx'}$$

then we see by the $T$-convolution theorem that $p_{k_1,k_2}(x')$ is simply the $T$-convolution:

$$p_{k_1,k_2}(x') = \mathrm{sub}_{k_1}(p_1(x')) * \mathrm{sub}_{k_2}(p_2(x')) \qquad (10.31)$$

where $\mathrm{sub}_k(p(x'))$ is the $k$-sub-Fourier series of $p(x')$, i.e., the periodic function (with period $\mathbf{T} = (1,0)$) whose Fourier series contains only every $k$-th coefficient from the Fourier series $p(x') = \sum_{n=-\infty}^{\infty} c_n e^{i2\pi nx'}$:

$$\mathrm{sub}_k(p(x')) = \sum_{n=-\infty}^{\infty} d_n e^{i2\pi nx'} \qquad \text{with:} \quad d_n = c_{kn}$$

and $x' = k_1 g_1(x,y) + k_2 g_2(x,y)$ is the bending function which brings $p_{k_1,k_2}(x')$ into the geometric layout of the $(k_1,k_2)$-moiré. In other words, we obtain the following generalization of Proposition 4.3:

---

[29] As for the minus signs in the $T$-convolution see Footnote 4 in Sec. 4.2.

**The fundamental moiré theorem (for the superposition of two curved gratings):** The $(k_1,k_2)$-moiré $m_{k_1,k_2}(\mathbf{x})$ in the superposition of two gratings $r_1(\mathbf{x}) = p_1(g_1(\mathbf{x}))$ and $r_2(\mathbf{x}) = p_2(g_2(\mathbf{x}))$ is given by $m_{k_1,k_2}(\mathbf{x}) = p_{k_1,k_2}(g_{k_1,k_2}(\mathbf{x}))$, where:

(1) $p_{k_1,k_2}(x')$, the normalized periodic-profile of the $(k_1,k_2)$-moiré, is given by:

$$p_{k_1,k_2}(x') = \mathrm{sub}_{k_1}(p_1(x')) * \mathrm{sub}_{k_2}(p_2(x'))$$

(2) $g_{k_1,k_2}(\mathbf{x})$, the bending function which brings $p_{k_1,k_2}(x')$ into the $(k_1,k_2)$-moiré $m_{k_1,k_2}(\mathbf{x})$, is given by:

$$g_{k_1,k_2}(\mathbf{x}) = k_1 g_1(\mathbf{x}) + k_2 g_2(\mathbf{x}).$$

In other words: the bending function $g_{k_1,k_2}(\mathbf{x})$ of the $(k_1,k_2)$-moiré is a weighted sum of the bending functions of the individual gratings, where the weighting coefficients are the moiré indices $k_i$. ■

Using less formal language, we may reformulate this theorem as follows:

**Proposition 10.4:** The $(k_1,k_2)$-moiré $m_{k_1,k_2}(x,y)$ generated in the superposition of the two curvilinear gratings $r_1(x,y) = p_1(g_1(x,y))$ and $r_2(x,y) = p_2(g_2(x,y))$ can be seen from the image-domain point of view as the result of a 3-stage process:

(1) Normalization of the original curvilinear gratings by, in each of them, replacing $g_i(x,y)$ with $x'$ (i.e., by undoing in each of them the effect of the bending function), in order to straighten them into uncurved, normalized periodic gratings having identical periods $\mathbf{T} = (1,0)$. (See also Sec. C.16 in Appendix C.)

(2) $T$-convolution of the $k_1$-sub-Fourier series of the first normalized grating with the $k_2$-sub-Fourier series of the second normalized grating. This gives the uncurved, normalized periodic-profile of the $(k_1,k_2)$-moiré, with the same period $\mathbf{T} = (1,0)$.

(3) Bending the normalized periodic-profile of the moiré into the actual curvilinear geometric layout of the moiré, by replacing $x'$ with $k_1 g_1(x,y) + k_2 g_2(x,y)$ (i.e., by applying the non-linear bending function $x' = k_1 g_1(x,y) + k_2 g_2(x,y)$). ■

Note that Propositions 4.2 and 4.3 of Chapter 4 are, indeed, particular cases of Propositions 10.3 and 10.4, in which $g_1(x,y)$ and $g_2(x,y)$ are linear functions, namely: $g_1(x,y) = u_1 x + v_1 y$, $g_2(x,y) = u_2 x + v_2 y$. In this case the gratings $r_1(x,y) = p_1(g_1(x,y))$ and $r_2(x,y) = p_2(g_2(x,y))$ are, indeed, straight, periodic gratings whose frequencies are given by $\mathbf{f}_1 = (u_1,v_1)$ and $\mathbf{f}_2 = (u_2,v_2)$, respectively.

**Example 10.19:** As an illustration to the fundamental moiré theorem, let us find the explicit expression of the (1,-1)-moiré $m_{1,-1}(x,y)$ in the superposition of a cosinusoidal straight grating and a cosinusoidal circular grating that are given by:

$$r_1(x,y) = \cos(2\pi f_1 x) \qquad \text{i.e.:} \quad p_1(x') = \cos(2\pi x'), \quad g_1(x,y) = f_1 x$$

$$r_2(x,y) = \cos(2\pi f_2 \sqrt{x^2+y^2}) \quad \text{i.e.:} \quad p_2(x') = \cos(2\pi x'), \quad g_2(x,y) = f_2\sqrt{x^2+y^2}$$

According to part (1) of the fundamental moiré theorem the normalized periodic-profile of the (1,-1)-moiré is given by:

$$p_{1,-1}(x') = p_1(x') * p_2(-x')$$
$$= \cos(2\pi x') * \cos(2\pi x')$$
$$= \tfrac{1}{2}\cos(2\pi x')$$

(This result is most easily obtained by using the $T$-convolution theorem, or by multiplying the spectra of $p_1(x')$ and $p_2(-x')$ and taking the inverse Fourier transform of the result.)

Now, according to part (2) of the fundamental moiré theorem the bending function of the (1,-1)-moiré is:

$$g_{1,-1}(x,y) = g_1(x,y) - g_2(x,y)$$
$$= f_1 x - f_2\sqrt{x^2 + y^2}$$

Therefore, the isolated (1,-1)-moiré $m_{1,-1}(x,y)$ is given by:

$$m_{1,-1}(x,y) = p_{1,-1}(g_{1,-1}(x,y))$$
$$= \tfrac{1}{2}\cos(2\pi[f_1 x - f_2\sqrt{x^2 + y^2}]) = \tfrac{1}{2}\cos(2\pi[f_2\sqrt{x^2 + y^2} - f_1 x])$$

Similarly, the isolated (1,1)-moiré $m_{1,1}(x,y)$ in the same superposition is given by:

$$m_{1,1}(x,y) = p_{1,1}(g_{1,1}(x,y))$$
$$= \tfrac{1}{2}\cos(2\pi[f_2\sqrt{x^2 + y^2} + f_1 x])$$

The moirés $m_{1,-1}(x,y)$ and $m_{1,1}(x,y)$ have been discussed at length in Sec. 10.7.5, and they are shown in Fig. 10.20(a),(b) for the case of $f_2 < f_1$. As we can see in the figure, the additive (1,1)-moiré $m_{1,1}(x,y)$ consists of the moiré curves to the left of the origin, while the subtractive (1,-1)-moiré $m_{1,-1}(x,y)$ consists of the curves to the right of the origin. Further explanation on these functions can be found in [Amidror98a pp. 911–912].

Note that the Fourier series development of the moirés $m_{1,-1}(x,y)$ and $m_{1,1}(x,y)$ could be also obtained directly from Eq. (10.18) or from its cosine analog. ■

An interesting consequence of part (2) of the fundamental moiré theorem is that in order to synthesize a $(k_1,k_2)$-moiré whose geometric layout is given by a certain desired function $g(x,y)$, all that we have to do is to choose two original layers whose bending functions $g_1(x,y)$ and $g_2(x,y)$ satisfy the condition:

$$k_1 g_1(x,y) + k_2 g_2(x,y) = g(x,y) \qquad (10.32)$$

In the case of a (1,-1)-moiré this condition is simplified into:[30]

$$g_1(x,y) - g_2(x,y) = g(x,y) \qquad (10.33)$$

[30] This particular case has already been given in [Lohman67]; further examples involving this case can be found in [Burch77].

In particular, as we have seen in Sec. 10.8, it may also happen that a superposition gives a periodic moiré even when the original layers are curved. Using condition (10.32) we can say now exactly when the $(k_1,k_2)$-moiré in the superposition of two curvilinear gratings is periodic: this occurs *iff* the bending function $g_{k_1,k_2}(x,y) = k_1 g_1(x,y) + k_2 g_2(x,y)$ is affine, namely:

$$k_1 g_1(x,y) + k_2 g_2(x,y) = ax + by + c \qquad (10.34)$$

In the case of a (1,-1)-moiré this condition becomes:

$$g_1(x,y) - g_2(x,y) = ax + by + c \qquad (10.35)$$

**Example 10.20:** A periodic (1,-1)-moiré that is generated by the lateral shift of two identical curvilinear layers on top of each other:

This kind of situation occurs when the bending function $g(x,y)$ (which is common to both layers) happens to have the property:

$$\forall x_1 \quad g(x+x_1,y) - g(x,y) = a_0 x + b_0 y + c_0 \qquad (10.36)$$

or its vertical analog:

$$\forall y_1 \quad g(x,y+y_1) - g(x,y) = a_0 x + b_0 y + c_0 \qquad (10.37)$$

or a combination of both. Note that if we consider the surfaces defined over the $x,y$ plane by $z = g(x,y)$ and by its shifted copy, conditions (10.36) or (10.37) mean that the difference between these two surfaces gives a plane, for any shift $x_1$ or $y_1$.

We have already met such cases in Examples 10.17 and 10.18 in Sec. 10.8. For instance, in the case of Example 10.18 (the superposition of two identical, laterally shifted zone gratings, whose bending function is $g(x,y) = ax^2 \pm by^2$) we have both:

$$\begin{aligned} \forall x_1 \quad g(x+x_1,y) - g(x,y) &= [a(x+x_1)^2 \pm by^2] - [ax^2 \pm by^2] \\ &= a(x^2 + 2xx_1 + x_1^2 - x^2) \\ &= 2ax_1x + ax_1^2 \end{aligned}$$

and:

$$\begin{aligned} \forall y_1 \quad g(x,y+y_1) - g(x,y) &= [ax^2 \pm b(y+y_1)^2] - [ax^2 \pm by^2] \\ &= \pm b(y^2 + 2yy_1 + y_1^2 - y^2) \\ &= \pm 2by_1y \pm by_1^2 \end{aligned}$$

This means that in this case both horizontal and vertical layer shifts (or any combinations thereof) give in the superposition a periodic (1,-1)-moiré.

Note that (1,-1)-moirés obtained in cases that satisfy Eqs. (10.36) or Eq. (10.37) remain periodic for *any* horizontal (or vertical) shift between the original layers. The period of the moiré bands increases as the shifts tend to 0, until a singular state is reached when the two layers precisely coincide. But when the shifts are increased the period of the moiré bands becomes smaller and smaller, until they finally completely disappear to the eye. ■

**Example 10.21:** A periodic (1,-1)-moiré which is generated by the rotation of two identical curvilinear layers on top of each other:

This kind of situation occurs when the bending function $g(x,y)$ (which is common to both layers) happens to have the following property:

$$\forall\theta \quad g(x\cos\theta + y\sin\theta,\ y\cos\theta - x\sin\theta) - g(x,y) = a_0x + b_0y + c_0 \tag{10.38}$$

or equivalently, in terms of polar coordinates:

$$\forall\theta \quad g(r,\theta) - g(r,0) = a_0r\cos\theta + b_0r\sin\theta + c_0 \tag{10.39}$$

Geometrically this condition means that the difference between the surface defined by $z = g(x,y)$ and the surface defined by its rotated copy gives a plane, for any rotation $\theta$.

What types of functions $g(x,y)$ satisfy this condition? Two trivial solutions are:

(a) All functions of the form $g(x,y) = ax + by + c$. In this case the difference surface is obviously a plane. However, such functions are not an interesting solution, since they do not correspond to curvilinear gratings but rather to straight, periodic gratings, whose moirés are periodic anyway.

(b) All the circular functions, like $g(x,y) = x^2+y^2$, $g(x,y) = e^{-(x^2+y^2)}$, etc. In this case the difference surface is, of course, the identical-zero plane, namely: the $x,y$ plane itself. However, these functions are not an interesting solution, either, since the rotation of a circular grating on top of itself does not produce a visible moiré (see, for example, Fig. 10.23).

More interesting solutions can be obtained by linear combinations of functions of types (a) and (b), like: $g(x,y) = e^{-(x^2+y^2)} + ax + by + c$, etc. In such cases the difference surface is a plane, and the curvilinear grating $r(x,y) = p(g(x,y))$ has, indeed, the required property: its rotation on top of a copy of itself gives a (1,-1)-moiré consisting of periodic, straight bands. This is illustrated in Fig. 10.34 for the case of $g(x,y) = x - e^{-(x^2+y^2)/4}$.

Note that (1,-1)-moirés obtained in such cases remain periodic for *any* rotation $\theta$ between the original gratings. The period of the moiré bands increases as $\theta$ tends to 0°, until a singular state is reached when the two layers precisely coincide. But when $\theta$ increases the period of the moiré bands becomes smaller, until they finally completely disappear to the eye. Note, however, that any shifts between the two superposed gratings may destroy the periodicity of the moiré, as shown in Figs. 10.34(c),(d). ■

**Example 10.22:** When the bending function $g(x,y)$ of the two superposed curvilinear gratings does not satisfy either of the conditions (10.36)–(10.39), we can still "force" the superposed layers to give a periodic moiré by slightly modifying one or even both of them. For example, if we slightly modify the bending function of the first layer into

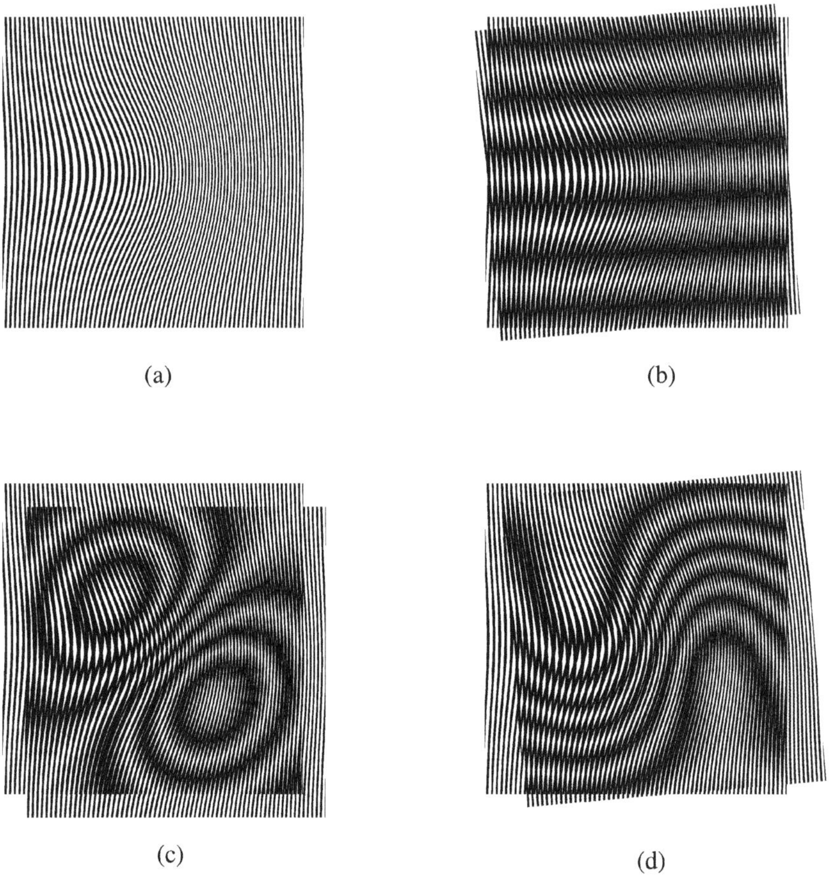


**Figure 10.34:** An example of a periodic (1,-1)-moiré which is generated by the rotation of two identical curvilinear gratings on top of one another. (a) A curvilinear grating whose bending function is $g(x,y) = x - e^{-(x^2+y^2)/4}$. (b) The superposition of this curvilinear grating with a rotated copy of itself gives a periodic (1,-1)-moiré, whose period and orientation depend on the rotation angle. Note that any shift between the two superposed gratings destroys the periodicity of the moiré; this is illustrated in (c): pure shift, and in (d): rotation and shift.

$g(x,y) + x/8$, the difference $g_1(x,y) - g_2(x,y)$ now becomes $x/8$ and therefore, according to condition (10.35), the (1,-1)-moiré obtained in their superposition consists of periodic vertical bands (see Fig. 10.35). However, unlike in the two previous examples, the (1,-1)-moiré in this case is periodic only when the two layers are superposed center on center, without any shifts or rotations, and its period is fixed and cannot be modified by layer shifts or rotations. The slightest shift or rotation between the two layers will destroy the periodicity of this moiré, as shown in Figs. 10.35(b),(c). ■

**Remark 10.9:** The geometric layout of the curvilinear $(k_1,k_2)$-moiré, i.e., the locus of the centerlines of its curvilinear corrugations in the $x,y$ plane, is determined by the bending function $g_{k_1,k_2}(\mathbf{x})$. In other words, the reflectance (or transmittance) function of the isolated $(k_1,k_2)$-moiré is constant along the curves $g_{k_1,k_2}(\mathbf{x}) = const.$ We will see in Sec. 11.2.2 that a general idea about the geometric layout of the $(k_1,k_2)$-moiré can be already obtained from the first harmonic term of the generalized Fourier series development of $m_{k_1,k_2}(x,y)$, $\cos(2\pi g_{k_1,k_2}(x,y))$, by drawing the locus of its maxima, namely, the curves where the cosine equals 1, $g_{k_1,k_2}(x,y) = m,\ m \in \mathbb{Z}$. ■

The generalization of Propositions 10.3 and 10.4 and of the fundamental moiré theorem to any $(k_1,...,k_m)$-moiré between $m$ superposed curvilinear gratings is straightforward.

### 10.9.2 The case of curved dot-screens

The $(k_1,k_2,k_3,k_4)$-moiré generated between two superposed dot-screens (see Sec. 4.3) is a case of particular interest. Because of its special importance, this case will now be analyzed separately, although it is already covered by the above generalization with $m = 4$.

We start by reformulating the 2D $T$-convolution theorem (see Sec. 4.3) in terms of Fourier series:

**2D $T$-convolution theorem:** Let $p_1(x,y)$ and $p_2(x,y)$ be doubly periodic functions of period $T_x$, $T_y$ integrable on a one-period interval $(0 \le x \le T_x,\ 0 \le y \le T_y)$, and let their Fourier series representations be:

$$p_1(x,y) = \sum_{m=-\infty}^{\infty} \sum_{n=-\infty}^{\infty} c^{(1)}{}_{m,n}\, e^{i2\pi(mx/T_x + ny/T_y)} \qquad p_2(x,y) = \sum_{m=-\infty}^{\infty} \sum_{n=-\infty}^{\infty} c^{(2)}{}_{m,n}\, e^{i2\pi(mx/T_x + ny/T_y)}$$

Then their $T$-convolution:

$$h(x,y) = p_1(x,y) ** p_2(x,y) = \frac{1}{T_xT_y} \iint_{T_xT_y} p_1(x-x',y-y')\, p_2(x',y')\, dx'dy'$$

is also periodic with the same periods $T_x$, $T_y$, and its Fourier series representation is:

$$h(x,y) = \sum_{m=-\infty}^{\infty} \sum_{n=-\infty}^{\infty} c^{(1)}{}_{m,n} c^{(2)}{}_{m,n}\, e^{i2\pi(mx/T_x + ny/T_y)} \qquad ■$$

Using this Fourier series formalism we now derive the counterpart of Proposition 4.5 for the case of curved screens:

Let $r_1(x,y)$ and $r_2(x,y)$ be curved dot-screens. We have, therefore:

$$r_1(x,y) = p_1(g_1(x,y),g_2(x,y)) = \sum_{n_1=-\infty}^{\infty} \sum_{n_2=-\infty}^{\infty} c^{(1)}{}_{n_1,n_2}\, e^{i2\pi(n_1g_1(x,y)+n_2g_2(x,y))}$$

$$r_2(x,y) = p_2(g_3(x,y),g_4(x,y)) = \sum_{n_3=-\infty}^{\infty} \sum_{n_4=-\infty}^{\infty} c^{(2)}{}_{n_3,n_4}\, e^{i2\pi(n_3g_3(x,y)+n_4g_4(x,y))}$$

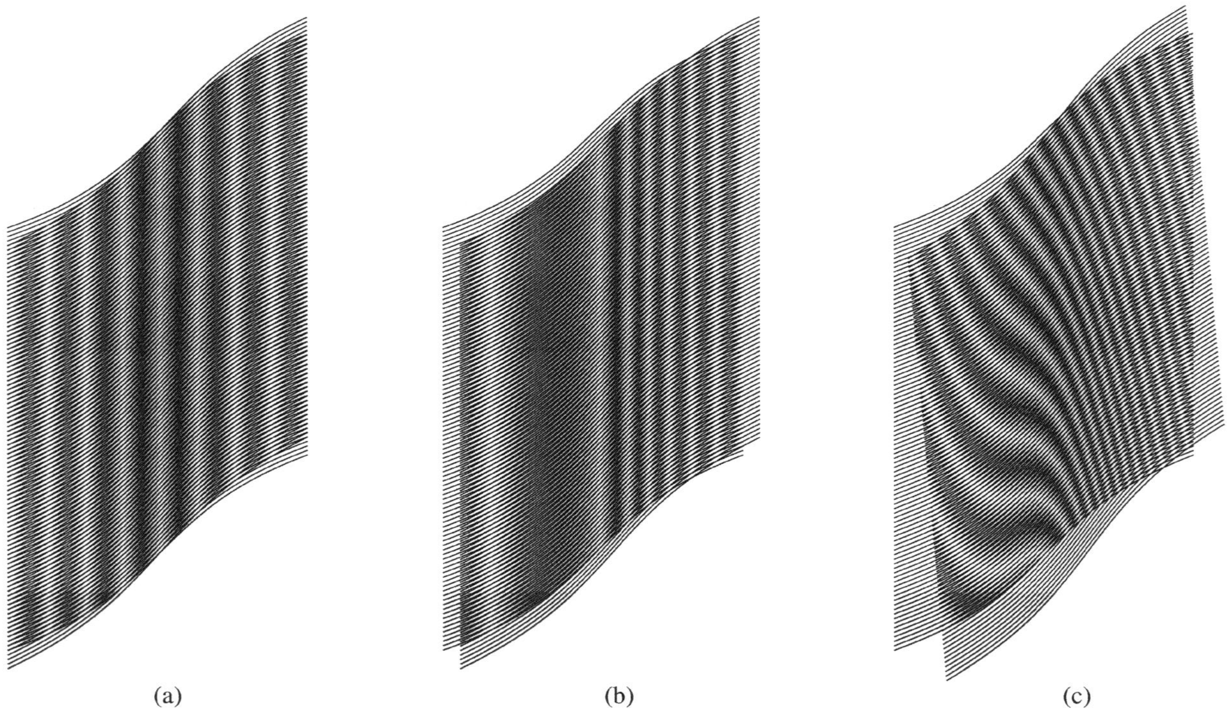

**Figure 10.35:** Superposition of two curvilinear gratings whose bending functions are $g_1(x,y) = \arg\sinh(x) + x/8$, $g_2(x,y) = \arg\sinh(x)$ so that $g_1(x,y) - g_2(x,y) = x/8$. The (1,-1)-moiré obtained in their superposition consists of periodic vertical bands as shown in (a). However, this moiré is periodic only when the two layers are superposed center on center, and the slightest shift or rotation between the two layers destroys the periodicity of the moiré, as shown in (b) and (c). Note that the moiré bands look darker in the center of each drawing; this happens because the curvilinear gratings were drawn here with a constant linewidth (compare with the correct, varying linewidths in Fig. 10.2(b); see also Remark 10.3 in Sec. 10.2).

where $p_1(x',y') = \sum_{n_1=-\infty}^{\infty} \sum_{n_2=-\infty}^{\infty} c^{(1)}_{n_1,n_2} e^{i2\pi(n_1x'+n_2y')}$ and $p_2(x',y') = \sum_{n_3=-\infty}^{\infty} \sum_{n_4=-\infty}^{\infty} c^{(2)}_{n_3,n_4} e^{i2\pi(n_3x'+n_4y')}$ are the periodic-profiles of the dot-screens $r_1(x,y)$ and $r_2(x,y)$, normalized to a 2D period of $T_{x'} = T_{y'} = 1$ (see Sec. C.16 in Appendix C), and where:

$$\begin{pmatrix} x' \\ y' \end{pmatrix} = \begin{pmatrix} g_1(x,y) \\ g_2(x,y) \end{pmatrix} \quad \text{and} \quad \begin{pmatrix} x' \\ y' \end{pmatrix} = \begin{pmatrix} g_3(x,y) \\ g_4(x,y) \end{pmatrix}$$

(or in vector notation: $\mathbf{x}' = \mathbf{g}_1(\mathbf{x})$ and $\mathbf{x}' = \mathbf{g}_2(\mathbf{x})$) are the non-linear coordinate transformations which transform $p_1(x',y')$ and $p_2(x',y')$ into the curved dot-screens $r_1(x,y)$ and $r_2(x,y)$, respectively.

The superposition of the two dot-screens is expressed by the product (10.20):

$$r_1(x,y)\, r_2(x,y)$$

$$= \sum_{n_1=-\infty}^{\infty} \sum_{n_2=-\infty}^{\infty} \sum_{n_3=-\infty}^{\infty} \sum_{n_4=-\infty}^{\infty} c^{(1)}_{n_1,n_2} c^{(2)}_{n_3,n_4} e^{i2\pi(n_1g_1(x,y)+n_2g_2(x,y)+n_3g_3(x,y)+n_4g_4(x,y))} \tag{10.40}$$

As we have seen in Sec. 10.7.1, the partial sum which corresponds to the (1,0,-1,0)-moiré consists of all the terms of this double sum whose indices are $m$, $n$, $-m$, $-n$, namely:

$$m_{1,0,-1,0}(x,y) = \sum_{m=-\infty}^{\infty} \sum_{n=-\infty}^{\infty} c^{(1)}_{m,n} c^{(2)}_{-m,-n} e^{i2\pi(m[g_1(x,y)-g_3(x,y)] + n[g_2(x,y)-g_4(x,y)])}$$

Let us denote by $p_{1,0,-1,0}(x',y')$ the periodic profile of $m_{1,0,-1,0}(x,y)$ (normalized to the 2D period $T_{x'} = T_{y'} = 1$; see the note about layer normalizations in Sec. C.16 of Appendix C):

$$p_{1,0,-1,0}(x',y') = \sum_{m=-\infty}^{\infty} \sum_{n=-\infty}^{\infty} c^{(1)}_{m,n} c^{(2)}_{-m,-n} e^{i2\pi(mx'+ny')}$$

We see, therefore, by the 2D $T$-convolution theorem that $p_{1,0,-1,0}(x',y')$ is simply the 2D $T$-convolution:

$$p_{1,0,-1,0}(x',y') = p_1(x',y') ** p_2(-x',-y') \tag{10.41}$$

where $\begin{pmatrix} x' \\ y' \end{pmatrix} = \begin{pmatrix} g_1(x,y) - g_3(x,y) \\ g_2(x,y) - g_4(x,y) \end{pmatrix}$, or in vector notation: $\mathbf{x}' = \mathbf{g}_1(\mathbf{x}) - \mathbf{g}_2(\mathbf{x})$, is the 2D coordinate transformation which brings $p_{1,0,-1,0}(x',y')$ into the geometric layout of the (1,0,-1,0)-moiré. In other words, we obtained the following result:

**Proposition 10.5:** The (1,0,-1,0)-moiré $m_{1,0,-1,0}(\mathbf{x})$ in the superposition of two curved dot-screens $r_1(\mathbf{x}) = p_1(\mathbf{g}_1(\mathbf{x}))$ and $r_2(\mathbf{x}) = p_2(\mathbf{g}_2(\mathbf{x}))$ is given by $m_{1,0,-1,0}(\mathbf{x}) = p_{1,0,-1,0}(\mathbf{g}_{1,0,-1,0}(\mathbf{x}))$, where:

(1) $p_{1,0,-1,0}(\mathbf{x}')$, the normalized periodic-profile of the (1,0,-1,0)-moiré, is the $T$-convolution of the normalized periodic-profiles of the original dot-screens:

$$p_{1,0,-1,0}(\mathbf{x}') = p_1(\mathbf{x}') ** p_2(-\mathbf{x}')$$

(2) $\mathbf{g}_{1,0,-1,0}(\mathbf{x})$, the bending transformation of the (1,0,-1,0)-moiré, is given by:

$$\mathbf{g}_{1,0,-1,0}(\mathbf{x}) = \mathbf{g}_1(\mathbf{x}) - \mathbf{g}_2(\mathbf{x}). \quad \blacksquare$$

More generally, we have seen in Sec. 10.7.1 that the partial sum of (10.40) which corresponds to the $(k_1,k_2,k_3,k_4)$-moiré consists of all the terms of the quadruple sum whose indices are: $mk_1-nk_2$, $mk_2+nk_1$, $mk_3-nk_4$ and $mk_4+nk_3$, namely:

$$m_{k_1,k_2,k_3,k_4}(x,y)$$

$$= \sum_{m=-\infty}^{\infty} \sum_{n=-\infty}^{\infty} c^{(1)}_{mk_1-nk_2,\, mk_2+nk_1}\, c^{(2)}_{mk_3-nk_4,\, mk_4+nk_3}$$
$$\times\ e^{i2\pi([mk_1-nk_2]g_1(x,y) + [mk_2+nk_1]g_2(x,y) + [mk_3-nk_4]g_3(x,y) + [mk_4+nk_3]g_4(x,y))}$$

$$= \sum_{m=-\infty}^{\infty} \sum_{n=-\infty}^{\infty} c^{(1)}_{mk_1-nk_2,\, mk_2+nk_1}\, c^{(2)}_{mk_3-nk_4,\, mk_4+nk_3}$$
$$\times\ e^{i2\pi(m[k_1g_1(x,y)+k_2g_2(x,y)+k_3g_3(x,y)+k_4g_4(x,y)] + n[-k_2g_1(x,y)+k_1g_2(x,y)-k_4g_3(x,y)+k_3g_4(x,y)])}$$

If we denote by $p_{k_1,k_2,k_3,k_4}(x',y')$ the periodic-profile of $m_{k_1,k_2,k_3,k_4}(x,y)$ (normalized to the 2D period $T_{x'} = T_{y'} = 1$; see Sec. C.16 in Appendix C):

$$p_{k_1,k_2,k_3,k_4}(x',y') = \sum_{m=-\infty}^{\infty} \sum_{n=-\infty}^{\infty} c^{(1)}_{mk_1-nk_2,\, mk_2+nk_1}\, c^{(2)}_{mk_3-nk_4,\, mk_4+nk_3}\, e^{i2\pi(mx'+ny')}$$

then we see by the 2D $T$-convolution theorem that $p_{k_1,k_2,k_3,k_4}(x',y')$ is simply the 2D $T$-convolution:

$$p_{k_1,k_2,k_3,k_4}(x',y') = \mathrm{sub}_{k_1,k_2}(p_1(x',y')) ** \mathrm{sub}_{k_3,k_4}(p_2(x',y')) \tag{10.42}$$

where $\mathrm{sub}_{r,s}(p(x',y'))$ is the $(r,s)$-sub-Fourier series of $p(x',y')$, i.e., the periodic function (with 2D period $T_{x'} = T_{y'} = 1$) whose Fourier series contains only every $(r,s)$-th coefficient from the 2D Fourier series $p(x',y') = \sum_{m=-\infty}^{\infty} \sum_{n=-\infty}^{\infty} c_{m,n}\, e^{i2\pi(mx'+ny')}$:[31]

$$\mathrm{sub}_{r,s}(p(x',y')) = \sum_{m=-\infty}^{\infty} \sum_{n=-\infty}^{\infty} d_{m,n}\, e^{i2\pi(mx'+ny')} \qquad \text{with:} \qquad d_{m,n} = c_{mr-ns,\, ms+nr}$$

and where $\begin{pmatrix} x' \\ y' \end{pmatrix} = \begin{pmatrix} k_1g_1(x,y) + k_2g_2(x,y) + k_3g_3(x,y) + k_4g_4(x,y) \\ -k_2g_1(x,y) + k_1g_2(x,y) - k_4g_3(x,y) + k_3g_4(x,y) \end{pmatrix}$ is the 2D coordinate transformation which brings $p_{k_1,k_2,k_3,k_4}(x',y')$ into the geometric layout of the $(k_1,k_2,k_3,k_4)$-moiré. Note that this bending transformation can be rearranged in the form:

$$\begin{pmatrix} x' \\ y' \end{pmatrix} = \begin{pmatrix} k_1 & k_2 \\ -k_2 & k_1 \end{pmatrix} \begin{pmatrix} g_1(x,y) \\ g_2(x,y) \end{pmatrix} + \begin{pmatrix} k_3 & k_4 \\ -k_4 & k_3 \end{pmatrix} \begin{pmatrix} g_3(x,y) \\ g_4(x,y) \end{pmatrix}$$

from which we obtain its vector form:

$$\mathbf{x}' = \mathrm{K}_1\mathbf{g}_1(\mathbf{x}) + \mathrm{K}_2\mathbf{g}_2(\mathbf{x})$$

where $\mathrm{K}_1$ and $\mathrm{K}_2$ denote the matrices $\begin{pmatrix} k_1 & k_2 \\ -k_2 & k_1 \end{pmatrix}$ and $\begin{pmatrix} k_3 & k_4 \\ -k_4 & k_3 \end{pmatrix}$, respectively.[32] We therefore obtain the following generalization of Proposition 4.5:

---

[31] Compare with the definition of a $(r,s)$-subnailbed in Sec. 4.3.

[32] Note that these matrices correspond to *similarity* transformations, i.e. linear transformations that consist of rotation and uniform scaling (see Proposition C.3 in Sec. C.15.3 of Appendix C).

**The fundamental moiré theorem (for the superposition of two curved screens):** The $(k_1,k_2,k_3,k_4)$-moiré $m_{k_1,k_2,k_3,k_4}(\mathbf{x})$ in the superposition of two curved dot-screens $r_1(\mathbf{x}) = p_1(\mathbf{g}_1(\mathbf{x}))$ and $r_2(\mathbf{x}) = p_2(\mathbf{g}_2(\mathbf{x}))$ is given by $m_{k_1,k_2,k_3,k_4}(\mathbf{x}) = p_{k_1,k_2,k_3,k_4}(\mathbf{g}_{k_1,k_2,k_3,k_4}(\mathbf{x}))$, where:

(1) $p_{k_1,k_2,k_3,k_4}(\mathbf{x}')$, the normalized periodic-profile of the $(k_1,k_2,k_3,k_4)$-moiré, is given by:

$$p_{k_1,k_2,k_3,k_4}(\mathbf{x}') = \text{sub}_{k_1,k_2}(p_1(\mathbf{x}')) ** \text{sub}_{k_3,k_4}(p_2(\mathbf{x}'))$$

(2) $\mathbf{g}_{k_1,k_2,k_3,k_4}(\mathbf{x})$, the bending transformation of the $(k_1,k_2,k_3,k_4)$-moiré, is given by:

$$\mathbf{g}_{k_1,k_2,k_3,k_4}(\mathbf{x}) = \mathrm{K}_1\mathbf{g}_1(\mathbf{x}) + \mathrm{K}_2\mathbf{g}_2(\mathbf{x}). \quad \blacksquare$$

The generalization of this theorem to any $(k_1,...,k_{2m})$-moiré in the superposition of $m$ curved screens is straightforward.

We obtain, therefore, as an immediate consequence of this theorem:

**Proposition 10.6:** The geometric layout of the moiré (which is determined by its bending transformation) and the periodic-profile of the moiré are completely independent of each other: the geometric layout of the moiré is influenced only by the geometric layout of the superposed layers, and the periodic-profile of the moiré depends only on the periodic-profiles of the superposed layers. ■

Using a less formal language than in the theorem, we can now state the counterpart of Proposition 4.5 for the superposition of two curved dot-screens as follows:

**Proposition 10.7:** Let $r_1(x,y)$ and $r_2(x,y)$ be two curved dot-screens, which are obtained from two normalized two-fold periodic dot-screens $p_1(x',y')$ and $p_2(x',y')$ by the non-linear coordinate transformations $\mathbf{g}_1(x,y)$ and $\mathbf{g}_2(x,y)$, namely:

$$\begin{pmatrix} x' \\ y' \end{pmatrix} = \begin{pmatrix} g_1(x,y) \\ g_2(x,y) \end{pmatrix} \quad \text{and} \quad \begin{pmatrix} x' \\ y' \end{pmatrix} = \begin{pmatrix} g_3(x,y) \\ g_4(x,y) \end{pmatrix}$$

respectively. The $(k_1,k_2,k_3,k_4)$-moiré $m_{k_1,k_2,k_3,k_4}(x,y)$ generated in the superposition of these curved dot-screens can be seen from the image-domain point of view as the result of a 3-stage process:

(1) Normalization of the original curved dot-screens by, in each of them, replacing $(g_i(x,y), g_{i+1}(x,y))$ with $(x',y')$ (i.e., by undoing in each of them the coordinate transformation), in order to straighten them into the uncurved, normalized 2D periodic dot-screens $p_1(x',y')$ and $p_2(x',y')$ having unit periods $T_{x'} = T_{y'} = 1$.

(2) $T$-convolution of the 2D $(k_1,k_2)$-sub-Fourier series of the first normalized dot-screen with the 2D $(k_3,k_4)$-sub-Fourier series of the second normalized dot-screen. This gives the uncurved, normalized periodic-profile of the $(k_1,k_2,k_3,k_4)$-moiré, with the same unit periods $T_{x'} = T_{y'} = 1$. (See also Sec. C.16 in Appendix C.)

(3) Bending the normalized periodic-profile of the moiré into the actual curved geometric layout of the moiré, by replacing $(x',y')$ with $(k_1g_1(x,y) + k_2g_2(x,y) + k_3g_3(x,y) + k_4g_4(x,y),\ -k_2g_1(x,y) + k_1g_2(x,y) - k_4g_3(x,y) + k_3g_4(x,y))$, i.e., by applying the non-linear coordinate transformation $\begin{pmatrix} x' \\ y' \end{pmatrix} = \begin{pmatrix} k_1g_1(x,y) + k_2g_2(x,y) + k_3g_3(x,y) + k_4g_4(x,y) \\ -k_2g_1(x,y) + k_1g_2(x,y) - k_4g_3(x,y) + k_3g_4(x,y) \end{pmatrix}$. ■

It is interesting to note, as we have already seen for curvilinear gratings, that in certain cases the coordinate transformation in step (3) may give a 2D periodic moiré even when the original layers are curved, i.e., when the coordinate transformations $g_i(x,y)$ of the individual layers are not linear. In the particular case of the (1,0,-1,0)-moiré (see Proposition 10.5), where the coordinate transformation of step (3) is simplified into:

$$\begin{pmatrix} x' \\ y' \end{pmatrix} = \begin{pmatrix} g_1(x,y) - g_3(x,y) \\ g_2(x,y) - g_4(x,y) \end{pmatrix} \tag{10.43}$$

this happens *iff* the coordinate transformation (10.43) is an affine transformation, namely:

$$\begin{aligned} g_1(x,y) - g_3(x,y) &= a_1x + b_1y + c_1 \\ g_2(x,y) - g_4(x,y) &= a_2x + b_2y + c_2 \end{aligned} \tag{10.44}$$

**Example 10.23:** Let $p_1(x',y')$ be a periodic dot-screen whose period consists of the digit "1", and let $r_1(x,y)$ be the curved dot-screen obtained by applying to $p_1(x',y')$ the coordinate transformation:

$$\begin{pmatrix} x' \\ y' \end{pmatrix} = \begin{pmatrix} 2xy \\ y^2 - x^2 \end{pmatrix}$$

If we laterally shift on top of this curved dot-screen a second dot-screen that was subject to the same coordinate transformation, we obtain a two-fold periodic moiré since:

$$\forall x_0 \qquad 2(x+x_0)y - 2xy = 2x_0y\ ,$$

$$[y^2-(x+x_0)^2] - [y^2-x^2] = -2x_0x - x_0^2$$

and similarly:

$$\forall y_0 \qquad 2x(y+y_0) - 2xy = 2y_0x\ ,$$

$$[(y+y_0)^2-x^2] - [y^2-x^2] = 2y_0y + y_0^2$$

(compare with conditions (10.36)–(10.37) above).

Now, if the second layer consists of small pinholes, we obtain in the superposition a periodic (1,0,-1,0)-moiré whose normalized periodic-profile is, according to Proposition 10.5, a $T$-convolution of the shape of "1" with the pinhole, which gives again a "1"-shaped periodic-profile (see Fig. 4.5). We obtain therefore a periodic (1,0,-1,0)-moiré whose period consists of a magnified digit "1", exactly as in Sec. 4.4, even though the two superposed screens are not periodic. This is illustrated in Fig. 10.36. ■

More generally, in order to synthesize a $(k_1,k_2,k_3,k_4)$-moiré whose geometric layout is given by the two independent functions $g^{(1)}(x,y)$ and $g^{(2)}(x,y)$, all that we have to do is to choose two original layers whose bending transformations $\mathbf{g}_1(x,y) = (g_1(x,y), g_2(x,y))$ and $\mathbf{g}_2(x,y) = (g_3(x,y), g_4(x,y))$ satisfy the condition:

$$\begin{aligned} k_1g_1(x,y) + k_2g_2(x,y) + k_3g_3(x,y) + k_4g_4(x,y) &= g^{(1)}(x,y) \\ -k_2g_1(x,y) + k_1g_2(x,y) - k_4g_3(x,y) + k_3g_4(x,y) &= g^{(2)}(x,y) \end{aligned}$$

In the case of a (1,0,-1,0)-moiré this condition is simplified into:

$$g_1(x,y) - g_3(x,y) = g^{(1)}(x,y)$$

$$g_2(x,y) - g_4(x,y) = g^{(2)}(x,y)$$

The periodic-profile of the synthesized moiré will be determined by the periodic-profiles of the superposed layers, in accordance with the first part of the fundamental moiré theorem.

**Proposition 10.8:** Suppose that we are given two periodic layers which have been obtained from their respective normalized unit-period counterparts by the layer transformations $\mathbf{x}' = \mathbf{g}_1(\mathbf{x})$ and $\mathbf{x}' = \mathbf{g}_2(\mathbf{x})$, respectively. The periodic $(k_1,k_2,k_3,k_4)$-moiré between these layers has, therefore, the geometric layout $\mathbf{g}_{k_1,k_2,k_3,k_4}(\mathbf{x}) = \mathrm{K}_1\mathbf{g}_1(\mathbf{x}) + \mathrm{K}_2\mathbf{g}_2(\mathbf{x})$. Now, if we apply to the two given non-normalized layers the transformations $\mathbf{x} = \mathbf{h}_1(\mathbf{s})$ and $\mathbf{x} = \mathbf{h}_2(\mathbf{s})$, respectively, the geometric layout of the $(k_1,k_2,k_3,k_4)$-moiré becomes $\mathrm{K}_1\mathbf{g}_1(\mathbf{h}_1(\mathbf{s})) + \mathrm{K}_2\mathbf{g}_2(\mathbf{h}_2(\mathbf{s}))$, because now the transformations of the individual layers with respect to their unit-period normalized counterparts are $\mathbf{g}_1(\mathbf{h}_1(\mathbf{s}))$ and $\mathbf{g}_2(\mathbf{h}_2(\mathbf{s}))$, respectively. Note that this moiré layout is *not* obtained by applying the transformation $\mathbf{x} = \mathbf{h}_{k_1,k_2,k_3,k_4}(\mathbf{s}) = \mathrm{K}_1\mathbf{h}_1(\mathbf{s}) + \mathrm{K}_2\mathbf{h}_2(\mathbf{s})$ to the original moiré layout $\mathbf{g}_{k_1,k_2,k_3,k_4}(\mathbf{x}) = \mathrm{K}_1\mathbf{g}_1(\mathbf{x}) + \mathrm{K}_2\mathbf{g}_2(\mathbf{x})$. ■

It should be mentioned, finally, that although our results have been presented here for the case of two superposed layers, their generalization to any $m$ superposed curved screens is straightforward.

## 10.10 Local frequencies and singular states in curved, repetitive layers

As we have already seen in Remarks 10.3 and 10.4 (Sec. 10.2), the application of a non-linear transformation (coordinate change) $\mathbf{g}(x,y)$ to a periodic function $p(x,y)$ such as a line-grating, line-grid, dot-screen, etc., causes local variations in the size or in the orientation of its period, depending on the location of the point $(x,y)$ in the $x,y$ plane. Thus,

**Figure 10.36:** (*See following pages.*) A dot-screen superposition illustrating Example 10.23. (a) The curved dot-screen $r_1(x,y)$ consisting of distorted "1"s. (b) The curved dot-screen $r_2(x,y)$ consisting of small pinholes. The two layers have been distorted by the same non-linear coordinate transformation $\mathbf{g}(x,y) = (2xy,\ y^2 - x^2)$. As shown in (c), the (1,0,-1,0)-moiré generated when $r_2(x,y)$ is laterally shifted on top of $r_1(x,y)$ is purely periodic, although both $r_1(x,y)$ and $r_2(x,y)$ are not periodic; this periodic moiré consists of a screen of magnified "1"s, whose period and orientation depend on the shift. Note that rotations destroy the periodicity of the moiré, as illustrated in (d).

(a)

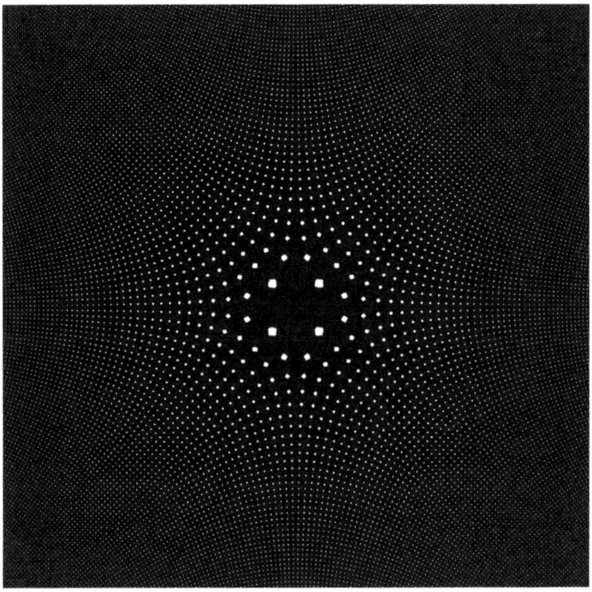

(b)

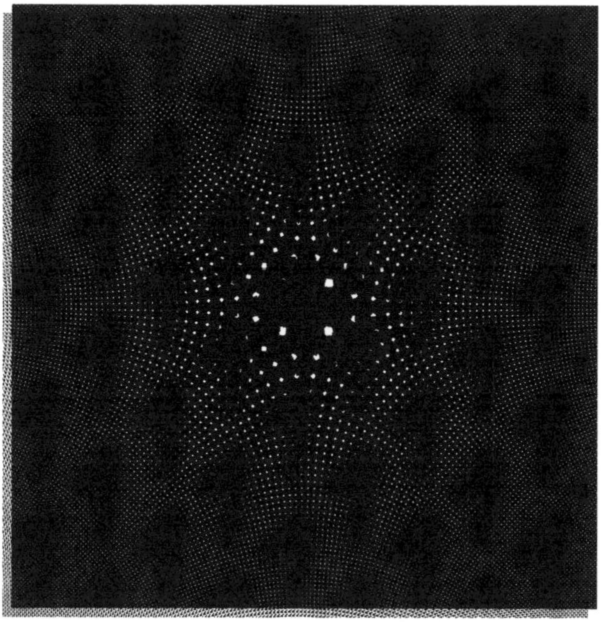

(c)

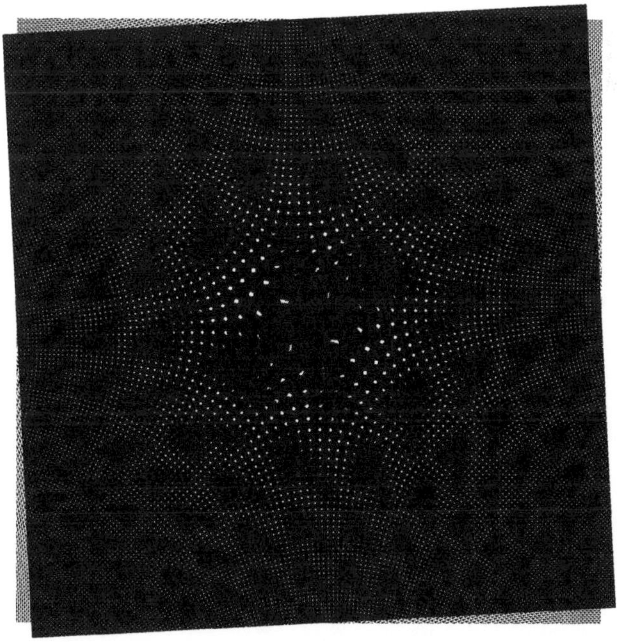

(d)

unlike in the periodic function $p(x,y)$, the *local period* (and the *local frequency*) of the transformed, curved function $r(x,y)$ are not constant and they vary according to the location in the $x,y$ plane.

It will be shown later, in Sec. 11.4, that the local frequency $\mathbf{f}(x,y)$ of a curvilinear grating $r(x,y) = p(g(x,y))$ is given by the *gradient* of the bending function $g(x,y)$:

$$\mathbf{f}(x,y) = (\frac{\partial}{\partial x}g(x,y), \frac{\partial}{\partial y}g(x,y))$$

Similarly, in the case of curved screens the local frequencies $\mathbf{f}_1(x,y)$ and $\mathbf{f}_2(x,y)$ of the curved screen $r(x,y) = p(g_1(x,y),g_2(x,y))$ are the gradients of the bending functions $g_1(x,y)$ and $g_2(x,y)$:

$$\mathbf{f}_1(x,y) = (\frac{\partial}{\partial x}g_1(x,y), \frac{\partial}{\partial y}g_1(x,y))$$
$$\mathbf{f}_2(x,y) = (\frac{\partial}{\partial x}g_2(x,y), \frac{\partial}{\partial y}g_2(x,y))$$

As we can see, the local frequency $\mathbf{f}(x,y)$ of a curvilinear grating $r(x,y)$ at the point $(x,y)$ is not a scalar function of $x$ and $y$, but a mapping of $\mathbb{R}^2$ onto itself: $\mathbf{f}$: $\mathbb{R}^2 \rightarrow \mathbb{R}^2$; we denote it by a boldface letter $\mathbf{f}$ since the value it returns, $\mathbf{f}(x,y)$, is a vector. This is a natural generalization of the notion of frequency in a periodic function $p(x,y)$, where the repetitivity properties of the function are constant throughout the $x,y$ plane, and are expressed by a constant vector $\mathbf{f}$, the frequency of $p(x,y)$.

The notion of local frequency will allow us to introduce here the generalization of singular moirés to repetitive, non-periodic cases. As we have seen in Sec. 2.9, a singular state of the $(k_1,...,k_m)$-moiré in the superposition of $m$ periodic gratings occurs when the frequency vectors of the superposed gratings satisfy:

$$k_1\mathbf{f}_1 + ... + k_m\mathbf{f}_m = \mathbf{0}$$

Similarly, we will see in Sec. 11.4 that a local singular state in the case of the curvilinear $(k_1,...,k_m)$-moiré occurs in the $x,y$ plane in the points $(x,y)$ where the *local* frequency vectors of the $m$ superposed curvilinear grating satisfy:

$$k_1\mathbf{f}_1(x,y) + ... + k_m\mathbf{f}_m(x,y) = \mathbf{0}$$

namely: $$(\frac{\partial}{\partial x}[k_1g_1(x,y) + ... + k_mg_m(x,y)] , \frac{\partial}{\partial y}[k_1g_1(x,y) + ... + k_mg_m(x,y)]) = \mathbf{0}$$

or: $$(\frac{\partial}{\partial x}g_{k_1,...,k_m}(x,y) , \frac{\partial}{\partial y}g_{k_1,...,k_m}(x,y)) = (0,0)$$

We have therefore:

**Proposition 10.8:** A singular state of the curvilinear $(k_1,...,k_m)$-moiré between $m$ curvilinear gratings occurs in the $x,y$ plane in the points $(x,y)$ where the gradient of the bending function of the $(k_1,...,k_m)$-moiré is zero. ■

This result can be generalized to the case of a superposition of curved screens as follows:

**Proposition 10.9:** A singular state of the curved $(k_1,...,k_{2m})$-moiré between $m$ curved screens occurs in the $x,y$ plane in the points $(x,y)$ where the Jacobian of the bending transformation $\mathbf{g}_{k_1,...,k_{2m}}(x,y) = \begin{pmatrix} g_{1,k_1,...,k_{2m}}(x,y) \\ g_{2,k_1,...,k_{2m}}(x,y) \end{pmatrix}$ of the $(k_1,...,k_{2m})$-moiré is zero:

$$\begin{vmatrix} \frac{\partial}{\partial x} g_{1,k_1,...,k_{2m}}(x,y) & \frac{\partial}{\partial y} g_{1,k_1,...,k_{2m}}(x,y) \\ \frac{\partial}{\partial x} g_{2,k_1,...,k_{2m}}(x,y) & \frac{\partial}{\partial y} g_{2,k_1,...,k_{2m}}(x,y) \end{vmatrix} = 0. \quad \blacksquare$$

## 10.11 Moirés in the superposition of screen gradations

In the preceding sections we assumed that each of the repetitive, non-periodic layers in the superposition belonged to the category of coordinate-transformed structures, i.e., that it has been obtained by a gradual variation in its *geometric layout*, while its *periodic profile* remained unchanged throughout. In other words: we were dealing with curved layers defined by $r(\mathbf{x}) = p(\mathbf{g}(\mathbf{x}))$, where $\mathbf{g}(\mathbf{x})$ is a bending transformation that is applied to the coordinates of the periodic-profile $p(\mathbf{x})$.

However, repetitive non-periodic layers can be also obtained by varying the periodic-profile of the layer, leaving its geometric layout unchanged throughout the $x,y$ plane (i.e., taking $\mathbf{g}(\mathbf{x}) = \mathbf{x}$). Such layers belong to the category of profile-transformed structures, which is represented by $r(\mathbf{x}) = t(p(\mathbf{x}))$ (see Remark 10.2). In the one-fold periodic case this corresponds to a *grating gradation*, i.e., a straight grating whose basic periodicity is constant, and only its periodic-profile varies (say, from a square-wave form to a triangular form, or from thinner to wider linewidths within the fixed period). In the two-fold periodic case this corresponds to a *screen gradation*, i.e., a dot-screen whose basic periodicity is constant, but its dot sizes (and/or its dot shapes) vary depending on the location in the $x,y$ plane (see Figs. 4.1, 4.4).

Screen gradations have the special property that their frequency remains unchanged throughout the image, and only the internal detail (the dot size and shape) within each "period" or cell is varied. But since the dot size and shape only affect in the spectrum of a dot-screen the impulse amplitudes and not the impulse locations, it is intuitively clear that the spectral representation of uniform screen gradations ("wedges") is still impulsive (its spectrum support being the lattice which corresponds to the cell periodicity), and only the complex *impulse amplitudes* are influenced. The same is true also for the case of uniform grating gradations, where the spectrum support is a comb. This result is explained in more detail in Sec. C.11 of Appendix C.

It follows, therefore, that uniform screen gradations and uniform grating gradations can be treated, both in the image domain and in the spectral domain, just like usual dot-screens or gratings. The moiré obtained in the superposition of uniform screen gradations has the same periodicity as the moiré in the superposition of uniform dot-screens with the same respective periods, and only the amplitude of the moiré (its intensity profile) varies throughout the superposed area: At each point of the superposition the moiré profile is locally identical to the uniform moiré profile obtained between uniform dot-screens having the "local screen dots" of the screen gradations at that point.[33] We have tacitly used this result in previous chapters where we illustrated our results by means of screen gradations in order to show in a single figure the intensity profile of the moiré between dot-screens of various dot-sizes or shapes (see, for example, Figs. 4.1, 4.4, etc. in Chapter 4).

Finally, it should be mentioned that there exists also a third category of repetitive, non-periodic layers which are combinations of curved layers and of gradations; they are obtained by letting both the geometric layout and the periodic-profile vary depending on the location in the $x,y$ plane. An example is given in Remark 10.3 of Sec. 10.2, and illustrated in Fig. 10.35. Such cases, too, are included within the scope of our generalized Fourier-based approach. Note that the fact that the fundamental moiré theorem remains valid in such cases has already been shown (for the case of the (1,0,-1,0)-moiré between two screens) in [Harthong81 pp. 68–71], using the theory of non-standard analysis.

## 10.12 Concluding remarks

In the present chapter we generalized our Fourier-based moiré theory to the case of moirés between repetitive, non-periodic layers. The fact that our spectral approach remains valid in non-periodic cases too is, indeed, quite significant: it confirms that the theory which has been developed in the preceding chapters is not just a nice coincidence that occurs when the structures involved happen to be periodic, but indeed, a more general theory which covers the full range of repetitive structures, periodic or not.

However, this generalized spectral approach does not provide the ultimate answer to all possible needs. For example, real-world moiré cases in various fields of engineering and technology may prove to be too complex to analyze by spectral means, and they may need to be treated using simpler methods (see Chapter 11). But the importance of the spectral approach cannot be overestimated, since understanding the fundamental properties of layer superpositions both in the image domain and in the spectral domain clearly provides a new, important insight into the nature of the various phenomena involved. In fact, as we will see in Chapter 11, the simple classical methods that are often used to investigate real-world cases are encompassed by the spectral approach, and the information they provide is therefore only partial to the full information that is provided by the spectral approach.

[33] Note the similarity of this reasoning with that of Sec. 11.4 below, where curved moirés are considered as locally straight and locally periodic at each point $(x,y)$ of the superposition.

# PROBLEMS

10-1. *Superposition of a parabolic grating and a straight grating.* Consider the superposition of a parabolic grating $r_1(x,y)$ and a straight periodic grating $r_2(x,y)$, with frequencies $f_1 = f_2$ and orientations $\theta_1 = 0$ and $\theta_2$ (see Fig. 10.9 and Example 10.13).

(a) What happens in the superposition while the straight grating is gradually rotated on top of the parabolic grating between $\theta_2 = 0...90°$? Within what range of angles is the (1,-1)-moiré visible? Within what range of angles is the (1,-2)-moiré visible? Can you also observe higher order moirés?

(b) At what angle $\theta_2$ is each of these moirés singular? Draw a sketch of the corresponding spectrum for each of these cases.

10-2. Repeat the previous problem but this time using a straight periodic grating $r_2(x,y)$ whose frequency is $f_2 > f_1$. How will your answers to (a) and (b) above be modified? (See also Fig. 10.12.)

10-3. Repeat the same problem once more, but this time using a straight periodic grating $r_2(x,y)$ whose frequency is $f_2 < f_1$. How will your answers to (a) and (b) above be modified? Can you find this time the singular state of the (1,-1)-moiré? Explain, based on your sketch of the spectrum.

10-4. *Superposition of two perpendicular parabolic gratings* (see Example 10.15 and Fig. 10.14). Show that the $(k_1,k_2)$-moiré eyelet in the superposition of two perpendicular parabolic gratings is a circular, elliptic or hyperbolic zone grating that is shifted from the origin by:

$$x_0 = \frac{k_1 f_1}{2k_2 f_2 a_2}, \quad y_0 = \frac{k_2 f_2}{2k_1 f_1 a_1}$$

For which $(k_1,k_2)$-moiré eyelets the zone grating is circular? For which $(k_1,k_2)$ it is elliptic? And for which $(k_1,k_2)$ it is hyperbolic?

10-5. What happens to the moiré eyelets of the previous problem when we shift the superposed layers on top of one another, without changing their angles or their periods? *Hint*: Unlike in the superposition of periodic layers (see Chapter 7), shifting the original parabolic gratings does not cause global shifts of the moiré effects in the superposition. It only causes relative phase shifts in the moiré bands which surround the singular locus of the moiré, but the *location* of the singular locus of the moiré in the superposition is not shifted.

10-6. Suppose that a regular periodic pinhole-screen with frequency $f_2$ in both directions is superposed on top of a parabolic grating, like the grating shown in Figs. 10.9(a) or 10.12(a), whose horizontal frequency is $f_1$. What do you expect to see in the superposition when $f_1 > f_2$, $f_1 = f_2$ or $f_1 < f_2$? How does this explain the sampling moirés that occur when a parabolic grating is displayed on a digital device? What do you expect to see when the pinhole-screen is rotated on top of the parabolic grating?

10-7. *Superposition of a zone grating and a square grid (or screen).* Consider the superposition of a zone grating $r_1(x,y)$ and a square grid (or screen) $r_2(x,y)$. What would you expect to see in this case in the image domain? and in the spectral domain? Can you identify the index of each of the moiré eyelets in the superposition? (Such superpositions are shown, for example, in [Walls75 p. 597] or in [Stecher64 pp. 254–255].) What happens when the superposed grid (or screen) is not orthogonal?

10-8. *Superposition of two mutually shifted zone gratings with different frequencies.* Consider the superposition of two mutually shifted zone gratings which have different frequencies. What would you expect to see in this case in the image domain? and in the spectral domain?

10-9. *Superposition of three mutually shifted zone gratings.* Consider the superposition of three identical but mutually shifted zone gratings. Depending on the shift configurations, what would you expect to see in the image domain? and in the spectral domain? What happens when the shift sizes are relatively small, and what happens when they become larger? Can you identify the index of each of the moiré eyelets in the superposition? (Such a superposition is shown, for example, in [Leifer73 p. 42].)

10-10. *Moiré effects in the spectrum?* In cases where the spectra of the original layers are continuous and oscillatory, there may occur between the individual elements of the spectrum an interference phenomenon which looks like a moiré pattern (see, for example, Figs. 10.28(f), 10.31(f), etc.). Obviously, this phenomenon cannot occur in spectra which only consist of impulses. Is this really a moiré effect in the spectral domain? What does it correspond to in the image domain?

10-11. Using the fundamental moiré theorem for the superposition of two curvilinear gratings (Sec. 10.9.1), find the explicit expression of the isolated (1,-1) moiré $m_{1,-1}(x,y)$ and the (1,1) moiré $m_{1,1}(x,y)$ in the superposition of a periodic straight grating and a circular grating (see Example 10.19 and Figs. 10.17–10.19):

(a) When the two original gratings have a raised cosinusoidal profile, i.e.:

$$r_1(x,y) = \tfrac{1}{2}\cos(2\pi f_1 x) + \tfrac{1}{2}$$

$$r_2(x,y) = \tfrac{1}{2}\cos(2\pi f_2\sqrt{x^2+y^2}) + \tfrac{1}{2}$$

(b) When the two original gratings have a square-wave profile, i.e.:

$$r_1(x,y) = \sum_{m=-\infty}^{\infty} a^{(1)}_m \cos(2\pi m f_1 x)$$

$$r_2(x,y) = a_0 + 2\sum_{n=1}^{\infty} a^{(2)}_n \cos(2\pi n f_2\sqrt{x^2+y^2})$$

with Fourier coefficients $a^{(1)}_m = (\tau_1 f_1)\text{sinc}(m\tau_1 f_1)$ and $a^{(2)}_n = (\tau_2 f_2)\text{sinc}(m\tau_2 f_2)$. (Note that $\tau_i f_i = \tau_i/T_i$ is the opening ratio of the square wave; for the sake of simplicity you may suppose that the opening ratios of both gratings are identical: $\tau_1 f_1 = \tau_2 f_2$.)

10-12. *A periodic moiré between two identical non-periodic gratings that are shifted on top of each other.* We have seen in Example 10.20 that the required conditions on a curvilinear grating in order that its superposition with any laterally shifted copy of itself give a (1,-1)-moiré consisting of periodic straight bands is that its bending function $g(x,y)$ (common to both layers) satisfy Eqs. (10.36) or (10.37). Clearly, this condition is satisfied by any second order polynomial function $g(x,y) = ax^2 + bxy + cy^2 + dx + ey + f$, since $g(x+x_1,y) - g(x,y)$ and $g(x,y+y_1) - g(x,y)$ only contain first order terms of $x$ and $y$. What is the condition on third order polynomial functions $g(x,y)$ in order that they satisfy Eqs. (10.36), (10.37)? Can you generalize this result to polynomial functions $g(x,y)$ of any order $n$?

10-13. *A periodic moiré between two identical non-periodic gratings that are scaled on top of each other.* We have seen in Examples 10.20 and 10.21 the required conditions on a curvilinear grating in order that its superposition with any laterally shifted copy of itself (or respectively, with any rotated copy of itself) give a (1,-1)-moiré consisting of periodic straight bands. What is the required condition on a curvilinear grating in order that its superposition with any scaled copy of itself give a (1,-1)-moiré consisting of periodic straight bands? Can you find such a curvilinear grating?

10-14. What is the required condition on a curvilinear grating in order that its superposition with any copy of itself which is both shifted and rotated give a (1,-1)-moiré consisting of periodic straight bands? Can you find a bending function $g(x,y)$ other than $g(x,y) = ax + by + c$ which satisfies this condition?

10-15. *A 2-fold periodic moiré between two non-periodic dot-screens that are rotated on top of each other.* Design two curved dot-screens that give, when they are rotated on top of each other, a 2-fold periodic (1,0,-1,0)-moiré consisting of a screen of magnified "1"s (compare with Fig. 10.36, where a similar moiré was obtained when the dot-screens were translated on top of each other). What is the non-linear coordinate transformation $\mathbf{g}(x,y)$ that you propose to use in order to obtain the two curved screens? Is it a conformal transformation?

10-16. *Coordinate transformations defined by implicit functions.* In the present chapter we have dealt with coordinate transformations that are defined by explicit functions: $x' = g_1(x,y)$, $y' = g_2(x,y)$. However, a curvilinear grating (or curved screen) may be also obtained by a coordinate transformation that is defined by an implicit function $G(x,y,z) = 0$, where $z$ cannot be expressed explicitly as a function of $x$ and $y$, i.e., by $z = g(x,y)$.

(a) Give simple examples of such curvilinear gratings.

(b) How can the $(k_1,k_2)$-moiré between two curvilinear gratings (Eq. (10.18)) be expressed when both of the curvilinear gratings are defined by implicit coordinate transformations? *Hint*: show that in this case we have:

$$m_{k_1,k_2}(x,y) = \sum_{n=-\infty}^{\infty} c^{(1)}{}_{nk_1}\, c^{(2)}{}_{nk_2}\, e^{i2\pi n(k_1 z_1 + k_2 z_2)}$$

where $z_1$ and $z_2$ are defined by the implicit functions $G_1(x,y,z_1) = 0$, $G_2(x,y,z_2) = 0$.

10-17. *Counterfeit deterrents.* A widely used method for protecting documents (cheques, banknotes, etc.) against counterfeiting consists of printing on the document some special repetitive elements that generate a strong moiré pattern when the document is counterfeited by means of halftone reproduction, by digital photocopying or by digital scanning. Can you design such *moiré-inducing patterns* (also known as *screen traps*) that are effective against all scanning angles and a wide range of frequencies? (See, for example, [Renesse05 pp. 147–151] or Figs. 14–15 in [CanadianBank68]).

10-18. *Halftone screen meter (*or *screen tester).* The moiré eyelets generated between a zone grating and a dot-screen can find a useful application in the measurement of the frequency of halftone screens. When a transparency of a binary zone grating is laid on the halftoned image whose dot frequency is to be determined, a number of moiré eyelets is generated in various locations around the zone grating's center. The displacement of the first-order (and hence strongest) moiré eyelet indicates the screen's frequency. Show that the displacement of each moiré eyelet is proportional to the measured halftone frequency. This allows us to draw on top of the transparency a linear scale of screen frequencies, so that the screen frequency being measured can be easily read off (see [Bracewell95 pp. 292–295]). Note that the same device can be also used to determine the screen angle; can you explain how? (see, for example, [Bann90 pp. 72–73]). Other types of moiré-based halftone screen meters use a grating in the form of a fan of straight lines radiating from a common center (see [Bann90 p. 72] or [Oster69 p. 39 and Figs 30, 44]); however, the scale in such devices is not linear (i.e., it is not proportional to the halftone frequency).

10-19. *Moiré effects as models of physical phenomena.* Moiré effects can be often used to illustrate physical wave phenomena. For example, the moiré patterns in the superposition of two identical circular gratings can be used as a model for the interference fringes created when two concentric circular waves are generated in phase from different centers (e.g., water waves on the surface of a pool). The same moiré effect can be also used as a model for the interference patterns produced when light waves from a common point source pass through two adjacent, narrow slits in a wall. Because waves of all kinds behave in much the same way, the analogy is valid for sound waves, water waves, electromagnetic waves, etc. As an example, design a set of circular gratings that can be used to illustrate the interference patterns between two spherical waves emanating from two point sources, and to show how they vary when:

(a) The distance between the two sources is changed;

(b) The phase between the two sources is changed;

(c) The frequency of one of the two sources is changed;

(d) The amplitude of one of the two sources is changed.

10-20. *Interference of two circular water waves.* The various properties of waves are very conveniently demonstrated by means of water waves. In order to produce regular and controllable water waves a *ripple tank* is often used. This is a tank of water, usually made of glass, with a small mechanical device mounted above the surface which drives a pointed vertical rod up and down into the water, using a motor with great regularity. The result is a continuous circular wave which propagates on the water surface from its point source, moving radially outward (see, for example, [Pssc65, Chapter 16]).

Suppose now that two pointed rods are simultaneously driven by the same motor in a ripple tank. A set of circular waves will spread out on the water surface from each of the point sources (see, for example, [Pssc65, Chapter 17]). These continuous waves will overlap everywhere and interact in an *additive* way. In other words, at any point on the water surface where there is simultaneously a crest from both sources, the water surface is lifted to a height equal to the sum of both crests; at any point where there is simultaneously a valley from both sources, the water surface is lowered to a deepness equal to the sum of the two negative valleys; and at any point where a crest from one source meets a valley from the other source, their effects are cancelled out and the water surface remains undisturbed. The two first cases are known as *constructive interference*, while the third case is called a *destructive interference*. At intermediate points there will exist an intermediate situation called *partial interference*. The interference between the two circular waves generates a hyperbolic interference pattern like in Fig. 10.24. (It is interesting to note that even when we consider the moving waves rather than the "frozen" waves at a given instant, this interference pattern remains stationary, whence the traditional name of *standing waves*. The lines where the surface altitude remains zero due to destructive interference are traditionally called "node lines".) Explain the analogy and the differences between this physical phenomenon and the moiré effect shown in Fig. 10.24.

10-21. *"Raised" vs. pure cosinusoidal gratings* (see Remark 2.2 and Problem 2-8). Figs. 10.24 and 10.26 show the superposition of two circular gratings with raised cosinusoidal periodic profiles. How would these superpositions and their spectra look like if the circular gratings had pure, unraised cosine profiles?

10-22. *Different layer superposition rules* (see Remarks 2.1 and 2.3 and Problem 2-9). How would the superpositions and spectra in Figs. 10.24 and 10.26 vary if the multiplicative superposition rule were replaced by an additive superposition rule? See also Problem 2-10 concerning pseudo moiré fringes in additive superpositions.

10-23. *3D moirés.* What is the form of the interference pattern generated between two spherical waves emanating from two point sources? Can you think of an example of a 3D moiré that is based on the *multiplicative* superposition rule?

# Chapter 11

# Other possible approaches for moiré analysis

## 11.1 Introduction

We have seen in Chapter 10 that the Fourier-based approach can be extended to superpositions of curvilinear gratings or non-periodic, repetitive layers in general (provided that their mathematical expressions are known). Although the spectrum in such cases may be only partially impulsive or even purely continuous, our basic theory remains valid in these cases as well.

However, the analytic expressions describing the spectral behaviour in non-periodic cases may become very complicated, and even the Fourier series approach, purely in the image domain (see Sec. 10.7.1), is not always simple to use. Although this does not affect the validity of the theory, its use may prove to be impractical in various real-world moiré applications. In such cases the moiré analysis can be done directly in the image domain using one of several possible simplified methods. In the present chapter we briefly review some of these alternative methods, compare them, and evaluate their advantages and their drawbacks. Furthermore, we also show how these methods are related to each other and how they are related to the spectral approach. As we will see, these alternative methods are, indeed, encompassed by the spectral approach, and hence the results they can provide are only partial to the full information which can be obtained by the spectral approach.

The alternative methods reviewed in the present chapter include the indicial equations method (Sec. 11.2), approximations using the first harmonic of the Fourier series development (Sec. 11.3), and the local frequency method (Sec. 11.4). Further details on these methods can be found in the references given in each section below.

## 11.2 The indicial equations method

The simplest and probably also the oldest method for analyzing the geometric shape of moiré patterns in the superposition of two given curvilinear gratings is the indicial (or the parametric) equations method [Oster64; Durelli70 pp. 64–78; Patorski93 pp. 14–21]. This method, which involves only the image domain, is based on the curve equations of the original curvilinear gratings: If each of the original layers is regarded as an indexed family of lines (or curves), the moiré pattern that results from their interaction forms a new indexed family of lines (or curves), whose equations can be deduced from the equations of the original layers. We illustrate this method by three simple examples; additional examples can be found, for instance, in the above mentioned references.

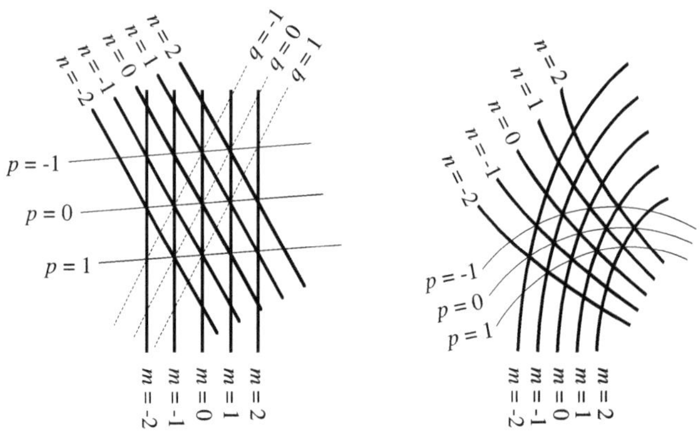


(a) (b)

**Figure 11.1:** Schematic illustration of a (1,-1)-moiré (a) between two straight periodic gratings; and (b) between two curvilinear gratings. The two line families enumerated by the indices $m$ and $n$, respectively, represent the centerlines of the superposed gratings, and the family of thin lines enumerated by the index $p$ indicates the centerlines of the bright bands of the resulting (1,-1)-moiré. Similarly, the family of dotted lines enumerated by the index $q$ indicates the centerlines of the bright bands of the (2,-1)-moiré. (For the sake of clarity the respective dotted lines have not been drawn in case (b)).

**Example 11.1:** Finding the line equations of the moiré between two periodic gratings:

Assume that we are given a periodic grating of vertical lines with period $T_1$, and a second periodic line grating with period $T_2$ which is rotated by angle $\theta$ (like in Figs. 2.5 and 11.1(a)). For the sake of simplicity we may assume that both gratings are centered about the origin. We consider each of the gratings as a family of lines, and we focus our attention to their centerlines, ignoring their real linewidths or their intensity profiles. If we enumerate the lines of the first grating by $m = ...,-2,-1,0,1,2,...$ then the equations of their centerlines in the $x,y$ plane are given by:

$$x = mT_1, \qquad m \in \mathbb{Z} \tag{11.1}$$

Similarly, the equations of the centerlines of the rotated grating are:

$$x\cos\theta + y\sin\theta = nT_2, \qquad n \in \mathbb{Z} \tag{11.2}$$

As we already know, moiré bands occur in a grating superposition since areas where black lines of the two gratings cross each other contain less black than areas where the grating lines fall between each other. Therefore, the bright bands of the most visible moiré run along the lines which connect closest crossing points in the superposition (see the thin lines in Fig. 11.1, which correspond here to the (1,-1)-moiré, like in Fig. 2.5). Note that in general, the eye automatically selects as the most prominent moiré the locus of intersection points in which the density of crossing points is the greatest; in the case of Fig. 2.6, for example, this would be the (2,-1)-moiré.

Let us find the line equations of the most prominent moiré shown in Fig. 11.1(a), i.e., the subtractive, (1,-1)-moiré.[1] In this case, the 0-th line of the moiré line family (i.e., the centerline of the 0-th bright band of the moiré) joins all the intersection points where $m - n = 0$, namely, the intersection points $(m,n) = ..., (-1,-1), (0,0), (1,1), ...$ . But since the moiré bands are continuous, the 0-th line of the moiré contains also all the intermediate points between these intersection points; clearly, it contains all the $(x,y)$ points in the plane for which equations (11.1) and (11.2) satisfy: $m - n = 0$.

Similarly, the $p$-th line of the (1,-1)-moiré consists of all the $(x,y)$ points in the plane for which equations (11.1) and (11.2) satisfy the condition:

$$m - n = p \tag{11.3}$$

In order to find the $p$-th line of the moiré, i.e., the locus of all the $(x,y)$ points which satisfy Eqs. (11.1)–(11.3), we have to solve these three simultaneous equations for $x$, $y$ and $p$. This can be done by solving for $m$ in Eq. (11.1) and for $n$ in Eq. (11.2), and inserting these results into the indicial equation (11.3). We obtain, therefore:

$$\frac{x}{T_1} - \frac{x\cos\theta + y\sin\theta}{T_2} = p$$

or after rearrangement:

$$x(T_2 - T_1\cos\theta) - yT_1\sin\theta = T_1T_2p$$

This is the equation of the centerline of the $p$-th bright moiré band, using the normal form of the line equation [Spiegel68 p. 35]. If we let the index $p$ vary through all integers, $p = ...,-2,-1,0,1,2,...,$ this equation represents the line family of the centerlines of the subtractive (1,-1)-moiré bands.[2]

More generally, the line equations of any $(k_1,k_2)$-moiré can be obtained in the same way, but this time using instead of condition (11.3) the general indicial equation:

$$k_1m + k_2n = p \tag{11.4}$$

where $k_1$ and $k_2$ are constant integers, and $m$, $n$ and $p$ are the indexing parameters of the three line families (the two original gratings and the $(k_1,k_2)$-moiré bands).[3]

---

[1] Note that in Chapters 10 and 11 we sometimes find it convenient to use the classical terms *subtractive moiré* and *additive moiré* (not to be confused with *additive superposition*!). These terms designate moirés which correspond, respectively, to frequency differences or frequency sums in the spectrum. For example, the (1,-1)-moiré is subtractive, while the (1,1)-moiré is an additive moiré.

[2] It can be shown that this equation leads, indeed, to the classical formulas (2.9) that give the period and the angle of the (1,-1)-moiré between two superposed line gratings (see [Oster64 p. 170]).

[3] Note that a moiré of order > 1, too, is the locus of points of intersection (see, for example, the dotted lines of the (2,-1)-moiré in Fig. 11.1(a)). But since usually the density of points of intersection along these loci is lower than in a first-order moiré, a higher order moiré is usually less clearly visible.

In the most general case of a $(k_1,...,k_s)$-moiré between $s$ superposed gratings we will have $s$ equations, one for each layer, plus the condition formulated by the general indicial equation:

$$k_1 n_1 + ... + k_s n_s = p , \qquad p \in \mathbb{Z} \tag{11.5}$$

The line equations of the $(k_1,...,k_s)$-moiré bands can be found, again, by elininating the indices $n_1,...,n_s$ in Eq. (11.5) using the line equations of the $s$ original gratings, i.e., by solving the set of $s+1$ equations for $x$, $y$ and $p$.

Note, however, that although the indicial equations method can give the equation family of each of the $(k_1,...,k_s)$-moirés in the superposition, it can not tell which of them is the most prominent and clearly visible. This question depends, of course, on the grating periods, on their superposition angles, and also on the grating profiles. This kind of information can be obtained in the spectral method from the locations and the amplitudes of the impulses in the spectrum of the superposition. ■

**Example 11.2:** Finding the curve equations of the different moirés between two perpendicular parabolic gratings (see Fig. 10.14(a)):

In this example we apply the indicial equations method to the case of two perpendicular parabolic gratings, which has already been analyzed using our spectral method in Sec. 10.7.4, in order to compare the results obtained by the two methods.

The centerlines of the two curve families in the $x,y$ plane are given in this case by:

$$x - a_1 y^2 = mT_1 , \qquad m \in \mathbb{Z}$$

$$y - a_2 x^2 = nT_2 , \qquad n \in \mathbb{Z}$$

where $T_1$ is the spacing between the centerlines of two consecutive parabolas of the first family, and $T_2$ is the spacing between the centerlines of two consecutive parabolas of the second family.

The equations of the indexed family of curves of the $(k_1,k_2)$-moiré are obtained by solving the above equations together with condition (11.4) for $x$, $y$ and $p$. This can be done by solving the two equations above for $m$ and $n$, respectively, and inserting the resulting expressions into the indicial equation (11.4) in order to eliminate there $m$ and $n$. This gives the curve family of the $(k_1,k_2)$-moiré:

$$\tfrac{k_1}{T_1}(x - a_1 y^2) + \tfrac{k_2}{T_2}(y - a_2 x^2) = p , \qquad p \in \mathbb{Z}$$

or, by putting $f_1 = 1/T_1$, $f_2 = 1/T_2$:

$$k_2 a_2 f_2 x^2 + k_1 a_1 f_1 y^2 - k_1 f_1 x - k_2 f_2 y = p , \qquad p \in \mathbb{Z} \tag{11.6}$$

Note that $p$ in the right hand side of the equation is *not* squared. This result means that the curve family of any additive moiré, where $k_1$ and $k_2$ have the same signs, is an elliptic (or circular) zone grating, while the curves of any subtractive moiré, where $k_1$ and $k_2$ have opposite signs, form a hyperbolic zone grating. Furthermore, the centers of these zone gratings are shifted from the origin by $(x_0,y_0) = (\frac{k_1 f_1}{2k_2 f_2 a_2}, \frac{k_2 f_2}{2k_1 f_1 a_1})$. This can be seen by comparing Eq. (11.6) with the general equation of a shifted ellipse (or hyperbola, if $a$ and $b$ have opposite signs), namely:

$$\tfrac{1}{a}(x-x_0)^2 + \tfrac{1}{b}(y-y_0)^2 = 1$$

or:

$$\tfrac{1}{a}x^2 + \tfrac{1}{b}y^2 - 2\tfrac{1}{a}x_0 x - 2\tfrac{1}{b}y_0 y + (\tfrac{1}{a}x_0{}^2 + \tfrac{1}{b}y_0{}^2 - 1) = 0$$

and equating the respective coefficients.

These results are, of course, in accordance with those that we have already obtained using the spectral approach. Note, however, that in spite of the remarkable simplicity of the derivation using the indicial equations method, the information this method provides is limited to the geometric layout of the centerlines of the moiré curves. ■

**Example 11.3:** Finding the curve equations of the moiré between two identical shifted circular gratings (see Figs. 10.24 and 10.26):

This is a case in which derivations in the spectral domain become quite complicated (see Remark 10.7 in Sec. 10.7.6). However, the derivation of the moiré shapes using the indicial equations method remains straightforward: The curve equations of the two horizontally shifted circular gratings in the $x,y$ plane are given by:

$$(x-x_0)^2 + y^2 = (mT)^2\,, \qquad m = 1,2,...$$

$$(x+x_0)^2 + y^2 = (nT)^2\,, \qquad n = 1,2,...$$

where $T$ is the radial period of both circular gratings, and $+x_0$ and $-x_0$ are their respective horizontal shifts. The equations of the curve families of the (1,-1)-moiré and of the (1,1)-moiré are obtained by solving the above equations for $m$ and $n$ and inserting the resulting expressions into the indicial equation $m \pm n = p$ in order to eliminate $m$ and $n$. After some rearrangements one obtains [Patorski93 p. 17]:

$$\frac{x^2}{(pT/2)^2} \pm \frac{y^2}{x_0^2 - (pT/2)^2} = 1\,, \qquad p = 1,2,...$$

This means that the curves of the subtractive (1,-1)-moiré form a family of hyperbolas, while the curves of the additive (1,1)-moiré form a family of ellipses. ■

The indicial equations method can be summarized, therefore, as follows: If the centerlines of the $s$ superposed curvilinear gratings are given by the curve families:

$$\begin{aligned} g_1(x,y) &= n_1\,, & n_1 &\in \mathbb{Z} \\ &\vdots \\ g_s(x,y) &= n_s\,, & n_s &\in \mathbb{Z} \end{aligned} \qquad (11.7)$$

(where the line spacing $T_i$ of each layer is already incorporated into $g_i(x,y)$) then the centerlines of the $(k_1,...,k_s)$-moiré curves can be obtained from the indicial equation (11.5) by eliminating the indices $n_1,...,n_s$ using Eqs. (11.7). We obtain, therefore:

$$k_1 g_1(x,y) + ... + k_s g_s(x,y) = p \,, \qquad p \in \mathbb{Z} \tag{11.8}$$

This is the relationship between $x$ and $y$ that describes the curve family of the $(k_1,...,k_s)$-moiré. As we can see, this result is equivalent to the second part of the fundamental moiré theorem for curvilinear gratings (see Sec. 10.9.1).

Further examples using the indicial equations method can be found in the references mentioned above. Particularly interesting examples can be also found in [Leifer73] and [Walls75], which analyze in depth the various $(k_1,k_2)$-moirés obtained between circular zone gratings or between a circular zone grating and a periodic straight grating.[4]

It is interesting to note that the indicial equations method can be given also a more visual interpretation, by regarding the indexed family of curves that describes a given layer as the level lines of a 2D curved surface which are orthogonally projected onto the $x,y$ plane, like in a topographic map. According to this interpretation, if the $i$-th indexed family of curves represents level lines of the surface $z = g_i(x,y)$, then the indexed curve family of the $s$-layers $(k_1,...,k_s)$-moiré consists of the level lines of the surface:

$$z = k_1 g_1(x,y) + ... + k_s g_s(x,y) \tag{11.9}$$

In particular, the indexed curve family of the (1,-1)-moiré between two curvilinear gratings consists of the level lines of the difference surface $z = g_1(x,y) - g_2(x,y)$; the level line $z = p$ corresponds therefore to the projection onto the $x,y$ plane of the intersection curve between the two surfaces $z = g_1(x,y)$ and $z = g_2(x,y) + p$, namely: $g_1(x,y) = g_2(x,y) + p$ (this is nicely illustrated in Fig. 31 in [Oster69 p. 25]). We will return to this interpretation in Sec. 11.4.1 below.

The indicial equations method can be also generalized to the 2D case of line grids (or dot screens) by regarding each of the superposed layers as *two* indexed families of line gratings (see Secs. 2.11–2.12). The resulting moiré, too, is given by two indexed families of curves, each of which can be obtained independently using Eq. (11.8) and the respective $k_i$ values. For example, in the case of a (1,0,-1,0)-moiré between two line grids the first indexed curve family of the moiré takes the indices $k_1 = 1$, $k_2 = 0$, $k_3 = -1$, $k_4 = 0$ and the second indexed curve family of the moiré takes the indices $k_1 = 0$, $k_2 = 1$, $k_3 = 0$, $k_4 = -1$; see Eq. (5.2) in *Vol. II*.

### 11.2.1 Evaluation of the method

The indicial equations method yields the curve equations of the different $(k_1,...,k_m)$-moirés that are generated in the superposition of $m$ given curve families. The original

[4] Their indicial equations even include the phases of the different gratings and of the resulting moirés. Note, however, that unlike in the spectral approach, analysis of the phase in the indicial equations method is rather limited; for example, it fails to discriminate between black and white zones of a zone grating (see [Leifer73 p. 40] and [Walls75 p. 596]).

layers, as well as the different moirés in the superposition, are expressed by indexed families of curves which represent the centerlines of the curvilinear gratings and of the moiré bands. This method involves only the image domain, and it takes into account only the geometric layout of the centerlines of the curvilinear gratings and of the resulting moirés, but it totally ignores their intensity profiles (i.e., their real linewidths and their intensity or gray-level variations). The biggest advantage of the indicial equations method is in its remarkable simplicity, which is particularly welcome in cases where the derivations in the spectral approach become complicated. The results it provides are, however, more restricted.

### 11.2.2 Comparison with the spectral approach

It is interesting to note that although the indicial equations method stands on its own right independently of any Fourier considerations, this method is in fact fully included in the Fourier-based approach. To see this, suppose that the superposed curvilinear layers are given by (see Secs. 10.5.1 and 10.7.1):[5]

$$r_1(x,y) = \sum_{m=-\infty}^{\infty} c^{(1)}_m \, e^{i2\pi m g_1(x,y)}$$

$$r_2(x,y) = \sum_{n=-\infty}^{\infty} c^{(2)}_n \, e^{i2\pi n g_2(x,y)}$$

Their superposition is expressed therefore by the product:

$$\begin{aligned} r_1(x,y)\, r_2(x,y) &= \Big(\sum_{m=-\infty}^{\infty} c^{(1)}_m \, e^{i2\pi m g_1(x,y)}\Big) \Big(\sum_{n=-\infty}^{\infty} c^{(2)}_n \, e^{i2\pi n g_2(x,y)}\Big) \\ &= \sum_{m=-\infty}^{\infty} \sum_{n=-\infty}^{\infty} c^{(1)}_m c^{(2)}_n \, e^{i2\pi (m g_1(x,y) + n g_2(x,y))} \end{aligned} \tag{11.10}$$

As we have seen in Sec. 10.7.1, the partial sum which corresponds to the (1,-1)-moiré consists of all the terms whose indices are $n$ and $-n$, namely:

$$m_{1,-1}(x,y) = \sum_{n=-\infty}^{\infty} c^{(1)}_n c^{(2)}_{-n} \, e^{i2\pi n (g_1(x,y) - g_2(x,y))}$$

And more generally, the partial sum which corresponds to the $(k_1,k_2)$-moiré consists of all the terms of the double sum whose indices are $nk_1$ and $nk_2$:

$$m_{k_1,k_2}(x,y) = \sum_{n=-\infty}^{\infty} c^{(1)}_{nk_1} c^{(2)}_{nk_2} \, e^{i2\pi n (k_1 g_1(x,y) + k_2 g_2(x,y))}$$

Now, the geometric layout of the corrugations of $r_1(x,y)$ is given by its first harmonic term (the term with $m = 1$, $e^{i2\pi g_1(x,y)}$), which determines the shape of the "curved periods" of $r_1(x,y)$ throughout the $x,y$ plane. Because $e^{i2\pi(...)} = \cos 2\pi(...) + i \sin 2\pi(...)$ it is clear that the geometric layout of the corrugations of $e^{i2\pi g_1(x,y)}$ is given by the locus of all the points in the $x,y$ plane where the argument is an integer multiple of $2\pi$:[6]

---

[5] Note that we adopt here the convention that the frequencies $f_1$ and $f_2$ (or their reciprocals $T_1$ and $T_2$) are already incorporated into the functions $g_1(x,y)$ and $g_2(x,y)$, respectively (see Sec. 10.2).

[6] In these points the cosine term equals 1 and the imaginary sine term vanishes. Note that when the periodic-profiles of the gratings are symmetric the sine term is identically zero anyway.

$$2\pi g_1(x,y) = 2\pi m\,, \qquad m \in \mathbb{Z}$$

or:

$$g_1(x,y) = m\,, \qquad m \in \mathbb{Z} \tag{11.11}$$

Similarly, the geometric layout of the corrugations of $r_2(x,y)$ is given by the locus of all the points in the $x,y$ plane where:

$$g_2(x,y) = n\,, \qquad n \in \mathbb{Z} \tag{11.12}$$

In the same manner, the geometric layout of the corrugations of $m_{1,-1}(x,y)$ is determined by its first harmonic term, $e^{i2\pi(g_1(x,y) - g_2(x,y))}$, and hence, applying the same reasoning, we find that the geometric layout of the corrugations of the (1,-1)-moiré is given by:

$$2\pi(g_1(x,y) - g_2(x,y)) = 2\pi p\,, \quad p \in \mathbb{Z}$$

or:

$$g_1(x,y) - g_2(x,y) = p\,, \qquad p \in \mathbb{Z} \tag{11.13}$$

By applying the same reasoning to $m_{k_1,k_2}(x,y)$ we find that the geometric layout of the corrugations of the $(k_1,k_2)$-moiré is given by:

$$2\pi(k_1 g_1(x,y) + k_2 g_2(x,y)) = 2\pi p\,, \qquad p \in \mathbb{Z}$$

or:

$$k_1\, g_1(x,y) + k_2\, g_2(x,y) = p\,, \qquad p \in \mathbb{Z} \tag{11.14}$$

This follows, of course, from the fact that the exponentials in the Fourier series product $r_1(x,y)r_2(x,y)$ combine *additively* via the identity $e^{ia}\, e^{ib} = e^{i(a+b)}$ (see Eq. (11.10)).

As we can immediately recognize, equation family (11.13) is simply the indicial equation $m - n = p$, where $m$ and $n$ have been replaced by (11.11) and (11.12). Similarly, the more general equation family (11.14) is precisely the indicial equation $k_1 m + k_2 n = p$. The generalization to the case of the $(k_1,\ldots,k_s)$-moiré between $s$-layers is straightforward.

As we can see, the indicial equations that represent the curve families of the original layers and of the resulting moirés are already incorporated into their respective Fourier series representations. This shows that the classical indicial equations method is indeed fully encompassed by the Fourier-based approach; in fact, it is equivalent to the *second part* of the fundamental moiré theorem for curvilinear gratings (see Sec. 10.9.1). Clearly, since it only uses a part of the full information included in the Fourier expressions, it is not surprising that the indicial equations method can only provide partial information about the moirés.

## 11.3 Approximation using the first harmonic

Until now we have only dealt with moiré effects between ideal structures, i.e., between original layers that are defined by mathematical formulas. However, important real-world applications of the moiré theory in various fields of technology often involve non-ideal structures. In a typical situation moiré patterns are generated in the superposition of a deformed specimen grating with an undeformed reference grating. Clearly, in such real-world cases the analytic expression of the deformation is generally unknown. How can such slight deformations in the specimen grating be detected and evaluated by means of the moiré theory?

Let us consider first the superposition of the two straight, periodic gratings before the application of the deformation. As a first order approximation we will only consider the fundamental harmonics and the DC, ignoring all higher order harmonics of the Fourier series development. We are therefore facing the same situation as in Sec. 2.3, namely, the superposition of two raised cosinusoidal gratings (see Fig. 2.2 and Eqs. (2.5) and (2.6)).

Assume now that the straight cosinusoidal grating of Eq. (2.5) undergoes a slight deformation $g(x,y)$.[7] The deformed grating is given, therefore, by:

$$r_1(x,y) = \tfrac{1}{2}\cos(2\pi f_1[x - g(x,y)]) + \tfrac{1}{2} \tag{11.15}$$

while the reference grating remains unchanged, as given in Eq. (2.6):

$$r_2(x,y) = \tfrac{1}{2}\cos(2\pi f_2[x\cos\theta + y\sin\theta]) + \tfrac{1}{2} \tag{11.16}$$

The superposition of these two gratings is given by their product:

$$\begin{aligned}
r_1(x,y)\, r_2(x,y) &= \left(\tfrac{1}{2}\cos(2\pi f_1[x - g(x,y)]) + \tfrac{1}{2}\right) \left(\tfrac{1}{2}\cos(2\pi f_2[x\cos\theta + y\sin\theta]) + \tfrac{1}{2}\right)\\
&= \tfrac{1}{4}\cos(2\pi f_1[x - g(x,y)])\cos(2\pi f_2[x\cos\theta + y\sin\theta]) + \tfrac{1}{4}\cos(2\pi f_1[x - g(x,y)])\\
&\quad + \tfrac{1}{4}\cos(2\pi f_2[x\cos\theta + y\sin\theta]) + \tfrac{1}{4}\\
&= \tfrac{1}{8}\cos 2\pi(f_1[x - g(x,y)] - f_2[x\cos\theta + y\sin\theta]) && \leftarrow \mathit{Term}\#1\\
&\quad + \tfrac{1}{8}\cos 2\pi(f_1[x - g(x,y)] + f_2[x\cos\theta + y\sin\theta]) && \leftarrow \mathit{Term}\#2\\
&\quad + \tfrac{1}{4}\cos(2\pi f_1[x - g(x,y)]) && \leftarrow \mathit{Term}\#3\\
&\quad + \tfrac{1}{4}\cos(2\pi f_2[x\cos\theta + y\sin\theta]) && \leftarrow \mathit{Term}\#4\\
&\quad + \tfrac{1}{4} && \leftarrow \mathit{Term}\#5
\end{aligned}$$

If we look back at Fig. 2.2(f), the spectrum of the product of the two ideal gratings *before* the deformation $g(x,y)$ has been applied, we see that:

* *Term*#1 corresponds in the spectrum to the impulse pair located at $\mathbf{f}_1-\mathbf{f}_2$, $\mathbf{f}_2-\mathbf{f}_1$;

* *Term*#2 corresponds in the spectrum to the impulse pair located at $\mathbf{f}_1+\mathbf{f}_2$, $-\mathbf{f}_1-\mathbf{f}_2$;

---

[7] The function $g(x,y)$ may represent any physical quantity which can be encoded in the departure of the grating lines from straightness: In strain analysis $g(x,y)$ is the in-plane deformation of the object under load, in moiré topography it is the out-of-plane deformation, etc. [Patorski93 p. 4]. The distortion $g(x,y)$ is considered as the information which is contained in the deformed grating.

* *Term*#3 corresponds in the spectrum to the impulse pair located at $\mathbf{f}_1$, $-\mathbf{f}_1$;

* *Term*#4 corresponds in the spectrum to the impulse pair located at $\mathbf{f}_2$, $-\mathbf{f}_2$;

* *Term*#5 corresponds in the spectrum to the DC impulse.

Now, as we have seen in Sec. 10.4, if $g(x,y)$ is only a small deformation, the spectrum of the deformed superposition will be approximately similar to the spectrum in the undeformed case. Therefore we may assume that the predominant moiré, the subtractive (1,-1)-moiré, is still represented in the spectrum by the impulses at $\mathbf{f}_1-\mathbf{f}_2$, $\mathbf{f}_2-\mathbf{f}_1$ (that only undergo some smearing deformations), namely, by *Term*#1 in the image domain. Since the impulses $\pm\mathbf{f}_1$ and $\pm\mathbf{f}_2$ of the original gratings, that correspond to *Term*#3 and *Term*#4, as well as the impulses $\mathbf{f}_1+\mathbf{f}_2$, $-\mathbf{f}_1-\mathbf{f}_2$ of the additive (1,1)-moiré, that correspond to *Term*#2, fall outside the visibility circle, the only terms which contribute to the visible moiré are *Term*#1 and *Term*#5. We can therefore extract an *approximate* version of the (1,-1)-moiré by considering only these two terms out of the five terms of the product:

$$m_{1,-1}(x,y) \approx \tfrac{1}{4} + \tfrac{1}{8}\cos 2\pi(f_1[x - g(x,y)] - f_2[x\cos\theta + y\sin\theta])$$

In other words: we have obtained here an approximate evaluation of the (1,-1)-moiré, ignoring all its harmonics higher than 1 (see Fig. 2.5(f)).

The geometric layout of this moiré is given by the locus of all the points in the $x,y$ plane where the cosine function in the first harmonic of $m_{1,-1}(x,y)$ has its peak values, namely, where the argument of the cosine function is an integer multiple of $2\pi$:

$$2\pi(f_1[x - g(x,y)] - f_2[x\cos\theta + y\sin\theta]) = 2\pi k \qquad k \in \mathbb{Z}$$

or:

$$(f_1 - f_2\cos\theta)x - (f_2\sin\theta)y - f_1\, g(x,y) = k \qquad k \in \mathbb{Z}$$

In particular, if the two original gratings before the application of the deformation $g(x,y)$ are both vertical gratings with identical frequencies (namely, if $\theta = 0$ and $f_2 = f_1$), then this expression, the geometric layout of the deformed (1,-1)-moiré, is reduced into:

$$g(x,y) = kT \qquad k \in \mathbb{Z}$$

where $T = 1/f$. This means that the deformed (1,-1)-moiré fringes are simply level lines of the unknown function $z = g(x,y)$ that describes the deformation applied to the specimen grating. We obtained, therefore, a simple moiré technique which generates the contour plot of the deformation function $g(x,y)$.

Different variants of this method can be found in many references, including [Patorski93 pp. 2–8], [Gåsvik95 pp. 161–166], [Yokozeki70] and others.

### 11.3.1 Evaluation of the method

Although this method still only involves the image domain, it makes one step forward with respect to the indicial equations method in that it *does* consider the development of the original layers into Fourier series. However, it only takes into account the fundamental harmonic of the Fourier series development. Its advantage is that it can be used even with non-ideal structures, such as slightly deformed gratings or grids, where the analytic expression of the deformation is generally unknown. This advantage is most important in real-world applications in which moiré effects are used to detect and to map deformations.

## 11.4 The local frequency method

Another method for analyzing moirés between curvilinear layers is based on considering each of the superposed structures as locally straight and locally periodic in every infinitesimally small region, and studying the locally straight, periodic moirés that are generated in each such region. In other words, each curvilinear grating is approximated in any infinitesimally small region by the straight, periodic grating which is tangential to it there and which has the same periodic-profile. Therefore the curvilinear moiré in any point $(x,y)$ is locally equal to the straight, periodic moiré bands obtained between the approximating tangential straight, periodic gratings at the point $(x,y)$ [Yokozeki74 p. 379]. Note that the *local frequency* of a curvilinear grating or a curvilinear moiré in any point $(x,y)$ of the $x,y$ plane is defined as the frequency of the approximating tangential periodic grating (or moiré) at the point $(x,y)$.

The local frequency method has been presented in [Bryngdahl74] and in [Rogers77] in slightly different ways. In [Bryngdahl74] it is proposed to express the local frequency variations around a point $(x_0,y_0)$ in a curvilinear structure by a Taylor series expansion of the frequency $\mathbf{f}(x,y)$ around the point $(x_0,y_0)$, neglecting derivatives higher than the first. This technique gives for any point $(x_0,y_0)$ in the $x,y$ plane the local frequency $\mathbf{f}(x_0,y_0)$ of the curvilinear structure in question at that point, in the form of a 2D vector emanating from the point $(x_0,y_0)$, whose length and direction correspond to the magnitude and direction of the local frequency there. In [Rogers77] a geometric construction is presented which allows one to find the local frequency vector of the moiré at any point $(x,y)$ based on the local frequency vectors of the original curvilinear layers at the same point.

Here we will present the local frequency method using a slightly different formulation, which puts it in line with our mathematical notations and allows an easier comparison between this method and the various other methods.

Let us start with the 1D case. Suppose that $g(x)$ represents a 1D coordinate transformation (for instance, $g(x) = ax^2$). The local frequency $f(x)$ of the cosinusoidal function $\cos(2\pi g(x))$ at any point $x$ is given by the slope of $g(x)$ at the point $x$: $f(x) = \frac{d}{dx}g(x)$. For instance, in the periodic function $\cos(2\pi f_1 x)$ the local frequency is constant and independent of $x$: $f(x) = \frac{d}{dx}f_1 x = f_1$; whereas in the function $\cos(2\pi f_1 x^2)$ (see Example 10.4 in Sec. 10.3) the local frequency is a linear function of the variable $x$: $f(x) = \frac{d}{dx}f_1 x^2 = 2f_1 x$. Now, if our 1D function has a periodic-profile which is more complex than a simple cosine (say, a square wave periodic-profile), we resort to its Fourier series development: Clearly, the frequency of our function is only determined by the first harmonic term of this Fourier series, since higher harmonic terms only contribute to the precise reconstruction of the periodic-profile shape, but not to the periodicity of the function. The local frequency of our 1D function is given, therefore, by the local frequency of its first harmonic term, which is, as we have just seen, the slope of $g(x)$ at the point $x$: $f(x) = \frac{d}{dx}g(x)$.[8]

The situation in the 2D case is similar, except that in this case the local frequency at any point $(x,y)$ is represented by a 2D vector, whose length and direction correspond to the magnitude and direction of the local frequency there. Let us consider first the 2D curvilinear cosinusoidal function $\cos(2\pi g(x,y))$. Its local frequency vector $\mathbf{f}(x,y) = (f_u(x,y), f_v(x,y))$ at any point $(x,y)$ is given by the gradient of $g(x,y)$ (i.e., the 2D slope of $g(x,y)$ in the direction of maximum increase) at the point $(x,y)$: $\mathbf{f}(x,y) = (\frac{\partial}{\partial x}g(x,y), \frac{\partial}{\partial y}g(x,y))$.[9] For example, in the 2D periodic function $\cos(2\pi(f_u x + f_v y))$ the frequency vector is constant and independent of the point $(x,y)$: $\mathbf{f}(x,y) = (f_u,f_v)$. Other cases are given in the examples which follow. Here, too, if our 2D curvilinear grating has a periodic-profile which is more complex than a simple cosine, we must consider its Fourier series development: the local frequency of the curvilinear grating is only determined by the first harmonic term of the Fourier series. We see, therefore, that if our curvilinear grating $r(x,y)$ is expressed by the Fourier series development:

$$r(x,y) = p(g_1(x,y)) = \sum_{n=-\infty}^{\infty} a_n \cos(2\pi n g_1(x,y))$$

or by the more general exponential notation:

$$r(x,y) = p(g_1(x,y)) = \sum_{n=-\infty}^{\infty} c_n \, e^{i2\pi n g_1(x,y)}$$

then the local frequency vector of $r(x,y)$ is given at any point $(x,y)$ by:

$$\mathbf{f}(x,y) = (\frac{\partial}{\partial x}g_1(x,y), \frac{\partial}{\partial y}g_1(x,y))$$

**Example 11.4:** In the case of the cosinusoidal parabolic grating $r(x,y) = \cos(2\pi f_1(y-ax^2))$ (see Example 10.5 in Sec. 10.3) the local frequency at the point $(x,y)$ is given by:

$$f_u(x,y) = f_1\frac{\partial}{\partial x}(y-ax^2) = -2f_1ax$$

$$f_v(x,y) = f_1\frac{\partial}{\partial y}(y-ax^2) = f_1$$

This means that the horizontal component of the local frequency $\mathbf{f}(x,y)$ is a linear function of the variable $x$, whereas the vertical component of the local frequency is the constant $f_1$ (as indeed expected, since our function is periodic in the vertical direction with frequency $f_1$). The local frequency remains the same even if the parabolic grating has any other

[8] Further information about the local frequency, the local magnitude and the local phase of a signal can be found in [Hahn96 pp. 44–46].

[9] Note that the direction of the local frequency vector is perpendicular to the local (tangential) direction of the corrugations.

periodic-profile form, such as a square wave periodic-profile. Note also that the local frequency $\mathbf{f}(x,y)$ is *not* directly related to the Fourier spectrum $R(u,v)$ of our parabolic grating $r(x,y)$. The Fourier spectrum of the parabolic grating has been presented in Example 10.5 of Sec. 10.3. ■

**Example 11.5:** Consider the cosinusoidal circular grating with radial frequency of $f_1$: $\cos(2\pi f_1\sqrt{x^2+y^2})$ (see Example 10.6 in Sec. 10.3). In this case we have:

$$f_u(x,y) = f_1\frac{\partial}{\partial x}\sqrt{x^2+y^2} = f_1 x/\sqrt{x^2+y^2}$$

$$f_v(x,y) = f_1\frac{\partial}{\partial y}\sqrt{x^2+y^2} = f_1 y/\sqrt{x^2+y^2}$$

It is easy to verify that in this case the local frequency vector $\mathbf{f}(x,y) = (f_u(x,y), f_v(x,y))$ has at each point $(x,y)$ of the $x,y$ plane a constant magnitude of $\sqrt{f_u(x,y)^2 + f_v(x,y)^2} = f_1$, and its direction is radial, as expected. ■

**Example 11.6:** Consider the cosinusoidal zone grating $\cos(2\pi f_1(x^2+y^2))$ (see Example 10.7 in Sec. 10.3). In this case we have:

$$f_u(x,y) = f_1\frac{\partial}{\partial x}(x^2+y^2) = 2f_1 x$$

$$f_v(x,y) = f_1\frac{\partial}{\partial y}(x^2+y^2) = 2f_1 y$$

Like in the previous example the local frequency vector $\mathbf{f}(x,y)$ is oriented at each point $(x,y)$ of the $x,y$ plane in the radial direction; its magnitude, however, is not constant but rather a linear function of the distance of the point $(x,y)$ from the origin: $\sqrt{f_u(x,y)^2 + f_v(x,y)^2} = 2f_1\sqrt{x^2+y^2} = 2f_1 r$. ■

Once we have understood how to find the local frequency vector of any curvilinear grating, we are ready to make the next step. We will show now that knowing the local frequency vector of each of the curvilinear layers in an $m$-layers superposition, one can deduce the local frequency vector of any $(k_1,\dots,k_m)$-moiré in their superposition. Suppose that the original curvilinear layers are given by the following Fourier series decompositions:

$$r_1(x,y) = \sum_{n_1=-\infty}^{\infty} c^{(1)}_{n_1}\, e^{i2\pi n_1 g_1(x,y)}$$

$$\vdots$$

$$r_m(x,y) = \sum_{n_m=-\infty}^{\infty} c^{(m)}_{n_m}\, e^{i2\pi n_m g_m(x,y)}$$

Their respective local frequency vectors are therefore:

$$\mathbf{f}_1(x,y) = \left(\frac{\partial}{\partial x}g_1(x,y), \frac{\partial}{\partial y}g_1(x,y)\right)$$

$$\vdots$$

$$\mathbf{f}_m(x,y) = \left(\frac{\partial}{\partial x}g_m(x,y), \frac{\partial}{\partial y}g_m(x,y)\right)$$

Now, the superposition of the $m$ layers $r_1(x,y),...,r_m(x,y)$ is expressed by the product:

$$r_1(x,y) \cdot ... \cdot r_m(x,y)$$
$$= \Big(\sum_{n_1=-\infty}^{\infty} c^{(1)}_{n_1}\, e^{i2\pi n_1 g_1(x,y)}\Big) \cdot ... \cdot \Big(\sum_{n_m=-\infty}^{\infty} c^{(m)}_{n_m}\, e^{i2\pi n_m g_m(x,y)}\Big)$$
$$= \sum_{n_1=-\infty}^{\infty} ... \sum_{n_m=-\infty}^{\infty} c^{(1)}_{n_1} \cdot ... \cdot c^{(m)}_{n_m}\; e^{i2\pi(n_1 g_1(x,y) + ... + n_m g_m(x,y))}$$

The partial sum which corresponds to the $(k_1,...,k_m)$-moiré consists of all the terms of this multiple sum whose indices are $nk_1,...,nk_m$, namely:

$$m_{k_1,...,k_m}(x,y) = \sum_{n=-\infty}^{\infty} c^{(1)}_{nk_1} \cdot ... \cdot c^{(m)}_{nk_m}\; e^{i2\pi n(k_1 g_1(x,y) + ... + k_m g_m(x,y))}$$

Therefore, the local frequency vector of $m_{k_1,...,k_m}(x,y)$ is:

$$\mathbf{f}_{k_1,...,k_m}(x,y) = \big(\tfrac{\partial}{\partial x}[k_1 g_1(x,y) + ... + k_m g_m(x,y)],\, \tfrac{\partial}{\partial y}[k_1 g_1(x,y) + ... + k_m g_m(x,y)]\big)$$
$$= \big(\tfrac{\partial}{\partial x}k_1 g_1(x,y) + ... + \tfrac{\partial}{\partial x}k_m g_m(x,y),\, \tfrac{\partial}{\partial y}k_1 g_1(x,y) + ... + \tfrac{\partial}{\partial y}k_m g_m(x,y)\big)$$
$$= \big(\tfrac{\partial}{\partial x}k_1 g_1(x,y),\, \tfrac{\partial}{\partial y}k_1 g_1(x,y)\big) + ... + \big(\tfrac{\partial}{\partial x}k_m g_m(x,y),\, \tfrac{\partial}{\partial y}k_m g_m(x,y)\big)$$
$$= k_1\mathbf{f}_1(x,y) + ... + k_m\mathbf{f}_m(x,y)$$

Hence we obtain the following remarkable extension of Eq. (2.26):

$$\mathbf{f}_{k_1,...,k_m}(x,y) = k_1\mathbf{f}_1(x,y) + ... + k_m\mathbf{f}_m(x,y) \qquad (11.17)$$

This means that the local frequency vector of the $(k_1,...,k_m)$-moiré at any point $(x,y)$ is a linear combination of the local frequency vectors of the $m$ individual layers at the same point $(x,y)$, with the integer coefficients $k_1,...,k_m$ as weighting factors.

The local frequency vector of a given $(k_1,...,k_m)$-moiré can give us useful information about the symmetry properties and the general shape of the moiré patterns. For example:

(a) Knowing the local frequency vector $\mathbf{f}_{k_1,...,k_m}(x,y)$ of the $(k_1,...,k_m)$-moiré at any point $(x,y)$ of the $x,y$ plane we can get an idea on the general direction of the moiré fringes in a close neighbourhood of the point $(x,y)$ from the fact that the contour lines are always perpendicular to the local frequency vector. In fact, the local frequency vector $\mathbf{f}(x,y)$ of any layer (and of any $(k_1,...,k_m)$-moiré) can be regarded as the gradient at the point $(x,y)$ of a 2D curved surface, whose equispaced contours represent the grating (or the moiré pattern) [Rogers77 p. 11]. The geometric layout of the $(k_1,...,k_m)$-moiré can be obtained, therefore, by finding the 2D curved surface whose gradient at any point $(x,y)$ is given by $\mathbf{f}_{k_1,...,k_m}(x,y)$ (see, for example, [Bronshtein97 p. 531–532]) and working out its contour lines.

(b) Furthermore, from the local frequency vector $\mathbf{f}_{k_1,...,k_m}(x,y)$ one may find out in which areas of the $x,y$ plane the $(k_1,...,k_m)$-moiré in the superposition can be visible: the moiré is only visible in points $(x,y)$ in which the local frequency $\mathbf{f}_{k_1,...,k_m}(x,y)$ is shorter than the radius of the visibility circle.

(c) Finally, the $(k_1,...,k_m)$-moiré becomes singular in all the points $(x,y)$ of the $x,y$ plane in which:

$$\mathbf{f}_{k_1,...,k_m}(x,y) = \mathbf{0}$$

These points form the singular locus of the $(k_1,...,k_m)$-moiré in question (see Sec. 10.7.3). This is illustrated in the following example.

**Example 11.7:** Let us find the singular loci of the different $(k_1,k_2)$-moirés in the superposition of two perpendicular parabolic gratings (see Example 10.15 in Sec. 10.7.4). The original curvilinear gratings are given here by:

$$r_1(x,y) = p_1(x-a_1y^2) = \sum_{n=-\infty}^{\infty} a^{(1)}_n \cos(2\pi n f_1 (x-a_1y^2))$$

$$r_2(x,y) = p_2(y-a_2x^2) = \sum_{n=-\infty}^{\infty} a^{(2)}_n \cos(2\pi n f_2 (y-a_2x^2))$$

Their local frequency vectors are, therefore:

$$\mathbf{f}_1(x,y) = (f_1 \frac{\partial}{\partial x}(x-a_1y^2)\,, f_1 \frac{\partial}{\partial y}(x-a_1y^2)) = (f_1\,, -2f_1a_1y)$$

$$\mathbf{f}_2(x,y) = (f_2 \frac{\partial}{\partial x}(y-a_2x^2)\,, f_2 \frac{\partial}{\partial y}(y-a_2x^2)) = (-2f_2a_2x\,, f_2)$$

Hence, according to Eq. (11.17) the local frequency vector of the $(k_1,k_2)$-moiré is:

$$\begin{aligned}\mathbf{f}_{k_1,k_2}(x,y) &= k_1\mathbf{f}_1(x,y) + k_2\mathbf{f}_2(x,y)\\ &= k_1(f_1\,, -2f_1a_1y) + k_2(-2f_2a_2x\,, f_2)\\ &= (k_1f_1 - 2k_2f_2a_2x\,, -2k_1f_1a_1y + k_2f_2)\end{aligned}$$

Now, the $(k_1,k_2)$-moiré is singular at any point $(x,y)$ in the $x,y$ plane where this local frequency vector is zero, i.e., where:

$$k_1f_1 - 2k_2f_2a_2x = 0$$

$$-2k_1f_1a_1y + k_2f_2 = 0$$

which gives:

$$x = \frac{k_1f_1}{2k_2f_2a_2}, \quad y = \frac{k_2f_2}{2k_1f_1a_1}$$

The same result can be also obtained using the spectral approach (see Sec. 10.7.4). ■

**Example 11.8:** The (1,-1)- and (1,1)-moirés between two shifted zone gratings:[10]

Let $r_1(x,y)$ and $r_2(x,y)$ be two identical zone gratings that are horizontally shifted by $+x_0$ and $-x_0$, respectively. Their Fourier series representations are (see Example 10.18 in Sec. 10.8):

[10] This example has been readapted from [Rogers77 pp. 4–5] to our present mathematical formulation; it may clearly illustrate the difference between our formulation of the local frequency method and that of the original paper, which uses purely geometric considerations.

$$r_1(x,y) = \sum_{n=-\infty}^{\infty} a^{(1)}_n \cos(2\pi nf[a(x-x_0)^2 + by^2])$$

$$r_2(x,y) = \sum_{n=-\infty}^{\infty} a^{(2)}_n \cos(2\pi nf[a(x+x_0)^2 + by^2])$$

The local frequency vector of $r_1(x,y)$ (i.e., of its first harmonic term) is, therefore:

$$\begin{aligned}\mathbf{f}_1(x,y) &= (f\tfrac{\partial}{\partial x}[a(x-x_0)^2 + by^2], f\tfrac{\partial}{\partial y}[a(x-x_0)^2 + by^2])\\ &= (2fax - 2fax_0, 2fby)\end{aligned}$$

Similarly, the local frequency vector of $r_2(x,y)$ is:

$$\begin{aligned}\mathbf{f}_2(x,y) &= (f\tfrac{\partial}{\partial x}[a(x+x_0)^2 + by^2], f\tfrac{\partial}{\partial y}[a(x+x_0)^2 + by^2])\\ &= (2fax + 2fax_0, 2fby)\end{aligned}$$

Therefore, according to Eq. (11.17) we find that the local frequency vector of the subtractive (1,-1)-moiré is:

$$\mathbf{f}_{1,-1}(x,y) = \mathbf{f}_1(x,y) - \mathbf{f}_2(x,y) = (-4fax_0, 0)$$

This local frequency vector is constant (independent of the point $(x,y)$), and it corresponds to a periodic structure of vertical bands having a constant horizontal frequency of $-4fax_0$. This result agrees, of course, with the results obtained in Example 10.18 of Sec. 10.8.

Similarly, according to Eq. (11.17) we find that the local frequency vector of the additive (1,1)-moiré is:

$$\mathbf{f}_{1,1}(x,y) = \mathbf{f}_1(x,y) + \mathbf{f}_2(x,y) = (4fax, 4fay)$$

This local frequency vector corresponds to a centered zone grating whose first harmonic term is $\cos(2\pi f2[ax^2+by^2])$ (see Example 11.6 above). The additive (1,1)-moiré in this case is therefore a zone grating having the same coefficients $a$ and $b$ as the original zone gratings but a double frequency, and which is centered about the origin. ■

### 11.4.1 Evaluation of the method

The local frequency method is basically equivalent to the indicial equations method; in fact, these two methods are simply different facets of the same mathematical reality. If we regard the indexed family of curves which describes a given layer in the indicial equations method as contour lines (level lines) of a 2D curved surface $z = g(x,y)$ over the $x,y$ plane, then the frequency vector $\mathbf{f}(x,y)$ at any point $(x,y)$ is simply the gradient of this curved surface at the point $(x,y)$ (i.e., the 2D slope of $z = g(x,y)$ at the point $(x,y)$ in the direction of maximum increase, which is, of course, perpendicular to the contour line $g(x,y) = c$ passing through that point [Courant88 p. 90]).[11] Therefore, switching from the indicial

[11] In particular, Eq. (11.17) is simply the gradient to the level-lines of the curved surface (11.9) which represents the $(k_1,...,k_m)$-moiré.

equations method to the local frequency method (i.e., from contours to gradients) simply involves finding the gradient of the 2D curved surface $z = g(x,y)$ which is defined by the given family of contour curves $g(x,y) = n$, $n \in \mathbb{Z}$; and going back from the second method to the first one (from gradients to contours) involves finding the 2D surface $z = g(x,y)$ back from its gradient $\mathbf{f}(x,y) = (\frac{\partial}{\partial x}g(x,y), \frac{\partial}{\partial y}g(x,y))$ (see, for example, [Bronshtein97 pp. 531–532]) and then finding the equispaced contour lines $z = n$, $n \in \mathbb{Z}$ of the surface.

In spite of their equivalence, each of the two methods illuminates the situation from a different angle (contours vs. gradients), and each of them has its own advantages. For example, the indicial equations method is better suited for tracing the general shape (the corrugations) of a curvilinear moiré, but the local frequency method is more convenient for finding the singular locus of the moiré. The two methods should be therefore considered as complementary, the choice between them in each case being basically a question of convenience.

### 11.4.2 Comparison with the spectral approach

It should be clearly noted that although we are dealing here with local frequency vectors, this technique does *not* give the Fourier spectrum of the curvilinear structure, but only the local frequency of each infinitesimal portion of the structure, considering it as locally periodic.[12] In fact, the local frequency method (just like the indicial equations method) only involves the image domain, and it only takes into account the geometric layout of the curvilinear gratings and of the resulting moirés, but it totally ignores their intensity profiles as well as their spectral properties. It is clear, therefore, that the information offered by this method is only partial to the full information that is provided by the spectral approach.

## 11.5 Concluding remarks

Thanks to the full duality between the image and the spectral domains via the Fourier transform, the information included in the spectral domain is completely equivalent to the information given in the image domain, and it is only represented in a different way. It is clear that any approach which uses partial information from the image domain or from the spectral domain (for example, only the geometric layout of the gratings, or only their first harmonic term) necessarily gives only partial information about the resulting moirés with respect to the full information provided by the spectral approach. However, as we have seen in the present chapter, such partial approaches can be still very welcome, since they often offer a simple way to determine the geometric properties of the moiré in question. This is particularly true in cases where the full analysis of the spectrum becomes impractical.

---

[12] The local frequency vector $\mathbf{f}(x_0,y_0)$ indicates the location $(u_0,v_0)$ of the fundamental impulse in the "local Fourier spectrum" which corresponds to the infinitesimal portion of the curvilinear grating at $(x_0,y_0)$.

## PROBLEMS

11-1. *Finding the curve equations of the* $(k_1,k_2)$*-moiré between two curvilinear gratings.* Using the indicial equations method find the curve equations of the $(k_1,k_2)$-moiré fringes in the following cases:

(a) The superposition of a periodic vertical line grating and a circular zone grating, both of which are centered on the origin.

(b) The superposition of two circular zone gratings whose centers are located at $+x_0$ and $-x_0$ on the $x$ axis.

(c) Same as in (a), but admitting that the original layers are horizontally shifted from the origin by $x_1$ and $x_2$, respectively.

In each case, start by formulating the equations of the curve families that represent the centerlines of the two original gratings; then, solve these equations for their respective indices, $m$ and $n$, and insert these results into the indicial equation (11.4) of the $(k_1,k_2)$-moiré. (You may compare your results with those obtained in [Leifer73] and [Walls75], or with our results in Secs. 10.7.5 and 10.7.6.)

11-2. *Geometric interpretation of the indicial equations method.* As we can see in Fig. 11.1, in the superposition of two gratings the loci of common solutions of the grating equations:

$$g_1(x,y) = m, \qquad m \in \mathbb{Z}$$

$$g_2(x,y) = n, \qquad n \in \mathbb{Z}$$

correspond to the set of all the intersection points $(m,n)$ in the superposition. The indicial equation of the $(k_1,k_2)$-moiré:

$$k_1 m + k_2 n = p, \qquad p \in \mathbb{Z}$$

simply selects from this set of intersection points those points which belong to each bright band of the particular $(k_1,k_2)$-moiré in question (see, for example, the thin and the dotted line families in Fig. 11.1, which represent the bright bands of the (1,-1)- and of the (2,-1)-moirés, respectively).

However, in superpositions of three gratings or more this interpretation does not hold any longer since in such cases there may exist in the superposition no common intersection points at all. And yet, as we have seen in Sec. 11.2, the indicial equations method does work for any $(k_1,\dots,k_s)$-moiré between $s$ superposed gratings (see Eqs. (11.7)–(11.8)). How can you explain this apparent contradiction?

*Hint*: The absence of common intersection points between three or more gratings can be remedied by considering the line indices $m$, $n$, etc. as continuously varying real numbers rather than as integers. This is equivalent to the alternative interpretation of the original $s$ layers and their $(k_1,\dots,k_s)$-moiré as continuous 2D surfaces over the $x,y$ plane (see Eq. (11.9), and compare $z$ with the integer $p$ in Eq. (11.8)).

11-3. How do you explain the fact that the line families of the two original gratings in Fig. 11.1 represent the centerlines of the black grating elements, while the line families of the moiré represent the centerlines of the bright moiré fringes? What will be the situation in a superposition of more than two gratings?

11-4. *A moiré technique for drawing the contour plot of a function* $g(x,y)$. Suppose that we want to visualize the contour plot (=level lines) of the surface defined by the function $z = g(x,y)$. A simple moiré-based technique has been proposed in [Marsh80]: First, a series of equally spaced vertical lines, called the reference grating, is drawn. We denote the period of this grating by $T$. On top of this reference grating we draw a copy of the first grating, called the function grating, whose lines $x = x_n$ have been laterally shifted by an amount equal to $g(x_n,y)$, the vertical section through the function $g(x,y)$ along the straight line $x = x_n$: $x = x_n + g(x_n,y)$. The line equations of the reference grating and of the function grating are given, respectively, by:

$$x = mT \qquad m \in \mathbb{Z}$$

$$x = nT + g(nT,y) \qquad n \in \mathbb{Z}$$

Show that the system of fringes selected by the indicial equation:

$$m - n = p \qquad p \in \mathbb{Z}$$

(i.e., the (1,-1)-moiré fringes obtained by eliminating $m$ and $n$) are level lines of the function $z = g(x-z, y)$:

$$g(x-pT, y) = pT \qquad p \in \mathbb{Z}$$

Note that these level lines are horizontally displaced by $\Delta x = z$ from their normal position in the contour plot of the required function $z = g(x,y)$. However, this is not necessarily a disadvantage, since as demonstrated in [Marsh80] this displacement of the contour lines may give an interesting appearance of perspective to the contour plot (see Figs. 1 and 2 in [Marsh80]).

11-5. *A moiré technique for drawing the contour plot of a function $g(x,y)$ (continued).* Suppose that instead of the function grating of the previous problem we use a copy of the reference grating whose lines are distorted by laterally shifting each point $(x,y)$ of the grating by an amount equal to $g(x,y)$. The line equations of the reference grating and of the function grating are given, therefore, by:

$$x = mT \qquad m \in \mathbb{Z}$$

$$x - g(x,y) = nT \qquad n \in \mathbb{Z}$$

In this case the lines of the distorted grating are, of course, level lines of the surface $z = x - g(x,y)$. An example of such a case is shown in Fig. 11.2. Show that the system of fringes selected by the indicial equation:

$$m - n = p \qquad p \in \mathbb{Z}$$

(i.e., the (1,-1)-moiré fringes obtained by eliminating $m$ and $n$) are level lines of the function $z = g(x,y)$:

$$g(x,y) = pT \qquad p \in \mathbb{Z}$$

What are the differences between this method and the previous one?

11-6. *Moiré fringe magnification.* Consider the superposition of a horizontal line grating and a $g(x)$-shaped curvilinear grating, having slightly different line spacings of $T_1$ and $T_2$, respectively (Fig. 11.3). The two gratings are given, therefore, by the line families:

$$y = mT_1 \qquad m \in \mathbb{Z}$$

and:

$$y - g(x) = nT_2 \qquad n \in \mathbb{Z}$$

Using the relation $\frac{1}{T_M} = \frac{1}{T_1} - \frac{1}{T_2}$ (i.e., $f_M = f_1 - f_2$; see Eq. (2.11) in Chapter 2) show that the line equations of the (1,-1)-moiré fringes are given by:

$$y + \frac{T_M}{T_2} g(x) = pT_M \qquad p \in \mathbb{Z}$$

This means that the visible moiré fringes are a vertically scaled version of the $g(x)$-shaped grating, where the magnification factor is given by $\frac{T_M}{T_2}$. How do the (1,-1)-moiré fringes look like when $T_1 > T_2$, when $T_1 < T_2$ and when $T_1 = T_2$? (Note that we have used here a variant of Eq. (2.11) without the absolute value; can you explain why?) Finally, when will the (2,-1)- and (1,-2)-moirés be visible in the superposition, and how will they look like? Show that the magnification factor for a $(k_1,k_2)$-moiré is $k_2\frac{T_M}{T_2}$.

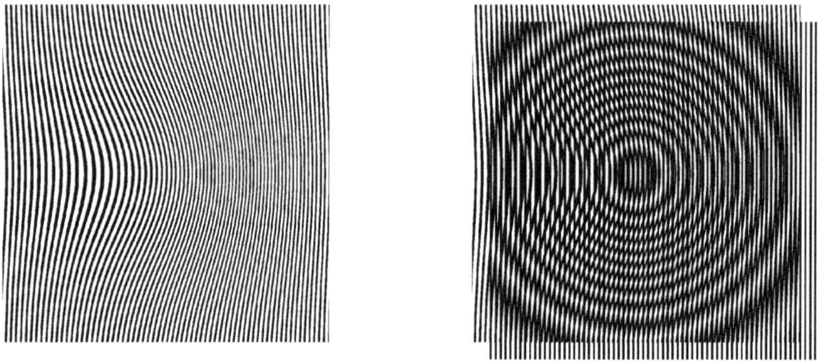

(a) (b)

**Figure 11.2:** A (1,-1)-moiré showing the contour plot of the function $g(x,y) = e^{-(x^2+y^2)/4}$. (a) A curvilinear grating whose lines are obtained by distorting a reference periodic grating of vertical lines by shifting each point by an amount equal to $g(x,y) = e^{-(x^2+y^2)/4}$. The bending function of this curvilinear grating is $g_1(x,y) = x - e^{-(x^2+y^2)/4}$ (see Fig. 10.34(a)). (b) The superposition of this curvilinear grating with the reference grating of periodic vertical lines gives a (1,-1)-moiré whose bending function is $g_{1,-1}(x,y) = e^{-(x^2+y^2)/4}$.

11-7. *A moiré technique for drawing the contour plot of derivatives of a function $g(x,y)$.* The following method has been proposed in literature (see, for example, [Giger86 p. 337]): Suppose that we want to visualize the contour plot of the derivative of the function $g(x,y)$ in a given direction, say, its partial derivative $\frac{\partial g}{\partial y}$ in the $y$ direction. We draw the curve family which corresponds to the level lines of $g(x,y)$, and superpose on top of it a second curve family, identical to the first one, which is slightly shifted in the desired direction (in our case: in the $y$ direction). The equations of the two line families are, therefore:

$$g(x,y) = nT \qquad n \in \mathbb{Z}$$

$$g(x, y+\Delta y) = mT \qquad m \in \mathbb{Z}$$

Consider the system of fringes selected by the indicial equation:

$$m - n = p \qquad p \in \mathbb{Z}$$

i.e., the (1,-1)-moiré fringes. Their equations are obtained by eliminating here $m$ and $n$:

$$g(x, y+\Delta y) - g(x,y) = pT \qquad p \in \mathbb{Z}$$

which means, if $\Delta y$ is small:

$$\frac{\partial g}{\partial y} \approx pT/\Delta y$$

As we can see, if the shift $\Delta y$ is small, the (1,-1)-moiré curves closely plot the contour map of the partial derivative of $g(x,y)$ in the $y$ direction, with increments of $T/\Delta y$.

Similarly, if the second curve family is shifted with respect to the first one in any arbitrary direction by a small shift $(\Delta x,\Delta y)$, the resulting (1,-1)-moiré curves closely approximate the contour plot of the directional derivative of $g(x,y)$ in that direction. Moreover, if the second copy of the level lines of $g(x,y)$ is rotated by a very small angle $\Delta\theta$ on top of the first copy, the resulting (1,-1)-moiré curves give a close approxi-

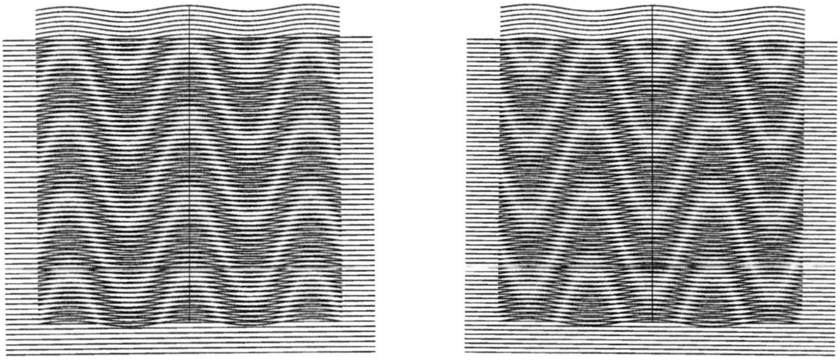

(a) (b)

**Figure 11.3:** The (1,-1)-moiré in the superposition of a horizontal line grating and a cosine-shaped grating having slightly different line spacings of $T_1$ and $T_2$, respectively: (a) when $T_1 > T_2$; (b) when $T_1 < T_2$. Note the magnification effect of the moiré, and the sign inversion in case (b).

mation of the contour plot of the rotational derivative $\frac{\partial g}{\partial \theta}$ of $g(r,\theta)$ (the polar coordinate expression of $g(x,y)$). And if the second copy of $g(x,y)$ is slightly stretched on top of the first copy, the resulting (1,-1)-moiré approximates the radial derivative $\frac{\partial g}{\partial r}$ of $g(r,\theta)$. (A good mathematical background on the directional, rotational and radial derivatives of a function $g(x,y)$ can be found, for example, in [Courant88 pp. 62–64, 74–76].)

What do you think of this method? How does it depend on the step $T$ of the two original curve families? Make a few tests with superposed gratings and evaluate the usefulness and the precision of this method.

11-8. *Cartesian ovals.* Example 11.3 shows, using the indicial equations method, that the curve equations of the subtractive (1,-1)-moiré between two shifted circular gratings with identical radial periods represent a family of hyperbolas (Fig. 11.4(a)).

(a) Using the same method, show that in the case of two circular gratings that are both centered on the origin but whose radial periods $T_1$ and $T_2$ are slightly different, the curve equations of the subtractive (1,-1)-moiré represent a family of circles that are centered on the origin, and whose radial period is $T_1T_2/|T_2-T_1|$ (see Fig. 11.4(b) and Sec. 10.7.6).

(b) Fig. 11.4(c) shows what happens to the (1,-1)-moiré between the same circular gratings when the second grating is horizontally shifted by $x_0$ from the origin. Can you find using the indicial equations method the equations of these lovely moiré curves?

*Hint*: Using polar coordinates, the curve equations of the two circular gratings are:

$$r = mT_1 \qquad m = 1,2,...$$

$$r^2 + x_0^2 - 2rx_0\cos\theta = (nT_2)^2 \qquad n = 1,2,...$$

By solving these equations for $m$ and $n$, respectively, and inserting the resulting expressions into the indicial equation $m - n = p$, one obtains the following (1,-1)-moiré curve equations in polar coordinates:

$$r^2(T_2^2 - T_1^2) + 2r(x_0T_1^2\cos\theta - T_1T_2^2p) + (p^2T_1^2T_2^2 - T_1^2x_0^2) = 0 \qquad p = 0,1,2,...$$

These equations represent a family of curves called *Cartesian ovals* ([Lockwood61 p. 188; Shikin95 pp. 102–103]). Note that in general the Cartesian oval consists of two closed curves, one enclosing the other: Since the Cartesian oval is given by a quadratic equation in $r$, its radius $r = f(\theta)$ may have for any angle $\theta$ two different values. It can be shown (see [Baudoin38 pp. 21, 129–131]) that in the particular cases where $p = x_0/T_1$ or $p = x_0/T_2$ the Cartesian ovals reduce into a simpler curve known as a *limaçon* [Shikin95 pp. 238–239]. This means that the moiré curves whose indices are $p = x_0/T_1$ or $p = x_0/T_2$ are limaçons. (What is the meaning of a non-integer $p$?) Assuming $T_1 = 3$, $T_2 = 4$, $x_0 = 5T_1$, plot the moiré curves for $p = 0,1,...,8$.

11-9. *Precision alignment.* In the process of fabricating integrated circuits photolithographic masks must be accurately aligned over a silicon wafer prior to contact printing. Propose a high precision moiré-based alignment method using circular gratings (see, for example, [King72]). Would you prefer to use instead circular zone gratings?

11-10. Can you think of other applications of moirés between circular gratings? (See, for example, [Post94 pp. 108–110].)

11-11. *Finding the singular locus of a moiré.* Use the local frequency method to find the singular locus of the moirés which are shown in: (a) Fig. 11.2(b); (b) Fig. 10.34(c).

11-12. *Finding the singular locus of a moiré (continued).* Use the local frequency method to find the singular locus of the (1,1)-moiré, the (1,-1)-moiré and the general $(k_1,k_2)$-moiré in the superposition of a vertical straight grating with a circular zone grating. Compare your results with those obtained in Sec. 10.7.7. (Note that the information obtained there was more complete: we obtained there the full expression of the moiré in question, including its intensity profile, its initial phase, etc.; while here, using the local frequency method, we only obtained the singular locus of the moiré.)

11-13. *Local frequency and the Fourier spectrum.* Discuss the relationship between the local frequency $\mathbf{f}(x,y)$ of a curvilinear grating $r(x,y)$ and its frequency spectrum $R(u,v)$. You may use Example 11.4 as a simple illustration. What is the situation when $r(x,y)$ is periodic, and what happens when $r(x,y)$ is slightly bent?

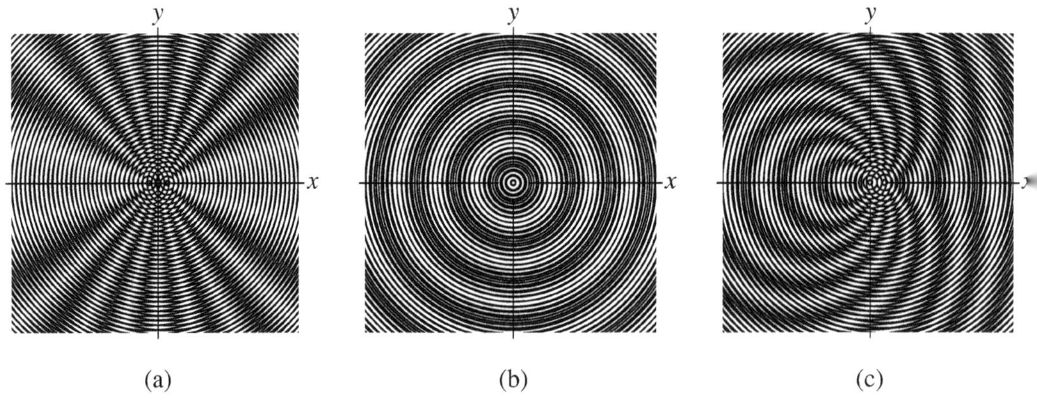


**Figure 11.4:** The (1,-1)-moiré in the superposition of two binary circular gratings, when: (a) both gratings have identical line spacings and are horizontally shifted by $\pm x_0$ from the origin; (b) both gratings are centered on the origin but their line spacings $T_1$ and $T_2$ are slightly different; (c) same as in (b) but the second grating is horizontally shifted by $x_0$ from the origin.

# Appendix A

## Periodic functions and their spectra

### A.1 Introduction

According to the Fourier theory, any periodic function (which satisfies certain conditions; see Sec. C.12 in Appendix C) can be represented by means of a Fourier series, and moreover, its Fourier series expansion is *unique*. In this appendix we briefly review periodic functions in one or two variables, along with their Fourier series expansions and their spectral representations. The main purpose of this appendix is to put together, using our own nomenclature, the main results that we need about periodic functions and their spectra, results which are normally scattered in literature among several different domains. Some of these results can be found in standard mathematic textbooks, while others are treated in textbooks on optics or crystallography (see the cited references). Some aspects of our approach are, however, original (notably Sec. A.6).

It should be emphasized here that the Fourier series expansion of a periodic function $p(x)$ is just an alternative representation of that function in the image domain. The importance of this representation is in that it explicitly gives the spectral decomposition of $p(x)$, i.e., the frequencies and the amplitudes of the impulses which make up the spectrum of $p(x)$. The Fourier series expansion of a periodic function will serve us therefore as a link between the original function $p(x)$ in the image domain and its spectrum $P(u)$ in the frequency domain.

### A.2 Periodic functions, their Fourier series and their spectra in the 1D case

A function $p(x)$ is called *periodic* if there exists a number $T \neq 0$ such that for all $x \in \mathbb{R}$:

$$p(x+T) = p(x)$$

The number $T$ is called a *period* of the function $p(x)$; note however that $T$ is not unique, since if $T$ is a period of $p(x)$, so is $nT$ for any integer $n$. The smallest period $T > 0$ is called the *fundamental period* of $p(x)$, and its reciprocal value, $f = 1/T$, is called the *fundamental frequency* of $p(x)$.[1] Note that the set of all periods of $p(x)$, i.e., the set of all integer multiples of the fundamental period of $p(x)$, forms a lattice in $\mathbb{R}$: $L_T = \{nT \mid n \in \mathbb{Z}\}$, whose basis is the fundamental period of $p(x)$.[2] Similarly, the set of all integer multiples of the fundamental frequency, i.e., all the harmonics of $f$, forms a lattice in the spectral domain:

---

[1] When no risk of confusion arises it is customary to omit the word "fundamental", and to use the terms *period* and *frequency* of $p(x)$ as abbreviations for the *fundamental period* and the *fundamental frequency* of $p(x)$.

[2] For the definition of a lattice, see Sec. 5.2.

$L_f = \{nf \mid n \in \mathbb{Z}\} = \{n/T \mid n \in \mathbb{Z}\}$, whose basis is the fundamental frequency of $p(x)$. As we will see later in this section, $L_f$ is the support of the spectrum of $p(x)$ in the frequency domain. We will also see below (in Sec. A.4) that the lattices $L_T$ in the image domain and $L_f$ in the frequency domain are said to be *reciprocal*; note, however, that the only member of $L_T$ whose reciprocal value is found in $L_f$ is the fundamental period itself. (For instance, although $2T$ is a period of $p(x)$, there is no corresponding reciprocal frequency $\frac{1}{2T} = \frac{f}{2}$ in $L_f$).

Suppose that $p(x)$ is a periodic function of period $T$ which satisfies the required convergence conditions (see Sec. C.12 in Appendix C). Then $p(x)$ can be expanded (or *developed*, or *decomposed*) into the form of a Fourier series, i.e., an infinite series of weighted cosine and sine functions at the fundamental frequency $f = 1/T$ and its harmonics $nf = n/T$ [Bracewell86 p. 205]:[3]

$$p(x) = a_0 + 2\sum_{n=1}^{\infty} a_n \cos(2\pi nx/T) + 2\sum_{n=1}^{\infty} b_n \sin(2\pi nx/T) \qquad \text{(A.1)}$$

where the weighting coefficients $a_n$ and $b_n$, which are called the *Fourier series coefficients* of $p(x)$, are real numbers given by:[4]

$$a_n = \frac{1}{T}\int_T p(x)\cos(2\pi nx/T)\,dx \qquad b_n = \frac{1}{T}\int_T p(x)\sin(2\pi nx/T)\,dx \qquad \text{(A.2)}$$

$\int_T$ means here that the integration may be done over any full period of $p(x)$, i.e., from $x_0$ to $x_0+T$ where $x_0$ is arbitrary; depending on the case it may be more convenient to integrate between $0...T$, between $-T/2...T/2$, etc.

We notice from (A.2) that for negative $n$ we have:

$$a_{-n} = a_n, \qquad b_{-n} = -b_n \qquad (n = 1,2,...)$$

and $b_0 = 0$. Therefore the Fourier series of $p(x)$ can be rewritten as a two-sided series, in a *symmetric* way, as follows (putting also $f = 1/T$):

$$p(x) = \sum_{n=-\infty}^{\infty} a_n \cos(2\pi nfx) + \sum_{n=-\infty}^{\infty} b_n \sin(2\pi nfx) \qquad \text{(A.3)}$$

with:

$$a_n = \frac{1}{T}\int_T p(x)\cos(2\pi nfx)\,dx \qquad b_n = \frac{1}{T}\int_T p(x)\sin(2\pi nfx)\,dx \qquad \text{(A.4)}$$

Note that if $p(x)$ is symmetric about the origin there are no sine components, and $b_n = 0$ for all $n$.

However, although in Chapter 2 we adopt, for didactic reasons, this trigonometric form of the Fourier series, in more advanced chapters we will usually prefer the *exponential* (or

[3] Note that depending on $p(x)$ some or even most of the weighting coefficients $a_n$ and $b_n$ may be zero, so that the Fourier series expansion may include only a finite number of non-zero terms; a trivial example of this type is the function $p(x) = \cos(2\pi x/T)$.

[4] Note that in most textbooks the factors 2 appear within the Fourier coefficients $a_n$ and $b_n$. We prefer, however, to put them before the summations in (A.1), in order to emphasize the correspondence between the Fourier series coefficients and the impulse amplitudes of the two-sided comb which extends to both directions in the spectrum.

*complex*) notation [Champeney73, p. 2], which is more compact and lends itself more easily to mathematical manipulations. This form is obtained from (A.3) and (A.4) by expressing the cosines and sines using the Euler identities: $2\cos\vartheta = e^{i\vartheta} + e^{-i\vartheta}$ and $2\sin\vartheta = -i(e^{i\vartheta} - e^{-i\vartheta})$, grouping the terms $e^{i2\pi nfx}$ and $e^{-i2\pi nfx}$ separately, and combining them into a single series. In the exponential notation the Fourier series expansion of $p(x)$ becomes:

$$p(x) = \sum_{n=-\infty}^{\infty} c_n \, e^{i2\pi nfx} \tag{A.5}$$

where the $n$-th Fourier series coefficient $c_n$ is given (as a single complex number instead of a pair of real numbers $a_n$, $b_n$ as in (A.4)) by:

$$c_n = \frac{1}{T}\int_T p(x) \, e^{-i2\pi nfx} \, dx \tag{A.6}$$

Note that the trigonometric and the exponential forms of the Fourier series are equivalent; by comparing expressions (A.5), (A.6) with expressions (A.3), (A.4) the following relations between their coefficients are obtained:

$$\begin{aligned} c_0 &= a_0 & & \\ c_n &= a_n - ib_n, & c_{-n} &= a_n + ib_n \quad (n \geq 1) \\ 2a_n &= (c_n + c_{-n}), & 2b_n &= i(c_n - c_{-n}) \quad (n \geq 1) \end{aligned} \tag{A.7}$$

(see, for example, [Champeney73 p. 3], with the required adaptations to our notation conventions).

As already mentioned in Sec. A.1, the representation of a periodic function $p(x)$ in the form of its Fourier series expansion will serve us as a link between the original function $p(x)$ and its spectrum, $P(u)$. This is based on the fact that the Fourier transform (the spectrum) of $g(x) = e^{i2\pi fx}$ is $G(u) = \delta(u-f)$, namely: an impulse at the frequency $f$ [Bracewell86 p. 101]. Consequently, the spectrum of each term in the Fourier series (A.5) consists of a single impulse at the frequency of $nf$, whose amplitude (real or complex) is given by the corresponding Fourier coefficient. It follows therefore (under the appropriate convergence conditions) that the spectrum $P(u)$ of the periodic function $p(x)$ is an impulse-comb of step $f$, whose $n$-th impulse is located at the frequency $u = nf$ and has the amplitude $c_n$. The general form of the spectrum of a periodic function $p(x)$ with period $T = 1/f$ is, therefore [Papoulis68 p. 107]:

$$P(u) = \sum_{n=-\infty}^{\infty} c_n \, \delta(u-n/T) = \sum_{n=-\infty}^{\infty} c_n \, \delta(u-nf) \tag{A.8}$$

where $\delta(u)$ is the impulse symbol (see Chapter 5 in [Bracewell86]). As we can see, the support of this spectrum is the lattice $L_f$, which contains all the integer multiples of the fundamental frequency $f$. Moreover, the converse is also true: given a spectrum $P(u)$ whose support is a lattice $L_f$, it follows that the original function $p(x)$ in the image domain, which can be represented by (A.5), is periodic with period $T = 1/f$ (of course, this is only meaningful if the convergence conditions on the Fourier series (A.5) are satisfied). We have, therefore, the following result, under the appropriate convergence conditions:

**Proposition A.1:** A function $p(x)$ in the image domain is periodic *iff* its spectrum support in the frequency domain is a lattice. ■

Finally, note that since we restrict ourselves to *real* functions $p(x)$ in the image domain, it follows from Eq. (A.7) that $c_{-n}$ is the complex conjugate of $c_n$, namely: $\text{Re}(c_{-n}) = \text{Re}(c_n)$ and $\text{Im}(c_{-n}) = -\text{Im}(c_n)$. This means that our impulse-comb in the spectrum is always *Hermitian*: its real part is symmetric while its imaginary part is antisymmetric. This is also in agreement with the properties of Fourier transforms of real functions; see [Bracewell86 pp. 14–15]. Note also that depending on $p(x)$ some or even most of the comb impulses may have amplitudes $c_n = 0$, as in the case of $p(x) = \cos(2\pi fx)$, where only the fundamental impulse pair, with indices $n = \pm 1$, has a non-zero amplitude.

## A.3 Periodic functions, their Fourier series and their spectra in the 2D case

The case of periodic functions with two variables is more diversified than the 1D case, and it can be divided into 4 subcases. A periodic function $p(x,y)$ may be either 1-fold periodic[5] (like a grating or a cosine over the plane) or 2-fold periodic (such as a dot-screen); and in each of these cases it may be periodic either in the direction of the main axes, or in any other direction. In the following subsections we briefly review each of these 4 subcases. The most general case of 2D periodicity is reviewed last, in Sec. A.3.4.

### A.3.1 1-fold periodic functions in the $x$ or $y$ direction

The simplest case of 2D periodic functions is that of a 1-fold periodic function $p(x,y)$ whose fundamental period (smallest period>0) is in the direction of one of the axes, say, in the $x$ direction. In this case $p(x,y)$ can be considered as a 1D periodic function $p(x)$, since it is constant in the $y$ direction, and its Fourier series expansion is identical to (A.5). The 2D spectrum of $p(x,y)$ consists in this case of a 1D impulse-comb, identical to the spectrum (A.8) of the 1D function $p(x)$, which is located in the $u,v$ plane on top of the $u$ axis.

### A.3.2 2-fold periodic functions in the $x$ and $y$ directions

Let us proceed now to the case of a 2-fold periodic function $p(x,y)$ which is periodic in the $x$ and in the $y$ directions with periods $T_x$ and $T_y$, namely: $p(x+T_x,y) = p(x,y)$ and $p(x,y+T_y) = p(x,y)$ for all $(x,y) \in \mathbb{R}^2$. If $p(x,y)$ satisfies certain convergence conditions (see Sec. C.12 in Appendix C), then it can be expanded into a 2D Fourier series as follows:

$$p(x,y) = \sum_{m=-\infty}^{\infty} \sum_{n=-\infty}^{\infty} a_{m,n} \cos 2\pi(mx/T_x + ny/T_y) + \sum_{m=-\infty}^{\infty} \sum_{n=-\infty}^{\infty} b_{m,n} \sin 2\pi(mx/T_x + ny/T_y) \quad \text{(A.9)}$$

[5] By the term *1-fold periodic* we understand that the function $p(x,y)$ is *constant* in the direction perpendicular to the direction of periodicity. Otherwise the Fourier development of $p(x,y)$ is no longer a pure 1D Fourier series, and it may even be a hybrid case having Fourier series in one direction and Fourier transform in the other [Korn68 p. 143].

where:
$$a_{m,n} = \frac{1}{T_xT_y}\iint_{T_xT_y} p(x,y)\cos 2\pi(mx/T_x + ny/T_y)\,dxdy$$
$$b_{m,n} = \frac{1}{T_xT_y}\iint_{T_xT_y} p(x,y)\sin 2\pi(mx/T_x + ny/T_y)\,dxdy \qquad \text{(A.10)}$$

$\iint_{T_xT_y}$ means an integration over any full period of $p(x,y)$, i.e., over a rectangle of sides $T_x$ and $T_y$ defined by the points $(x_0,y_0)$, $(x_0+T_x,y_0)$, $(x_0,y_0+T_y)$ and $(x_0+T_x,y_0+T_y)$ where $x_0$ and $y_0$ are arbitrary. Clearly, this is a 2D extension of the 1D trigonometric expressions (A.3) and (A.4). Note that here, too, if $p(x,y)$ is symmetric about the origin then there are no sine components, and $b_{m,n} = 0$ for all $m,n$.

However, for the sake of convenience and conciseness we will adopt here, too, as in the 1D case, the exponential form of the Fourier series, which is obtained from the trigonometric form by the Euler identities. In the exponential notation the 2D Fourier series expansions of $p(x,y)$ becomes (using also $u_0 = 1/T_x$ and $v_0 = 1/T_y$):

$$p(x,y) = \sum_{m=-\infty}^{\infty}\sum_{n=-\infty}^{\infty} c_{m,n}\, e^{i2\pi(mu_0x + nv_0y)} \qquad \text{(A.11)}$$

where the Fourier coefficients $c_{m,n}$ are given by:

$$c_{m,n} = \frac{1}{T_xT_y}\iint_{T_xT_y} p(x,y)\, e^{-i2\pi(mu_0x + nv_0y)}\,dxdy \qquad \text{(A.12)}$$

Expressions (A.11) and (A.12) are clearly 2D extensions of (A.5) and (A.6). These 2D expressions can be rendered more compact by using the vector notation:

$$p(\mathbf{x}) = \sum_{m=-\infty}^{\infty}\sum_{n=-\infty}^{\infty} c_{m,n}\, e^{i2\pi\mathbf{f}_{m,n}\cdot\mathbf{x}} \qquad \text{(A.13)}$$

with the Fourier series coefficients:

$$c_{m,n} = \frac{1}{T_xT_y}\iint_{T_xT_y} p(\mathbf{x})\, e^{-i2\pi\mathbf{f}_{m,n}\cdot\mathbf{x}}\,d\mathbf{x} \qquad \text{(A.14)}$$

where $\mathbf{x} = (x,y)$, $\mathbf{f}_{m,n} = (mu_0,nv_0)$ is the $(m,n)$-th frequency harmonic, and $\mathbf{f}_{m,n}\cdot\mathbf{x}$ denotes the scalar product of the vectors $\mathbf{f}_{m,n}$ and $\mathbf{x}$. Note that an even more compact notation can be obtained by substituting $\mathbf{n}$ for $(m,n)$ in the indices:

$$p(\mathbf{x}) = \sum_{\mathbf{n}} c_{\mathbf{n}}\, e^{i2\pi\mathbf{f}_{\mathbf{n}}\cdot\mathbf{x}} \qquad \text{(A.15)}$$

However, for the sake of clarity we will usually prefer the more explicit form of (A.13).

Like in the 1D case, the spectrum of the 2-fold periodic function $p(x,y)$ is readily obtained from the Fourier series representation of $p(x,y)$, (A.11): it is an impulse-nailbed, whose $(m,n)$-th impulse is located in the $u,v$ plane at the point $\mathbf{f}_{m,n} = (mu_0,nv_0)$ and has the amplitude $c_{m,n}$ (see Fig. 2.12(f)). Using the 2D impulse symbol $\delta(u,v)$ the spectrum of $p(x,y)$ is, therefore [Papoulis68 p. 114]:

$$P(u,v) = \sum_{m=-\infty}^{\infty}\sum_{n=-\infty}^{\infty} c_{m,n}\, \delta(u-mu_0, v-nv_0) \qquad \text{(A.16)}$$

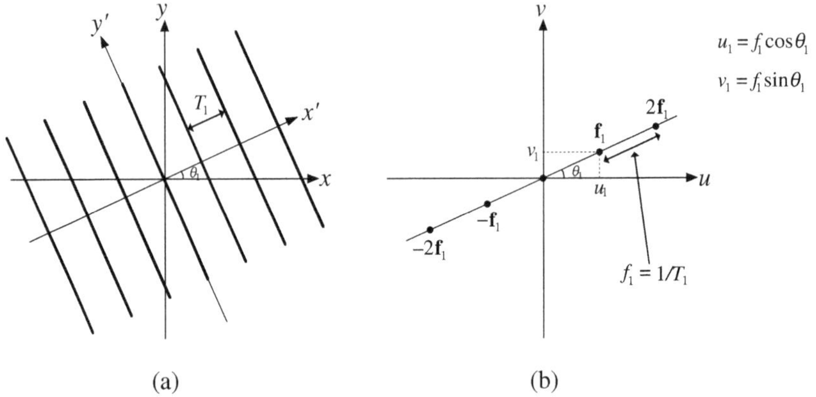


**Figure A.1:** A schematic plot of the 1-fold periodic function $p_1(x,y)$ in the image domain (a), and its impulse-comb in the spectral domain (b).

or in a vector form, where $\mathbf{f} = (u,v)$ and $\mathbf{f}_{m,n} = (mu_0,nv_0)$:

$$P(\mathbf{f}) = \sum_{m=-\infty}^{\infty} \sum_{n=-\infty}^{\infty} c_{m,n}\, \delta(\mathbf{f}-\mathbf{f}_{m,n}) \tag{A.17}$$

This means that the support of the spectrum of $p(x,y)$ is the 2D lattice, oriented in the axes directions, which is spanned by the fundamental frequencies $\mathbf{f}_{1,0} = (u_0,0)$ and $\mathbf{f}_{0,1} = (0,v_0)$. Moreover, like in the 1D case we have the following result (under the appropriate convergence conditions):

**Proposition A.2:** A function $p(x,y)$ in the image domain is periodic in the $x$ and $y$ directions *iff* its spectrum support in the frequency domain is a 2D lattice in the axes directions. ■

### A.3.3 1-fold periodic functions in an arbitrary direction

Unlike in the 1D case, in functions of two variables periodicity is not necessarily limited to the direction of the main axes. Let us consider first the case of a 1-fold periodic function, $p_1(x,y)$, whose fundamental period (i.e., smallest period > 0) is $T_1 = 1/f_1$ in the direction $\theta_1$ (for example: a rotated grating). This is, in fact, a function of the single variable $x'$ in the rotated coordinate system $x',y'$, and along the $y'$ direction it is constant (see Fig. A.1). Therefore its Fourier series expansion is given in the rotated coordinate system $x',y'$ by expressions (A.5) and (A.6) of the 1D case:

$$p_1(x',y') = \sum_{n=-\infty}^{\infty} c_n\, e^{i2\pi n f_1 x'} \qquad \text{where:} \qquad c_n = \frac{1}{T_1}\int_{T_1} p_1(x',y')\, e^{-i2\pi n f_1 x'}\, dx' \tag{A.18}$$

Changing back to the $x,y$ coordinate system by putting: $x' = x\cos\theta_1 + y\sin\theta_1$ [Spiegel68 p. 36] and hence: $f_1 x' = u_1 x + v_1 y$ (see Fig. A.1(b)) we obtain:

$$p_1(x,y) = \sum_{n=-\infty}^{\infty} c_n\, e^{i2\pi n(u_1x+v_1y)} \qquad \text{(A.19)}$$

Note that $(u_1,v_1)$ are the Cartesian coordinates, in the $u,v$ plane, of the *frequency vector* $\mathbf{f}_1$ in the spectrum of $p_1(x,y)$, whose polar coordinates are $(f_1,\theta_1)$. Therefore the expression $(u_1x + v_1y)$ in the exponential part of (A.19) can be rewritten in a more compact vector form as the scalar product $\mathbf{f}_1\boldsymbol{\cdot}\mathbf{x}$ of the two vectors $\mathbf{f}_1 = (u_1,v_1)$ and $\mathbf{x} = (x,y)$, and the Fourier series becomes:

$$p_1(\mathbf{x}) = \sum_{n=-\infty}^{\infty} c_n\, e^{i2\pi n\mathbf{f}_1\cdot\mathbf{x}} \qquad \text{(A.20)}$$

The 2D spectrum of $p_1(x,y)$ consists of a 1D impulse-comb, identical to the comb (A.8) of the 1D case, which is rotated in the $u,v$ plane by the angle of $\theta_1$. Using the 2D impulse symbol $\delta(u,v)$ the spectrum of $p_1(x,y)$ is given by (see Fig. A.1(b)):

$$P_1(u,v) = \sum_{n=-\infty}^{\infty} c_n\, \delta(u-nu_1, v-nv_1) \qquad \text{(A.21)}$$

or in a vector form, where $\mathbf{f} = (u,v)$ and $\mathbf{f}_1 = (u_1,v_1)$:

$$P_1(\mathbf{f}) = \sum_{n=-\infty}^{\infty} c_n\, \delta(\mathbf{f}-n\mathbf{f}_1) \qquad \text{(A.22)}$$

The support of the spectrum $P_1(u,v)$ is a 1D lattice in the $\theta_1$ direction, which contains all the integer multiples of the fundamental frequency $\mathbf{f}_1$. And moreover, under the appropriate convergence conditions we obtain, here too:

**Proposition A.3:** A function $p_1(x,y)$ is 1-fold periodic in the $\theta_1$ direction *iff* its spectrum support in the frequency domain is a 1D lattice in the $\theta_1$ direction. ■

### A.3.4 2-fold periodic functions in arbitrary directions (skew-periodic functions)

Let us proceed now to the most general 2D case: that of a 2-fold periodic function having two arbitrary periods, which are not necessarily oriented in the $x,y$ directions, not necessarily orthogonal to each other, and do not necessarily have the same period length in both directions. Such functions are called *2-fold periodic* or *skew-periodic* functions [Papoulis68 p. 116]. An example of such a function is shown in Fig. A.2(a).

Formally, a function $p(x,y)$ is called skew-periodic if there exist two non-zero and non-collinear vectors $\mathbf{P}_1 = (x_1,y_1)$ and $\mathbf{P}_2 = (x_2,y_2)$ so that for all $(x,y) \in \mathbb{R}^2$:

$$p(x+x_1,y+y_1) = p(x,y) \quad \text{and} \quad p(x+x_2,y+y_2) = p(x,y).$$[6]

The vectors $\mathbf{P}_1$ and $\mathbf{P}_2$ are called *periods* or *period-vectors* of $p(x,y)$;[7] note however that they are not unique, since for any integers $m,n$ the vector $m\mathbf{P}_1+n\mathbf{P}_2$ is also a period of $p(x,y)$ (in vector notation: $p(\mathbf{x}+m\mathbf{P}_1+n\mathbf{P}_2) = p(\mathbf{x})$ for all $\mathbf{x} \in \mathbb{R}^2$; see Fig. A.2(a)). Note that $p(x,y)$ is completely determined from its values in the period-parallelogram $A$ defined by

---

[6] An alternative definition, based on the *period-parallelogram* of $p(x,y)$ rather than on its *period-vectors*, is given in the Glossary (see under the term "period").

[7] Like in the case of frequency-vectors, we always consider period-vectors as radius-vectors emanating from the origin (and hence, the period-parallelograms which they define are attached to the origin).

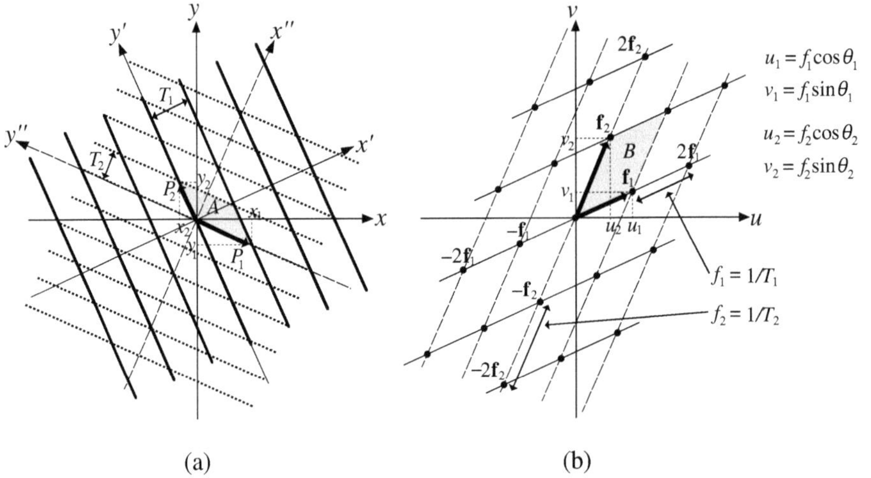


**Figure A.2:** A schematic plot of the 2-fold periodic (skew-periodic) function $p(x,y)$ in the image domain (a), and its skewed impulse-nailbed in the spectral domain (b). The gray parallelogram $A$ in the image domain represents a one-period element (tile) of $p(x,y)$. $P_1$ and $P_2$ are segments of this parallelogram which coincide with the period-vectors $\mathbf{P}_1$ and $\mathbf{P}_2$.

$\mathbf{P}_1$ and $\mathbf{P}_2$, which repeats itself throughout the $x,y$ plane. Clearly, each pair of non-zero and non-collinear period-vectors of $p(x,y)$ defines a different period-parallelogram $A$. A period parallelogram is called a *fundamental period-parallelogram* of $p(x,y)$ if it has the smallest area > 0 (the area of the smallest period). However, unlike in the 1D case in which there existed a single shortest period $T > 0$ attached to the origin, in the 2D case there exist infinitely many different period-parallelograms attached to the origin which all have the smallest period area (see, for example, parallelograms $A$ in Fig. A.2 and $A'$ in Fig. A.4). Each of these parallelograms is a fundamental period-parallelogram of $p(x,y)$, and each pair of period-vectors $\mathbf{P}_1$ and $\mathbf{P}_2$ of $p(x,y)$ which defines such a fundamental period-parallelogram is called a pair of *fundamental period-vectors* of $p(x,y)$. Their reciprocal vector pair (in a sense to be defined below in Sec. A.4) in the spectral domain, $\mathbf{f}_1$ and $\mathbf{f}_2$, is called a pair of *fundamental frequency-vectors* of $p(x,y)$ (see Fig. A.2(b)). For now, we just mention that $\mathbf{f}_1$ and $\mathbf{f}_2$ are oriented perpendicularly to the corresponding period-vectors $\mathbf{P}_2$ and $\mathbf{P}_1$, respectively: $\mathbf{f}_1 \perp \mathbf{P}_2$ and $\mathbf{f}_2 \perp \mathbf{P}_1$. Since each of the different fundamental period-parallelograms fully defines $p(x,y)$, we may freely choose one of them as a representative fundamental period-parallelogram, and consider its associated period-vectors $\mathbf{P}_1$, $\mathbf{P}_2$ and frequency vectors $\mathbf{f}_1$, $\mathbf{f}_2$ as the representative period and frequency vectors of $p(x,y)$.

Note that the set of all the periods of $p(x,y)$, i.e., the set of all integer linear combinations of the fundamental period-vectors $\mathbf{P}_1$ and $\mathbf{P}_2$, forms a skewed (=oblique) 2D lattice in the $x,y$ plane: $L_\mathbf{P} = \{m\mathbf{P}_1 + n\mathbf{P}_2 \mid m,n \in \mathbb{Z}\}$. And similarly, the set of all integer linear combinations of the fundamental frequency-vectors $\mathbf{f}_1$ and $\mathbf{f}_2$ forms a skewed 2D lattice in the

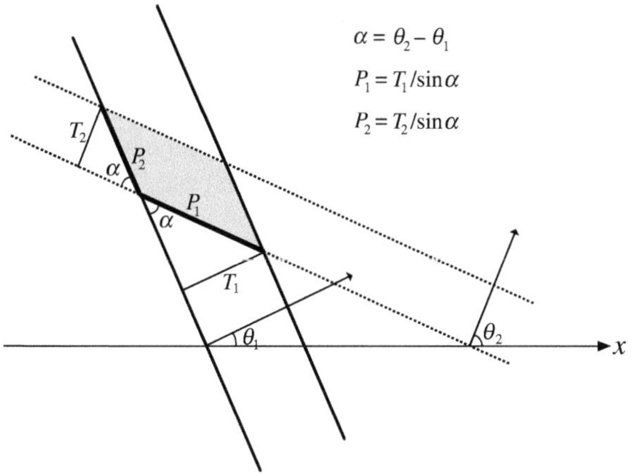


**Figure A.3:** A magnified view of a single period-parallelogram from Fig. A.2(a), showing the relations between $P_i$ and $T_i$.

spectral $u,v$ plane: $L_\mathbf{f} = \{m\mathbf{f}_1+n\mathbf{f}_2 \mid m,n \in \mathbb{Z}\}$. Both of these lattices are spanned by any of the possible fundamental vector pairs $\mathbf{P}_1$, $\mathbf{P}_2$ (or $\mathbf{f}_1$, $\mathbf{f}_2$) of $p(x,y)$. We will see later in this section that like in the 1D case the lattice $L_\mathbf{f}$ is the support of the spectrum of $p(x,y)$ in the frequency domain. The lattice $L_\mathbf{P}$ in the image domain and the lattice $L_\mathbf{f}$ in the frequency domain are said to be *reciprocal* (in a sense to be defined below, in Sec. A.4).

It would be instructive at this point to illustrate the case of a skew-periodic function $p(x,y)$ by an example:

**Example A.1:** Let $p_1(x,y)$ be a 1-fold periodic function in the $\theta_1$ direction as in Sec. A.3.3 above (see Fig. A.1, or the solid grating in Fig. A.2(a)), and let $p_2(x,y)$ be a similar 1-fold periodic function with the period $T_2 = 1/f_2$ in the $\theta_2$ direction (see the dotted grating in Fig. A.2(a)). Then, the superposed function $p(x,y) = p_1(x,y)\,p_2(x,y)$ is skew-periodic. However, as it can be seen in Fig. A.2, the periodicity of $p(x,y)$ is *not* given by the periods $T_1$ and $T_2$ of the original functions $p_1(x,y)$ and $p_2(x,y)$, but rather by the periods $P_1$ and $P_2$, where $P_1 \perp T_2$, $P_2 \perp T_1$. This fact, although surprising at first sight, is explained as follows:

Consider the original function $p_1(x,y)$ (Fig. A.1). Clearly, this function is periodic in the direction of $x'$ with the period $T_1$; but at the same time it is also periodic in the direction of $x$ (with the period $T_1/\cos\theta_1$), in the direction of $y$ (with the period $T_1/\sin\theta_1$), or in any other direction in the plane — *with only one exception*: the direction perpendicular to $x'$, i.e., the direction of $y'$, in which the function $p_1(x,y)$ is *constant*.

Similarly, $p_2(x,y)$ is periodic in all directions in the plane, except for the direction of $y''$, in which it is constant. Now, when $p_1(x,y)$ and $p_2(x,y)$ are superposed, their periodicities are destroyed in all directions — except in the directions in which $p_1(x,y)$ or $p_2(x,y)$ are constant and do not intervene, namely: except in the directions of $y'$ and $y''$. And indeed, in the direction of $y'$ (in which $p_1(x,y)$ is constant) $p(x,y)$ behaves like $p_2(x,y)$ and has the period $P_2$, and in the direction of $y''$ (in which $p_2(x,y)$ is constant) $p(x,y)$ behaves like $p_1(x,y)$ and has the period $P_1$. These two exceptional directions define therefore the periodicity of the composite function $p(x,y)$. If we denote the angle difference $\theta_2-\theta_1$ between the two gratings by $\alpha$ (see Fig. A.3), we find that $P_1 = T_1/\sin\alpha$ and $P_2 = T_2/\sin\alpha$.

In the frequency domain, however, there are no such "surprises", and the spectrum of $p(x,y)$ is indeed the convolution of the 1D comb of $p_1(x,y)$ in the $\theta_1$ direction and the 1D comb of $p_2(x,y)$ in the $\theta_2$ direction (see Fig. A.2(b)). The result of this convolution is a skewed nailbed whose support is spanned by $\mathbf{f}_1$ and $\mathbf{f}_2$, the fundamental frequencies of the two original combs. Note that only when the directions $\theta_1$ and $\theta_2$ are perpendicular to each other ($\alpha = \pm 90°$) do $P_1$ and $P_2$ coincide with $T_1$ and $T_2$, respectively, and $\mathbf{f}_1 \perp \mathbf{P}_2$ and $\mathbf{f}_2 \perp \mathbf{P}_1$ mean, then, as we would expect: $\mathbf{f}_1 \| \mathbf{P}_1$ and $\mathbf{f}_2 \| \mathbf{P}_2$.

Note that Example A.1 illustrates also another typical phenomenon which is related to non-orthogonal lattices: Let $L_\mathbf{f}$ be the 2D frequency lattice of $p(x,y)$; this lattice is obtained in the spectrum by the convolution of the two 1D lattices (combs) of the superposed 1-fold periodic functions $p_1(x,y)$ and $p_2(x,y)$. Clearly, $L_\mathbf{f}$ is spanned by the vectors $\mathbf{f}_1$ and $\mathbf{f}_2$, the fundamental frequency vectors of $p_1(x,y)$ and $p_2(x,y)$; but unless these two functions are superposed perpendicularly, $\mathbf{f}_1$ and $\mathbf{f}_2$ are not necessarily the *shortest* frequency vectors which span $L_\mathbf{f}$. In the case of Fig. A.2(b), for example, the shortest fundamental frequency vectors which span $L_\mathbf{f}$ are $\mathbf{f}'_1 = \mathbf{f}_1$ and $\mathbf{f}'_2 = \mathbf{f}_2 - \mathbf{f}_1$, and not $\mathbf{f}_1$ and $\mathbf{f}_2$. This is clearly shown in Fig. A.4(b). Similarly, in the image domain (Fig. A.2(a)) the period-vectors $\mathbf{P}_1$ and $\mathbf{P}_2$ (the reciprocal vector pair of $\mathbf{f}_1$, $\mathbf{f}_2$) are not the shortest fundamental period-vectors of the 2-fold periodic function $p(x,y)$: the shortest period-vectors are $\mathbf{P}'_1 = \mathbf{P}_1 + \mathbf{P}_2$ and $\mathbf{P}'_2 = \mathbf{P}_2$, the reciprocal vector pair of $\mathbf{f}'_1$ and $\mathbf{f}'_2$ (see Fig. A.4(a)).

It is interesting to note that the question of finding the shortest basis vectors of a given lattice of dimension $n > 1$ is not trivial, and it is one of the subjects which are dealt in geometry of numbers [Kannan87 pp. 4*ff*, Gruber93 pp. 742, 751*ff*]. However, in our study of the moiré effect we will always choose the fundamental frequency vectors $\mathbf{f}_i$ of the superposition to be the fundamental frequency vectors of the superposed layers, like in Fig. A.2, even when they are not the shortest possible frequency vectors in the superposition. ■

Suppose now that $p(x,y)$ is a skew-periodic function which satisfies the required convergence conditions, and suppose that $\mathbf{P}_1$, $\mathbf{P}_2$ and $\mathbf{f}_1$, $\mathbf{f}_2$ are representative fundamental period and frequency vectors of $p(x,y)$ (see Fig. A.2). Then its Fourier series expansion in terms of the $x'$ and $x''$ coordinates is given by:

$$p(x',x'') = \sum_{m=-\infty}^{\infty} \sum_{n=-\infty}^{\infty} c_{m,n}\, e^{i2\pi[mf_1x' + nf_2x'']}$$

Changing back to the $x,y$ coordinate system by putting: $x' = x\cos\theta_1 + y\sin\theta_1$, $x'' = x\cos\theta_2 + y\sin\theta_2$ [Spiegel68 p. 36] and hence: $f_1x' = u_1x + v_1y$, $f_2x' = u_2x + v_2y$ (see Fig. A.2(b)) we obtain:

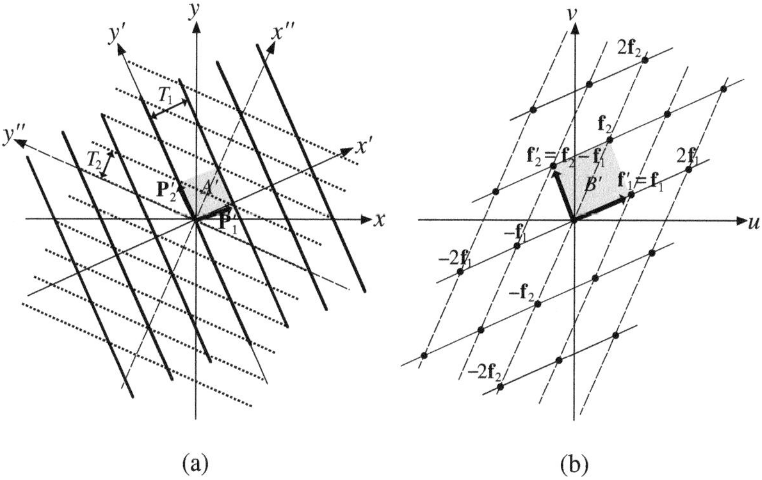


**Figure A.4:** A different set of fundamental period-vectors $\mathbf{P}'_1$, $\mathbf{P}'_2$ and fundamental frequency vectors $\mathbf{f}'_1$, $\mathbf{f}'_2$ which span the same period-lattice $L_\mathbf{P}$ and frequency-lattice $L_\mathbf{f}$ as in Fig. A.2. Note that $\mathbf{P}'_1$, $\mathbf{P}'_2$ and $\mathbf{f}'_1$, $\mathbf{f}'_2$ are the shortest vector pairs which span the lattices $L_\mathbf{P}$ and $L_\mathbf{f}$; the parallelograms $A'$ and $B'$ they define have the same respective areas as the parallelograms $A$ and $B$ in Fig. A.2, but they are almost square.

$$p(x,y) = \sum_{m=-\infty}^{\infty} \sum_{n=-\infty}^{\infty} c_{m,n}\, e^{i2\pi[m(u_1x + v_1y) + n(u_2x + v_2y)]}$$

(compare with the 1D case of Eq. (A.19)), or in other words [Papoulis68 p. 117]:

$$p(x,y) = \sum_{m=-\infty}^{\infty} \sum_{n=-\infty}^{\infty} c_{m,n}\, e^{i2\pi[(mu_1 + nu_2)x + (mv_1 + nv_2)y]} \tag{A.23}$$

where $u_1$, $u_2$, $v_1$, $v_2$ are defined in Fig. A.2, and the Fourier coefficients $c_{m,n}$ are:

$$c_{m,n} = \frac{1}{\bar{A}} \iint_A p(x,y)\, e^{-i2\pi[(mu_1 + nu_2)x + (mv_1 + nv_2)y]}\, dxdy \tag{A.24}$$

$\iint_A$ means here an integration over the parallelogram $A$ defined by $\mathbf{P}_1$ and $\mathbf{P}_2$ (see Fig. A.2(a)), whose area $\bar{A}$ is given by the cross product:

$$\bar{A} = \mathbf{P}_1 \times \mathbf{P}_2 = x_1y_2 - y_1x_2 \tag{A.25}$$

This can be expressed more compactly in the vector notation:

$$p(\mathbf{x}) = \sum_{m=-\infty}^{\infty} \sum_{n=-\infty}^{\infty} c_{m,n}\, e^{i2\pi(m\mathbf{f}_1+n\mathbf{f}_2)\cdot\mathbf{x}} \tag{A.26}$$

with the Fourier series coefficients:

$$c_{m,n} = \frac{1}{\bar{A}} \iint_A p(\mathbf{x})\, e^{-i2\pi(m\mathbf{f}_1+n\mathbf{f}_2)\cdot\mathbf{x}}\, d\mathbf{x} \tag{A.27}$$

where $\mathbf{f}_1 = (u_1,v_1)$ and $\mathbf{f}_2 = (u_2,v_2)$ are the corresponding fundamental frequency vectors, and $\mathbf{x} = (x,y)$.

Note that Eqs. (A.26) and (A.27) can be further simplified into the vector form of Eqs. (A.13) and (A.14) (with the integration being done over a parallelogram $A$ rather than over a rectangle $T_xT_y$) if we take:

$$\mathbf{f}_{m,n} = m\mathbf{f}_1+n\mathbf{f}_2 = (mu_1+nu_2,\ mv_1+nv_2) \tag{A.28}$$

as the $(m,n)$-th frequency harmonic. In this notation each frequency vector takes the indices of its impulse in the spectrum convolution; for example, the fundamental frequency vectors $\mathbf{f}_1$ and $\mathbf{f}_2$ are denoted by $\mathbf{f}_{1,0}$ and $\mathbf{f}_{0,1}$, respectively.

The spectrum of the skew-periodic function $p(x,y)$ is a skewed impulse-nailbed (see Fig. A.2(b)), whose $(m,n)$-th impulse is located in the $u,v$ plane at the point given by Eq. (A.28), and has the amplitude $c_{m,n}$. Using the 2D impulse symbol $\delta(u,v)$ the spectrum of $p(x,y)$ is given by [Papoulis68 p. 117]:

$$P(u,v) = \sum_{m=-\infty}^{\infty} \sum_{n=-\infty}^{\infty} c_{m,n}\, \delta(u-mu_1-nu_2,\ v-mv_1-nv_2) \tag{A.29}$$

or in a vector form, with $\mathbf{f} = (u,v)$:

$$P(\mathbf{f}) = \sum_{m=-\infty}^{\infty} \sum_{n=-\infty}^{\infty} c_{m,n}\, \delta(\mathbf{f}-(m\mathbf{f}_1+n\mathbf{f}_2)) \tag{A.30}$$

Note that using the notation of (A.28), the spectrum (A.30), too, can be further simplified, and it coincides with the vector form of (A.17).

As we can see, the support of the spectrum $P(u,v)$ is the skewed 2D lattice $L_\mathbf{f}$, which contains all the integer linear combinations of the fundamental frequency vectors $\mathbf{f}_1$ and $\mathbf{f}_2$.[8] Furthermore, under the appropriate convergence conditions we have the following result:

**Proposition A.4:** A function $p(x,y)$ in the image domain is 2-fold periodic (skew-periodic) *iff* its spectrum support in the frequency domain is a 2D lattice. ■

The relationship between the periodicity of $p(x,y)$ in the image domain and its spectrum support in the frequency domain is the subject of the following section.

## A.4 The period-lattice and the frequency-lattice (= spectrum support)

In this section we summarize the reciprocity relationship between the fundamental period(s) of a periodic function in the image domain and the corresponding fundamental frequency(ies) in the spectral domain. We have seen above, both in the 1D and in the 2D

[8] Note that the lattice $L_\mathbf{f}$ is independent of the choice of the fundamental frequency vectors, as illustrated in Figs. A.2(b) and A.4(b).

cases, that the set of all the periods of a periodic function (i.e., the set of all integer linear combinations of its fundamental period(s)) is a lattice $L_\mathbf{P}$ in the image domain, while the set of all the integer linear combinations of the fundamental frequency(ies) of the periodic function forms a lattice $L_\mathbf{f}$ in the spectral domain. This last lattice, $L_\mathbf{f}$, is in fact the *support* of the spectrum of the periodic function, i.e., the set of the geometric locations of all the impulses of the comb or the nailbed in the spectrum of the periodic function (including those impulses whose amplitudes happen to be zero).

For example, in the 1-fold periodic case of Fig. A.1 the 1D period-lattice $L_\mathbf{P}$ in the image domain (left) is represented by all integer multiples of the fundamental period along the $x'$ axis; and the 1D frequency-lattice $L_\mathbf{f}$ in the frequency domain (right) is the set of all the integer multiples of the fundamental frequency $\mathbf{f}_1$. Both lattices are oriented in the same direction, $\theta_1$, but their steps are reciprocal: the step of the period-lattice is $T_1$, while the step of the frequency-lattice is $f_1 = 1/T_1$. In the 2-fold periodic case shown in Fig. A.2 the 2D period-lattice $L_\mathbf{P}$ in the image domain (left) is represented by all the integer linear combinations of the fundamental periods $\mathbf{P}_1$ and $\mathbf{P}_2$; and the 2D frequency-lattice $L_\mathbf{f}$ in the spectral domain (right) is represented by all the linear combinations of the fundamental frequencies $\mathbf{f}_1$ and $\mathbf{f}_2$ (marked by dots in the figure). However, while the reciprocity between the lattices $L_\mathbf{P}$ and $L_\mathbf{f}$ is straightforward in the 1D case, in the 2D case some further explanation is required.

Let us see what is the relationship in the 2D case between the fundamental periods $\mathbf{P}_1$ and $\mathbf{P}_2$ (which are a basis of the period-lattice $L_\mathbf{P}$ in the image domain) and the corresponding fundamental frequencies $\mathbf{f}_1$ and $\mathbf{f}_2$ (which are a basis of the frequency-lattice $L_\mathbf{f}$ in the spectrum). According to Fig. A.2 we have:

in the image domain:
$$\begin{aligned}\mathbf{P}_1 &= (x_1,y_1)\\ \mathbf{P}_2 &= (x_2,y_2)\end{aligned} \qquad \text{(A.31)}$$

and in the spectrum:
$$\begin{aligned}\mathbf{f}_1 &= (u_1,v_1)\\ \mathbf{f}_2 &= (u_2,v_2)\end{aligned} \qquad \text{(A.32)}$$

In order to find how $\mathbf{f}_1$ and $\mathbf{f}_2$ are related to $\mathbf{P}_1$ and $\mathbf{P}_2$ we would like to express the coordinates $u_i$ and $v_i$ in terms of $x_i$ and $y_i$. And indeed, we have (see Fig. A.1(b)):

$$\begin{aligned}u_1 &= f_1 \cos\theta_1\\ &= \tfrac{1}{T_1}\cos\theta_1\\ &= \frac{y_2}{T_1 P_2} && \text{(since } \cos\theta_1 = \tfrac{y_2}{P_2}\text{; see Fig. A.2(a))}\\ &= \frac{y_2}{\bar{A}}\end{aligned}$$

where $\bar{A}$ is the area of the parallelogram $A$, as given by Eq. (A.25): $\bar{A} = \mathbf{P}_1 \times \mathbf{P}_2 = x_1 y_2 - y_1 x_2$. We can also express in a similar way $v_1$, $u_2$ and $v_2$; and by substituting them into (A.32) we obtain, therefore, the required expressions for $\mathbf{f}_1$, $\mathbf{f}_2$ in terms of $x_i$ and $y_i$:

$$\begin{aligned} \mathbf{f}_1 &= \tfrac{1}{A}(y_2, -x_2) \\ \mathbf{f}_2 &= \tfrac{1}{A}(-y_1, x_1) \end{aligned} \qquad \text{(A.33)}$$

From Eqs. (A.31) and (A.33) we immediately obtain the following properties:

(1) As we have already seen above: $\mathbf{f}_1 \perp \mathbf{P}_2$ and $\mathbf{f}_2 \perp \mathbf{P}_1$.

(2) Concerning the vector lengths we obtain:

from Eq. (A.31):
$$|\mathbf{P}_1| = \sqrt{x_1^2 + y_1^2}$$
$$|\mathbf{P}_2| = \sqrt{x_2^2 + y_2^2}$$

and from Eq. (A.33):
$$|\mathbf{f}_1| = \tfrac{1}{A}\sqrt{x_2^2 + y_2^2} = \tfrac{1}{A}|\mathbf{P}_2|$$
$$|\mathbf{f}_2| = \tfrac{1}{A}\sqrt{x_1^2 + y_1^2} = \tfrac{1}{A}|\mathbf{P}_1|$$

In other words, the length ratio between the two pairs of fundamental vectors is preserved reciprocally:

$$\frac{|\mathbf{f}_1|}{|\mathbf{f}_2|} = \frac{|\mathbf{P}_2|}{|\mathbf{P}_1|}$$

(For example, if the length of $\mathbf{P}_1$ is twice the length of $\mathbf{P}_2$ in the image domain, then in the spectrum the length of $\mathbf{f}_1$ is half the length of $\mathbf{f}_2$).

(3) Another interesting result concerns the areas of the fundamental period-parallelogram $A$ in the image domain and the fundamental frequency-parallelogram $B$ in the spectrum:

$$\bar{A} = \mathbf{P}_1 \times \mathbf{P}_2 = x_1 y_2 - y_1 x_2$$
$$\bar{B} = \mathbf{f}_1 \times \mathbf{f}_2 = u_1 v_2 - v_1 u_2 = \tfrac{1}{A^2}(x_1 y_2 - y_1 x_2) = \tfrac{1}{A}$$

This means that the areas of the parallelograms $A$ and $B$ are, indeed, reciprocal.

Due to these reciprocity relations which prevail between the period-lattice $L_{\mathbf{P}}$ and the frequency lattice $L_{\mathbf{f}}$ these two lattices are said to be *reciprocal*. Similarly, the vector pair $\mathbf{f}_1$, $\mathbf{f}_2$ which spans the lattice $L_{\mathbf{f}}$ is said to be reciprocal to the vector pair $\mathbf{P}_1$, $\mathbf{P}_2$ which spans the lattice $L_{\mathbf{P}}$.

Finally, it is interesting to mention the following relations regarding the mixed scalar products of the basis vectors of the 2D reciprocal lattices $L_{\mathbf{P}}$ and $L_{\mathbf{f}}$, which are also obtained from Eqs. (A.31) and (A.33):

$$\mathbf{P}_1 \cdot \mathbf{f}_1 = x_1 \tfrac{y_2}{A} - y_1 \tfrac{x_2}{A} = \tfrac{1}{A}\bar{A} = 1 \qquad \mathbf{P}_1 \cdot \mathbf{f}_2 = -x_1 \tfrac{y_1}{A} + y_1 \tfrac{x_1}{A} = 0$$
$$\mathbf{P}_2 \cdot \mathbf{f}_1 = x_2 \tfrac{y_2}{A} - y_2 \tfrac{x_2}{A} = 0 \qquad \mathbf{P}_2 \cdot \mathbf{f}_2 = -x_2 \tfrac{y_1}{A} + y_2 \tfrac{x_1}{A} = \tfrac{1}{A}\bar{A} = 1$$

This property of the two reciprocal lattices is most useful, since it can be readily generalized to lattices in 3D (such as in the case of crystallography) or any other dimension $n \geq 1$. In fact, this property is often used to define reciprocal lattices (see, for example, [Rosenfeld82 p. 75] or [Jarić89 p. 15]):

**Definition A.1:** Let $L$ be an $n$-dimensional lattice and let the vectors $\mathbf{P}_1,\dots,\mathbf{P}_n \in \mathbb{R}^n$ be a basis of $L$. Then its reciprocal lattice $L^*$ is a lattice of the same dimension, whose basis vectors $\mathbf{f}_1,\dots,\mathbf{f}_n \in \mathbb{R}^n$ are defined by:

$$\mathbf{P}_i \cdot \mathbf{f}_j = \begin{cases} 1 & \text{if } i = j \\ 0 & \text{if } i \neq j \end{cases} \quad \blacksquare \tag{A.34}$$

The geometric interpretation of this definition is as follows:

(a) The second condition in (A.34) means that the vector $\mathbf{f}_j$ is perpendicular to all the vectors $\mathbf{P}_i$ with $i \neq j$; this determines the direction of the line through the origin of $\mathbb{R}^n$ on which $\mathbf{f}_j$ is situated. (For example, in the 3D case $\mathbf{f}_1$ is situated on the line emanating from the origin of $\mathbb{R}^3$ perpendicularly to the plane spanned by $\mathbf{P}_2$ and $\mathbf{P}_3$).

(b) The first condition in (A.34) determines the precise length and direction of the vector $\mathbf{f}_j$ on the line defined in (a). $\mathbf{P}_j \cdot \mathbf{f}_j = 1$ means that the length of $\mathbf{f}_j$ on this line is reciprocal to the length of the projection of $\mathbf{P}_j$ on the same line: Since by [Vygodski73 p. 142] $\mathbf{P}\cdot\mathbf{f} = |\mathbf{f}|\,|\text{proj}(\mathbf{P})_\mathbf{f}|$ (where $\text{proj}(\mathbf{P})_\mathbf{f}$ denotes the projection of $\mathbf{P}$ on $\mathbf{f}$), and since we have here $\mathbf{P}\cdot\mathbf{f} = 1$, it follows that $|\mathbf{f}| = 1/|\text{proj}(\mathbf{P})_\mathbf{f}|$. And furthermore, since we have $\mathbf{P}_j \cdot \mathbf{f}_j = 1 > 0$ the direction of the vector $\mathbf{f}_j$ on the same line is determined such that the angle between $\mathbf{f}_j$ and $\mathbf{P}_j$ is sharp [*ibid.*].

Hence, the two conditions of (A.34) fully determine each of the $n$ vectors $\mathbf{f}_j$.

Although for our present work we will only need the 2D or 1D cases, it is interesting to mention that this definition permits us to generalize Propositions A.1–A.4 above to the $n$-dimensional case (again, under the appropriate convergence conditions), as follows:

**Proposition A.5:** A function $p(\mathbf{x})$ in the $n$-dimensional image domain is periodic with period-lattice $L$ *iff* its $n$-dimensional spectrum $P(\mathbf{u})$ has as its support in the frequency domain the reciprocal lattice $L^*$. $\blacksquare$

## A.5 The matrix notation, its appeal, and its limitations for our needs

It is interesting to note that based on Eq. (A.34) the reciprocity between the vectors $\mathbf{P}_i$ (which are a basis of the period-lattice $L$) and the vectors $\mathbf{f}_i$ (which are a basis of the frequency-lattice $L^*$, i.e., of the spectrum support) can be also expressed in matrix notation. Since from (A.34) we have:[9]

$$\begin{pmatrix} \mathbf{P}_1 \\ \vdots \\ \mathbf{P}_n \end{pmatrix} (\mathbf{f}_1,\dots,\mathbf{f}_n) = \begin{pmatrix} \mathbf{P}_1\cdot\mathbf{f}_1 & \cdots & \mathbf{P}_1\cdot\mathbf{f}_n \\ \cdots & \cdots & \cdots \\ \mathbf{P}_n\cdot\mathbf{f}_1 & \cdots & \mathbf{P}_n\cdot\mathbf{f}_n \end{pmatrix} = \begin{pmatrix} 1 & & 0 \\ & \ddots & \\ 0 & & 1 \end{pmatrix}$$

---

[9] Note that each vector in the following expressions represents an $n$-tuple of coordinates, and therefore each entity in parentheses is, in fact, a *matrix*.

it follows that if the matrix $(\mathbf{f}_1,...,\mathbf{f}_n)$ is invertible, i.e., non-singular (which is true *iff* the vectors $\mathbf{f}_1,...,\mathbf{f}_n$ are linearly independent over $\mathbb{R}$ [Birkhoff77 pp. 237–238]), then:

$$\begin{pmatrix} \mathbf{P}_1 \\ \vdots \\ \mathbf{P}_n \end{pmatrix} = (\mathbf{f}_1,...,\mathbf{f}_n)^{-1}$$

and by writing both sides as columns we obtain:

$$\begin{pmatrix} \mathbf{P}_1 \\ \vdots \\ \mathbf{P}_n \end{pmatrix} = \begin{pmatrix} \mathbf{f}_1 \\ \vdots \\ \mathbf{f}_n \end{pmatrix}^{-\mathrm{T}} \tag{A.35}$$

(where "–T" means the transpose of the inverse matrix). We will henceforth denote these two matrices in short by P and F, and hence Eq. (A.35) becomes: $\mathrm{P} = \mathrm{F}^{-\mathrm{T}}$.

For example, in the 2D case we have, indeed, by Eqs. (A.33) and (A.31):

$$\mathrm{F} = \begin{pmatrix} \mathbf{f}_1 \\ \mathbf{f}_2 \end{pmatrix} = \begin{pmatrix} u_1 & v_1 \\ u_2 & v_2 \end{pmatrix} = \frac{1}{x_1y_2 - y_1x_2}\begin{pmatrix} y_2 & -x_2 \\ -y_1 & x_1 \end{pmatrix} = \begin{pmatrix} x_1 & y_1 \\ x_2 & y_2 \end{pmatrix}^{-\mathrm{T}} = \begin{pmatrix} \mathbf{P}_1 \\ \mathbf{P}_2 \end{pmatrix}^{-\mathrm{T}} = \mathrm{P}^{-\mathrm{T}} \tag{A.36}$$

We note that $\bar{A} = x_1y_2 - y_1x_2$ is, in fact, the determinant of the matrix $\mathrm{P} = \begin{pmatrix} \mathbf{P}_1 \\ \mathbf{P}_2 \end{pmatrix} = \begin{pmatrix} x_1 & y_1 \\ x_2 & y_2 \end{pmatrix}$.

This matrix representation of (A.34) leads us also to a matrix notation for the periodic function $p(x,y)$ and for its spectrum (=Fourier transform) $P(u,v)$:

We start first with the spectrum $P(u,v)$. As we have seen earlier, $P(u,v)$ is given in vector notation by Eq. (A.30), as follows:

$$P(\mathbf{f}) = \sum_{m=-\infty}^{\infty} \sum_{n=-\infty}^{\infty} c_{m,n}\, \delta(\mathbf{f} - (m\mathbf{f}_1 + n\mathbf{f}_2))$$

We note that:

$$\begin{aligned} m\mathbf{f}_1 + n\mathbf{f}_2 &= m(u_1,v_1) + n(u_2,v_2) \\ &= (mu_1+nu_2,\ mv_1+nv_2) \\ &= (m,n)\begin{pmatrix} u_1 & v_1 \\ u_2 & v_2 \end{pmatrix} \\ &= \mathbf{n}\mathrm{F} \end{aligned}$$

where $\mathrm{F} = \begin{pmatrix} \mathbf{f}_1 \\ \mathbf{f}_2 \end{pmatrix} = \begin{pmatrix} u_1 & v_1 \\ u_2 & v_2 \end{pmatrix}$ and $\mathbf{n} = (m,n)$.

Hence we obtain the following matrix notation for the spectrum $P(u,v)$:

$$P(\mathbf{f}) = \sum_{\mathbf{n}} c_{\mathbf{n}}\, \delta(\mathbf{f} - \mathbf{n}\mathrm{F}) \tag{A.37}$$

We proceed now to the periodic function $p(x,y)$ itself, in the image domain: Let $\mathbf{P}_1$, $\mathbf{P}_2$ be the fundamental period-vectors of $p(x,y)$ and let $d(x,y)$ be its restriction to the fundamental period-parallelogram defined by $\mathbf{P}_1$, $\mathbf{P}_2$, i.e., a single period-element of $p(x,y)$. Therefore, we can rewrite $p(x,y)$ in vector notation as follows:

$$p(\mathbf{x}) = \sum_{m=-\infty}^{\infty} \sum_{n=-\infty}^{\infty} d(\mathbf{x}-(m\mathbf{P}_1+n\mathbf{P}_2)) \tag{A.38}$$

Again, we note that:

$$\begin{aligned} m\mathbf{P}_1+n\mathbf{P}_2 &= m(x_1,y_1) + n(x_2,y_2) \\ &= (mx_1+nx_2,\ my_1+ny_2) \\ &= (m,n)\begin{pmatrix} x_1 & y_1 \\ x_2 & y_2 \end{pmatrix} \\ &= \mathbf{n}\mathrm{P} \end{aligned}$$

where $\mathrm{P} = \begin{pmatrix} \mathbf{P}_1 \\ \mathbf{P}_2 \end{pmatrix} = \begin{pmatrix} x_1 & y_1 \\ x_2 & y_2 \end{pmatrix}$ and $\mathbf{n} = (m,n)$.

Hence we obtain the following matrix notation for the periodic function $p(x,y)$:

$$p(\mathbf{x}) = \sum_{\mathbf{n}} d(\mathbf{x}-\mathbf{n}\mathrm{P}) \tag{A.39}$$

where, as we have already seen in Eqs. (A.35) and (A.36): $\mathrm{P} = \mathrm{F}^{-\mathrm{T}}$.[10]

The matrix notation of Eqs. (A.37) and (A.39) is frequently used in literature, due to its concise and appealing form;[11] see, for example, [Ulichney88 pp. 17–19, 44–47] and [Cartwright90 pp. 123–126]. However, although useful for expressing 2D or multi-dimensional periodic functions and their spectra (=Fourier transforms), we will not make use of this notation in our study on the superposition of periodic functions, and this for two main reasons:

First, in spite of its concise, appealing form, the matrix notation tends to obscure the detailed structure of the spectrum support — while the explicit vector notations that we introduced earlier, like in Eqs. (A.26) or (A.30), clearly reflect this structure, and are therefore particularly well adapted for our needs (see also Sec. 6.7 in Chapter 6).

But even more importantly, it must be noted that Eq. (A.39) in the image domain, which is based on the matrix $\mathrm{P} = \mathrm{F}^{-\mathrm{T}}$, is valid only for non-singular cases, where the matrix F is invertible (i.e., where the frequency-vectors $\mathbf{f}_1,...,\mathbf{f}_n$ are linearly independent; see earlier in this section). However, in our present study not only do such singular cases occur, but they even constitute some of the most interesting cases of our research. In fact, there exist two possible types of failure due to the singularity of the matrix F:

(a) If the vectors $\mathbf{f}_1,...,\mathbf{f}_n$ have $\mathrm{rank}_{\mathbb{Z}}\ \mathrm{Md}(\mathbf{f}_1,...,\mathbf{f}_m) = \dim \mathrm{Sp}(\mathbf{f}_1,...,\mathbf{f}_m) = k$ where $k < n$, meaning that the spectrum support is still a *lattice* (see Proposition 5.1) — but of a lower dimension $k < n$, then the failure is curable and Eq. (A.39) can still be used, after a manual adaptation to the new, lower dimension $k$ (based on a new basis of $k$ rather than $n$ vectors $\mathbf{f}_i$, and on new $k \times k$ matrices F and P). For instance, in the case of $n = 2$

---

[10] Note that the multiplication of the integer index-vector $\mathbf{n} \in \mathbb{Z}^2$ by the matrix F in Eq. (A.37) (or by the matrix P, in Eq. (A.39)) can be interpreted as an application of a linear mapping on the integer lattice $\mathbb{Z}^2$ which transforms it into the skewed frequency-lattice $L_{\mathbf{f}}$, like in Fig. A.2(b) (or, respectively, into the reciprocal skewed period-lattice $L_{\mathbf{P}}$ in the image domain).

[11] Note, for instance, the similarity between Eq. (A.37) and its 1D counterpart, Eq. (A.8).

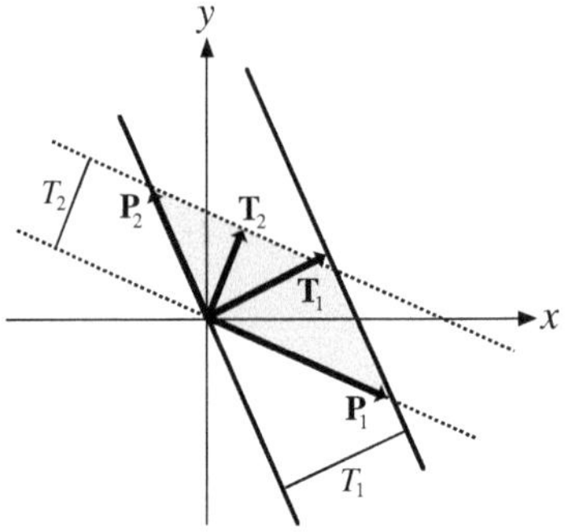


**Figure A.5:** A magnified view of the main period-parallelogram of Fig. A.2(a), showing the period-vectors $\mathbf{P}_i$ and the step-vectors $\mathbf{T}_i$.

this occurs when $\mathbf{f}_1$ and $\mathbf{f}_2$ are collinear and commensurable, i.e., linearly dependent over $\mathbb{R}$ and over $\mathbb{Z}$ (see Example 2 in Sec. 5.2).

(b) If, however, the vectors $\mathbf{f}_1,...,\mathbf{f}_n$ have $\text{rank}_{\mathbb{Z}}\ \text{Md}(\mathbf{f}_1,...,\mathbf{f}_m) > \dim \text{Sp}(\mathbf{f}_1,...,\mathbf{f}_m)$, meaning that the spectrum support is not a lattice but rather a *module* (so that our superposition in the image domain is an almost-periodic case; see Proposition 6.1), then there is no basis to the spectrum support (see Sec. 5.2) and therefore the singularity of matrix F cannot be remedied. In this case notation (A.39) fails altogether and cannot be used.

These reasons make the matrix notation unsuitable for our study of superposed layers and their spectra, and they favour, instead, our vector notation, based on the layer frequencies $\mathbf{f}_1,...,\mathbf{f}_n$.

## A.6 The period-vectors $\mathbf{P}_i$ vs. the step-vectors $\mathbf{T}_i$

One further remark is due at this point. The failure of matrix P whenever matrix F is singular is, in fact, just an indication to a more fundamental failure: Since the period-vectors $\mathbf{P}_i$ in the image domain only exist when the frequency-vectors $\mathbf{f}_1,...,\mathbf{f}_n$ (the individual layer frequencies) in the spectrum are linearly independent over $\mathbb{R}$ (see Sec. A.5), the vectors $\mathbf{P}_i$ are inappropriate for our study of layer superpositions, because superpositions may well occur even when $\mathbf{f}_1,...,\mathbf{f}_n$ are linearly dependent. (And in fact, in the case of our 2D spectrum this necessarily occurs whenever $n > 2$). Therefore, instead of the vectors $\mathbf{P}_i$ we will only consider in the image domain the periods $T_i$ of the $n$ individual gratings (see Figs. A.2, A.3). And if a 2-fold periodic layer $p(x,y)$ appears in the superposition (say, a dot-screen), it will contribute *two* values $T_i$, which are the periods of the two *virtual gratings* defined by the borders of the fundamental period-parallelogram of that layer (see the solid and the dotted gratings in Fig. A.2(a) and the magnified view in

Fig. A.3).[12] And indeed, even the classical moiré formulas, such as Eqs. (2.9) – (2.11), are only based on the grating periods $T_i$. If we consider these grating periods as vectors $\mathbf{T}_i$ emanating from the origin of the image domain at the grating directions $\theta_i$ (see Fig. A.5), and call them *step-vectors*, we find that the relationship between each step-vector $\mathbf{T}_i$ and its counterpart $\mathbf{f}_i$ in the spectral domain is straightforward:

(1) The vectors $\mathbf{T}_i$ and $\mathbf{f}_i$ (for every $i$) are collinear, i.e., they have the same angle $\theta_i$;

(2) Their lengths are reciprocal, namely: for every $i$, $|\mathbf{T}_i| = 1/|\mathbf{f}_i|$.

This means that for every $i$ we have $\mathbf{T}_i = \frac{|\mathbf{T}_i|}{|\mathbf{f}_i|}\mathbf{f}_i = \frac{1}{|\mathbf{f}_i|^2}\mathbf{f}_i$. In other words, we obtain:

$$\begin{aligned} \mathbf{T}_1 &= \frac{1}{|\mathbf{f}_1|^2}\mathbf{f}_1 \\ &\vdots \\ \mathbf{T}_n &= \frac{1}{|\mathbf{f}_n|^2}\mathbf{f}_n \end{aligned} \qquad \text{(A.40)}$$

As we can see, this relationship holds between any pair $\mathbf{T}_i$, $\mathbf{f}_i$, individually; and moreover, it exists in *all* cases and for any number $n$ of superposed layers — even if $\mathbf{f}_1,...,\mathbf{f}_n$ are linearly dependent.

In order to formulate better the relationship between the vector pair $\mathbf{T}_i$, $\mathbf{f}_i$ we introduce here the following definition:

**Definition A.2:** For any vector $\mathbf{v} \neq 0$, the *reciprocal vector* of $\mathbf{v}$ (with respect to scalar product) is defined as:

$$\mathbf{v}^{-1} = \frac{1}{|\mathbf{v}|^2}\mathbf{v} \qquad \blacksquare \qquad \text{(A.41)}$$

This definition requires a short explanation. Although it is clear that the vector $\mathbf{v}^{-1}$ is reciprocal to $\mathbf{v}$ with respect to scalar product:

$$\mathbf{v}^{-1}\cdot\mathbf{v} = \mathbf{v}\cdot\mathbf{v}^{-1} = \frac{1}{|\mathbf{v}|^2}\mathbf{v}\cdot\mathbf{v} = 1 \qquad \text{(A.42)}$$

$\mathbf{v}^{-1}$ is *not* the unique vector with this property. In fact, the locus of all the vectors $\mathbf{x}$ which satisfy $\mathbf{x}\cdot\mathbf{v} = 1$ contains (in the case of $\mathbb{R}^3$) the entire plane perpendicular to the vector $\mathbf{v}$, whose distance from the origin (along the line spanned by $\mathbf{v}$) is $\frac{1}{|\mathbf{v}|}$. Therefore, the uniqueness of $\mathbf{v}^{-1}$ in this definition is obtained only through the requirement, implicit in Eq. (A.41), that the reciprocal vector $\mathbf{v}^{-1}$ is collinear with the vector $\mathbf{v}$.

Using this definition and in view of (1), (2) and Eq. (A.40) above, we can now reformulate the relationship between the vector pair $\mathbf{T}_i$ and $\mathbf{f}_i$ as follows:

[12] Each of the different possible choices of fundamental frequency-vectors $\mathbf{f}_1,\mathbf{f}_2$ for $p(x,y)$ automatically determines a corresponding pair of fundamental period-vectors $\mathbf{P}_1,\mathbf{P}_2$ (by Eq. (A.34)), and hence it determines also the fundamental period-parallelogram they define, and the corresponding virtual-grating periods $T_1,T_2$. Note that all the different possible choices represent the same 2D lattices $L_\mathbf{f}$ and $L_\mathbf{P}$ of $p(x,y)$.

(3) For every $i$, the vector $\mathbf{T}_i$ in the image domain is the reciprocal vector (with respect to scalar product) of the frequency-vector $\mathbf{f}_i$ in the spectral domain:

$$\mathbf{T}_i = \frac{1}{|\mathbf{f}_i|^2}\,\mathbf{f}_i = \mathbf{f}_i^{-1} \tag{A.43}$$

This is, in fact, the 2D vectorial generalization of the relation $T = 1/f$ between period and frequency in the 1D case. This vector notation proves to be particularly useful in Chapter 7.

Returning now to our vector comparison, the main difference between the step-vectors $\mathbf{T}_i$ and the period-vectors $\mathbf{P}_i$ (all of which subsist in the image domain) is that each vector $\mathbf{T}_i$ depends only on a *single* frequency-vector $\mathbf{f}_i$, while as we have seen in Eqs. (A.34) and (A.35), each of the vectors $\mathbf{P}_i$ depends on *all* of the $n$ frequency-vectors $\mathbf{f}_1,...,\mathbf{f}_n$. In the first case we speak about reciprocity (with respect to scalar product) between the individual vectors $\mathbf{T}_i$ and $\mathbf{f}_i$, but in the second case we speak about reciprocity between vector $n$-tuples, $\mathbf{P}_1,...,\mathbf{P}_n$ and $\mathbf{f}_1,...,\mathbf{f}_n$, or between the lattices $L_\mathbf{P}$ and $L_\mathbf{f}$ spanned by them. And while the vectors $\mathbf{P}_i$ exist only when the vectors $\mathbf{f}_1,...,\mathbf{f}_n$ are linearly independent, the vectors $\mathbf{T}_i$, on the contrary, exist for any vectors $\mathbf{f}_1,...,\mathbf{f}_n$ with no restrictions, since every vector $\mathbf{T}_i$ is only dependent on its own counterpart $\mathbf{f}_i$, and no matrix inversion is involved.[13]

Finally, it should be emphasized that in spite of the apparent symmetry between the frequency-lattice $L_\mathbf{f}$ and the period-lattice $L_\mathbf{P}$ (or between the frequency-vectors $\mathbf{f}_1,...,\mathbf{f}_n$ and the period-vectors $\mathbf{P}_1,...,\mathbf{P}_n$) due to Eqs. (A.34) and (A.35), there exists a substantial difference between them: While the spectrum support $L_\mathbf{f}$ is a fundamental property of any superposition of $n$ periodic functions, the period-lattice $L_\mathbf{P}$ (and the period-vectors $\mathbf{P}_1,...,\mathbf{P}_n$) are only derived properties, and they exist only conditionally: *iff* the vectors $\mathbf{f}_1,...,\mathbf{f}_n$ are linearly independent (i.e., *iff* the superposition in the image domain is periodic in $n$ dimensions).

[13] Note that when the vector frequencies $\mathbf{f}_1,...,\mathbf{f}_n$ are all orthogonal to each other (in the $n$-dimensional spectrum), the vectors $\mathbf{T}_i$ and $\mathbf{P}_i$ ($i = 1,...,n$) coincide in the image domain. The vectors $\mathbf{T}$ and $\mathbf{P}$ also coincide in any 1-fold periodic function.

# Appendix B

## Almost-periodic functions and their spectra

### B.1 Introduction

Almost-periodic functions constitute an important generalization of the class of periodic functions. The theory of almost-periodic functions was founded in 1923 by the Danish mathematician Harald Bohr [Bohr23] and was further developed by A. S. Besicovitch and by others. The main importance of this theory is in the discovery of the tight relationship (or duality) between extensions of the structural concept of periodicity of functions on the one hand and generalizations of the analytic concept of the Fourier series representation of these functions on the other hand. This, in turn, further extends the scope of the reciprocity between functions in the image domain and their spectra in the frequency domain to a larger range of functions. In fact, this generalization introduces a new class of functions situated between periodic functions and aperiodic functions, whose spectral representation, too, is intermediate between the two: its spectrum is no longer composed of a comb or a nailbed of impulses, as in the periodic case, but it is not yet a continuous Fourier transform as in the aperiodic case, either. It is also distinct from the diffuse spectrum of random or pseudo-random functions. Rather, it is still composed of a denumerable set of impulses, but these impulses are freely located in the spectrum, and may be even everywhere dense.[1]

In this appendix we shortly review the concept of almost-periodic functions in one or two variables, along with their main properties. We mainly concentrate here on results which are needed for our purposes; additional details as well as a rigorous mathematic development of the subject can be found in the references cited throughout.

### B.2 A simple illustrative example

Let $p_1(x)$ and $p_2(x)$ be periodic functions with periods $T_1$, $T_2$. What can be said about the functions $f(x) = p_1(x) + p_2(x)$, $g(x) = p_1(x)\, p_2(x)$ ? We must distinguish here between two cases:

---

[1] It is interesting to note that in 1984 almost-periodicity was brought to the center of scientific interest by a very remarkable discovery, made by Dan Shechtman *et al.* [Shechtman84]. They discovered solid materials with a new kind of microstructure, which is intermediate between the periodic structure of crystalline materials and the structure of amorphous solids. Further research of these materials, which were named *quasi-crystals*, has shown through spectral and other evidences that this new kind of physical structure corresponds to the mathematical properties of almost-periodicity [Nelson86], [Katz86]. This recent discovery, which has shaken the very foundations of the science of crystallography, shows that almost-periodic structures do not only belong to the wild imagination of mathematicians, but well to the contrary, they do correspond to the physical reality of our world.

(1) If the periods $T_1$ and $T_2$ are *commensurable*, i.e., if $T_2/T_1$ is rational, then there exist non-zero integers $m$, $n$ such that $mT_1 = nT_2$. Since $mT_1$ is a period of $p_1(x)$ and $nT_2$ is a period of $p_2(x)$ it follows that $T = mT_1 = nT_2$ is a common period of both, and therefore also of their sum $f(x)$ and of their product $g(x)$. This means that the functions $f(x)$ and $g(x)$ are periodic with period $T$ (obviously, $T \geq T_1, T_2$).

(2) If the periods $T_1$ and $T_2$ are *incommensurable*, i.e., if $T_2/T_1$ is irrational, then except at the origin the periods of the functions $p_1(x)$ and $p_2(x)$ never meet again after any integer numbers $m$, $n$ of full periods $T_1$ and $T_2$. This means that there exists no common period to $p_1(x)$ and $p_2(x)$, and therefore $f(x)$ and $g(x)$ never exactly repeat themselves, and they are not periodic. As we will see below, $f(x)$ and $g(x)$ belong to a wider class of functions called: *almost-periodic functions*. This situation is illustrated in the following example.

**Example B.1:** Consider the functions $p_1(x) = \cos(2\pi x/20)$ and $p_2(x) = \frac{1}{4}\cos(2\pi x/\sqrt{2})$, whose periods are respectively $T_1 = 20$ and $T_2 = \sqrt{2}$. Since $T_1$ and $T_2$ are *incommensurable*, it is clear that except at the origin the periods of these two functions will never meet again after any integer numbers $m$, $n$ of full periods $T_1$ and $T_2$. This means that the function $f(x) = \cos(2\pi x/20) + \frac{1}{4}\cos(2\pi x/\sqrt{2})$ is not periodic since no value of $T$, no matter how large it is, gives *exact* repetitions of $f(x)$ (see Fig. B.1). For instance, $f(x)$ gets its maximum value $f(x) = 1\frac{1}{4}$ only at $x = 0$, and for no other value of $x$. However, as it can be seen in the figure, $f(x)$ looks "almost" as a periodic function. And indeed, if we admit a certain small error $\varepsilon$, say $\varepsilon = 0.3$, then $\tau_\varepsilon = 20$ can be considered as an $\varepsilon$-almost-period of $f(x)$, which gives a repetition of $f(x)$ up to an error of $\varepsilon$. If we require a smaller error, such as $\varepsilon = 0.01$, then there can be found a larger number, such as $\tau_\varepsilon = 140$, which may be taken as an $\varepsilon$-almost-period. In fact, for any $\varepsilon > 0$ we can find an arbitrarily large $\varepsilon$-almost-period $\tau_\varepsilon$ for $f(x)$ such that for any $-\infty < x < \infty$, $|f(x+\tau_\varepsilon) - f(x)| < \varepsilon$ (this is demonstrated, for a similar function, in [Besicovitch32 p. ix]). A slightly stronger version of this property will serve as a basis for the definition of an almost-periodic function (see Sec. B.3). ■

As we will see below, any periodic function is almost-periodic, but clearly not every almost-periodic function is periodic. It should also be emphasized that not every non-periodic function is almost-periodic; an obvious counter-example would be $f(x) = x$. This means that almost-periodic functions form an intermediate class of functions between periodic functions and aperiodic functions (see Fig. B.3 in Sec. B.5 below).

### B.3 Definitions and main properties

The definition of almost-periodic functions relies on the definition of the concept of $\varepsilon$-almost-period as a generalization of the concept of a period. Let us start by rephrasing the definitions of a period and of a periodic function which are given in Sec. A.2.

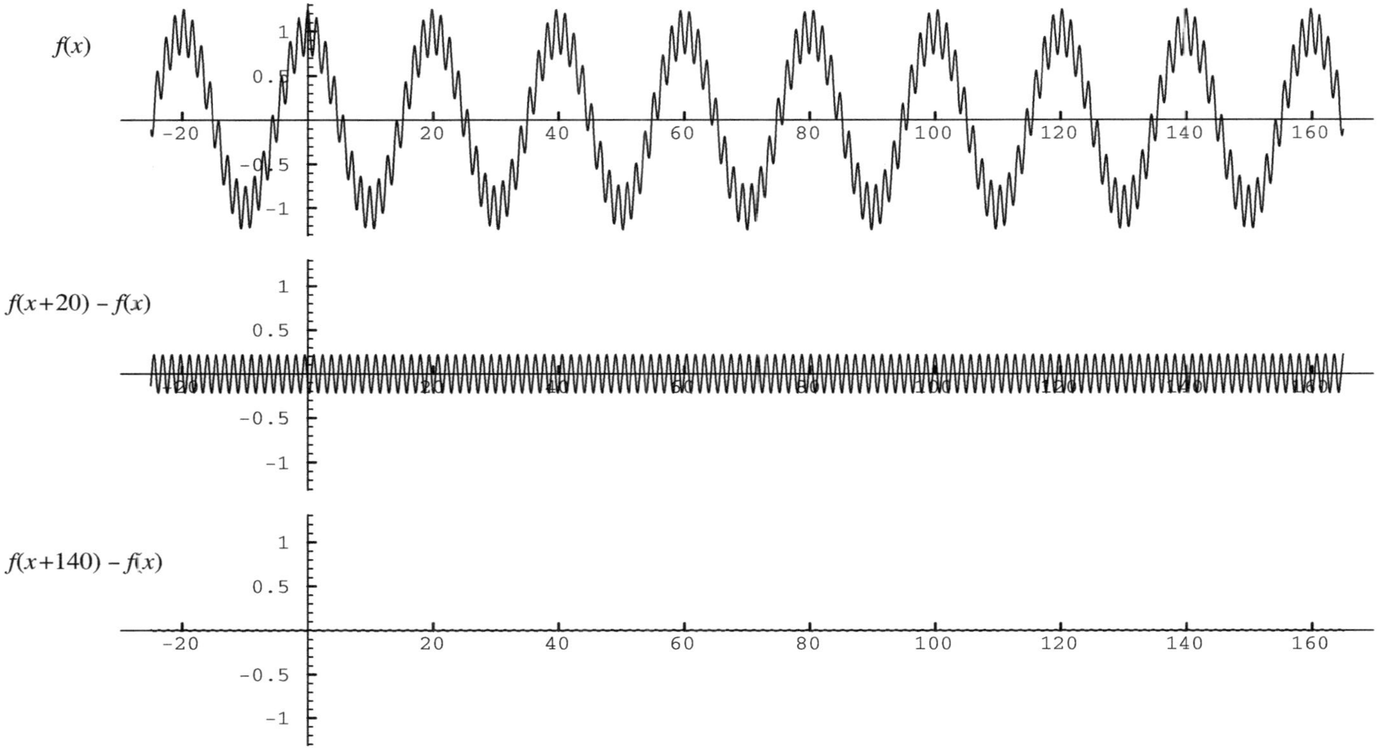


**Figure B.1:** Top: almost periodicity is clearly illustrated by the graphic plot of the function: $f(x) = \cos(2\pi x/20) + \frac{1}{4}\cos(2\pi x/\sqrt{2})$. In this case, a high-frequency oscillation (with period $T_2 = \sqrt{2}$) is added to a cosine with period of $T_1 = 20$. The resulting function $f(x)$ is not periodic, but still it "almost" repeats itself with $\tau = 20$, with just a tiny error; it repeats itself with an even smaller error if we take $\tau = 140$, and so forth. Center and bottom: the error $f(x+\tau) - f(x)$ for $\tau = 20$ and for $\tau = 140$, shown at the same scale as $f(x)$. Note that for $\tau = 20$ the error is smaller than 0.3 everywhere; for $\tau = 140$ the error is smaller than 0.01.

**Definition B.1:** A number $T$ is called a *period* of $f(x)$ if for all $x \in \mathbb{R}$:

$$|f(x+T) - f(x)| = 0$$

If there exists such a number $T \neq 0$, then $f(x)$ is called *periodic*. As was noted in Sec. A.2 $T$ is not unique, since if $T$ is a period of $f(x)$, so is $nT$ for any integer $n$. In fact, the set of all periods of $f(x)$ is denumerably infinite, and it forms a lattice in $\mathbb{R}$: $L_T = \{nT \mid n \in \mathbb{Z}\}$. Normally one considers the *fundamental period* of $f(x)$, i.e., the minimal positive number in $L_T$. ■

Using this definition for inspiration, we give now the following definitions [Bohr51 p. 32]:

**Definition B.2:** An infinite set $S$ of real numbers $\tau$ is called *relatively dense* if there exists some number $l > 0$ (called *inclusion length*) such that every interval of length $l$, $(x, x+l) \subset \mathbb{R}$, contains at least one number $\tau$ of the set $S$. (This means that there are no arbitrarily large gaps between the numbers $\tau$. As an example, the set $L_T$ defined above is relatively dense since every interval $(x, x+2T)$ contains at least one number of $L_T$; but the set $\{n^2 \mid n \in \mathbb{Z}\}$ is not relatively dense.) ■

**Definition B.3:** Given $\varepsilon > 0$, a number $\tau_\varepsilon$ is called an *$\varepsilon$-almost-period*[2] of $f(x)$ if for all $x \in \mathbb{R}$:

$$|f(x+\tau_\varepsilon) - f(x)| < \varepsilon \tag{B.1}$$

A continuous function $f(x)$ is called *Bohr-almost-periodic*[3] (or *uniformly-almost-periodic*) if for *any* $\varepsilon > 0$, no matter how small, there exists a relatively dense set of $\varepsilon$-almost-periods.[4] ■

In other words, $f(x)$ is Bohr-almost-periodic if the equation $f(x+\tau_\varepsilon) = f(x)$ is satisfied with an arbitrary degree of accuracy $\varepsilon > 0$ by infinitely many values of $\tau_\varepsilon$, these values being spread over the whole range from $-\infty$ to $+\infty$ in such a way as not to leave empty intervals of arbitrarily great length [Besicovitch32 p. x].

It should be noted that every periodic function is almost-periodic (since for any $\varepsilon$ we may take the periods $nT$ for all $n \in \mathbb{Z}$ as a relatively dense set of $\varepsilon$-almost-periods). But as we have seen in Example B.1 above there exist almost-periodic functions which are not periodic.

---

[2] This term is current in modern literature; the term originally used by Bohr (and still being used by others) is: "a translation number of $f(x)$ corresponding to $\varepsilon$" [Bohr51 p. 31].

[3] As we will see in Sec. B.5, this definition, initially given by Bohr, has been extended later in various different ways (for example, in order to lift the restriction of the continuity of $f(x)$). Each of these extensions defines a different class of almost-periodic functions, which is a superset of the class of Bohr-almost-periodic functions. We will use the general term "almost-periodic function" in the widest sense, and whenever a particular class of almost-periodic functions is intended we will refer to it specifically, such as: "Bohr-almost-periodic function" etc.

[4] As already pointed out in Example B.1 above, the property required by this definition is stronger than just having for any $\varepsilon > 0$ an arbitrarily large $\varepsilon$-almost-period $\tau_\varepsilon$. The reasons for this choice are explained in [Bohr51 p. 32].

Almost-periodicity is not always as easy to identify visually as in Fig. B.1, and the graphic behaviour of an almost-periodic function may be quite complex. However, almost-periodic functions have the following structural properties:

(1) A (Bohr) almost-periodic function is bounded and uniformly continuous throughout $-\infty < x < \infty$ [Bohr51 pp. 33–35].

(2) A non-constant almost-periodic function does not tend to any limit when $x \to \pm\infty$; in fact, throughout $-\infty < x < \infty$ it oscillates irregularly between two finite extreme values without damping [Bass71 p. 345, 367].

Other important properties of almost-periodic functions include the following:

(3) If $f(x)$ is almost-periodic, so are $|f(x)|$, $cf(x)$ and $f(x+b)$ for any $a,b,c \in \mathbb{R}$, with the same almost-periods [Bass71 p. 344]. More generally, $cf(ax+b)$ is also almost-periodic [Corduneanu68 p. 11].

(4) If $f(x)$ and $g(x)$ are almost-periodic, so are $f(x) \pm g(x)$ and $f(x)g(x)$ [Bohr51 pp. 36–38]. If in addition $\inf_{-\infty<x<\infty} |g(x)| = m > 0$ then $1/g(x)$ and $f(x)/g(x)$ are also almost-periodic [Bohr51 pp. 34–35].

(5) Every almost-periodic function is representable as a generalized Fourier series and consequently has an impulsive spectrum.

This last property, which is probably the most remarkable of all, will be discussed in the following section.

## B.4 The spectrum of almost-periodic functions

As we have seen in Appendix A, according to the Fourier theory if $p(x)$ is a periodic function of period $T$, then it can be uniquely developed into the form of a Fourier series:

$$p(x) \sim \sum_{n=-\infty}^{\infty} c_n \, e^{i2\pi nfx} \tag{B.2}$$

where the Fourier coefficients $c_n$ are given by:

$$c_n = \frac{1}{T} \int_T p(x) \, e^{-i2\pi nfx} \, dx \tag{B.3}$$

(In all the above expressions $f = 1/T$ is the fundamental frequency of $p(x)$, and $nf$ are its harmonics). Furthermore, if $p(x)$ satisfies certain convergence conditions (see Sec. C.12 in Appendix C), then the Fourier series indeed converges, and the '~' sign can be replaced by an equality: '='.

This implies that the spectrum of $p(x)$, $P(u)$, consists of impulses which are all included in a common impulse comb of step $f$, whose $n$-th impulse is located at the frequency $u = nf$

and its amplitude is $c_n$. Note that some, or even most, of the values $c_n$ may be zero, as in the case of $p(x) = \cos(2\pi fx)$, where only the fundamental impulse pair, with indices $n = \pm 1$, has a non-zero amplitude.

One of the outstanding consequences of the theory of almost-periodic functions is that this result, namely, the correspondence between a periodic function and its Fourier series (and hence also with its spectrum), can be extended also to the case of almost-periodic functions [Bohr51 p. 50, 60]:

If $f(x)$ is an almost-periodic function then it can be uniquely[5] developed into the form of a *generalized Fourier series*:

$$f(x) \sim \sum_n c_n \, e^{i2\pi f_n x} \tag{B.4}$$

where $f_n$ are arbitrary real numbers, called the *Fourier exponents* of $f(x)$, and the complex numbers $c_n$, called the *Fourier coefficients* of $f(x)$, are given by:

$$c_n = \lim_{T\to\infty} \frac{1}{T} \int_T f(x) \, e^{-i2\pi f_n x} \, dx \tag{B.5}$$

$\int_T$ means here an integration over any interval of length $T$, i.e., from $x_0$ to $x_0+T$ where $x_0$ is arbitrary.[6]

Note that the infinite sum in (B.4) is a *generalized* Fourier series, in the sense that the frequencies $f_n$ are no longer integer multiples (harmonics) of a fundamental frequency $f$, as was the case in the Fourier series (B.3) of a periodic function $p(x)$. In fact, the frequencies $f_n$ may be here any denumerable set of arbitrary real numbers, which may have finite accumulation points or even be everywhere dense [EncMath88 Vol. 1 p. 155].

This means that the spectrum of an almost-periodic function consists of a denumerable set of impulses, which are located at the frequencies $f_n$ and whose amplitudes are $c_n$:

$$P(u) = \sum_{n=-\infty}^{\infty} c_n \, \delta(u-f_n) \tag{B.6}$$

where $\delta(u)$ is the impulse symbol. In fact, almost-periodic functions are characterized by having *impulsive spectra* (also called *line-spectra*). This is a generalization of the spectrum of a periodic function: although it still consists of a denumerable set of impulses, it is no longer limited to a *comb* of impulses with a fixed step, as in Eq. (A.8), and it may consist of any set of impulses with arbitrary frequencies $f_n$, which may even be everywhere dense.

---

[5] Uniqueness is guaranteed for the Bohr-almost-periodic functions by the uniqueness theorem [Bohr51 p. 60]. In the case of generalized almost-periodic functions (see Sec. B.5 below) uniqueness is true up to a null function [Bohr51 p. 98].

[6] This integral is guaranteed to exist for every Bohr-almost-periodic function $f(x)$ by the mean value theorem [Bohr51 p. 39, 44]. However, it is interesting to note that the series (B.4) does not always converge to $f(x)$ in the sense of *uniform convergence* (not even in the case of Bohr-almost-periodic functions), but it does converge to $f(x)$ in the sense of *convergence in the square mean* (see, for example, [Bass71 p. 366]). In this sense, the symbol '~' can be replaced by an equality: '='.

The following converse statement is also true: if $\{f_n\}$ is a denumerable set of arbitrary real numbers and $\{c_n\}$ are complex numbers for which $\sum|c_n| < \infty$, then there exist a Bohr-almost-periodic function $f(x)$ which has (B.4) as its generalized Fourier series [Bohr51 p. 52].[7] Note that in contrast to the frequencies $\{f_n\}$, the amplitudes $\{c_n\}$ can not be completely arbitrary, and they must satisfy some conditions in order that the Fourier series (B.4) converge.

**Example B.2:** Consider the almost-periodic function $f(x)$ defined in Example B.1. Clearly, the spectrum of $f(x)$ is the sum of the spectra of its two terms, namely: it consists of a pair of impulses with amplitude 1/2 at the frequency of $\pm f_1 = 1/20$, and a pair of impulses with amplitude 1/8 at the frequency of $\pm f_2 = 1/\sqrt{2}$. And indeed, since $f_1$ and $f_2$ are incommensurable, these four impulses on the $x$ axis are not located on any common comb. This is an almost-periodic case in which the number of impulses in the spectrum is finite. As we will see in Sec. B.5 below, this function belongs in fact to the subclass of quasi-periodic functions. ■

We conclude this section with one further remark concerning the connection between the almost-periods of an almost-periodic function and its Fourier exponents. In the periodic case, the fundamental period of a function $p(x)$ uniquely defines the set of its Fourier exponents: if $T > 0$ is the fundamental period of $p(x)$, then all its Fourier exponents are integer multiples of $f = 1/T$. It is interesting to note that there exists also a similar connection between the *almost-periods* of an almost-periodic function and its *Fourier exponents*, though it is not as simple as in the periodic case. More details on this subject can be found, for example, in [Levitan82 pp. 40–41].

## B.5 The different classes of almost-periodic functions and their spectra

As we have seen in Sec. B.3, the definition of almost-periodic functions given by Bohr is only valid for *continuous* functions. This is however a severe restriction in our case, since it excludes all discontinuous functions such as square waves, gratings and their superpositions. Moreover, mathematically speaking, this class is, in a way, not fully complete, since it is only closed[8] under the strongest limiting process (uniform convergence for all $-\infty < x < \infty$), but not under other limiting processes. These considerations gave rise to many efforts to extend the definition of almost-periodic functions in various different ways (see: "Almost-periodic function", "Generalized almost-periodic functions" in [EncMath88]). A beautiful account on these different extensions and on the reciprocity between their definitions in terms of almost-periods and in terms of their Fourier series is given in [Bohr51 pp. 91–99]. Without going here into detail we will only mention that the widest (and most complete) class of almost-periodic

[7] For the wider class of Besicovitch-almost-periodic functions (see Sec. B.5 below) the weaker condition of $\sum|c_n|^2 < \infty$ is enough [Besicovitch32 p. xii].

[8] A set $S$ is *closed* under a certain limiting process if it includes all the possible elements which can be obtained by applying this limiting process on sequences of members of $S$.

functions, known as the class of *Besicovitch almost-periodic functions*, contains all the Bohr almost-periodic functions plus other functions, including all "reasonably" discontinuous functions (and in particular all the discontinuous functions which may occur in our case, such as square waves, gratings, etc., and their superpositions; see Fig. B.2). In fact, in this class the requirement of continuity of $f(x)$ is replaced by the requirement of integrability in the Lebesgue sense [Bohr51 pp. 92–94].

It is interesting to note that the class of Bohr-almost-periodic functions (which is itself a subset of the Besicovitch class) contains, in turn, several subclasses of interest. These include:

(1) The subclass of *quasi-periodic functions* (also called *Bohl-almost-periodic functions*). This subclass contains all continuous functions which can be represented as:

$$\sum_{n_1=-\infty}^{\infty} \cdots \sum_{n_m=-\infty}^{\infty} c_{n_1,\dots,n_m} \; e^{i2\pi(n_1 f_1+\dots+n_m f_m)x} \tag{B.7}$$

or in other words, all continuous almost-periodic functions whose Fourier exponents $f_n$ in (B.4) are not completely arbitrary, but rather can be generated as a linear combination with integer coefficients of a finite number $m$ of arbitrary (possibly incommensurable) frequencies $f_1,\dots,f_m$. Using our terminology, the Fourier exponents $f_n$ in this case are simply the members of the module $\{n_1 f_1 + \dots + n_m f_m \mid n_i \in \mathbb{Z}\}$. It is said therefore that the support of the spectrum of $f(x)$ is *finitely generated*. Any sum or product of $m$ continuous periodic functions belongs to this class [EncMath88 Vol. 1 p. 414]. And indeed, expression (B.7) has exactly the same form as Eqs. (6.3) and (6.5) that we obtained in Chapter 6 for the product of $m$ periodic functions $p_1(x)\cdot\ldots\cdot p_m(x)$ with periods $f_1,\dots,f_m$. The only detail which disqualifies the class of quasi-periodic functions for our needs in the treatment of layer superpositions is the fact that this class is restricted only to the case where $p_1(x),\dots,p_m(x)$ are *continuous* (which obviously excludes square waves, binary gratings etc.).

(2) The subclass of *limit-periodic functions* [Besicovitch32 pp. 32–34]. A function is called limit-periodic if it is the limit of a uniformly convergent sequence $\{f_n(x)\}$, $n = 1,2,\dots$ of continuous periodic functions. It can be shown [*ibid.*] that this subclass contains all the Bohr-almost-periodic functions whose Fourier exponents are rational multiples of one arbitrary number $f \in \mathbb{R}$, i.e., all functions which can be represented in the form:

$$f(x) \sim \sum_n c_n \, e^{i2\pi r_n f x} \qquad \text{where } r_n \in \mathbb{Q}. \tag{B.8}$$

(3) The subclass of periodic functions (which is a subclass of both the quasi-periodic and the limit-periodic classes).

Fig. B.3 gives a schematic description of the hierarchy of the different classes and subclasses of almost-periodic functions that are mentioned in this section.

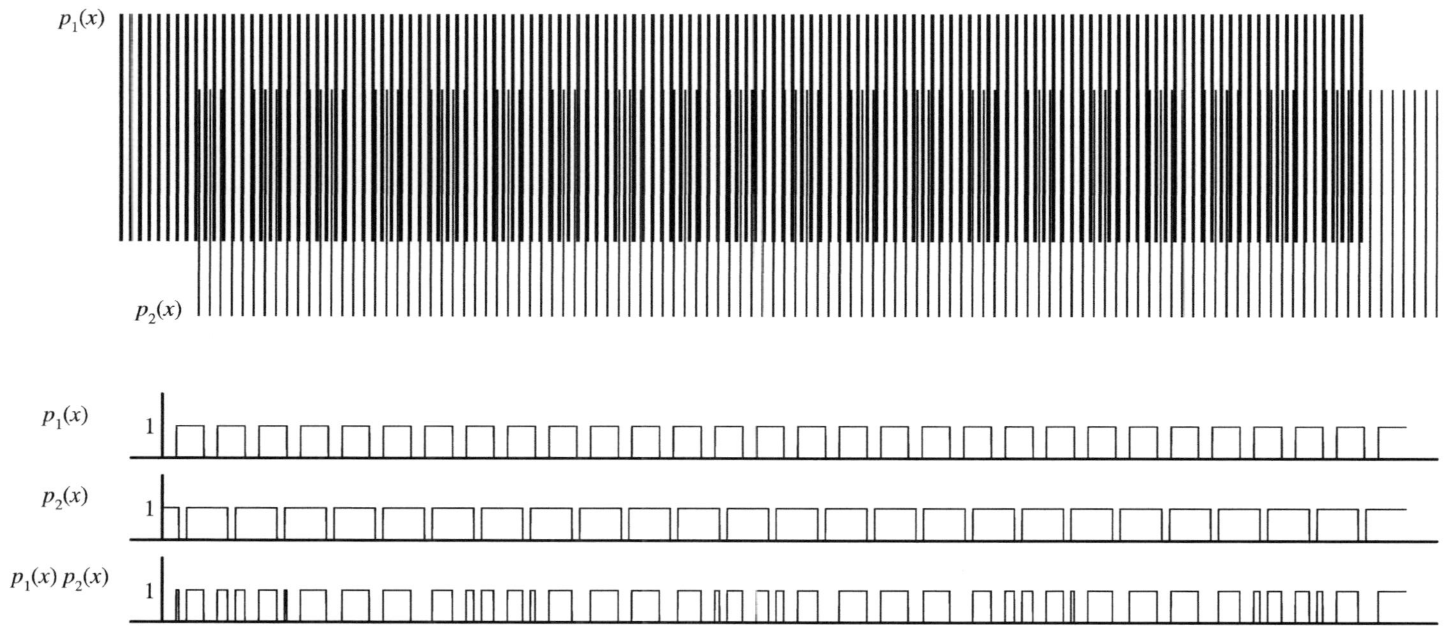


**Figure B.2:** Top: an example of a discontinuous (and hence non-Bohr) almost-periodic function, obtained as a superposition (= multiplication) of two periodic gratings with incommensurable frequencies $f_1$ and $f_2$. Bottom: a magnified section through the two gratings and their superposition (1 = white, 0 = black). Note the irregular microstructure of the superposition; both the pulse widths and their locations vary irregularly. Although the microstructure detail never repeats precisely, the global visual impression is that the superposition looks "almost" periodic with a low frequency of $f_2 - f_1$. In fact, the extracted intensity profile of this moiré, which eliminates all microstructure detail and only preserves the macrostructure of the moiré (see Fig. 4.2(a)), *is* indeed a periodic function with the frequency $f = f_2 - f_1$.

## B.6 Characterization of functions according to their spectrum support

It will be interesting to conclude this discussion by a short review of the different classes of functions through the perspective of their spectral characteristics. Let us denote the support of the spectrum of the function $f(x)$ by $F$. Then:

(1) If $f(x)$ is a "pure vibration" [Bohr51 p. 2], namely: $\cos 2\pi fx$, $\sin 2\pi fx$, or more generally $a\cos(2\pi fx + b)$, then its spectrum simply consists of one pair of impulses:

$$F = \{\pm f\}$$

(2) If $f(x)$ is periodic with period $T = 1/f$, then its spectrum consists of a denumerable (finite or infinite) set of impulses, which are located on a common lattice (i.e., included in a common comb) with a fixed step of $f$:

$$F = \{nf \mid n \in \mathbb{Z}\}$$

(Note that some or even most of the impulses on the lattice may have a zero amplitude, as is case (1) above. But if $f(x)$ is discontinuous, like a square wave, then the number of non-zero impulses on the lattice is denumerably infinite.)

(3) If $f(x)$ is quasi-periodic, then its spectrum consists of a denumerable set of impulses which are located on a common module spanned by a finite number $m$ of arbitrary (possibly incommensurable) frequencies $f_1,\dots,f_m$:

$$F = \{n_1 f_1 + \dots + n_m f_m \mid n_i \in \mathbb{Z}\}$$

(Here, too, some or even most of the impulses may have a zero amplitude, as in the case of Example B.1, where $F = \{\pm f_1, \pm f_2\}$. However, if the number of non-zero impulses in $F$ is infinite, then $F$ may be everywhere dense.)

(4) If $f(x)$ is limit-periodic, then its spectrum consists of a finite or denumerable set of impulses which are located on rational multiples of one arbitrary frequency $f$:

$$F = \{r_n f \mid r_n \in \mathbb{Q}\}$$

Here, too, if the number of non-zero impulses is infinite, then $F$ is everywhere dense.

(5) If $f(x)$ is almost-periodic (in the sense of Bohr, Besicovitch, etc.), then its spectrum consists of a denumerable set of impulses with any arbitrary frequencies:

$$F = \{f_n \in \mathbb{R} \mid n \in \mathbb{Z}\}$$

This is the most general case of impulsive spectrum, in which no restrictions exist on the values of the frequencies $f_n$, and they may have finite accumulation points, or even be everywhere dense.[9]

[9] Note that the inverse question of determining to which set of impulses in the spectrum there actually corresponds an almost-periodic function is more delicate, since it requires also conditions on the

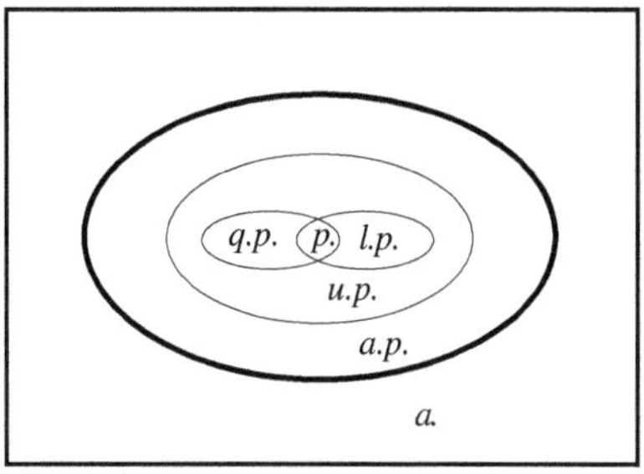


*p.* = *periodic* functions
*q.p.* = *quasi-periodic* functions
*l.p.* = *limit-periodic* functions
*u.p.* = *Bohr (uniformly) a. p.* funcs.
*a.p.* = *almost-periodic* functions
*a.* = *aperiodic* functions

(*q.p.*, *l.p.*, *u.p.*, *a.p.*, *a.*: *non-periodic* functions)

**Figure B.3:** A schematic diagram showing the classification of functions according to their periodicity properties, and the inclusion relationships between these different function-classes. The outside rectangle represents the universe of all functions; thin lines represent the borders of classes which are contained in each other, while thick borders separate between mutually exclusive classes.

It is interesting to note, just in order to complete the picture, that if we go one (big) step further and admit a non-denumerable set of frequencies $f$ in the spectrum [Bohr51 pp. 2–3], then the summation in (B.4) is no longer denumerable and should be understood in the sense of integration. We arrive then to the realm of the functions $f(x)$ which are representable by:[10]

$$f(x) \sim \int_{-\infty}^{\infty} F(u)\, e^{i2\pi ux}\, du \qquad \text{(B.9)}$$

rather than by (B.4): $$f(x) \sim \sum_n c_n\, e^{i2\pi f_n x}$$

$F(u)$ plays in (B.9) the same role as the Fourier coefficients $c_n$ in (B.4), namely: assigning the proper amplitudes (weights) to the various frequencies. It can be said, therefore, that in this case the spectrum of the function $f(x)$ is a function of the continuous frequency $u$, which is given by:

$$F(u) \sim \int_{-\infty}^{\infty} f(x)\, e^{-i2\pi ux}\, dx \qquad \text{(B.10)}$$

Visibly, this is the continuous counterpart of the impulsive spectrum of almost-periodic functions: as we have seen, in the almost-periodic case the spectrum is only defined on a denumerable set of frequencies $\{f_n\}$, and its impulse amplitudes at these frequencies are given by Eq. (B.5):

impulse amplitudes in order to guarantee that their Fourier series indeed converges and "makes sense" [Besicovitch32 p. xii–xiii]. The same is true also for the previous function classes having infinitely many impulses in their spectra, such as periodic functions, etc.

[10] By convention we use here the letter '$u$' rather than '$f$' to denote the frequency values.

$$c_n = \lim_{T\to\infty} \frac{1}{T}\int_T f(x)\, e^{-i2\pi f_n x}\, dx$$

Note, however, that these amplitudes can be also regarded as a function of the continuous frequency $u$ $(-\infty < u < \infty)$:

$$c(u) = \lim_{T\to\infty} \frac{1}{T}\int_T f(x)\, e^{-2\pi u x}\, dx \tag{B.11}$$

And indeed, a fundamental theorem in the theory of almost-periodic functions states that for any almost-periodic function $f(x)$, the function (B.11) is zero for all values of $u$ with the exception of an at most denumerable set of numbers $u = f_n$ [Bohr51 p. 48–50; the equivalent theorem for the periodic case is given on pp. 50–51]. These numbers $\{f_n\}$ are the Fourier exponents of $f(x)$, i.e., the frequencies which appear in the Fourier series (B.4), and the values of $c(u)$ at these points, $c_n = c(f_n)$, are the Fourier coefficients of $f(x)$.

We recognize, of course, that (B.10) and (B.9) above are simply the formulas of the (continuous) Fourier transform and inverse Fourier transform of the (aperiodic) function $f(x)$; $F(u)$ is the continuous-frequency spectrum of $f(x)$.

It can be said, therefore, roughly speaking, that making the step from denumerable to non-denumerable spectra is the spectral-domain equivalent of proceeding from almost-periodic to aperiodic functions in the image domain. More on the subject of impulsive and continuous spectra can be found in Chapter 11 of [Champeney87], especially on pp. 109–114.[11]

## B.7 Almost-periodic functions in two variables

The theory of almost-periodic functions can be also extended to functions of two or more variables (see, for instance, [Besicovitch32 pp. 59–66]); for our needs, however, we are only concerned with the 2D case, i.e., the case of two variables.

The 2D generalized Fourier series representation of an almost-periodic function $f(x,y)$, namely, the 2D extension of (B.4), is given by:

$$f(x,y) \sim \sum_m \sum_n c_{m,n}\, e^{i2\pi(u_{m,n}x + v_{m,n}y)} \tag{B.12}$$

or in the more compact vector notation:

$$f(\mathbf{x}) \sim \sum_m \sum_n c_{m,n}\, e^{i2\pi \mathbf{f}_{m,n}\cdot\mathbf{x}} \tag{B.13}$$

[11] It should be noted, however, that many authors present the transition between impulsive and continuous spectra as a limiting process directly between the periodic and aperiodic cases, ignoring the intermediate case of almost-periodic functions. Such a transition from Fourier series to the Fourier transform as a limit case is given, for example, in [Cartwright90 pp. 101–103], [Bracewell86 pp. 208–209] and [Gaskill78 pp. 111–112]. For the inverse direction, i.e., obtaining the Fourier series as a limit case of the Fourier transform, see [Cartwright90 pp. 99–101]; [Bracewell86 pp. 205–208].

where $\mathbf{f}_{m,n} = (u_{m,n}, v_{m,n})$, the $(m,n)$-th Fourier exponents, are arbitrary points in $\mathbb{R}^2$ (i.e., in the $(u,v)$ frequency plane of the spectrum), and $\mathbf{x} = (x,y)$. Note that an even more compact notation can be obtained by substituting $\mathbf{n}$ for $(m,n)$ in the indices:

$$f(\mathbf{x}) \sim \sum_{\mathbf{n}} c_{\mathbf{n}}\, e^{i2\pi \mathbf{f}_{\mathbf{n}} \cdot \mathbf{x}} \tag{B.14}$$

However, for the sake of clarity we will usually prefer the form of (B.13). The Fourier coefficients $c_{m,n}$ of $f(x,y)$, i.e., the 2D extension of (B.5), are given by:

$$c_{m,n} = \lim_{T\to\infty} \frac{1}{T^2} \iint_{TT} f(\mathbf{x})\, e^{-i2\pi \mathbf{f}_{m,n} \cdot \mathbf{x}}\, d\mathbf{x} \tag{B.15}$$

where $\iint_{TT}$ means an integration over any square interval of side $T$, i.e., over the area defined by the points $(x_0,y_0)$, $(x_0+T,y_0)$, $(x_0,y_0+T)$ and $(x_0+T,y_0+T)$ where $x_0$ and $y_0$ are arbitrary.

The spectrum of the 2D almost-periodic function (B.12) is given by:

$$P(u,v) = \sum_{m=-\infty}^{\infty} \sum_{n=-\infty}^{\infty} c_{m,n}\, \delta(u-u_{m,n},\, v-v_{m,n}) \tag{B.16}$$

which is the 2D extension of Eq. (B.6).

Examples of 2D almost-periodic functions appear throughout this book; let us mention here, for instance, the superposition of three identical dot-screens with angle differences of 30° that is traditionally used in colour printing (see Example 5.13 in Sec. 5.7).

For the sake of comparison, we recall from Sec. A.3 in Appendix A that the Fourier series representation of the 2-fold periodic function $p(x,y)$ whose periods are $T_x = 1/u_0$ and $T_y = 1/v_0$ is given by (A.11):

$$p(x,y) \sim \sum_{m=-\infty}^{\infty} \sum_{n=-\infty}^{\infty} c_{m,n}\, e^{i2\pi(mu_0 x + nv_0 y)}$$

Note that the vector notation of this expression is identical to (B.13), with only $\mathbf{f}_{m,n} = (mu_0, nv_0)$ replacing the more general $\mathbf{f}_{m,n} = (u_{m,n}, v_{m,n})$. The Fourier coefficients $c_{m,n}$ in the 2-fold periodic case are given by Eq. (A.14):

$$c_{m,n} = \frac{1}{T_x T_y} \iint_{T_x T_y} p(\mathbf{x})\, e^{-i2\pi \mathbf{f}_{m,n} \cdot \mathbf{x}}\, d\mathbf{x}$$

instead of Eq. (B.15) of the almost-periodic case; and the spectrum of the 2-fold periodic function $p(x,y)$ is given by Eq. (A.16):

$$P(u,v) = \sum_{m=-\infty}^{\infty} \sum_{n=-\infty}^{\infty} c_{m,n}\, \delta(u-mu_0,\, v-nv_0)$$

instead of the more general expression (B.16) in the almost-periodic case.

Note that there exist also "hybrid" functions which are periodic in one direction and almost-periodic in the other direction (see Sec. 6.2). These hybrid cases are still considered as almost-periodic functions.

# Appendix C

## Miscellaneous issues and derivations

### C.1 Derivation of the classical moiré formula (2.9) of Sec. 2.4

We show here that the classical formula (2.9) is simply a special case of Eqs. (2.28) and (2.8), that is obtained when the number of superposed gratings is $m = 2$, and the moiré in question is the (1,-1)-moiré, namely: $k_1=1$, $k_2=-1$. In this particular case (2.28) reduces to:

$$u = f_1 \cos\theta_1 - f_2 \cos\theta_2$$

$$v = f_1 \sin\theta_1 - f_2 \sin\theta_2$$

Therefore we have by Eq. (2.8):

$$\begin{aligned} f^2 &= u^2 + v^2 \\ &= f_1^2 \cos^2\theta_1 - 2f_1f_2 \cos\theta_1 \cos\theta_2 + f_2^2 \cos^2\theta_2 \\ &\qquad + f_1^2 \sin^2\theta_1 - 2f_1f_2 \sin\theta_1 \sin\theta_2 + f_2^2 \sin^2\theta_2 \\ &= f_1^2 (\cos^2\theta_1 + \sin^2\theta_1) - 2f_1f_2 (\cos\theta_1 \cos\theta_2 + \sin\theta_1 \sin\theta_2) + f_2^2 (\cos^2\theta_2 + \sin^2\theta_2) \\ &= f_1^2 - 2f_1f_2 \cos(\theta_2 - \theta_1) + f_2^2 \end{aligned}$$

In terms of periods rather than frequencies we have, therefore (where $\alpha = \theta_2 - \theta_1$):

$$\frac{1}{T_M^2} = \frac{1}{T_1^2} - 2\frac{1}{T_1}\frac{1}{T_2}\cos\alpha + \frac{1}{T_2^2}$$

$$= \frac{T_2^2 - 2T_1T_2\cos\alpha + T_1^2}{(T_1T_2)^2}$$

which finally gives, indeed, the period $T_M$ of the moiré, as predicted by formula (2.9):

$$T_M = \frac{T_1T_2}{\sqrt{T_1^2 + T_2^2 - 2T_1T_2\cos\alpha}}$$

As for the angle $\varphi_M$ of the moiré, we obtain from Eqs. (2.8) and (2.28):

$$\varphi_M = \arctan(v/u) = \arctan\left(\frac{f_1\sin\theta_1 - f_2\sin\theta_2}{f_1\cos\theta_1 - f_2\cos\theta_2}\right)$$

$$= \arctan\left(\frac{\frac{\sin\theta_1}{T_1} - \frac{\sin\theta_2}{T_2}}{\frac{\cos\theta_1}{T_1} - \frac{\cos\theta_2}{T_2}}\right) = \arctan\left(\frac{T_2\sin\theta_1 - T_1\sin\theta_2}{T_2\cos\theta_1 - T_1\cos\theta_2}\right) \quad ■$$

## C.2 Derivation of the first part of Proposition 2.1 of Sec. 2.5

We show in this section that if the values of the periodic function $p(x)$ are bounded between 0 and 1 then its Fourier series coefficients (impulse amplitudes) given by (A.2) satisfy: $0 \le a_0 \le 1$, and for any $n \ne 0$: $|a_n| \le 1/\pi$, $|b_n| \le 1/\pi$.

**Proof:** According to (A.4) we have:

$$a_0 = \frac{1}{T}\int_T p(x)\,dx$$

$a_0$ is, therefore, the average value of $p(x)$ on a single period; and since $0 \le p(x) \le 1$ it is obvious that $0 \le a_0 \le 1$.

Now, according to (A.4) we have for any $n \ne 0$:

$$a_n = \frac{1}{T}\int_T p(x)\cos(2\pi nx/T)\,dx \tag{C.1}$$

The function $\cos(2\pi nx/T)$ is periodic with the period $T/n$, and therefore it makes exactly $n$ full cosinusoidal cycles (periods) within the interval $T$, oscillating between 1 and –1. Its total area within the interval $T$ is therefore 0 (see Fig. C.1).

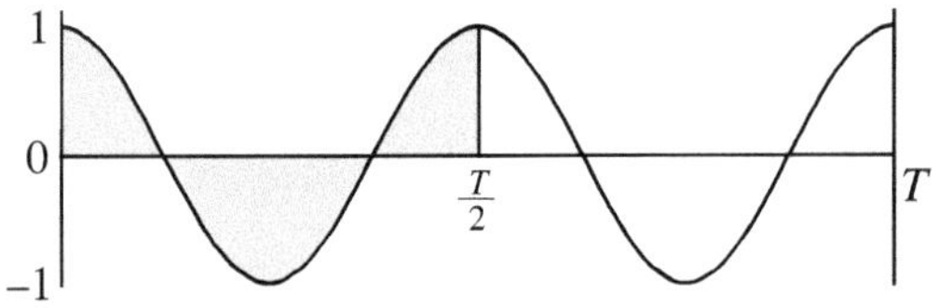


**Figure C.1:** The function $\cos(2\pi nx/T)$ for $n = 2$.

Now, since the value of the function $p(x)$ is bounded between 0 and 1, it is clear that the most it can do to increase the area defined by the integral (C.1) is to become 0 at all the negative areas of $\cos(2\pi nx/T)$ in order to mask them out. But even then, the maximum possible area left within the interval $T$ is the positive area of $\cos(2\pi nx/T)$, which equals $2n$ times the area of half a lobe, namely:

$$2n\int_0^{\frac{T}{4n}} \cos(2\pi nx/T)\,dx = 2n\left[\frac{\sin(2\pi nx/T)}{2\pi n/T}\right]_0^{\frac{T}{4n}} = \frac{\sin(\pi/2)}{\pi/T} = \frac{T}{\pi}$$

which means by (C.1) that $a_n \le 1/\pi$. In a similar way, by masking out the positive areas we obtain $a_n \ge -1/\pi$, and hence, by combining both results we get, indeed, $|a_n| \le 1/\pi$.

$|b_n| \le 1/\pi$ is obtained in a similar way. ■

## C.3 Invariance of the impulse amplitudes under rotations and $x,y$ scalings

We used in Secs. 4.2 and 4.3 the fact that stretching and rotating a periodic (or doubly periodic) image does not affect the *impulse amplitudes* of its comb (or nailbed), but only their *impulse locations* in the spectrum. In this section we show how these properties can be derived, based on well-known results in the Fourier theory. We will use here the letters $p$ and $P$ to denote a periodic function and its spectrum, and the letters $f$ and $F$ to denote an arbitrary function (not necessarily periodic) and its spectrum.

### C.3.1 Invariance of the 2D Fourier transform under rotations

The rotation-invariance property of the 2D Fourier transform means that a rotation of $f(x,y)$ in the image domain by angle $\theta$ has no effect on its spectrum $F(u,v)$ other than a rotation by the same angle $\theta$. This property of the 2D Fourier transform, which is valid for any function, is guaranteed by the rotation theorem [Bracewell95 p. 157]. This invariance of the 2D Fourier transform under rotation is a special case of its more complex behaviour under a general linear transformation, which is given for instance in [Bracewell95 p. 160].

### C.3.2 Invariance of the impulse amplitudes under $x, y$ scalings

According to the similarity theorem [Bracewell86 p. 244], for any function $f(x,y)$ with Fourier transform $F(u,v)$ we have:

$$f(ax,by) \leftrightarrow \frac{1}{|ab|}F(u/a,v/b) \tag{C.2}$$

However, a special case of interest occurs with periodic functions, where the spectrum is impulsive. In this case, thanks to the particular scaling property of $\delta(u,v)$ [Bracewell86 p. 85]: $\delta(u/a,v/b) = |ab|\,\delta(u,v)$, the factor $1/|ab|$ in Eq. (C.2) is cancelled out, and we obtain [Bracewell86 p. 103]:

$$p(ax,by) \leftrightarrow P(u/a,v/b)$$

This is usually formulated in the following way (see, for example, the 1D equivalent in [Cartwright90 pp. 59–61]):

Let the function $p(x,y)$ be periodic with periods $T_x$, $T_y$ and generate the Fourier series:

$$p(x,y) \sim \sum_{m=-\infty}^{\infty}\sum_{n=-\infty}^{\infty} a_{m,n}\cos 2\pi\left(\frac{mx}{T_x}+\frac{ny}{T_y}\right) + \sum_{m=-\infty}^{\infty}\sum_{n=-\infty}^{\infty} b_{m,n}\sin 2\pi\left(\frac{mx}{T_x}+\frac{ny}{T_y}\right)$$

Then $p(ax,by)$ is periodic with periods $T_x/a$, $T_y/b$, and the Fourier series it generates preserves the original coefficients (impulse amplitudes) $a_{m,n}$ and $b_{m,n}$:

$$p(ax,by) \sim \sum_{m=-\infty}^{\infty}\sum_{n=-\infty}^{\infty} a_{m,n}\cos 2\pi\left(\frac{mx}{T_x/a}+\frac{ny}{T_y/b}\right) + \sum_{m=-\infty}^{\infty}\sum_{n=-\infty}^{\infty} b_{m,n}\sin 2\pi\left(\frac{mx}{T_x/a}+\frac{ny}{T_y/b}\right)$$

## C.4 Shift and phase

This section, which complements the introduction to Chapter 7, provides a more detailed explanation on the connection between the phase in the context of complex number theory and the phase in periodic functions.

### C.4.1 The shift theorem

Let $f(x)$ be a 1D function (periodic or not) in the image domain, and let $F(u)$ be its spectrum. As already mentioned in Sec. 2.2, if $f(x)$ is symmetric about the origin then its spectrum $F(u)$ is purely real; and if $f(x)$ is non-symmetric or non-centered about the origin then $F(u)$ is complex-valued (its imaginary part is non-zero).

Assume now that we shift $f(x)$ in the image domain by $a$. The shift theorem [Bracewell86 p. 104] states that if the spectrum of $f(x)$ is $F(u)$, then the spectrum of the shifted function $f(x-a)$ is $F_a(u) = e^{-i2\pi ua} \cdot F(u)$. This means that a shift of $a$ in the image domain multiplies the spectrum at each frequency $u$ by the complex factor $e^{-i2\pi ua}$. Therefore, even if the spectrum $F(u)$ of the unshifted function $f(x)$ is purely real, the spectrum $F_a(u)$ of the shifted function $f(x-a)$ is a complex-valued function of the real variable $u$, namely: $F_a : \mathbb{R} \to \mathbb{C}$.

The situation in the 2D case is similar: Assume that $f(x,y)$ is a 2D function (periodic or not) in the image domain and that $F(u,v)$ is its spectrum. Then, according to the 2D shift theorem [Bracewell95 p. 156], the spectrum of the shifted function $f(x-a, y-b)$ is $F_{a,b}(u,v) = e^{-i2\pi(ua+vb)} \cdot F(u,v)$. In other words, a shift of $\mathbf{a} = (a,b)$ in the image domain multiplies the spectrum at each frequency $\mathbf{f} = (u,v)$ by the complex factor $e^{-i2\pi \mathbf{f} \cdot \mathbf{a}}$. We see, therefore, that as in the 1D case, even if the spectrum $F(u,v)$ of the unshifted function $f(x,y)$ is purely real, the spectrum $F_{a,b}(u,v)$ of the shifted function $f(x-a, y-b)$ is a complex-valued function of the real variables $u,v$, namely: $F_{a,b} : \mathbb{R}^2 \to \mathbb{C}$.

It may be in order, therefore, to review here some of the properties of complex-valued functions. We will do it here for complex-valued functions of two real variables, but the situation in the case of a single real variable is completely analogous.

A complex-valued function $F(u,v)$ can be represented either by its real and its imaginary parts:

$$F(u,v) = \mathrm{Re}[F(u,v)] + i\,\mathrm{Im}[F(u,v)]$$

or, using the polar notation, by its *magnitude* (also called *absolute value* or *modulus*) and its *phase* (also called *argument*):

$$F(u,v) = \mathrm{Abs}[F(u,v)] \cdot e^{i\,\mathrm{Arg}[F(u,v)]}$$

where:

$$\mathrm{Abs}[F(u,v)] = \sqrt{\mathrm{Re}[\mathrm{F}(u,v)]^2 + \mathrm{Im}[\mathrm{F}(u,v)]^2}$$

$$\mathrm{Arg}[F(u,v)] = \arctan \frac{\mathrm{Im}[\mathrm{F}(u,v)]}{\mathrm{Re}[\mathrm{F}(u,v)]}$$

Note that Re[$F(u,v)$], Im[$F(u,v)$], Abs[$F(u,v)$] and Arg[$F(u,v)$] are all real-valued functions of the real variables $u,v$. The functions Abs[$F(u,v)$] and Arg[$F(u,v)$] represent the local magnitude and the local phase of the complex-valued function $F(u,v)$ at the point $(u,v)$. Therefore, by analogy with the polar notation of a complex number, the complex-valued function $F(u,v)$ can be interpreted as a phasor (a varying radius-vector) rotating in the complex plane, whose length and angle at any point $(u,v)$ are given, respectively, by Abs[$F(u,v)$] and Arg[$F(u,v)$].[1] As Abs[$F(u,v)$] varies with $u$ and $v$, the length of the vector varies, and as Arg[$F(u,v)$] varies with $u$ and $v$, the direction of the vector varies. (A good, concise introduction on phasors can be found, for example, in [Gaskill78 pp. 18–29].) Therefore, in the context of complex number theory the term "the phase of the function $F(u,v)$" refers to the argument Arg[$F(u,v)$], which represents the angle of the phasor of $F(u,v)$ in the complex plane at each point $(u,v)$ of the $u,v$ plane.

Now, if $F(u,v)$ is the spectrum of $f(x,y)$, then according to the 2D shift theorem the spectrum of the shifted function $f(x-a, y-b)$ is $F_{a,b}(u,v) = e^{-i2\pi(ua+vb)} \cdot F(u,v)$, or, using the polar notation: $F_{a,b}(u,v) = \text{Abs}[F(u,v)] \cdot e^{i[\text{Arg}[F(u,v)] - 2\pi(ua+vb)]}$. We obtain, therefore, that:

$$\text{Abs}[F_{a,b}(u,v)] = \text{Abs}[F(u,v)]$$

$$\text{Arg}[F_{a,b}(u,v)] = \text{Arg}[F(u,v)] - 2\pi(ua + vb)$$

This means that a shift of $(a,b)$ in $f(x,y)$ in the image domain does not influence the magnitude of the spectrum, but it does decrement its phase at any point $(u,v)$ by $2\pi(ua + vb)$. We obtain, therefore, the following corollary:

**Corollary of the 2D shift theorem:** When a function $f(x,y)$ in the image domain is shifted by $(a,b)$, its *magnitude-spectrum* (i.e., the magnitude of its complex spectrum) remains unchanged, but its *phase-spectrum* (i.e., the phase, or the argument, of its complex spectrum) is linearly decremented at any point $(u,v)$ of the $u,v$ plane by the linear function $2\pi(ua + vb)$. ■

We see, therefore, that the increment generated in the phase spectrum due to a shift of **a** in the image domain is a linear function of the frequency, i.e., it is a continuous linear plane through the origin, whose slopes are determined by $\mathbf{a} = (a,b)$.[2] Denoting this phase increment by $\varphi$ we have, therefore:

$$\varphi(\mathbf{f}) = -2\pi\mathbf{f}\cdot\mathbf{a}$$

namely: $$\varphi(u,v) = -2\pi(ua + vb) \qquad \text{(C.3)}$$

Similarly, we can obtain from the 1D shift theorem the 1D counterpart of the above corollary:

[1] Note that the complex plane should not be confused here with the $u,v$ plane: the complex-valued function $F(u,v)$ is defined on the $u,v$ plane, but its image is located in the complex plane $\mathbb{C}$.

[2] Note that the converse is also true: it follows from the shift theorem that a linear increment occurs in the phase-spectrum *iff* the original function has undergone a shift in the image domain.

**Corollary of the 1D shift theorem:** When a function $f(x)$ in the image domain is shifted by $a$, its *magnitude-spectrum* remains unchanged, but its *phase-spectrum* is linearly decremented at any frequency $u$ by $2\pi ua$. ■

As we can see, the 1D case is indeed a straightforward simplification of the 2D case: The increment generated in the phase spectrum due to a shift of $a$ in the image domain is a linear function of the frequency $u$, i.e., it is a straight line through the origin, whose slope is determined by $a$. Denoting this phase increment by $\varphi$, the 1D equivalent of Eq. (C.3) is given therefore by:

$$\varphi(u) = -2\pi ua \qquad \text{(C.4)}$$

### C.4.2 The particular case of periodic functions

Let us return now from the general shift theorem to the particular case of periodic functions. Let $p(x,y)$ (or in short, $p(\mathbf{x})$) be a 2-fold periodic function with fundamental frequency vectors $\mathbf{f}_1 = (u_1,v_1)$, $\mathbf{f}_2 = (u_2,v_2)$ (see Sec. A.3.4 in Appendix A). As we have seen above, Eq. (C.3) says that for any given $f(x,y)$ the increment generated in the phase-spectrum due to a shift of $\mathbf{a}$ in the image domain is a linear function of the frequency, i.e., a continuous plane whose slopes are determined by $\mathbf{a} = (a,b)$. In our case, however, the spectrum of $p(\mathbf{x})$ is an impulse nailbed, whose $(m,n)$-th impulse has the frequency $\mathbf{f} = m\mathbf{f}_1+n\mathbf{f}_2$, or in other words: $(u,v) = m(u_1,v_1) + n(u_2,v_2) = (mu_1+nu_2\ ,\ mv_1+nv_2)$. This spectrum is given in vector form by Eq. (A.30) in Appendix A. The phase increment generated at the $(m,n)$-th impulse in the spectrum as a result of the shift of $\mathbf{a}$ in the image domain is, therefore:

$$\varphi(m\mathbf{f}_1+n\mathbf{f}_2) = -2\pi(m\mathbf{f}_1+n\mathbf{f}_2)\cdot\mathbf{a}$$

namely:

$$\varphi(mu_1+nu_2\ ,\ mv_1+nv_2) = -2\pi[(mu_1+nu_2)a + (mv_1+nv_2)b] \qquad \text{(C.5)}$$

which is simply the restriction of the linear plane (C.3) to the points of our nailbed. In other words, Eq. (C.5) samples the continuous plane (C.3) of the phase-spectrum increment, which is owed to the shift theorem, at all the impulse locations $m\mathbf{f}_1+n\mathbf{f}_2$. This is clearly seen in the spectrum of the shifted function $p(\mathbf{x}-\mathbf{a})$ (see Eq. (7.4)).

Similarly, in the case of a 1-fold periodic function $p(x)$, the spectrum is an impulse comb whose $n$-th impulse has the frequency $u = nf$ (this spectrum is given by Eq. (A.8) in Appendix A). The phase increment generated at the $n$-th impulse in the spectrum as a result of the shift of $a$ in the image domain is, therefore:

$$\varphi(nf) = -2\pi nfa \qquad \text{(C.6)}$$

which is simply the restriction of the straight line (C.4) to the points of our comb. In other words, Eq. (C.6) samples the continuous line (C.4) of the phase-spectrum increment, which is due to the shift theorem, at all the impulse locations $nf$.

### C.4.3 The phase of a periodic function: the $\varphi$ and the $\phi$ notations

We have seen until now how shifts of a function in the image domain are related to phase changes in the spectrum, where the term "phase" is understood as the argument of the complex spectrum.

However, in the particular case of periodic functions the term "phase" can be also used in a different sense, which is related, this time, to the image domain. Consider, as a simple example, the 1D periodic function $p(x) = \cos(2\pi fx)$, whose period is $T = 1/f$. Its counterpart that is shifted in the image domain by $a$ is given by: $p(x-a) = \cos(2\pi f(x-a))$. As explained in detail in Sec. 7.3, an alternative way to specify the amount of shift in a periodic function is to state it as a fraction of the period $T$: $\phi = \frac{a}{T} = fa$. This value is often called in literature the *phase* of the shifted function $p(x-a)$; for example, when $\phi = n$, $n \in \mathbb{Z}$, it is said that $p(x)$ and $p(x-a)$ are "in phase", and when $\phi = n+\frac{1}{2}$ it is said that they are "in counter-phase". In order to avoid confusion between the two meanings of the term "phase" we prefer to call $\phi$ the *period-shift* of the periodic function; this name will clearly distinguish it from the phase $\text{Arg}[P_a(u)]$ and the phase increment $\varphi$ which were defined in the context of complex number theory. Obviously, the period-shift $\phi$ is only meaningful in periodic functions, while the phase increment $\varphi$ in the sense of complex numbers is meaningful in the spectrum of any function. Therefore, in the case of a periodic function both $\phi$ and $\varphi$ can be used as a measure of its shifts in the image domain.

Let us first illustrate this for the case of a 1D periodic function $p(x)$ and its shifted copy $p(x-a)$. As we have seen above, the shift of $a$ in the image domain is expressed, in terms of the phase increment of the complex spectrum, by the linear function (C.4):

$$\varphi(u) = -2\pi ua$$

But in our particular case in which the shifted function is periodic this continuous line is sampled (and is only meaningful) at the frequencies $u = nf$ ($n \in \mathbb{Z}$), since the impulsive spectrum is only defined at these points. The phase increment of the $n$-th impulse in the spectrum of $p(x-a)$ owed to the shift $a$ is given, therefore, by:

$$\varphi(nf) = -2\pi nfa$$

or:

$$\varphi(nf) = -2\pi n\phi$$

where the period-shift $\phi = fa = \frac{a}{T}$ expresses the shift of the periodic function $p(x-a)$ in the image domain in terms of its period. This equation gives, indeed, the connection between $\varphi$ and $\phi$, which are both measures of the shift $a$ in the periodic function $p(x)$: While $\phi$ describes the shift in terms of the period of $p(x)$, $\varphi$ describes the shift as an angle difference in the complex plane (where for the $n$-th impulse, $\varphi = -2\pi n$ is equivalent to a shift of one full period, $\phi = 1$).

This connection between $\varphi$ and $\phi$ can be also generalized to the case of a 2-fold periodic function $p(\mathbf{x})$. As explained in detail in Sec. 7.5.2, an alternative way to specify the amount

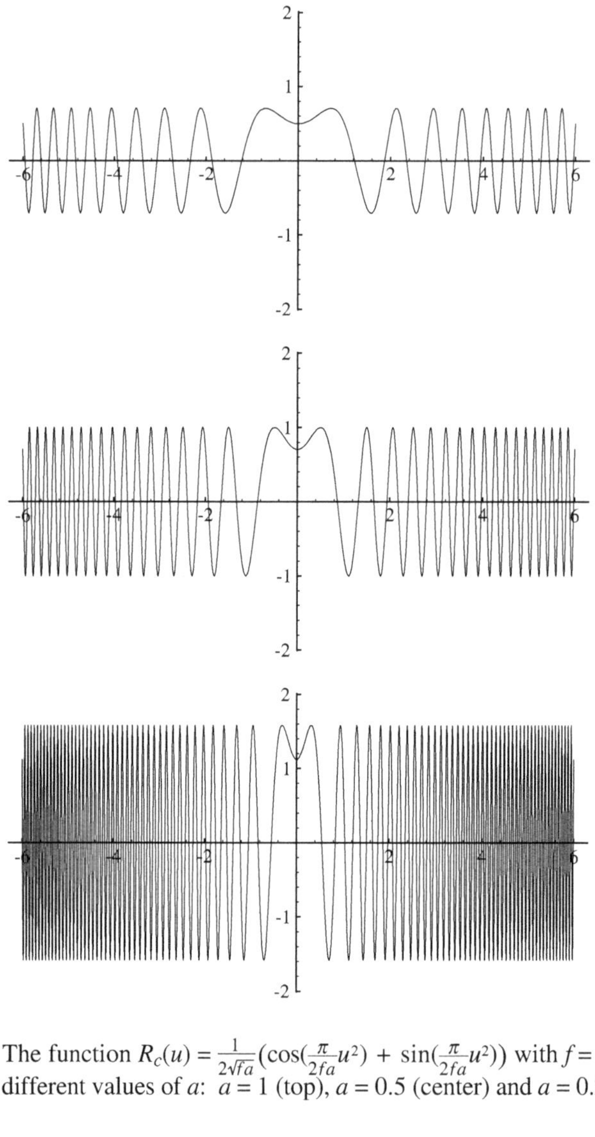

**Figure C.2:** The function $R_C(u) = \frac{1}{2\sqrt{fa}}(\cos(\frac{\pi}{2fa}u^2) + \sin(\frac{\pi}{2fa}u^2))$ with $f = 1$ and three different values of $a$: $a = 1$ (top), $a = 0.5$ (center) and $a = 0.2$ (bottom).

of shift in a 2-fold periodic function is to state it using the period-shifts $\phi_1$ and $\phi_2$, i.e., as fractions of the two periods $\mathbf{T}_1$ and $\mathbf{T}_2$. And indeed, as we can see in Eq. (7.12) at the end of Sec. 7.5.2, the phase increment of the $(m,n)$-th impulse in the spectrum of $p(\mathbf{x}-\mathbf{a})$ owing to the shift $\mathbf{a}$ can be expressed by:

$$\varphi(m\mathbf{f}_1+n\mathbf{f}_2) = -2\pi(m\phi_1+n\phi_2) \tag{C.7}$$

This equation is, indeed, the connection between $\varphi$ and $\phi$ in the 2D case.

It should be noted, however, that in Chapters 7 and 8 we usually use the period-shifts $\phi$, while the phase increment $\varphi$ is only occasionally mentioned.

## C.5 The function $R_c(u)$ converges to $\delta(u)$ as $a\to 0$

We derive here the rather surprising fact (see Sec. 10.4.1) that in spite of its undamped oscillatory nature, the function $R_c(u) = \frac{1}{2\sqrt{fa}}(\cos(\frac{\pi}{2fa}u^2) + \sin(\frac{\pi}{2fa}u^2))$ tends when $a\to 0$ to the impulse $\delta(u)$ (see Fig. C.2). Following [Gelfand64 pp. 36–38], we have to show for this end that: (a) the total area under $R_c(u)$ is 1 independently of $a$; and (b) the area under $R_c(u)$ in the ranges $(u_1,u_2)$ and $(-u_2,-u_1)$ for any $u_2 > u_1 > 0$ tends to zero when $a\to 0$.

We start by showing part (a). Since we have [Spiegel68 p. 97]:

$$\int_0^\infty \sin cx^2\, dx = \int_0^\infty \cos cx^2\, dx = \tfrac{1}{2}\sqrt{\tfrac{\pi}{2c}}$$

it follows that:

$$\int_0^\infty R_c(u)\, du = \frac{1}{2\sqrt{fa}}\int_0^\infty (\cos(\tfrac{\pi}{2fa}u^2) + \sin(\tfrac{\pi}{2fa}u^2))\, du$$
$$= \frac{1}{2\sqrt{fa}}(\tfrac{1}{2}\sqrt{fa} + \tfrac{1}{2}\sqrt{fa}) = \tfrac{1}{2}$$

and since $R_c(u)$ is symmetric we obtain, as required:

$$\int_{-\infty}^\infty R_c(u)\, du = 2\int_0^\infty R_c(u)\, du = 1.$$

Proceeding now to part (b), we wish to show that for any $u_2 > u_1 > 0$ we have:

$$\lim_{a\to\infty} \int_{u_1}^{u_2} R_c(u)\, du = 0.$$

And indeed, using the formulas [Gradshteyn94 pp. 178–179]:

$$\int \sin cx^2\, dx = \sqrt{\tfrac{\pi}{2c}}\, S(\sqrt{c}\,x)$$

$$\int \cos cx^2\, dx = \sqrt{\tfrac{\pi}{2c}}\, C(\sqrt{c}\,x)$$

where $S(x)$ and $C(x)$ are the Fresnel sine and cosine integrals, defined as:

$$S(x) = \sqrt{\tfrac{2}{\pi}}\int_0^x \sin t^2\, dt$$

$$C(x) = \sqrt{\tfrac{2}{\pi}}\int_0^x \cos t^2\, dt$$

we obtain:

$$\int_{u_1}^{u_2} R_c(u)\,du = \frac{1}{2\sqrt{fa}}\int_{u_1}^{u_2} (\cos(\tfrac{\pi}{2fa}u^2) + \sin(\tfrac{\pi}{2fa}u^2))\,du$$

$$= \frac{1}{2\sqrt{fa}}\left(\left[\sqrt{fa}\,C(\sqrt{\tfrac{\pi}{2fa}}x)\right]_{u_1}^{u_2} + \left[\sqrt{fa}\,S(\sqrt{\tfrac{\pi}{2fa}}x)\right]_{u_1}^{u_2}\right)$$

$$= \tfrac{1}{2}\left([C(\sqrt{\tfrac{\pi}{2fa}}u_2) - C(\sqrt{\tfrac{\pi}{2fa}}u_1)] + [S(\sqrt{\tfrac{\pi}{2fa}}u_2) - S(\sqrt{\tfrac{\pi}{2fa}}u_1)]\right)$$

and hence, when $a\to 0$ we get:

$$\lim_{a\to\infty}\int_{u_1}^{u_2} R_c(u)\,du = \tfrac{1}{2}([C(\infty)-C(\infty)] + [S(\infty)-S(\infty)])$$

but since $S(\infty) = C(\infty) = \frac{1}{2}$ [Spiegel68 p. 184] we obtain, as required:

$$\lim_{a\to\infty}\int_{u_1}^{u_2} R_c(u)\,du = 0.$$

The proof of (b) for the range $(-u_2,-u_1)$ is similar. ■

## C.6 The 2D spectrum of a cosinusoidal zone grating

As shown in Example 10.7 of Sec. 10.3, a cosinusoidal zone grating is defined by:

$$r^+(x,y) = p(x^2+y^2) = \cos(2\pi f(x^2+y^2))$$

and its hyperbolic counterpart is given by:

$$r^-(x,y) = p(x^2-y^2) = \cos(2\pi f(x^2-y^2))$$

We find now the spectra $R^+(u,v)$ and $R^-(u,v)$ of these two functions. According to the trigonometric identity $\cos(\alpha\pm\beta) = \cos\alpha\cos\beta \mp \sin\alpha\sin\beta$ we have:

$$r^\pm(x,y) = \cos(2\pi fax^2)\cos(2\pi fby^2) \mp \sin(2\pi fax^2)\sin(2\pi fby^2)$$

Thanks to the separable-product theorem [Bracewell95 p. 166] we obtain:

$$\cos(2\pi fax^2)\cos(2\pi fby^2) \leftrightarrow \frac{1}{2\sqrt{f}}(\cos(\tfrac{\pi}{2f}u^2) + \sin(\tfrac{\pi}{2f}u^2))\frac{1}{2\sqrt{f}}(\cos(\tfrac{\pi}{2f}v^2) + \sin(\tfrac{\pi}{2f}v^2))$$

$$= \frac{1}{4f}[\cos(\tfrac{\pi}{2f}u^2)\cos(\tfrac{\pi}{2f}v^2) + \sin(\tfrac{\pi}{2f}u^2)\cos(\tfrac{\pi}{2f}v^2)$$

$$+ \cos(\tfrac{\pi}{2f}u^2)\sin(\tfrac{\pi}{2f}v^2) + \sin(\tfrac{\pi}{2f}u^2)\sin(\tfrac{\pi}{2f}v^2)]$$

Now, using the trigonometric identities

$$\cos\alpha\cos\beta = \tfrac{1}{2}[\cos(\alpha-\beta) + \cos(\alpha+\beta)]$$

$$\sin\alpha\sin\beta = \tfrac{1}{2}[\cos(\alpha-\beta) - \cos(\alpha+\beta)]$$

$$\sin\alpha\cos\beta = \tfrac{1}{2}[\sin(\alpha-\beta) + \sin(\alpha+\beta)]$$

we obtain:

$$= \frac{1}{8f}\Big[\cos\tfrac{\pi}{2f}(u^2-v^2) + \cos\tfrac{\pi}{2f}(u^2+v^2) + \sin\tfrac{\pi}{2f}(u^2-v^2) + \sin\tfrac{\pi}{2f}(u^2+v^2)$$

$$+ \sin\tfrac{\pi}{2f}(v^2-u^2) + \sin\tfrac{\pi}{2f}(u^2+v^2) + \cos\tfrac{\pi}{2f}(u^2-v^2) - \cos\tfrac{\pi}{2f}(u^2+v^2)\Big]$$

$$= \frac{1}{4f}\Big[\cos\tfrac{\pi}{2f}(u^2-v^2) + \sin\tfrac{\pi}{2f}(u^2+v^2)\Big]$$

Similarly, thanks to the separable-product theorem we obtain:

$$\sin(2\pi fax^2)\sin(2\pi fby^2) \leftrightarrow \frac{1}{2\sqrt{f}}\big(\cos(\tfrac{\pi}{2f}u^2) - \sin(\tfrac{\pi}{2f}u^2)\big)\frac{1}{2\sqrt{f}}\big(\cos(\tfrac{\pi}{2f}v^2) - \sin(\tfrac{\pi}{2f}v^2)\big)$$

$$= \frac{1}{4f}\Big[\cos(\tfrac{\pi}{2f}u^2)\cos(\tfrac{\pi}{2f}v^2) - \sin(\tfrac{\pi}{2f}u^2)\cos(\tfrac{\pi}{2f}v^2)$$

$$- \cos(\tfrac{\pi}{2f}u^2)\sin(\tfrac{\pi}{2f}v^2) + \sin(\tfrac{\pi}{2f}u^2)\sin(\tfrac{\pi}{2f}v^2)\Big]$$

$$= \frac{1}{4f}\Big[\cos\tfrac{\pi}{2f}(u^2-v^2) - \sin\tfrac{\pi}{2f}(u^2+v^2)\Big]$$

We have, therefore:

$$R^{\pm}(u,v) = \frac{1}{4f}\Big[\cos\tfrac{\pi}{2f}(u^2-v^2) + \sin\tfrac{\pi}{2f}(u^2+v^2) \mp \cos\tfrac{\pi}{2f}(u^2-v^2) \pm \sin\tfrac{\pi}{2f}(u^2+v^2)\Big]$$

and hence:

$$R^{+}(u,v) = \frac{1}{2f}\sin(\tfrac{\pi}{2f}(u^2+v^2))$$

$$R^{-}(u,v) = \frac{1}{2f}\cos(\tfrac{\pi}{2f}(u^2-v^2)). \quad \blacksquare$$

## C.7 The convolution of two orthogonal line-impulses

Suppose that we are given a horizontal line-impulse $f(x)\delta(y)$, whose amplitude is defined by $f(x)$, and a vertical 1D line-impulse $g(y)\delta(x)$, whose amplitude is defined by $g(y)$. We assume that both line-impulses are centered on the origin. We want to show (see Sec. 10.7.3) that their 2D convolution is given by the 2D function $f(x)g(y)$, namely:

$$f(x)\delta(y) ** g(y)\delta(x) = f(x)g(y)$$

And indeed, according to the definition of 2D convolution [Bracewell86 p. 243] we have:

$$f(x)\delta(y) ** g(y)\delta(x) = \int_{-\infty}^{\infty}\int_{-\infty}^{\infty} f(x')\delta(y')\, g(y-y')\delta(x-x')\, dx'\, dy'$$

$$= \int_{-\infty}^{\infty} f(x')\delta(x-x')\, dx' \int_{-\infty}^{\infty} \delta(y')g(y-y')\, dy'$$

but since each of these two integrals is simply a 1D convolution of two 1D functions:

$$= [f(x)*\delta(x)][g(y)*\delta(y)]$$

$$= f(x)g(y)$$

This is a 2D surface which is centered about the origin. Similarly, if the centers of the original line-impulses are shifted from the origin to the points $(x_1,y_1)$ and $(x_2,y_2)$, respectively, so that the line-impulses are given by $f(x-x_1)\delta(y-y_1)$ and $g(y-y_2)\delta(x-x_2)$, their 2D convolution gives:

$$f(x-x_1)\delta(y-y_1) ** g(y-y_2)\delta(x-x_2) = f(x-x_1-x_2)g(y-y_1-y_2)$$

This is the same 2D surface as before, but as expected its center is shifted to the point $(x_1+x_2, y_1+y_2)$. ■

## C.8 The compound line-impulse of the singular $(k_1,k_2)$-line-impulse cluster

We have seen in Sec. 10.7.3 that a $(k_1,k_2)$-moiré in the superposition of a parabolic grating and a periodic straight grating becomes singular when the line-impulses of the $(k_1,k_2)$-cluster in the spectrum convolution fall on a single line through the spectrum origin. This gives us in the spectrum a compound line-impulse, whose amplitude is the sum of all the individual line-impulses of the $(k_1,k_2)$-cluster. Note that the collapsed line-impulses do not necessarily fall center-on-center: in the general case the distance between the centers of consecutive line-impulses of the collapsed $(k_1,k_2)$-cluster is not zero but some other constant, so that the individual line-impulses are summed up along the compound line-impulse with a constant shift between each other. We call this shift (the distance between the DC and the center of the first line-impulse of the collapsed cluster) the *internal discrepancy* of the compound line-impulse and we denote it by $f_0$; the distance between the DC and the $n$-th impulse center is therefore $nf_0$. The role of the internal discrepancy $f_0$ will become clear below; we will see that when $f_0 = 0$ (so that all the collapsed line-impulses are centered on the DC) the compound line-impulse corresponds in the image domain to a singular moiré which is centered on the origin, but when $f_0 \neq 0$ the center of the singular moiré is shifted away from the center of the image domain.

Let us now concentrate on the $n$-th line-impulse pair of the $(k_1,k_2)$-cluster. For the sake of simplicity we assume, like in Example 10.13, that the parabolic grating $r_1(x,y)$ is oriented horizontally, so that all the line-impulses in the spectrum are vertical and the $(k_1,k_2)$-cluster may only collapse on the vertical $v$ axis. Since the shift $f_0$ in this case takes place vertically along the $v$ axis, we prefer to denote it henceforth by $v_0$ (see Fig. C.3).

As a generalization of Eq. (10.23), the $n$-th line-impulse pair of the $(k_1,k_2)$-cluster, namely: the $(nk_1,nk_2)$- and $(-nk_1,-nk_2)$-line-impulses, are given (for $n \neq 0$) by:

$$a_{nk_1,nk_2}(u,v) = \tfrac{1}{2}[R_c(v-nv_0) + iR_s(v-nv_0)]\ \delta(u)\ a^{(1)}{}_{nk_1}\ a^{(2)}{}_{nk_2}$$

$$a_{-nk_1,-nk_2}(u,v) = \tfrac{1}{2}[R_c(v+nv_0) - iR_s(v+nv_0)]\ \delta(u)\ a^{(1)}{}_{-nk_1}\ a^{(2)}{}_{-nk_2}$$

with:

$$R_c(v) = \frac{1}{2\sqrt{nk_1f_1a}}\left(\cos\left(\frac{\pi}{2nk_1f_1a}v^2\right) + \sin\left(\frac{\pi}{2nk_1f_1a}v^2\right)\right)$$

$$R_s(v) = \frac{1}{2\sqrt{nk_1f_1a}}\left(\cos\left(\frac{\pi}{2nk_1f_1a}v^2\right) - \sin\left(\frac{\pi}{2nk_1f_1a}v^2\right)\right)$$

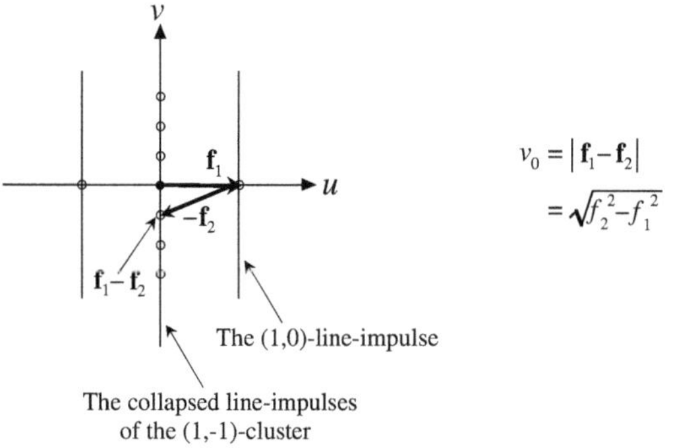


**Figure C.3:** The spectrum convolution showing the compound line-impulse of the singular (1,-1)-moiré, which is collapsed on the $v$ axis, and its internal discrepancy $v_0$. Note that it is assumed here that $f_2 \geq f_1$. When $f_2 < f_1$ the line-impulses of the (1,-1)-cluster cannot fall on the $v$ axis, and hence no rotation angle $\theta$ of $\mathbf{f}_2$ (i.e., of the straight grating $r_2(x,y)$) can bring the (1,-1)-moiré to a singular state.

(the terms $\delta(u)$ indicate that both line-impulses are located now on the vertical $v$ axis).

We prefer to use here the exponential notation, which is more compact and easier to handle. We have, therefore:

$$R_c(v) + iR_s(v) = \frac{1}{2\sqrt{nk_1f_1a}}[\cos(\tfrac{\pi}{2nk_1f_1a}v^2) + \sin(\tfrac{\pi}{2nk_1f_1a}v^2) + i\cos(\tfrac{\pi}{2nk_1f_1a}v^2) - i\sin(\tfrac{\pi}{2nk_1f_1a}v^2)]$$

$$= \sqrt{\tfrac{c}{2\pi}}\,[e^{-icv^2} + ie^{-icv^2}]$$

where $c = \frac{\pi}{2nk_1f_1a}$; and similarly:

$$R_c(v) - iR_s(v) = \frac{1}{2\sqrt{nk_1f_1a}}[\cos(\tfrac{\pi}{2nk_1f_1a}v^2) + \sin(\tfrac{\pi}{2nk_1f_1a}v^2) - i\cos(\tfrac{\pi}{2nk_1f_1a}v^2) + i\sin(\tfrac{\pi}{2nk_1f_1a}v^2)]$$

$$= \sqrt{\tfrac{c}{2\pi}}\,[e^{icv^2} - ie^{icv^2}]$$

Therefore we can rewrite the line-impulses $a_{nk_1,nk_2}(u,v)$ and $a_{-nk_1,-nk_2}(u,v)$ using the exponential notation as follows:

$$a_{nk_1,nk_2}(u,v) = \tfrac{1}{2}\sqrt{\tfrac{c}{2\pi}}\,[e^{-ic(v-nv_0)^2} + ie^{-ic(v-nv_0)^2}]\;\delta(u)\;a^{(1)}_{nk_1}\,a^{(2)}_{nk_2}$$

$$a_{-nk_1,-nk_2}(u,v) = \tfrac{1}{2}\sqrt{\tfrac{c}{2\pi}}\,[e^{ic(v+nv_0)^2} - ie^{ic(v+nv_0)^2}]\;\delta(u)\;a^{(1)}_{-nk_1}\,a^{(2)}_{-nk_2}$$

Now, since at the singular state the amplitudes of these two line-impulses are summed together on the $v$ axis we have (note that for the sake of simplicity we assume that the grating profiles are symmetric, so that $a^{(1)}_{-nk_1} = a^{(1)}_{nk_1}$ and $a^{(2)}_{-nk_2} = a^{(2)}_{nk_2}$):

$$a_{nk_1,nk_2}(u,v) + a_{-nk_1,-nk_2}(u,v) =$$

$$= \tfrac{1}{2}\sqrt{\tfrac{c}{2\pi}}\,[e^{-ic(v-nv_0)^2} + ie^{-ic(v-nv_0)^2} + e^{ic(v+nv_0)^2} - ie^{ic(v+nv_0)^2}]\ \delta(u)\ a^{(1)}_{nk_1}\ a^{(2)}_{nk_2}$$

$$= \tfrac{1}{2}\sqrt{\tfrac{c}{2\pi}}\,[e^{-ic(v^2-2nv_0v+n^2v_0^2)} + ie^{-ic(v^2-2nv_0v+n^2v_0^2)}$$
$$+ e^{ic(v^2+2nv_0v+n^2v_0^2)} - ie^{ic(v^2+2nv_0v+n^2v_0^2)}]\ \delta(u)\ a^{(1)}_{nk_1}\ a^{(2)}_{nk_2}$$

$$= \tfrac{1}{2}\sqrt{\tfrac{c}{2\pi}}\,e^{i2cnv_0v}\,[e^{-ic(v^2+n^2v_0^2)} + ie^{-ic(v^2+n^2v_0^2)}$$
$$+ e^{ic(v^2+n^2v_0^2)} - ie^{ic(v^2+n^2v_0^2)}]\ \delta(u)\ a^{(1)}_{nk_1}\ a^{(2)}_{nk_2}$$

$$= \sqrt{\tfrac{c}{2\pi}}\,e^{i2cnv_0v}\,[\cos(c(v^2 + n^2v_0^2)) + \sin(c(v^2 + n^2v_0^2))]\ \delta(u)\ a^{(1)}_{nk_1}\ a^{(2)}_{nk_2}$$

and using the trigonometric identity $\cos\alpha + \sin\alpha = \sqrt{2}\cos(\alpha - \frac{\pi}{4})$:

$$= \sqrt{\tfrac{c}{\pi}}\,e^{i2cnv_0v}\,[\cos(cv^2 + cn^2v_0^2 - \tfrac{\pi}{4})]\ \delta(u)\ a^{(1)}_{nk_1}\ a^{(2)}_{nk_2} \tag{C.8}$$

Now, using the 1D Fourier transform pair (see Sec. C.9 below):

$$\cos(ax^2 - b) \leftrightarrow \sqrt{\tfrac{\pi}{a}}\cos(\tfrac{\pi^2}{a}u^2 + b - \tfrac{\pi}{4})$$

or rather its vertical 2D counterpart:

$$\cos(ay^2 - b) \leftrightarrow \sqrt{\tfrac{\pi}{a}}\cos(\tfrac{\pi^2}{a}v^2 + b - \tfrac{\pi}{4})\ \delta(u)$$

we find that:

$$\cos(\tfrac{\pi^2}{c}y^2 - cn^2v_0^2) \leftrightarrow \sqrt{\tfrac{c}{\pi}}\cos(cv^2 + cn^2v_0^2 - \tfrac{\pi}{4})\ \delta(u)$$

Denoting this 2D Fourier pair by $h_n(x,y) \leftrightarrow H_n(u,v)$, we see that expression (C.8) is simply $e^{i2cnv_0v}H_n(u,v)$ (multiplied by the constants $a^{(1)}_{nk_1}$ and $a^{(2)}_{nk_2}$). Therefore, according to the 2D shift theorem [Bracewell86 p. 244], expression (C.8) is the Fourier transform of a vertically shifted version of $h_n(x,y)$, where $h_n(x,y)$ is a centered, horizontal linear zone grating.[3] Denoting this vertical shift by $y_0$, we have by the shift theorem:

$$y_0 = \frac{v_0}{2k_1f_1a}$$

We see, therefore, that the internal discrepancy $v_0$ of the collapsed $(k_1,k_2)$-cluster in the spectrum convolution causes a vertical shift of $y_0 = \frac{v_0}{2k_1f_1a}$ in the horizontal linear zone grating of the singular $(k_1,k_2)$-moiré in the image domain. And indeed, if $v_0 = 0$ (so that all the collapsed line-impulses of the $(k_1,k_2)$-cluster are centered on the DC) then $y_0 = 0$ and the singular moiré linear zone grating is centered on the $x$ axis, i.e., unshifted; this occurs in the (1,-1)-moiré when $f_1 = f_2$ and $\theta = 0$ (see Fig. C.3).[4]

[3] Note that although the 2D functions $\cos(ay^2 - b)$ and $\cos ay^2$ differ in their vertical relative phase at any value of $y$, both of them are horizontal, cosinusoidal linear zone gratings which are centered along the $x$ axis.

[4] Note that when the shifted singular $(k_1,k_2)$-moiré slightly moves away from its singular state, for example when the grating $r_2(x,y)$ is slightly rotated on top of the parabolic grating $r_1(x,y)$, then the shifted horizontal linear zone grating of the singular moiré turns into a parabolic moiré grating with the same shift $y_0$. Only when the moiré is getting farther away from its singular state, the shift of the parabolic moiré grating visibly diverges from $y_0$. As we have seen in Sec. 10.7.3, the shift of the parabolic moiré grating in the superposition can be explained by the shear theorem.

As we can see, the amount of the shift $y_0$ in the image domain is independent of $n$ and hence it is identical for all harmonics of the $(k_1,k_2)$-moiré. This means that all harmonics build up together a moiré grating (in the form of a horizontal linear zone grating) which is vertically shifted by $y_0$ from the $x$ axis.

Finally, it is important to emphasize that in contrast to the moiré shifts in the periodic case, which were discussed in Chapter 7, we are dealing here with moiré shifts of a different type, namely: shifts of the *singular locus* of the moiré. Moreover, the moiré shifts in the present case are *not* generated by shifting the original layers in the superposition, and they are even *invariant* under such shifts. In fact, shifts of the original layers in the present case will only cause relative phase shifts in the moiré bands which surround the singular locus of the moiré, but they will not influence the *location* of the singular locus of the moiré in the superposition.

## C.9 The 1D Fourier transform of the chirp $\cos(ax^2+b)$

Knowing that for $a>0$ we have (see Example 10.5 in Sec. 10.3):

$$\cos ax^2 \leftrightarrow \sqrt{\tfrac{\pi}{a}}\cos(\tfrac{\pi^2}{a}u^2-\tfrac{\pi}{4}) = \sqrt{\tfrac{\pi}{2a}}(\cos\tfrac{\pi^2}{a}u^2+\sin\tfrac{\pi^2}{a}v^2)$$

$$\sin ax^2 \leftrightarrow \sqrt{\tfrac{\pi}{a}}\sin(\tfrac{\pi^2}{a}u^2+\tfrac{\pi}{4}) = \sqrt{\tfrac{\pi}{2a}}(\cos\tfrac{\pi^2}{a}u^2-\sin\tfrac{\pi^2}{a}v^2)$$

we show here that:

$$\cos(ax^2+b) \leftrightarrow \sqrt{\tfrac{\pi}{a}}\cos(\tfrac{\pi^2}{a}u^2-b-\tfrac{\pi}{4})$$

(Note that although $\cos(ax^2+b)$ and $\cos ax^2$ differ in their relative phase at any value of $x$, both of them, as well as their spectra, are centered on the origin.)

Using the trigonometric identity $\cos(\alpha+\beta)=\cos\alpha\cos\beta-\sin\alpha\sin\beta$ we have:

$$\cos(ax^2+b) = \cos ax^2\cos b - \sin ax^2 \sin b$$

and therefore:

$$\cos(ax^2+b) \leftrightarrow \cos b\sqrt{\tfrac{\pi}{2a}}(\cos\tfrac{\pi^2}{a}u^2+\sin\tfrac{\pi^2}{a}v^2) - \sin b\sqrt{\tfrac{\pi}{2a}}(\cos\tfrac{\pi^2}{a}u^2-\sin\tfrac{\pi^2}{a}v^2)$$

$$= \sqrt{\tfrac{\pi}{2a}}\,[(\cos b-\sin b)\cos\tfrac{\pi^2}{a}u^2+(\cos b+\sin b)\sin\tfrac{\pi^2}{a}u^2]$$

and using the known trigonometric identity for $c_1\sin\alpha+c_2\cos\alpha$ [Bronstein p. 273]:

$$= \sqrt{\tfrac{\pi}{2a}}\sqrt{(\cos b+\sin b)^2+(\cos b-\sin b)^2}\,\cos(\tfrac{\pi^2}{a}u^2-\varphi)$$

where: $$\varphi = \operatorname{arctg}\frac{\cos b+\sin b}{\cos b-\sin b} = \operatorname{arctg}\frac{\sqrt{2}\sin(b+\pi/4)}{\sqrt{2}\cos(b+\pi/4)} = b+\frac{\pi}{4}$$

and: $$\sqrt{(\cos b+\sin b)^2+(\cos b-\sin b)^2} = \sqrt{2}$$

so that we obtain, as required:

$$= \sqrt{\tfrac{\pi}{a}}\, \cos(\tfrac{\pi^2}{a}u^2 - b - \tfrac{\pi}{4}) \quad \blacksquare$$

## C.10 The 2D Fourier transform of the 2D chirp $\cos(ax^2 + by^2 + c)$

Knowing that for $a,b > 0$ we have (see Example 10.7 in Sec. 10.3):

$$\cos(ax^2 + by^2) \leftrightarrow \tfrac{\pi}{\sqrt{ab}} \sin(\tfrac{\pi^2}{a}u^2 + \tfrac{\pi^2}{b}v^2)$$

and hence also (by reading the same Fourier pair the other way around):

$$\sin(ax^2 + by^2) \leftrightarrow \tfrac{\pi}{\sqrt{ab}} \cos(\tfrac{\pi^2}{a}u^2 + \tfrac{\pi^2}{b}v^2)$$

we show here that:

$$\cos(ax^2 + by^2 + c) \leftrightarrow \tfrac{\pi}{\sqrt{ab}} \sin(\tfrac{\pi^2}{a}u^2 + \tfrac{\pi^2}{b}v^2 - c - \tfrac{\pi}{2})$$

and also that:

$$\sin(ax^2 + by^2 + c) \leftrightarrow \tfrac{\pi}{\sqrt{ab}} \cos(\tfrac{\pi^2}{a}u^2 + \tfrac{\pi^2}{b}v^2 - c - \tfrac{\pi}{2})$$

(Note that although $\cos(ax^2 + by^2 + c)$ and $\cos(ax^2 + by^2)$ differ in their relative phase at any point $(x,y)$, both of them, as well as their spectra, are centered on the origin. The same goes also for their sine counterparts.)

Using the trigonometric identity $\cos(\alpha+\beta) = \cos\alpha \cos\beta - \sin\alpha \sin\beta$ we have:

$$\cos(ax^2 + by^2 + c) = \cos(ax^2 + by^2) \cos c - \sin(ax^2 + by^2) \sin c$$

and therefore:

$$\cos(ax^2 + by^2 + c) \leftrightarrow \cos c \tfrac{\pi}{\sqrt{ab}} \sin(\tfrac{\pi^2}{a}u^2 + \tfrac{\pi^2}{b}v^2) - \sin c \tfrac{\pi}{\sqrt{ab}} \cos(\tfrac{\pi^2}{a}u^2 + \tfrac{\pi^2}{b}v^2)$$

Using the known trigonometric identity for $c_1\sin\alpha + c_2\cos\alpha$ [Bronstein p. 273]:

$$= \tfrac{\pi}{\sqrt{ab}} \sqrt{\cos^2 c + \sin^2 c}\; \sin(\tfrac{\pi^2}{a}u^2 + \tfrac{\pi^2}{b}v^2 - \varphi)$$

where: $\quad \varphi = \tfrac{\pi}{2} - \operatorname{arctg} \dfrac{-\sin c}{\cos c} = \tfrac{\pi}{2} + c$

and: $\quad \sqrt{\cos^2 c + \sin^2 c} = 1$

so that we obtain, as required:

$$= \tfrac{\pi}{\sqrt{ab}} \sin(\tfrac{\pi^2}{a}u^2 + \tfrac{\pi^2}{b}v^2 - c - \tfrac{\pi}{2})$$

Furthermore, if we denote $a' = \tfrac{\pi^2}{a}$, $b' = \tfrac{\pi^2}{b}$ and $c' = -c - \tfrac{\pi}{2}$ we obtain, by reading the same Fourier pair the other way around:

$$\sin(a'u^2 + b'v^2 + c') \leftrightarrow \tfrac{\pi}{\sqrt{a'b'}} \cos(\tfrac{\pi^2}{a'}x^2 + \tfrac{\pi^2}{b'}y^2 - c' - \tfrac{\pi}{2})$$

and by renaming the variables we obtain:

$$\sin(ax^2 + by^2 + c) \leftrightarrow \frac{\pi}{\sqrt{ab}}\cos(\frac{\pi^2}{a}u^2 + \frac{\pi^2}{b}v^2 - c - \frac{\pi}{2}) \quad \blacksquare$$

## C.11 The spectrum of screen gradations

A screen gradation is a profile-transformed dot-screen (see Sec. 10.2) in which the "period" remains constant throughout the image, but the dot size and/or the dot shape within each such "period" (or cell) vary smoothly according to a certain given rule. Halftoned images (printed with a clustered dot halftoning method) are in fact screen gradations in which the dot size varies according to the tone values in the original, continuous-tone image to be reproduced (see Sec. 3.2). A halftoned image can be seen therefore as a *surface-area modulation* of an underlying dot-screen (the "*carrier*") by the tone values of the original image (the "*modulator*"). This is in fact a 2D extension of the modulation method known in communication theory as *pulse-width modulation*, in which a train of square pulses with period $T$ is modulated by varying the duration $\tau$ of each of its pulses within the limits permitted by the period $T$ [Black53 pp. 32, 263–281].

The influence of surface-area modulation on the nailbed spectrum of the underlying periodic dot-screen (the carrier) is quite complex. It is therefore instructive to start by studying the simplest case, in which the modulating function is a raised cosine function so that its spectrum contains only the DC and the fundamental impulse pair at the cosine frequency. Since in this case both of the functions involved (the carrier and the modulator) are periodic, the modulated function is either periodic or almost-periodic (see Appendix B), depending on whether their periods are commensurable or not. This means that the spectrum of the modulated screen is impulsive; in fact, its impulses are located:

(1) At the fundamental frequency $f_c$ of the underlying dot-screen and all its harmonics (i.e., at the impulse locations of the nailbed spectrum of the unmodulated screen);

(2) At the frequency $f_m$ of the modulating cosine and all its harmonics; and

(3) In all their sums and differences (the intermodulation frequencies).

More formally speaking, the impulse locations in the resulting spectrum form the module $\{mf_c + nf_m \mid m,n \in \mathbb{Z}\}$. In the 1D case this spectrum looks as if a decaying comb of period $f_m$ has been placed on top of each of the impulses of the comb of the underlying square wave, all the combs being intermingled together (see Fig. C.4). In the 2D case all the above mentioned frequencies become frequency-vectors (which represent the impulse locations in the 2D spectrum), and the combs are replaced by nailbeds. The explicit expression for the impulse amplitudes is quite complicated; an example for the 1D case is given in [Black53 p. 275].[5] In general, the strongest impulses in the spectrum are those of

[5] The formula given there is in fact the Fourier series development of the modulated function, from which the impulse frequencies and amplitudes can be readily found. This formula was developed there for the case of a modulating function with a triangular profile.

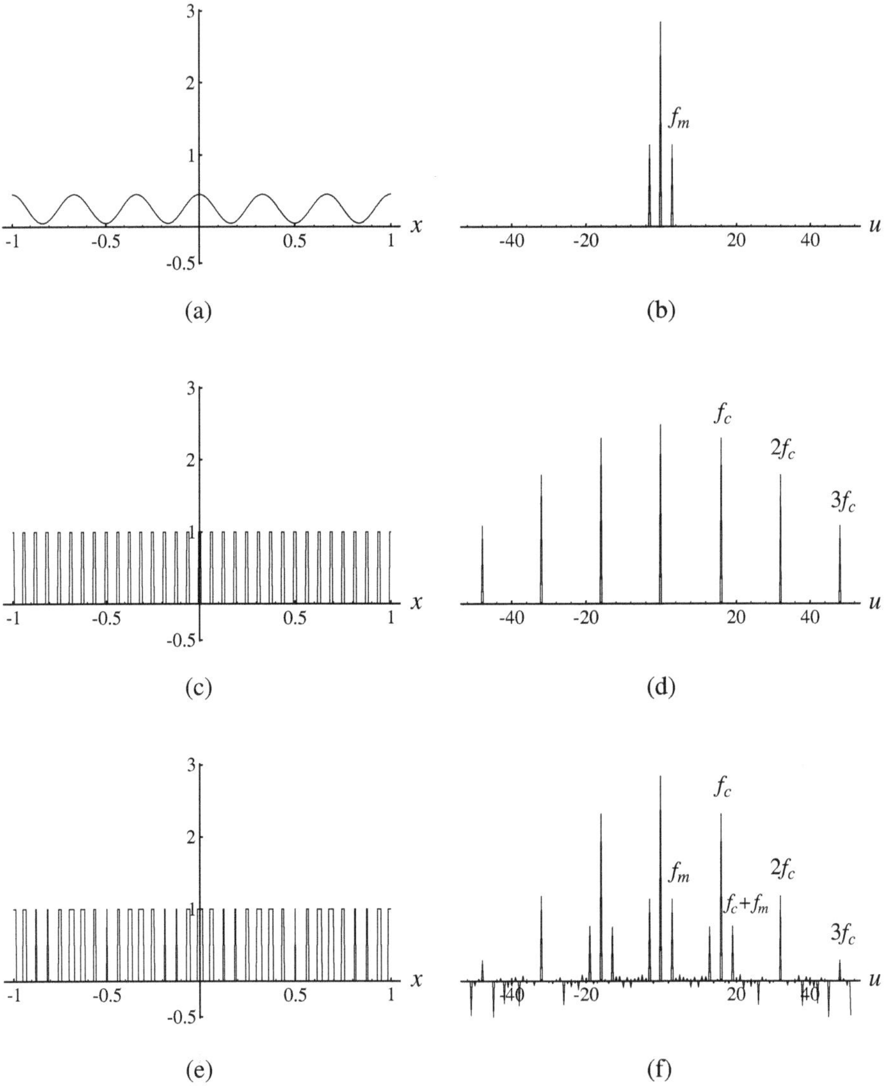


**Figure C.4:** A simple case of 1D pulse-width modulation. The modulating function (a) is a raised cosine $0.2\cos(2\pi f_m x) + 0.25$ with frequency $f_m = 3$; its spectrum is shown in (b). The carrier (c) is a pulse-train of frequency $f_c = 16$; its spectrum is shown in (d). The result of the pulse-width modulation is shown in (e), and its spectrum is shown in (f). Note that all the spectra have been obtained by FFT, which explains the slightly visible noise.

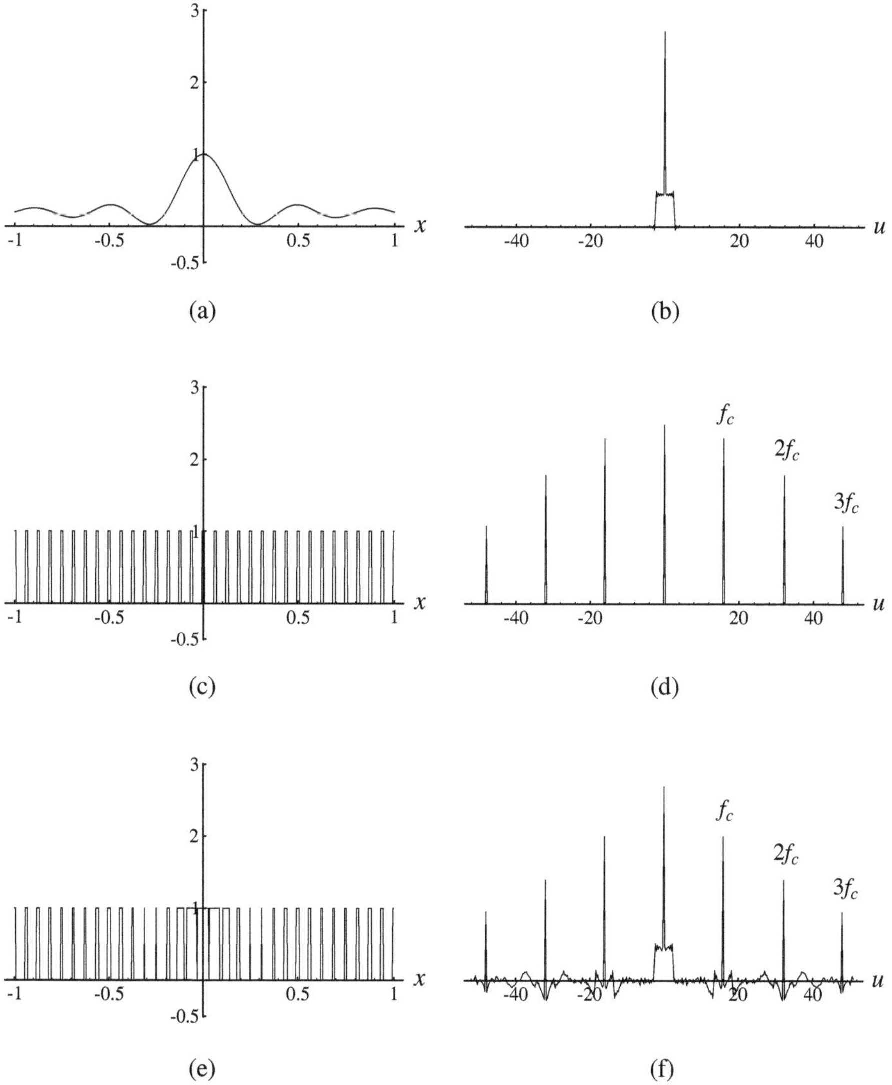


**Figure C.5:** A 1D pulse-width modulation simulating a halftoned image and its spectral interpretation. The original continuous-tone image (a), i.e., the modulating function, is $0.8\text{sinc}(5x) + 0.2$; its spectrum (b) is a square pulse of width 5 (plus a DC impulse). The carrier (c) is a pulse-train of frequency $f_c = 16$; its spectrum is shown in (d). The result of the pulse-width modulation is shown in (e), and its spectrum in (f). It is clearly seen that the spectrum (f) contains around its DC a faithful replica of the spectrum (b) of the original image; farther away it contains also the impulses of the carrier at the halftoning frequency $f_c$ and its harmonics $nf_c$, as well as the distorted replicas of (b) around each of them (the *intermodulation distortions*).

the DC, the underlying dot lattice frequency $f_c$ and its first few higher harmonics, and the modulation frequency $f_m$; the impulses at higher harmonics and at intermodulation frequencies die out quite rapidly. Note that $f_c$ is in fact the frequency of the halftone cells which constitute the halftone image; it is therefore often called the *halftone frequency*.

Now, if the modulating function is not a pure cosine but rather consists of several frequencies, then all of these frequencies are present in the spectrum, and they participate in the generation of the new intermodulation impulses. The same principle is also true for the general case where the modulating function is a real-world image, whose original spectrum may consist of infinitely many frequencies $f_m$ of various amplitudes (obviously, $f_m < f_c$). It has been shown in [Kermisch75 p. 723] that even in this case the spectrum of the modulated function (the halftoned image) can be written in the form of a series, where the first term represents the spectrum of the original continuous tone image, and the other terms represent the distortions introduced by the halftoning process. Similar results have been obtained in [Allebach79] for the case of discrete images. In general, the spectrum of a halftoned image contains a replica of the spectrum of the original continuous tone image (the modulating function), which is centered on the DC, as well as distorted replicas which are centered on each of the carrier's harmonic frequencies $nf_c$. This is illustrated in Fig. C.5, where the modulating function is sinc($5x$), a function whose own spectrum (a square pulse of width 5) is easily recognizable. As we can see in the figure, the spectrum of the modulated function (the halftoned image) shows a clear replica of this square pulse around the DC, as well as distorted replicas of this pulse which surround the halftone frequency $f_c$ and its harmonics $nf_c$.[6]

However, in our case here we are not interested in general halftoned images, but rather in the particular case of uniform screen gradations ("wedges") which are modulated by a uniform slope. We use such gradations for demonstrating within a single image the moirés which are generated at various possible tone levels of each of the superposed screens (see, for example, Figs. 4.1, 4.4, etc.). In the 1D case the modulating function of the gradation is $f(x) = ax + b$, whose Fourier transform is $F(u) = (ai/2\pi)\delta'(u) + b\delta(u)$ [Champeney87 p. 138], where $\delta'(u)$ is the first derivative of the impulse $\delta(u)$ (see [Bracewell86 p. 80–82]). Therefore the spectrum of the modulating function $f(x)$ consists of a single impulsive entity at the spectrum origin; this guarantees that no new frequencies appear in the spectrum of a dot-screen when it is modulated by this function. A similar result is obtained in the 2D case, where the modulating function is $f(x,y) = ax + by + c$. This confirms that the profile-transformation which transforms the original dot-screen into such a screen gradation only affects the amplitudes and the nature of the impulses in the spectrum of the original dot-screen, but it does not modify the impulse locations nor does it introduce any new impulses. In other words: The spectrum support of such a screen

[6] Note that when looking at the halftoned image from a normal viewing distance the halftone frequency $f_c$ and its harmonics $nf_c$ are already beyond the border of the visibility circle, and only the frequencies of the main replica of the spectrum of the original image around the DC are still located within the visibility circle. This explains why when looking at a halftoned image the eye normally perceives an almost identical version of the original, continuous-tone image.

gradation remains the same lattice as in the spectrum of the underlying dot-screen and it corresponds to the cell periodicity of the gradation.

As a consequence, the moiré effects in the superposition of such screen gradations vary locally in their *profiles* according to the profile variations in the superposed layers; but their angles and "periods" remain unchanged throughout the superposition (see Figs. 4.1, 4.4, etc.).

## C.12 Convergence issues related to Fourier series

In this section we briefly review questions concerning the convergence of Fourier series (proper and generalized), and the order of summation in multiple Fourier series. Additional details as well as more rigorous mathematical development of the subjects may be found in the cited references.

### C.12.1 On the convergence of Fourier series

Suppose that $p(x)$ is a periodic function of period $T$; $p(x)$ is therefore fully defined by any 1-period interval of length $T$, such as $0...T$.[7] If $p(x)$ is integrable on a 1-period interval, then a set of complex numbers $c_n$ for $n = 0, \pm 1, \pm 2,...$ may be defined by:

$$c_n = \frac{1}{T}\int_T p(x)\, e^{-i2\pi nfx}\, dx \qquad \text{(C.9)}$$

where $f = 1/T$, and $\int_T$ means integration over any full period of $p(x)$, i.e., from $x_0$ to $x_0+T$ where $x_0$ is arbitrary. The complex numbers $c_n$ are called the *Fourier series coefficients* of $p(x)$, and their values are independent of the choice of $x_0$.

The following series, involving the Fourier series coefficients $c_n$ of the function $p(x)$, is called the *(formal) Fourier series belonging to* $p(x)$:

$$p(x) \sim \sum_{n=-\infty}^{\infty} c_n\, e^{i2\pi nfx} \qquad \text{(C.10)}$$

(where the symbol '~' is used to denote the relation "belongs to"). Note that at any value of $x$ the formal Fourier series in (C.10) may or may not converge, depending on the choice of $p(x)$ and of $x$. Many periodic functions $p(x)$ possess a Fourier series that converges to the function $p(x)$ itself at all $x$; in such cases the symbol '~' in (C.10) can be replaced by an equality '=', and the term "formal Fourier series belonging to $p(x)$" can be replaced by the term "*Fourier series expansion of* $p(x)$". But in other cases the formal Fourier series may fail to converge at some or even all values of $x$ (see for example [Zygmund68 pp. 298–315]). One of the main aims of the theory of Fourier series has been to determine

[7] If $f(x)$ is not periodic (e.g., $f(x) = x^2$) but its behaviour is of interest only within a given interval of $x$, it can still be artificially made periodic by simply defining $p(x) = f(x)$ within this interval, and redefining $p(x)$ outside this interval to be a periodic repetition of its values within the interval.

in what sense (what type of convergence) and under what conditions the formal Fourier series tends to $p(x)$; see, for instance, [Zygmund68; Katzenelson68].

Several convergence criteria for Fourier series can be stated, depending on the class of functions considered and on the way "convergence" is understood (see, for example, [Champeney87 pp. 156–164], [Gaskill78 pp. 107–108]). Some convergence criteria are also known for the case of almost-periodic functions, i.e., for generalized Fourier series (see [EncMath88 Vol. 1 pp. 154–156: "Almost-periodic functions"; Vol. 4 p. 79: "Fourier series of an almost-periodic function"]). However, this problem is more difficult than its periodic counterpart and is still far from a complete solution [Corduneanu68 pp. 31–38].

It is beyond the scope of this work to define rigorously the precise class of periodic or almost-periodic functions which satisfy the required convergence conditions for our needs. Instead, we simply restrict ourselves to those functions which do satisfy these conditions. It should be noted that practically all functions which represent real physical quantities satisfy the convergence conditions [Gaskill78 p. 108]. Therefore we may adopt for our purposes the pragmatic approach which says, roughly speaking: if a function describes a physically realizable phenomenon (e.g., if a theoretic reflectance function can be realized and demonstrated, for instance reproduced on film[8]), then the required convergence conditions are mathematically met. In particular, common functions such as square waves, gratings, etc. present no convergence problems, and their Fourier series developments can be found tabulated in literature.

### C.12.2 Multiplication of infinite series

As we have seen in Sec. 2.2, the superposition of periodic layers in the image domain means, mathematically, their multiplication. Therefore, the Fourier series of the superposition is the product of the Fourier series of each of the individual layers. Intuitively, as an extension of the finite case, the product of infinite series $\Sigma a_n$ and $\Sigma b_k$ should be given by the infinite series $\Sigma\Sigma a_n b_k$. However, handling infinite series must be done with care, since even simple operations which are obvious in the finite case, such as reordering the terms within the series, may affect its convergence.

If we are lucky to have original functions (individual layers in the superposition) whose Fourier series are absolutely convergent,[9] then we are in a happy situation, thanks to the following theorem:

**Theorem 1:** If the infinite series $\Sigma a_n = A$ and $\Sigma b_k = B$ are *absolutely convergent*, then the double series $\Sigma\Sigma a_n b_k$ is absolutely convergent, and has the sum *AB however the terms are arranged* [Hardy73 pp. 227–228]. ■

---

[8] Unlike various "pathological" functions such as $\sin(1/x)$, the Cantor function, etc.

[9] An infinite series $\Sigma a_n$ is *absolutely convergent* if $\Sigma|a_n|$ is convergent (this implies that $\Sigma a_n$ is also convergent).

This means that the double series is indeed the product of the two given series, and furthermore, any order of summation in this double series is permitted.

The most familiar multiplication rule (order of terms) is Cauchy's, in which the double series is summed "diagonally", by associating together the terms in which $n+k$ has a fixed value. We then define the product series $C = AB$ as:

$$C = \sum_{m=0}^{\infty} \sum_{n+k=m} a_n b_k = \sum_{m=0}^{\infty} \sum_{n=0}^{m} a_n b_{m-n} = \sum_{m=0}^{\infty} \sum_{k=0}^{m} a_{m-k} b_k \tag{C.11}$$

However, this diagonal summation is by no means the only possible summation order for the double series. In fact, by the above theorem, if the series being multiplied are absolutely convergent, any arrangement of the terms is permitted (provided that it is exhaustive, and that no term is taken more than once). This can be generalized to the multiplication of any number of absolutely convergent series, i.e., to the ordering of the terms in any such multiple series.

The above theorem for absolutely convergent series stands unchanged, for any rule of multiplication, also for two-sided series (i.e., series infinite in both directions); for example, if $\sum_{n=-\infty}^{\infty} a_n = A$ and $\sum_{k=-\infty}^{\infty} b_k = B$, then Cauchy's diagonal summation:

$$C = \sum_{m=-\infty}^{\infty} \sum_{n+k=m} a_n b_k = \sum_{m=-\infty}^{\infty} \sum_{n=-\infty}^{\infty} a_n b_{m-n} = \sum_{m=-\infty}^{\infty} \sum_{k=-\infty}^{\infty} a_{m-k} b_k$$

satisfies $C = AB$ [Hardy73 pp. 239–241].

If, however, the Fourier series of our superposed functions are not absolutely convergent, the following theorem (Abel's product theorem) may come to the rescue:

**Theorem 2:** If the series $\Sigma a_n = A$, $\Sigma b_k = B$ and (C.11) are all convergent, then $AB = C$ [Hardy73 p. 228]. ■

In other words, if we know in advance that the original series $\Sigma a_n = A$, $\Sigma b_k = B$ and also their Cauchy series (C.11) are all convergent, then the Cauchy series (C.11) is indeed the product of the original series.[10] It can be shown that other linear variants of the Cauchy diagonal ordering are also permitted, such as: $\sum_{m=0}^{\infty} \sum_{qn+pk=m} a_n b_k = C$, where $p$ and $q$ are constant integers (see, for example, Fig. 5.1(a)).

Clearly, this is not as general as in the case of Theorem 1, where virtually any order of summation is permitted. But for our needs, when our original superposed functions do not have absolutely convergent Fourier series as required by Theorem 1, Theorem 2 will also do: since we do know — due to the physical realizability of the superposition — that its Fourier series is indeed convergent.

---

[10] Note that Abel's theorem is not always true for two-sided series [Hardy73 p. 244]; however, since in our case the Fourier series is symmetric (or Hermitian, in the complex case), we have no problem to represent the two-sided Fourier series as a one-sided infinite series, i.e., in the form (A.1) rather than (A.3).

## C.13 Moiré effects in image reproduction

Undesired moiré patterns may appear in the printing process for several possible reasons, which can be classified into the following main categories:

(a) *Screening moirés*. These are moiré patterns which occur as an interaction between repetitive patterns in the original image (such as striped clothes, fences, etc.) and the halftone screen which is used to print the image. An example containing such moirés is shown in [Bann90 p. 65]. Screening moirés are also called *subject* or *content moirés*, since they depend on the details of the original image [Blatner98 p. 282].

(b) *Auto moirés*. These are moiré patterns which occur owing to an interaction between the halftone screen and the pixel grid of the output device [Jones94 pp. 267–268]. These moirés are also called *internal moiré artifacts* [Levien93] because they are generated internally to the screening process, and do not depend on an "external" source.

(c) *Superposition moirés*. These are moiré patterns which occur in the superposition of two or more repetitive structures, such as the halftone screens of the different process colours which are used in colour printing [Blatner98 p. 279]. See Chapter 3.

(d) *Sampling moirés*. These are moiré patterns which occur in the analog to digital conversion of an image between the sampling grid (the device resolution) and repetitive patterns in the original image. Sampling moirés most frequently occur when attempting to scan an already halftoned image (such as a photo from a book or a newspaper), or an image which includes repetitive patterns (see also Sec. 2.13, and several of the problems at the end of Chapter 3).

All of these types of moiré are generated as interactions between repetitive structures (which may be, depending on the case, the halftone screen, the pixel grid of the input or output device, or repetitive details within the original image). These different types of moirés are all included within the framework of our superposition moiré theory if we consider sampling grids, pixel grids, etc., as particular cases of dot-screens.

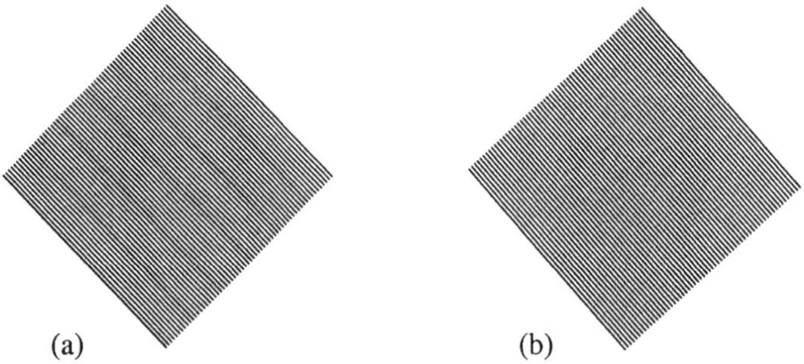

(a) (b)

**Figure C.6:** (a) An auto moiré may occur when printing a single line-grating (or dot-screen) on a PostScript printer, owing to periodic pixel rounding errors. Is this really a *single layer moiré*? (b) This printer artifact may disappear by slightly changing the angle or the frequency of the grating (or dot-screen).

## C.14 Hybrid (1,-1)-moiré effects whose moiré bands have 2D intensity profiles

In this section we describe an interesting moiré effect which may carry 2D information although it is based on the 1D case of the (1,-1)-moiré between two line gratings. This 1D moiré effect could be also considered, therefore, as a "$1\frac{1}{2}$D case". This phenomenon was known to Joe Huck who used it in his artistic work [Huck03], but it was first investigated in depth and fully explored in [Hersch04] and in [Chosson06]. Interesting applications of this phenomenon to the field of document protection have been described in detail in [Hersch04a], [Hersch06], [Hersch06a] and [Schilling06].

It should be noted that the terminology, the notations and the conventions used in the above mentioned references differ from the ones that we use throughout this book. But for the sake of consistensy and uniformity within this volume we will use here the same conventions and notation standards as in the rest of the book. Note also that although the original publications were only based on geometric considerations and on the indicial equations approach, we will base our discussion here on the Fourier-based approach, which provides a much deeper insight into this phenomenon. In particular, the Fourier-based approach can also explain the intensity profiles of the phenomena in question, and it is not only limited to their planar geometric properties as the other approaches.

### C.14.1 Preliminary considerations

Suppose that we are given a (1,-1)-moiré between two periodic line gratings, as shown in Figs. 1.1 or 2.5. The period and the angle of the moiré bands are given, as we have seen in Chapter 2, by Eq. (2.9) or by its particular cases, Eqs. (2.10) or (2.11).

Assume, now, that we replace the simple black lines of the first grating by lines that incorporate along their main direction some predefined information (such as tiny letters, digits or symbols), while the other grating is replaced by a series of linear slits (narrow white or rather transparent lines) on a black background (see Fig. C.7). The resulting effect may remind us of the 2D case that we have already studied in Sec. 4.3 (see, for example, Figs. 4.4 or C.23). In that 2D case, when the first layer is a periodic dot screen whose individual dots have some predefined dot shapes and the second layer is a periodic pinhole screen having similar periods, we obtain in the superposition a moiré effect whose intensity profile is a largely magnified (and possibly rotated) version of the information that is incorporated in the dot shapes of the first screen. And indeed, as we can see in Fig. C.7, a similar phenomenon occurs also in our 1D case: The superposition of our two line gratings gives moiré bands whose intensity profile is a largely magnified (and possibly slanted) version of the information that is incorporated along the lines of the first grating. However, the periodicity and the magnification of this moiré effect only occur along *one* direction, and not along two directions as in the 2D case (compare Figs. C.7 and C.23). The same goes also for the dynamic behaviour of the moiré effect under layer shifts.

Obviously, because the baseline direction of the incorporated text coincides with the line direction of our grating, the period and the orientation of the baselines of our moiré bands

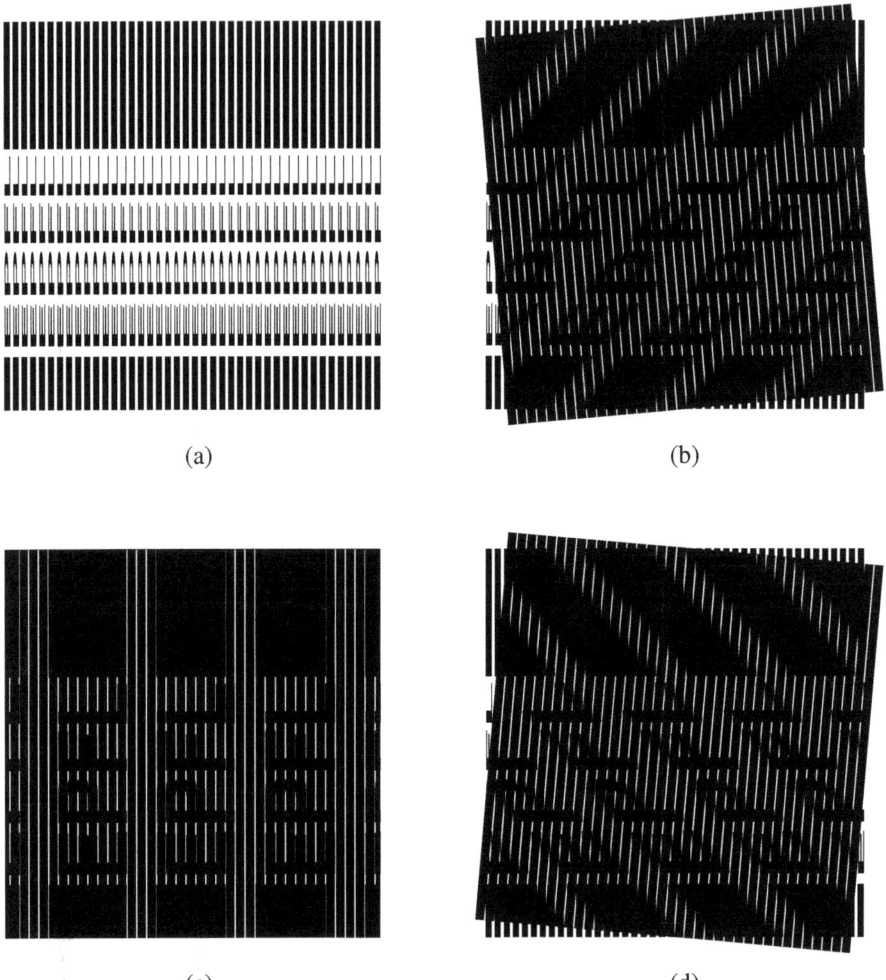


**Figure C.7:** (a) A periodic line grating whose individual lines contain a flattened version of the letters "EPFL". (b)–(d) The superposition of this line grating with a second periodic line grating having a slightly larger period $T_2 > T_1$, which consists of narrow slits on a black background. The angle difference between the two superposed gratings is 5° in (b), 0° in (c) and –5° in (d). Each of these superpositions gives a periodic (1,-1)-moiré effect whose individual bands contain a largely stretched-out (and possibly sheared) version of the 2D information that is embedded in the individual lines of the first grating. Note that we have intentionally left the upper and the lower parts of the grating (a) non-modulated; this allows us to compare in each of the superpositions our modulated moiré bands with the simple moiré bands that are obtained in the classical, non-modulated case.

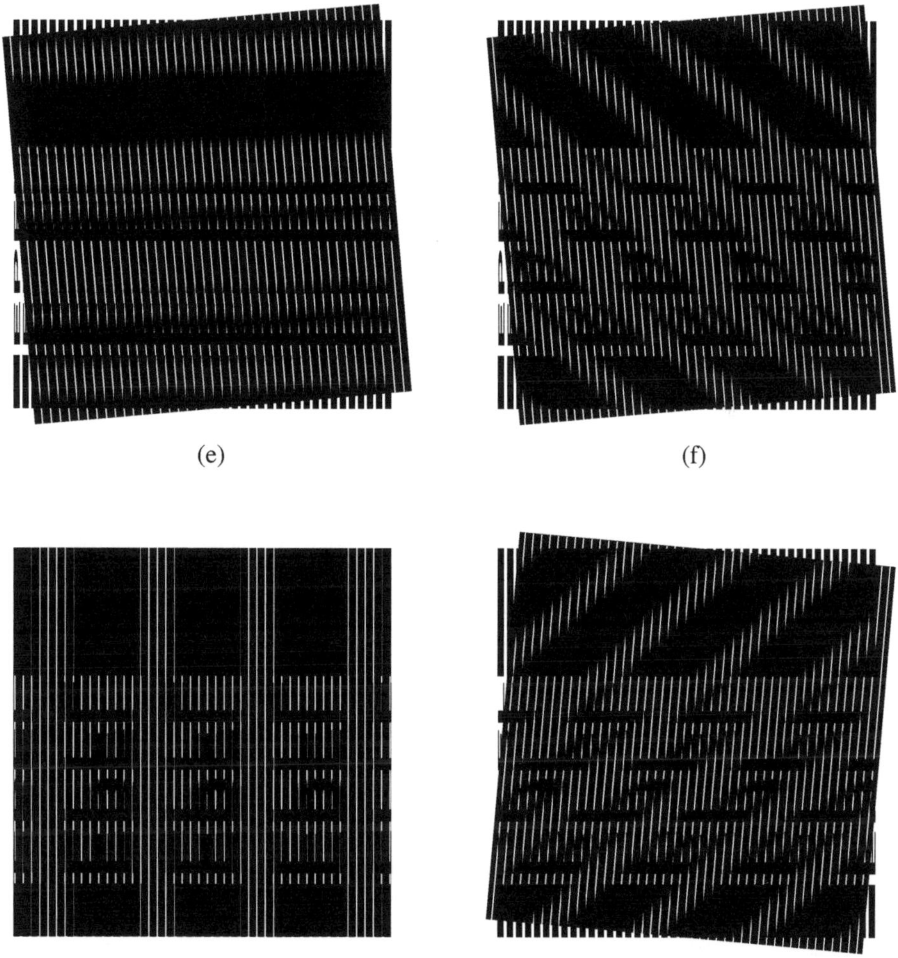

(e) (f) (g) (h)

**Figure C.7:** (*continued.*) (e) Same as in (b), but with periods $T_2 = T_1$; this case is explained in greater detail in Fig. C.8. (f)–(h) Same as in (b)–(d), but this time with periods $T_2 < T_1$; in this case the letters "EPFL" appear in the moiré bands mirror-imaged, and under layer shifts they move in the opposite direction than in (b)–(d), respectively.

are still given by the same equations as in the case consisting of simple, black lines (Eqs. (2.9)–(2.11)). However, as we can see in Fig. C.7(b), this periodicity does not correspond to the periodicity of the moiré text that is incorporated within the moiré bands. The reason is that unlike in the simple case of black lines, in our present case two of the three structures involved — the first grating and the resulting moiré bands — also have a *secondary orientation*, i.e. the height direction of the letters.[11] And indeed, it turns out that the secondary direction of the moiré letters is not necessarily orthogonal to their baseline direction. For instance, in the case shown in Fig. C.7 the height direction of the moiré letters remains horizontal, like in the original text of the first grating, so that the resulting moiré text can be significantly slanted, depending on the direction of the moiré bands.

This interesting phenomenon can be explained using geometric considerations, as done, indeed, in the original publications mentioned above. However, by also considering the intensity information, i.e. the intensity profiles of the two given gratings and the intensity profile of the resulting moiré effect, in addition to their planar geometric layouts, we can obtain a much deeper understanding of this phenomenon. This deeper insight is provided by the Fourier-based approach, as we will see below.

In fact, the (1,-1)-moiré that we discuss here differs from the classical case that is shown in Fig. 2.5 in that it is based on lines having 2D profile variations rather than on lines having only 1D profile variations across the line width, as in the classical case. This type of moiré will be called here a *hybrid* (1,-1)-moiré, due to the fact that it still carries 2D information although it is based on line gratings and not on dot screens. As shown in the references mentioned above, the hybrid (1,-1)-moiré has some nice properties which make it very useful in applications. These include, notably, the larger amount of light that passes through a grating made of line slits (as compared to a 2D pinhole screen), which makes the resulting moiré more easily visible than its 2D counterpart even in difficult light conditions; and the fact that it can carry along its moiré bands information of practically any desired length. But on the other hand, the hybrid (1,-1)-moiré only provides 1D rather than 2D magnification (compare Figs. C.7(a),(c) with Fig. C.24(a)). Moreover, the hybrid (1,-1)-moiré is more sensitive to layer rotations than its 2D counterpart, since such rotations do not cause a *rotation* of the resulting moiré as in the 2D case (as shown, for example, in Fig. C.23(a)) but rather a strong *shearing* effect (as shown in Figs. C.7(b) or C.8), which may distort the carried information and make it harder to recognize.

### C.14.2 The Fourier-based approach

Let us start with the familiar superposition of two simple straight periodic gratings having similar periods and angles, as already described in Chapters 2 and 4 (see Fig. 2.5). This time, however, we allow each individual line of our first grating $r_1(x,y)$ to carry some given information, not necessarily periodic, along its main direction. In other words, the intensity of each of the lines of our grating $r_1(x,y)$ may be modulated by any given information, such as letters, digits or symbols, provided that these objects are sufficiently narrow or flattened to fit within each individual line. For example, in the case shown in Fig. C.9(a) each of the grating lines contains a very flattened version of the letters "EPFL". Because the line grating $r_1(x,y)$ is periodic along its main direction (in our figure, along the $x$ axis),[12] it is clear that its Fourier spectrum $R_1(u,v)$ is impulsive.

[11] The *height direction* of the letters is understood here as the direction defined by letters such as "I".
[12] Note that the main direction of a line grating is perpendicular to its individual lines.

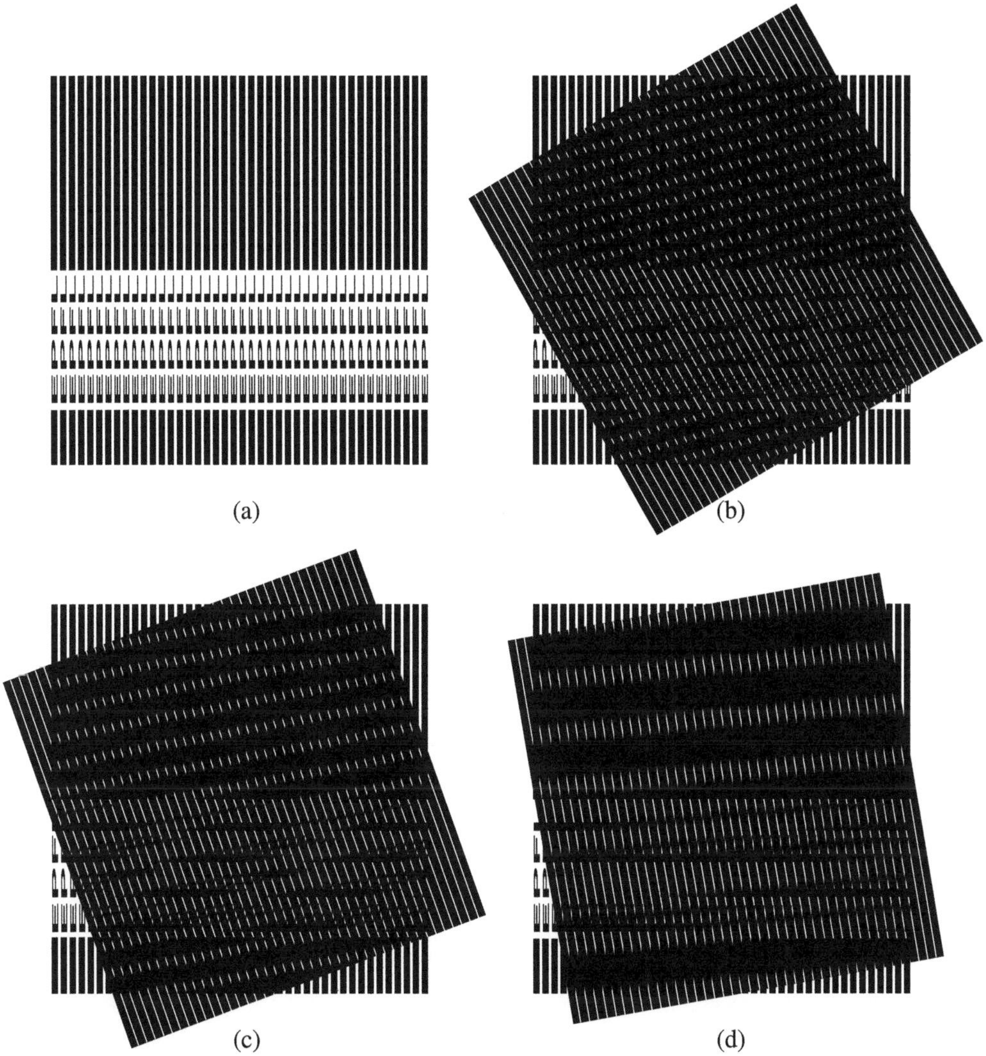

(a) (b) (c) (d)

**Figure C.8:** A more detailed explanation of the case shown in Fig. C.7(e), in which both of the superposed gratings have identical periods, $T_2 = T_1$. (a) A periodic line grating whose individual lines contain a flattened version of the letters "EPFL". Note that the black, unmodulated part of the individual lines at the top of this grating is intentionally left longer than in Fig. C.7(a), in order to leave more room for the black, unmodulated moiré bands at the top of the resulting superpositions. (b)–(d) The superposition of this line grating with a second periodic line grating which consists of narrow slits on a black background. The periods of the two superposed gratings are identical, and the angle difference between them is 30° in (b), 20° in (c) and 10° in (d). As we can clearly see, the letters "EPFL" are still present within each of the moiré bands even though $T_2 = T_1$. But in this case, when the angle difference between the gratings is lower than about 20°, the letter shapes become so elongated and slanted that they are no longer recognizable.

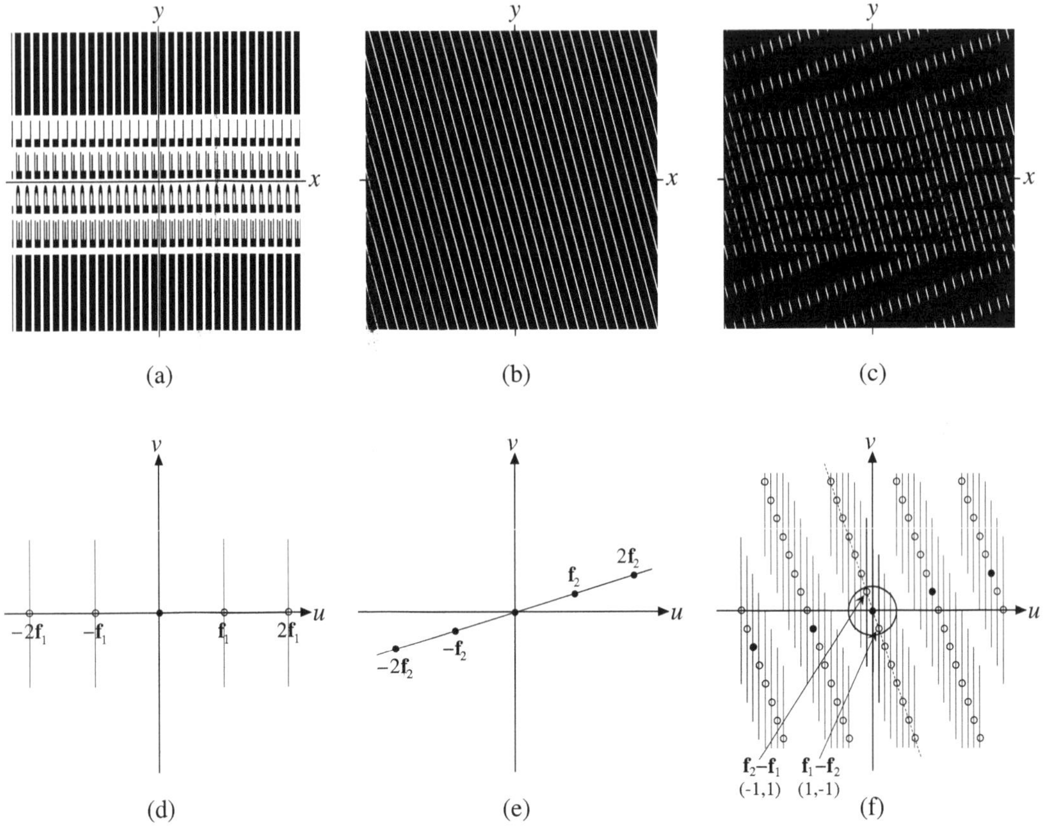


**Figure C.9:** Periodic grating (a) whose individual lines are modulated by the flattened letters "EPFL", periodic line grating (b), and their superposition (c) in the image domain. Their respective spectra are the infinite line-impulse comb (d), the infinite impulse comb (e) and their convolution (f). Black dots in the spectra represent impulses, while open dots indicate the skeleton locations of the line-impulses. Notice the hybrid (1,-1)-moiré which appears in the superposition. Compare with the classical (1,-1)-moiré shown in Fig. 2.5. (Note that for the sake of clarity the periods in the upper row of the present figure have been magnified with respect to those shown in Fig. 2.5.)

However, unlike the simple non-modulated line gratings that we considered so far, the intensity along each line within our grating $r_1(x,y)$ is not constant, and it needs not even be periodic. Therefore the Fourier spectrum of such a grating is no longer an impulse comb like in Fig. 2.5(d), but rather a line-impulse comb, i.e. a sequence of equispaced parallel line impulses that are orthogonal to the grating's main direction (see Fig. C.9(d) and the illustrative explanation that is provided in Fig. C.11). These continuous line impulses are centered about their *skeleton locations*, which are the locations of the corresponding

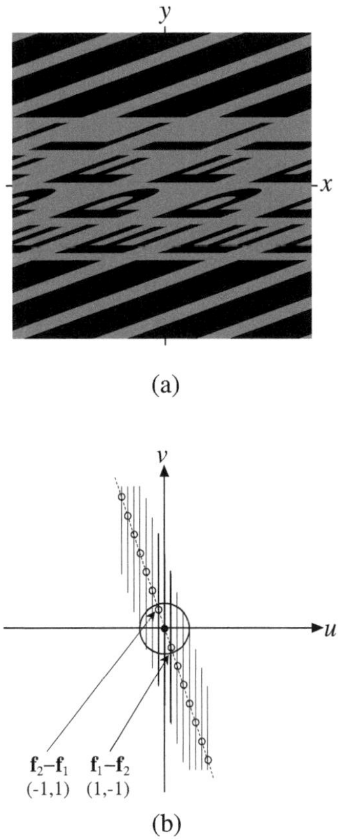


**Figure C.10:** Extraction of the (1,-1)-moiré of Fig. C.9: (b) shows the isolated line-impulse comb of the (1,-1)-moiré after its extraction from the full spectrum of Fig. C.9(f). (a) shows the image domain function which corresponds to the spectrum (b). This is the intensity profile of the modulated (1,-1)-moiré shown in Fig. C.9(c); note that it only contains the extracted moiré (i.e., its isolated contribution to the superposition), but not the microstructure details of the original gratings and of the superposition. The orientation of the *skeleton* of the line-impulse comb corresponds to the direction of the 1-fold periodicity of the non-modulated moiré bands, while the *shearing* of this line-impulse comb with respect to the original line-impulse comb of the first grating (Fig. C.9(d)) determines the slanting angle of the letters within each moiré band (see Fig. C.12 below). Compare with the classical (1,-1)-moiré shown in Fig. 4.2.

impulses in the spectrum of the equivalent non-modulated line grating (see Fig. 2.5(d)), i.e. the points $n\mathbf{f}_1$, $n \in \mathbb{Z}$, where $\mathbf{f}_1$ is the frequency vector of our grating. The information that is modulated along each of the grating lines is encoded in the spectrum by the

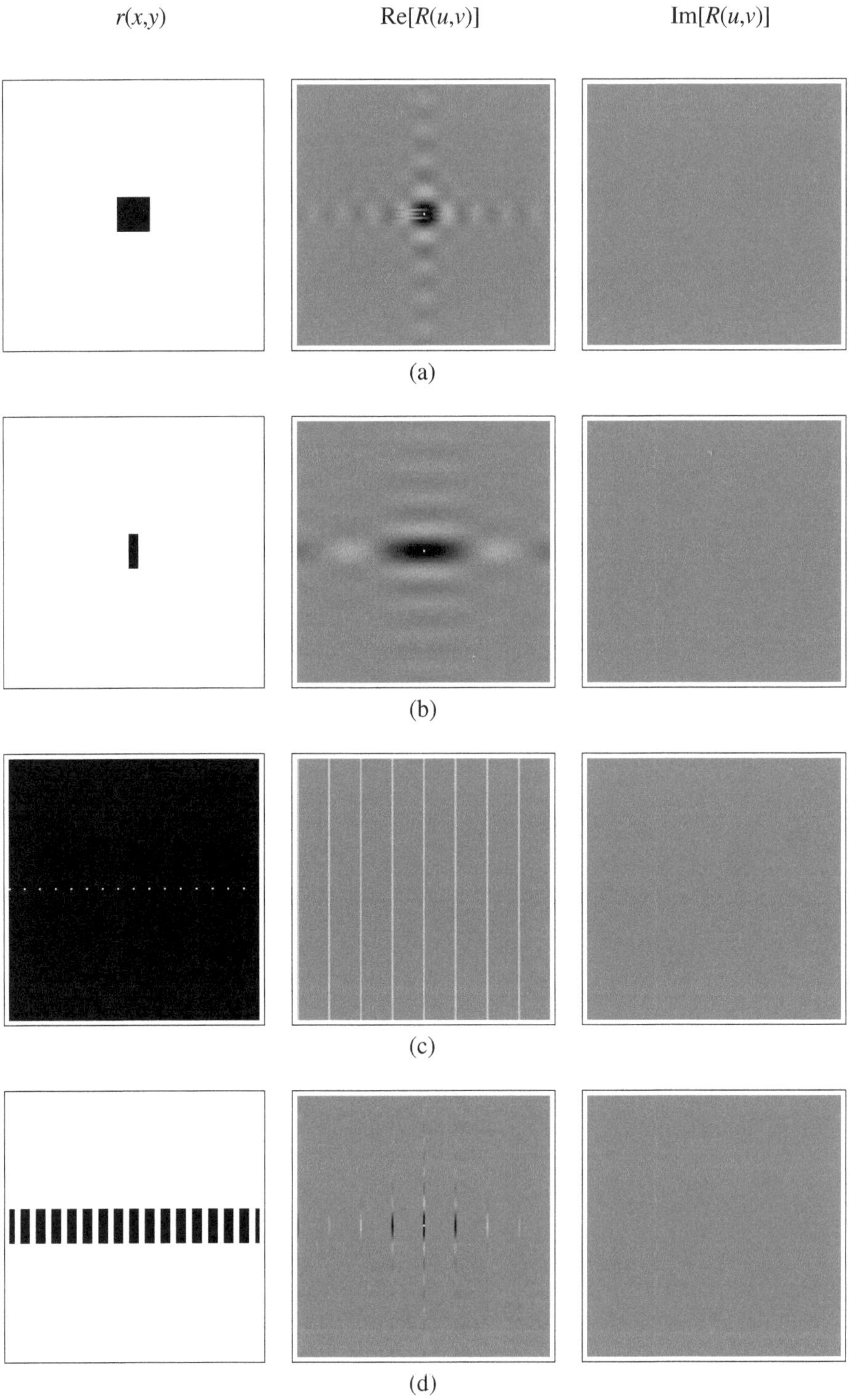
r(x,y)
Re[R(u,v)]
Im[R(u,v)]
(a)
(b)
(c)
(d)

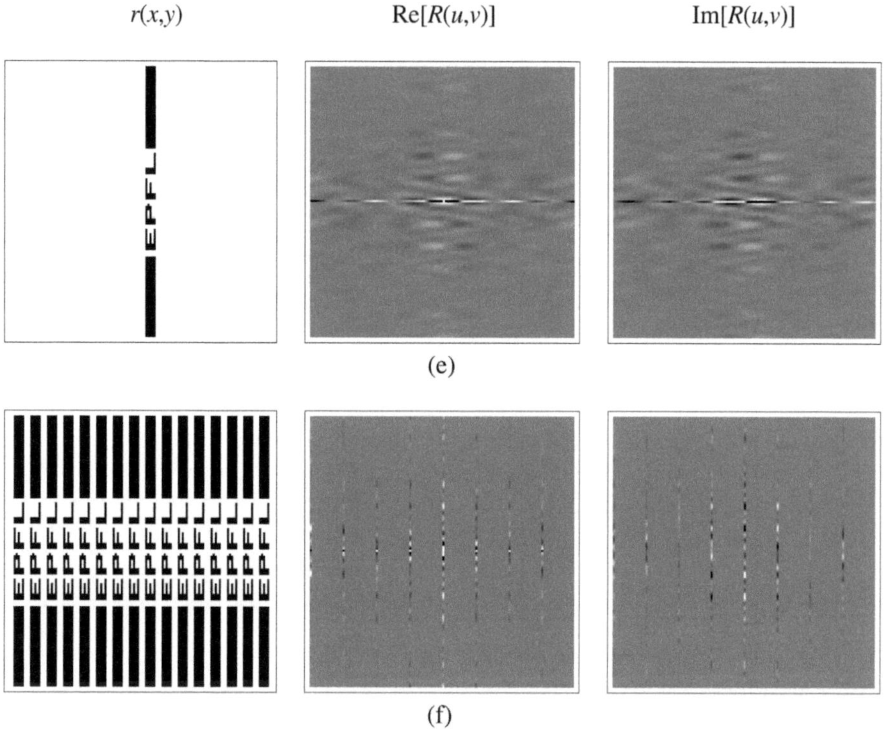


**Figure C.11:** Explanation of the spectrum of a line grating whose individual lines are modulated by 2D information; two simple examples are shown in (d) and (f). For each image $r(x,y)$ in this figure, Re[$R(u,v)$] and Im[$R(u,v)$] show the real and the imaginary parts of the spectrum $R(u,v)$ as obtained on computer by 2D DFT. We start with the simple modulated line grating shown in (d): (a) Since the Fourier spectrum of a centered white square of width $\tau$ on black background, $d(x,y) = \text{rect}(x/\tau, y/\tau)$, is given by $D(u,v) = \tau^2 \, \text{sinc}(\tau u)\, \text{sinc}(\tau v)$ [Bracewell95 p. 150], it follows that the Fourier spectrum of a centered *black* square of width $\tau$ on *white* background, $1 - d(x,y)$, is given by $\mathcal{F}[1 - d(x,y)] = \mathcal{F}[1] - \mathcal{F}[d(x,y)] = \delta(u,v) - D(u,v)$, where $\delta(u,v)$ is an impulse at the origin (note that this impulse is clearly visible in the center of the spectrum). (b) Compressing the black square (a) horizontally results in a horizontal expansion of its spectrum (due to the similarity theorem [Bracewell86 p. 244]). (c) A horizontal impulse comb of constant amplitude; its spectrum is a comb of constant vertical line impulses [Bracewell86 p. 246]). (d) Periodic repetition of the line segment (b) (i.e. its convolution with the constant horizontal impulse comb (c)) gives in the spectral domain, according to the convolution theorem, the product of the spectrum of (b) with the spectrum of (c), namely, a line-impulse comb which is a sequence of vertical slices through the spectrum of (b). The spectrum of the modulated line grating (f) is obtained in the same way: (e) A single modulated line and its spectrum. (f) A periodic repetition of the modulated line (e) gives in the spectral domain a line-impulse comb which is a sequence of vertical slices through the spectrum of (e). Note that the spectra in (e) and (f) are complex-valued.

amplitude variations along the line impulses.[13] It is interesting to note, however, that if the grating lines are only modulated across the line width and their profile remains constant along the line, then the spectrum of the grating reduces into a simple impulse comb in the grating's main direction, as described in Sec. 4.2 and in Fig. 2.5. Another interesting special case occurs when the information that is modulated within each of the lines *is* periodic along the line's direction. In this case each of the parallel continuous line impulses in the spectrum reduces into an impulse comb (which is a discrete subset of the continuous line impulse), and the entire spectrum turns into a 2D nailbed, reflecting the fact that our grating is now a 2-fold periodic structure. Note, however, that in the discussions which follow we do not require periodicity along the grating's individual lines.

Let us now consider the second line grating, $r_2(x,y)$, that is superposed on top of our modulated grating. In the most general case this grating may be modulated by some given information, just as the first grating. However, in our case of interest (see Fig. C.7) the lines of the second grating are only modulated in one direction, across the line widths; typically, our second grating consists of black lines on a white background, or of white lines (or transparent slits) on a black background. The spectrum $R_2(u,v)$ of this grating (see Fig. C.9(e)) is, therefore, a classical impulse comb whose impulses are located in the $u,v$ plane at integer multiples of the grating frequency $\mathbf{f}_2$, exactly as in Fig. 2.5(e).

What happens, now, when we superpose both of our gratings on top of each other with a small angle difference? As we already know, the superposition of the two layers $r_1(x,y)$ and $r_2(x,y)$ is given in the image domain by the product of the two individual layers, $r_1(x,y)r_2(x,y)$; therefore, according to the convolution theorem, the spectrum of the layer superposition is given by the convolution of the two individual spectra, $R_1(u,v)**R_2(u,v)$. In the simple case where none of the grating lines contains information, which is shown in Fig. 2.5, this convolution gives an oblique impulse nailbed (see Fig. 2.5(f)). However, in our present case the spectrum $R_1(u,v)$ of the first grating is no longer an impulse comb but rather a line-impulse comb. And indeed, as already shown in Sec. 10.7.3 for a similar case (see Figs. 10.9(d)–(f)), the convolution of such a line-impulse comb with the impulse comb $R_2(u,v)$ of the second spectrum consists of an infinite number of replicas of the spectrum $R_1(u,v)$, each of which being centered on top of an impulse of the comb $R_2(u,v)$.[14] The resulting spectrum convolution $R_1(u,v)**R_2(u,v)$ is shown in Fig. C.9(f).

At this point it may be helpful to recall the intuitive reasoning that we have presented in Sec. 10.7.3, and to momentarily think of $R_1(u,v)$ as a comb of impulses that "leaked out" perpendicularly to the comb direction to form our parallel continuous line-impulses. Before the impulses "leak out", i.e., when our first grating is still a non-modulated periodic grating $p_1(x,y)$, its spectrum $P_1(u,v)$ is an impulse comb, just like $R_2(u,v)$, and

[13] Obviously, if our grating is not symmetric about the origin, then each of the line impulses in the spectrum may have a complex-valued amplitude. But even in this case the spectrum still remains Hermitian (because our grating is always real valued; see [Bracewell86 pp. 14–15]).

[14] Remember that the convolution of any object with a comb of impulses places a centered replica of that object on top of each impulse of the comb (after properly scaling its amplitude).

therefore the convolution $P_1(u,v)**R_2(u,v)$ is an oblique lattice of impulses like in Fig. 2.5(f). We call this oblique lattice the *skeleton* of our line-impulse spectrum $R_1(u,v)**R_2(u,v)$. As the first grating starts being modulated, all the impulses of this skeleton lattice (except for the impulses of the comb $R_2(u,v)$) start "leaking out" to both directions, forming the line-impulse spectrum of Fig. C.9(d). This point of view allows us to identify each line-impulse in the spectrum of the superposition with the corresponding impulse in the skeleton lattice. Each line-impulse may thus "inherit" the properties of its original skeleton-impulse; the *skeleton location* (or the *center*) of a line-impulse in the spectrum will be defined as the location of its skeleton-impulse, and the *index* of a line-impulse will be defined as the index of its skeleton impulse. This allows us to carry over the important notions of *impulse location* and *impulse index* to the case of continuous line-impulses, too. Hence, a $(k_1,k_2)$-moiré in the superposition is the moiré which is caused by the $(k_1,k_2)$-line-impulse in the spectrum convolution (or, in fact, by the $(k_1,k_2)$-comb of line-impulses). This moiré becomes visible if the center of the $(k_1,k_2)$-line-impulse is located inside the visibility circle, close to the spectrum origin.

Using this terminology we can say, therefore, that the visible moiré effect in our case is represented in the spectrum convolution (Fig. C.9(f)) by the (1,-1)-line-impulse (whose center is located inside the visibility circle), or, rather, by the entire (1,-1)-comb of line-impulses that it spans.

Thus, by extracting from the spectrum convolution only this line-impulse comb (see Fig. C.10(b)) and taking its inverse Fourier transform, we obtain, back in the image domain, the isolated contribution of the (1,-1)-moiré in question to the image superposition. And indeed, as shown in Fig. C.10(a), this gives us precisely the macroscopic intensity profile of the moiré bands. Note that although this moiré is visible both in the layer superposition (Fig. C.9(c)) and in the extracted moiré intensity profile (Fig. C.10(a)), the latter does not contain the fine structure of the original layers $r_1(x,y)$ and $r_2(x,y)$ but only the pure contribution of the extracted moiré itself.

Now, following the same reasoning as in Sec. 4.2, we may ask ourselves how the intensity profile of the resulting (1,-1)-moiré (Fig. C.10(a)) is related to the gratings $r_1(x,y)$ and $r_2(x,y)$ themselves, in terms of image domain considerations only. To see this, let us first briefly review the simpler case that we have already studied in Chapter 4, in which none of the superposed line gratings contains modulated information.

In the case of Chapter 4 no line combs are involved in the spectral domain, and the spectra of the two given gratings as well as the spectrum of the extracted (1,-1)-moiré are all simple impulse combs. As we have seen in Sec. 4.2, the $n$-th impulse of the resulting moiré-comb is located in the spectrum at the point:

$$\mathbf{f}_{n,-n} = n\mathbf{f}_1 - n\mathbf{f}_2 \tag{C.12}$$

and its amplitude is given by:

$$d_n = a^{(1)}{}_n \, a^{(2)}{}_{-n} \tag{C.13}$$

where $a^{(1)}_i$ and $a^{(2)}_i$ are the amplitudes of the $i$-th impulses in the combs of the first and of the second line gratings, respectively. This means, as we have seen in Proposition 4.1, that the impulse amplitudes of the comb of the (1,-1)-moiré in the spectrum convolution are obtained by a term-by-term multiplication of the combs of the original superposed gratings, one of which is being inverted (rotated by 180°) before the multiplication.

Proceeding with this simple non-modulated case we see, therefore, that the moiré comb (Fig. 4.2(b)) can be considered in the spectral domain as a product of the two original combs (Figs. 2.5(d) and 2.5(e)), after they have been normalized (rotated and stretched) to fit the impulse locations of the resulting moiré comb. However, thanks to the 1D $T$-convolution theorem (see Sec. 4.2), this term-by-term multiplication of the original combs, as defined by Eq. (C.13), can be also represented as a 1D convolution in the image domain. Thus, the (normalized) 1D intensity profile of the extracted (1,-1)-moiré bands is simply the 1D $T$-convolution of the (normalized) 1D intensity profiles of the two original gratings (see Proposition 4.2 for a more detailed formulation). For example, if one of the superposed gratings (say, the second one) consists of narrow slits on a black background, the 1D intensity profile $p_2(x)$ of its period may be approximated by the 1D impulse $\delta(x)$, and therefore, the normalized 1D intensity profile of the resulting (1,-1)-moiré bands is almost identical to the normalized 1D intensity profile of the first grating.[15] In other words, the 1D intensity profile of the resulting (1,-1)-moiré is simply a magnified and rotated version of the 1D intensity profile of the first grating, where the magnification rate is controlled by the periods of the individual gratings and their angle difference, according to Eqs. (2.9). As we have seen in Sec. 4.3, this reasoning extends easily to the 2D case, too, where both of the superposed layers are dot screens.

Let us now return to our present case, the superposition of a *modulated* line grating with a simple line grating. In this case the situation is slightly different, since the spectra of the first grating and of the resulting (1,-1)-moiré consist of line-impulses, and are no longer simple impulse combs. And indeed, in this case the line-impulse comb in the spectrum convolution that corresponds to our modulated (1,-1)-moiré bands is not a term-by-term product of the (normalized) line-impulse comb of $R_1(u,v)$ with the (normalized) impulse comb of $R_2(u,v)$, since such a term-by-term product would yield a simple *impulse comb* rather than the expected *line-impulse comb*. So how is the line-impulse spectrum of our (1,-1)-moiré effect related to the spectra of the two original gratings? Since the spectrum of Fig. C.9(f) is obtained as a convolution of the line-impulse comb $R_1(u,v)$ with a simple impulse comb, $R_2(u,v)$, it follows from the properties of a convolution with an impulse comb that the line-impulse comb of the (1,-1)-moiré that is generated in the spectrum convolution of Fig. C.9(f) still consists of the same line impulses as the spectrum $R_1(u,v)$ of the modulated grating, where the amplitude of each of these continuous line impulses has only been scaled (during the convolution process) by the amplitude of the corresponding impulse of the simple impulse comb $R_2(u,v)$. Thus, the line-impulse comb of our modulated (1,-1)-moiré is not a normalized term-by-term product of the line-impulse

[15] Remember that the convolution of any object with an impulse $\delta(x)$ simply gives the original object.

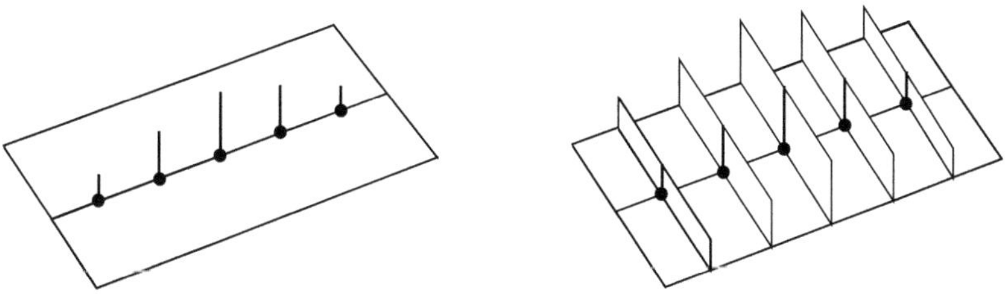

(a) (b)

**Figure C.12:** (a) The original impulse comb $R_2(u,v)$ of Fig. C.9(e). (b) The line-impulse comb $V_2(u,v)$ which is the constant perpendicular extension of $R_2(u,v)$.

comb $R_1(u,v)$ with the impulse comb $R_2(u,v)$, but rather a normalized term-by-term product of the line-impulse comb $R_1(u,v)$ with the new *line-impulse comb* $V_2(u,v)$ that results from the constant perpendicular extension of the simple impulse comb $R_2(u,v)$ (see Fig. C.12), after having inversed its elements order, as required by Eq. (C.13). The normalization step, which consists of rotation, scaling and *shearing*, guarantees that both of the line-impulse combs being multiplied have the same line-impulse locations and orientations.

Now, returning to the image domain, what is the inverse Fourier transform of the new line-impulse comb $V_2(u,v)$? It turns out that this inverse Fourier transform is simply the 1D section through our second grating along its main direction.[16] Therefore, we may distinguish here between 3 possible cases, just as we did in Sec. 4.4.1 for the classical 2D (1,0,-1,0)-moiré (for the effects of *shearing* in the normalization see Remark C.1 below):

**Case 1:** If our second grating consists of narrow slits on a black background, as in Fig. C.9(b), its spectrum $R_2(u,v)$ consists of a 1D impulse comb along the grating direction. In this case the inverse Fourier transform of the corresponding extended line-impulse comb

---

[16] Because the horizontal line impulse $\delta(y)$ over the $x,y$ plane and the vertical line impulse $\delta(u)$ over the $u,v$ plane are a 2D Fourier pair (see [Bracewell86p. 247] or [Bracewell95 pp. 152–153]), it follows from the 2D convolution theorem that if $f(x,y)$ and $F(u,v)$ are a 2D Fourier pair, then the horizontal slice $f(x,y)\delta(y)$ and the vertical extension $F(u,v)**\delta(u)$ are also a 2D Fourier pair [Gaskill78 pp. 307–308]. Note that this result is not limited to horizontal and vertical line impulses, and it can be generalized using the rotation theorem [Bracewell95 p. 157] to perpendicular pairs of line impulses at any other directions. Some graphical examples that may illustrate this rule are shown in [Bracewell86 pp. 246–247]. For example, because the 2D spectrum of a vertical cosinusoidal grating $p(x,y) = \cos(2\pi fx)$ consists of two impulses that are located along the $u$ axis at $u = f$ and $u = -f$, it follows that the 2D spectrum of the horizontal section through this cosinusoidal grating, $p(x,y)\delta(y)$, consists of two vertical line impulses having a constant amplitude, that cross the $u$ axis at the points $u = f$ and $u = -f$. These line impulses are the constant perpendicular extension of the original impulse pair. Note that $\delta(x)$ and $\delta(y)$ are considered here as 2D functions of the two variables $x,y$, just like the function $p(x,y) = \cos(2\pi fx)$, and they denote here, therefore, the vertical and horizontal line impulses over the $x,y$ plane through its origin [Gaskill78 pp. 85–86]; they should not be confused with the simple impulse $\delta(x)$ in the 1D case.

$V_2(u,v)$ gives back in the image domain a 1D section through our grating's slits, i.e. a train of 1D narrow pulses which can be closely approximated as a simple impulse comb. Therefore, the profile of the resulting (1,-1)-moiré bands in the image domain is simply a 2D *T*-convolution of one period of the modulated grating $r_1(x,y)$ (i.e. an entire 2D modulated line of the grating $r_1(x,y)$) with a simple impulse. This means that the 2D intensity profile of the resulting modulated (1,-1)-moiré band is basically a magnified and rotated (or rather *sheared*, as explained in Remark C.1 below) version of the modulated lines of our first grating $r_1(x,y)$.[17] But because our slits and their 1D section are not really perfect impulses the moiré shapes obtained in the convolution are slightly blurred and rounded, just like in Fig. 4.5(a).

**Case 2:** Similarly, if our second grating consists of narrow black lines on a white (or rather transparent) background, the profile of the resulting (1,-1)-moiré bands is simply a 2D *T*-convolution of a modulated line of the grating $r_1(x,y)$ with an "inverse" 1D pulse of 0-amplitude on a constant background of amplitude 1. This gives, just as in Case 2 of Sec. 4.4.1, an *inverse-video* version of the moiré bands that are obtained in Case 1. However, for reasons similar to those given in Case 2 of Sec. 4.4.1, the perceived contrast of the moiré in Case 2 appears to the eye much weaker than in Case 1.

**Case 3:** If our second grating consists of lines having any other profile, the profile form of the resulting (1,-1)-moiré is still a magnified version of the *T*-convolution between a modulated line of the grating $r_1(x,y)$ and a period of the 1D section through the grating $r_2(x,y)$. This *T*-convolution gives, again, some kind of blending between the two original profile shapes, but this time the resulting shape has a rather blurred or smoothed-out appearance and the moiré looks less attractive to the eye. This explains, in particular, the triangular or trapezoidal profile shape of the (1,-1)-moiré bands in the superposition of two simple line gratings as shown, for example, in Fig. 2.5(c) or in Fig. 2.9.

Note that in all of these cases the direction and the frequency of the line-comb of our modulated (1,-1)-moiré, i.e., the direction and the frequency of its (1,-1)-skeleton comb, are given by the direction $\varphi_M$ and the frequency $f_M$ of the vector $\mathbf{f}_1 - \mathbf{f}_2$; see Eqs. (2.9)–(2.11).

**Remark C.1:** As we can see in Figs. C.9(f) and C.10(b), the line-impulse comb of our modulated (1,-1)-moiré differs from the spectrum of a modulated grating (Fig. C.9(d)) in that its parallel line-impulses are not orthogonal to the comb direction. Let us try to understand the meaning of this fact.

---

[17] Remember that the inverse Fourier transform of the impulse comb $R_2(u,v)$ gives the entire 2D grating $r_2(x,y)$, while the inverse Fourier transform of $V_2(u,v)$, the constant perpendicular extension of the impulse comb $R_2(u,v)$, gives the 1D section through the 2D grating $r_2(x,y)$ along its main direction. Therefore, if we "forget" to take the constant perpendicular extension of the impulse comb $R_2(u,v)$, the 2D *T*-convolution in the image domain would be performed with the 2D profile of an entire slit of the grating $r_2(x,y)$, i.e. with a *line impulse* rather than with a simple impulse. This, however, does not give the expected result, since the convolution of any given object with a line impulse gives a blurred, continuous replication (or averaging) of the given object along the entire line impulse (see [Gaskill78 p. 308]).

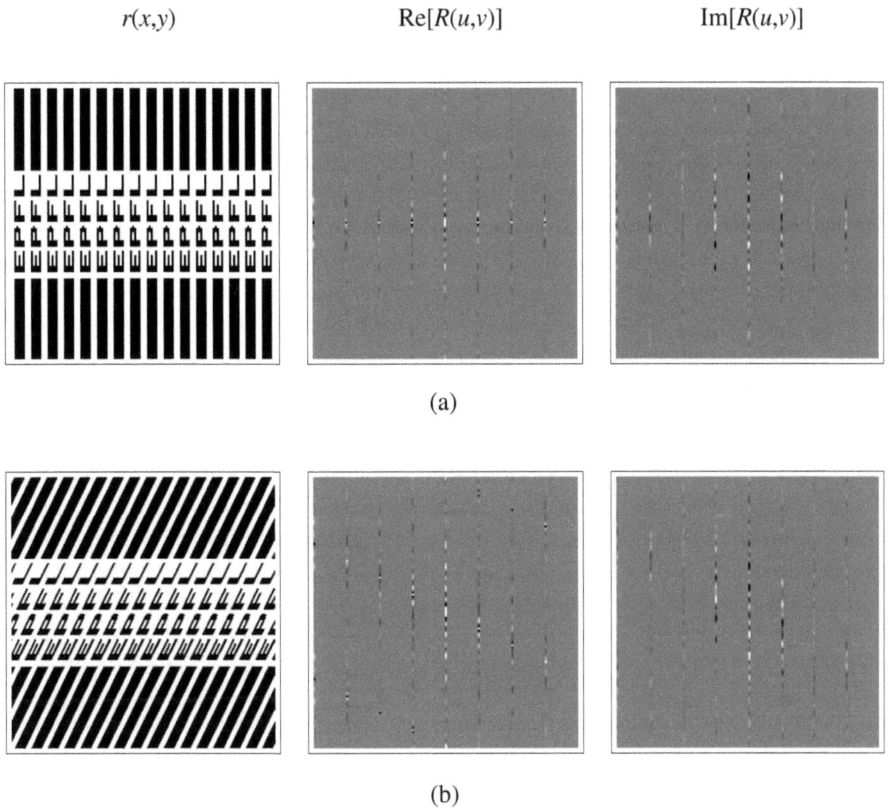


**Figure C.13:** A vertical shear transformation in the spectrum causes a horizontal shear transformation in the image domain, and vice versa. This is explained by the shear theorem [Bracewell95 p. 158].

As shown in Fig. C.13 (and as we have already seen in Sec. 10.7.3), the non-orthogonal line-impulse comb of Fig. C.10(b) can be obtained from an orthogonal line-impulse comb whose skeleton is located on the $u$ axis by applying a vertical shear transformation, namely, by replacing the vertical coordinate $v$ with $v+bu$ (where the coefficient $b$, in our case negative, is given by $b = \tan\varphi_M$, $\varphi_M$ being the direction of the skeleton comb). Now, according to the well-known shear theorem (see, for example, [Bracewell95 p. 158]), a vertical shear in the spectral domain corresponds to a horizontal shear in the image domain, namely: if $f(x,y) \leftrightarrow F(u,v)$ then $f(x+by,y) \leftrightarrow F(u,v-bu)$. This means that our vertically sheared (1,-1)-line-impulse comb corresponds in the image domain to a horizontally sheared modulated moiré-grating. And indeed, as it can be seen in Figs. C.9(c) and C.10(a), the (1,-1)-moiré that we obtain in the present example forms a horizontally sheared periodic grating pattern.

We see, therefore, that while in the simple superposition of non-modulated line gratings (Fig. 2.5) the resulting moiré effect is a magnified and rotated version of the lines of the original grating, in our present case the resulting moiré effect is a magnified and *sheared* version of the modulated lines of the original grating. The reason is, in fact, that the rotation is applied here only to one dimension of the moiré bands (the baseline direction of the modulated text), while the other direction of the modulated text, the letter's height, remains unchanged. This results in a *shearing* effect on the modulated moiré text rather than a *rotation* effect (that would have influenced *both* directions of the text). But when the moiré bands are not modulated and they contain no text (see, for example, the upper or the lower parts of Fig. C.7) this shearing effect can be also considered as a rotation of the moiré bands (with an appropriate scaling of their periods), and both interpretations are in fact equivalent and give the same results. ■

**Remark C.2:** If the letters that are embedded (or modulated) within the individual lines of the grating $r_1(x,y)$ are not upright but rather slanted (i.e. sheared), while they still keep their original baseline direction, the resulting text within the moiré bands will also be sheared, without affecting its baseline direction. The resulting shear transformation of the moiré text (and hence, its slanting angle, too) can be determined by using the shear theorem. We will return to this point in Remark C.4 at the end of Sec. C.14.4. ■

**Remark C.3:** It is interesting to note that just like in the similar case discussed in Sec. 10.7.3, when $f_2 > f_1$ (i.e. when the period of the slit grating is smaller than the period of the modulated grating), there exists a critical superposition angle at which all the vertical line impulses of the line-impulse comb of the hybrid (1,-1)-moiré collapse into a single vertical line impulse that coincides with the $y$ axis. This is a particularly interesting situation, because in this case the hybrid (1,-1)-moiré is singular, while its simple (1,-1)-moiré counterpart between non-modulated gratings is not. This singularity of the hybrid moiré can be easily understood in the image domain, too: at this particular superposition angle the resulting moiré bands are perfectly horizontal, which means that the letters embedded in them become infinitely elongated and slanted and hence they are no longer visible. An analogous spectral situation is depicted, for the case described in Sec. 10.7.3, in Fig. 10.12. Note that just like in that analogous case, such a singular state can only occur when $f_2 \geq f_1$, i.e. when $T_2 \leq T_1$; the equality here corresponds to the trivial singular case in which the two superposed layers have identical periods and angles. ■

Finally, it should be noted that just as in Sec. 4.2, our reasoning here can be also generalized to any $(k_1,k_2)$-moiré. However, the reason we have limited ourselves here to the simplest substractive case, the (1,-1)-moiré, is that this is the only case in which the resulting moiré effect can clearly preserve the letter shapes that are embedded in the individual lines of our first grating. As we have already seen in Chapter 4 (both in the 1D case of line gratings and in the 2D case of dot screens), when higher order moirés are considered the $T$-convolution theorem is no longer applicable in the image domain, and the relationship between the intensity profile of the resulting moiré and the intensity profiles of the two original layers becomes much more complex.

### C.14.3 Generalization to curvilinear gratings

As we have seen in Chapter 10, the results obtained by the Fourier-based approach can be also generalized to cases in which the original periodic layers undergo geometric transformations, either linear or non-linear. The fundamental moiré theorem for the superposition of two curved gratings (see Sec. 10.9.1) determines the effects of such layer transformations on the resulting profile and geometric layout of the $(k_1,k_2)$-moiré between two gratings. Similarly, the fundamental moiré theorem for the superposition of two curved screens (see Sec. 10.9.2) determines the effects of such layer transformations on the resulting profile and geometric layout of the $(k_1,k_2,k_3,k_4)$-moiré between two screens. It may be asked, therefore, if a similar rule can be also formulated for our hybrid case, allowing us to see how the application of linear or non-linear transformations to the original gratings will affect the resulting modulated moiré bands.

Because we are only interested here in the simplest, subtractive first-order moiré, let us first reformulate the two fundamental moiré theorems of Chapter 10 specifically for such first-order moirés (see Propositions 10.2 and 10.5):

**The fundamental (1,-1)-moiré theorem for line gratings:** Let $r_1(x,y)$ and $r_2(x,y)$ be two curvilinear line gratings that are obtained by applying the bending functions (linear or not) $g_1(x,y)$ and $g_2(x,y)$, respectively, to two periodic line gratings having the intensity profiles $p_1(x')$ and $p_2(x')$:

$$r_1(x,y) = p_1(g_1(x,y)), \qquad r_2(x,y) = p_2(g_2(x,y))$$

Then, the (1,-1)-moiré $m(x,y)$ in the superposition of $r_1(x,y)$ and $r_2(x,y)$ is given by:

$$m(x,y) = p(g(x,y))$$

where:

(1) $p(x')$, the normalized intensity profile of the (1,-1)-moiré, is the 1D $T$-convolution of the normalized intensity profiles of the original gratings:

$$p(x') = p_1(x') * p_2(-x') \tag{C.13}$$

(2) $g(x,y)$, the bending function which brings $p(x')$ back into the actual moiré pattern $m(x,y)$ as it appears in the superposition of the two *transformed* layers, is given by:

$$g(x,y) = g_1(x,y) - g_2(x,y) \quad \blacksquare \tag{C.14}$$

**The fundamental (1,0,-1,0)-moiré theorem for dot screens:** Let $r_1(\mathbf{x})$ and $r_2(\mathbf{x})$ be two curvilinear screens that are obtained by applying the mappings (linear or not) $\mathbf{g}_1(\mathbf{x})$ and $\mathbf{g}_2(\mathbf{x})$, respectively, to two periodic screens having the intensity profiles $p_1(\mathbf{x}')$ and $p_2(\mathbf{x}')$:

$$r_1(\mathbf{x}) = p_1(\mathbf{g}_1(\mathbf{x})), \qquad r_2(\mathbf{x}) = p_2(\mathbf{g}_2(\mathbf{x}))$$

Then, the (1,0,-1,0)-moiré $m(\mathbf{x})$ in the superposition of $r_1(\mathbf{x})$ and $r_2(\mathbf{x})$ is given by:

$$m(\mathbf{x}) = p(\mathbf{g}(\mathbf{x}))$$

where:

(1) $p(\mathbf{x}')$, the normalized intensity profile of the (1,0,-1,0)-moiré, is the 2D *T*-convolution of the normalized intensity profiles of the original, untransformed screens:

$$p(\mathbf{x}') = p_1(\mathbf{x}') ** p_2(-\mathbf{x}') \tag{C.15}$$

(2) $\mathbf{g}(\mathbf{x})$, the transformation which brings $p(\mathbf{x}')$ back into the actual moiré pattern $m(\mathbf{x})$ as it appears in the superposition of the two *transformed* layers, is given by:

$$\mathbf{g}(\mathbf{x}) = \mathbf{g}_1(\mathbf{x}) - \mathbf{g}_2(\mathbf{x}) \quad \blacksquare \tag{C.16}$$

We now return to the new moiré theorem that we wish to establish for our hybrid case. We have already seen in Sec. C.14.2 above how to formulate its first part, which concerns the intensity profiles: The intensity profile of our hybrid (1,-1)-moiré is the 2D *T*-convolution of the profile of the first, modulated grating with a section through the second, unmodulated grating along its main direction. How can we now formulate the second part of the theorem, which concerns the geometric transformations?

To see this, consider the full componentwise notation of the two layer transformations $\mathbf{g}_1(\mathbf{x})$ and $\mathbf{g}_2(\mathbf{x})$ and of the moiré transformation $\mathbf{g}(\mathbf{x})$:

$$\mathbf{g}_1(\mathbf{x}) = \begin{pmatrix} g_{1,1}(x,y) \\ g_{1,2}(x,y) \end{pmatrix}, \quad \mathbf{g}_2(\mathbf{x}) = \begin{pmatrix} g_{2,1}(x,y) \\ g_{2,2}(x,y) \end{pmatrix}, \quad \mathbf{g}(\mathbf{x}) = \begin{pmatrix} g_1(x,y) \\ g_2(x,y) \end{pmatrix}$$

According to Eq. (C.16) of the fundamental (1,0,-1,0)-moiré theorem, the transformation $\mathbf{g}(\mathbf{x})$ undergone by the moiré is given by $\mathbf{g}(\mathbf{x}) = \mathbf{g}_1(\mathbf{x}) - \mathbf{g}_2(\mathbf{x})$. And indeed, in the 2D case, where both of the original, untransformed layers are dot screens, this simply means:

$$\begin{pmatrix} g_1(x,y) \\ g_2(x,y) \end{pmatrix} = \begin{pmatrix} g_{1,1}(x,y) \\ g_{1,2}(x,y) \end{pmatrix} - \begin{pmatrix} g_{2,1}(x,y) \\ g_{2,2}(x,y) \end{pmatrix} \tag{C.17}$$

Now, what happens in the 1D case, i.e. when both of the original, untransformed layers consist of vertical lines with a purely 1D profile? In this case the second component in each of the above transformations becomes irrelevant, and we obtain:

$$\mathbf{g}_1(\mathbf{x}) = \begin{pmatrix} g_{1,1}(x,y) \\ 0 \end{pmatrix}, \quad \mathbf{g}_2(\mathbf{x}) = \begin{pmatrix} g_{2,1}(x,y) \\ 0 \end{pmatrix}, \quad \mathbf{g}(\mathbf{x}) = \begin{pmatrix} g_1(x,y) \\ 0 \end{pmatrix}$$

or, more simply, by dropping the unused components and indices:

$$\mathbf{g}_1(\mathbf{x}) = g_1(x,y), \qquad \mathbf{g}_2(\mathbf{x}) = g_2(x,y), \qquad \mathbf{g}(\mathbf{x}) = g(x,y)$$

Therefore in this case Eq. (C.16) reduces into its single-component counterpart:

$$g(x,y) = g_1(x,y) - g_2(x,y) \tag{C.18}$$

which is, indeed, Eq. (C.14) of the fundamental moiré theorem for *line gratings*.

We now return to our present hybrid case. In this case, only one of the two original, untransformed gratings (the second one) has a purely 1D profile, and therefore we have:

$$\mathbf{g}_1(\mathbf{x}) = \begin{pmatrix} g_{1,1}(x,y) \\ g_{1,2}(x,y) \end{pmatrix}, \quad \mathbf{g}_2(\mathbf{x}) = \begin{pmatrix} g_{2,1}(x,y) \\ 0 \end{pmatrix}, \quad \mathbf{g}(\mathbf{x}) = \begin{pmatrix} g_1(x,y) \\ g_2(x,y) \end{pmatrix}$$

Hence, Eq. (C.16) of the fundamental moiré theorem simply becomes here:[18]

$$\begin{pmatrix} g_1(x,y) \\ g_2(x,y) \end{pmatrix} = \begin{pmatrix} g_{1,1}(x,y) \\ g_{1,2}(x,y) \end{pmatrix} - \begin{pmatrix} g_{2,1}(x,y) \\ 0 \end{pmatrix} \tag{C.19}$$

which is, indeed, intermediate between the 2D case of Eq. (C.17) and the 1D case of Eq. (C.18). This suggests, once again, that our hybrid superposition could be considered in fact as a "$1\frac{1}{2}$D case". Our new fundamental moiré theorem for the (1,-1)-hybrid case can be therefore formulated as follows:[19]

**The fundamental moiré theorem for the hybrid (1,-1)-moiré:** Let $r_1(\mathbf{x})$ and $r_2(\mathbf{x})$ be two curvilinear gratings that are obtained by applying the mappings (linear or not) $\mathbf{g}_1(\mathbf{x})$ and $\mathbf{g}_2(\mathbf{x})$, respectively, to a first, modulated periodic grating having the 2D intensity profile $p_1(\mathbf{x}')$, and to a second, unmodulated periodic grating having the intensity profile $p_2(\mathbf{x}')$:

$$r_1\begin{pmatrix} x \\ y \end{pmatrix} = p_1\begin{pmatrix} g_{1,1}(x,y) \\ g_{1,2}(x,y) \end{pmatrix}, \qquad r_2\begin{pmatrix} x \\ y \end{pmatrix} = p_2\begin{pmatrix} g_{2,1}(x,y) \\ g_{2,2}(x,y) \end{pmatrix}$$

Then, the hybrid (1,-1)-moiré $m(\mathbf{x})$ in the superposition of $r_1(\mathbf{x})$ and $r_2(\mathbf{x})$ is given by:

$$m\begin{pmatrix} x \\ y \end{pmatrix} = p\begin{pmatrix} g_1(x,y) \\ g_2(x,y) \end{pmatrix}$$

where:

(1) $p(\mathbf{x}')$, the normalized intensity profile of our hybrid (1,-1)-moiré, is the 2D *T*-convolution of the normalized intensity profile of the first, original modulated (but untransformed) grating, with a 1D section through the normalized intensity profile of the second, unmodulated grating along its main direction:

$$p\begin{pmatrix} x' \\ y' \end{pmatrix} = p_1\begin{pmatrix} x' \\ y' \end{pmatrix} ** \; p_2\begin{pmatrix} -x' \\ 0 \end{pmatrix} \tag{C.20}$$

(2) $\mathbf{g}(\mathbf{x})$, the transformation which brings $p(\mathbf{x}')$ back into the actual moiré pattern $m(\mathbf{x})$ as it appears in the superposition of the two *transformed* layers, is given by:

$$\begin{pmatrix} g_1(x,y) \\ g_2(x,y) \end{pmatrix} = \begin{pmatrix} g_{1,1}(x,y) \\ g_{1,2}(x,y) \end{pmatrix} - \begin{pmatrix} g_{2,1}(x,y) \\ 0 \end{pmatrix} \quad \blacksquare \tag{C.21}$$

---

[18] The fact that we have chosen here the *second* component of $\mathbf{g}_2(\mathbf{x})$ to be zero is just a matter of convention. We could equally well start our discussion on the hybrid (1,-1)-moiré with gratings made of horizontal rather than vertical lines, in which case the *first* component of $\mathbf{g}_2(\mathbf{x})$ would have been zero. We have chosen to present the hybrid (1,-1)-moiré using vertical gratings in order to remain consistent with our discussions in Chapter 2 (see, for example, Fig. 2.5).

[19] Note that the mathematical development that leads to this result is basically the same as in Sec. 10.9.2, except that $g_4(x,y) = 0$. The slight difference between the indexing conventions that are used here and in Sec. 10.9.2 for the components of the transformations is just a matter of convenience in the notations.

**Example C.1:** As a simple illustration to this theorem consider, once again, the hybrid (1,-1)-moiré shown in Fig. C.7(c). In this case, the normalized intensity profile of the moiré effect is, indeed, equal to the normalized intensity profile of the first layer, as predicted by the first part of the theorem (see Case 1 in Sec. C.14.2). In order to see how the second part of the theorem works, we start with three normalized vertical gratings having identical periods, which represent the initial, untransformed version of the two original gratings and of the resulting moiré bands. The transformations $\mathbf{g}_1$, $\mathbf{g}_2$ and $\mathbf{g}$ are given, with respect to these initial gratings, as follows: The first grating (the modulated one) does not undergo any layer transformations, so that $\mathbf{g}_1(x,y) = (x,y)$. The second, 1D grating undergoes the 1D transformation $\mathbf{g}_2(x,y) = (0.9x, 0)$, meaning that its horizontal period is slightly stretched out by $1/0.9 = 1.111$. Consequently, according to the second part of our theorem, the moiré bands are determined by the transformation:

$$\mathbf{g}\begin{pmatrix} x \\ y \end{pmatrix} = \begin{pmatrix} x \\ y \end{pmatrix} - \begin{pmatrix} 0.9x \\ 0 \end{pmatrix} = \begin{pmatrix} 0.1x \\ y \end{pmatrix} \tag{C.22}$$

meaning that the modulated moiré bands are horizontally stretched by the factor $1/0.1 = 10$, while in the vertical direction they remain unchanged.[20] Note that both the first layer and the moiré effect are 2D entities (because they carry 2D information), while the second layer is only of 1D nature, so that the second component of the transformation $\mathbf{g}_2(x,y) = (g_1(x,y), g_2(x,y))$ is $g_2(x,y) = 0$.

As a second simple example, consider the hybrid (1,-1)-moiré shown in Fig. C.7(b). In this case the first part of the theorem remains exactly as in the previous case, but in the second part of the theorem we have:

$$\mathbf{g}\begin{pmatrix} x \\ y \end{pmatrix} = \begin{pmatrix} x \\ y \end{pmatrix} - \begin{pmatrix} x\cos\theta + y\sin\theta \\ 0 \end{pmatrix} = \begin{pmatrix} (1 - \cos\theta)x - \sin\theta y \\ y \end{pmatrix} \tag{C.23}$$

This means that the moiré bands have been horizontally stretched out by the factor $1/(1 - \cos\theta)$ and horizontally sheared by $\sin\theta y$.[21] ■

Finally, it should be always remembered that the transformations $\mathbf{g}_1$, $\mathbf{g}_2$ and $\mathbf{g}$ are applied to the initially normalized and untransformed layers as *domain* (and hence, *inverse*) transformations. For example, $\mathbf{g}(x,y) = (0.1x, y)$ corresponds to a 10-fold *magnification* in the $x$ direction, and $\mathbf{g}(x,y) = (x - ay, y)$ represents a horizontal shearing effect to the *right*.

---

[20] In fact, if the three original normalized gratings have the period 1, then after the application of the transformations $\mathbf{g}_1$, $\mathbf{g}_2$ and $\mathbf{g}$ the horizontal periods of the two superposed gratings and of the resulting moiré bands are given, respectively, by 1, $1/0.9 = 1.111$ and $1/0.1 = 10$.

[21] It is interesting to note that $\mathbf{g}_2(x,y)$ only consists here of the first component of the rotation transformation $\mathbf{f}(x,y) = (x\cos\theta + y\sin\theta, -x\sin\theta + y\cos\theta)$, since its second component has no influence on our vertical slit grating and is set to zero (see also Fig. 10.1(b) in Chapter 10). But if the rotation transformation were applied to the *first* layer (the 2D modulated line grating), then $\mathbf{g}_1(x,y)$ would consist of both components of $\mathbf{f}(x,y)$, and the second component of the resulting moiré transformation $\mathbf{g}(x,y)$ would no longer be simply $y$. And indeed, in this case the resulting moiré text would not only be horizontally scaled and sheared, as in Fig. C.7(b), but also rotated (by the same angle $\theta$ as the first, modulated grating).

### C.14.4 Synthesis of hybrid (1,-1)-moiré effects

The fundamental moiré theorem for the hybrid (1,-1) case tells us also how we can design layers that will give in their superposition a hybrid (1,-1)-moiré effect having any desired geometric layout, and any predefined information running along each of its moiré bands. In order to obtain hybrid moiré bands whose intensity profiles are modulated by some predefined information, we simply need to embed a flattened version of the information in question inside each of the lines of the first grating, while the second grating should consist of linear slits on a black background. Then, in order to impose on the resulting moiré bands a certain desired geometric layout, we simply have to apply to our original straight gratings transformations $\mathbf{g}_1(\mathbf{x})$ and $\mathbf{g}_2(\mathbf{x})$ whose difference gives the desired moiré-band transformation $\mathbf{g}(\mathbf{x})$, in accordance with Eq. (C.21). Note that $\mathbf{g}_1(\mathbf{x})$, $\mathbf{g}_2(\mathbf{x})$ and $\mathbf{g}(\mathbf{x})$ are understood here as domain transformations that are applied to three untransformed, normalized periodic structures, all of which have the same initial periods and orientations; after the application of these domain transformations the first two structures give the geometric layouts of the two transformed layers, and the third one gives the geometric layout of the resulting moiré pattern.

To illustrate the synthesis of such a hybrid (1,-1)-moiré, let us consider the following interesting example which involves non-linear transformations.

**Example C.2:** Suppose we wish to generate two layers that give in their superposition a circular moiré effect whose intensity profile consists of repeated occurrences of the digit "1". For didactic reasons we will do this exercise twice, once using the approach based on the fundamental (1,0,-1,0)-moiré theorem for dot screens (as described in Sec. 10.9.2), and then using our new approach based on the fundamental moiré theorem for the hybrid (1,-1) case. Note that in the latter case we could have incorporated into the moiré bands any aperiodic text, but we have chosen to use here the same repetitive pattern of "1"s in order to be able to compare the two approaches using the very same underlying data.

In order to obtain our desired moiré effect using the first approach, as described in Sec. 10.9.2, we start with two original periodic dot screens having identical frequencies and orientations, one of which consists of dots having the shape of tiny "1"s, while the other consists of tiny pinholes on a black background. In order to obtain the desired moiré geometric layout, we may define the moiré transformation $\mathbf{g}(\mathbf{x})$ as follows:

$$\mathbf{g}\begin{pmatrix} x \\ y \end{pmatrix} = \begin{pmatrix} \varepsilon \log(\sqrt{x^2+y^2}) \\ \varepsilon \arctan(y/x) \end{pmatrix} \tag{C.24}$$

where $\varepsilon$ is a small positive constant. Note that by using here the logarithm of the radius rather than the radius itself we obtain gradually increasing elements along the radial direction, which is more visually pleasing than keeping fixed sized elements along the radial direction.

(a)

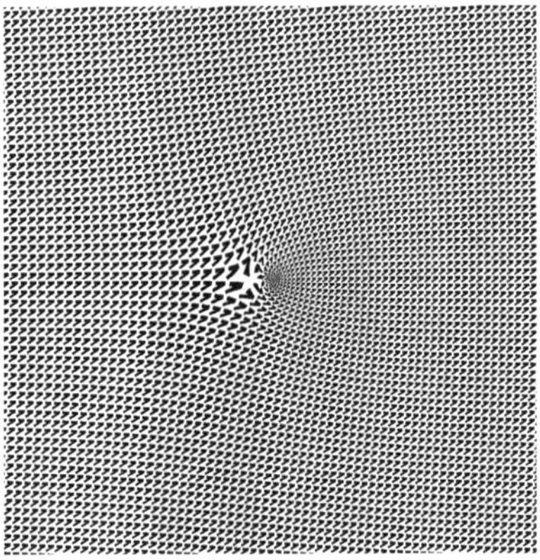

(b)

(c)

(d)

**Figure C.14:** Synthesis of a circular moiré effect whose intensity profile consists of repeated "1"s using the fundamental (1,0,-1,0)-moiré theorem (see Example C.2). (a) A periodic dot screen whose individual dots have the shape of "1". (b) The same dot screen after having undergone the transformation $\mathbf{g}_1(x,y)$ of Eq. (C.25). (c) A periodic pinhole screen having the same periodicity as the screen (a). (d) The superposition of the screens (b) and (c) gives the desired moiré effect. Note that shifting the second layer horizontally causes the moiré effect to move in the radial direction, while shifting it vertically causes the moiré effect to rotate clockwise or counterclockwise. The dynamics of the moiré effect can be best appreciated by using transparencies of the layers (b) and (c) and sliding them on top of each other.

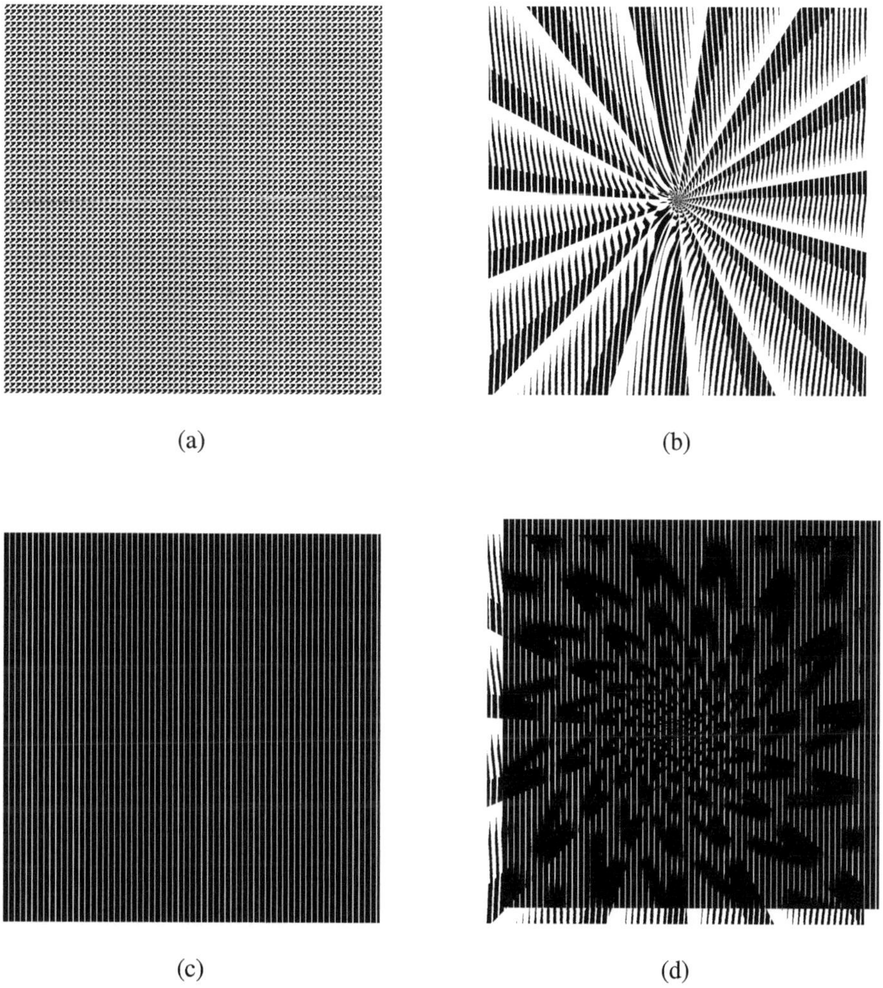


**Figure C.15:** Synthesis of the same macroscopic moiré effect as in Fig. C.14 using, this time, the fundamental moiré theorem for the hybrid (1,-1)-moiré (see Example C.2). (a) The same periodic dot screen as in Fig. C.14(a). (b) The same dot screen after having undergone the transformation $\mathbf{g}_1(x,y)$ of Eq. (C.26). (c) A periodic grating consisting of vertical slits having the same horizontal periodicity as the screen (a). (d) The superposition of the screens (b) and (c) gives the desired moiré effect. Note that shifting the second layer horizontally causes the moiré effect to move in the radial direction, but unlike in Fig. C.14, shifting it vertically has no effect on the resulting moiré. The dynamics of the moiré effect can be best appreciated by using transparencies of the layers (b) and (c) and sliding them on top of each other.

Now, according to Eq. (C.16), all that we have to do is to apply to our original periodic layers two layer transformations $\mathbf{g}_1(\mathbf{x})$ and $\mathbf{g}_2(\mathbf{x})$ such that $\mathbf{g}(\mathbf{x}) = \mathbf{g}_1(\mathbf{x}) - \mathbf{g}_2(\mathbf{x})$. Obviously, there exist infinitely many ways to define transformations $\mathbf{g}_1(\mathbf{x})$ and $\mathbf{g}_2(\mathbf{x})$ that satisfy this condition, but if we wish to transform only one of the two original layers, say, the first one, we may choose the transformations:

$$\mathbf{g}_1\begin{pmatrix} x \\ y \end{pmatrix} = \begin{pmatrix} x \\ y \end{pmatrix} + \begin{pmatrix} \varepsilon\log(\sqrt{x^2+y^2}) \\ \varepsilon\arctan(y/x) \end{pmatrix}, \qquad \mathbf{g}_2\begin{pmatrix} x \\ y \end{pmatrix} = \begin{pmatrix} x \\ y \end{pmatrix} \tag{C.25}$$

And indeed, as shown in Fig. C.14, this choice of layers and geometric transformations gives us in the superposition the desired moiré effect, using the first approach.

Now, if we wish to obtain a similar result using the second approach, i.e. using the hybrid (1,-1)-moiré effect, we have to start with two original periodic line gratings having identical frequencies and orientations: One of the gratings incorporates along each of its lines a sequence of tiny "1"s (which, in principle, could be replaced, if we so desired, by any aperiodic text running along the line), while the other grating consists of straight narrow slits on a black background.[22] Assuming that we want to obtain precisely the same moiré transformation, as given by Eq. (C.24), all that we have to do now is to apply to our two original periodic gratings two layer transformations $\mathbf{g}_1(\mathbf{x})$ and $\mathbf{g}_2(\mathbf{x})$ that satisfy Eq. (C.21) (rather than Eq. (C.16) in the first approach). If, once again, we wish to transform only one of the two original layers, we may choose this time the following layer transformations:[23]

$$\mathbf{g}_1\begin{pmatrix} x \\ y \end{pmatrix} = \begin{pmatrix} x \\ 0 \end{pmatrix} + \begin{pmatrix} \varepsilon\log(\sqrt{x^2+y^2}) \\ \varepsilon\arctan(y/x) \end{pmatrix}, \qquad \mathbf{g}_2\begin{pmatrix} x \\ y \end{pmatrix} = \begin{pmatrix} x \\ 0 \end{pmatrix} \tag{C.26}$$

Note that the only difference between the geometric transformations (C.25) and (C.26) that we apply to the layers in our two approaches is that in the latter approach the second component of $\mathbf{g}_2$ remains zero, so that we must replace the term $(x,y)$ in both $\mathbf{g}_1(\mathbf{x})$ and $\mathbf{g}_2(\mathbf{x})$ by $(x,0)$. And indeed, as shown in Fig. C.15, this choice of layers and geometric transformations gives us in the superposition a moiré effect having the same intensity profile as in Fig. C.14, but, this time using the hybrid (1,-1)-moiré effect. Note, however, that only the macroscopic properties of the moiré (its intensity profile, its geometric layout, etc.) are common to both cases; the microstructure details, on their part, are obviously different.

Finally, it is interesting to note that the two similar moiré effects shown in Figs. C.14 and C.15 also have a similar dynamic behaviour: In Fig. C.14, when the second layer is slowly shifted on top of the first layer horizontally, the resulting moiré effect moves

---

[22] Note that in practice we can use here the same first layer as in the first approach, since a dot screen consisting of periodic "1"s can be also viewed as a grating composed of lines that consist of a periodic sequence of "1"s.

[23] Note that in this case the *second* component of our desired transformation (C.24) must be entirely taken care of by $\mathbf{g}_1(\mathbf{x})$, since the second component of $\mathbf{g}_2(\mathbf{x})$ has no influence and is set to zero. However, the *first* component of our desired transformation (C.24) can be distributed between the two layer transformations $\mathbf{g}_1(\mathbf{x})$ and $\mathbf{g}_2(\mathbf{x})$ as we may wish.

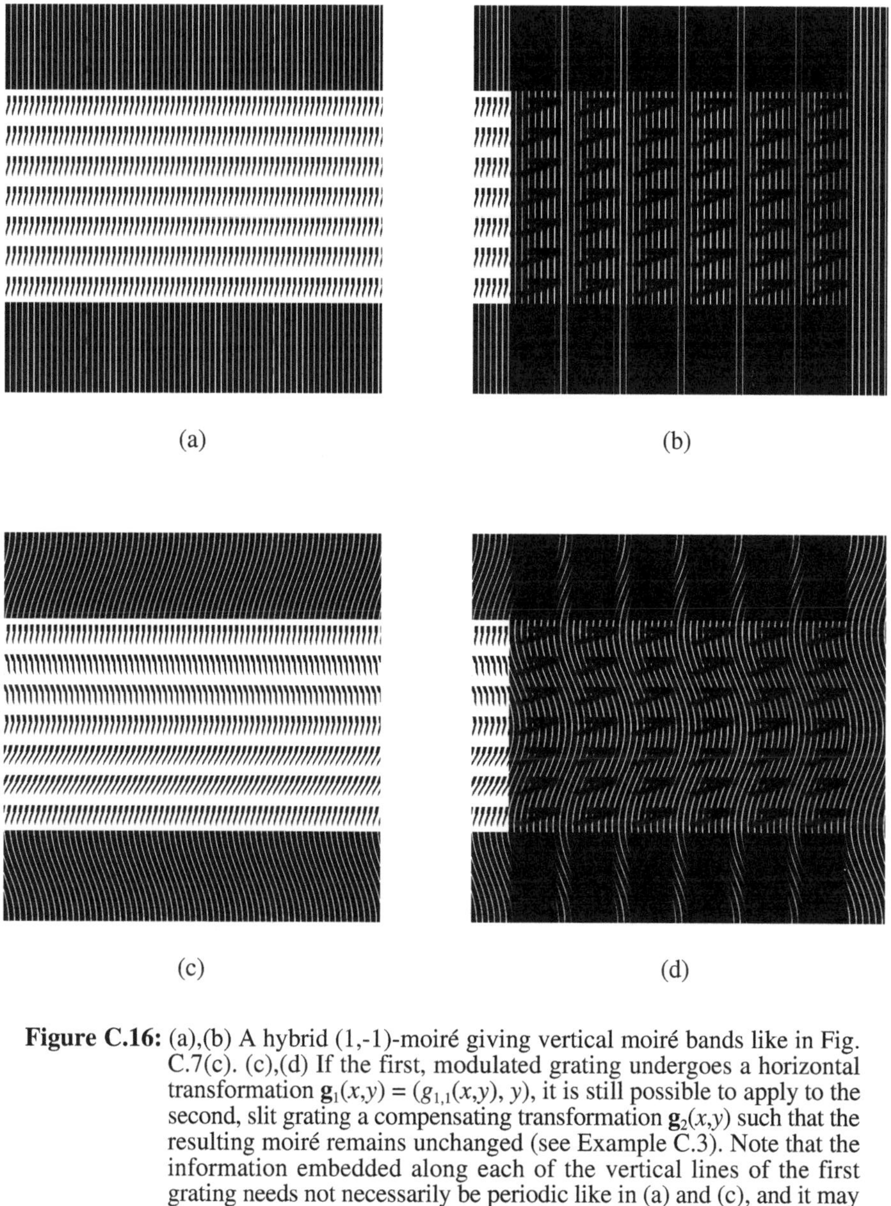


**Figure C.16:** (a),(b) A hybrid (1,-1)-moiré giving vertical moiré bands like in Fig. C.7(c). (c),(d) If the first, modulated grating undergoes a horizontal transformation $\mathbf{g}_1(x,y) = (g_{1,1}(x,y), y)$, it is still possible to apply to the second, slit grating a compensating transformation $\mathbf{g}_2(x,y)$ such that the resulting moiré remains unchanged (see Example C.3). Note that the information embedded along each of the vertical lines of the first grating needs not necessarily be periodic like in (a) and (c), and it may include any aperiodic text.

rapidly in the radial direction outward or inward, depending on the direction of the horizontal shift; and when applying a vertical shift, the resulting moiré effect rotates clockwise or counterclockwise, depending on the direction of the vertical shift. The hybrid case of Fig. C.15, however, only inherits the dynamic behaviour of Fig. C.14 in the *horizontal* direction, giving a radial motion of the moiré, while vertical shifts have no effect on the resulting moiré. Can you think of a similar hybrid (1,-1)-moiré that only inherits the dynamic behaviour of Fig. C.14 in the *vertical* direction, giving a circular motion of the moiré, while horizontal shifts have no effect on the resulting moiré? (*Hint*: in Eqs. (C.25), zero the $x$ components rather than the $y$ components as we did in Eqs. (C.26).) Note that the dynamic effects of the moiré can be best observed by printing the two layers in question on transparencies and shifting them on top of each other. ■

**Example C.3:** Suppose that we are given a vertical hybrid (1,-1)-moiré like in Fig. C.7(c) (see Figs. C.16(a),(b)), and that we apply to its first, modulated grating (Fig. C.16(a)) a geometric transformation $\mathbf{g}_1(x,y)$. For example, we may bend the straight vertical lines of this grating into cosinusoidal lines, as shown in Fig. C.16(c), by using the transformation:

$$\mathbf{g}_1\begin{pmatrix} x \\ y \end{pmatrix} = \begin{pmatrix} x - \varepsilon\cos(2\pi f y) \\ y \end{pmatrix} \tag{C.27}$$

where $\varepsilon$ is a small positive constant. What geometric transformation $\mathbf{g}_2(x,y)$ should be applied to the second layer, the slit grating, in order to keep the moiré bands straight as before?

Following the same reasoning as in Sec. 10.9 (see, for instance, Eqs. (10.35) and (10.44) and Examples 10.20–10.23), it is easy to see that vertical, periodic moiré bands occur in our superposition *iff* the moiré transformation has the form:

$$\mathbf{g}\begin{pmatrix} x \\ y \end{pmatrix} = \begin{pmatrix} x/T_x + c \\ y \end{pmatrix}$$

where $T_x$ is the desired horizontal magnification of the vertical moiré bands,[24] and $c$ is their horizontal displacement (that we assume here to be zero).

Now, since the transformations $\mathbf{g}_1(x,y)$ and $\mathbf{g}(x,y)$ are already known, it follows from Eq. (C.21) that the transformation $\mathbf{g}_2(x,y)$ is given by their difference:

$$\mathbf{g}_2\begin{pmatrix} x \\ y \end{pmatrix} = \mathbf{g}_1\begin{pmatrix} x \\ y \end{pmatrix} - \mathbf{g}\begin{pmatrix} x \\ y \end{pmatrix} = \begin{pmatrix} (1 - 1/T_x)x - \varepsilon\cos(2\pi f y) \\ 0 \end{pmatrix}$$

And indeed, as illustrated in Fig. C.16(d), the application of this transformation to our slit grating gives in the superposition exactly the same straight moiré bands as in Fig. C.16(b).

Now, is it also possible to expand the resulting moiré text vertically, say, by a factor of two? In order to obtain this result, the moiré transformation should obviously be:

$$\mathbf{g}\begin{pmatrix} x \\ y \end{pmatrix} = \begin{pmatrix} x/T_x \\ y/2 \end{pmatrix}$$

[24] In fact, if the original gratings are normalized and have the period 1, then $T_x$ simply indicates the horizontal period of the moiré bands.

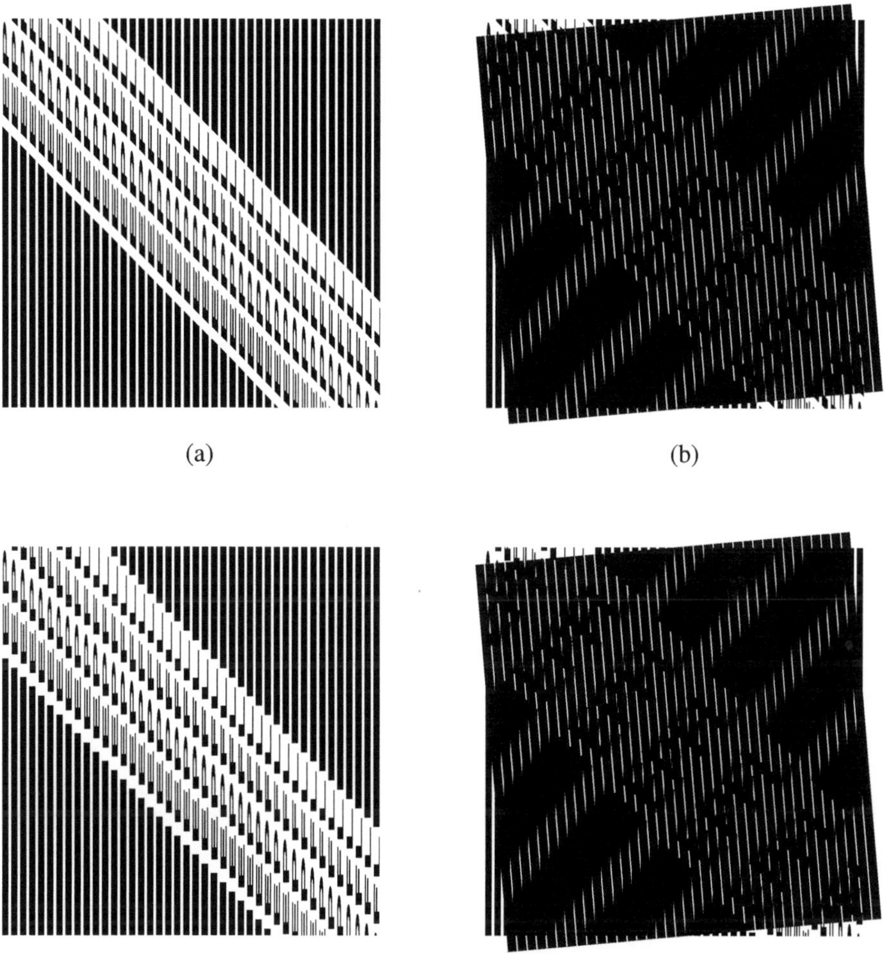

(a) (b) (c) (d)

**Figure C.17:** Rectification of the slanted text within the moiré bands can be obtained in two different ways (compare with Fig. C.7(a),(b)): By applying to the original modulated grating a shearing transformation, as shown in (a) and (b); or by staggering, i.e. by gradually shifting (advancing or retracting) the text periods within the original modulated grating without slanting the letters themselves, as shown in (c) and (d). In the latter case the resulting moiré periods are both slanted and shifted with respect to each other (staggered). The slit grating used in (b) and (d) is identical to the one used in Fig. C.7(b). Note that in both (b) and (d) the non-modulated parts of the moiré bands remain identical to those of Fig. C.7(b), and only the text within the bands is affected. (The quality of the resulting moiré letters can be significantly improved by increasing the resolution and the frequency of the original layers or by modifying their rotation angles.)

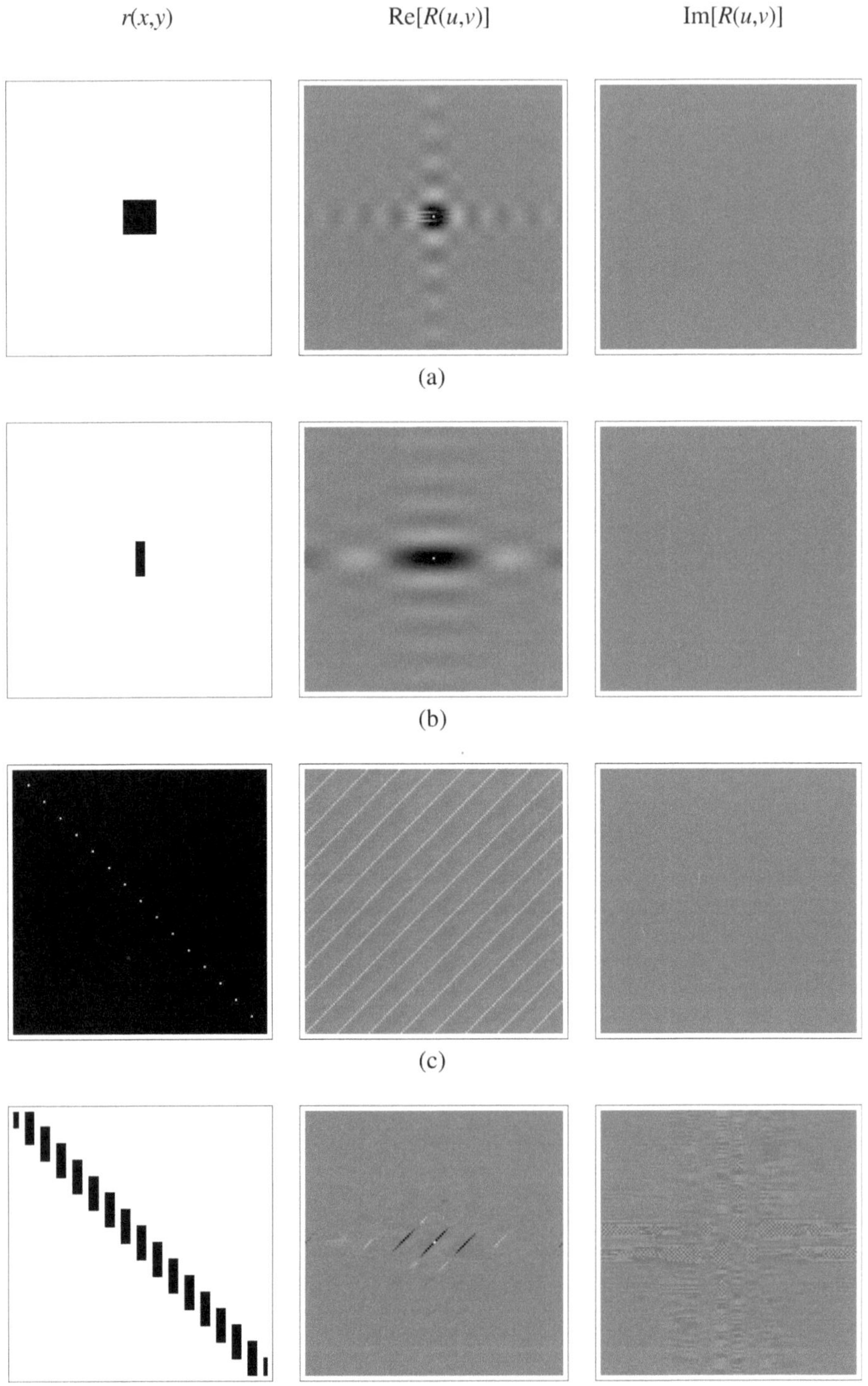
$r(x,y)$
$\mathrm{Re}[R(u,v)]$
$\mathrm{Im}[R(u,v)]$
(a)
(b)
(c)
(d)

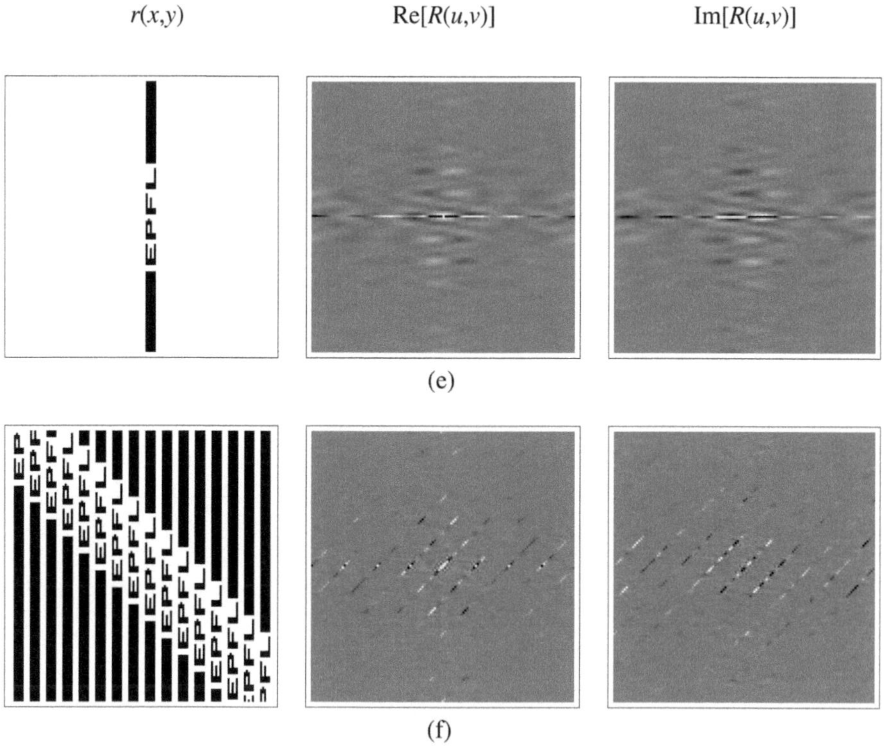


**Figure C.18:** Explanation of the spectrum of a staggered line grating whose individual modulated lines are gradually shifted along the line direction; two simple examples are shown in (d) and (f). The explanation here is the same as in Fig. C.11, except that this time the single modulated lines, (b) or (e), are convolved with a *slanted* impulse comb (c). In the resulting line gratings, shown respectively in (d) and (f), the modulated periods are gradually shifted along the lines of the grating. The spectrum of such staggered gratings consists of slanted (rather than vertical) blades that sample the same spectrum as in Fig. C.11, i.e. the spectrum of the corresponding single line, (b) or (e). Note that according to the shearing theorem the spectrum of a vertically *sheared* version of the gratings of Figs. C.11(d) and C.11(f) would consist of similar slanted blades, but these blades would sample the horizontally sheared spectra that correspond to the vertically sheared version of the single lines (b) and (e). In other words, in both cases the sampling blades are the same, and the difference is only in the *envelope* of the sampled spectra, i.e. in the amplitude of the resulting line impulses.

If we were using here the (1,0,-1,0)-moiré between two dot screens we could have distributed this transformation between the two layer transformations $\mathbf{g}_1(x,y)$ and $\mathbf{g}_2(x,y)$ at will, provided that their difference gives $\mathbf{g}(x,y)$ (see Eq. (C.16)). However, because we are using here the hybrid (1,-1)-moiré we are slightly more limited, since the second component of $\mathbf{g}_2(x,y)$ has no influence and is set to zero (see Eq. (C.21)). This means that the *second* component of $\mathbf{g}(x,y)$ must be entirely taken care of by $\mathbf{g}_1(x,y)$, although the *first* component of $\mathbf{g}(x,y)$ can be distributed between $\mathbf{g}_1(x,y)$ and $\mathbf{g}_2(x,y)$ at will. In other words, using the hybrid (1,-1)-moiré the only way to expand our resulting moiré text *vertically* is by applying this vertical expansion to the first layer (the modulated grating). For example, we may choose the layer transformations:

$$\mathbf{g}_1\begin{pmatrix} x \\ y \end{pmatrix} = \begin{pmatrix} x - \varepsilon\cos(2\pi f y) \\ y/2 \end{pmatrix}, \qquad \mathbf{g}_2\begin{pmatrix} x \\ y \end{pmatrix} = \begin{pmatrix} (1 - 1/T_x)x - \varepsilon\cos(2\pi f y) \\ 0 \end{pmatrix}$$

But if we insist on using the same modulated grating as before (namely, Eq. (C.27); see Fig. C.16(c)), then no transformation $\mathbf{g}_2(x,y)$ to be applied to the slit grating will be able to provide a vertical magnification of the resulting moiré bands. ■

**Remark C.4:** Due to the shearing effect that is inherent to the hybrid (1,-1)-moiré the resulting text in the moiré bands may sometimes appear too slanted and therefore hardly recognizable (see, for example, Fig. C.8). In such cases it may be advantageous, in order to straighten up the text which appears within the resulting moiré bands, to design the original, modulated grating with already back-slanted (or over-slanted) text. This initial slanting of the text can be seen mathematically as a vertical shearing transformation that has been applied beforehand to the first, modulated grating (see Fig. C.17(a)):

$$\mathbf{v}\begin{pmatrix} x \\ y \end{pmatrix} = \begin{pmatrix} x \\ y + ax \end{pmatrix} \tag{C.28}$$

To see the effect of this layer transformation on the resulting moiré bands compare Figs. C.17(a),(b) with Figs. C.7(a),(b); note that in both cases the same slit grating is being used. As we can see in Fig. C.17(b), this straightening method only affects the text slanting within the moiré bands, but the moiré bands themselves remain unchanged, as in Fig. C.7(b). This shearing effect can be easily explained by applying the shear theorem to Figs. C.9(a),(d) and observing its effects on Figs. C.9(c),(f) and C.10.

If, in addition to the shearing $\mathbf{v}(x,y)$, we wish to distort our modulated grating by a geometric transformation:

$$\mathbf{f}\begin{pmatrix} x \\ y \end{pmatrix} = \begin{pmatrix} f_1(x,y) \\ f_2(x,y) \end{pmatrix} \tag{C.29}$$

then we have to apply this transformation to the already sheared (back-slanted or over-slanted) layer, meaning that the global transformation undergone by the original, modulated layer is given by $\mathbf{g}_1(x,y) = \mathbf{f}(\mathbf{v}(x,y))$, namely:

$$\mathbf{g}_1\begin{pmatrix} x \\ y \end{pmatrix} = \mathbf{f}\begin{pmatrix} x \\ y + ax \end{pmatrix} = \begin{pmatrix} f_1(x,\ y + ax) \\ f_2(x,\ y + ax) \end{pmatrix}$$

The resulting transformation of the moiré bands can be obtained, as usual, using Eq. (C.21).

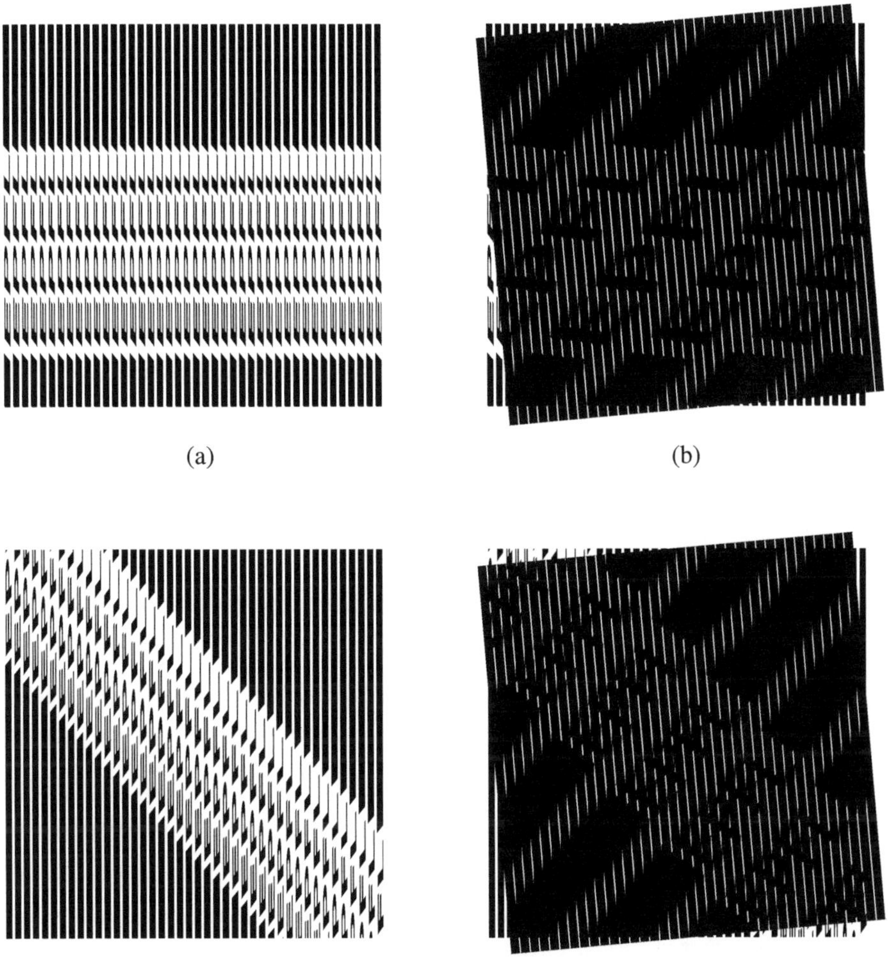

(a) (b) (c) (d)

**Figure C.19:** The effect of combined shearing and staggering in the first, modulated grating on the resulting moiré bands. (a),(b) Vertical shearing like in Fig. C.17(a) which is compensated by staggering in the opposite direction. (c),(d) Vertical shearing in the opposite direction followed by staggering; compare with Figs. C.17(c),(d). The slit grating used in (b) and (d) is identical to the one used in Fig. C.7(b) and in Figs. C.17(b),(d).

Another straightening effect can be also obtained by gradually shifting (advancing or retracting) the text positioning along consecutive lines of the first grating, without slanting the letters themselves (see Fig. C.17(c)). In this case the period of the first, modulated grating is no longer a scalar $T_1$, as in Fig. C.7(a), but rather a vector $\mathbf{T}_1 = (T_x,T_y)$, where

$T_x = T_1$ and $T_y$ is the new vertical shift increment between two consecutive periods of the first, modulated grating (Fig. C.17(c)). To see how this staggering effect affects the resulting moiré bands, compare Figs. C.17(c),(d) with Figs. C.7(a),(b); note that in both cases the same slit grating is being used. As we can see in Fig. C.17(d), this staggering in the first, modulated grating results in a shearing of the moiré text of Fig. C.7(b), which is also accompanied by staggering. But here, too, the non-modulated parts of the moiré bands remain unchanged, as they were in Fig. C.7(b). This effect can be explained once we understand the spectrum of the staggered grating, which is explained in detail in Fig. C.18. As we can see, the spectrum of a staggered grating is similar to the spectrum of a vertically sheared grating that is obtained by the application of the shear theorem, and both consist of slanted line impulses that sample the spectrum of a single modulated line. The difference between these spectra is only in their envelopes (i.e. in the amplitudes of the continuous spectra that are being sampled by the line impulses): In the case of a sheared grating the spectrum of the single modulated line that is being sampled is sheared, too, while in the case of a staggered grating the spectrum of the single modulated line is *not* sheared. Now, if the staggering effect and its spectral counterpart are applied to Figs. C.9(a),(d), they also influence accordingly Figs. C.9(c),(f) and the extracted moiré in Fig. C.10. Note, in particular, that the resulting moiré periods are not only staggered, but also sheared. The *shearing* effect in the resulting moiré is due to the slanting of the line impulses in the spectrum of the staggered grating (see Figs. C.18(d) and (f)), which also affects the slanting of the line impulses in the spectrum convolution (Fig. C.9(f)) and in the spectrum of the resulting moiré (Fig. C.10(b)). The *staggering* effect in the resulting moiré is due to the difference, in the spectrum of the resulting moiré, between the shearing that applies to the sampling blades and the shearing that applies to the underlying continuous spectrum of a single moiré period. This can be best understood by considering a version of Fig. C.18 in which the single period element, (b) or (e), is in itself vertically sheared, independently of the staggering effect.

Obviously, the two straightening effects mentioned above (shearing and staggering) can be also used together in various different combinations, as shown, for example, in Figs. 19(a),(b) and in Figs. 19(c),(d).

A detailed formulation of the effects of shearing, staggering and geometric transformations on the resulting moiré periods (using a different approach that is based on geometric considerations and indicial equations) can be found in [Hersch04] and in [Chosson06]. ■

## C.15 Moiré effects between general 2-fold periodic layers

Moiré effects that are generated between 2-fold periodic layers such as line grids or dot screens have a particular importance in the moiré theory and in many applications. Such moiré effects have been discussed in the second part of Chapter 4 (in Secs. 4.3–4.5), following the discussion on the moiré effect between 1-fold periodic line gratings in Sec.

4.2. But although the results obtained in Chapter 4 are fully general, we have only illustrated them there for the particular case in which each of the given 2-fold periodic layers (line grid or dot screen) is *regular*, i.e. periodic in two orthogonal directions with an identical period length to both directions. This simple 2-fold periodic case is, indeed, a rather straightforward generalization of the 1-fold periodic case, because each regular grid (or screen) can be considered as a composition of two orthogonal gratings (or virtual gratings), who generate in the superposition the two orthogonal directions of the resulting 2-fold periodic moiré (see Secs. 2.11, 2.12 and Fig. 2.10).

However, while the simple case consisting of regular layers is certainly the most basic 2-fold periodic case, both for didactic reasons and for many practical purposes, it is not yet sufficiently general. For example, it does not include the moiré effects that are generated between oblique screens or between hexagonal screens (see Fig. 2.13(b)), cases which may turn to be quite advantageous in certain applications.[25] Although such cases are covered by our results in Chapter 4, their particular behaviour is not explicitly obvious from these general results.

In the present section we provide an alternative, yet completely equivalent approach that more explicitly illustrates all of the 2-fold periodic cases, either regular or not. This approach is based on results that we obtained in Chapter 10 for the superposition of curved screens (see Sec. 10.9.2); but instead of considering general non-linear layer transformations $\mathbf{g}_1(\mathbf{x})$ and $\mathbf{g}_2(\mathbf{x})$, that yield curved layers, we will limit ourselves here to the case where the layer transformations $\mathbf{g}_1(\mathbf{x})$ and $\mathbf{g}_2(\mathbf{x})$ are *linear* (scalings, rotations, shearings, combinations thereof, or any other non-degenerate linear transformations[26]).

### C.15.1 Examples of general 2-fold periodic layers

As we have seen in Chapter 10, any curved dot screen or line grid $r(x,y)$ can be obtained by applying a certain transformation $(x',y') = \mathbf{g}(x,y)$ to $p(x',y')$, the original uncurved regular counterpart of $r(x,y)$ having a unit period to both directions:

$$r(x,y) = p(\mathbf{g}(x,y)) = p(g_1(x,y),g_2(x,y)) \qquad \text{(C.30)}$$

The general 2-fold periodic layer (dot screen or line grid) can be seen as a particular case of Eq. (C.30) in which $\mathbf{g}(x,y)$ is a *linear* transformation:

$$(x',y') = \mathbf{g}(x,y) = \begin{pmatrix} g_1(x,y) \\ g_2(x,y) \end{pmatrix} = \begin{pmatrix} u_1x + v_1y \\ u_2x + v_2y \end{pmatrix} = \begin{pmatrix} u_1 & v_1 \\ u_2 & v_2 \end{pmatrix}\begin{pmatrix} x \\ y \end{pmatrix} = \begin{pmatrix} \mathbf{f}_1 \\ \mathbf{f}_2 \end{pmatrix}\begin{pmatrix} x \\ y \end{pmatrix} \qquad \text{(C.31)}$$

namely:

$$r(x,y) = p(\mathbf{g}(x,y)) = p(u_1x + v_1y,\ u_2x + v_2y) = p(\mathbf{f}_1\cdot\mathbf{x},\ \mathbf{f}_2\cdot\mathbf{x}) \qquad \text{(C.32)}$$

---

[25] For instance, hexagonal microlens arrays have the advantage of better filling the plane than their orthogonal counterparts, and consequently they may let more light pass through. Indeed, a hexagonal pattern is known to be the most effective way to pack the largest number of similar objects (circles, etc.) in a minimum area [Weisstein99 pp. 254–255].

[26] Degenerate linear transformations, such as the null transformation $\mathbf{g}(\mathbf{x}) = \mathbf{0}$ or transformations that project the entire 2D plane onto a 1D line, will obviously not interest us here.

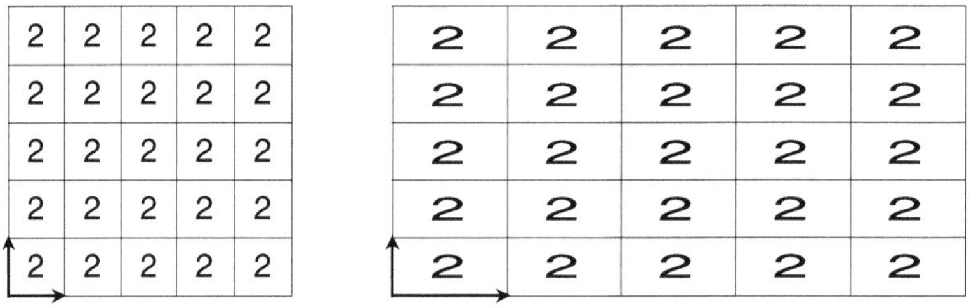

(a) (b)

**Figure C.20:** (a) A regular dot screen $p(x',y')$ having a unit period to both directions. (b) Its horizontally stretched version $r(x,y) = p(x/2,y)$ is a non-regular 2-fold periodic dot screen. See Example C.4.

where $\mathbf{x} = (x,y)$. Note that $\mathbf{f}_1 = (u_1,v_1)$ and $\mathbf{f}_2 = (u_2,v_2)$ are the frequency vectors of the 2-fold periodic layer $r(x,y)$; in the general case these two vectors are not necessarily orthogonal, and their lengths may be different. As we have seen in Secs. A.4 and A.5 of Appendix A, the frequency vectors $\mathbf{f}_1$ and $\mathbf{f}_2$ are related to the period vectors $\mathbf{P}_1 = (x_1,y_1)$ and $\mathbf{P}_2 = (x_2,y_2)$ of the same layer by Eq. (A.33), meaning that these two vector pairs are reciprocal to each other. According to Eq. (A.36) this also means that the two matrices

$$\mathrm{F} = \begin{pmatrix} \mathbf{f}_1 \\ \mathbf{f}_2 \end{pmatrix} = \begin{pmatrix} u_1 & v_1 \\ u_2 & v_2 \end{pmatrix} \quad \text{and} \quad \mathrm{P} = \begin{pmatrix} \mathbf{P}_1 \\ \mathbf{P}_2 \end{pmatrix} = \begin{pmatrix} x_1 & y_1 \\ x_2 & y_2 \end{pmatrix}$$

are the transpose inverse of each other: $\mathrm{F} = \mathrm{P}^{-\mathrm{T}}$.

**Remark C.5:** As shown above, the linear transformation $\mathbf{g}(x,y)$ is always expressed in terms of the *frequencies* of the 2-fold periodic layer $r(x,y)$ in the spectral domain, and not in terms of its *periodicities* in the image domain.[27] This fact will accompany us throughout our discussion below. ■

Let us now see a few concrete examples of 2-fold periodic layers that are not regular:

**Example C.4:** Consider the horizontally stretched dot screen $r(x,y)$ shown in Fig. C.20(b). The period vectors of this screen are $\mathbf{P}_1 = (2,0)$ and $\mathbf{P}_2 = (0,1)$, and its frequency vectors are therefore, according to Eq. (A.33), $\mathbf{f}_1 = (\frac{1}{2},0)$ and $\mathbf{f}_2 = (0,1)$. This horizontally stretched dot screen is obtained from its normalized, regular counterpart $p(x',y')$ (see Fig. C.20(a)) by applying to $p(x',y')$ the linear domain transformation:

$$(x',y') = \mathbf{g}(x,y) = \begin{pmatrix} \mathbf{f}_1 \\ \mathbf{f}_2 \end{pmatrix} \begin{pmatrix} x \\ y \end{pmatrix} = \begin{pmatrix} \frac{1}{2} & 0 \\ 0 & 1 \end{pmatrix} \begin{pmatrix} x \\ y \end{pmatrix} = \begin{pmatrix} \frac{1}{2}x \\ y \end{pmatrix}$$

[27] This follows from the fact that the transformation $(x',y') = \mathbf{g}(x,y)$ is applied to $p(x',y')$ as a *domain transformation*, and thus its effect on $p(x',y')$ is indeed that of the inverse transformation. For example, applying the transformation $(x',y') = (x/2,y/2)$ gives $p(x/2,y/2)$, a two-fold *magnification* of $p(x',y')$, and applying the transformation $(x',y') = (x\cos\theta + y\sin\theta, -x\sin\theta + y\cos\theta)$ gives a *counterclockwise* rotation of $p(x',y')$ by the angle $\theta$. This is explained in more detail in Sec. D.6 of *Vol. II*.

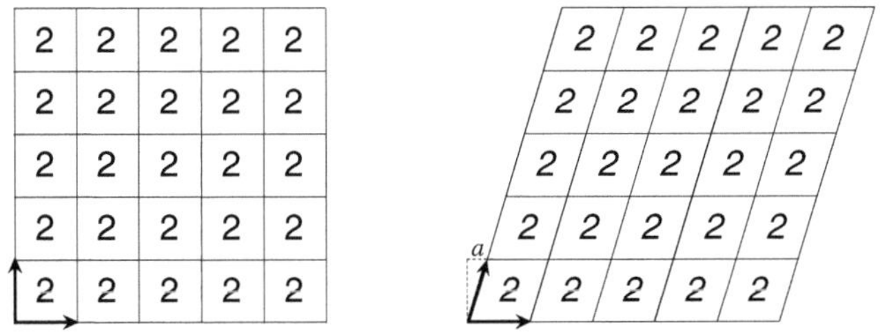

(a) (b)

**Figure C.21:** (a) A regular dot screen $p(x',y')$ having a unit period to both directions. (b) Its horizontally slanted version $r(x,y) = p(x-ay,y)$ is a non-regular 2-fold periodic dot screen. See Example C.5.

And indeed, it is easy to see that the resulting transformed screen:

$$r(x,y) = p(\mathbf{g}(x,y)) = p(x/2,y)$$

is a horizontally stretched version of the original screen $p(x',y')$. Note that, in accordance with Remark C.5, $\mathbf{g}(x,y)$ expresses the *frequencies* of $r(x,y)$ in the spectral domain, and not its *periodicities* in the image domain. ■

**Example C.5:** An *oblique* (also called *slanted* or *skew-periodic*) dot screen is obtained from a regular dot screen by a shearing transformation. Consider, for example, the horizontally sheared dot screen $r(x,y)$ shown in Fig. C.21(b). The period vectors of this screen are $\mathbf{P}_1 = (1,0)$ and $\mathbf{P}_2 = (a,1)$, and its frequency vectors are therefore, according to Eq. (A.33), $\mathbf{f}_1 = (1,-a)$ and $\mathbf{f}_2 = (0,1)$. This horizontally slanted dot screen is obtained from its normalized, regular counterpart $p(x',y')$ (see Fig. C.21(a)) by applying to $p(x',y')$ the linear domain transformation:

$$(x',y') = \mathbf{g}(x,y) = \begin{pmatrix} \mathbf{f}_1 \\ \mathbf{f}_2 \end{pmatrix}\begin{pmatrix} x \\ y \end{pmatrix} = \begin{pmatrix} 1 & -a \\ 0 & 1 \end{pmatrix}\begin{pmatrix} x \\ y \end{pmatrix} = \begin{pmatrix} x-ay \\ y \end{pmatrix}$$

And indeed, it is easy to see that the resulting transformed screen:

$$r(x,y) = p(\mathbf{g}(x,y)) = p(x-ay,y)$$

is a horizontally slanted version of the original screen $p(x',y')$. ■

**Example C.6:** A *hexagonal* dot screen. Consider the hexagonal screen shown in Fig. C.22(b). The period vectors of this screen are $\mathbf{P}_1 = (1,0)$ and $\mathbf{P}_2 = (\frac{1}{2}, \sin 60°) = (\frac{1}{2}, \frac{\sqrt{3}}{2})$, and its frequency vectors are therefore, according to Eq. (A.33), $\mathbf{f}_1 = (1,-\frac{1}{\sqrt{3}})$ and $\mathbf{f}_2 = (0,\frac{2}{\sqrt{3}})$. This hexagonal dot screen is obtained from its normalized, regular counterpart $p(x',y')$ (see Fig. C.22(a)) by applying to $p(x',y')$ the linear domain transformation:

$$(x',y') = \mathbf{g}(x,y) = \begin{pmatrix} \mathbf{f}_1 \\ \mathbf{f}_2 \end{pmatrix}\begin{pmatrix} x \\ y \end{pmatrix} = \begin{pmatrix} 1 & -\frac{1}{\sqrt{3}} \\ 0 & \frac{2}{\sqrt{3}} \end{pmatrix}\begin{pmatrix} x \\ y \end{pmatrix} = \begin{pmatrix} x-\frac{1}{\sqrt{3}}y \\ \frac{2}{\sqrt{3}}y \end{pmatrix}$$

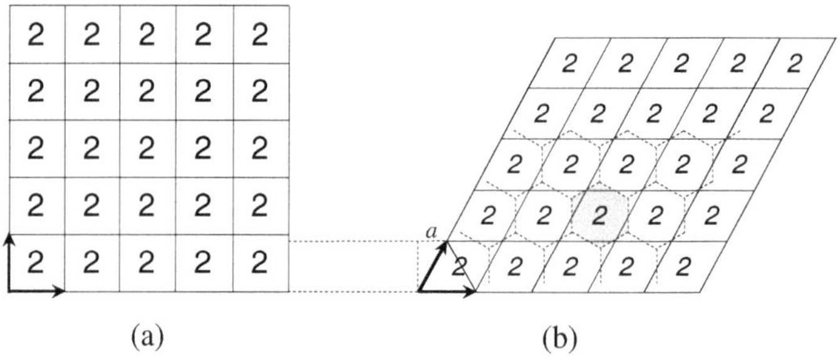

**Figure C.22:** (a) A regular dot screen $p(x',y')$ having a unit period to both directions. (b) Its hexagonal version $r(x,y) = p(x-\frac{1}{\sqrt{3}}y, \frac{2}{\sqrt{3}}y)$ is a non-regular 2-fold periodic dot screen. Note that in this case the transformation $\mathbf{g}(x,y)$ only modifies the angle between the two period vectors, but not their lengths; consequently, unlike in Fig. C.21, the cell height in (b) is smaller than in (a). See Example C.6.

We will henceforth call this transformation the *hexagonality transformation* and its matrix the *hexagonality matrix.* And indeed, it is easy to see that the resulting transformed screen:

$$r(x,y) = p(\mathbf{g}(x,y)) = p(x-\tfrac{1}{\sqrt{3}}y, \tfrac{2}{\sqrt{3}}y)$$

is the hexagonal version of the original screen $p(x',y')$.

Note that the hexagonal screen can be also seen as a slanted screen with $a = \frac{1}{2}$ and a further vertical scaling of $\sin 60° = \frac{\sqrt{3}}{2}$. Its two vector periods $\mathbf{P}_1$ and $\mathbf{P}_2$ have the same length, and they form an angle of 60°. ■

**Remark C.6:** A more general form of the hexagonal screen is obtained by applying to the above hexagonal screen a similarity transformation, i.e. a linear transformation that consists of *rotation* and *uniform scaling* (scaling with identical scaling factors to both directions). This more general form of the hexagonal screen is obtained by applying to $p(x',y')$ the linear domain transformation:

$$(x',y') = \mathbf{g}(x,y) = \begin{pmatrix} s & 0 \\ 0 & s \end{pmatrix} \begin{pmatrix} \cos\theta & \sin\theta \\ -\sin\theta & \cos\theta \end{pmatrix} \begin{pmatrix} 1 & -\frac{1}{\sqrt{3}} \\ 0 & \frac{2}{\sqrt{3}} \end{pmatrix} \begin{pmatrix} x \\ y \end{pmatrix} \tag{C.33}$$

where the two first matrices represent a similarity transformation: a uniform scaling (magnification by the factor $1/s$) and a counterclockwise rotation by angle $\theta$; see the footnote in Remark C.5 above. ■

**Remark C.7:** Note that the contents of the period cell in a hexagonal screen is not necessarily slanted as in Fig. C.22(b). If the desired hexagonal cell contains, for example, an upright "2", it simply means that in the original regular screen, before the application of the transformation **g**, we had a back-slanted version of "2", so that after the application of the transformation we obtain, indeed, an upright "2" in each of the period cells. ■

### C.15.2 Adaptation of results from Chapter 10 to our particular case

Let us now return to Sec. 10.9.2 and reformulate some of its main results for the particular case in which the layer transformations $\mathbf{g}_1(x,y)$ and $\mathbf{g}_2(x,y)$ are linear.

We start by reformulating Proposition 10.5 for the linear case:

**Proposition C.1:** The (1,0,-1,0)-moiré $m_{1,0,-1,0}(\mathbf{x})$ in the superposition of the two 2-fold periodic layers $r_1(\mathbf{x}) = p_1(\mathbf{g}_1(\mathbf{x}))$ and $r_2(\mathbf{x}) = p_2(\mathbf{g}_2(\mathbf{x}))$, where $\mathbf{g}_1(\mathbf{x})$ and $\mathbf{g}_2(\mathbf{x})$ are linear transformations, is given by $m_{1,0,-1,0}(\mathbf{x}) = p_{1,0,-1,0}(\mathbf{g}_{1,0,-1,0}(\mathbf{x}))$, where:

(1) $p_{1,0,-1,0}(\mathbf{x}')$, the normalized periodic-profile of the (1,0,-1,0)-moiré, is the *T*-convolution of the normalized periodic-profiles of the original layers:

$$p_{1,0,-1,0}(\mathbf{x}') = p_1(\mathbf{x}') ** p_2(-\mathbf{x}')$$

(2) $\mathbf{g}_{1,0,-1,0}(\mathbf{x})$, the linear transformation of the (1,0,-1,0)-moiré, is given by:

$$\mathbf{g}_{1,0,-1,0}(\mathbf{x}) = \mathbf{g}_1(\mathbf{x}) - \mathbf{g}_2(\mathbf{x}). \quad \blacksquare$$

Using a less formal language we can now state the counterpart of Proposition 4.5 for the superposition of two 2-fold periodic layers as follows:

**Proposition C.2:** Let $r_1(x,y)$ and $r_2(x,y)$ be two 2-fold periodic layers, which are obtained from two normalized regular 2-fold periodic layers $p_1(x',y')$ and $p_2(x',y')$ by the linear coordinate transformations $\mathbf{g}_1(x,y)$ and $\mathbf{g}_2(x,y)$, namely:

$$\begin{pmatrix} x' \\ y' \end{pmatrix} = \begin{pmatrix} g_1(x,y) \\ g_2(x,y) \end{pmatrix} \quad \text{and} \quad \begin{pmatrix} x' \\ y' \end{pmatrix} = \begin{pmatrix} g_3(x,y) \\ g_4(x,y) \end{pmatrix}$$

respectively. The (1,0,-1,0)-moiré $m_{1,0,-1,0}(x,y)$ generated in the superposition of these 2-fold periodic layers can be seen from the image-domain point of view as the result of a 3-stage process:

(1) Normalization of the original 2-fold periodic layers by, in each of them, replacing $(g_i(x,y), g_{i+1}(x,y))$ with $(x',y')$ (i.e., by undoing in each of them the coordinate transformation), in order to straighten them into the normalized regular 2-fold periodic layers $p_1(x',y')$ and $p_2(x',y')$ having unit periods $T_{x'} = T_{y'} = 1$ (see Sec. C.16 below).

(2) *T*-convolution of these normalized layers. This gives the normalized periodic-profile of the (1,0,-1,0)-moiré, with the same unit periods $T_{x'} = T_{y'} = 1$.

(3) Bending the normalized periodic-profile of the moiré into the actual geometric layout of the moiré, by replacing $(x',y')$ with $\mathbf{g}_1(x,y) - \mathbf{g}_2(x,y)$, i.e., by applying the linear coordinate transformation $\begin{pmatrix} x' \\ y' \end{pmatrix} = \begin{pmatrix} g_1(x,y) - g_3(x,y) \\ g_2(x,y) - g_4(x,y) \end{pmatrix}$. $\blacksquare$

It follows, therefore, that in order to synthesize between two 2-fold periodic layers a (1,0,-1,0)-moiré whose geometric layout is given by the two independent linear functions $g^{(1)}(x,y)$ and $g^{(2)}(x,y)$, all that we have to do is to choose two original 2-fold periodic layers whose linear transformations $\mathbf{g}_1(x,y) = (g_1(x,y), g_2(x,y))$ and $\mathbf{g}_2(x,y) = (g_3(x,y), g_4(x,y))$ satisfy the condition:

$$g_1(x,y) - g_3(x,y) = g^{(1)}(x,y)$$

$$g_2(x,y) - g_4(x,y) = g^{(2)}(x,y)$$

The periodic-profile of the synthesized moiré will be determined by the periodic-profiles of the superposed layers, in accordance with the first part of Proposition C.1.

Similar results for the general $(k_1,k_2,k_3,k_4)$-moiré between any two 2-fold periodic layers can be obtained from the fundamental moiré theorem and Proposition 10.7 by restricting them to the particular case in which the layer transformations $\mathbf{g}_1(x,y)$ and $\mathbf{g}_2(x,y)$ are linear.

### C.15.3 The (1,0,-1,0)-moiré between two regular screens or grids

In Chapter 4 we have seen that the (1,0,-1,0)-moiré between any two regular screens or grids is itself regular (i.e. orthogonal with identical periods to both directions), and we determined the angle and the period of the resulting (1,0,-1,0)-moiré using the formulas that we have obtained previously for the (1,-1)-moiré between two line gratings (see Sec. 4.4.2).

Let us now show, based on Proposition C.1, how this result can be obtained directly, using a simple 2D matrix formulation, without having to resort to the (1,-1)-moiré between line gratings. The most general case with non-regular layers will be treated in Sec. C.15.5 below.

Because in our present case both of the layers $r_1(x,y)$ and $r_2(x,y)$ are regular, it follows that the linear transformations $\mathbf{g}_1(x,y)$ and $\mathbf{g}_2(x,y)$ that generate these layers from their normalized counterparts $p_1(x',y')$ and $p_2(x',y')$ having a unit period to both directions are similarity transformations, i.e. they only consist of rotations and uniform scalings:

$$\mathbf{g}_1(x,y) = \begin{pmatrix} f_1 & 0 \\ 0 & f_1 \end{pmatrix} \begin{pmatrix} \cos\theta_1 & \sin\theta_1 \\ -\sin\theta_1 & \cos\theta_1 \end{pmatrix} \begin{pmatrix} x \\ y \end{pmatrix} \tag{C.34}$$

$$\mathbf{g}_2(x,y) = \begin{pmatrix} f_2 & 0 \\ 0 & f_2 \end{pmatrix} \begin{pmatrix} \cos\theta_2 & \sin\theta_2 \\ -\sin\theta_2 & \cos\theta_2 \end{pmatrix} \begin{pmatrix} x \\ y \end{pmatrix} \tag{C.35}$$

where $\theta_i$ is the rotation angle of the regular layer $r_i(x,y)$ and $f_i$ is its frequency (i.e. the scaling ratio between the frequency of $r_i(x,y)$ and the unit-frequency of its counterpart $p_i(x',y')$). Note that the first matrix in $\mathbf{g}_i(x,y)$ corresponds to a magnification of $p_i(x',y')$ by the factor $T_i = 1/f_i$ and the second matrix corresponds to a rotation by $\theta_i$ counterclockwise (see Remark C.5 above and the footnote therein).

Now, according to Proposition C.1, the linear transformation of the (1,0.-1,0)-moiré between two regular layers $r_1(x,y)$ and $r_2(x,y)$ is given by:

$$\mathbf{g}_{1,0,-1,0}(\mathbf{x}) = \mathbf{g}_1(\mathbf{x}) - \mathbf{g}_2(\mathbf{x})$$

$$= \begin{pmatrix} f_1 & 0 \\ 0 & f_1 \end{pmatrix} \begin{pmatrix} \cos\theta_1 & \sin\theta_1 \\ -\sin\theta_1 & \cos\theta_1 \end{pmatrix} \begin{pmatrix} x \\ y \end{pmatrix} - \begin{pmatrix} f_2 & 0 \\ 0 & f_2 \end{pmatrix} \begin{pmatrix} \cos\theta_2 & \sin\theta_2 \\ -\sin\theta_2 & \cos\theta_2 \end{pmatrix} \begin{pmatrix} x \\ y \end{pmatrix}$$

$$= \left[\begin{pmatrix} f_1 & 0 \\ 0 & f_1 \end{pmatrix}\begin{pmatrix} \cos\theta_1 & \sin\theta_1 \\ -\sin\theta_1 & \cos\theta_1 \end{pmatrix} - \begin{pmatrix} f_2 & 0 \\ 0 & f_2 \end{pmatrix}\begin{pmatrix} \cos\theta_2 & \sin\theta_2 \\ -\sin\theta_2 & \cos\theta_2 \end{pmatrix}\right]\begin{pmatrix} x \\ y \end{pmatrix}$$

$$= \begin{pmatrix} f_1\cos\theta_1 - f_2\cos\theta_2 & f_1\sin\theta_1 - f_2\sin\theta_2 \\ -(f_1\sin\theta_1 - f_2\sin\theta_2) & f_1\cos\theta_1 - f_2\cos\theta_2 \end{pmatrix}\begin{pmatrix} x \\ y \end{pmatrix} \tag{C.36}$$

Let us recall at this point the following result from linear algebra (see, for example, [Lay03 p. 339]):

**Proposition C.3:** Any matrix of the form $\left(\begin{smallmatrix} a & -b \\ b & a \end{smallmatrix}\right)$ represents a similarity transformation, i.e. a linear transformation consisting of rotation by angle $\theta$ and uniform scaling by a scaling factor $f$. Furthermore, the angle and the scaling factor of this similarity transformation are given by:

$$\theta = \arctan(b/a)$$

$$f = \sqrt{a^2 + b^2} \quad \blacksquare$$

Returning now to Eq. (C.36), we see that the matrix it contains (i.e. the matrix which represents the moiré transformation $\mathbf{g}_{1,0,-1,0}(\mathbf{x})$) has indeed the form $\left(\begin{smallmatrix} a & -b \\ b & a \end{smallmatrix}\right)$, meaning that it represents a similarity transformation. This means in turn that the (1,0,-1,0)-moiré $m_{1,0,-1,0}(\mathbf{x}) = p_{1,0,-1,0}(\mathbf{g}_{1,0,-1,0}(\mathbf{x}))$ is itself regular, since it is obtained by the application of a similarity transformation (rotation and uniform scaling) to its normalized counterpart $p_{1,0,-1,0}(\mathbf{x}')$ which is obviously regular.

Furthermore, the angle $\varphi_M$ of this (1,0,-1,0)-moiré and its scaling factor $f_M$ are given according to Proposition C.3 by:[28]

$$\varphi_M = \arctan(b/a) = \arctan\left(\frac{f_1\sin\theta_1 - f_2\sin\theta_2}{f_1\cos\theta_1 - f_2\cos\theta_2}\right) \tag{C.37}$$

$$f_M{}^2 = a^2 + b^2 = (f_1\cos\theta_1 - f_2\cos\theta_2)^2 + (f_1\sin\theta_1 - f_2\sin\theta_2)^2$$

$$= f_1{}^2(\cos^2\theta_1 + \sin^2\theta_1) - 2f_1f_2(\cos\theta_1\cos\theta_2 + \sin\theta_1\sin\theta_2) + f_2{}^2(\cos^2\theta_2 + \sin^2\theta_2)$$

$$= f_1{}^2 - 2f_1f_2\cos(\theta_2 - \theta_1) + f_2{}^2 \tag{C.38}$$

As we can see, this is precisely the result that we have obtained in Sec. C.1 for the case of the (1,-1)-moiré between two line gratings. Note that in accordance with Remark C.5, all our results are formulated here in terms of the spectral domain, i.e. in terms of *frequencies* rather than in terms of *periods* (see Eqs. (C.34)–(C.38)). We can, however, express Eq. (C.38) in terms of periods by rewriting it as follows (denoting the angle difference between the two layers by $\alpha = \theta_2 - \theta_1$):

[28] Note that Proposition C.3 assumes that the transformation belonging to the matrix $\left(\begin{smallmatrix} a & -b \\ b & a \end{smallmatrix}\right)$ is applied as a direct transformation. When it is applied as a domain (and hence inverse) transformation, like in our case (see Remark C.5 and the footnote therein), the rotation is applied to the inverse direction. But since in our case the matrix has the form $\left(\begin{smallmatrix} a & b \\ -b & a \end{smallmatrix}\right)$, i.e. $b$ is negative, the sign inversion in the rotation angle is finally cancelled out.

$$\frac{1}{T_M^2} = \frac{1}{T_1^2} - 2\frac{1}{T_1}\frac{1}{T_2}\cos\alpha + \frac{1}{T_2^2}$$

$$= \frac{T_2^2 - 2T_1T_2\cos\alpha + T_1^2}{(T_1T_2)^2}$$

This finally gives, indeed, the period $T_M$ of the moiré (to both directions) as predicted by Eq. (2.9):

$$T_M = \frac{T_1T_2}{\sqrt{T_1^2 + T_2^2 - 2T_1T_2\cos\alpha}} \tag{C.39}$$

Similarly, we can also express the angle $\varphi_M$ of the moiré in terms of periods, by rewriting Eq. (C.37) as follows:

$$\varphi_M = \arctan\left(\frac{\frac{\sin\theta_1}{T_1} - \frac{\sin\theta_2}{T_2}}{\frac{\cos\theta_1}{T_1} - \frac{\cos\theta_2}{T_2}}\right) = \arctan\left(\frac{T_2\sin\theta_1 - T_1\sin\theta_2}{T_2\cos\theta_1 - T_1\cos\theta_2}\right) \tag{C.40}$$

As we can see, the period and the angle of the (1,0,-1,0)-moiré between two regular screens or grids are, indeed, identical to those of the (1,-1)-moiré between two line gratings (see Eq. (2.9) and Sec. C.1).

**Example C.7:** Suppose that we are given two regular dot screens $r_1(x,y)$ and $r_2(x,y)$, the first consisting of tiny "2"-shaped periods and the second consisting of tiny pinholes (see Fig. C.23(a)). Suppose that the period lengths and the orientations of our two regular screens are, respectively, $T_1 = 5$, $\theta_1 = 0°$ and $T_2 = 5.2$, $\theta_2 = 5°$. What are the period and the orientation of the (1,0,-1,0)-moiré in the superposition of these two screens?

According to Proposition C.1 (see also Sec. 4.4), the resulting (1,0,-1,0)-moiré consists of a periodic pattern of magnified "2"-shaped elements. The periodicity of this moiré (in both directions) is given according to Eq. (C.39) by:

$$T_M = \frac{5\cdot 5.2}{\sqrt{5^2 + 5.2^2 - 2\cdot 5\cdot 5.2\cos 5°}} = 53.28$$

meaning that its magnification factor is 53.28/5 = 10.66. The direction of this moiré is given according to Eq. (C.40) by:

$$\varphi_M = \arctan\left(\frac{5.2\sin 0° - 5\sin 5°}{5.2\cos 0° - 5\cos 5°}\right) = \arctan(-1.991) = -63.33°$$

meaning that the "2"-shaped moiré is rotated by 66.77° clockwise with respect to the $x$ axis. Note that the moiré angle $\varphi_M$ is independent of the length units in which the periods $T_1$ and $T_2$ are expressed, since if we express $T_1$ and $T_2$ in terms of another length unit we simply multiply their values by a constant factor which is then cancelled out in Eq. (C.40). The moiré period $T_M$, on its part, is expressed in terms of the same length units as the periods $T_1$ and $T_2$. ■

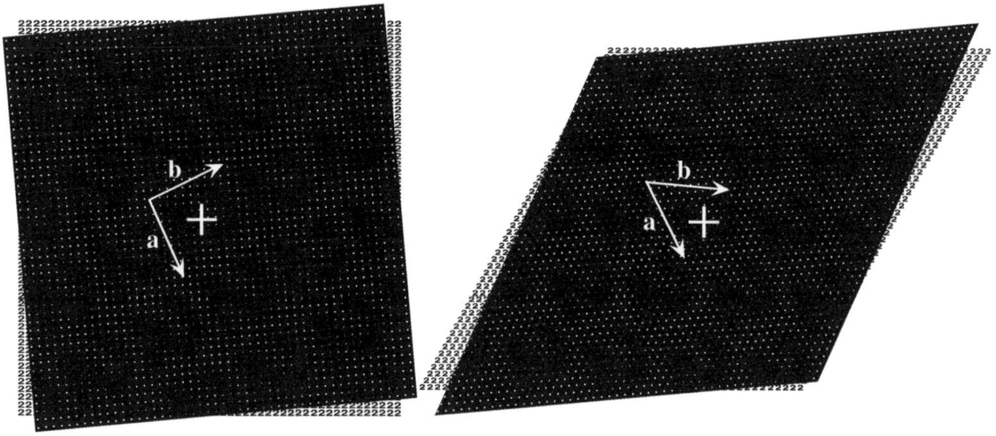

(a) (b)

**Figure C.23:** (a) The moiré effect between the two regular dot screens of Example C.7. (b) The moiré effect between the two hexagonal dot screens of Example C.8. Each of the two layers in (b) is obtained by applying the hexagonality transformation to the corresponding layer in (a); the resulting moiré in (b) is the hexagonal counterpart of the regular moiré in (a), which is obtained by applying the hexagonality transformation to the regular moiré. Note, however, that for the sake of clarity the symbol "2" within the oblique periods in (b) has been kept upright, as explained in Remark C.7. The two white arrows in (a) and (b) show the period-vectors **a** and **b** of the respective moiré effects.[29] Note that the angle $\varphi_M$ of the moiré, i.e. the baseline direction of the "2"-shaped moiré,[30] as well as the length $T_M$ of the two moiré period-vectors, are exactly the same in (a) and (b), and only the internal angle between the two period vectors is different (90° in (a) and 60° in (b)). Although hexagonal screens are usually clipped to have square or rectangular external borders, just like their regular counterparts, we have chosen for didactic reasons not to clip them in the present figure, in order to clearly show their oblique nature.

Obviously, these results apply only to the particular case in which both layers are regular grids or screens (note that we have assumed that our original normalized layers $p_1(x',y')$ and $p_2(x',y')$ are regular, and that the linear transformations $\mathbf{g}_1(x,y)$ and $\mathbf{g}_2(x,y)$ that are applied to them consist of rotations and uniform scalings alone, i.e. they are similarity transformations). In the general case $\mathbf{g}_1(x,y)$ and $\mathbf{g}_2(x,y)$ may be any linear transformations, that may include shearing, scalings with different factors to both directions, etc., so that the transformed layers (as well as the resulting moiré) are no longer regular (orthogonal with identical periods to both directions). This general case is treated in Sec. C.15.5 below.

---

[29] The moiré period-vectors **a** and **b** can be found by applying the moiré transformation $\mathbf{g}_{1,0,-1,0}(x,y)$ to the normalized layer period-vectors (1,0) and (0,1), respectively, using Eq. (C.36) in the regular case and Eq. (C.43) in the hexagonal case. Note that the hexagonality matrix in Eq. (C.43) does not modify the vector (1,0), meaning that the vector **a** is precisely the same in (a) and (b).

[30] The baseline direction of the moiré is indicated by the vector **a**.

And yet, interestingly, we can still use the above considerations as a useful basis for some other particular cases, for example, the case in which both layers are hexagonal screens, as shown below.

### C.15.4 The (1,0,-1,0)-moiré between two hexagonal screens or grids

In this subsection we discuss the particular case of hexagonal layers, which is probably the most widely used 2-fold periodic case after regular layers. We will show that a (1,0,-1,0)-moiré between two hexagonal layers is always hexagonal, and we will determine the periodicities and the orientation of this moiré.

Because in this case both of the layers $r_1(x,y)$ and $r_2(x,y)$ are hexagonal, it follows from Remark C.6 that the linear transformations $\mathbf{g}_1(x,y)$ and $\mathbf{g}_2(x,y)$ that generate these layers from their normalized counterparts $p_1(x',y')$ and $p_2(x',y')$ having a unit period in both directions are given by:

$$\mathbf{g}_1(x,y) = \begin{pmatrix} f_1 & 0 \\ 0 & f_1 \end{pmatrix} \begin{pmatrix} \cos\theta_1 & \sin\theta_1 \\ -\sin\theta_1 & \cos\theta_1 \end{pmatrix} \begin{pmatrix} 1 & -\frac{1}{\sqrt{3}} \\ 0 & \frac{2}{\sqrt{3}} \end{pmatrix} \begin{pmatrix} x \\ y \end{pmatrix} \tag{C.41}$$

$$\mathbf{g}_2(x,y) = \begin{pmatrix} f_2 & 0 \\ 0 & f_2 \end{pmatrix} \begin{pmatrix} \cos\theta_2 & \sin\theta_2 \\ -\sin\theta_2 & \cos\theta_2 \end{pmatrix} \begin{pmatrix} 1 & -\frac{1}{\sqrt{3}} \\ 0 & \frac{2}{\sqrt{3}} \end{pmatrix} \begin{pmatrix} x \\ y \end{pmatrix} \tag{C.42}$$

Note that Eqs. (C.41) and (C.42) only differ from their counterparts (C.34) and (C.35) of the regular case by the presence of a new, third matrix. This additional matrix is, indeed, the *hexagonality matrix* (see Example C.6), and its corresponding linear transformation is the *hexagonality transformation*.

Now, according to Proposition C.1, the linear transformation of the (1,0.-1,0)-moiré between the two hexagonal layers $r_1(x,y)$ and $r_2(x,y)$ is given by:

$$\mathbf{g}_{1,0,-1,0}(\mathbf{x}) = \mathbf{g}_1(\mathbf{x}) - \mathbf{g}_2(\mathbf{x})$$

$$= \begin{pmatrix} f_1 & 0 \\ 0 & f_1 \end{pmatrix} \begin{pmatrix} \cos\theta_1 & \sin\theta_1 \\ -\sin\theta_1 & \cos\theta_1 \end{pmatrix} \begin{pmatrix} 1 & -\frac{1}{\sqrt{3}} \\ 0 & \frac{2}{\sqrt{3}} \end{pmatrix} \begin{pmatrix} x \\ y \end{pmatrix}$$

$$- \begin{pmatrix} f_2 & 0 \\ 0 & f_2 \end{pmatrix} \begin{pmatrix} \cos\theta_2 & \sin\theta_2 \\ -\sin\theta_2 & \cos\theta_2 \end{pmatrix} \begin{pmatrix} 1 & -\frac{1}{\sqrt{3}} \\ 0 & \frac{2}{\sqrt{3}} \end{pmatrix} \begin{pmatrix} x \\ y \end{pmatrix}$$

$$= \left[\begin{pmatrix} f_1 & 0 \\ 0 & f_1 \end{pmatrix} \begin{pmatrix} \cos\theta_1 & \sin\theta_1 \\ -\sin\theta_1 & \cos\theta_1 \end{pmatrix} - \begin{pmatrix} f_2 & 0 \\ 0 & f_2 \end{pmatrix} \begin{pmatrix} \cos\theta_2 & \sin\theta_2 \\ -\sin\theta_2 & \cos\theta_2 \end{pmatrix}\right] \begin{pmatrix} 1 & -\frac{1}{\sqrt{3}} \\ 0 & \frac{2}{\sqrt{3}} \end{pmatrix} \begin{pmatrix} x \\ y \end{pmatrix}$$

$$= \begin{pmatrix} f_1\cos\theta_1 - f_2\cos\theta_2 & f_1\sin\theta_1 - f_2\sin\theta_2 \\ -(f_1\sin\theta_1 - f_2\sin\theta_2) & f_1\cos\theta_1 - f_2\cos\theta_2 \end{pmatrix} \begin{pmatrix} 1 & -\frac{1}{\sqrt{3}} \\ 0 & \frac{2}{\sqrt{3}} \end{pmatrix} \begin{pmatrix} x \\ y \end{pmatrix} \tag{C.43}$$

A comparison between Eqs. (C.43) and (C.36) shows that the only difference between the transformations $\mathbf{g}_{1,0,-1,0}(\mathbf{x})$ in the hexagonal case and in the regular case is the presence of the hexagonality matrix in the hexagonal case. And indeed, because the first matrix in

Eq. (C.43) is a similarity matrix, it follows from Remark C.6 that the (1,0,-1,0)-moiré $m_{1,0,-1,0}(\mathbf{x}) = p_{1,0,-1,0}(\mathbf{g}_{1,0,-1,0}(\mathbf{x}))$ in the hexagonal case is hexagonal in itself, since it is obtained by the application of the hexagonality transformation (plus a certain rotation and uniform scaling due to the similarity transformation) to its regular, normalized counterpart $p_{1,0,-1,0}(\mathbf{x}')$.

Furthermore, the additional hexagonality transformation in Eq. (C.43) only modifies the internal angle between the two period vectors of the 2-fold periodic moiré but not the length of these period vectors or the baseline direction of the moiré (see Figs. C.22, C.23). Therefore the periodicity and the rotation angle (i.e. the baseline direction) of the hexagonal moiré are still determined by the same similarity transformation as in the regular case, and hence they are given by the same formulas as in the regular case, i.e. by Eqs. (C.39) and (C.40). And if both layers have the same periods these formulas reduce again into Eqs. (2.10).

**Example C.8:** Suppose that we are given two hexagonal dot screens $r_1(x,y)$ and $r_2(x,y)$, the first consisting of tiny "2"-shaped periods and the second consisting of tiny pinholes (see Fig. C.23(b)). Suppose that the period lengths and the orientations of our two hexagonal screens are, respectively, $T_1 = 5$, $\theta_1 = 0°$ and $T_2 = 5.2$, $\theta_2 = 5°$, like in Example C.7 with the regular screens. What are the period and the orientation of the (1,0,-1,0)-moiré in the superposition of these two screens?

We note that the period lengths and the rotation angles of the two hexagonal screens in this case are identical to those of the two regular screens in Example C.7, the only difference between the two cases being in the internal angle that is formed between the two period vectors of each layer (90° in a regular screen, and 60° in a hexagonal screen). Therefore, as we have seen above, the period length $T_M$ and the rotation angle $\varphi_M$ of the moiré in the hexagonal case are identical to those of the regular moiré of Example C.7, the only difference between them being, once again, in the internal angle between the two period vectors of the moiré (see Fig. C.23).

Note, however, that because the two individual layers (as well as the moiré effect) in this hexagonal case have been obtained by applying the hexagonality transformation to their respective counterparts of Example C.7, it follows that the "2"-shaped elements in the hexagonal case (both in the original layer $r_1(x,y)$ and in the moiré effect) are slightly smaller than in the regular case (compare Figs. C.23(a) and C.23(b)), although the period vectors in both cases have the same length. The reason is that the height of the oblique period cell in the hexagonal case is only $T\sin 60° = 0.866T$ rather than $T$, the height of the square period cell in the corresponding regular case. In principle, the "2"-shaped elements in the hexagonal case should be slanted versions of their counterparts in the regular case; but according to Remark C.7 it is always possible to "rectify" them, if so desired (as we have done, indeed, in Fig. C.23(b)). ■

Similar considerations can be also devised for other layer types, e.g. the hexagonal screen variant obtained by *vertical* rather than horizontal shearing, etc.

### C.15.5 The (1,0,-1,0)-moiré between two general 2-fold periodic screens or grids

We now proceed to the case involving general 2-fold periodic layers.

**Proposition C.4:** Suppose we are given two 2-fold periodic layers $r_1(\mathbf{x}) = p_1(\mathbf{g}_1(\mathbf{x}))$ and $r_2(\mathbf{x}) = p_2(\mathbf{g}_2(\mathbf{x}))$ that are obtained from their original, unit-period normalized counterparts $p_1(\mathbf{x})$ and $p_2(\mathbf{x})$ by the linear transformations:

$$\mathbf{g}_1(\mathbf{x}) = \begin{pmatrix} u_1x + v_1y \\ u_2x + v_2y \end{pmatrix} = \mathrm{F}_1\boldsymbol{\cdot}\mathbf{x} \quad \text{with} \quad \mathrm{F}_1 = \begin{pmatrix} u_1 & v_1 \\ u_2 & v_2 \end{pmatrix} = \begin{pmatrix} \mathbf{f}_1 \\ \mathbf{f}_2 \end{pmatrix}, \quad \mathbf{x} = (x,y)$$

$$\mathbf{g}_2(\mathbf{x}) = \begin{pmatrix} u_3x + v_3y \\ u_4x + v_4y \end{pmatrix} = \mathrm{F}_2\boldsymbol{\cdot}\mathbf{x} \quad \text{with} \quad \mathrm{F}_2 = \begin{pmatrix} u_3 & v_3 \\ u_4 & v_4 \end{pmatrix} = \begin{pmatrix} \mathbf{f}_3 \\ \mathbf{f}_4 \end{pmatrix}, \quad \mathbf{x} = (x,y)$$

It therefore follows from Proposition C.1 that the transformation $\mathbf{g}_{1,0,-1,0}(\mathbf{x})$ of the (1,0,-1,0)-moiré between the two layers, i.e. the transformation which brings the unit-period normalized moiré profile $p_{1,0,-1,0}(\mathbf{x})$ into the final (1,0,-1,0)-moiré $p_{1,0,-1,0}(\mathbf{g}_{1,0,-1,0}(\mathbf{x}))$ between the two layers $r_1(\mathbf{x})$ and $r_2(\mathbf{x})$, is given by:

$$\begin{aligned}\mathbf{g}_{1,0,-1,0}(\mathbf{x}) &= \mathbf{g}_1(\mathbf{x}) - \mathbf{g}_2(\mathbf{x}) \\ &= \begin{pmatrix} (u_1-u_3)x + (v_1-v_3)y \\ (u_2-u_4)x + (v_2-v_4)y \end{pmatrix} = \mathrm{F}_M\boldsymbol{\cdot}\mathbf{x} \quad \text{with} \quad \mathrm{F}_M = \begin{pmatrix} u_1-u_3 & v_1-v_3 \\ u_2-u_4 & v_2-v_4 \end{pmatrix} = \begin{pmatrix} \mathbf{a} \\ \mathbf{b} \end{pmatrix}\end{aligned}$$

$$\text{where} \quad \mathrm{F}_M = \mathrm{F}_1 - \mathrm{F}_2, \quad \text{namely:} \quad \begin{pmatrix} \mathbf{a} \\ \mathbf{b} \end{pmatrix} = \begin{pmatrix} \mathbf{f}_1 - \mathbf{f}_3 \\ \mathbf{f}_2 - \mathbf{f}_4 \end{pmatrix}. \qquad \blacksquare$$

This gives us the connection between the frequency vectors of the two individual layers ($\mathbf{f}_1$, $\mathbf{f}_2$ and $\mathbf{f}_3$, $\mathbf{f}_4$) and the frequency vectors **a** and **b** of the resulting (1,0,-1,0)-moiré. This result is, indeed, identical to the result that we have already obtained in Chapter 4 (see Eq. (4.17)).

Note that the explicit results that we have obtained in the previous subsections for the cases of regular or hexagonal screens are simply particular cases of this general result, although their particularities are not explicitly obvious from the general result.

### C.15.6 Allowing for layer shifts

So far we have considered here the (1,0,-1,0)-moiré effects between 2-fold periodic layers $r_1(\mathbf{x})$ and $r_2(\mathbf{x})$ that were obtained from their original, unit-period normalized counterparts $p_1(\mathbf{x})$ and $p_2(\mathbf{x})$ by purely linear transformations such as rotations, scalings, shearings, or any combinations thereof. Note, however, that all linear transformations leave the layer's origin at the point (0,0). Therefore, if we want to take into consideration layer shifts as well, we must consider rather than purely linear transformations the slightly larger family of *affine* transformations. The general 1D affine transformation has the form:

$$x' = ax + x_0$$

and the general 2D affine transformation has the form:

$$x' = a_1x + b_1y + x_0$$

$$y' = a_2x + b_2y + y_0$$

or in vector notation, denoting $\mathbf{x} = (x,y)$, $\mathbf{x}' = (x',y')$, $\mathbf{x}_0 = (x_0,y_0)$ and $A = \begin{pmatrix} a_1 & b_1 \\ a_2 & b_2 \end{pmatrix}$:

$$\mathbf{x}' = A{\cdot}\mathbf{x} + \mathbf{x}_0$$

where $x_0$, $y_0$ and $\mathbf{x}_0 = (x_0,y_0)$ are arbitrary constants that represent a given displacement. Note that linear transformations are a particular case of affine transformations in which $x_0 = 0$ and $y_0 = 0$.

Now, in order to be able to consider layer shifts, as we already did in Chapter 7, we must use here a slightly wider variant of Proposition C.1 in which the transformations $\mathbf{g}_1(\mathbf{x})$, $\mathbf{g}_2(\mathbf{x})$ and hence $\mathbf{g}_{1,0,-1,0}(\mathbf{x}) = \mathbf{g}_1(\mathbf{x}) - \mathbf{g}_2(\mathbf{x})$ are *affine* rather than *linear*. We thus obtain the following generalization of Proposition C.4:

**Proposition C.5:** Suppose we are given two 2-fold periodic layers $r_1(\mathbf{x}) = p_1(\mathbf{g}_1(\mathbf{x}))$ and $r_2(\mathbf{x}) = p_2(\mathbf{g}_2(\mathbf{x}))$ that are obtained from their original, unit-period normalized counterparts $p_1(\mathbf{x})$ and $p_2(\mathbf{x})$ by the affine transformations $\mathbf{g}_1(\mathbf{x})$ and $\mathbf{g}_2(\mathbf{x})$:

$$\mathbf{g}_1(\mathbf{x}) = \begin{pmatrix} u_1x + v_1y + x_1 \\ u_2x + v_2y + y_1 \end{pmatrix} = F_1{\cdot}\mathbf{x} + \mathbf{x}_1 \quad \text{with} \quad F_1 = \begin{pmatrix} u_1 & v_1 \\ u_2 & v_2 \end{pmatrix} = \begin{pmatrix} \mathbf{f}_1 \\ \mathbf{f}_2 \end{pmatrix}, \quad \mathbf{x} = (x,y), \ \mathbf{x}_1 = (x_1,y_1)$$

$$\mathbf{g}_2(\mathbf{x}) = \begin{pmatrix} u_3x + v_3y + x_2 \\ u_4x + v_4y + y_2 \end{pmatrix} = F_2{\cdot}\mathbf{x} + \mathbf{x}_2 \quad \text{with} \quad F_2 = \begin{pmatrix} u_3 & v_3 \\ u_4 & v_4 \end{pmatrix} = \begin{pmatrix} \mathbf{f}_3 \\ \mathbf{f}_4 \end{pmatrix}, \quad \mathbf{x} = (x,y), \ \mathbf{x}_2 = (x_2,y_2)$$

It follows, therefore, from the affine variant of Proposition C.1, that the transformation $\mathbf{g}_{1,0,-1,0}(\mathbf{x})$ which brings the unit-period normalized moiré profile $p_{1,0,-1,0}(\mathbf{x})$ into the final (1,0,-1,0)-moiré $p_{1,0,-1,0}(\mathbf{g}_{1,0,-1,0}(\mathbf{x}))$ between the two layers $r_1(\mathbf{x})$ and $r_2(\mathbf{x})$ is given by:

$$\mathbf{g}_{1,0,-1,0}(\mathbf{x}) = \mathbf{g}_1(\mathbf{x}) - \mathbf{g}_2(\mathbf{x})$$

$$= \begin{pmatrix} (u_1-u_3)x + (v_1-v_3)y + (x_1-x_2) \\ (u_2-u_4)x + (v_2-v_4)y + (y_1-y_2) \end{pmatrix} = F_M{\cdot}\mathbf{x} + (\mathbf{x}_1 - \mathbf{x}_2) \quad \text{with} \quad F_M = \begin{pmatrix} u_1-u_3 & v_1-v_3 \\ u_2-u_4 & v_2-v_4 \end{pmatrix} = \begin{pmatrix} \mathbf{a} \\ \mathbf{b} \end{pmatrix}$$

where $\quad F_M = F_1 - F_2, \quad$ namely: $\begin{pmatrix} \mathbf{a} \\ \mathbf{b} \end{pmatrix} = \begin{pmatrix} \mathbf{f}_1 - \mathbf{f}_3 \\ \mathbf{f}_2 - \mathbf{f}_4 \end{pmatrix}$. ■

This slight generalization of Proposition C.4 allows us now to consider the most general case of the (1,0,-1,0)-moiré between periodic layers, in which the individual layers are also shifted by $\mathbf{x}_1 = (x_1,y_1)$ and $\mathbf{x}_2 = (x_2,y_2)$, respectively.

**Remark C.8:** Obviously, the behaviour of the 2-fold periodic (1,0,-1,0)-moiré effect under layer shifts, as predicted by Proposition C.5, must be consistent with the results that we have already obtained in Chapter 7. Therefore, the following interesting question may be asked at this point: We already know from Chapter 7 that when the original periodic layers undergo slight layer shifts, the resulting periodic moiré usually undergoes a much larger shift (see, for example, Fig. 7.6). How is this fact reflected in Proposition C.5, given that the resulting shift $\mathbf{x}_1 - \mathbf{x}_2$ in $\mathbf{g}_{1,0,-1,0}(\mathbf{x})$ is not necessarily greater than the individual layer shifts $\mathbf{x}_1$ and $\mathbf{x}_2$?

To answer this question we must remember, once again, that the layer transformations $(x',y') = \mathbf{g}_1(x,y)$, $(x',y') = \mathbf{g}_2(x,y)$ as well as the resulting moiré transformation $(x',y') = \mathbf{g}_{1,0,-1,0}(x,y)$ are applied to the respective unit-period normalized layers $p_1(x',y')$, $p_2(x',y')$ and $p_{1,0,-1,0}(x',y')$ as *domain* transformations, and thus their effects are indeed those of the *inverse* transformations (see Remark C.5 above and the footnote therein). To better illustrate this point, consider the simple 1-fold periodic (1,-1)-moiré that is obtained in the superposition of two parallel periodic line gratings, one of which has been slightly shifted to the *right* while the other has been slightly *scaled up* (see Fig. 7.9 and Problem 7-18 in Chapter 7). In this case the layer transformations with respect to the original unit-period normalized layers are given by:

$$g_1(x) = x - x_0$$

$$g_2(x) = 0.9x$$

and therefore the moiré transformation with respect to the unit-period normalized layer is:

$$g_{1,-1}(x) = g_1(x) - g_2(x) = 0.1x - x_0 \tag{C.44}$$

At first sight this seems to indicate that the moiré effect is shifted by the same amount $x_0$ as the first layer. But if we remember that the actual effect of the domain transformation $x' = g_{1,-1}(x)$ is expressed by the *inverse* transformation, $x = g^{-1}_{1,-1}(x')$:[31]

$$x = 10x' + 10x_0 \tag{C.45}$$

we see immediately that Eq. (C.44) represents, in fact, a 10-fold magnification and a shift that is 10 times bigger than the shift $x_0$ in $g_1(x)$ (in the same direction). ■

**Example C.9:** As a further example, let us consider here the case illustrated in Fig. 7.6 of Chapter 7. This figure shows the superposition of two regular line grids, one of which is simply shifted by $(x_1,y_1)$:

$$\mathbf{g}_1(x,y) = \begin{pmatrix} 1 & 0 \\ 0 & 1 \end{pmatrix}\begin{pmatrix} x \\ y \end{pmatrix} - \begin{pmatrix} x_1 \\ y_1 \end{pmatrix}$$

while the second is rotated by angle $\alpha$ counterclockwise:

$$\mathbf{g}_2(x,y) = \begin{pmatrix} \cos\alpha & \sin\alpha \\ -\sin\alpha & \cos\alpha \end{pmatrix}\begin{pmatrix} x \\ y \end{pmatrix}$$

It follows, therefore, that the transformation undergone by the moiré effect is given by:

$$\begin{aligned}\mathbf{g}_{1,0,-1,0}(x,y) &= \mathbf{g}_1(x,y) - \mathbf{g}_2(x,y)\\ &= \begin{pmatrix} 1 & 0 \\ 0 & 1 \end{pmatrix}\begin{pmatrix} x \\ y \end{pmatrix} - \begin{pmatrix} x_1 \\ y_1 \end{pmatrix} - \begin{pmatrix} \cos\alpha & \sin\alpha \\ -\sin\alpha & \cos\alpha \end{pmatrix}\begin{pmatrix} x \\ y \end{pmatrix}\end{aligned}$$

[31] Note that being the inverse of $g_{1,-1}(x)$, Eq. (C.45) expresses the transformation undergone by the moiré effect in terms of a *direct* transformation. Similarly, the transformations undergone by the two original layers can be expressed as direct transformations by $x = x' + x_0$ and $x = \frac{10}{9}x'$, respectively. But the relationship $g_{1,-1}(x) = g_1(x) - g_2(x)$ is only valid for the domain (and hence, inverse) transformations.

$$= \begin{pmatrix} 1-\cos\alpha & -\sin\alpha \\ \sin\alpha & 1-\cos\alpha \end{pmatrix} \begin{pmatrix} x \\ y \end{pmatrix} - \begin{pmatrix} x_1 \\ y_1 \end{pmatrix}$$

Note, again, that the layer transdormations $(x',y') = \mathbf{g}_1(x,y)$, $(x',y') = \mathbf{g}_2(x,y)$ as well as the resulting moiré transformation $(x',y') = \mathbf{g}_{1,0,-1,0}(x,y)$ are applied to the respective unit-period normalized layers $p_1(x',y')$, $p_2(x',y')$ and $p_{1,0,-1,0}(x',y')$ as *domain* transformations, and hence their effects are indeed those of the *inverse* transformations. Thus, the effect of $\mathbf{g}_1(x,y)$ is a shift of the first layer by $(x_1,y_1)$ to the *right*, and the effect of $\mathbf{g}_2(x,y)$ is a rotation of the second layer by angle $\alpha$ *counterclockwise*. In order to see the actual effect of the transformation $(x',y') = \mathbf{g}_{1,0,-1,0}(x,y)$, namely, how it transforms the unit-period normalized moiré profile $p_{1,0,-1,0}(x',y')$ into the final (1,0,-1,0)-moiré $p_{1,0,-1,0}(\mathbf{g}_{1,0,-1,0}(x,y))$ between the two given layers $r_1(x',y')$ and $r_2(x',y')$, let us find the inverse transformation $(x,y) = \mathbf{g}^{-1}_{1,0,-1,0}(x',y')$:

$$\begin{pmatrix} x \\ y \end{pmatrix} = \begin{pmatrix} 1-\cos\alpha & -\sin\alpha \\ \sin\alpha & 1-\cos\alpha \end{pmatrix}^{-1} \left[\begin{pmatrix} x' \\ y' \end{pmatrix} + \begin{pmatrix} x_1 \\ y_1 \end{pmatrix}\right]$$

$$= \frac{1}{(1-\cos\alpha)^2 + (\sin\alpha)^2} \begin{pmatrix} 1-\cos\alpha & \sin\alpha \\ -\sin\alpha & 1-\cos\alpha \end{pmatrix} \left[\begin{pmatrix} x' \\ y' \end{pmatrix} + \begin{pmatrix} x_1 \\ y_1 \end{pmatrix}\right]$$

$$= \tfrac{1}{2} \begin{pmatrix} 1 & \frac{\sin\alpha}{1-\cos\alpha} \\ -\frac{\sin\alpha}{1-\cos\alpha} & 1 \end{pmatrix} \begin{pmatrix} x' \\ y' \end{pmatrix} + \tfrac{1}{2} \begin{pmatrix} 1 & \frac{\sin\alpha}{1-\cos\alpha} \\ -\frac{\sin\alpha}{1-\cos\alpha} & 1 \end{pmatrix} \begin{pmatrix} x_1 \\ y_1 \end{pmatrix}$$

$$= \tfrac{1}{2} \begin{pmatrix} 1 & \cot(\alpha/2) \\ -\cot(\alpha/2) & 1 \end{pmatrix} \begin{pmatrix} x' \\ y' \end{pmatrix} + \tfrac{1}{2} \begin{pmatrix} 1 & \cot(\alpha/2) \\ -\cot(\alpha/2) & 1 \end{pmatrix} \begin{pmatrix} x_1 \\ y_1 \end{pmatrix} \qquad \text{(C.46)}$$

We note that the matrix in this transformation has the form $\frac{1}{2}\begin{pmatrix} 1 & -b \\ b & 1 \end{pmatrix}$ with $b = -\cot(\alpha/2)$. This matrix is therefore a particular case of the similarity matrix $\begin{pmatrix} a & -b \\ b & a \end{pmatrix}$. As we know from Proposition C.3 this matrix corresponds to a linear transformation that is composed of:

- a rotation by angle $\theta$, where: $\quad \theta = \arctan(b/a)$
- and a scaling by: $\quad s = \sqrt{a^2 + b^2}$

By inserting here $a = \frac{1}{2}$ and $b = -\frac{1}{2}\cot(\alpha/2)$ (taking into account the factor $\frac{1}{2}$ before the matrix) and using the identities $\arctan x = \frac{\pi}{2} - \operatorname{arccot} x$ [Bronstein90 p. 280] and $1/\sqrt{1+\cot^2 x} = \sin x$ [Spiegel68 p. 15] we see from Eq. (C.46) that the moiré shift is obtained from the layer shift $(x_1,y_1)$ by:

- a rotation by angle $\theta$: $\quad \theta = \arctan(-\cot\frac{\alpha}{2}) = -\arctan(\cot\frac{\alpha}{2}) = \operatorname{arccot}(\cot\frac{\alpha}{2}) - \frac{\pi}{2}$

  namely: $\quad \theta = \frac{\alpha}{2} - \frac{\pi}{2} \qquad$ (C.47)

- and a scaling by: $\quad s = \frac{1}{2}\sqrt{1 + \cot^2(\alpha/2)} = \frac{1}{2\sin(\alpha/2)} \qquad$ (C.48)

And indeed, as expected, this result fully agrees with the moiré shift as explained in Chapter 7, which is obtained as follows:

(a) As predicted by proposition 7.2, the moiré effect is shifted in its own direction, which is, as shown in Fig. 4.8, exactly $\frac{\alpha}{2} - \frac{\pi}{2}$.

(b) Furthermore, the extent of the shift of the moiré effect is given, in terms of periods, by Eqs. (7.26) and (2.10), namely:

$$b_M = T_M(\phi_1 - \phi_2) = \frac{T}{2\sin(\alpha/2)}(\phi_1 - \phi_2) \tag{C.49}$$

where $b_M$ is the resulting shift of the moiré, $T_M$ is the period length of the moiré, $T$ is the period length of the two individual layers, and $\phi_1$ and $\phi_2$ are the shifts of the individual layers in terms of the period $T$. Noting that in our case $\phi_2 = 0$ (the second layer is not shifted), we see that for a shift of $d$ in the first layer (i.e., $\phi_1 = d/T$ periods), the extent of the resulting shift of the moiré is:

$$b_M = \frac{1}{2\sin(\alpha/2)}\, d \tag{C.50}$$

Hence, the shift of the periodic moiré is obtained by scaling up the shift $d$ of the first layer by the factor $s = \frac{1}{2\sin(\alpha/2)}$, exactly as predicted by Eq. (C.48) above.

This shows us, indeed, that the moiré shifts obtained by Proposition C.5 are in full agreement with our previous results from Chapter 7. ■

**Example C.10:** Finally, returning to the general case of Proposition C.5, what is the actual shift undergone by the moiré effect in the superposition?

According to Proposition C.5, the moiré transformation $\mathbf{g}_{1,0,-1,0}(\mathbf{x})$ is expressed by:

$$\mathbf{x}' = \mathrm{F}_M \cdot \mathbf{x} + (\mathbf{x}_1 - \mathbf{x}_2)$$

where $\mathrm{F}_M = \mathrm{F}_1 - \mathrm{F}_2$. Remembering that the actual effect of the domain transformation $\mathbf{x}' = \mathbf{g}_{1,0,-1,0}(\mathbf{x})$ is expressed by its *inverse*, $\mathbf{x} = \mathbf{g}^{-1}{}_{1,0,-1,0}(\mathbf{x}')$:

$$\mathbf{x} = \mathrm{F}_M^{-1} \cdot (\mathbf{x}' - (\mathbf{x}_1 - \mathbf{x}_2)) = \mathrm{F}_M^{-1} \cdot \mathbf{x}' \; - \; \mathrm{F}_M^{-1} \cdot (\mathbf{x}_1 - \mathbf{x}_2)$$

we see that the actual shift of the moiré effect in the superposition is given here by the second term, $-\mathrm{F}_M^{-1} \cdot (\mathbf{x}_1 - \mathbf{x}_2)$. Denoting this vector by $\mathbf{x}_M = (x_M, y_M)$, it follows that the direction and the extent of the moiré shift are given by $\arctan(y_M/x_M)$ and $\sqrt{x_M^2 + y_M^2}$, respectively. ■

### C.15.7 The order of the superposed layers

At this point, the following interesting question may be asked: Suppose that we change the numbering of the superposed layers, so that the layer transformations $\mathbf{g}_1$ and $\mathbf{g}_2$ are interchanged. Physically, this may correspond to changing the order of the superposed layers, such that the top layer in the original superposition becomes now the bottom layer and vice versa. Clearly, this has no influence whatsoever on the resulting moiré effect. And yet, because $\mathbf{g}_1$ and $\mathbf{g}_2$ have been interchanged, it turns out that the sign of the resulting moiré transformation $\mathbf{g}_{1,0,-1,0}$ has been inverted (since $\mathbf{g}_{1,0,-1,0}$ is the difference between the two individual layer transformations). This should mean that the resulting moiré effect has

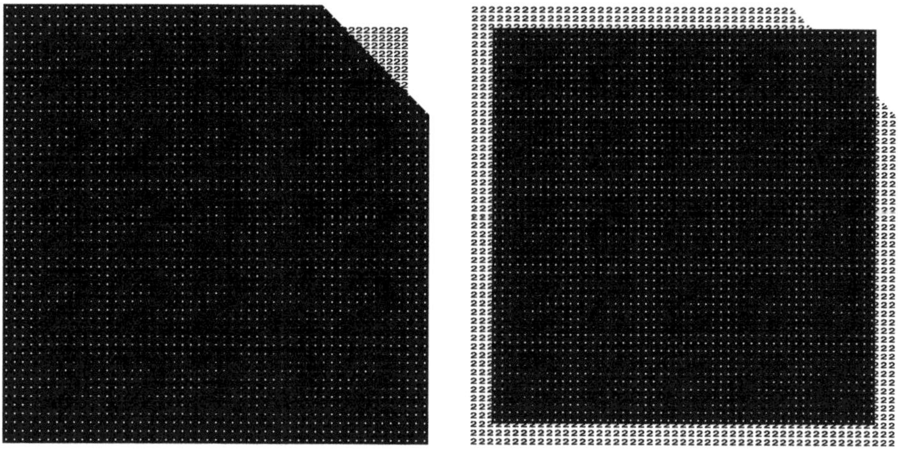

(a) (b)

**Figure C.24:** (a) The superposition of two regular dot screens $r_1(x,y)$ and $r_2(x,y)$, the first consisting of tiny "2"-shaped periods and the second consisting of tiny pinholes, where the first layer has been slightly scaled down: $\mathbf{g}_1(\mathbf{x}) = 1.1\mathbf{x}$, $\mathbf{g}_2(\mathbf{x}) = \mathbf{x}$. (b) The superposition of the same dot screens, but this time it is the second layer (the pinhole screen) that has been slightly scaled down: $\mathbf{g}_1(\mathbf{x}) = \mathbf{x}$, $\mathbf{g}_2(\mathbf{x}) = 1.1\mathbf{x}$. Note that the multiplication of a layer transformation by 1.1 corresponds, indeed, to a scale *down* effect in that layer (see Remark C.5 and the footnote therein).

been rotated by 180° — but as mentioned above, we know that the moiré effect does not depend on the order in which the two layers are superposed (or on the way we choose to number the layers). How can we explain this contradiction? The same question applies also to the sign of the frequency or the period of the moiré effect in Eqs. (2.11), (4.17), etc. The answer is, of course, that the orientation of the moiré is not only determined by the moiré transformation $\mathbf{g}_{1,0,-1,0}$, but also by its intensity profile $p_{1,0,-1,0}$, as clearly indicated by the two parts of Proposition C.1 (see also Propositions 10.2 and 10.5). Indeed, Propositions C.4 and C.5 only concentrate on the second part of Proposition C.1, the geometric transformations, but we should never forget that the first part of Proposition C.1 is also involved in the determination of the orientation of the moiré. To illustrate this more clearly, consider the following example.

**Example C.11:** Fig. C.24(a) shows a superposition of two regular dot screens, the first consisting of tiny "2"-shaped periods and the second consisting of tiny pinholes, where the first layer has been slightly scaled down: $\mathbf{g}_1(\mathbf{x}) = 1.1\mathbf{x}$, $\mathbf{g}_2(\mathbf{x}) = \mathbf{x}$, so that the layer frequencies (to both directions) are $f_1 = 1.1$, $f_2 = 1$ units. In this case we have, therefore:

$$\mathbf{g}_{1,0,-1,0}(\mathbf{x}) = 1.1\mathbf{x} - \mathbf{x} = 0.1\mathbf{x} \qquad \text{(or: } f_M = f_1 - f_2 = 0.1\text{)}$$

$$p_{1,0,-1,0}(\mathbf{x}') = p_1(\mathbf{x}') ** p_2(-\mathbf{x}') = \text{“2”} ** \text{“pinhole”} = \text{“2”}$$

Note that in this case the sign inversion in the second term of the convolution has no effect, since a pinhole is invariant under a 180° rotation.

What happens, now, if we change the order of the layers? In this case we obtain:

$$\mathbf{g}_{1,0,-1,0}(\mathbf{x}) = \mathbf{x} - 1.1\mathbf{x} = -0.1\mathbf{x} \qquad \text{(or: } f_M = 1 - 1.1 = -0.1\text{)}$$

which may lead us to the conclusion that the moiré effect has been inverted (rotated by 180°). But if we also consider the intensity profiles, we see that in this case the moiré profile $p_{1,0,-1,0}(\mathbf{x}')$ itself has been inverted, due to the sign inversion in the second term of the convolution:

$$p_{1,0,-1,0}(\mathbf{x}') = \text{"pinhole"} ** \text{"inverted 2"} = \text{"inverted 2"}$$

This means that the inversion in the moiré transformation rectifies the inversion in the moiré intensity profile, so that both inversions are cancelled out.

Consider now Fig. C.24(b). In this case, it is the second layer (the pinhole screen) that has been slightly scaled down: $\mathbf{g}_1(\mathbf{x}) = \mathbf{x}$, $\mathbf{g}_2(\mathbf{x}) = 1.1\mathbf{x}$, so that the layer frequencies are $f_1 = 1$, $f_2 = 1.1$ frequency units. As we can see in the figure, in this case the moiré effect is indeed rotated by 180°. To understand this, we note that in the present case we have:

$$\mathbf{g}_{1,0,-1,0}(\mathbf{x}) = \mathbf{x} - 1.1\mathbf{x} = -0.1\mathbf{x} \qquad \text{(or: } f_M = 1 - 1.1 = -0.1\text{)}$$

$$p_{1,0,-1,0}(\mathbf{x}') = p_1(\mathbf{x}') ** p_2(-\mathbf{x}') = \text{"2"} ** \text{"pinhole"} = \text{"2"}$$

This means that unlike in the previous case, the moiré profile itself has not been rotated by 180°, so that the sign inversion in the moiré transformation is not cancelled out, and it does, indeed, cause an inversion in the resulting moiré effect.

Now, if we change the order of the layers we obtain:

$$\mathbf{g}_{1,0,-1,0}(\mathbf{x}) = 1.1\mathbf{x} - \mathbf{x} = 0.1\mathbf{x} \qquad \text{(or: } f_M = 1.1 - 1 = 0.1\text{)}$$

$$p_{1,0,-1,0}(\mathbf{x}') = \text{"pinhole"} ** \text{"inverted 2"} = \text{"inverted 2"}$$

meaning that the moiré profile has been rotated by 180°, but the moiré transformation is not sign-inverted, and hence it does not rectify this inversion. The resulting moiré remains, therefore, rotated by 180°, as expected. ■

**Remark C.9:** The 1D counterpart of the explanation above gives a more precise formulation for the orientation of the 1D moiré in Eq. (2.11), which is no longer dependent on the order of the layers or on their numbering. This formulation could not be given in Chapter 2 since it depends on notions from Chapter 10. Indeed, the orientation of the moiré — just like its intensity profile — cannot be deduced from geometric considerations only. ■

## C.16 Layer normalization issues

As we have seen in Chapter 4, the $T$-convolution of the two given periodic layers (or, equivalently, the multiplication of their impulse combs or nailbeds in the spectral domain) requires that the two layers be first *normalized* (see the explanation preceding Proposition 4.2 in Sec. 4.2, as well as Propositions 4.2, 4.3 and 4.5). This normalization is required in order that the periods of the two layers to be convolved coincide (or, equivalently, in terms of the spectral domain, in order that the two combs or nailbeds to be multiplied have a common support). This normalization allows us, therefore, to apply the $T$-convolution theorem even though the two original layers do not necessarily have the same period sizes and orientations.

A similar consideration is also used in Chapter 10, which deals with the more general case in which the superposed layers are *repetitive* (i.e. non-linear transformations of periodic layers); see, for example, the explanation that precedes Proposition 10.2, as well as Propositions 10.3, 10.4, 10.5 and 10.7 in Sec. 10.9.

However, as the reader may have noticed, the normalizations being used in Chapter 4 and in Chapter 10 are not the same: In Chapter 4 the original layers are normalized to the period size and orientation of the final moiré; but in Chapter 10 the original curvilinear layers are normalized to their straight (uncurved) periodic counterparts having a unit period and angle 0° (i.e. the unit grid), and after the $T$-convolution the resulting normalized moiré is then transformed back into its actual geometric layout.

The reason for this difference is as follows. In fact, in order to perform the $T$-convolution of the two given layers (or, equivalently, the multiplication of their respective spectra in the Fourier domain), it is enough to normalize both of the layers into *any* common periodicity and orientation, be it the periodicity of the final moiré, a unit-grid periodicity, or any other arbitrary periodicity. In the case of Chapter 4, the moiré layout considerations are only *qualitative*,[32] and therefore it is more natural there to normalize the two original layers directly into the geometric layout of the final moiré, rather than to make first a normalization of the original layers into a unit-period structure and then transform the resulting unit-period moiré back into its actual geometric layout. But in Chapter 10, where we provide the full *quantitative* moiré layout considerarions (such as $g_{1,-1}(x,y) = g_1(x,y) - g_2(x,y)$, etc.), it is most natural to use a unit-period normalization, since here $g_1(x,y)$, $g_2(x,y)$ and $g_{1,-1}(x,y)$ express, indeed, the transformations undergone by the two original layers and by the resulting moiré, respectively, *with respect to this underlying unit-period grid or coordinate system.* Any other choice would imply a more complex interpretation of the transformations $g_1(x,y)$, $g_2(x,y)$ and $g_{1,-1}(x,y)$, which would be less intuitive and less useful. For example, if we choose to normalize our layers with respect to a two-units grid, or with respect to the final moiré layout (as we did in Chapter 4), the intuitive meaning of $g_1(x,y)$, $g_2(x,y)$ and $g_{1,-1}(x,y)$ would be lost.

**Remark C.10:** Note that in terms of the moiré layout relationships ($g_{1,-1}(x,y) = g_1(x,y) - g_2(x,y)$, $\mathbf{g}_{1,0,-1,0}(x,y) = \mathbf{g}_1(x,y) - \mathbf{g}_2(x,y)$, etc.) choosing a different normalization simply means a coordinate change in the already unit-period normalized layers. For example, if we choose to normalize our layers with respect to the coordinate grid having a

[32] In the sense that they do not yet explicitly provide the quantitative moiré layout relationships $g_{1,-1}(x,y) = g_1(x,y) - g_2(x,y)$, $\mathbf{g}_{1,0,-1,0}(x,y) = \mathbf{g}_1(x,y) - \mathbf{g}_2(x,y)$, etc. that are given later in Chapter 10.

two units periodicity, it simply means that we now use the new $s,t$ coordinate system with $s = 2x$, $t = 2y$, i.e. $x = s/2$, $y = t/2$, and our layout relationship then becomes $g_{1,-1}(s/2,t/2) = g_1(s/2,t/2) - g_2(s/2,t/2)$. In the general case, if $(s,t) = \mathbf{h}(x,y)$, and the inverse coordinate transformation is given by $(x,y) = \mathbf{h}^{-1}(s,t)$, then our layout relationship becomes $g_{1,-1}(\mathbf{h}^{-1}(s,t)) = g_1(\mathbf{h}^{-1}(s,t)) - g_2(\mathbf{h}^{-1}(s,t))$. Similar considerations can be devised for all the other moiré layout relationships, such as the (1,0,-1,0)-moiré between two dot screens $\mathbf{g}_{1,0,-1,0}(x,y) = \mathbf{g}_1(x,y) - \mathbf{g}_2(x,y)$, or any higher order moirés. This simply reflects the fact that the function identities expressing the moiré layout relationships are invariant under coordinate changes in the plane. ■

**Example C.12:** Let us see what happens to the layout relationship of the (1,0,-1,0)-moiré between two periodic dot screens, $\mathbf{g}_{1,0,-1,0}(x,y) = \mathbf{g}_1(x,y) - \mathbf{g}_2(x,y)$, if we choose to normalize our layers with respect to the final moiré layout, as we did in Chapter 4. As shown in Remark C.10, this normalization is obtained, in fact, by applying the coordinate change $(x,y) = \mathbf{g}^{-1}{}_{1,0,-1,0}(s,t)$ to the layers that are already normalized with respect to the unit grid.

Suppose that we are given two periodic dot screens $r_1(x,y)$ and $r_2(x,y)$ that have been obtained from their respective unit-period normalized counterparts $p_1(x',y')$ and $p_2(x',y')$ by the linear transformations $(x',y') = \mathbf{g}_1(x,y)$ and $(x',y') = \mathbf{g}_2(x,y)$, respectively. The periodic (1,0,-1,0)-moiré between these layers has, therefore, the geometric layout $\mathbf{g}_{1,0,-1,0}(x,y) = \mathbf{g}_1(x,y) - \mathbf{g}_2(x,y)$, where $\mathbf{g}_1$, $\mathbf{g}_2$ and $\mathbf{g}_{1,0,-1,0}$ are expressed with respect to the unit grid. Now, if we apply to the unit-period normalized layers the coordinate change $(x,y) = \mathbf{g}^{-1}{}_{1,0,-1,0}(s,t)$, the layout relationship of the moiré becomes, in terms of this new coordinate system:

$$\mathbf{g}_{1,0,-1,0}(\mathbf{g}^{-1}{}_{1,0,-1,0}(s,t)) = \mathbf{g}_1(\mathbf{g}^{-1}{}_{1,0,-1,0}(s,t)) - \mathbf{g}_2(\mathbf{g}^{-1}{}_{1,0,-1,0}(s,t))$$

namely:

$$(s,t) = \mathbf{g}_1(\mathbf{g}^{-1}{}_{1,0,-1,0}(s,t)) - \mathbf{g}_2(\mathbf{g}^{-1}{}_{1,0,-1,0}(s,t))$$

As we can see, under this normalization (coordinate change) the moiré layout is simply $(x',y') = (s,t)$ (i.e. it coincides with the new coordinate lines), and the layouts of the two dot screens become, respectively, $(x',y') = \mathbf{g}_1(\mathbf{g}^{-1}{}_{1,0,-1,0}(s,t))$ and $(x',y') = \mathbf{g}_2(\mathbf{g}^{-1}{}_{1,0,-1,0}(s,t))$. But this result is obviously less useful for quantitative calculations than its counterpart that is normalized with respect to the unit grid, $\mathbf{g}_{1,0,-1,0}(x,y) = \mathbf{g}_1(x,y) - \mathbf{g}_2(x,y)$. ■

**Example C.13:** Suppose that we are given two periodic gratings $r_1(x,y)$ and $r_2(x,y)$ that have been obtained from their respective unit-period normalized counterparts $p_1(x')$ and $p_2(x')$ by the linear bending transformations $x' = g_1(x,y)$ and $x' = g_2(x,y)$, respectively. The periodic (1,-1)-moiré between these layers has, therefore, the geometric layout $g_{1,-1}(x,y) = g_1(x,y) - g_2(x,y)$. Now, if we apply to the two given non-normalized gratings new layer transformations $(x,y) = \mathbf{h}_1(s,t)$ and $(x,y) = \mathbf{h}_2(s,t)$, respectively, the geometric layout of the (1,-1)-moiré becomes $g_1(\mathbf{h}_1(s,t)) - g_2(\mathbf{h}_2(s,t))$, because now the transformations of the individual gratings with respect to their unit-period normalized counterparts are $g_1(\mathbf{h}_1(s,t))$ and $g_2(\mathbf{h}_2(s,t))$, respectively. Note that this moiré layout is *not* obtained by applying the transformation $(x,y) = \mathbf{h}_{1,-1}(s,t) = \mathbf{h}_1(s,t) - \mathbf{h}_2(s,t)$ to the original moiré layout $g_{1,-1}(x,y) = g_1(x,y) - g_2(x,y)$. The counterpart of this example for the case of the (1,0,-1,0)-moiré between two dot screens can be obtained in a similar way; its generalization to the $(k_1,k_2,k_3,k_4)$-moiré is given in Chapter 10 in Proposition 10.8. ■

# Appendix D

## Glossary of the main terms

### D.1 About the glossary

Several thousands of publications on the moiré phenomenon have appeared during the last decades, in many different fields and applications. However, the terminology used in this vast literature is very far from being consistent and uniform. Different authors use different terms for the same entities, and what is even worse, the same terms are often used in different meanings by different authors. As a few examples among many others, let us cite here some of the many terms used in literature for what we call here *gratings*: line gratings, rulings [Nishijima64], sets of parallel lines [Fink92 p. 44], grids [Jarić89 pp. 29–31], parallel-line grids [Stecher64], grilles [Post67], and even line-screens [Tollenaar64]. For what we call here *screens* one can find the terms: lattices, grids (again!), meshes, masks, etc. Even the moiré patterns are often called fringes, beats, interferences, aliasing effects, and so forth.

Obviously, in such an interdisciplinary domain as the moiré theory it would be quite impossible to adopt a universally acceptable standardization of the terms, because of the different needs and traditions in the various fields involved (optics, mechanics, mathematics, printing, etc.).[1] Nevertheless, even without having any far-reaching pretensions, we were obliged to make our own terminological choices in a systematic and coherent way, in order to prevent confusion and ambiguity in our own work. We tried to be consistent in our terminology throughout this work, even if it forced us to assign to some terms a somewhat different meaning than one would expect (depending on his own background, of course).

In the present glossary we included all the terms for which we felt a clear definition was desirable to avoid any risk of ambiguity. Note, however, that this glossary is not ordered alphabetically; rather, we preferred to group the various terms according to subjects. We hope this should help the reader not only to clearly see the meaning of each individual term by itself, but also to put it in relation with other closely related terms (which would be completely dispersed throughout the glossary if an alphabetical order were preferred). Note that terms in the glossary can be found alphabetically through the general index at the end of the book.

---

[1] For example, the term *density* has very different meanings in almost any imaginable field of science or technology. A non-exhaustive list, just for the sake of illustration, may include: density of matter (in physics), density of population (in statistics), probability density (in probability), spectral density (in spectral analysis), density of a set (in mathematic topology), ink or colour density (in printing and colorimetry), etc. Another similar example is the term *phase*. It should be also noted that terminology sometimes tends to change over the years or according to fashions. As an example, the term *function convolution* in modern literature appears in older publications as *function composition* [Zygmund68 p. 36], the *resultant* of two functions, or even using the German term *Faltung* [Hardy68 p. 10].

## D.2 Terms in the image domain

**grating** (or *line-grating*) —

A pattern consisting of parallel lines. Unless otherwise mentioned it will be assumed that a grating is periodic and consists of equally wide parallel, straight lines that are separated by equal spaces. For example, a *binary grating* is a grating with a square-wave intensity profile consisting of white lines (with a constant value of 1) on a black background (whose value is 0).

**curvilinear grating** —

A repetitive pattern consisting of parallel curvilinear lines. A curvilinear grating can be seen as a non-linear transformation of an initial uncurved periodic grating of straight lines (see Sec. 10.2). Examples of curvilinear gratings with a cosinusoidal periodic profile are shown in Fig. 10.1; examples with a square-wave periodic profile are shown in the left side of Fig. 10.8.

**cosinusoidal grating** (not to be confused with *cosine-shaped grating*) —

A grating with a cosinusoidal periodic-profile; for example, a cosinusoidal circular grating is a circular grating with a cosinusoidal periodic-profile. Note, however, that since reflectance and transmittance functions always take values ranging between 0 and 1, the cosinusoidal grating is normally "raised" and rescaled into this range of values. For example, a reflectance function in the form of a vertical straight cosinusoidal grating is expressed by: $r(x,y) = \frac{1}{2}\cos(2\pi fx) + \frac{1}{2}$.

**cosine-shaped grating** (not to be confused with *cosinusoidal grating*) —

A grating (with any periodic-profile form) whose corrugations in the $x,y$ plane are bent into a cosinusoidal shape, like in Fig. 10.1(*l*).

**grid** (or *line-grid*; also called in literature *cross-line grating*) —

A pattern consisting of two superposed line-gratings, crossing each other at a non-zero angle. Unless otherwise mentioned it will be assumed that a grid is 2-fold periodic, and consists of two binary straight line gratings. Note that every grid can be also seen as a screen (whose dot-elements are the spaces left between the lines of the grid).

**regular grid** (or *square grid*) —

A 2-fold periodic grid composed of two superposed straight line-gratings that are identical but perpendicular to each other.

**curved grid** —

A repetitive pattern obtained by applying a non-linear transformation on a periodic grid (see Sec. 10.2). An example of a curved grid is shown in Fig. 10.2(b).

**screen** (or *dot-screen*; not to be confused with *lattice* or *dot-lattice*) —

A pattern consisting of dots. In most cases it will be assumed that a screen is 2-fold periodic, with a parallel, equally spaced arrangement of identical dots.

**regular screen** —

A 2-fold periodic screen whose dot arrangement is orthogonal and whose periods (or frequencies) to both orthogonal directions are equal.

**curved screen** —

A repetitive pattern obtained by applying a non-linear transformation on a periodic screen (see Sec. 10.2). An example of a curved screen is shown in Fig. 10.2(b).

**halftone screen** —

A more relaxed case of a binary screen in which the size (and the shape) of the screen dots may vary (typically, according to the gray level of a given original continuous-tone image), while its frequency and direction remain fixed throughout. Halftone screens are used in the printing world for the reproduction of continuous-tone images on bilevel printing devices.

**screen gradation** (or *wedge*) —

A halftone screen whose screen dots vary gradually (in their size and possibly also in their shape) across the image, generating a halftoned image with a smooth and uniform tone gradation.

**image** (has nothing to do with the image of a transformation) —

The most general term we use to cover "anything" in the image domain. It may be periodic or not, binary or continuous, etc. In principle, a monochrome (black-and-white) image has reflectance (or transmittance) values that vary between 0 (black) and 1 (white); similarly, a colour image has reflectance (or transmittance) values varying between 0 and 1 for each wavelength $\lambda$ of its colour spectrum.

**opening** (of a periodic binary grating or square wave) —

The length of the white span within the period of a periodic binary grating or square wave (see Fig. 2.4).

**opening ratio** (of a periodic binary grating or square wave) —

The ratio $\tau/T$ between the opening $\tau$ and the period $T$ of a periodic binary grating or square wave (see Fig. 2.9(a)–(c)). Note that the opening ratio remains fixed even when the periodic binary grating undergoes a non-linear coordinate transformation.

**spot function** —

A function which defines the way the size and the shape of the dots in a given halftone screen change as they grow from 0 to 100 percent coverage.

**period** (or *repetition-period* of a function $p$) —

A number $T \neq 0$ such that for any $x \in \mathbb{R}$, $p(x+T) = p(x)$. Note that the set of all the periods of $p(x)$ forms a lattice in $\mathbb{R}$. In the case of a 2-fold periodic function $p(x,y)$, a double period (or period parallelogram) of $p(x,y)$ is any parallelogram $A$ which tiles the $x,y$ plane so that $p(x,y)$ repeats itself identically on any of these tiles (see also Sec. A.3.4 in Appendix A).

**period-vector** (of a periodic function $p(x,y)$) —
A non-zero vector $\mathbf{P} = (x_0,y_0)$ such that for any $(x,y)\in\mathbb{R}^2$, $p(x+x_0,y+y_0) = p(x,y)$. The set $L_\mathbf{P}$ of all the period-vectors of $p(x,y)$ forms a lattice in the $x,y$ plane, which is 2D if $p(x,y)$ is 2-fold periodic, 1D if $p(x,y)$ is 1-fold periodic, and 0D, i.e., $L_\mathbf{P} = \{(0,0)\}$, if $p(x,y)$ is not periodic. If $\mathbf{P}_1$, $\mathbf{P}_2$ are two non-collinear period-vectors of a 2-fold periodic function $p(x,y)$, then for any point $\mathbf{x}\in\mathbb{R}^2$ the points $\mathbf{x}$, $\mathbf{x}+\mathbf{P}_1$, $\mathbf{x}+\mathbf{P}_2$, $\mathbf{x}+\mathbf{P}_1+\mathbf{P}_2$ define a period parallelogram of $p(x,y)$.

**period-lattice** (of a periodic function) —
The lattice formed by the set of all periods (or period-vectors) of the periodic function (see Secs. A.2, A.3.4 in Appendix A).

**step-vector** (of a periodic function $p(x,y)$) —
See Sec. A.6 in Appendix A, and the use of step-vectors in Sec. 7.5.2.

**periodic function** —
A function having a period. Note that a 2D function $p(x,y)$ can be 2-fold periodic (such as $p(x,y) = \cos x + \cos y$) or only 1-fold periodic (such as $p(x,y) = \cos x$).

**almost-periodic function** —
See Secs. B.3, B.5 in Appendix B.

**aperiodic function** (not to be confused with *non-periodic function*) —
A function which is not included in the class of almost-periodic functions. See Fig. B.3 in Appendix B.

**non-periodic function** (not to be confused with *aperiodic function*) —
A function which is not included in the class of periodic functions. See Fig. B.3 in Appendix B.

**repetitive structure** (or *repetitive function*) —
A structure (or a function) which is repetitive according to a certain rule, but which is not necessarily periodic (or almost-periodic). For example: concentric circles; gratings with logarithmic line-distances; screen gradations; etc. Note that such structures are sometimes called in literature *quasi-periodic* (like in [Bryngdahl74 p. 1290]); however, we reserve the term quasi-periodic only to its meaning in the context of the theory of almost-periodic functions (see Sec. B.5 in Appendix B).

**coordinate-transformed structure** —
A repetitive structure $r(x,y)$ which is obtained by the application of a non-linear coordinate transformation $\mathbf{g}(x,y)$ on a certain initial periodic structure $p(x,y)$. More formally, using vector notation: $r(\mathbf{x}) = p(\mathbf{g}(\mathbf{x}))$. Curvilinear gratings (such as parabolic or circular gratings) and gratings with a varying frequency (such as a

grating with logarithmic line-distances) are examples of coordinate-transformed structures.

**profile-transformed structure** —
A repetitive structure $r(x,y)$ which is obtained by the application of a non-linear transformation $t(z)$ on the profile of a certain initial periodic structure $p(x,y)$. More formally, using vector notation: $r(\mathbf{x}) = t(p(\mathbf{x}))$. Screen gradations are an example of profile-transformed structures.

**coordinate-and-profile transformed structure** —
A repetitive structure $r(x,y)$ which is obtained from a certain initial periodic structure $p(x,y)$ by the application of both a non-linear coordinate-transformation $\mathbf{g}(x,y)$ and a non-linear profile-transformation $t(z)$. More formally, using vector notation: $r(\mathbf{x}) = t(p(\mathbf{g}(\mathbf{x})))$. An example of a coordinate-and-profile-transformed structure is given in Remark 2 of Sec. 10.2.

**intensity profile** (of a structure $r(x,y)$) —
A function (surface) over the $x,y$ plane that gives at any point $(x,y)$ the intensity of the structure $r(x,y)$.

**periodic profile** (of a curvilinear grating, curved grid, etc.) —
The periodic profile of a curvilinear grating or a curved screen $r(x,y)$ is defined as the intensity profile of the original, uncurved periodic grating (or screen), before the non-linear transformation has been applied to it (see Sec. 10.2). Examples of curvilinear gratings with a cosinusoidal periodic profile are shown in Fig. 10.1; examples with a square-wave periodic profile are shown in Fig. 10.8(left). Note that in periodic structures the periodic profile coincides with the intensity profile.

**normalized periodic profile** (of a curvilinear grating, curved grid, etc.) —
See Sec. 10.2.

**geometric layout** (of a curvilinear grating, curved grid, etc.) —
The geometric layout of a curvilinear grating $r(x,y)$ is the locus of the centers of its curvilinear corrugations in the $x,y$ plane; it is defined by the bending function of the curvilinear grating (see Sec. 10.2). Similarly, the geometric layout of a curved grid or a curved screen is defined by its two bending functions. Fig. 10.1 shows curvilinear gratings with various geometric layouts; Fig. 10.2(b) shows a curved screen whose geometric layout is given by two inverse hyperbolic sine functions.

**bending function** (of a curvilinear grating, curved grid, etc.) —
The bending function of a curvilinear grating $r(x,y) = p(g(x,y))$ is the non-linear function $x' = g(x,y)$ which bends the original, uncurved periodic grating $p(x')$ into the curvilinear grating $r(x,y)$. The bending functions of a curved grid or a curved screen $r(x,y) = p(g_1(x,y),g_2(x,y))$ are the two components $x' = g_1(x,y)$, $y' = g_2(x,y)$ of the non-linear coordinate transformation $\mathbf{g}(x,y) = (g_1(x,y),g_2(x,y))$ which bends the original, uncurved periodic grid or screen $p(x',y')$ into the curved structure $r(x,y)$.

We usually assume that the bending transformation is smooth (a diffeomorphism), so that it has no abrupt jumps or other troublesome singularities.

**local period** (of a curvilinear grating, curved grid, etc.) —

The local period of a curvilinear grating (or curved grid) at a point $(x,y)$ is the period of the straight, periodic grating (or grid) which is defined by the tangents to the curvilinear grating (or curved grid) at a small neighbourhood around the point $(x,y)$. In a periodic structure the local period is constant throughout the $x,y$ plane, and it equals the period of the structure.

**zone grating** (or *zone plate*) —

According to the classical definition, a zone grating is a circularly symmetric grating which is obtained by drawing a family of concentric circles $x^2+y^2 = nr_1^2$, $n = 1,2,...$ (where the radius of the $n$-th circle is proportional to $\sqrt{n}$: $r_n = r_1\sqrt{n}$, $r_1$ being the radius of the central circle), and blackening alternate rings (zones) between these circles. This construction implies that the surface areas of the central circle and of each of the black or white rings which surround it are all equal. According to this classical definition the periodic-profile of a zone grating is a binary (black-and-white) square wave with opening ratio $\tau/T = 1/2$. Such classical zone gratings are called *Fresnel zone plates*, and they are often used in optics as focusing devices which are based on diffraction [Baez61, Myers51] (just like focusing lenses, that are based on refraction, or focusing mirrors, that are based on reflectance). However, we prefer to use the term zone grating in a wider sense, where the periodic-profile may have an opening ratio other than 1/2, or even where the periodic-profile is not at all a binary square wave. For example, a zone grating may have a cosinusoidal periodic-profile; such a zone grating is sometimes called a *Gabor zone plate* [Chau69]. In fact, we extend the definition of a zone grating even further, allowing also elliptic, hyperbolic and linear zone gratings [Welberry76]; see Example 10.7 in Sec. 10.3.

## D.3 Terms in the spectral domain

**spectrum** (or *frequency spectrum*; not to be confused with *colour spectrum*) —

The frequency decomposition of a given function, which specifies the contribution of each frequency to the function in question. The frequency spectrum is obtained by taking the Fourier transform of the given function.

**visibility circle** —

A circle around the spectrum origin whose radius represents the cutoff frequency, i.e., the threshold frequency beyond which fine detail is no longer detected by the eye. Obviously, its radius depends on several factors such as the viewing distance, the light conditions, etc. It should be noted that the visibility circle is just a first-order approximation. In fact, the sensitivity of the human eye is a continuous 2D bell-shaped function [Daly92 p. 6], with a steep "crater" in its center (representing frequencies which are too small to be perceived), and "notches" in the diagonal directions (owing to the drop in the eye sensibility in the diagonal directions [Ulichney88 pp.79–84]).

**frequency vector** —

A vector in the $u,v$ plane of the spectrum which represents the geometric location of an impulse in the spectrum (see Sec. 2.2 and Fig. 2.1).

**DC impulse** —

The impulse that is located on the spectrum origin. This impulse represents the frequency of zero, which corresponds to the constant component in the Fourier series decomposition of the periodic image; the amplitude of the DC impulse corresponds to the intensity of this constant component. This impulse is traditionally called the *DC impulse* because it represents in electrical transmission theory the direct current component, i.e., the constant term in the frequency decomposition of an electric wave; we are following here this naming convention.

**comb** (or *impulse-comb, Dirac-comb, impulse-train*) —

An infinite train of equally spaced impulses located on a straight line in the spectrum. Any 1-fold periodic function is represented in the spectrum by a comb centered on the spectrum origin. The step and the direction of this comb represent the frequency and the orientation of the periodic function; its impulse amplitudes, which are given by the Fourier series development of the periodic function, determine its intensity profile.

**nailbed** (or *impulse-nailbed*) —

An infinite 2D train of equally spaced impulses located in the spectrum on a dot-lattice (either square-angled or skewed). Any 2-fold periodic function is represented in the spectrum by an impulse nailbed centered on the spectrum origin. The steps and the two main directions of this nailbed represent the frequency and the orientation of the two main directions of the function's 2D periodicity; the impulse amplitudes, which are given by the 2D Fourier series development of the periodic function, determine its intensity profile.

**compound impulse** —

An impulse in the spectrum which is composed of several distinct impulses that happen to fall on the same location and hence "fuse down" into a single impulse. The amplitude of a compound impulse is the sum of the amplitudes (real or complex) of the individual impulses from which it is composed. See Sec. 6.4 in Chapter 6.

**compound comb** —

A comb of compound impulses.

**compound nailbed** —
A nailbed of compound impulses.

**support** (of a *comb*, a *nailbed*, a *spectrum*, etc.) —
The set of the geometric locations on the $u,v$ plane of all the impulses of the specified comb, nailbed, or spectrum.

**lattice** (or *dot-lattice*; not to be confused with *screen* or *dot-screen*) —
An algebraic structure, subset of $\mathbb{R}^n$; see definition 5.1 in Sec. 5.2.1. Note that our definition is narrower than the classical definition of a lattice in algebra textbooks such as [Jacobson85 pp. 457–459]; it rather corresponds to a lattice (or a lattice of points) in geometry of numbers [Cassels71 p. 9], or a lattice in crystallography. However, any lattice by our definition is indeed a lattice also in the larger sense.

**frequency lattice** (of a periodic function) —
The lattice formed by the set of all the integer linear combinations of the fundamental frequency(ies) of a periodic function, in the spectral domain. It is the reciprocal lattice of the period-lattice in the image domain (see Sec. A.4 in Appendix A).

**reciprocal lattice** (or *dual lattice*) —
See Appendix A, Sec. A.4.

**module** (or $\mathbb{Z}$-*module*; has nothing to do with the module of a complex number) —
An algebraic structure, subset of $\mathbb{R}^n$; see definition 5.2 in Sec. 5.2.1. Note that our definition is narrower than the classical definition of a module in algebra textbooks such as [Lang78 pp. 127–128]; however, any module by our definition is indeed a module also in the larger sense.

**cluster** (or *impulse-cluster*; has nothing to do with clusters in halftoning) —
A subset (either a lattice or a module) of impulse-locations in the spectrum support which collapse down, at a given singular state, into a single point in the spectrum. Algebraically, each cluster contains all points (impulses) which belong to one equivalence class in the singular state, and to each equivalence class there corresponds a cluster in the spectrum (see Chapter 5). As the superposed layers move away from the singular state, each cluster spreads-out in the spectrum. The impulse cluster which is generated around the spectrum origin (which belongs to the equivalence class of $\mathbf{0}$) has a particular significance, since it represents the spectrum of the isolated (extracted) moiré in question.

**line-impulse** —
A generalized function which is impulsive along a 1D line through the plane, and null everywhere else. A line-impulse can be graphically illustrated as a "blade" whose behaviour is continuous along its 1D line support but impulsive in the perpendicular direction. As an example, the spectrum of a parabolic cosinusoidal grating consists of two parallel line-impulses (see Example 10.5 in Sec. 10.3).

Note that the amplitude of a line-impulse does not necessarily die out away from its center, and it may even rapidly oscillate between two constant values.

**compound line-impulse** —

A line-impulse in the spectrum which is composed of several distinct line-impulses that happen to fall on the same 1D line and hence "fuse down" into a single line-impulse. Note that the center (skeleton location) of each of the individual line-impulses may be found in a different point along their common support. The amplitude of a compound line-impulse is the sum of the amplitudes (real or complex) of the individual line-impulses from which it is composed.

**curvilinear impulse** —

A generalized function which is impulsive along a 1D curvilinear path through the plane, and null everywhere else. A curvilinear impulse can be graphically illustrated as a curvilinear "blade" whose behaviour is continuous along its 1D curvilinear support but impulsive in the perpendicular direction.

**hump** —

A 2D continuous surface, often bell-shaped, elliptic or hyperbolic, which is defined around a given center on the plane. For example, the convolution of two non-parallel line-impulses gives a hump (see Sec. 10.7.3 and Fig. 10.13). Note that the amplitude of a hump does not necessarily die out away from its center, and in some cases it may even rapidly oscillate between two constant values.

**wake** —

A 2D continuous surface which trails off from an impulsive element (line-impulse, curvilinear impulse, etc.), gradually dying out as it goes away from the impulsive element in question. The amplitude of the wake may be considered as negligible with respect to that of its generating impulsive element. As an example, the spectrum of the cosinusoidal circular grating $\cos(2\pi f\sqrt{x^2+y^2})$ is a circular impulse ring of radius $f$, with a particular dipole-like impulsive behaviour on the perimeter of the circle and a negative, continuous wake which gradually trails off toward the center (see Example 10.6 in Sec. 10.3 and Fig. 10.4(d)).

**local frequency** (of a curvilinear grating, curved grid, etc.) —

The local frequency of a curvilinear grating (or curved grid) at a point $(x,y)$ is the frequency of the straight, periodic grating (or grid) which is defined by the tangents to the curvilinear grating (or curved grid) at a small neighbourhood around the point $(x,y)$. In a periodic structure the local frequency is constant throughout the $x,y$ plane, and it equals the frequency of the structure. See also Remark 10.4 in Sec. 10.2, and Sec. 11.4.

**internal discrepancy** —

The distance between the centers (skeleton locations) of two consecutive line-impulses in a cluster of line-impulses which has collapsed on a common line to form a compound line-impulse; see Sec. C.8 in Appendix C. If all the individual line-impulses are collapsed with a common center, the internal discrepancy of the compound line-impulse is zero. One may also define by analogy the internal discrepancy of a cluster of humps as the distance between the centers (skeleton locations) of two consecutive humps of the cluster.

**skeleton** —
According to the gradual transition approach (see Sec. 10.4.1) we may often think of a line-impulse cluster, a hump-cluster etc. as originating from an impulse-cluster whose impulses have "leaked out" or "melted down" to form the line-impulses or humps in question. This impulse-cluster is called the *skeleton* of the line-impulse cluster (or of the hump cluster). See, for example, Sec. 10.7.3.

**skeleton location** (or *center* of a line-impulse, a hump, etc.) —
The location in the spectrum of the skeleton-impulse which "leaked out" to give the line-impulse or hump in question.

**impulsive spectrum** —
A spectrum which only consists of impulses, i.e., whose support consists of a finite or at most denumerably infinite number of points. All periodic and almost-periodic functions have impulsive spectra.

**line-spectrum** —
A spectrum which consists of line impulses (see, for example, Fig. 10.11).

**hybrid spectrum** —
A spectrum which contains any combination of impulses, line-impulses and continuous humps (as opposed to a purely impulsive spectrum, a purely line-spectrum or a purely continuous spectrum). See, for instance, Example 10.14 of Sec. 10.7.4 and Fig. 10.13.

**singular support** (of a *spectrum*, etc.; distinguish from a *singular locus of a moiré*) —
The subset of the spectrum support over which the spectrum is impulsive. The singular support of a given spectrum includes the support of all the impulsive elements which are included in the spectrum (impulses, line-impulses, etc.), but not the support of continuous elements such as humps or wakes.

## D.4 Terms related to moiré

**moiré effect** (or *moiré phenomenon*) —
A visible phenomenon which occurs when repetitive structures (such as line-gratings, dot-screens, etc.) are superposed. It consists of a new pattern which is clearly observed in the superposition, although it does not appear in any of the original structures.

**$(k_1,...,k_m)$-moiré** —

The 1-fold periodic structure in the image domain which corresponds to the $(k_1,...,k_m)$-comb in the spectrum convolution (the spectrum of the superposition); see Sec. 2.8. In other words, this is the moiré which is generated due to the interaction between the $k_i$ harmonic frequencies of the respective layers in the superposition. This moiré may be visible if at least its fundamental impulse, the $(k_1,...,k_m)$-impulse, is located inside the visibility circle.

**singular moiré** (or *singular state, singular superposition*) —

A configuration of the superposed layers in which the period of the moiré in question becomes infinitely large (i.e., its frequency becomes 0), and hence it can no longer be seen in the superposition. Singular moirés are unstable moiré-free states, since any slight deviation in the angle or frequency of any of the superposed layers may cause the moiré in question to "come back from infinity" and to reappear with a large, visible period. Formally, we say that a $(k_1,...,k_m)$-moiré reaches a singular state whenever the geometric location of its fundamental impulse, the $(k_1,...,k_m)$-impulse in the spectrum convolution, falls on the spectrum origin (0,0) (i.e., whenever the frequency vectors $\mathbf{f}_1,...,\mathbf{f}_m$ which define the superposed layers are linearly dependent over $\mathbb{Z}$: $\sum k_i\mathbf{f}_i = \mathbf{0}$). Note that every $(k_1,...,k_m)$-moiré can be made singular by sliding the vector-sum $\sum k_i\mathbf{f}_i$ to the spectrum origin, namely: by appropriately modifying the vectors $\mathbf{f}_1,...,\mathbf{f}_m$ (or the frequencies and angles of the superposed layers).

**stable moiré-free state** —

A moiré-free configuration of the superposed layers in which no moiré becomes visible even when small deviations occur in the angle or frequency of any of the layers (see Fig. 2.8).

**unstable moiré-free state** —

A moiré-free configuration of the superposed layers in which any slight deviation in the angle or frequency of any of the layers causes the reappearance of a moiré with a large, visible period (see Fig. 2.8). Any singular state is an unstable moiré-free state.

**combined-moiré** (or *composite-moiré*) —

A 2D moiré which is combined of two or more $(k_1,...,k_m)$-moirés. See Sec. 2.8.

**moiré profile** (or *moiré intensity profile*; *moiré intensity surface*) —

A function (surface) which defines the intensity level of the moiré at any point of the image (see Secs. 2.10 and 4.1).

**macrostructure, microstructure** (within a superposition) —

The superposition of two or more layers (gratings, screens, etc.) may generate new structures which appear in the superposition but not in the original layers. These new structures can be classified into two categories: the *macrostructures*, i.e., the moiré effects proper, which are much coarser than the detail of the original layers; and the *microstructures*, i.e., the tiny geometric forms which are almost as small as the periods of the original layers, and are normally visible only from a close distance or through a magnifying glass (see Chapter 8).

**rosettes** —
The various tiny flower-like shapes which are often present in the microstructure of dot-screen or grid superpositions.

**singular locus of a moiré** (not to be confused with *singular support* in the spectrum) —
The locus in the image domain (a point, a straight line, a curve etc.) along which a moiré between curved, repetitive layers is singular. Along this singular locus the moiré locally disappears to the eye, its local period there being infinitely large. But at any other point of the $x,y$ plane outside this singular locus the moiré in question has a finite local period, which gradually decreases as one moves away from the singular locus. For example, the singular locus of the moiré shown in Fig. 10.18(c) is a straight line which coincides with the $x$ axis. Note that the singular locus of any moiré between periodic layers consists of the entire $x,y$ plane.

**moiré eyelet** (or *eye-shaped moiré*) —
A moiré effect whose singular locus consists of a single point. The moiré eyelet is centered on its singular locus point, where its local period is infinitely large, and around this point its frequency gradually increases in all directions until it exceeds the resolving power of the eye and disappears. See, for example, the moiré eyelets in Fig. 10.14(a).

**additive / subtractive moiré** (not to be confused with *additive superposition*) —
Classical terms often used in literature to designate moirés which are generated by frequency sums or frequency differences, respectively, in the spectrum. For example, the (1,-1)-moiré is subtractive, while the (1,1)-moiré is additive. Note, however, that these terms cannot be generalized to more complex cases such as the (1,1,-1)-moiré between three gratings. These terms are mostly useful in the superposition of two curvilinear gratings, where both the additive and the subtractive moiré are often observed simultaneously, each of them having a different shape and location (see, for example, Fig. 10.31). In this case the most convenient way to define them is based on their indicial equations (see Sec. 11.2): the additive moiré is the system of moiré fringes which corresponds to the indicial equation $m+n = p$, while the subtractive moiré is the system of moiré fringes which is selected by the indicial equation $m-n = p$.

## D.5 Terms related to light and colour

**colour spectrum** (not to be confused with *frequency spectrum*) —
The wavelength decomposition of a given light, which specifies the contribution of

each visible light wavelength $\lambda$ (approximately between $\lambda = 380$ nm for violet and $\lambda = 750$ nm for red) to the given light. The colour spectrum determines the visible colour of the light in question.

**monochrome** (or *black-and-white*; not to be confused with *monochromatic*) —
Achromatic light, image, etc. involving only black, white and all the intermediate gray levels. The colour spectrum of an ideal monochrome light is flat, i.e., it has a constant value (between 0 and 1) for all wavelengths $\lambda$ of the visible light.

**monochromatic** (not to be confused with *monochrome*) —
Chromatic light, image, etc. involving only a single pure wavelength $\lambda$ of the visible light. The colour spectrum of an ideal monochromatic light consists of a single impulse of intensity 1 at the wavelength $\lambda$.

**reflectance** (or *reflectance function*) —
A function $r(x,y)$ which assigns to any point $(x,y)$ of a monochrome image viewed by reflection a value between 0 and 1 representing its light reflection: 0 for black (or no reflected light), 1 for white (or full light reflection), and intermediate values for in-between shades. More formally, reflectance is defined at any point $(x,y)$ as the ratio of reflected to incident radiant power [Wyszecki82 p. 463].

**transmittance** (or *transmittance function*) —
A function $r(x,y)$ which assigns to any point $(x,y)$ of a monochrome image viewed by transmission (such as a transparency, a film, etc.) a value between 0 and 1 representing its light transmission: 0 for black (or no transmitted light), 1 for white (or full light transmission), and intermediate values for in-between shades. More formally, transmittance is defined at any point $(x,y)$ as the ratio of transmitted to incident radiant power [Wyszecki82 p. 463].

**chromatic reflectance** (or *chromatic reflectance function*) —
A function $r(x,y;\lambda)$ which assigns to any point $(x,y)$ of a colour image viewed by reflection its full colour spectrum. In other words, it gives for every wavelength $\lambda$ of the visible light (approximately between $\lambda = 380$ nm and $\lambda = 750$ nm) a value between 0 and 1, which represents the reflectance of light of wavelength $\lambda$ at the point $(x,y)$ of the image. This is a straightforward generalization of the reflectance function $r(x,y)$ in the monochrome case.

**chromatic transmittance** (or *chromatic transmittance function*) —
A function $r(x,y;\lambda)$ which assigns to any point $(x,y)$ of a colour image viewed by transmission its full colour spectrum. In other words, it gives for every wavelength $\lambda$ of the visible light (approximately between $\lambda = 380$ nm and $\lambda = 750$ nm) a value between 0 and 1, which represents the transmittance of light of wavelength $\lambda$ at the point $(x,y)$ of the image. This is a straightforward generalization of the transmittance function $r(x,y)$ in the monochrome case.

## D.6 Miscellaneous terms

**binary** (grating, etc.) —
A structure which contains only two transmittance (or reflectance) levels: 0 and 1.

**discrete** —
A subset $D$ of $\mathbb{R}^n$ is called *discrete* if there exists a number $d > 0$ so that for any points $a,b \in D$ the distance between $a$ and $b$ is larger than $d$. Note, however, that the term *discrete* is also used (often carelessly) as the opposite of *continuous*. For example: *discrete spectra* (including everywhere dense spectra of almost-periodic functions!) vs. *continuous spectra* in [Champeney87 pp. 109–114]; or in our own case, the *discrete* mapping $\Psi$ vs. the *continuous* mapping $\Phi$, in Sec. 5.5.

**dense** (or *everywhere dense*; has nothing to do with the term *density* below) —
A subset $S$ of $\mathbb{R}^n$ is called *dense* or *everywhere dense* in $\mathbb{R}^n$ if $[S] = \mathbb{R}^n$, where $[S]$ denotes the closure of $S$, i.e., the set containing $S$ and all its limit points [EncMath88 Vol. 3 p. 434]. Note that a dense subset of $\mathbb{R}^n$ is not necessarily continuous; for example, the set $\mathbb{Q}$ of all rational numbers is everywhere dense in $\mathbb{R}$ but nowhere continuous.

**density** (has nothing to do with the term *dense* above) —
A representation of reflectance (or transmittance) values in logarithmic terms (see Fig. 2.9). This representation corresponds better to human visual perception due to the rather logarithmic nature of the eye's sensibility to light intensity [Pratt91 pp. 27–29]. Also called *reflection density* (or *transmission density*), or more generally: *optical density* [Rosenfeld82 p. 4].

**scaling** (of a *comb*, *nailbed*, etc.; not to be confused with *spreading-out / squeezing*) —
We distinguish between *amplitude* scalings, and *period* or *frequency* scalings (in which the expansion or contraction occurs along the $x,y$ axes in the image, or the $u,v$ axes in the spectrum).

**squeezing / spreading-out** (of an *impulse cluster*; not to be confused with *scaling*) —
We reserve these terms only to lateral (or spatial) contractions or expansions in the geometric locations of the cluster impulses when the superposed layers approach or move away from a singular state (see Sec. 5.6.3). A cluster can be squeezed towards its "center", or spread-out from the "center" outwards.

**commensurable** (or *commensurate*) —
Two vectors $\mathbf{v}_1,\mathbf{v}_2 \in \mathbb{R}^n$ (or real numbers in $\mathbb{R}$) are called *commensurable* if there exist non-zero integers $m,n$ for which $\mathbf{v}_2 = \frac{m}{n}\mathbf{v}_1$ (so that both $\mathbf{v}_1$ and $\mathbf{v}_2$ can be measured as integer multiples of the same length unit, say $\frac{1}{n}\mathbf{v}_1$). Note that two numbers $x,y \in \mathbb{R}$ are commensurable *iff* their ratio $x/y$ is rational. More generally, $k$ vectors $\mathbf{v}_1,...,\mathbf{v}_k \in \mathbb{R}^n$ (or real numbers in $\mathbb{R}$) are called commensurable if they are linearly dependent over $\mathbb{Q}$ (which is identical to linear dependence over $\mathbb{Z}$).

**incommensurable** (or *incommensurate*) —

Two vectors $\mathbf{v}_1,\mathbf{v}_2 \in \mathbb{R}^n$ (or real numbers in $\mathbb{R}$) are called *incommensurable* if there do not exist non-zero integers $m,n$ so that $\mathbf{v}_2 = \frac{m}{n}\mathbf{v}_1$. Note that two numbers $x,y \in \mathbb{R}$ are incommensurable *iff* their ratio $x/y$ is irrational. More generally, $k$ vectors $\mathbf{v}_1,...,\mathbf{v}_k \in \mathbb{R}^n$ (or real numbers in $\mathbb{R}$) are called incommensurable if they are linearly independent over $\mathbb{Q}$ (which is identical to linear independence over $\mathbb{Z}$).

**vector** (or *point* in a vector space) —

An element of the vector space in question ($\mathbb{R}^n$, the $u,v$ plane, etc.). We always consider vectors as radius-vectors attached to the origin, and we do not distinguish between a *vector* and a *point* in the vector space (= the end point or the head of the vector).

**phase** (of a periodic function, a moiré, etc.) —

See Appendix C Sec. C.4 and Chapter 7 Secs. 7.1–7.5.

**geometry of numbers** (or *geometric number theory*) —

A branch of number theory, initiated by Minkowski in 1896, that studies number-theoretical problems by the use of geometric methods [EncMath88 Vol. 4 pp. 267–271]. One of the distinctive characteristics of geometry of numbers is that it combines concepts from both continuous and discrete mathematics: it studies properties which come from the realm of continuous mathematics (like volume, area, etc.) in relation to lattices, which are discrete objects [Kannan87 p. 2]. A typical task of this theory is the problem of finding the minimum of some real function $f(x_1,...,x_m)$ where $(x_1,...,x_m)$ are restricted to integral points (i.e., to points of the lattice $\mathbb{Z}^m$), normally with some supplementary condition, such as $(x_1,...,x_m) \neq \mathbf{0}$. The main concepts we use from the theory of geometry of numbers (see Chapter 5) are the restriction of the continuous linear mapping $\Phi$ (which is defined on $\mathbb{R}^m$) into its counterpart $\Psi$ which is only defined on integral points, i.e., on $\mathbb{Z}^m$ (see Secs. 5.3–5.5); the investigation of their kernels and images and the interrelations between their ranks over $\mathbb{R}$ and over $\mathbb{Z}$; and the notions of lattice and module (which are already bordering on algebra).

**diffeomorphism** —

A diffeomorphism (in our case, on $\mathbb{R}^2$) is a one-to-one continuously differentiable mapping of $\mathbb{R}^2$ onto itself whose inverse mapping is also continuously differentiable.

**chirp** (or *chirp signal*) —

An oscillatory signal with an increasing (or decreasing) oscillation rate. For example, $\cos(ax^2)$ is a 1D chirp signal, and $\cos(ax^2 + by^2)$ is a 2D chirp signal.

**equivalent grating number** —

The number $m$ of virtual gratings in a given superposition (where each 2D dot-screen or line-grid contributes two virtual gratings). See Sec. 2.12 in Chapter 2.

**virtual gratings** (or *grating equivalents*) —
See Sec. 2.12 and Sec. 7.5.2.

**separable function** (of two variables) —
A function $f(x,y)$ is said to be *separable* if it can be presented as (or separated into) a product of a function of $x$ and a function of $y$: $f(x,y) = g(x) \cdot h(y)$ [Gaskill78 pp. 16–17; Cartwright90 p. 117]. Note, however, that we use this term in a slightly larger sense: A 2D function $f(x,y)$ is separable if it can be presented as a product of two independent 1D functions. Therefore, although $f(x,y) = g(x) \cdot h(y)$ may no longer be separable (in the narrower sense) after it has undergone a rotation or a skewing transformation, we will still consider it as separable (with respect to the rotated or skewed axes $x'$ and $y'$: $f(x',y') = g(x') \cdot h(y')$).

**inseparable function** (of two variables) —
A function that is not *separable*. For example, the function representing a square white dot is separable: $\text{rect}(x,y) = \text{rect}(x) \cdot \text{rect}(y)$, while the function representing a circular white dot is inseparable. See also Sec. 2.12.

**spatially separable** (not to be confused with a *separable function*) —
Two functions $F(u,v)$ and $G(u,v)$ in the spectrum are called spatially separable if their supports in the $u,v$ plane are not overlapping. Spatially separable elements in the spectrum can be separated and extracted by means of filtering, i.e., by multiplying the spectrum with an appropriate 2D low-pass or band-pass filter (see, for example, Fig 10.22).

**spatially inseparable** (not to be confused with an *inseparable function*) —
Two functions $F(u,v)$ and $G(u,v)$ in the spectrum are called spatially inseparable if their supports in the $u,v$ plane are at least partially overlapping. Spatially inseparable elements in the spectrum cannot be separated or extracted by multiplying the spectrum with 2D low-pass or band-pass filters (see, for example, Fig 10.27).

**dots per inch (dpi)** (or *dots per centimeter*; not to be confused with *lines per inch*) —
A term used to specify the resolution of a digital device such as a printer, a scanner, etc. For example, a device whose resolution is 300 dpi can only address points on an underlying pixel-grid whose period is 1/300 of an inch, and no in-between points or pixel-fractions can be addressed. Some devices have different resolutions in the horizontal and in the vertical directions.

**lines per inch (lpi)** (or *lines per centimeter*; not to be confused with *dots per inch*) —
A term used to specify the frequency of gratings, dot-screens, etc. This term specifies the number of periods per inch. For example: the finest grating that can be produced on a 300 dpi device, namely: a sequence of alternating one-pixel wide black and white lines, is a grating of 150 lpi (since one period consists here of two device pixels).

**1D** —

One dimensional (layer, periodicity, spectrum, impulse comb, etc.). Strictly, a one dimensional entity is an entity that has only one dimension (such as a straight line). However, by abuse of language we often use this term to designate a 2D entity that only varies along one dimension, while its other dimension is constant. For example, we say that an impulse comb such as in Fig. 2.5(d) or 2.5(e) is 1D, even though it subsists in the 2D $u,v$ spectrum, because it varies only in one dimension, while in the orthogonal direction it remains constantly zero. Similarly, we often say that a 1-fold periodic structure such as a line grating (see, for example, Fig. 2.5(a) or 2.5(b)) is a 1D structure, even though it actually spreads in the 2D $x,y$ space, because it only varies along one dimension but remains constant along the orthogonal direction (i.e. along the individual grating lines). Such a 1D periodic layer is, in fact, a constant extension of a really one-dimensional structure (such as a square wave) into the second dimension. Note that we sometimes use the terms "1D periodic" and "1-fold periodic" interchangeably as synonyms.

**2D** —

Two dimensional (layer, periodicity, spectrum, impulse nailbed, etc.). Strictly, any entity that has two dimensions, including a 1-fold periodic structure such as a line grating, is a 2D layer. However, by abuse of language we often use this term to designate a 2D entity that indeed varies along two dimensions. For example, we say that a structure such as a line grid or a dot screen is a 2D layer, because it varies along two dimensions. Note that we sometimes use the terms "2D periodic" and "2-fold periodic" interchangeably as synonyms.

**domain / range transformation** —

Any image $r(x,y)$ (or function $r$: $\mathbb{R}^2 \to \mathbb{R}$) can undergo two types of coordinate transformations: Either a transformation of its *domain*, $r(x,y) \mapsto r(\mathbf{g}(x,y))$, or a transformation of its *range*, $r(x,y) \mapsto t(r(x,y))$. As explained in Sec. D.6 of Appendix D in *Vol. II*, in the first case $\mathbf{g}(x,y)$ is applied as an inverse transformation, while in the second case $t(x)$ is used as a direct transformation. Similarly, any mapping $\mathbf{f}(x,y)$, $\mathbf{f}$: $\mathbb{R}^2 \to \mathbb{R}^2$, can undergo two types of coordinate transformations: Either a transformation of its *domain*, $\mathbf{f}(x,y) \mapsto \mathbf{f}(\mathbf{g}(x,y))$, or a transformation of its *range*, $\mathbf{f}(x,y) \mapsto \mathbf{g}(\mathbf{f}(x,y))$. Again, in the first case $\mathbf{g}(x,y)$ is applied as an inverse transformation, while in the second case it is used as a direct transformation.

**direct / inverse transformation** (or *direct / inverse mapping*) —

The mathematical terms used to designate a mapping (geometric transformation) $\mathbf{g}$ and its inverse $\mathbf{g}^{-1}$. We have, therefore, $\mathbf{g}(\mathbf{g}^{-1}(x,y)) = \mathbf{g}^{-1}(\mathbf{g}(x,y)) = (x,y)$. Note that the designations *direct* and *inverse* are interchangeable, and they depend on our point of view; thus, if we focus our attention to the mapping $\mathbf{h} = \mathbf{g}^{-1}$, we may consider $\mathbf{h}$ as the direct mapping and $\mathbf{h}^{-1} = (\mathbf{g}^{-1})^{-1} = \mathbf{g}$ as its inverse. For example, if $\mathbf{g}(x,y) = (2x,2y)$ then $\mathbf{g}^{-1}(x,y) = (x/2,y/2)$; and if $\mathbf{g}(x,y) = (x/2,y/2)$ then

$\mathbf{g}^{-1}(x,y) = (2x,2y)$. Note that the similar terms *direct / inverse transforms* are reserved to operators that act on functions, such as the Fourier transform.

# List of notations and symbols

This list consists of the main symbols used in the text. They appear with a very brief description and a reference to the page in which they are first used or defined. Obvious symbols such as '+', '–', etc. have not been included.

| Symbol | Short description | Page |
|---|---|---|
| $x, y$ | The coordinates (axes) of the image plane | 10 |
| $u, v$ | The coordinates (axes) of the spectral plane | 10 |
| $x', y'$ | Rotated coordinates (axes) in the image plane | 380 |
| $p(x)$ | A 1D periodic function | 375 |
| $p(x,y)$ | A 2D periodic function | 378 |
| $p(\mathbf{x})$ | The vector notation for $p(x,y)$ | 379 |
| $P(u)$ | The spectrum of $p(x)$ | 377 |
| $P(u,v)$ | The spectrum of $p(x,y)$ | 379 |
| $P(\mathbf{u})$, $P(\mathbf{f})$ | The vector notation for $P(u,v)$ | 380 |
| $r(x)$ | A 1D reflectance (or transmittance) function | 21 |
| $r(x,y)$ | A 2D reflectance (or transmittance) function | 10 |
| $R(u)$ | The spectrum of $r(x)$ | 22 |
| $R(u,v)$ | The spectrum of $r(x,y)$ | 11 |
| $d(x,y)$ | A single dot (of a dot-screen) | 44 |
| $D(u,v)$ | The spectrum of $d(x,y)$ | 44 |
| $*$ | 1D convolution (or $T$-convolution) | 86 |
| $**$ | 2D convolution (or $T$-convolution) | 11, 95 |
| $\theta, \theta_1, \ldots$ | Angles of superposed layers | 12, 18 |
| $\alpha, \beta, \gamma$ | Angle differences between superposed layers | 20, 68–69 |

| | | |
|---|---|---|
| $\varphi_M$ | Angle of a moiré effect | 18, 20 |
| $\alpha \to 0°$ | The angle difference $\alpha$ tends to 0° | 156 |
| $f, f_1, \ldots$ | Frequencies of 1D periodic functions $p(x), p_1(x), \ldots$ | 12 |
| $f_M$ | Frequency of a moiré effect | 20 |
| $T, T_1, \ldots$ | Periods of 1D periodic functions $p(x), p_1(x), \ldots$ | 18 |
| $T_M$ | Period of a moiré effect | 20 |
| $\mathbf{P}_1, \mathbf{P}_2$ | Period-vectors | 381 |
| $\mathbf{T}_1, \mathbf{T}_2$ | Step-vectors | 393 |
| P, F | Matrices | 390 |
| $\mathrm{P}^{-\mathrm{T}}$ | The inverse transpose of matrix P | 390 |
| $a, b, r, s, t$ | Real numbers (sometimes also used as integer numbers) | 169 |
| $\mathbf{a}, \mathbf{b}, \mathbf{r}, \mathbf{t}, \mathbf{w}$ | Vectors | 171 |
| $i, j, k, l, m, n$ | Integer numbers | 33 |
| $i$ | (In complex numbers): the imaginary unit, $\sqrt{-1}$ | 10 |
| $a_n, b_n, c_n, d_n$ | Fourier series coefficients; impulse amplitudes | 21, 23, 376 |
| $\mathbb{Z}$ | The set of all integer numbers (positive, negative, and 0) | 110 |
| $\mathbb{Q}$ | The set of all rational numbers | 113 |
| $\mathbb{R}$ | The set of all real numbers | 110 |
| $\mathbb{C}$ | The set of all complex numbers | 412 |
| $\mathbb{Z}^n$ | The $n$-dimensional integer lattice | 110 |
| $\mathbb{R}^n$ | The $n$-dimensional Euclidean space | 110 |
| $L$ | A lattice | 110 |
| $M$ | A module | 110 |
| $\mathrm{rank}_{\mathbb{R}} M$ | The rank of $M$ (also denoted: rank $M$) | 111 |
| $\mathrm{rank}_{\mathbb{Z}} M$ | The integral rank of $M$ | 111 |
| $\mathbf{f}_1, \ldots, \mathbf{f}_m$ | Frequency vectors in the $u,v$ plane of the spectrum | 12, 28 |

| | | |
|---|---|---|
| $\mathbf{v}_1,\ldots,\mathbf{v}_m$ | Vectors in $\mathbb{R}^n$ | 110 |
| $\mathrm{Sp}(\mathbf{v}_1,\ldots,\mathbf{v}_m)$ | The set of all linear combinations of $\mathbf{v}_1,\ldots,\mathbf{v}_m$ | 112 |
| $\mathrm{Md}(\mathbf{v}_1,\ldots,\mathbf{v}_m)$ | The set of all integral linear combinations of $\mathbf{v}_1,\ldots,\mathbf{v}_m$ | 112 |
| $V$, $W$ | Vector spaces | 115 |
| $\dim V$ | Dimension of vector space $V$ | 117 |
| $\Phi : V \to W$ | A linear transformation from vector space $V$ to vector space $W$ | 115 |
| $\mathrm{Im}\Phi$ | The image of transformation $\Phi$ | 115 |
| $\mathrm{Ker}\Phi$ | The kernel of transformation $\Phi$ | 115 |
| $\Psi_{\mathbf{f}_1,\ldots,\mathbf{f}_m}$ | The discrete linear transformation from $\mathbb{Z}^m$ to $\mathrm{Md}(\mathbf{f}_1,\ldots,\mathbf{f}_m)$ | 114–115 |
| $\Phi_{\mathbf{f}_1,\ldots,\mathbf{f}_m}$ | The continuous extension of $\Psi_{\mathbf{f}_1,\ldots,\mathbf{f}_m}$ | 115 |
| Re[ ] | The real-valued part of a complex entity | 253, 412 |
| Im[ ] | The imaginary-valued part of a complex entity | 253, 412 |
| Abs[ ] | The magnitude of a complex entity | 412 |
| Arg[ ] | The phase of a complex entity | 412 |
| $U$ | A subspace of the vector space $V$ | 116 |
| $V/U$ | The quotient space of $V$ modulo $U$ | 117 |
| 0 | The number zero | 37 |
| $\mathbf{0}$ | The zero vector | 37 |
| $(u,v)$ | The Cartesian coordinates of the frequency vector $\mathbf{f}$ | 12 |
| $(f,\theta)$ | The polar coordinates of the frequency vector $\mathbf{f}$ | 12 |
| $a{\cdot}b$ | Multiplication | 20 |
| $\mathbf{v}\cdot\mathbf{w}$ | Scalar product of two vectors | 181 |
| $\mathbf{v}\times\mathbf{w}$ | Vector product of two vectors | 387–388 |
| $\mathbf{v}^{-1}$ | The reciprocal vector of $\mathbf{v}$ | 393 |
| $\tau$ | Opening (white width) of a binary grating or a square wave | 21 |
| $\tau/T$ | Opening ratio of a binary grating or a square wave | 23 |

| | | |
|---|---|---|
| $a \bmod b$ | The remainder of the integer division of $a$ by $b$ | 169 |
| $\phi$ | The period-shift of $p(x)$ | 169 |
| $\boldsymbol{\phi}$ | The period-shift of $p(\mathbf{x})$ | 172 |
| $\varphi$ | Phase increment (in the sense of complex numbers) | 413–417 |
| $\xi$ | Period-coordinate | 211 |
| $\Xi(x,y)$ | Period-coordinate function | 215 |
| $\mathbf{g}(x,y)$ | A 2D coordinate transformation | 256 |
| $\mathbf{f}(x,y)$ | Local frequency vector | 364 |
| $\forall$ | For all ... | 21 |
| ■ | End of example, proof, etc. | 22 |

# List of abbreviations

| Symbol | Short description | Page |
|---|---|---|
| 1D | 1-dimensional | 123 |
| 2D | 2-dimensional | 10, 123 |
| 1D-L | A 1D lattice | 122–123 |
| 1D-M | A 1D module | 122–123 |
| 2D-L | A 2D lattice | 122–123 |
| 2D-M | A 2D module | 122–123 |
| CMYK | The four process ink colours: Cyan, Magenta, Yellow, blacK | 60 |
| RGB | The three colour display primaries: Red, Green, Blue | 235, 243 |
| DC | The impulse at the spectrum origin (i.e., at frequency zero) | 12 |

| | | |
|---|---|---|
| DFT | Discrete Fourier transform | 99 |
| FFT | Fast Fourier transform | 84, 100 |
| dpi | Dots per inch (printer resolution) | 448 |
| lpi | Lines per inch (frequency of a grating or a screen) | 448 |
| *iff* | If and only if | 111 |

# References


[Adkins92] W. A. Adkins and S. H. Weintraub, *Algebra, an Approach via Module Theory.* Springer-Verlag, NY, 1992.

[Adobe90] Adobe Systems Inc., *PostScript language Reference Manual.* Addison-Wesely, Reading, Massachusetts, 1990 (second edition).

[Ahumada83] A. J. Ahumada, Jr., D. C. Nagel and A. B. Watson, "Reduction of display artifacts by random sampling," in *Applications of Digital Image Processing VI*, Proceedings of the SPIE, Vol. 432, 1983, pp. 216–221.

[Alasia98] A. Alasia, "Digital anti-counterfeiting software methos and apparatus," US Patent No. 5,708,717, 1998.

[Allebach78] J. P. Allebach, "Random nucleated halftone screen," *Photographic Science and Engineering*, Vol. 22, No. 2, 1978, pp. 89–91.

[Allebach79] J. P. Allebach, "Aliasing and quantization in the efficient display of images," *Jour. of the Optical Society of America*, Vol. 69, June 1979, pp. 869–877.

[Amidror91] I. Amidror, "The moiré phenomenon in colour separation," in *Raster Imaging and Digital Typography II*, Proceedings of the 2nd International Conference on Raster Imaging and Digital Typography, R. A. Morris and J. André (Eds.), Cambridge University Press, 1991, pp. 98–119.

[Amidror94] I. Amidror, R. D. Hersch and V. Ostromoukhov, "Spectral analysis and minimization of moiré patterns in colour separation," *Journal of Electronic Imaging*, Vol. 3, No. 3, July 1994, pp. 295–317.

[Amidror94a] I. Amidror, "A generalized Fourier-based method for the analysis of 2D moiré envelope forms in screen superpositions," *Journal of Modern Optics*, Vol. 41, No. 9, September 1994, pp. 1837–1862.

[Amidror95] I. Amidror, *Analysis of Moiré Patterns in Multi-Layer Superpositions.* Ph.D. Thesis No. 1341, EPFL, Lausanne, 1995.

[Amidror96] I. Amidror, "Fourier-based analysis of phase shifts in the superposition of periodic layers and their moiré effects," *Jour. of the Optical Society of America A*, Vol. 13, No. 5, 1996, pp. 974–987.

[Amidror97] I. Amidror, "Fourier spectrum of radially periodic images," *Jour. of the Optical Society of America A*, Vol. 14, No. 4, 1997, pp. 816–826.

[Amidror97a] I. Amidror and R. D. Hersch, "Quantitative analysis of multichromatic moiré effects in the superposition of coloured periodic layers," *Jour. of Modern Optics*, Vol. 44, No. 5, 1997, pp. 883–899.

[Amidror98] I. Amidror and R. D. Hersch, "Analysis of the superposition of periodic layers and their moiré effects through the algebraic structure of their Fourier spectrum," *Jour. of Mathematical Imaging and Vision*, Vol. 8, 1998, pp. 99–130.

[Amidror98a] I. Amidror, "Fourier spectrum of curvilinear gratings of the second order," *Jour. of the Optical Society of America A*, Vol. 15, No. 4, 1998, pp. 900–913.

[Amidror98b] I. Amidror and R. D. Hersch, "Fourier-based analysis and synthesis of moirés in the superposition of geometrically transformed periodic structures," *Jour. of the Optical Society of America A*, Vol. 15, No. 5, 1998, pp. 1100–1113.

[Amidror98c] I. Amidror, "The Fourier-spectrum of circular sine and cosine gratings with arbitrary radial phases," *Optics Communications*, Vol. 149, 1998, pp. 127–134.

[Amidror99] I. Amidror and R. D. Hersch, "Methods and apparatus for authentication of documents by using the intensity profile of moire patterns," US Patent No. 5,995,638, 1999.

[Amidror01] I. Amidror and R. D. Hersch, "Method and apparatus for authentication of documents by using the intensity profile of moire patterns," US Patent No. 6,249,588, 2001.

[Amidror02] I. Amidror, "A new print-based security strategy for the protection of valuable documents and products using moiré intensity profiles," in *Optical Security and Counterfeit Deterrence Techniques IV*, R. L. Van Renesse (Ed.), Proceedings of SPIE Vol. 4677, SPIE, Bellingham, 2002, pp. 89–100.

[Amidror04] I. Amidror and R. D. Hersch, "Authentication of documents and valuable articles by using moire intensity profiles," US Patent No. 6,819,775, 2004.

[Artin91] M. Artin, *Algebra*. Prentice Hall, NJ, 1991.

[Asundi93] A. K. Asundi, "Moiré methods using computer-generated gratings," *Optical Engineering*, Vol. 32, No. 1, January 1993, pp. 107–116.

[Badcock85] D. R. Badcock and A. M. Derrington, "Detecting the displacement of periodic patterns," *Vision Research*, Vol. 25, No. 9, 1985, pp. 1253–1258.

[Badcock89] D. R. Badcock and A. M. Derrington, "Detecting the displacement of spacial beats: no role for distortion products," *Vision Research*, Vol. 29, No. 6, 1989, pp. 731–739.

[Baez61] A. V. Baez, "Fresnel zone plate for optical image formation using extreme ultraviolet and soft X radiation," *Jour. of the Optical Society of America*, Vol. 51, No. 4, April 1961, pp. 405–412.

[Bann90] D. Bann and J. Gargan, *How to Check and Correct Colour Proofs*. North Light Books, USA, 1990.

[Bass71] J. Bass, *Cours de Mathématiques (Tome 3)*. Masson, Paris, 1971 (in French).

[Baudoin38] P. Baudoin, *Les Ovales de Descartes et le Limaçon de Pascal*. Librairie Vuibert, Paris, 1938 (In French).

[Besicovitch32] A. S. Besicovitch, *Almost Periodic Functions*. Cambridge University Press, Cambridge, 1932.

[Birkhoff77] G. Birkhoff and S. Mac Lane, *A Survey of Modern Algebra*. Macmillan Publishing, NY, 1977 (fourth edition).

[Black53] H. S. Black, *Modulation Theory*. D. van Nostrand, NY, 1953.

[Blatner98] D. Blatner, G. Fleishman and S. Roth, *Real World Scanning and Halftones*. Peachpit Press, Berkeley, 1998 (second edition).

[Boff86] K. R. Boff, L. Kaufman and J. P. Thomas, *Handbook of Perception and Human Performance (Vol. 2)*, John Wiley & sons, NY, 1986.

[Bohr23] H. Bohr, "Sur les fonctions presque périodiques," *Comptes Rendus Hebd. de L'academie des Sciences*, Paris, Vol. 177, 1923, pp. 737–739; "Sur l'approximation des fonctions presque périodiques par des sommes trigonométriques," *ibid.*, pp. 1090–1092 (in French).

[Bohr51] H. Bohr, *Almost Periodic Functions.* Chelsea Publishing Company, NY, 1951.

[Bracewell86] R. N. Bracewell, *The Fourier Transform and its Applications.* McGraw-Hill Publishing Company, Reading, NY, 1986 (second edition).

[Bracewell95] R. N. Bracewell, *Two Dimensional Imaging.* Prentice Hall, NJ, 1995.

[Brigham88] E. O. Brigham, *The Fast Fourier Transform and Its Applications.* Prentice-Hall, NJ, 1988.

[Bronshtein97] I. N. Bronshtein and K. A. Semendyayev, *Handbook of Mathematics.* Springer, Berlin, 1997 (Reprint of the third edition; English translation).

[Bronstein90] I. N. Bronstein and K. A. Semendiaev, *Aide-Mémoire de Mathématiques.* Eyrolles, Paris, 1990 (9th edition; French translation).

[Bryngdahl74] O. Bryngdahl, "Moiré: formation and interpretation," *Jour. of the Optical Society of America,* Vol. 64, No. 10, 1974, pp. 1287–1294.

[Bryngdahl75] O. Bryngdahl, "Moiré and higher grating harmonics," *Jour. of the Optical Society of America,* Vol. 65, No. 6, 1975, pp. 685–694.

[Bryngdahl76] O. Bryngdahl, "Characteristics of superposed patterns in optics," *Jour. of the Optical Society of America,* Vol. 66, No. 2, 1976, pp. 87–94.

[Bryngdahl81] O. Bryngdahl, "Beat pattern selection — multi-color-grating moiré," *Optics Communications,* Vol. 39, No. 3, 1981, pp. 127–131.

[Burch77] J. M. Burch and D. C. Williams, "Varifocal moiré zone plates for straightness measurement," *Applied Optics*, Vol. 16, No. 9, September 1977, pp. 2445–2450.

[CanadianBank68] Canadian Bank Note Company Ltd., "Improvements in printed matter for the purpose of rendering counterfeiting more difficult," UK Patent No. 1,138,011, 1968.

[Cartwright90] M. Cartwright, *Fourier Methods for Mathematicians, Scientists and Engineers.* Ellis Horwood, UK, 1990.

[Cassels71] J. W. S. Cassels, *An Introduction to the Geometry of Numbers.* Springer-Verlag, NY, 1971 (second printing, corrected).

[Champeney73] D. C. Champeney, *Fourier Transforms and their Physical Applications.* Academic Press, London, 1973.

[Champeney87] D. C. Champeney, *Fourier Theorems.* Cambridge University Press, UK, 1987.

[Chau69] H. H. M. Chau, "Zone plates produced optically," *Applied Optics*, Vol. 8, No. 6, June 1969, pp. 1209–1211.

[Chosson06] S. Chosson, *Synthèse d'Images Moiré.* Ph.D. Thesis No. 3434, EPFL, Lausanne, 2006 (in French).

[Corduneanu68] C. Corduneanu, *Almost Periodic Functions.* Interscience Publishers, US, 1968.

[Cornsweet70] T. N. Cornsweet, *Visual Perception.* Harcourt Brace Jovanovich, Florida, 1970.

[Coudray91] M. A. Coudray, "Understanding halftone moiré," *Screen Printing,* Vol. 81, No. 11, 1991, pp. 142–147, 212.

[Courant88] R. Courant, *Differential and Integral Calculus (Vol. II)*. Wiley-Interscience, USA, 1988.

[Daels94] K. Daels and P. Delabastita, "Color balance in conventional halftoning," Proceedings of the TAGA Conference, Baltimore, May 1994, pp. 1–18.

[Daly92] S. Daly, "The visible differences predictor: an algorithm for the assessment of image fidelity," *Human Vision, Visual Processing and Digital Display III*, SPIE proceedings, Vol. 1666, USA, 1992, pp. 2–15.

[Delabastita92] P. A. Delabastita, "Screening techniques, moiré in four color printing," Proceedings of the TAGA Conference, Vancouver, April 1992, pp. 44–65.

[Durelli70] A. J. Durelli and V. J. Parks, *Moiré Analysis of strain*. Prentice-Hall, Englewood Cliffs, New Jersey, 1970.

[Ebbeni70] J. Ebbeni, "Nouvaux aspects du phénomène de moiré," *Nouvelle Revue d'Optique Appliquée*, Vol. 1, 1970, pp. 333–342, 353–358 (in French).

[Egorov93] Yu. V. Egorov and M. A. Shubin (Eds.), *Partial Differential Equations IV: Microlocal Analysis and Hyperbolic Equations*. Springer Verlag, Berlin, 1993.

[EncMath88] *Encyclopaedia of Mathematics*, Vols. 1–10. Kluwer Academic Publishers, Dordrecht, 1988–1994.

[Erdélyi54] A. Erdélyi (Ed.), *Tables of Integral Transforms (Vol. 1)*. McGraw-Hill, NY, 1954.

[Eschbach88] R. Eschbach, "Generation of moiré by nonlinear transfer characteristics," *Jour. of the Optical Society of America A*, Vol. 5, No. 11, 1988, pp. 1828–1835.

[Fink92] P. Fink, *PostScript Screening: Adobe Accurate Screens*. Adobe Press, California, 1992.

[Firby84] P. A. Firby and D. J. Stone, "Interference patterns and two-dimensional potential theory," *IMA Journal of Applied Mathematics*, Vol. 33, 1984, pp. 291–301.

[Foley90] J. D. Foley, A. van Dam, S. K. Feiner and J. F. Hughes, *Computer Graphics: Principles and Practice*. Addison-Wesely, Reading, Massachusetts, 1990 (second edition).

[Foster72] A. M. Foster, "The moiré interference in digital halftone generation," Proceedings of the TAGA Conference, 1972, pp. 130–143.

[Gaskill78] J. D. Gaskill, *Linear Systems, Fourier Transforms, and Optics*. John Wiley & Sons, NY, 1978.

[Gåsvik95] K. J. Gåsvik, *Optical Metrology*. John Wiley & Sons, NY, 1995 (second edition).

[Gelfand64] I. M. Gelfand and G. E. Shilov, *Generalized Functions (Vol. 1)*. Academic Press, CA, 1964.

[Giger86] H. Giger, "Moirés," *Computers & Mathematics with Applications*, Vol. 12B, No. 1/2, 1986, pp. 329–361.

[Glass73] L. Glass and R. Pérez, "Perception of random dot interference patterns," *Nature*, Vol. 246, December 1973, pp. 360–362.

[Gradshteyn94] I. S. Gradshteyn and I. M. Ryzhik, *Table of Integrals, Series, and Products*. Academic Press, Boston, 1994 (fifth edition).

[Gruber93] P. M. Gruber and J. M. Wills (Eds.), *Handbook of Convex Geometry (Vol. B)*. North-Holland, Amsterdam, 1993.

[Hahn96] S. L. Hahn, *Hilbert Transforms in Signal Processing*. Artech House, Norwood, MA, 1996.

[Harburn75] G. Harburn, T. R. Welberry and R. P. Williams, "A Fourier-series approach to moiré patterns with special reference to those produced by overlapping zone plates," *Optica Acta*, Vol. 22, No. 5, 1975, pp. 409–420.

[Harding83] K. G. Harding and J. S. Harris, "Projection moiré interferometer for vibration analysis," *Applied Optics*, Vol. 22, No. 6, March 1983, pp. 856–861; reprinted in [Indebetouw92, pp. 269–274].

[Hardy68] G. H. Hardy and W. W. Rogosinski, *Fourier Series*. Cambridge University Press, London, 1968 (reprint of third edition of 1956).

[Hardy73] G. H. Hardy, *Divergent Series*. University Press, Oxford, 1973 (corrected reprint of first edition of 1949).

[Harthong81] J. Harthong, "Le Moiré," *Advances in Applied Mathematics,* Vol. 2, 1981, pp. 24–75 (in French).

[Heaviside71] O. Heaviside, *Electromagnetic Theory (Vol. II)*. Chelsea, NY, 1971 (third edition).

[Hersch04] R. D. Hersch and S. Chosson, "Band moiré images," Proceedings of SIGGRAPH 2004, *ACM Transactions on Graphics*, Vol. 23, No. 3, 2004, pp. 239–248.

[Hersch04a] R. D. Hersch and S. Chosson, "Authentication of documents and articles by moire patterns," Published US Patent Application No. 2004/0076310, 2004.

[Hersch06] R. D. Hersch and S. Chosson, "Model-based synthesis of band moire images for authenticating security documents and valuable products," Published US Patent Application No. 2006/0003295, 2006.

[Hersch06a] R. D. Hersch and S. Chosson, "Model-based synthesis of band moire images for authentication purposes," Published US Patent Application No. 2006/0129489, 2006.

[Hoy92] D. E. P. Hoy, "Color digital imaging system for bicolor moiré analysis," *Experimental Techniques,* Vol. 16, No. 3, 1992, pp. 25–27.

[Huang74] T. S. Huang, "Digital transmission of halftone pictures," *Computer Graphics and Image Processing,* Vol. 3, 1974, pp. 195–202.

[Huck03] J. Huck, *Mastering Moirés: Investigating Some of the Fascinating Properties of Interference Patterns*. Private publication by J. Huck, 2003. `http://pages.sbcglobal.net/joehuck`

[Huck04] J. Huck, "Moiré patterns and the illusion of depth," Proceedings of the fifth Interdisciplinary Conference of the International Society of the Arts, Mathematics and Architecture (ISAMA 2004), Chicago, June 2004.

[Hunt87] R. W. G. Hunt, *The Reproduction of Colour in Photography, Printing and Television.* Fountain Press, England, 1987 (fourth edition).

[Hutley91] M. Hutley, R. Stevens and D. Daly, "Microlens arrays," *Physics World,* Vol. 4, July 1991, pp. 27–32.

[Indebetouw92] G. Indebetouw and R. Czarnek (Eds.), *Selected Papers on Optical Moiré and Applications.* SPIE Milestone Series, Vol. MS64, SPIE Optical Engineering Press, Washington, 1992.

[Jacobson85] N. Jacobson, *Basic Algebra (Vol. 1).* W. H. Freeman and Co., NY, 1985 (second edition).

[Jansson97] P. A. Jansson (Ed.), *Deconvolution of Images and Spectra.* Academic Press, US, 1997 (second edition).

[Jarić89] M. V. Jarić (Ed.), *Introduction to the Mathematics of Quasicrystals (Vol. 2)*. Academic Press, US, 1989.

[Jones94] P. R. Jones, "Evolution of halftoning technology in the United States patent literature," *Journal of Electronic Imaging,* Vol. 3, No. 3, July 1994, pp. 257–275.

[Kafri89] O. Kafri and I. Glatt, *The Physics of Moiré Metrology*. John Wiley & Sons, NY, 1989.

[Kang97] H. R. Kang, *Color Technology for Electronic Imaging Devices*. SPIE Optical Engineering Press, Bellingham, Washington, 1997.

[Kannan87] R. Kannan, *Algorithmic Geometry of Numbers*. Report No. 87453-OR, Rheinische Friedrich-Wilhelm Universität, Bonn, 1987.

[Katz86] A. Katz and M. Duneau, "Quasiperiodic patterns and icosahedral symmetry," *Journal de Physique*, Vol. 47, 1986, pp. 181–196.

[Katzenelson68] Y. Katzenelson, *An Introduction to Harmonic Analysis*. John Wiley & Sons, US, 1968.

[Kendig80] K. M. Kendig, "Moiré phenomena in algebraic geometry: polynomial alterations in $\mathbb{R}^n$," *Pacific Journal of Mathematics*, Vol. 89, No. 2, 1980, pp. 327–349.

[Kermisch75] D. Kermisch and P. G. Roetling, "Fourier spectrum of halftone images," *Jour. of the Optical Society of America*, Vol. 65, June 1975, pp. 716–723.

[King72] M. C. King and D. H. Berry, "Photolithographic mask alignment using moiré techniques," *Applied Optics*, Vol. 11, No. 11, November 1972, pp. 2455–2459.

[Kipphan01] H. Kipphan, *Handbook of Print Media*. Springer, Berlin, 2001.

[Konno94] H. Konno, H. Ohara, K. Murata and Y. Nakano, "Measurement of focusing properties of axially asymmetric lenses using moiré technique," *Optical Review*, Vol. 1, No. 1, 1994, pp. 107–109.

[Korn68] G. A. Korn and T. M. Korn, *Mathematical Handbook for Scientific and Engineers*. McGraw-Hill, NY, 1968 (second edition).

[Kumar83] R. Kumar and M. M. Bindal, "Measurement of small angles and small angular rotations by moiré method," *Atti della Fondazione Giorgio Ronchi*, Vol. 38, No. 4, July-August 1983, pp. 399–410.

[Kunt86] M. Kunt, *Digital Signal Processing*. Artech House, MA, 1986.

[Lang78] S. Lang, *Undergraduate Algebra*. Springer-Verlag, NY, 1978.

[Latimer93] P. Latimer, "Talbot plane patterns: grating images or interference effects?" *Applied Optics*, Vol. 32, No. 7, March 1993, pp. 1078–1083.

[Lay03] D. C. Lay, *Linear Algebra and its Applications*. Addison-Wesley, Boston, 2003 (third edition).

[Legault73] R. Legault, "The aliasing problems in two-dimensional sampled imagery," in *Perception of Displayed Information*, L. M. Biberman (Ed.), Plenum Press, NY, 1973, pp. 279–312.

[Leifer73] I. Leifer, J. M. Walls and H. N. Southworth, "The moiré pattern produced by overlapping zone plates," *Optica Acta*, Vol. 20, No. 1, 1973, pp. 33–47.

[Levien93] R. Levien, "Well tempered screening technology," IS&T Third Technical Symposium on Prepress, Proofing and Printing, Chicago, 1993, pp. 98–101; also reprinted in *Recent Progress in Digital Halftoning,* R. Eschbach (Ed.), IS&T, 1994, pp. 83–86.

[Levitan82] B. M. Levitan and V. V. Zhikov, *Almost Periodic Functions and Differential Equations*. Cambridge University Press, Cambridge, 1982.

[Lipschutz91] S. Lipschutz, *Linear Algebra*. Schaum's Outline Series, McGraw-Hill, NY, 1991 (second edition).

[Lockwood61] E. H. Lockwood, *A Book of Curves*. Cambridge University Press, Cambridge, 1961.

[Lohmann67] A. W. Lohmann and D. P. Paris, "Variable Fresnel zone pattern," *Applied Optics*, Vol. 6, September 1967, pp. 1567–1570.

[Luo98] J. Luo, R. de Queiroz and Z. Fan, "Robust technique for image descreening based on the wavelet transform," *IEEE Transactions on Signal Processing*, Vol. 46, No. 4, April 1998, pp. 1179–1184.

[Marsh80] J. S. Marsh, "Contour plots using a moiré technique," *American Jour. of Physics*, Vol. 48, January 1980, pp. 39–40.

[Matteson04] R. Matteson, *Scanning for the SOHO — Small Office and Home Office*. Virtualbookworm Publishing, Texas, 2004.

[McDonald97] R. McDonald (Ed.), *Colour Physics for Industry*. Society of Dyers and Colourists, Bradford, England, 1997 (second edition).

[Meadows70] D. M. Meadows, W. O. Johnson and J. B. Allen, "Generation of surface contours by moiré patterns," *Applied Optics*, Vol. 9, No. 4, April 1970, pp. 942–947.

[Miceli92] C. M. Miceli and K. J. Parker, "Inverse Halftoning," *Journal of Electronic Imaging*, Vol. 1, No. 2, April 1992, pp. 143–151.

[Morimoto88] Y. Morimoto, Y. Seguchi and T. Higashi, "Application of moiré analysis of strain using Fourier transform," *Optical Engineering*, Vol. 27, No. 8, 1988, pp. 650–656.

[Morimoto90] Y. Morimoto, Y. Seguchi and M. Okada, "Screening and moiré suppression in printing and its analysis by Fourier transform," *Systems and Computers in Japan*, Vol. 21, No. 2, 1990, pp. 98–106.

[Morimura75] M. Morimura and K. Nakagawa, "Small-angle measuring device utilizing moiré fringe," *Japan Journal of Applied Physics*, Vol. 14, Supplement 14-1, 1975, pp. 461–464.

[Myers51] O. E. Myers, Jr., "Studies of transmission zone plates," *American Journal of Physics*, Vol. 19, No. 6, 1951, pp. 359–365.

[Nakano85] Y. Nakano and K. Murata, "Talbot interferometry for measuring the focal length of a lens," *Applied Optics*, Vol. 24, No. 19, October 1985, pp. 3162–3166.

[Nelson86] R. D. Nelson, "Quasicrystals," *Scientific American*, Vol. 255, August 1986, pp. 33–41.

[Nishijima64] Y. Nishijima and G. Oster, "Moiré patterns: their application to refractive index and refractive index gradient measurements," *Jour. of the Optical Society of America*, Vol. 54, No. 1, January 1964, pp. 1–5.

[Ohyama86] N. Ohyama, M. Yamaguchi, J. Tsujiuchi, T. Honda and S. Hiratsuka, "Suppression of moiré fringes due to sampling of halftone screened images," *Optics Communications*, Vol. 60, No. 6, December 1986, pp. 364–368.

[Oster63] G. Oster and Y. Nishijima, "Moiré patterns," *Scientific American*, Vol. 208, May 1963, pp. 54–63.

[Oster64] G. Oster, M. Wasserman and C. Zwerling, "Theoretical interpretation of moiré patterns," *Jour. of the Optical Society of America*, Vol. 54, No. 2, 1964, pp. 169–175.

[Oster65] G. Oster, "Optical art," *Applied Optics,* Vol. 4, No. 11, 1965, pp. 1359–1369.

[Oster69] G. Oster, *The Science of Moiré Patterns.* Edmund Scientific Co., Barrington, NJ, 1969 (second edition).

[Ostromoukhov93] V. Ostromoukhov, "Pseudo-random halftone screening for color and black&white printing," Proceedings of the 9-th International Congress on Advances in Non-impact Printing Technologies, Yokohama, October 1993, pp. 579–582; also reprinted in *Recent Progress in Digital Halftoning,* R. Eschbach (Ed.), IS&T, 1994, pp. 130–134.

[Ostromoukhov95] V. Ostromoukhov and R. D. Hersch, "Artistic screening," Proceedings SIGGRAPH 95, Computer Graphics Proceedings, Annual Conference Series, 1995, pp. 219–228.

[Papoulis68] A. Papoulis, *Systems and Transforms with Applications in Optics.* McGraw-Hill, NY, 1968.

[Patorski76] K. Patorski, S. Yokozeki and T. Suzuki, "Moiré profile prediction by using Fourier series formalism," *Japanese Jour. of Applied Physics,* Vol. 15, No. 3, 1976, pp. 443–456.

[Patorski93] K. Patorski, *Handbook of the Moiré Fringe Technique.* Elsevier, Amsterdam, 1993.

[Post67] D. Post, "Sharpening and multiplication of moiré fringes," *Experimental Mechanics,* Vol. 7, April 1967, pp. 154–159.

[Post94] D. Post, B. Han and P. Ifju, *High Sensitivity Moiré: Experimental Analysis for Mechanics and Materials.* Springer-Verlag, NY, 1994.

[Pratt91] W. K. Pratt, *Digital Image Processing.* John Wiley & Sons, NY, 1991 (second edition).

[Pssc65] Physical Science Study Committee, *Physics.* D.C. Heath and Company, Boston, 1965 (second edition).

[Rayleigh74] Lord Rayleigh, "On the manufacture and theory of diffraction-gratings," *London, Edinburgh, and Dublin Philosophical Magazine*, Series 4, Vol. 47, 1874, pp. 81–93 and 193–205.

[Renesse05] R. L. van Renesse, *Optical Document Security.* Artech House, Boston, 2005 (third edition).

[Réveillès91] J.-P. Réveillès, *Géométrie Discrète, Calcul en Nombres Entiers et Algorithmique.* Thèse d'Etat, Université Louis-Pasteur, Strasbourg, 1991 (in French).

[Righi87] A. Righi, "Sui fenomeni che si producono colla sovrapposizione di due reticoli e sopra alcune loro applicazioni," *Il Nuovo Cimento*, Series 3, Vol. 21, 1887, pp. 203–229 (in Italian).

[Rodriguez94] M. Rodriguez, "Promises and pitfalls of stochastic screening in the graphics arts industry," Proceedings of the 47th Annual Conference of the IS&T, May 1994; also reprinted in *Recent Progress in Digital Halftoning,* R. Eschbach (Ed.), IS&T, 1994, pp. 34–37.

[Rogers77] G. L. Rogers, "A geometrical approach to moiré pattern calculations," *Optica Acta*, Vol. 24, No. 1, 1977, pp. 1–13.

[Rosenfeld82] A. Rosenfeld and A. C. Kak, *Digital Picture Processing (Vol.1).* Academic Press, Florida, USA, 1982 (second edition).

[Russ95] J. C. Russ, *The Image Processing Handbook.* CRC Press, Florida, 1995 (second edition).

[Saichev97] A. I. Saichev and W. A. Woyczyński, *Distributions in the Physical and Engineering Sciences (Vol. 1)*. Birkhäuser, Boston, 1997.

[Scantips97] http://www.scantips.com/basics06.html

[Schilling06] A. Schilling, W. R. Tompkin, R. Staub, R. D. Hersch, S. Chosson and I. Amidror, "Diffractive moiré features for optically variable devices," in *Optical Security and Counterfeit Deterrence Techniques VI*, R. L. Van Renesse (Ed.), Proceedings of SPIE Vol. 6075, SPIE, Bellingham, 2006, pp. 60750V-1–60750V-12.

[Schoppmeyer85] J. Schoppmeyer, "Screen systems for multicolor printing," US Patent No. 4,537,470, 1985.

[Schreiber93] W. F. Schreiber, *Fundamentals of Electronic Imaging Systems*. Springer-Verlag, Berlin, 1993 (third edition).

[Shamir73] J. Shamir, "Moiré gauging by projected interference fringes," *Optics and Laser Technology*, Vol. 5, April 1973, pp. 78–86.

[Shechtman84] D. Shechtman, I. Blech, D. Gratias and J. W. Cahn, "Metallic phase with long-range orientational order and no translational symmetry," *Physical Review Letters*, Vol. 53, 1984, pp. 1951–1954.

[Shepherd79] A. T. Shepherd, "25 years of moiré fringe measurement," *Precision Engineering*, Vol. 1, No. 2, 1979, pp. 61–69.

[Shikin95] E. V. Shikin, *Handbook and Atlas of Curves*. CRC Press, Boca Raton, 1995.

[Shu89] J. S. Shu, R. Springer and C. L. Yeh, "Moiré factors and visibility in scanned and printed halftone images," *Optical Engineering*, Vol. 28, No. 7, July 1989, pp. 805–812.

[Siegel89] C. L. Siegel, *Lectures on the Geometry of Numbers*. Springer-Verlag, Berlin, 1989.

[Spiegel68] M. R. Spiegel, *Mathematical Handbook of Formulas and Tables*. Schaum's Outline Series, McGraw-Hill, NY, 1968.

[Stecher64] M. Stecher, "The moiré phenomenon," *American Journal of Physics*, Vol. 32, No. 4, 1964, pp. 247–257.

[Steinbach82] A. Steinbach and K. Y. Wong, "Moiré patterns in scanned halftone pictures," *Journal of the Optical Society of America*, Vol. 72, No. 9, 1982, pp. 1190–1198.

[Steinhardt90] P. J. Steinhardt and P. Taylor, "Methods and apparatus for eliminating moiré interference using quasiperiodic patterns," US Patent No. 4,894,726, 1990.

[Stevenson97] R. L. Stevenson, "Inverse halftoning via MAP estimation," *IEEE Transactions on Image Processing*, Vol. 6, No. 4, April 1997, pp. 574–583.

[Swift74] D. W. Swift, "A simple moiré fringe technique for magnification checking," *Journal of Physics E*, Vol. 7, No. 3, 1974, pp. 164–166.

[Takasaki70] H. Takasaki, "Moiré topography," *Applied Optics*, Vol. 9, No. 6, 1970, pp. 1467–1472.

[Takeda82] M. Takeda, H. Ina and S. Kobayahi, "Fourier-transform method of fringe-pattern analysis for computer-based topography and interferometry," *Journal of the Optical Society of America*, Vol. 72, No. 1, 1982, pp. 156–160.

[Taub86] H. Taub and D. L. Schilling, *Principles of Communication Systems*. McGraw-Hill, NY, 1986 (second edition).

[Theocaris69] P. S. Theocaris, *Moiré Fringes in Strain Analysis*. Pergamon Press, UK, 1969.

[Theocaris73] P. S. Theocaris and S. A. Paipetis, "Plane kinematics of moiré fringes," *Journal of Physics E*, Vol. 6, 1973, pp. 987–990.

[Theocaris75] P. S. Theocaris, S. A. Paipetis, and C. Liakopoulos, "Some aspects on the formation of spatial moiré fringes," *Optik*, Vol. 42, No. 3, 1975, pp. 269–286.

[Tollenaar64] D. Tollenaar, "Moiré in halftone printing interference phenomena." English translation published in 1964; reprinted in [Indebetouw92, pp. 618–633].

[Tolstov62] G. P. Tolstov, *Fourier Series.* Prentice-Hall, New Jersey, 1962.

[Ulichney88] R. Ulichney, *Digital Halftoning.* MIT Press, USA, 1988.

[Vygodski73] M. Vygodski, *Aide-Mémoire de Mathématiques Supérieures.* Mir, Moscow, 1973 (French translation).

[Walls75] J. M. Walls and H. N. Southworth, "The moiré pattern formed on superposing a zone plate with a grating or grid," *Optica Acta*, Vol. 22, No. 7, 1975, pp. 591–601.

[Wandell95] B. A. Wandell, *Foundations of Vision.* Sinauer, Massachusetts, 1995.

[Weber73] R. L. Weber, *A Random Walk in Science*. The Institute of Physics, London, 1973, pp. 199–200. Also published in *Applied Optics*, Vol. 7, No. 4, 1968, p. 625.

[Weisstein99] E. W. Weisstein, *CRC Concise Encyclopedia of Mathematics.* CRC, Boca Raton, 1999.

[Welberry76] T. R. Welberry and R. P. Williams, "On certain non-circular zone plates," *Optica Acta*, Vol. 23, No. 3, 1976, pp. 237–244.

[Widmer92] E. Widmer, K. Schläpfer, V. Humbel and S. Persiev, "The benefits of frequency modulation screening," Proceedings of the TAGA Conference, Vancouver, April 1992, pp. 28–43.

[Witschi86] W. Witschi, "Moirés," *Computers & Mathematics with Applications*, Vol. 12B, Nos. 1/2, 1986, pp. 363–378.

[Wurzburg61] F. L. Wurzburg, Jr., "Understanding moiré screen pattern," *Gravure*, Vol. 7, 1961, pp. 17–21.

[Wyszecki82] G. Wyszecki and W. S. Stiles, *Color Science: Concepts and Methods, Quantitative Data and Formulae*. John Wiley & Sons, NY, 1982 (second edition).

[Yokozeki70] S. Yokozeki and T. Suzuki, "Theoretical interpretation of moiré pattern for two grids of parallel lines," *Japan Jour. of Applied Physics*, Vol. 9, 1970, pp. 1011–1012.

[Yokozeki74] S. Yokozeki, "Theoretical interpretation of the moiré pattern," *Optics Communications*, Vol. 11, No. 4, August 1974, pp. 378–381.

[Yule67] J. A. C. Yule, *Principles of Color Reproduction.* John Wiley & Sons, NY, 1967, Chapter 13: "Moiré Patterns and Screen Angles", pp. 328–345.

[Zunino03] L. Zunino and M. Garavaglia, "Moiré by fractal structures," *Jour. of Modern Optics*, Vol. 50, No. 9, 2003, pp. 1477–1486.

[Zygmund68] A. Zygmund, *Trigonometric Series (Vol. 1).* Cambridge University Press, US, 1968 (corrected reprint of second edition of 1959).

# Index

Page numbers followed by the letter "g" indicate entries in the glossary (Appendix D).

**T**



**U**

MIX
Papier aus verantwortungsvollen Quellen
Paper from responsible sources
FSC® C105338

If you have any concerns about our products,
you can contact us on
**ProductSafety@springernature.com**

In case Publisher is established outside the EU,
the EU authorized representative is:
**Springer Nature Customer Service Center GmbH**
**Europaplatz 3, 69115 Heidelberg, Germany**

Printed by Libri Plureos GmbH
in Hamburg, Germany